Solutions to Exercises

Roxy Wilson
University of Illinois

Sixth Edition

CHEMISTRY
The Central Science

Theodore L. Brown
University of Illinois

H. Eugene LeMay, Jr.
University of Nevada

Bruce E. Bursten
The Ohio State University

Prentice Hall, Englewood Cliffs, NJ 07632

A Paramount Communications Company
Englewood Cliffs, N.J. 07632

10 9 8 7 6 5 4 3 2

ISBN 0-13-338724-0

Printed in the United States of America

Contents

Introduction

Chemistry: The Central Science, 6th edition, contains nearly 2000 end-of-chapter exercises. Much attention has been given to these exercises because one of the best ways for students to master chemistry is by solving problems. Grouping the exercises according to subject matter is intended to aid the student in selecting and recognizing particular types of problems. Within each subject matter group, similar problems are arranged in pairs. This provides the student with an opportunity to reinforce a particular kind of problem. There are also a substantial number of general exercises in each chapter to supplement those grouped by topic. Answers to the odd numbered topical exercises and selected general exercises, about 900 in all, are provided in the text. These appendix answers help to make the text a useful self-contained vehicle for learning.

This manual, **Solutions to Exercises in Chemistry: The Central Science, 6th edition**, was written to enhance the end-of-chapter exercises by providing documented solutions. The manual assists the instructor by saving time spent generating solutions for assigned problem sets and aids the student by offering an independent source to check their understanding of the material. Most solutions have been worked in the same detail as the in-chapter sample exercises to help guide the students in their studies.

Extraordinary efforts have been made to keep this manual as error-free as possible. All exercises were worked by at least two chemists and proofread by a third to ensure clarity in methods and accuracy in mathematics. However, the typo virus seems to find its way into all written work and notification of sightings would be appreciated. We hope that both instructors and students will find this manual accurate, helpful and instructive.

Roxy B. Wilson
School of Chemical Sciences
University of Illinois
Urbana 61801

CHAPTER ***1***

Introduction: Some Basic Concepts

Introduction to Matter

1.1 (a) liquid (b) solid (c) gas (d) gas

1.2 (a) gas (b) solid (c) liquid (d) solid

1.3 (a) chemical - a different substance with different physical properties is formed
(b) physical (c) chemical (d) physical

1.4 (a) chemical (b) physical (c) physical (The production of H_2O is a chemical change, but its **condensation** is a physical change.) (d) physical (The production of soot is a chemical change, but its **deposition** is a physical change.)

1.5 Physical properties: silvery white (color); lustrous; melting point = 649°C; boiling point = 1105°C; density at 20°C = 1.738 g/mL; pounded into sheets (maleable); drawn into wires (ductile); good conductor. Chemical properties: burns in air to give intense white light; reacts with Cl_2 to produce brittle white solid.

1.6 (a) physical (b) physical (c) chemical (d) physical (e) physical
(f) chemical (g) physical (h) physical (i) chemical

1.7 (a) heterogeneous mixture (b) pure substance (c) homogeneous mixture
(d) pure substance

1.8 (a) homogeneous mixture (b) homogeneous mixture (c) pure substance
(d) heterogeneous mixture (This applies to dressings such as Italian, Thousand Island, etc.; homogeneous looking dressings such as French are really emulsions which are classified somewhere between homogeneous and heterogeneous mixtures.)

Elements and Compounds

1.9 (a) C (b) Na (c) Fe (d) P (e) K (f) Cl (g) N (h) Ag

1.10 (a) helium (b) magnesium (c) lead (d) sulfur
(e) fluorine (f) zinc (g) copper (h) argon

1.11 Before modern instrumentation, the classification of a pure substance as an element was determined by whether it could be broken down into component elements. Scientists subjected the substance to all known chemical means of decomposition, and if the results were negative, the substance was an element. Classification by negative results was somewhat ambiguous, since an effective decomposition technique might exist, but not yet have been discovered.

1.12 $A(s) \xrightarrow{\text{heat}} B(s) + C(g)$

When carbon(s) is burned in excess oxygen the two elements combine to form a gaseous compound, carbon dioxide. Clearly substance C is this compound.

Since C is produced when A is heated in the absence of oxygen (from air), both the carbon and oxygen in C must have been present in A originally. A is, therefore, a compound composed of two or more elements chemically combined. Without more information on the chemical or physical properties of B, we cannot determine absolutely whether it is an element or compound. However, few if any elements exist as white solids, so B is probably also a compound.

Units and Measurement

1.13 (a) meters (b) $(\text{meters})^2$ (c) $(\text{meters})^3$ (d) kilograms
(e) meters/second (f) Kelvins

1.14 (a) $(\text{meters})^2$ (b) $(\text{meters})^3$ (c) kilograms (d) seconds (e) meters (f) Kelvins

1.15 (a) 1×10^{-1} (b) 1×10^{-2} (c) 1×10^{-15} (d) 1×10^{-6} (e) 1×10^{6} (f) 1×10^{3}
(g) 1×10^{-9} (h) 1×10^{-3} (i) 1×10^{-12}

1.16 (a) 3.4 $\underline{\text{p}}$m (b) 4.8 $\underline{\text{n}}$L (c) 7.23 $\underline{\text{k}}$g (d) 2.35 cm^3 ($1 \times 10^{-6}\ \text{m}^3 = (1 \times 10^{-2})^3\ \text{cm}^3$)
(e) 5.8 $\underline{\text{n}}$s (f) 3.45 $\underline{\text{m}}$mol

1.17 (a) volume (b) area (c) volume (d) density
(e) time (f) length (g) temperature

1.18 (a) time (b) density (c) length (d) area
(e) temperature (f) volume (g) temperature

1.19. (a) $$454\ \text{mg} \times \frac{1 \times 10^{-3}\ \text{g}}{1\ \text{mg}} = 0.454\ \text{g}$$

(b) $$5.0 \times 10^{-8}\ \text{m} \times \frac{1\ \text{nm}}{1 \times 10^{-9}\ \text{m}} = 5.0 \times 10^{1}\ \text{nm}$$ (50. nm, the zero is significant)

(c) $0.076\ mL \times \frac{1 \times 10^{-3}\ L}{1\ mL} \times \frac{1\ \mu L}{1 \times 10^{-6}\ L} = 76\ \mu L$

(d) $\frac{1.55\ kg}{m^3} \times \frac{1000\ g}{1\ kg} \times \frac{(1 \times 10^{-1}\ m)^3}{1\ dm^3} \times \frac{1\ dm^3}{1\ L} = 1.55\ \frac{g}{L}$

1.20 (a) $3.05 \times 10^5\ g \times \frac{1\ kg}{1000\ g} = 3.05 \times 10^2\ kg$ (305 kg)

(b) $0.0025\ \mu m \times \frac{1 \times 10^{-6}\ m}{1\ \mu m} \times \frac{1\ pm}{1 \times 10^{-12}\ m} = 2.5 \times 10^3\ pm$

(c) $3.45 \times 10^{-8}\ s \times \frac{1\ ns}{1 \times 10^{-9}\ s} = 34.5\ ns$

(d) $4.5 \times 10^8\ pm^3 \times \frac{(1 \times 10^{-12}\ m)^3}{1\ pm^3} = 4.5 \times 10^{-28}\ m^3$

1.21 (a) $\text{density} = \frac{\text{mass}}{\text{volume}} = \frac{37.25\ g}{25.0\ mL} \times \frac{1\ mL}{1\ cm^3} = 1.49\ \frac{g}{cm^3}$

(b) $50.0\ cm^3 \times \frac{23.4\ g}{cm^3} = 1.17 \times 10^3\ g$ (1.17 kg)

(c) $175\ g \times \frac{1\ cm^3}{1.74\ g} = 101\ cm^3$ (101 mL)

1.22 (a) Strategy: change length to cm, volume = length^3 (cm^3), density = mass/volume (g/cm^3)

$1.2 \times 10^{-5}\ km \times \frac{1000\ m}{1\ km} \times \frac{1\ cm}{1 \times 10^{-2}\ m} = 1.2\ cm$

volume = $(1.2\ cm)^3 = 1.7\ cm^3$

$\text{density} = \frac{1.1\ g}{1.7\ cm^3} = \frac{0.64\ g}{cm^3}$

The plastic is less dense than water (1.0 g/cm^3), so the object will float.

(b) $3.12\ \frac{g}{mL} \times \frac{1\ mL}{1 \times 10^{-3}\ L} \times 0.500\ L = 1.56 \times 10^3\ g = 1.56\ kg$

(c) $8.74\ kg \times \frac{1000\ g}{kg} \times \frac{1\ cm^3}{1.20\ g} = 7.28 \times 10^3\ cm^3 = 7.28\ L$

1.23 (a) °C = 5/9 (°F - 32°); 5/9 (68-32) = 20°C

(b) °F = 9/5 (°C) + 32°; 9/5 (-36.7) + 32° = -34.1°F

(c) K = °C + 273.15; -15°C + 273.15 = 258 K

(d) 415 K - 273 = 142°C; 9/5 (142°C) + 32 = 288°F

(e) °C = 5/9 (°F - 32°); 5/9 (1500 - 32) = 815.6°C; 815.6°C + 273.15 = 1088.7 K (assuming 1500°F has 4 sig. figs.)

1.24 (a) °C = 5/9 (92°F - 32°) = 33°C (b) 804°C + 273.15 = 1077 K
(c) 234.28 K - 273.15 = -38.87°C; °F = 9/5 (-38.87°C) + 32 = 37.97°F

Uncertainty in Measurement

1.25 Exact: (a), (d), (e), (f) (2.54 is considered an exact number when changing centimeters to inches.)

1.26 Exact: (b), (e), (f)

1.27 (a) 3 (b) 4 (c) 3 (d) 4 (e) 5

1.28 (a) 4 (b) 2 (c) 5, 6 or 7 (the trailing zeros may or may not be significant)
(d) 4 (e) 5

1.29 (a) 1.235×10^7 (b) 2.355 (c) 4.565×10^5 (d) 3.218×10^3
(e) 6.570×10^{-4} (f) 1.005×10^5

1.30 (a) 1.00×10^1 (b) 5.00×10^{-2} (c) 2.30×10^4 (d) 1.56×10^1 (e) 9.83×10^3
(f) -1.24×10^3

1.31 (a) 69.040 (b) -476 (c) 1.09×10^4 (d) 3917.0

1.32 (a) 51 (the intermediate quotient has three significant figures and zero decimal places, so the answer also has zero decimal places)
(b) 6.532×10^9 (the intermediate result has one decimal place and four significant figures)
(c) 140 (two significant figures) + 3400 (two significant figures) = 3500 (only the third digit to the left of the implied decimal point is significant) or 3.5×10^3
(d) -1.91×10^6 (the intermediate result has two decimal places and three significant figures)

Dimensional Analysis

1.33 (a) $$3.60 \text{ mi} \times \frac{1.609 \text{ km}}{1 \text{ mi}} \times \frac{1000 \text{ m}}{1 \text{ km}} = 5.79 \times 10^3 \text{ m}$$

(b) $$2.00 \text{ days} \times \frac{24 \text{ hr}}{1 \text{ day}} \times \frac{60 \text{ min}}{1 \text{ hr}} \times \frac{60 \text{ s}}{1 \text{ min}} = 1.73 \times 10^5 \text{ s}$$

(c) $$\frac{\$1.95}{\text{gal}} \times \frac{1 \text{ gal}}{4 \text{ qt}} \times \frac{1 \text{ qt}}{0.946 \text{ L}} = \frac{\$0.515}{\text{L}}$$

(d) $$\frac{5.0 \text{ pm}}{\mu\text{s}} \times \frac{1 \times 10^{-12} \text{ m}}{1 \text{ pm}} \times \frac{1 \text{ }\mu\text{s}}{1 \times 10^{-6} \text{ s}} = \frac{5.0 \times 10^{-6} \text{ m}}{\text{s}}$$

(e) $$\frac{85.00 \text{ mi}}{\text{hr}} \times \frac{1.609 \text{ km}}{1 \text{ mi}} \times \frac{1000 \text{ m}}{1 \text{ km}} \times \frac{1 \text{ hr}}{60 \text{ min}} \times \frac{1 \text{ min}}{60 \text{ s}} = \frac{37.99 \text{ m}}{\text{s}}$$

(f) $$33.35 \text{ ft}^3 \times \frac{(12)^3 \text{ in}^3}{1 \text{ ft}^3} \times \frac{(2.54)^3 \text{ cm}^3}{1 \text{ in}^3} = 9.444 \times 10^5 \text{ cm}^3$$

1.34 (a) $7.5\text{ ft} \times \frac{12\text{ in}}{1\text{ ft}} \times \frac{2.54\text{ cm}}{1\text{ in}} = 2.3 \times 10^2\text{ cm}$

(b) $4.45\text{ qt} \times \frac{0.946\text{ L}}{1\text{ qt}} \times \frac{1000\text{ mL}}{1\text{ L}} = 4.21 \times 10^3\text{ mL}$

(c) $\frac{35.7\text{ in}}{\text{hr}} \times \frac{2.54\text{ cm}}{1\text{ in}} \times \frac{10\text{ mm}}{1\text{ cm}} \times \frac{1\text{ hr}}{60\text{ min}} \times \frac{1\text{ min}}{60\text{ s}} = \frac{0.252\text{ mm}}{\text{s}}$

(d) $2.00\text{ yd}^3 \times \frac{1\text{ m}^3}{(1.094)^3\text{ yd}^3} = 1.53\text{ m}^3$

(e) $\frac{\$3.99}{\text{lb}} \times \frac{100\text{ ¢}}{1\text{ \$}} \times \frac{1\text{ lb}}{453.6\text{ g}} = \frac{0.880\text{ ¢}}{\text{g}}$

(f) $\frac{1.57\text{ g}}{\text{mL}} \times \frac{1\text{ kg}}{1000\text{ g}} \times \frac{1000\text{ mL}}{1\text{ L}} \times \frac{1\text{ L}}{1\text{ dm}^3} \times \frac{10^3\text{ dm}^3}{1\text{ m}^3} = \frac{1.57 \times 10^3\text{ kg}}{\text{m}^3}$

1.35 (a) $12\text{ gal} \times \frac{4\text{ qt}}{1\text{ gal}} \times \frac{0.946\text{ L}}{1\text{ qt}} = 45\text{ L}$

(b) $\frac{3.4\text{ m}}{\text{s}} \times \frac{1.094\text{ yd}}{1\text{ m}} \times \frac{1\text{ mi}}{1760\text{ yd}} \times \frac{60\text{ s}}{1\text{ min}} \times \frac{60\text{ min}}{1\text{ hr}} = \frac{7.6\text{ mi}}{\text{hr}}$

(c) $320\text{ in}^3 \times \frac{(2.54)^3\text{ cm}^3}{1\text{ in}^3} \times \frac{1\text{ dm}^3}{(10)^3\text{ cm}^3} \times \frac{1\text{ L}}{1\text{ dm}^3} = 5.24\text{ L}$

1.36 (a) $31\text{ gal} \times \frac{4\text{ qt}}{1\text{ gal}} \times \frac{0.946\text{ L}}{1\text{ qt}} = 1.2 \times 10^2\text{ L}$

(b) $\frac{6\text{ mg}}{\text{kg}} \times \frac{1\text{ kg}}{2.205\text{ lb}} \times 170\text{ lb} = 5 \times 10^2\text{ mg}$ (500 mg, the zeros are not significant)

(c) $\frac{244\text{ mi}}{11.2\text{ gal}} \times \frac{1.609\text{ km}}{1\text{ mi}} \times \frac{1\text{ gal}}{4\text{ qt}} \times \frac{1.057\text{ qt}}{1\text{ L}} = \frac{9.26\text{ km}}{\text{L}}$

1.37 $(8.32 \times 13.5 \times 2.75)\text{ m}^3 \times \frac{10^6\text{ cm}^3}{1\text{ m}^3} \times \frac{1\text{ L}}{1000\text{ cm}^3} \times \frac{1.19\text{ g}}{1\text{ L}} \times \frac{1\text{ kg}}{1000\text{ g}}$

$= 3.6 \times 10^2\text{ kg}$

1.38 $8\text{ ft} \times 12\text{ ft} \times 20\text{ ft} = 1920\text{ ft}^3$

$1920\text{ ft}^3 \times \frac{(1\text{ yd})^3}{(3\text{ ft})^3} \times \frac{(1\text{ m})^3}{(1.094\text{ yd})^3} \times \frac{10\text{ mg CO}}{1\text{ m}^3} \times \frac{1\text{ g}}{1000\text{ mg}} = 0.54\text{ g CO}$

(rounding to one significant figure gives 0.5 g CO)

1.39 $34.5\text{ kg} \times \frac{1000\text{ g}}{1\text{ kg}} \times \frac{1\text{ mL}}{13.6\text{ g}} \times \frac{1\text{ L}}{1000\text{ mL}} = 2.54\text{ L}$

1.40 (a) $26.73 \text{ g total} \times \frac{0.90 \text{ g Ag}}{1 \text{ g total}} \times \frac{1 \text{ lb}}{453.59 \text{ g}} \times \frac{16 \text{ oz}}{1 \text{ lb}} \times \frac{\$1.18}{1 \text{ oz}} = \$1.0$

(b) $\$25.00 \times \frac{1 \text{ oz}}{\$6.10} \times \frac{1 \text{ lb}}{16 \text{ oz}} \times \frac{453.59 \text{ g}}{1 \text{ lb}} \times \frac{1 \text{ g total}}{0.90 \text{ g Ag}} \times \frac{1 \text{ coin}}{26.73\text{g}} = 4.8 \text{ coins}$

1.41 $2.4 \times 10^5 \text{ bbl} \times \frac{42\text{gal}}{1 \text{ bbl}} \times \frac{4 \text{ qt}}{1 \text{ gal}} \times \frac{1000 \text{ mL}}{1.057 \text{ qt}} = 3.8 \times 10^{10} \text{ mL}$

1.42 $\frac{50 \text{ cups}}{1\text{lb}} \times \frac{8 \text{ oz}}{1 \text{ cup}} \times \frac{1 \text{ qt}}{32 \text{ oz}} \times \frac{1000 \text{ mL}}{1.057 \text{ qt}} \times \frac{1 \text{ lb}}{453.59 \text{ g}} = \frac{26 \text{ mL}}{\text{g}}$

Additional Exercises

1.43 Behavior when pressure is applied: if the volume changes significantly, the substance is a gas; if the volume doesn't change, the substance is liquid. Density: the densities of gases have units of g/L, while the densities of liquids are on the order of g/mL. Boiling point: if the boiling point is below room temperature, the substance is a gas. If it is above room temperature, the substance is a liquid.

1.44 Intensive: (b), (c) and (e). Density (a ratio), temperature and color do not depend on amount.

1.45 Physical: reddish-brown liquid at room temperature; vaporizes readily; red vapor; b.p. = 58.8°C; f.p. = -7.2°C; d(g) = 7.59 g/L; d(l) = 3.12 g/mL (at 20°C). Chemical: reacts readily with many metals including iron and aluminum.

1.46 (a) A **hypothesis** is a possible explanation for certain phenomena based on preliminary experimental data. A **theory** may be more general, and has a significant body of experimental evidence to support it; a theory has withstood the test of experimentation.
(b) A scientific **law** is a summary or statement of natural behavior; it tells how matter behaves. A **theory** is an explanation of natural behavior, it attempts to explain why matter behaves the way it does.

1.47 Any sample of vitamin C has the same relative amount of carbon and oxygen; the ratio of oxygen to carbon in the isolated sample is the same as the ratio in synthesized vitamin C.

$$\frac{2.00 \text{ g O}}{1.50 \text{ g C}} = \frac{x \text{ g O}}{6.35 \text{ g C}}; \quad x = \frac{(2.00 \text{ g O})(6.35 \text{ g C})}{1.50 \text{ g C}} = 8.47 \text{ g O}$$

This illustrates the *law of constant composition.*

1.48 (a) $\frac{\text{m}}{\text{s}^2}$ (b) $\frac{\text{kg} \cdot \text{m}}{\text{s}^2}$ (c) $\frac{\text{kg} \cdot \text{m}}{\text{s}^2} \times \text{m} = \frac{\text{kg} \cdot \text{m}^2}{\text{s}^2}$

(d) $\frac{\text{kg m}}{\text{s}^2} \times \frac{1}{\text{m}^2} = \frac{\text{kg}}{\text{m} \cdot \text{s}^2}$ (e) $\frac{\text{kg} \cdot \text{m}^2}{\text{s}^2} \times \frac{1}{\text{s}} = \frac{\text{kg} \cdot \text{m}^2}{\text{s}^3}$

1.49 Magnesium is *less dense* than steel. That is, a unit volume of magnesium weighs less than a unit volume of steel.

1.50 $K = {}^{\circ}C + 273.15$; $K = -268.9^{\circ}C + 273.15 = 4.3\ K$
${}^{\circ}F = 9/5\ ({}^{\circ}C) + 32$; ${}^{\circ}F = 9/5\ (-268.9) + 32 = -452.0^{\circ}F$

(We consider 32 to be exact, so the result has 4 significant figures, as does the data.)

1.51 (a) Inappropriate - The circulation of a widely read publication would vary over a year's time, and could simply not be counted to the nearest single subscriber. Probably about four significant figures would be appropriate. (b) Appropriate - It might be possible to do better than two significant figures, but an estimate to two significant figures should be easily possible. (c) Rainfall can be measured to within 0.02 in., but it is probably not possible to record an entire year's rainfall to the nearest 0.01 in. Further, the variation from year to year is sufficiently large that it does not make much sense to report the average to this significant number of figures. Probably two significant figures would be appropriate. (d) Inappropriate - The population of a city is not constant during a year, and probably cannot be counted at any one time to an accuracy of one person. An appropriate estimate would be 51,000.

1.52 (a) $$24.39 \times 10^{9}\ \text{lb} \times \frac{453.59\ \text{g}}{1\ \text{lb}} = 1.106 \times 10^{13}\ \text{g}$$

(b) $$1.106 \times 10^{13}\ \text{g} \times \frac{1\ \text{cm}^3}{2.130\ \text{g}} \times \frac{1\ \text{m}^3}{(100)^3\ \text{cm}^3} \times \frac{1\ \text{km}^3}{(1000)^3\ \text{m}^3} = 5.194 \times 10^{-3}\ \text{km}^3$$

1.53 density = (5.26 g - 3.01 g)/ 2.36 mL = 0.953 g/mL

1.54 mass of benzene = 24.54 g - 8.47 g = 16.07 g

$$\text{volume of benzene} = 16.07\ \text{g} \times \frac{1\ \text{mL}}{0.879\ \text{g}} = 18.3\ \text{mL}$$

volume of solid = 25.00 mL - 18.3 mL = 6.7 mL

$$\text{density of solid} = \frac{8.47\ \text{g}}{6.7\ \text{mL}} = 1.3\ \text{g/mL}$$

1.55 (a) $$2.4 \times 10^{5}\ \text{mi} \times \frac{1.609\ \text{km}}{1\ \text{mi}} \times \frac{1000\ \text{m}}{1\ \text{km}} \times \frac{1000\ \text{mm}}{1\ \text{m}} = 3.9 \times 10^{11}\ \text{mm}$$

(b) $$2.4 \times 10^{5}\ \text{mi} \times \frac{1.609\ \text{km}}{1\ \text{mi}} \times \frac{1\ \text{hr}}{2.4 \times 10^{3}\ \text{km}} \times \frac{60\ \text{min}}{1\ \text{hr}} \times \frac{60\ \text{sec}}{1\ \text{min}}$$

$$\times \frac{1\ \text{Ms}}{1 \times 10^{6}\ \text{s}} = 0.58\ \text{Ms}$$

1.56 (a) $$575\ \text{ft} \times \frac{12\ \text{in}}{1\ \text{ft}} \times \frac{2.54\ \text{cm}}{1\ \text{in}} \times \frac{10\ \text{mm}}{1\ \text{cm}} \times \frac{1\ \text{quarter}}{1.55\ \text{mm}} = 1.13 \times 10^{5}\ \text{quarters}$$

(b) $$1.13 \times 10^{5}\ \text{quarters} \times \frac{5.67\ \text{g}}{1\ \text{quarter}} = 6.41 \times 10^{5}\ \text{g}\ (641\ \text{kg})$$

(c) $1.13 \times 10^5 \text{ quarters} \times \dfrac{1 \text{ dollar}}{4 \text{ quarters}} = \$2.83 \times 10^4$
($28,250 but the result has only 3 significant figures)

(d) $\$4.2 \times 10^{12} \times \dfrac{1 \text{ stack}}{\$2.83 \times 10^4} = 1.5 \times 10^8 \text{ stacks}$
(approximately 150 million stacks)

1.57 (a) $\dfrac{\$2000}{\text{acre-ft}} \times \dfrac{1 \text{ acre}}{4840 \text{ yd}^2} \times \dfrac{3 \text{ ft}}{1 \text{ yd}} \times \dfrac{(1.094 \text{ yd})^3}{(1 \text{ m})^3} \times \dfrac{(1 \text{ m})^3}{(10 \text{ dm})^3} \times \dfrac{(1 \text{ dm})^3}{1 \text{ L}} =$

$\$1.6 \times 10^{-3}/\text{L}$ or 0.16 ¢ / L

(b) $\dfrac{\$2000}{\text{acre-ft}} \times \dfrac{1 \text{ acre-ft}}{2 \text{ households} \cdot \text{year}} \times 1 \text{ household} = \dfrac{\$1000}{\text{yr}}$

1.58 Calculate the volume of the aluminum foil:

$5.175\text{g} \times \dfrac{1 \text{ cm}^3}{2.70 \text{ g}} = 1.92 \text{ cm}^3$

Divide volume by area to get thickness

$1.92 \text{ cm}^3 \times \dfrac{1}{12.0 \text{ in}} \times \dfrac{1}{15.5 \text{ in}} \times \dfrac{1 \text{ in}^2}{(2.54)^2 \text{ cm}^2} \times \dfrac{10 \text{ mm}}{1 \text{ cm}} = 1.60 \times 10^{-2} \text{ mm}$

1.59 (a) $\text{volume} = \pi r^2 h = \pi \times (16.5 \text{ cm})^2 \times 22.3 \text{ cm} = 1.91 \times 10^4 \text{ cm}^3$

(b) $r = d/2 = 2.0 \text{ ft}/2 = 1.0 \text{ ft} = 12 \text{ in}$

$v = \pi \times (12 \text{ in})^2 \times \dfrac{(2.54 \text{ cm})^2}{1 \text{ in}^2} \times \dfrac{(1 \text{ m})^2}{(100 \text{ cm})^2} \times 6.3 \text{ ft} \times \dfrac{1 \text{ yd}}{3 \text{ ft}}$

$\times \dfrac{1 \text{ m}}{1.094 \text{ yd}} = 0.56 \text{ m}^3$

(c) $0.56 \text{ m}^3 \times \dfrac{(100 \text{ cm})^3}{(1 \text{ m})^3} \times \dfrac{13.6 \text{ g Hg}}{1 \text{ cm}^3} \times \dfrac{1 \text{ kg}}{1000 \text{ g}} = 7.6 \times 10^3 \text{ kg } H_2O$

1.60 $9.64 \text{ g ethanol} \times \dfrac{1 \text{ cm}^3}{0.789 \text{ g ethanol}} = 12.2 \text{ cm}^3$, volume of cylinder

$V = \pi r^2 h;\ r = (V/\pi h)^{\frac{1}{2}} \times \left[\dfrac{12.2 \text{ cm}^3}{\pi \times 15.0 \text{ cm}}\right]^{\frac{1}{2}} = 0.509 \text{ cm}$

$d = 2r = 1.02 \text{ cm}$

1.61 (a) Let x = mass of Au in jewelry

9.85 - x = mass of Ag in jewelry

The total volume of jewelry = volume of Au + volume of Ag

$$0.675\ \text{cm}^3 = x\ \text{g} \times \frac{1\ \text{cm}^3}{19.3\ \text{g}} + (9.85 - x)\ \text{g} \times \frac{1\ \text{cm}^3}{10.5\ \text{g}}$$

$$0.675 = \frac{x}{19.3} + \frac{9.85 - x}{10.5}$$ (To solve, multiply both sides by (19.3)(10.5))

$$0.675\,(19.3)(10.5) = 10.5\,x + (9.85 - x)(19.3)$$

$$136.79 = 10.5\,x + 190.105 - 19.3\,x$$

$$-53.315 = -8.8\,x$$

x = 6.06 g Au; 9.85 g total - 6.06 g Au = 3.79 g Ag

$$\text{mass \% Au} = \frac{6.06\ \text{g Au}}{9.85\ \text{g jewelry}} \times 100 = 61.5\ \%\ \text{Au}$$

(b) 24 karats × 0.615 = 15 karat gold

1.62 The separation is successful if two distinct spots are seen on the paper. To quantify the characteristics of the separation, calculate a reference value for each spot that is

$$\frac{\text{distance travelled by spot}}{\text{distance travelled by solvent}}$$

If the values for the two spots are fairly different, the separation is successful. (One could measure the distance between the spots, but this would depend on the length of paper used and be different for each experiment. The values suggested above are independent of the length of paper.)

1.63 A solution could be separated into components by physical means, so separation would be attempted. If the liquid is a solution, the solute could be a solid or a liquid; these two kinds of solutions would be separated differently. Therefore, divide the liquid into several samples and do different tests on each. Try evaporating the solvent from one sample. If a solid remains, the liquid is a solution and the solute is a solid. If the result is negative, try distilling a sample to see if two or more liquids with different boiling points are present. If this result is negative, the liquid is probably a pure substance, but negative results are never entirely conclusive. We might not have tried the appropriate separation technique.

1.64 (a) osmium, density = 22.6 g/cm^3
(b) tungsten, m.p. = 3410°C
(c) helium, b.p. = -268.9°C
(d) mercury and bromine - both have freezing points below room temperature and boiling points above it

CHAPTER 2

Atoms, Molecules, and Ions

Atomic Theory

2.1 (a) In nitrogen dioxide, nitrogen atoms and oxygen atoms have chemically combined to form molecules in which the relative number of nitrogen and oxygen atoms is the same in each molecule. Nitrogen dioxide molecules have properties different than independent nitrogen and oxygen molecules. NO_2 molecules cannot be broken down by physical means into elemental nitrogen and oxygen.

(b) In air, nitrogen and oxygen molecules exist independently. The physical properties of air are an average of the physical properties of oxygen and nitrogen. The component elements in air can be separated by physical means such as selective condensation, taking advantage of the difference in boiling points of nitrogen and oxygen.

2.2 Postulate 3 of the atomic theory is the *law of constant composition*. It states that the relative number and kinds of atoms in a compound are constant, regardless of the source. Therefore, 1.0 g of pure water should always contain the same relative amounts of hydrogen and oxygen, no matter where or how the sample is obtained.

2.3 (a) A : $\dfrac{55.0 \text{ g fluorine}}{23.2 \text{ g sulfur}} = 2.37 \text{ g fluorine/1 g sulfur}$

B : $\dfrac{9.8 \text{ g fluorine}}{16.6 \text{ g sulfur}} = 0.59 \text{ g fluorine/1 g sulfur}$

C : $\dfrac{68.6 \text{ g fluorine}}{19.3 \text{ g sulfur}} = 3.55 \text{ g fluorine/1 g sulfur}$

(b) Dividing through by 0.59 to look for integer relationships between these values, the ratios are A:B:C as 4:1:6. Yes, the different masses of fluorine that will combine with one gram of sulfur as calculated in (a) are in the ratio of small whole numbers and, therefore, obey the *law of multiple proportions*.

2.4 (a) $\frac{17.37 \text{ g oxygen}}{15.20 \text{ g nitrogen}} = \frac{1.143 \text{ g O}}{1 \text{ g N}} \div 1.143 = 1.0 \times 2 = 2$

$\frac{34.74 \text{ g oxygen}}{15.20 \text{ g nitrogen}} = \frac{2.286 \text{ g O}}{1 \text{ g N}} \div 1.143 = 2.0 \times 2 = 4$

$\frac{43.43 \text{ g oxygen}}{15.20 \text{ g nitrogen}} = \frac{2.857 \text{ g O}}{1 \text{ g N}} \div 1.143 = 2.5 \times 2 = 5$

(b) These masses of oxygen per one gram nitrogen are in the ratio of 2:4:5 and thus obey the *law of multiple proportions*.

2.5 According to postulate 4 of Dalton's Atomic Theory and the *law of conservation of mass*, the total mass of the products of a reaction must equal the total mass of the reactants. Thus (4.0 g natural gas + 16.0 g oxygen) = (9.0 g water + ? g carbon dioxide); the heat produced has no measureable mass. 11.0 g carbon dioxide are produced.

2.6 In the first case, all reactants and products are enclosed in the glass flash bulb. Thus, all reactants are weighed before reaction and all products weighed after reaction. In the second case, only one reactant, Mg ribbon, is weighed before reaction. The second observation does not violate the law of conservation of mass because the law requires that we consider the sum of the masses of all reactants and compare it with the sum of the masses of all products.

Atomic Structure

2.7 Evidence that cathode rays were negatively charged particles was (1) that electric and magnetic fields deflected the rays in the same way they would deflect negatively charged particles and (2) that a metal plate exposed to cathode rays acquired a negative charge.

2.8 (a) The electron itself has a negative charge so it is repelled by the negatively charged plate and attracted to the positively charged plate.

(b) As the charge on the plates is increased, the respective repulsion and attraction of the electron increases (Coloumb's Law) and the amount of bend should increase.

(c) As the mass of the particle increases, a greater force is required to deflect it. If the strength of the magnetic field is constant, the bend should decrease as the mass increases.

2.9 $3.24 \times 10^{-6} \text{ C} \times \frac{1e^-}{1.60 \times 10^{-19} \text{ C}} = 2.02 \times 10^{13}$ electrons

2.10 The droplets contain different charges because there may be 1, 2, 3 or more excess electrons on the droplet. The electronic charge is likely to be the lowest common factor in all the observed charges. Assuming this is so, we calculate the apparent electronic charge from each drop as follows:

$$A: 1.60 \times 10^{-19} \div 1 = 1.60 \times 10^{-19}\ C$$

$$B: 3.15 \times 10^{-19} \div 2 = 1.58 \times 10^{-19}\ C$$

$$C: 4.81 \times 10^{-19} \div 3 = 1.60 \times 10^{-19}\ C$$

$$D: 6.31 \times 10^{-19} \div 4 = 1.58 \times 10^{-19}\ C$$

The reported value is the average of these four values. Since each calculated charge has three significant figures, the average will also have three significant figures.

$$(1.60 \times 10^{-19}\ C + 1.58 \times 10^{-19}\ C + 1.60 \times 10^{-19}\ C + 1.58 \times 10^{-19}\ C) \div 4 = 1.59 \times 10^{-19}\ C$$

2.11 In the scattering experiment, most alpha particles, massive positively charged helium nuclei, passed directly through the foil, but a few were deflected at large angles. The diffuse positive charge in the "plum pudding" model would not have produced a repulsion strong enough to deflect the alpha particles at large angles. In Rutherford's model, the positive charge in the atom is concentrated in the small nucleus. If an alpha particle strickes a gold nucleus directly, it is deflected at a large angle.

2.12 The Mg nuclei have a much smaller volume and positive charge than the Au nuclei; the charge repulsion between the alpha particles and the Mg nuclei will be less, and there will be fewer direct hits because the Mg nuclei have even a smaller volume than the Au nuclei. Fewer alpha particles will be scattered in general and fewer will be strongly back scattered.

2.13 (a) $$4.7\ \text{Å} \times \frac{1 \times 10^{-10}\ m}{1\ \text{Å}} \times \frac{1\ nm}{1 \times 10^{-9}\ m} = 0.47\ nm$$

$$4.7\ \text{Å} \times \frac{1 \times 10^{-10}\ m}{1\ \text{Å}} \times \frac{1\ pm}{1 \times 10^{-12}\ m} = 470\ pm \quad (1\ \text{Å} = 100\ pm)$$

(b) $$1.0\ cm \times \frac{1 \times 10^{-2}\ m}{1\ cm} \times \frac{1\ \text{Å}}{1 \times 10^{-10}\ m} \times \frac{1\ \text{Cs atom}}{4.7\ \text{Å}} = 2.1 \times 10^{7}\ \text{Cs atoms}$$

2.14 $$\frac{3.1\ cm}{1.0 \times 10^{8}\ \text{Na atoms}} = \frac{3.1 \times 10^{-8}\ cm}{\text{Na atom}}$$

$$3.1 \times 10^{-8}\ cm \times \frac{1\ m}{100\ cm} \times \frac{1\ \text{Å}}{1 \times 10^{-10}\ m} = 3.1\ \text{Å}\ \text{(or 310 pm)}$$

2.15 p = protons, n = neutrons, e = electrons

(a) ^{20}Ne has 10 p, 10 n, 10 e
(b) ^{39}K has 19 p, 20 n, 19 e
(c) ^{48}Ti has 22 p, 26 n, 22 e
(d) ^{80}Br has 35 p, 45 n, 35 e
(e) ^{109}Ag has 47 p, 62 n, 47e
(f) ^{137}Ba has 56 p, 81 n, 56 e

2.16 (a) ^{60}Co, 27 p, 33 n (b) ^{131}I, 53 p, 78 n (c) ^{99}Tc, 43 p, 56 n
(d) ^{32}P, 15 p, 17 n (e) ^{51}Cr, 24 p, 27 n (f) ^{59}Fe, 26 p, 33 n

2.17

Symbol	^{19}F	^{74}As	^{137}Ba	^{122}Sb	^{196}Pt
Protons	9	33	**56**	51	78
Neutrons	10	41	**81**	**71**	118
Electrons	9	33	56	**51**	**78**
Mass No.	19	74	137	122	**196**

2.18

Symbol	^{31}P	^{56}Fe	^{119}Sn	^{127}I	^{201}Hg
Protons	15	**26**	50	53	**80**
Neutrons	16	**30**	69	74	121
Electrons	15	26	50	**53**	80
Atomic No.	15	26	**50**	53	80
Mass No.	31	56	**119**	**127**	**201**

2.19 (a) $^{23}_{11}Na$ (b) $^{51}_{23}V$ (c) $^{4}_{2}He$ (d) $^{37}_{17}Cl$ (e) $^{24}_{12}Mg$

2.20 $^{235}_{92}U$, $^{238}_{92}U$

The Periodic Table; Molecules and Ions

2.21 (a) Mn (metal) (b) P (nonmetal) (c) Al (metal) See Figure 2.15; although Al lies along the metalloid diagonal of the periodic chart, it is considered to be a metal because of its physical properties. (d) Ar (nonmetal) (e) I (nonmetal) (f) Cr (metal) (g) Ge (metalloid)

2.22 (a) sulfur (nonmetal) (b) selenium (nonmetal) (c) mercury (metal)
(d) calcium (metal) (e) neon (nonmetal) (f) beryllium (metal)
(g) zinc (metal)

2.23 (a) Cl, halogens (nonmetal) (b) Xe, noble gases (nonmetal)
(c) Li, alkali metals (metal) (d) Ba, alkline earth metals (metal)
(e) S, chalcogens (nonmetal)

2.24 O (nonmetal); S (nonmetal); Se (nonmetal); Te (metalloid); Po (metal); although Po is in the position of a metalloid, its properties are those of a metal.

2.25 The molecular formula conveys more information because it indicates **both** the combining ratio of elements in the compound and the exact number of atoms of each element in a molecule of the compound (and thus the molecular weight).

2.26 No. Two molecules with the same empirical formula can have different molecular formulas if the integer number of empirical formula units in the two molecules is different. For example, CH_2O is the empirical formula for both formaldehyde, CH_2O, and glucose, $C_6H_{12}O_6$.

2.27 (a) NO_2 (b) CH_2 (c) C_2HO_2 (d) P_2O_5 (e) CH_2O (f) SO_3

2.28 **CH:** C_2H_2, C_6H_6
CH_2: C_2H_4, C_3H_6, C_4H_8
NO_2: N_2O_4, NO_2

2.29

Symbol	$^{37}Cl^-$	$^{88}Sr^{2+}$	$^{44}Sc^{3+}$	$^{59}Ni^{2+}$	$^{31}P^{3-}$
Protons	17	38	21	28	15
Neutrons	20	50	23	31	16
Electrons	18	36	18	26	18
Net Charge	-1	2+	3+	2+	3-

2.30

Symbol	$^{17}O^{2-}$	$^{52}Cr^{3+}$	$^{88}Sr^{2+}$	$^{79}Se^{2-}$	$^{127}I^-$
Protons	8	24	38	34	53
Neutrons	9	28	50	45	74
Electrons	10	21	36	36	54
Net Charge	2-	3+	2+	2-	1-

2.31 (a) Mg^{2+} (b) S^{2-} (c) Al^{3+} (d) K^+ (e) Br^- (f) Y^{3+}

2.32 Unlikely ions: (a) F^+, (c) Be^-, (f) K^{2+}

2.33 (a) $CaBr_2$ (b) NH_4Cl (c) $Al(C_2H_3O_2)_3$ (d) K_2SO_4 (e) $Mg_3(PO_4)_2$
(f) $Ba(OH)_2$

2.34 (a) Na_2S (b) CaF_2 (c) MgO (d) Al_2O_3 (e) BeS (f) Li_3N

2.35 Molecular (all elements are nonmetals): (a) NO_2, (b) BF_3, (f) PF_5, (g) NF_3
Ionic (formed by a metal and a nonmetal): (c) Li_2O (d) Sc_2O_3, (e) CsBr, (h) LaP

2.36 Molecular (all elements are nonmetals or metalloids): (b) $SiCl_4$, (d) NOCl, (e) B_2H_6, (f) CH_3OH

Ionic (formed from ions, usually contain a metal cation): (a) CaO, (c) $Mg(NO_3)_2$, (g) Ag_2SO_4

Naming Ionic Compounds

2.37 sulfate - SO_4^{2-}; sulfite - SO_3^{2-}; sulfite has one less oxygen atom.
sulfide - S^{2-}; sulfide has no oxygen atoms;
hydrogen sulfate - HSO_4^-, H^+ added to SO_4^{2-} increases the net charge by +1.

2.38 (a) ClO_2^- (b) Cl^- (c) ClO_3^- (d) ClO_4^- (e) ClO^-

2.39 (a) zinc chloride (b) calcium cyanide (c) copper (II) hydroxide (or cupric hydroxide)
(d) barium nitrate (e) potassium phosphate (f) mercury (I) sulfide (or mercurous sulfide)
(g) ammonium sulfate (h) iron (III) fluoride (or ferric fluoride) (i) sodium chromate
(j) chromium (III) carbonate (or chromic carbonate)

2.40 (a) aluminum oxide (b) copper (II) perchlorate (or cupric perchlorate)
(c) nickel carbonate (d) tin (II) bromide (or stannous bromide)
(e) iron (II) hydroxide (or ferrous hydroxide) (f) potassium permanganate
(g) lead (II) acetate (or plumbous acetate) (h) zinc dihydrogenphosphate
(i) lithium sulfite (j) ammonium dichromate

2.41 (a) Ca_3N_2 (b) FeS (c) $Cr_2(SO_4)_3$ (d) $Cu(C_2H_3O_2)_2$ (e) $Ca(HCO_3)_2$ (f) KClO

2.42 (a) Fe_2O_3 (b) $Mg_3(PO_4)_2$ (c) Na_2O_2 (d) $Fe(NO_3)_2$ (e) CaH_2 (f) $Zn(HSO_4)_2$

2.43 (a) HNO_3 (b) HI (c) H_2SO_3 (d) chlorous acid (e) phosphoric acid
(f) carbonic acid

2.44 (a) hydrobromic acid (b) perbromic acid (c) nitrous acid (d) HClO
(e) HIO_3 (f) H_2SO_4

2.45 (a) dinitrogen pentoxide (b) iodine heptafluoride (c) xenon trioxide (d) $SiCl_4$
(e) H_2Se (f) P_4O_6

2.46 (a) sulfur hexafluoride (b) dichlorine heptoxide (c) iodine trichloride (d) CS_2
(e) HCN (f) N_2O_4

2.47 (a) $ZnCO_3$, ZnO, CO_2 (b) HF, SiO_2, SiF_4, H_2O (c) SO_2, H_2O, H_2SO_3
(d) H_3P (or PH_3) (e) $HClO_4$, $Cd(ClO_4)_2$ (f) VBr_3

2.48 (a) $KClO_3(s)$, $O_2(g)$ (b) $NaClO(aq)$ (c) $NH_3(g)$, $NH_4NO_3(s)$ (d) $HF(aq)$
(e) $H_2S(g)$ (f) $HCl(aq)$, $NaHCO_3(s)$, $CO_2(g)$

Additional Exercises

2.49 (a) Based on data accumulated in the late eighteenth century on how substances react with one another, **Dalton** postulated the atomic theory. Dalton's theory is based on the indivisible atom as the smallest unit of an element that can combine with other elements.

(b) By determining the effects of electric and magnetic fields on cathode rays, **Thomson** measured the mass-to-charge ratio of the electron. He also proposed the "plum pudding" model of the atom in which most of the space in an atom is occupied by a diffuse positive charge in which the tiny negatively charged electrons are imbedded.

(c) By observing the rate of fall of oil drops in and out of an electric field, Millikin measured the charge of an electron.

(d) After observing the scattering of alpha particles at large angles when the particles struck gold fail, **Rutherford** postulated the nuclear atom. In Rutherford's atom, most of the mass of the atom is concentrated in a small dense region called the nucleus and the tiny negatively charged electrons are moving through empty space around the nucleus.

2.50 (a) 5 significant figures. $^1H^+$ is a bare proton with mass 1.0073 amu. ^1H is a hydrogen atom, with 1 proton and 1 electron. The mass of the electron is 5.486×10^{-4} or 0.0005486 amu. Thus the mass of the electron is significant in the fourth decimal place or fifth significant figure in the mass of ^1H.

(b) Mass of ^1H = 1.0073 amu (proton)
0.0005486 amu (electron)
1.0078 amu (We have not rounded up to 1.0079 since 49 < 50 in the final sum.)

$$\text{Mass \% of electron} = \frac{\text{mass of } e^-}{\text{mass of } ^1H} \times 100 = \frac{5.486 \times 10^{-4} \text{ amu}}{1.0078 \text{ amu}} \times 100 = 0.05444\%$$

2.51

Symbol	^{35}Cl	$^{55}Mn^{2+}$	^{33}S	$^{76}As^{3-}$	$^{87}Sr^{2+}$
Protons	17	25	**16**	33	**38**
Neutrons	18	30	**17**	**43**	**49**
Electrons	17	23	16	**36**	**36**
Net Charge	0	2+	**0**	**3-**	2+

2.52 (a) deuterium - ${}^{2}_{1}H$, tritium - ${}^{3}_{1}H$

(b) Both deuterium and tritium have 1 proton and 1electron; their atomic number is 1. They differ in their number of neutrons and, therefore, have different mass numbers and nuclear composition.

2.53 The ionic compound XCl_2 is composed of X^{2+} and 2 Cl^-. If X^{2+} has 18 e^-, it must have 20 p^+; X is **Ca.**

2.54 (a) K (b) Ca (c) Ar (d) Br (e) Ge (f) H (g) Al (h) O (i) Ga

2.55 (a) $\frac{0.077 \text{ g H}}{0.923 \text{ g C}} = \frac{x \text{ g H}}{1.000 \text{ g C}}$; 0.083 g H/1 g C

(b) methane: $\frac{0.250 \text{ g H}}{0.750 \text{ g C}} = \frac{x \text{ g H}}{1.000 \text{ g C}}$; 0.333 g H/1 g C

ethylene: $\frac{0.143 \text{ g H}}{0.857 \text{ g C}} = \frac{x \text{ g H}}{1.000 \text{ g C}}$; 0.167 g H/1 g C

ethane: $\frac{0.200 \text{ g H}}{0.800 \text{ g C}} = \frac{x \text{ g H}}{1.000 \text{ g C}}$; 0.250 g H/1 g C

propane: $\frac{0.182 \text{ g H}}{0.818 \text{ g C}} = \frac{x \text{ g H}}{1.000 \text{ g C}}$; 0.222 g H/1 g C

butane: $\frac{0.172 \text{ g H}}{0.828 \text{ g C}} = \frac{x \text{ g H}}{1.000 \text{ g C}}$; 0.208 g H/1 g C

(c) If 0.083 g H/1 g C is a 1:1 combining ratio, dividing the g H/1 g C obtained in (b) by 0.083 should indicate the ratio of H:1C in the other compounds. (d) Empirical formulas follow:

methane: $\frac{0.333 \text{ g H/1 g C}}{0.083 \text{ g H/1 g C}}$ = 4H : 1C ; CH_4

ethylene: $\frac{0.167 \text{ g H/1 g C}}{0.083 \text{ g H/1 g C}}$ = 2H : 1C ; CH_2

ethane: $\frac{0.250 \text{ g H/1 g C}}{0.083 \text{ g H/1 g C}}$ = 3H : 1C ; CH_3

propane: $\frac{0.222 \text{ g H/1 g C}}{0.083 \text{ g H/1 g C}}$ = 2.67H : 1C = 8/3 H : 1C = 8H : 3C ; C_3H_8

butane: $\frac{0.208 \text{ g H/1 g C}}{0.083 \text{ g H/1 g C}}$ = 2.5 H : 1C = 5/2 H : 1C = 5H : 2C ; C_2H_5

2.56 (a) NH_4^+ , SO_4^{2-} , $(NH_4)_2SO_4$, ammonium sulfate

(b) Ca^{2+} , HCO_3^- , $Ca(HCO_3)_2$, calcium hydrogen carbonate (or calcium bicarbonate)

(c) Al^{3+} , HSO_4^- , $Al(HSO_4)_3$, aluminum hydrogen sulfate (or aluminum bisulfate)

(d) K^+ , CrO_4^{2-} , K_2CrO_4 , potassium chromate

(e) Na^+ , $C_2H_3O_2^-$, $NaC_2H_3O_2$, sodium acetate

(f) Ag^+ , S^{2-} , Ag_2S , silver sulfide

(g) Be^{2+} , Br^- , $BeBr_2$, beryllium bromide

(h) K^+ , Te^{2-} , K_2Te , potassium telluride

2.57 (a) sodium chloride (b) sodium bicarbonate (or sodium hydrogen carbonate)
(c) sodium hypochlorite (d) sodium hydroxide (e) ammonium carbonate
(f) calcium sulfate

2.58 (a) potassium nitrate (b) sodium carbonate (c) calcium oxide
(d) hydrochloric acid (e) magnesium sulfate (f) magnesium hydroxide

2.59 (a) CaS, $Ca(HS)_2$ (b) HBr, $HBrO_3$ (c) AlN, $Al(NO_2)_3$ (d) FeO, Fe_2O_3
(e) NH_3, NH_4^+ (f) K_2SO_3, $KHSO_3$ (g) Hg_2Cl_2, $HgCl_2$ (h) $HClO_3$, $HClO_4$

2.60 (a) IO_3^- (b) IO_4^- (c) IO^- (d) HIO (e) HIO_4 or (H_5IO_6)

2.61 (a) There are 10 known isotopes of sulfur, ranging from ^{29}S to ^{38}S.

(b) The four most abundant are: $^{32}_{16}S$ - 95.0%, $^{34}_{16}S$ - 4.22%,

$^{33}_{16}S$ - 0.76%, $^{36}_{16}S$ - 0.14%

2.62 PF_3 - phosphorus trifluoride; density = 3.907 g/mL; m.p. = -151.5°C; b. p. = -101.5°C

CHAPTER 3

Stoichiometry: Calculations with Chemical Formulas and Equations

Balancing Chemical Equations

3.1 (a) In balancing chemical equations, the **law of conservation of mass**, that atoms are neither created nor destroyed during the course of a reaction, is observed. This means that the **number** and **kinds** of atoms on both sides of the chemical equation must be the same.

(b) gases - (g) ; liquids - (l) ; solids - (s) ; aqueous solutions - (aq)

(c) P_4 indicates that there are four phosphorus atoms bound together by chemical bonds into a single molecule; 4P denotes four separate phosphorus atoms.

3.2 (a) Reactants are the substances we start with and products are what we end up with. Reactants appear to the left of the arrow in a chemical equation and products are written on the right.

(b) Subscripts in chemical formulas should not be changed when balancing equations because changing the subscript changes the identity of the compound (law of constant composition).

(c) This equation is **inconsistent** with the law of conservation of mass because the reactant's side has two more H atoms and one more O atom than the products side; $\underline{2}H_2O(l)$ are needed.

3.3 (a) $\mathbf{1}\ N_2O_5(g) + \mathbf{1}\ H_2O(l) \rightarrow \mathbf{2}\ HNO_3(aq)$

(b) $\mathbf{1}\ (NH_4)_2Cr_2O_7(s) \rightarrow \mathbf{1}\ N_2(g) + Cr_2O_3(s) + 4\ H_2O(l)$

(c) $\mathbf{1}\ PCl_3(l) + \mathbf{3}\ H_2O(l) \rightarrow \mathbf{1}\ H_3PO_3(aq) + \mathbf{3}\ HCl(aq)$

(d) $\mathbf{1}\ Mg_3N_2(s) + \mathbf{8}\ HCl(aq) \rightarrow \mathbf{3}\ MgCl_2(aq) + \mathbf{2}\ NH_4Cl(aq)$

(e) $2\ C_6H_6(l) + 15\ O_2(g) \rightarrow 12\ CO_2(g) + 6\ H_2O(l)$

(f) $4\ C_3H_5NO(g) + 19\ O_2(g) \rightarrow 12\ CO_2(g) + 4\ NO_2(g) + 10\ H_2O(l)$

3.4 (a) $1\ La_2O_3(s) + 3\ H_2O(l) \rightarrow 2\ La(OH)_3(aq)$

(b) $4\ FeO(s) + 1\ O_2(g) \rightarrow 2\ Fe_2O_3(s)$

(c) $1\ NCl_3(aq) + 3\ H_2O(l) \rightarrow 1\ NH_3(aq) + 3\ HOCl(aq)$

(d) $1\ C_5H_{10}(l) + 7\ O_2(g) \rightarrow 5\ CO_2(g) + 5\ H_2O(g)$

(e) $3\ NO_2(g) + 1\ H_2O(l) \rightarrow 2\ HNO_3(aq) + 1\ NO(g)$

(f) $2\ Fe(OH)_3(s) + 3\ H_2SO_4(aq) \rightarrow 1\ Fe_2(SO_4)_3(aq) + 6\ H_2O(l)$

3.5 (a) $SO_3(g) + H_2O(l) \rightarrow H_2SO_4(aq)$

(b) $B_2S_3(s) + 6H_2O(l) \rightarrow 2H_3BO_3(aq) + 3H_2S(g)$

(c) $4PH_3(g) + 8O_2(g) \rightarrow 6H_2O(g) + P_4O_{10}(s)$

(d) $2Hg(NO_3)_2(s) \rightarrow 2HgO(s) + 4NO_2(g) + O_2(g)$

(e) $3H_2S(g) + 2Fe(OH)_3(s) \rightarrow Fe_2S_3(s) + 6H_2O(g)$

3.6 (a) $2NH_3(g) + 2Na(l) \rightarrow 2NaNH_2(s) + H_2(g)$

(b) $Zn(s) + H_2SO_4(aq) \rightarrow ZnSO_4(aq) + H_2(g)$

(c) $2KNO_3(s) \xrightarrow{\Delta} 2KNO_2(s) + O_2(g)$

(d) $PCl_3(l) + 3H_2O(l) \rightarrow H_3PO_3(aq) + 3HCl(aq)$

(e) $Cu(s) + 2H_2SO_4(aq) \rightarrow CuSO_4(aq) + SO_2(g) + 2H_2O(l)$

Patterns of Chemical Reactivity

3.7 (a) $C_7H_{16}(l) + 11O_2(g) \rightarrow 7CO_2(g) + 8H_2O(l)$

(b) $2CH_3OC_2H_5(l) + 9O_2(g) \rightarrow 6CO_2(g) + 8H_2O(l)$

(c) $2Rb(s) + 2HOH(l) \rightarrow 2RbOH(aq) + H_2(g)$

(d) $Mg(s) + Cl_2(g) \rightarrow MgCl_2(s)$

CHAPTER 3
Stoichiometry: Calculations with Chemical Formulas and Equations

---

3.8 (a) $C_8H_8(l) + 10O_2(g) \rightarrow 8CO_2(g) + 4H_2O(l)$

(b) $2C_5H_{12}O(l) + 15O_2(g) \rightarrow 10CO_2(g) + 12H_2O(l)$

(c) $2Li(s) + 2HOH(l) \rightarrow 2LiOH(aq) + H_2(g)$

(d) $BaCO_3(s) \xrightarrow{\Delta} BaO(s) + CO_2(g)$

3.9 (a) $2KClO_3(s) \rightarrow 2KCl(s) + 3O_2(g)$ decomposition

(b) $C_7H_8O_2(l) + 8O_2(g) \rightarrow 7CO_2(g) + 4H_2O(l)$ combustion

(c) $2Cr(s) + 3Cl_2(g) \rightarrow 2CrCl_3(s)$ combination

(d) $2SO_3(g) \rightarrow 2SO_2(g) + O_2(g)$ decomposition

3.10 (a) $C_5H_6O(l) + 6O_2(g) \rightarrow 5CO_2(g) + 3H_2O(l)$ combustion

(b) $2H_2O_2(l) \rightarrow 2H_2O(l) + O_2(g)$ decomposition

(c) $CaO(s) + H_2O(l) \rightarrow Ca(OH)_2(aq)$ combination

(d) $6Li(s) + N_2(g) \rightarrow 2Li_3N(s)$ combination

3.11 (a) $2K(s) + 2NH_3(l) \rightarrow 2KNH_2(solv) + H_2(g)$

(b) $4CH_3NO_2(g) + 7O_2(g) \rightarrow 4NO_2(g) + 4CO_2(g) + 6H_2O(l)$

(c) $CH_4(g) + 4F_2(g) \rightarrow CF_4(g) + 4HF(g)$

3.12 (a) $2Si_2H_6(g) + 7O_2(g) \rightarrow 4SiO_2(s) + 6H_2O(l)$

(b) $CH_3SH(g) + 3O_2(g) \rightarrow SO_2(g) + CO_2(g) + 2H_2O(l)$

(c) $Mg(s) + Cl_2(g) \rightarrow MgCl_2(s)$

Atomic and Molecular Weights

3.13 (a) ${}^{12}_{6}C$ (b) An atomic mass unit is exactly 1/12 of the mass of one atom of ${}^{12}_{6}C$, or 1.66056×10^{-24} g.

3.14 (a) 12 amu (b) The mass of an average C atom is 12.011 amu.

If the mass of a carbon-12 atom were redefined to 20 nmu, the mass of an average H atom would increase proportionally.

$$\frac{20\text{ amu}}{12\text{ amu}} = \frac{x\text{ nmu}}{1.0079\text{ amu}}\text{ ; } x = 1.6798\text{ nmu}$$

Stoichiometry: Calculations with Chemical Formulas and Equations

3.15 The **average** mass of a boron atom is given by the sum of the mass of each isotope present in the sample times its **fractional** abundance:

average atomic mass = 0.1978(10.013 amu) + 0.8022(11.009 amu)
(atomic weight, AW) = 1.981 amu + 8.831 amu = 10.812 amu

3.16 Average atomic mass =19.99(0.9092) + 20.99(0.0025) + 21.99(0.0883) = 20.17 amu

3.17 GaAs is 48.20% Ga, 51.80% As

Since the mole ratio is 1:1, the mass ratio of Ga : As is 48.20 : 51.80.

$$\frac{48.20 \text{ Ga}}{51.80\text{As}} = \frac{x}{100}; \quad x = \frac{(48.20)(100)}{51.80} = 93.05$$

If the mass of As is 100 (arbitrary units), the relative atomic mass of Ga is 93.05 (arbitrary units).

3.18 Assuming that the formula is ZnO_2, let x = Berzelius's atomic weight of Zn

$$\frac{x}{2(16.00)} = 4.032 \; ; \; x = 129.0 \text{ amu}$$

Because the accepted AW of Zn is 64.5 amu, Berzelius's assumption must have been wrong. The correct formula is ZnO, leading to the correct AW for Zn.

3.19 FW in amu to 1 decimal place (see Sample Exercise 3.4).

(a) SO_3: 1(32.1) + 3(16.0) = 80.1 amu

(b) C_2F_6: 2(12.0) + 6(19.0) = 138.0 amu

(c) $(NH_4)_2Cr_2O_7$: 2(14.0) + 8(1.0) + 2(52.0) + 7(16.0) = 252.0 amu

(d) $Mg(C_2H_3O_2)_2$: 1(24.3) + 4(12.0) + 6(1.0) + 4(16.0) = 142.3 amu

(e) $CH_3OC_2H_5$: 3(12.0) + 8(1.0) + 1(16.0) = 60.0 amu

3.20 FW in amu to 1 decimal place

(a) C_2H_2: 2(12.0) + 2(1.0) = 26.0 amu

(b) $(NH_4)_2SO_4$: 2(14.0) + 8(1.0) + 1(32.1) + 4(16.0) = 132.1 amu

(c) $C_6H_8O_6$: 6(12.0) + 8(1.0) + 6(16.0) = 176.0 amu

(d) $PtCl_2(NH_3)_2$: 1(195.1) + 2(35.5) + 2(14.0) + 6(1.0) = 300.1 amu

(e) $C_{19}H_{28}O_2$: 19(12.0) + 28(1.0) + 2(16.0) = 288.0 amu

3.21 Calculate the formua weight, then the mass % of each element in the compound.

(a) SO_3: FW = 80.1 amu (Exercise 3.19(a))

$$\% \text{ S} = \frac{32.1 \text{ amu}}{80.1 \text{ amu}} \times 100 = 40.1\%; \ \%\text{O} = \frac{3(16.0) \text{ amu}}{80.1 \text{ amu}} \times 100 = 59.9\%$$

(b) CCl_4: FW = 1(12.0) + 4(35.5) = 154.0 amu

$$\% \text{ C} = \frac{12.0 \text{ amu}}{154.0 \text{ amu}} \times 100 = 7.79\%; \ \% \text{ Cl} = \frac{4(35.5) \text{ amu}}{154.0 \text{ amu}} \times 100 = 92.2\%$$

(c) CH_3OH: FW = 1(12.0) + 4(1.0) + 1(16.0) = 32.0 amu

$$\% \text{ C} = \frac{12.0 \text{ amu}}{32.0 \text{ amu}} \times 100 = 37.5\%; \ \%\text{H} = \frac{4(1.0) \text{ amu}}{32.0 \text{ amu}} \times 100 = 12.5\%$$

$$\% \text{ O} = \frac{16.0 \text{ amu}}{32.0 \text{ amu}} \times 100 = 50.0\%$$

(d) $Ca(NO_3)_2$: 1(40.1) + 2(14.0) + 6(16.0) = 164.1 amu

$$\% \text{ Ca} = \frac{40.1 \text{ amu}}{164.1 \text{ amu}} \times 100 = 24.4\%; \ \% \text{ N} = \frac{2(14.0) \text{ amu}}{164.1 \text{ amu}} \times 100 = 17.1\%$$

$$\% \text{ O} = \frac{6(16.0) \text{ amu}}{164.1 \text{ amu}} \times 100 = 58.5\%$$

(e) $(NH_4)_2SO_4$: FW = 132.1 (Exercise 3.20(b))

$$\% \text{ N} = \frac{2(14.0)}{132.1} \times 100 = 21.2\%; \ \% \text{ H} = \frac{8(1.0)}{132.1} \times 100 = 6.06\%$$

$$\% \text{ S} = \frac{32.1}{132.1} \times 100 = 24.3\%; \ \% \text{ O} = \frac{4(16.0)}{132.1} \times 100 = 48.4\%$$

3.22 (a) N_2O: FW = 2(14.0) + 1(16.0) = 44.0 amu

$$\% \text{ N} = \frac{28.0 \text{ amu}}{44.0 \text{ amu}} \times 100 = 63.6\%$$

(b) $C_7H_6O_2$: 7(12.0) + 6(1.0) + 2(16.0) = 122.0 amu

$$\% \text{ C} = \frac{7(12.0) \text{ amu}}{122.0 \text{ amu}} \times 100 = 68.9\%$$

(c) $Mg(OH)_2$: 1(24.3) + 2(16.0) + 2(1.0) = 58.3 amu

$$\% \text{ Mg} = \frac{24.3 \text{ amu}}{58.3 \text{ amu}} \times 100 = 41.7\%$$

(d) $(NH_2)_2CO$: 2(14.0) + 4(1.0) + 1(12.0) + 1(16.0) = 60.0 amu

$$\% \text{ N} = \frac{2(14.0) \text{ amu}}{60.0 \text{ amu}} \times 100 = 46.7\%$$

(e) $CH_3CO_2C_5H_{11}$: 7(12.0) + 14(1.0) + 2(16.0) = 130.0 amu

$$\% \text{ H} = \frac{14(1.0) \text{ amu}}{130.0 \text{ amu}} \times 100 = 10.8\%$$

3.23 (a) Three peaks: $^1H - {}^1H$, $^1H - {}^2H$, $^2H - {}^2H$

(b) $^1H - {}^1H$ = 2(1.00783) = 2.01566 amu
$^1H - {}^2H$ = 1.00783 + 2.01411 = 3.02194 amu
$^2H - {}^2H$ = 2(2.01411) = 4.02822 amu
[The mass ratios are 1 : 1.49923 : 1.99846 or 1 : 1.5 : 2]

(c) $^1H - {}^1H$ is largest, because there is the greatest chance that two atoms of the more abundant isotope will combine.

$^2H - {}^2H$ is the smallest, because there is the least chance that two atoms of the less abundant isotope will combine.

3.24 (a) A Br_2 molecule could consist of two atoms of the same isotope or one atom of each of the two different isotopes. This second possibility is twice as likely as the first. Therefore, the second peak (twice as large as peaks 1 and 3) represents a Br_2 molecule containing different isotopes. The mass numbers of the two isotopes are determined from the masses of the two smaller peaks. Since 157.84 ≈ 158, the first peak represents a $^{79}Br - {}^{79}Br$ molecule. Peak 3, 161.81 ≈ 162, represents a $^{81}Br - {}^{81}Br$ molecule. Peak 2 then contains one atom of each isotope, $^{79}Br - {}^{81}Br$, with an approximate mass of 160 amu.

(b) The mass of the lighter isotope is 157.84 amu/2 atoms, or 78.92 amu/atom. For the heavier one, 161.84 amu/2 atoms = 80.92 amu/atom.

(c) The relative size of the three peaks in the mass spectrum of Br_2 indicates their relative abundance. The average mass of a Br_2 molecule is

$$0.2534(157.84) + 0.5000(159.84) + 0.2466(161.84) = 159.83$$

(Each product has four significant figures and two decimal places, so the answer has two decimal places.)

(d) $$\frac{159.83 \text{ amu}}{\text{avg. } Br_2 \text{ molecule}} \times \frac{1\ Br_2 \text{ molecule}}{2 \text{ Br atoms}} = 79.915 \text{ amu}$$

(e) Let x = the abundance of ^{79}Br, 1-x = abundance of ^{81}Br. From (b), the masses of the two isotopes are 78.92 amu and 89.92 amu, respectively. From (d), the mass of an average Br atom is 79.915 amu.

$$x(78.92) + (1-x)(80.92) = 79.915, \ x = 0.502$$

$$^{79}Br = 50.2\%, \quad ^{81}Br = 49.8\%$$

Stoichiometry: Calculations with Chemical Formulas and Equations

The Mole

3.25 (a) A mole is the amount of matter that contains as many objects as the number of atoms in exactly 12 g of ^{12}C. (b) 6.022×10^{23} (c) The formula weight of a substance has the same numerical value as the molar mass expressed in grams.

3.26 (a) exactly 12 g (b) 6.0221367×10^{23}, Avogadro's number

(c) $$\frac{12 \text{ g } ^{12}C}{1 \text{ mol } ^{12}C} \times \frac{1 \text{ mol } ^{12}C}{6.0221367 \times 10^{23} \text{ atoms}} = \frac{1.9926482 \times 10^{-23} \text{ g}}{^{12}C \text{ atoms}}$$

3.27 (a) $$\frac{12 \text{ H atoms}}{6 \text{ C atoms}} = \frac{2H}{1C} \times 2.0 \times 10^{22} \text{ C atoms} = 4.0 \times 10^{22} \text{ H atoms}$$

(b) $$\frac{1\ C_6H_{12}O_6 \text{ molecule}}{6 \text{ C atoms}} \times 2.0 \times 10^{22} \text{ C atoms} = 3.3 \times 10^{21}\ C_6H_{12}O_6 \text{ molecules}$$

(c) $$3.3 \times 10^{21}\ C_6H_{12}O_6 \text{ molecules} \times \frac{1 \text{ mole}}{6.022 \times 10^{23} \text{ molecules}} = 5.5 \times 10^{-3} \text{ moles}$$

(d) 1 mole of $C_6H_{12}O_6$ weighs 180 g (Sample Exercise 3.7)

$$5.5 \times 10^{-3} \text{ mol } C_6H_{12}O_6 \times \frac{180 \text{ g } C_6H_{12}O_6}{1 \text{ mol}} = 0.99 \text{ g } C_6H_{12}O_6$$

(In the progressive calculation from (a) to (d), if the intermediate results are not rounded, the mass obtained (to 2 sig. figs.) is 1.0 g.)

3.28 (a) $$1.0 \times 10^{20} \text{ H atoms} \times \frac{18 \text{ C atoms}}{24 \text{ H atoms}} = 7.5 \times 10^{19} \text{ C atoms}$$

(b) $$1.0 \times 10^{20} \text{ H atoms} \times \frac{1\ C_{18}H_{24}O_2 \text{ molecule}}{24 \text{ H atoms}} = 4.2 \times 10^{18}\ C_{18}H_{24}O_2 \text{ molecules}$$

(c) $$4.2 \times 10^{18}\ C_{18}H_{24}O_2 \text{ molecules} \times \frac{1 \text{ mol}}{6.022 \times 10^{23} \text{ molecules}} = 6.9 \times 10^{-6} \text{ mol}$$

(d) molar mass = 18(12.0) + 24(1.01) + 2(16.0) = 272 g

$$6.9 \times 10^{-6} \text{ mol} \times \frac{272 \text{ g}}{\text{mol}} = 1.9 \times 10^{-3} \text{ g estradiol}$$

3.29 (a) molar mass: 63.55 + 2(14.01) + 6(16.00) = 187.57 g

(b) $$0.320 \text{ mol } Cu(NO_3)_2 \times \frac{187.57 \text{ g}}{1 \text{ mol}} = 60.0 \text{ g } Cu(NO_3)_2$$

(c) $$5.20 \text{ g } Cu(NO_3)_2 \times \frac{1 \text{ mol}}{187.57 \text{ g}} = 2.77 \times 10^{-2} \text{ mol}$$

(d) $$5.25 \text{ mg} \times \frac{1 \times 10^{-3} \text{ g}}{1 \text{ mg}} \times \frac{1 \text{ mol}}{187.57 \text{ g}} \times \frac{6.022 \times 10^{23} \text{ molecules}}{\text{mol}} \times \frac{2 \text{ N atoms}}{1 \text{ molecule}}$$

$$= 3.37 \times 10^{19} \text{ N atoms}$$

CHAPTER 3
Stoichiometry: Calculations with Chemical Formulas and Equations
Solutions to Exercises

3.30 (a) molar mass = 14(12.01) + 18(1.008) + 2(14.01) + 5(16.00) = 294.3 g

(b) $75.8 \text{ g aspartame} \times \frac{1 \text{ mol}}{294.3 \text{ g}} = 0.258 \text{ mol aspartame}$

(c) $0.736 \text{ mol aspartame} \times \frac{294.3 \text{ g}}{\text{mol}} = 217 \text{ g aspartame}$

(d) $$8.22 \text{ mg aspartame} \times \frac{1 \times 10^{-3} \text{ g}}{1 \text{ mg}} \times \frac{1 \text{ mol}}{294.3 \text{ g}} \times \frac{18 \text{ mol H}}{1 \text{ mol aspartame}}$$

$$\times \frac{6.022 \times 10^{23} \text{ H atoms}}{1 \text{ mol H}} = 3.03 \times 10^{20} \text{ aspartame molecules}$$

3.31 (a) $0.00650 \text{ mol } SO_2 \times \frac{64.06 \text{ g } SO_2}{1 \text{ mol } SO_2} = 0.416 \text{ g } SO_2$

(b) $4.58 \times 10^{22} \text{ Ar atom} \times \frac{1 \text{ mol}}{6.022 \times 10^{23} \text{ atoms}} \times 39.95 \frac{\text{g Ar}}{\text{mol Ar}} = 3.04 \text{ g Ar}$

(c) $$1.250 \times 10^{20} \text{ molecules} \times \frac{1 \text{ mol}}{6.022 \times 10^{23} \text{ molecules}} \times \frac{194.2 \text{ g } C_8H_{10}N_4O_2}{1 \text{ mol caffeine}}$$

$$= 4.03 \times 10^{-2} \text{ g } C_8H_{10}N_4O_2$$

3.32 (a) $0.0795 \text{ mol aspirin} \times \frac{180.2 \text{ g}}{1 \text{ mol}} = 14.3 \text{ g } C_9H_8O_4$

(b) $$3.25 \times 10^{20} \text{ molecules } O_3 \times \frac{1 \text{ mol } O_3}{6.022 \times 10^{23} \text{ molecules}} \times \frac{48.00\text{g } O_3}{1 \text{ mol } O_3}$$

$$= 2.59 \times 10^{-2} \text{ g } O_3$$

(c) $$5.64 \times 10^{24} \text{ molecules} \times \frac{1 \text{ mol}}{6.022 \times 10^{23} \text{ molecules}} \times \frac{386.7 \text{ g } C_{27}H_{46}O}{1 \text{ mol}}$$

$$= 3.62 \times 10^{3} \text{ g} = 3.62 \text{ kg } C_{27}H_{46}O$$

3.33 (a) $0.350 \text{ mol } C_2H_2 \times \frac{6.022 \times 10^{23} \text{ molecules}}{1 \text{ mol}} = 2.11 \times 10^{23} \text{ } C_2H_3 \text{ molecules}$

(b) $$500 \text{ mg } C_6H_8O_6 \times \frac{1 \times 10^{-3} \text{ g}}{1 \text{ mg}} \times \frac{1 \text{ mol } C_6H_8O_6}{176.1 \text{ g } C_6H_8O_6} \times \frac{6.022 \times 10^{23} \text{ molecules}}{1 \text{ mol}}$$

$$= 1.71 \times 10^{21} \text{ } C_6H_8O_6 \text{ molecules}$$

(c) $$5.0 \times 10^{-5} \text{g } H_2O \times \frac{1 \text{ mol } H_2O}{18.02 \text{ g } H_2O} \times \frac{6.022 \times 10^{23} \text{ molecules}}{1 \text{ mol}}$$

$$= 1.7 \times 10^{18} \text{ } H_2O \text{ molecules}$$

3.34 (a) $0.0350 \text{ mol } C_3H_8 \times \frac{6.022 \times 10^{23} \text{ molecules}}{1 \text{ mol}} = 2.11 \times 10^{22} \text{ } C_3H_8 \text{ molecules}$

(b) $$100 \text{ mg } C_8H_9O_2N \times \frac{1 \times 10^{-3} \text{ g}}{1 \text{ mg}} \times \frac{1 \text{ mol } C_8H_9O_2N}{151.2 \text{ g } C_8H_9O_2N} \times \frac{6.022 \times 10^{23} \text{ molecules}}{1 \text{ mol}}$$

$$= 3.98 \times 10^{20}\ C_8H_9O_2N \text{ molecules}$$

(c) $$12.6 \text{ g } C_{12}H_{22}O_{11} \times \frac{1 \text{ mol } C_{12}H_{22}O_{11}}{342.3 \text{ g } C_{12}H_{22}O_{11}} \times \frac{6.022 \times 10^{23} \text{ molecules}}{1 \text{ mol}}$$

$$= 2.22 \times 10^{22}\ C_{12}H_{22}O_{11} \text{ molecules}$$

3.35 $$\frac{2.05 \times 10^{-6} \text{ g } C_2H_3Cl}{1 \text{ L}} \times \frac{1 \text{ mol } C_2H_3Cl}{62.5 \text{ g } C_2H_3Cl} = \frac{3.28 \times 10^{-8} \text{ mol } C_2H_3Cl}{\text{L}}$$

$$\frac{3.28 \times 10^{-8} \text{ mol } C_2H_3Cl}{1 \text{ L}} \times \frac{6.022 \times 10^{23} \text{ molecules}}{1 \text{ mol}} = 1.97 \times 10^{16} \frac{\text{molecules}}{\text{L}}$$

3.36 $$25 \times 10^{-6} \text{ g } C_{21}H_{30}O_2 \times \frac{1 \text{ mol } C_{21}H_{30}O_2}{314 \text{ g } C_{21}H_{30}O_2} = 8.0 \times 10^{-8} \text{ mol } C_{21}H_{30}O_2$$

$$8.0 \times 10^{-8} \text{ mol } C_{21}H_{30}O_2 \times \frac{6.022 \times 10^{23} \text{ molecules}}{1 \text{ mol}} = 4.8 \text{ x } 10^{16}\ C_{21}H_{30}O_2 \text{ molecules}$$

Empirical Formulas

3.37 An empirical formula gives the relative number and kind of each atom in a compound, but a molecular formula gives the actual number of each kind of atom, and thus the molecular weight.

3.38 Percent composition gives the **relative** mass of each element in a compound. These percents yield **relative** numbers of each atom in the molecule, the empirical formula.

3.39 Find the **simplest ratio of moles** by dividing by the smallest number of moles present.

(a) 0.0230 mol C ÷ 0.0230 = 1.000
0.0615 mol H ÷ 0.0230 = 2.674

The ratio for H is too far from 3.0 to round up.
Multiplying by 3 gives an integer.

C : H = 3(1 : 2.674) = 3 : 8 The empirical formula is $\mathbf{C_3H_8}$.

(b) Calculate moles of each element present, then the simplest ratio of moles.

$$5.28 \text{ g Sn} \times \frac{1 \text{ mol Sn}}{118.7 \text{ g Sn}} = 0.04448 \text{ mol Sn};\ 0.04448 \div 0.04448 = 1$$

$$3.37 \text{ g F} \times \frac{1 \text{ mol F}}{19.00 \text{ g F}} = 0.1774 \text{ mol F};\ 0.1774 \div 0.04448 \approx 4$$

The integer ratio is 1 Sn : 4 F; SnF_4.

(c) Assume 100 g sample, calculate moles of each element, find the simplest ratio of moles.

$87.5\%\ N = 87.5\ g\ N \times \frac{1\ mol\ N}{14.01\ g} = 6.25\ mol\ N;\quad 6.25 \div 6.25 = 1$

$12.5\%\ H = 12.5\ g\ H \times \frac{1\ mol}{1.008\ g} = 12.4\ mol\ H;\quad 12.4 \div 6.25 \approx 2$

The empirical formula is NH_2.

3.40 (a) $0.104\ mol\ Na \div 0.052 = 2$

$0.052\ mol\ S \div 0.052 = 1$

$0.156\ mol\ O \div 0.052 = 3$

The empirical formula is Na_2SO_3.

(b) $11.66\ g\ Fe \times \frac{1\ mol\ Fe}{55.85\ g\ Fe} = 0.2088\ mol\ Fe;\ 0.2088 \div 0.2088 = 1$

$5.01\ g\ O \times \frac{1\ mol\ O}{16.00\ g\ O} = 0.3131\ mol\ O;\ 0.3131 \div 0.2088 \approx 1.5$

The integer ratio is 2 Fe : 3 O; the empirical formula is **Fe_2O_3**.

(c) Assume 100 g sample.

$40.0\ g\ C \times \frac{1\ mol\ C}{12.01\ g\ C} = 3.33\ mol\ C;\ 3.33 \div 3.33 = 1$

$6.7\ g\ H \times \frac{1\ mol\ H}{1.008\ mol\ H} = 6.65\ mol\ H;\ 6.65 \div 3.33 \approx 2$

$53.3\ g\ O \times \frac{1\ mol\ O}{16.00\ mol\ O} = 3.33\ mol\ O;\ 3.33 \div 3.33 = 1$

The empirical formula is CH_2O.

3.41 Assume 100 g sample in the following problems.

(a) $10.4\ g\ C \times \frac{1\ mol\ C}{12.01\ g\ C} = 0.866\ mol\ C;\quad 0.866 \div 0.866 = 1$

$27.8\ g\ S \times \frac{1\ mol\ S}{32.06\ g\ S} = 0.867\ mol\ S;\quad 0.867 \div 0.866 \approx 1$

$61.7\ g\ Cl \times \frac{1\ mol\ Cl}{35.45\ g\ Cl} = 1.74\ mol\ Cl;\quad 1.74 \div 0.866 \approx 2;$

The empirical formula is $CSCl_2$.

(b) $21.7\text{ g C} \times \frac{1\text{ mol C}}{12.01\text{ g C}} = 1.81\text{ mol C}$; $1.81 \div 0.600 \approx 3$

$9.6\text{ g O} \times \frac{1\text{ mol O}}{16.00\text{ g O}} = 0.600\text{ mol O}$; $0.600 \div 0.600 = 1$

$68.7\text{ g F} \times \frac{1\text{ mol F}}{19.00\text{ g F}} = 3.62\text{ mol F}$; $3.62 \div 0.600 \approx 6$

The empirical formula is C_3OF_6.

3.42 The procedure in all these cases is to assume 100 g of sample, calculate the number of moles of each element present in that 100 g, then obtain the ratios of the numbers of moles as smallest whole numbers.

(a) $32.79\text{ g Na} \times \frac{1\text{ mol Na}}{22.99\text{ g Na}} = 1.426\text{ mol Na O}$; $1.426 \div 0.4826 \approx 3$

$13.02\text{ g Al} \times \frac{1\text{ mol Al}}{26.98\text{ g Al}} = 0.4826\text{ mol Al}$; $0.4826 \div 0.4826 = 1$

$54.19\text{ g F} \times \frac{1\text{ mol F}}{19.00\text{ g F}} = 2.852\text{ mol F}$; $2.852 \div 0.4826 \approx 6$

The empirical formula is Na_3AlF_6.

(b) $62.1\text{ g C} \times \frac{1\text{ mol C}}{12.01\text{ g C}} = 5.17\text{ mol C}$; $5.17 \div 0.864 \approx 6$

$5.21\text{ g H} \times \frac{1\text{ mol H}}{1.008\text{ g H}} = 5.17\text{ mol O}$; $5.17 \div 0.864 \approx 6$

$12.1\text{ g N} \times \frac{1\text{ mol N}}{14.01\text{ g N}} = 0.864\text{ mol N}$; $0.864 \div 0.864 = 1$

$20.7\text{ g O} \times \frac{1\text{ mol O}}{16.00\text{ g O}} = 1.29\text{ mol O}$; $1.29 \div 0.864 \approx 1.5$

Multiplying by two, the formula is $C_{12}H_{12}N_2O_3$.

3.43 Assume 100 g in the following problems.

(a) $59.0\text{ g C} \times \frac{1\text{ mol C}}{12.01\text{ g C}} = 4.91\text{ mol C}$; $4.91 \div 0.550 \approx 9$

$7.1\text{ g H} \times \frac{1\text{ mol H}}{1.008\text{ g H}} = 7.04\text{ mol H}$; $7.04 \div 0.550 \approx 13$

$26.2\text{ g O} \times \frac{1\text{ mol O}}{16.00\text{ g O}} = 1.64\text{ mol O}$; $1.64 \div 0.550 \approx 3$

$7.7\text{ g N} \times \frac{1\text{ mol N}}{14.01\text{ g N}} = 0.550\text{ mol N}$; $0.550 \div 0.550 = 1$

Thus, the empirical formula is $C_9H_{13}O_3N$. This corresponds to a formula weight of 183, so it is also the molecular formula.

(b) $74.1 \text{ g C} \times 74.1 \text{ g C} \times \frac{1 \text{ mol C}}{12.01 \text{ g C}} = 6.17 \text{ mol C}$; $6.17 \div 1.23 \approx 5$

$8.6 \text{ g H} \times \frac{1 \text{ mol H}}{1.008 \text{ g H}} = 8.53 \text{ mol H}$; $8.53 \div 1.23 \approx 7$

$17.3 \text{ g N} \times \frac{1 \text{ mol N}}{14.01 \text{ g N}} = 1.23 \text{ mol N}$; $1.23 \div 1.23 \approx 1$

Thus, C_5H_7N. This corresponds to a formula weight of 81. If the molar mass is 160 ± 5 g, the mole ratios must be multiplied by a factor of 2 to obtain the molecular formula, $C_{10}H_{14}N_2$.

3.44 (a) $38.7 \text{ g C} \times \frac{1 \text{ mol C}}{12.01 \text{ g C}} = 3.22 \text{ mol C}$; $3.22 \div 3.22 = 1$

$9.7 \text{ g H} \times \frac{1 \text{ mol C}}{1.008 \text{ g H}} = 9.62 \text{ mol H}$; $9.62 \div 3.22 \approx 3$

$51.6 \text{ g O} \times \frac{1 \text{ mol O}}{16.00 \text{ g O}} = 3.22 \text{ mol O}$; $3.22 \div 3.22 = 1$

Thus, CH_3O, formula weight = 31. If the molar mass is 62.1 amu, a factor of 2 gives $C_2H_6O_2$ as the molecular formula.

(b) $49.5 \text{ g C} \times \frac{1 \text{ mol C}}{12.01 \text{ g C}} = 4.12 \text{ mol C}$; $4.12 \div 1.03 \approx 4$

$5.15 \text{ g H} \times \frac{1 \text{ mol H}}{1.008 \text{ g H}} = 5.11 \text{ mol H}$; $5.11 \div 1.03 \approx 5$

$28.9 \text{ g N} \times \frac{1 \text{ mol N}}{14.01 \text{ g N}} = 2.06 \text{ mol N}$; $2.06 \div 1.03 \approx 2$

$16.5 \text{ g O} \times \frac{1 \text{ mol O}}{16.00 \text{ g O}} = 1.03 \text{ mol O}$; $1.03 \div 1.03 = 1$

Thus, $C_4H_5N_2O$, FW = 97. If the molar mass is about 195, the molecular formula is twice the empirical formula, $C_8H_{10}N_4O_2$.

3.45 $3.14 \text{ g } CO_2 \times \frac{1 \text{ mol } CO_2}{44.01 \text{ g } CO_2} \times \frac{1 \text{ mol C}}{1 \text{ mol } CO_2} = 0.0713 \text{ mol C}$; $0.0713 \div 0.0713 = 1$

$1.29 \text{ g } H_2O \times \frac{1 \text{ mol } H_2O}{18.02 \text{ g } H_2O} \times \frac{2 \text{ mol H}}{1 \text{ mol } H_2O} = 0.143 \text{ mol H}$; $0.143 \div 0.0713 \approx 2$

We see that the molar ratio of C to H is 1:2. The simplest formula is CH_2.

3.46 We can calculate the mass of O by subtraction, but first the masses of C and H in the sample must be found.

$$6.32 \times 10^{-3}\ g\ CO_2 \times \frac{12.01\ g\ C}{44.01\ g\ CO_2} = 1.72 \times 10^{-3}\ g\ C = 1.72\ mg\ C$$

$$2.58 \times 10^{-3}\ g\ H_2O \times \frac{2.02\ g\ H}{18.02\ g\ H_2O} = 2.89 \times 10^{-4}\ g\ H = 0.289\ mg\ H$$

mass of O = 2.78 mg sample - (1.72 mg C + 0.289 mg H) = 0.77 mg O

$$1.72 \times 10^{-3}\ g\ C \times \frac{1\ mol\ C}{12.01\ g\ C} = 1.43 \times 10^{-4}\ mol\ C;\quad 1.43 \times 10^{-4} \div 4.81 \times 10^{-5} \approx 3$$

$$2.89 \times 10^{-4}\ g\ C \times \frac{1\ mol\ H}{1.008\ g\ H} = 2.87 \times 10^{-4}\ mol\ H;\quad 2.87 \times 10^{-4} \div 4.81 \times 10^{-5} \approx 6$$

$$7.7 \times 10^{-4}\ g\ O \times \frac{1\ mol\ O}{16.00\ g\ H} = 4.81 \times 10^{-5}\ mol\ O;\quad 4.81 \times 10^{-5} \div 4.81 \times 10^{-5} = 1$$

The empirical formula is C_3H_6O.

3.47 The reaction involved is $MgSO_4 \bullet xH_2O(s) \rightarrow MgSO_4(s) + xH_2O(g)$. First, calculate the number of moles of product $Mg(OH)_2$; this is the same as the number of moles of starting hydrate.

$$2.472\ g\ Mg(OH)_2 \times \frac{1\ mol\ MgSO_4}{120.4\ g\ MgSO_4} \times \frac{1\ mol\ MgSO_4 \bullet xH_2O}{1\ mol\ MgSO_4} = 0.02053\ mol\ MgSO_4{\bullet}xH_2O$$

$$\text{Thus, } \frac{5.061\ g\ MgSO_4{\bullet}xH_2O}{0.02053} = 246.5\ g/mol = \text{FW of } MgSO_4{\bullet}xH_2O.$$

FW of $MgSO_4{\bullet}xH_2O$ = FW of $MgSO_4$ + x(FW of H_2O)

246.5 = 120.4 + x(18.02). x = 6.998. The hydrate formula is $MgSO_4{\bullet}\underline{7}H_2O$.

Alternatively, we could calculate the number of moles of water represented by weight loss: (5.061 - 2.472) = 2.589 g H_2O lost.

$$2.589\ g\ H_2O \times \frac{1\ mol\ H_2O}{18.02\ g\ H_2O} = 0.1437\ mol\ H_2O;\quad \frac{mol\ H_2O}{mol\ Sr(OH)_2} = \frac{0.1437}{0.02053} = 7.000$$

Again the correct formula is $MgSO_4{\bullet}\underline{7}H_2O$.

3.48 Using the second method described in Exercise 3.47,

g H_2O lost = 2.558 g sample - 0.948 g Na_2CO_3 = 1.610 g H_2O

$$0.948\ g\ Na_2CO_3 \times \frac{1\ mol\ Na_2CO_3}{106.0\ g\ Na_2CO_3} = 0.00894\ mol\ Na_2CO_3$$

$$1.610\ g\ H_2O \times \frac{1\ mol\ H_2O}{18.02\ g\ H_2O} = 0.08935\ mol\ H_2O$$

$$\frac{\text{mol } H_2O}{\text{mol } Na_2CO_3} = \frac{0.08935}{0.00894} = 9.99 \text{ ; } x = 10.$$

The formula is $Na_2CO_3 \cdot \underline{\mathbf{10}}\ H_2O$.

3.49 Laccase is 0.39% Cu by mass. One gram of laccase contains 0.0039 g Cu.

$$\frac{3.9 \times 10^{-3} \text{ g Cu}}{1 \text{ g laccase}} \times \frac{1 \text{ mol Cu}}{63.55 \text{ Cu}} \times \frac{1 \text{ mol laccase}}{4 \text{ mol Cu}} = \frac{1.5 \times 10^{-5} \text{ mol laccase}}{1 \text{ g laccase}}$$

The molar mass is the reciprocal of this number: 6.5×10^4 g laccase/1 mol laccase.

3.50 1 gram of hemoglobin (hem) contains 3.4×10^{-3} g Fe.

$$\frac{3.40 \times 10^{-3} \text{ g Fe}}{1 \text{ g hem}} \times \frac{1 \text{ mol Fe}}{55.85 \text{ Fe}} \times \frac{1 \text{ mol hem}}{4 \text{ mol Fe}} = \frac{1.52 \times 10^{-5} \text{ mol hem}}{1 \text{ g hem}}$$

The reciprocal of this quantity is: 6.57×10^4 g hemoglobin/1 mol hemoglobin.

Calculations Based on Chemical Equations

3.51 The mole ratios implicit in the coefficients of a balanced chemical equation are essential for solving stoichiometry problems. If the equation is not balanced, the mole ratios will be incorrect and lead to erroneous calculated amounts of reactants and/or products.

3.52 The **integer coefficients** immediately preceeding each molecular formula in a chemical equation give information about relative numbers of moles of reactants and products involved in a reaction.

3.53 **Apply the mole ratio** of moles CO_2/moles C_2H_5OH to calculate moles CO_2 produced. This is the heart of every stoichiometry problem.

$$C_2H_5OH(l) + O_2(g) \rightarrow 2CO_2(g) + 3H_2O(l)$$

(a) $5.00 \text{ mol } C_2H_5OH \times \frac{2 \text{ mol } CO_2}{1 \text{ mol } C_2H_5OH} = 10.0 \text{ mol } CO_2$

(b) $\text{g } C_2H_5OH \longrightarrow \text{mol } C_2H_5OH \xrightarrow[\text{ratio}]{\text{mole}} \text{mol } CO_2 \longrightarrow \text{g } CO_2$

$$5.00 \text{ g } C_2H_5OH \times \frac{1 \text{ mol } C_2H_5OH}{46.07 \text{ } C_2H_5OH} \times \frac{2 \text{ mol } CO_2}{1 \text{ mol } C_2H_5OH} \times \frac{44.01 \text{ g } CO_2}{1 \text{ mol } CO_2} = 9.55 \text{ g } CO_2$$

3.54 $2C_8H_{18}(l) + 25O_2(g) \longrightarrow 16CO_2(g) + 18H_2O(l)$

(a) $5.00 \text{ mol } C_8H_{18} \times \frac{25 \text{ mol } O_2}{2 \text{ mol } C_8H_{18}} = 62.5 \text{ mol } O_2$

(b) The molar mass of C_8H_{18} = 8(12.01) + 18(1.008) = 114.2 g

$$5.00 \text{ g } C_8H_{18} \times \frac{1 \text{ mol } C_8H_{18}}{114.2 \text{ g } C_8H_{18}} \times \frac{25 \text{ mol } O_2}{2 \text{ mol } C_8H_{18}} \times \frac{32.00 \text{ g } O_2}{1 \text{ mol } O_2} = 17.5 \text{ g } O_2$$

(c) $$5.00 \text{ mL } C_8H_{18} \times \frac{0.692 \text{ g } C_8H_{18}}{1 \text{ mL } C_8H_{18}} \times \frac{1 \text{ mol } C_8H_{18}}{114.2 \text{ g } C_8H_{18}} \times \frac{25 \text{ mol } O_2}{2 \text{ mol } C_8H_{18}}$$

$$\times \frac{32.00 \text{ g } O_2}{1 \text{ mol } O_2} = 12.1 \text{ g } O_2$$

3.55 (a) $$1.50 \text{ mol } Na_2SiO_3 \times \frac{8 \text{ mol HF}}{1 \text{ mol } Na_2SiO_3} = 12.0 \text{ mol HF}$$

(b) $$3.00 \text{ mol HF} \times \frac{2 \text{ mol NaF}}{8 \text{ mol HF}} \times \frac{41.99 \text{ g NaF}}{1 \text{ mol NaF}} = 31.5 \text{ g NaF}$$

(c) $$3.00 \text{ g HF} \times \frac{1 \text{ mol HF}}{20.01 \text{ g HF}} \times \frac{1 \text{ mol } Na_2SiO_3}{8 \text{ mol HF}} \times \frac{122.1 \text{ g } Na_2SiO_3}{1 \text{ mol } Na_2SiO_3} = 2.29 \text{ g } Na_2SiO_3$$

3.56 $$C_6H_{12}O_6(aq) \rightarrow 2C_2H_5OH(aq) + 2CO_2(g)$$

(a) $$0.450 \text{ mol } C_6H_{12}O_6 \times \frac{2 \text{ mol } CO_2}{1 \text{ mol } C_6H_{12}O_6} = 0.900 \text{ mol } CO_2$$

(b) $$5.00 \text{ mol } C_2H_5OH \times \frac{1 \text{ mol } C_6H_{12}O_6}{2 \text{ mol } C_2H_5OH} \times \frac{180.0 \text{ g } C_6H_{12}O_6}{1 \text{ mol } C_6H_{12}O_6} = 450 \text{ g } C_6H_{12}O_6$$

(c) $$5.00 \text{ g } C_2H_5OH \times \frac{1 \text{ mol } C_2H_5OH}{46.07 \text{ g } C_2H_5OH} \times \frac{2 \text{ mol } CO_2}{2 \text{ mol } C_2H_5OH} \times \frac{44.01 \text{ g } CO_2}{1 \text{ mol } CO_2} = 4.78 \text{ g } CO_2$$

3.57 $$1.00 \text{ kg Al} = 1000 \text{ g Al} \times \frac{1 \text{ mol Al}}{26.98 \text{ g Al}} \times \frac{3 \text{ mol } NH_4ClO_4}{3 \text{ mol Al}}$$

$$\times \frac{117.5 \text{ g } NH_4ClO_4}{1 \text{ mol } NH_4ClO_4} = 4.36 \times 10^3 \text{ g or } 4.36 \text{ kg } NH_4ClO_4$$

3.58 $$2NaCl(aq) + 2H_2O(l) \longrightarrow 2NaOH(aq) + H_2(g) + Cl_2(g)$$

Calculate mol Cl_2 and relate to mol H_2, mol NaOH.

$$1.4 \times 10^6 \text{ kg} \times \frac{1000 \text{ g}}{1 \text{ kg}} \times \frac{1 \text{ mol } Cl_2}{70.9 \text{ g } Cl_2} = 2.0 \times 10^7 \text{ mol } Cl_2$$

$$2.0 \times 10^7 \text{ mol } Cl_2 \times \frac{1 \text{ mol } H_2}{1 \text{ mol } Cl_2} \times \frac{2.02 \text{ g } H_2}{1 \text{ mol } H_2} = 4.0 \times 10^7 \text{ g } H_2 = 4.0 \times 10^4 \text{ kg } H_2$$

$$4.0 \times 10^7 \text{ g} \times \frac{1 \text{ metric ton}}{1 \times 10^6 \text{ g (1 Mg)}} = \mathbf{40 \text{ metric tons } H_2}$$

$$2.0 \times 10^7 \text{ mol } Cl_2 \times \frac{2 \text{ mol NaOH}}{1 \text{ mol } Cl_2} \times \frac{40.0 \text{ g NaOH}}{1 \text{ mol NaOH}} = 1.6 \times 10^9 \text{ g NaOH}$$

1.6×10^9 g NaOH = 1.6×10^6 kg NaOH = **1.6×10^3 metric tons NaOH**

3.59 (a) $Al(OH)_3(s) + 3HCl(aq) \rightarrow AlCl_3(aq) + 3H_2O(l)$

(b) See 3.53 (b) for method.

$$2.50 \text{ g } Al(OH)_3 \times \frac{1 \text{ mol } Al(OH)_3}{78.00 \text{ g } Al(OH)_3} \times \frac{3 \text{ mol HCl}}{1 \text{ mol } Al(OH)_3} \times \frac{36.46 \text{ g HCl}}{1 \text{ mol HCl}} = 3.51 \text{ g HCl}$$

3.60 (a) $2NaN_3(s) \rightarrow 2Na(s) + 3N_2(g)$

(b) $$1.00 \text{ g } N_2 \times \frac{1 \text{ mol } N_2}{28.01 \text{ g } N_2} \times \frac{2 \text{ mol } NaN_3}{3 \text{ mol } N_2} \times \frac{65.01 \text{ g } NaN_3}{1 \text{ mol } NaN_3} = 1.55 \text{ g } NaN_3$$

(c) First determine how many g N_2 are in 12.0 ft^3, using the density of N_2.

$$\frac{1.25 \text{ g}}{1 \text{ L}} \times \frac{1 \text{ L}}{1000 \text{ cm}^3} \times \frac{(2.54)^3 \text{ cm}^3}{1 \text{ in}^3} \times \frac{(12)^3 \text{ in}^3}{1 \text{ ft}^3} \times 12.0 \text{ ft}^3 = 425 \text{ g } N_2$$

$$425 \text{ g } N_2 \times \frac{1 \text{ mol } N_2}{28.01 \text{ g } N_2} \times \frac{2 \text{ mol } NaN_3}{3 \text{ mol } N_2} \times \frac{65.01 \text{ g } NaN_3}{1 \text{ mol } NaN_3} = 658 \text{ g } NaN_3$$

Limiting Reagents; Theoretical Yields

3.61 (a) The **limiting reagent** determines the maximum number of product moles resulting from a chemical reaction; any other reactant is an **excess reagent**.

(b) The limiting reagent regulates the amount of products because it is completely used up during the reaction; no more product can be made when one of the reactants is unavailable.

3.62 (a) **Theoretical yield** is the maximum amount of product possible, as predicted by stoichiometry, assuming that the limiting reagent is converted entirely to product. **Actual yield** is the amount of product actually obtained, less than or equal to the theoretical yield. **Percent yield** is the ratio of (actual yield to theoretical yield) × 100.

(b) No reaction is perfect. Not all reactant molecules come together effectively to form products, alternative reaction pathways may produce secondary products and reduce the amount of desired product actually obtained, or it might not be possible to completely isolate the desired product from the reaction mixture. In any case, these factors reduce the actual yield of a reaction.

3.63 (a) Each bicycle needs 2 wheels, 1 frame and 1 set of handlebars. A total of 5350 wheels corresponds to 2675 pairs of wheels. This is fewer than the number of frames and handlebars, so only 2675 bicycles can be manufactured.

(b) 3133 available frames - 2675 bicycles = 458 frames leftover

2785 handlebars - 2675 bicycles = 110 handlebars leftover

(c) The wheels are the "limiting reactant" in that they determine the number of bicycles that can be produced.

3.64 (a) $53{,}575 \text{ L beverage} \times \frac{1 \text{ bottle}}{0.355 \text{ L}} = 150{,}915$ portions of beverage

121,550 bottles; 125,000 caps

121,550 bottles can be filled and capped.

(b) 125,000 caps - 121,550 bottles = 3,450 caps remain

$121{,}550 \text{ bottles} \times \frac{0.355 \text{ L}}{1 \text{ bottle}} = 43{,}150 \text{ L used}$

53,575 L total - 43,150 L used = 10,425 L beverage leftover

(c) The empty bottles limit production.

3.65 (a) $5.00 \text{ g } SiO_2 \times \frac{1 \text{ mol } SiO_2}{60.09 \text{ g } SiO_2} \times \frac{1 \text{ mol SiC}}{1 \text{ mol } SiO_2} \times \frac{40.10 \text{ g SiC}}{1 \text{ mol SiC}} = 3.34 \text{ g SiC}$

(b) $5.00 \text{ g } SiO_2 \times \frac{1 \text{ mol } SiO_2}{60.09 \text{ g } SiO_2} \times \frac{3 \text{ mol C}}{1 \text{ mol } SiO_2} \times \frac{12.01 \text{ g C}}{1 \text{ mol C}} = 3.00 \text{ g C}$

(c) Follow the approach in Sample Exercise 3.16:

$2.50 \text{ g } SiO_2 \times \frac{1 \text{ mol } SiO_2}{60.09 \text{ g } SiO_2} \times \frac{1 \text{ mol SiC}}{1 \text{ mol } SiO_2} \times \frac{40.10 \text{ g SiC}}{1 \text{ mol SiC}} = 1.67 \text{ g SiC}$

(This is exactly 1/2 the amount of SiO_2 reacted and 1/2 the amount of SiC produced in part (a).)

$2.50 \text{ g C} \times \frac{1 \text{ mol C}}{12.01 \text{ g}} \times \frac{1 \text{ mol SiC}}{3 \text{ mol C}} \times \frac{40.10 \text{ g SiC}}{1 \text{ mol SiC}} = 2.78 \text{ g SiC}$

The lesser amount, **1.67 g SiC** is produced.

(d) SiO_2 is the limiting reactant (it leads to the smaller amount of product) and C is present in excess.

(e) $2.50 \text{ g } SiO_2 \times \frac{1 \text{ mol } SiO_2}{60.09 \text{ g } SiO_2} \times \frac{3 \text{ mol C}}{1 \text{ mol } SiO_2} \times \frac{12.01 \text{ g C}}{1 \text{ mol C}} = 1.50 \text{ g C consumed}$

2.50 g C initially present - 1.50 g C consumed = **1.00 g C remain**

3.66 $4NH_3(g) + 5O_2(g) \rightarrow 4NO(g) + 6H_2O(g)$

(a) $2.50 \text{ g } NH_3 \times \frac{1 \text{ mol } NH_3}{17.03 \text{ g } NH_3} \times \frac{4 \text{ mol NO}}{4 \text{ mol } NH_3} \times \frac{30.01 \text{ g NO}}{1 \text{ mol NO}} = 4.41 \text{ g NO}$

(b) $2.50 \text{ g } NH_3 \times \frac{1 \text{ mol } NH_3}{17.03 \text{ g } NH_3} \times \frac{5 \text{ mol } O_2}{4 \text{ mol } NH_3} \times \frac{32.0 \text{ g } O_2}{1 \text{ mol } O_2} = 5.87 \text{ g } O_2$

(c) Following the approach in Sample Exercise 3.16:

$$1.50 \text{ g NH}_3 \times \frac{1 \text{ mol NH}_3}{17.03 \text{ g NH}_3} \times \frac{4 \text{ mol NO}}{4 \text{ mol NH}_3} \times \frac{30.01 \text{ g NO}}{1 \text{ mol NO}} = 2.64 \text{ g NO}$$

$$1.00 \text{ g O}_2 \times \frac{1 \text{ mol O}_2}{32.0 \text{ g O}_2} \times \frac{4 \text{ mol NO}}{5 \text{ mol O}_2} \times \frac{30.01 \text{ g NO}}{1 \text{ mol NO}} = 0.750 \text{ g NO}$$

The **lesser** amount, 0.750 g NO, is produced.

(d) $\mathbf{O_2}$ is the limiting reactant and $\mathbf{NH_3}$ is present in excess.

(e) $$1.00 \text{ g O}_2 \times \frac{1 \text{ mol O}_2}{32.0 \text{ g O}_2} \times \frac{4 \text{ mol NH}_3}{5 \text{ mol O}_2} \times \frac{17.03 \text{ g NH}_3}{1 \text{ mol NH}_3} = 0.426 \text{ g NH}_3 \text{ consumed}$$

1.50 g NH_3 initial - 0.43 g NH_3 consumed = 1.07 g NH_3 remain.

3.67 $$3.05 \text{ g H}_2\text{S} \times \frac{1 \text{ mol H}_2\text{S}}{34.08 \text{ g}} \times \frac{1 \text{ mol Na}_2\text{S}}{1 \text{ mol H}_2\text{S}} = 0.0895 \text{ mol Na}_2\text{S}$$

$$1.84 \text{ g NaOH} \times \frac{1 \text{ mol NaOH}}{40.00 \text{ g NaOH}} \times \frac{1 \text{ mol Na}_2\text{S}}{2 \text{ mol NaOH}} = 0.0230 \text{ mol Na}_2\text{S}$$

NaOH is the limiting reagent;

$$0.230 \text{ mol Na}_2\text{S} \times \frac{78.04 \text{ g Na}_2\text{S}}{1 \text{ mol Na}_2\text{S}} = 1.80 \text{ g Na}_2\text{S}$$

3.68 $$2.93 \text{ g C}_2\text{H}_4 \times \frac{1 \text{ mol C}_2\text{H}_4}{28.05 \text{ g C}_2\text{H}_4} \times \frac{2 \text{ mol CO}_2}{1 \text{ mol C}_2\text{H}_4} \times \frac{44.01 \text{ g CO}_2}{1 \text{ mol CO}_2} = 9.19 \text{ g CO}_2$$

$$4.29 \text{ g O}_2 \times \frac{1 \text{ mol O}_2}{32.00 \text{ g O}_2} \times \frac{2 \text{ mol CO}_2}{3 \text{ mol O}_2} \times \frac{44.01 \text{ g CO}_2}{1 \text{ mol CO}_2} = 3.93 \text{ g CO}_2$$

O_2 is the limiting reagent; thus, 3.93 g CO_2 are formed.

3.69 (a) $$30.0 \text{ g C}_6\text{H}_6 \times \frac{1 \text{ mol C}_6\text{H}_6}{78.11 \text{ g C}_6\text{H}_6} \times \frac{1 \text{ mol C}_6\text{H}_5\text{Br}}{1 \text{ mol C}_6\text{H}_6} \times \frac{157.0 \text{ g C}_6\text{H}_5\text{Br}}{1 \text{ mol C}_6\text{H}_5\text{Br}}$$

$$= 60.3 \text{ g C}_6\text{H}_5\text{Br}$$

$$65.0 \text{ g Br}_2 \times \frac{1 \text{ mol Br}_2}{159.8 \text{ g Br}_2} \times \frac{1 \text{ mol C}_6\text{H}_5\text{Br}}{1 \text{ mol Br}_2} \times \frac{157.0 \text{ g C}_6\text{H}_5\text{Br}}{1 \text{ mol C}_6\text{H}_5\text{Br}} = 63.9 \text{ g C}_6\text{H}_5\text{Br}$$

C_6H_6 is the limiting reagent, and the theoretical yield is 60.3 g C_6H_5Br.

(b) $$\text{The \% yield} = \frac{\text{actual yield}}{\text{theoretical yield}} \times 100\% = \frac{56.7 \text{ g C}_6\text{H}_5\text{Br}}{60.3 \text{ g C}_6\text{H}_5\text{Br}} \times 100 = 94.0\% \text{ yield}$$

3.70 (a) $$115 \text{ g C}_6\text{H}_5\text{NO}_2 \times \frac{1 \text{ mol C}_6\text{H}_5\text{NO}}{123.1 \text{ g C}_6\text{H}_{10}\text{NO}_2} \times \frac{1 \text{ mol C}_{12}\text{H}_{10}\text{N}_2}{2 \text{ mol C}_6\text{H}_5\text{NO}_2} \times \frac{182.2 \text{ g C}_{12}\text{H}_{10}\text{N}_2}{1 \text{ mol C}_{12}\text{H}_{10}\text{N}_2}$$

$$= 85.1 \text{ g C}_{12}\text{H}_{10}\text{N}_2$$

$$327 \text{ g } C_6H_{14}O_4 \times \frac{1 \text{ mol } C_6H_{14}O_4}{150.2 \text{ g } C_6H_{14}O_4} \times \frac{1 \text{ mol } C_{12}H_{10}N_2}{4 \text{ mol } C_6H_{14}O_4} \times \frac{182.2 \text{ g } C_{12}H_{10}N_2}{1 \text{ mol } C_{12}H_{10}N_2}$$

$$= 99.2 \text{ g } C_{12}H_{10}N_2$$

The theoretical yield is the smaller of these two quantities, 85.1 g $C_{12}H_{10}N_2$. Nitrobenzene is the limiting reagent.

(b) $\text{percent yield} = \frac{55 \text{ g } C_{12}H_{10}N_2}{85.1 \text{ g } C_{12}H_{10}N_2} \times 100 = 65\%$

Additional Exercises

3.71 (a) $PBr_5(l) + \mathbf{4}H_2O(l) \rightarrow H_3PO_4(aq) + \mathbf{5}HBr(aq)$

(b) $Li_3N(s) + \mathbf{3}H_2O(l) \rightarrow NH_3(g) + \mathbf{3}LiOH(aq)$

(c) $C_4H_9OH(l) + \mathbf{6}O_2(g) \rightarrow \mathbf{4}O_2(g) + \mathbf{5}H_2O(l)$

3.72 $C_4H_8O_2(s) + 5O_2(g) \rightarrow 4CO_2(g) + 4H_2O(l)$

3.73 In the current system, the mass of ^9Be = 9.01218 amu (its atomic weight, since ^9Be is the only naturally occuring isotope).

If we call the newly defined mass unit an nmu, the amu/nmu ratio is:

$\frac{9.01218 \text{ amu}}{9 \text{ nmu}} = \frac{1.00135 \text{ amu}}{1 \text{ nmu}}$ or 1 nmu = 1.00135 amu.

The nmu is a larger unit of mass than the amu.

3.74 (a) The 68.926 amu isotope has a mass number of 69, with 31 protons, 38 neutrons and the symbol $^{69}_{31}Ga$. The 70.926 amu isotope has a mass number of 71, 31 protons, 40 neutrons, and symbol $^{71}_{31}Ga$. (All Ga atoms have 31 protons.)

(b) The average mass of a Ga atom (given on the inside cover of the text) is 69.72 amu. Let x = abundance of the lighter isotope, 1-x = abundance of the heavier isotope Then x(68.926) + (1-x)(70.926) = 69.72; x = 0.603, ^{69}Ga = 60.3%, ^{71}Ga = 39.7%

3.75 (a) $\%Sm = \frac{150.4 \text{ g Sm}}{445.3 \text{ g } Co_5Sm} = 33.81\%$

(b) $\%N = \frac{28.02 \text{ g N}}{44.02 \text{ g } N_2O} = 63.65\%$

(c) $\%C = \frac{96.08 \text{ g C}}{153.2 \text{ g } C_8H_{11}O_2N} = 45.09\%$

(d) $\%Pt = \frac{195.1 \text{ g Pt}}{300.1 \text{ g } Pt(NH_3)_2Cl_2} = 65.01\%$

Stoichiometry: Calculations with Chemical Formulas and Equations

3.76 $10.0\ cm^3 \times \frac{3.97\ g\ Al_2O_3}{1\ cm^3} \times \frac{1\ mol\ Al_2O_3}{102.0\ g\ Al_2O_3} \times \frac{2\ mol\ Al}{1\ mol\ Al_2O_3} \times \frac{6.022 \times 10^{23}\ Al\ atoms}{1\ mol}$

$= 4.69 \times 10^{23}$ Al atoms

3.77 (a) $0.068\ g\ C_5H_5N \times \frac{1\ mol\ C_5H_5N}{79.1\ g\ C_5H_5N} \times \frac{6.02 \times 10^{23}\ molecules}{1\ mol}$

$= 5.2 \times 10^{20}$ C_5H_5N molecules

(b) $5.0\ g\ ZnO \times \frac{1\ mol\ ZnO}{81.4\ g\ ZnO} \times \frac{6.022 \times 10^{23}\ molecules}{1\ mol} = 3.7 \times 10^{22}$ ZnO formula units

There is one C_5H_5N molecule for each $3.7 \times 10^{22}/5.2 \times 10^{20} = 71$ ZnO units

(c) $5.0\ g\ ZnO \times \frac{48\ m^2}{1\ g\ ZnO} \times \frac{1}{5.2 \times 10^{20}\ C_5H_5N\ molecules}$

$= 4.6 \times 10^{-19}\ m^2/$ molecule

$\frac{4.6 \times 10^{-19}\ m^2}{C_5H_5N\ molecule} \times \frac{1 \times 10^{10}\ Å}{1\ m} = 46\ Å^2\ /\ C_5H_5N$ molecule

3.78 (a) Let AW = the atomic weight of X.
According to the chemical reaction moles, XI_3 reacted = moles XCl_3 produced

$0.5000\ g\ XI_3 \times \frac{1\ mol\ XI_3}{(AW + 380.70)\ g\ XI_3} = 0.2360\ g\ XCl_3 \times \frac{1\ mol\ XCl_3}{(AW + 106.35)\ g\ XCl_3}$

$0.5000\ (AW + 106.35) = 0.2360\ (AW + 380.70)$

$0.5000\ AW + 53.18 = 0.2360\ AW + 89.84$

$0.2640\ AW = 36.66;\ AW = 138.9$ g

(b) X is lanthanum, La, atomic number 57.

3.79 The ratio of the masses $KClO_3/KCl$ is experimentally determined to be 1.64382.

Let x = the mass of KCl. Then $\frac{[x + 3(15.9994)]}{x} = 1.64382$

Solving for x we obtain $x + 47.9982 = 1.64382\ x$

$x = 47.9982/0.6482 = 74.552$

The sum of the presently accepted atomic weights of K and Cl is 74.552.

3.80 We can proceed by writing the ratio of masses of Ag to $AgNO_3$, and call the unknown quantity, the atomic mass of nitrogen, y:

$$\frac{Ag}{AgNO_3} = 0.634985 = \frac{107.870}{107.870 + 3(15.9994) + y}$$

Solve for y to obtain y = 14.010. This is to be compared with the currently accepted value of 14.007.

3.81 (a) $0.7787 \text{ g C} \times \frac{1 \text{ mol C}}{12.01 \text{ g C}} = 0.06484 \text{ mol C}$

$0.1176 \text{ g H} \times \frac{1 \text{ mol H}}{1.008 \text{ g H}} = 0.1167 \text{ mol H}$

$0.1037 \text{ g O} \times \frac{1 \text{ mol C}}{16.00 \text{ g O}} = 0.006481 \text{ mol O}$

Dividing through by the smallest of these values we obtain $C_{10}H_{18}O$.

(b) The formula weight of this formula is 154. Thus, the empirical formula is also the molecular formula.

3.82 Since all the C in the vanillin must be present in the CO_2 produced, get g C from g CO_2.

$$2.43 \text{ g } CO_2 \times \frac{1 \text{ mol } CO_2}{44.01 \text{ g } CO_2} \times \frac{12.01 \text{ g C}}{1 \text{ mol C}} = 0.663 \text{ g C}$$

Since all the H in vanillin must be present in the H_2O produced, get g H from g H_2O.

$$0.50 \text{ g } H_2O \times \frac{1 \text{ mol } H_2O}{18.02 \text{ g } H_2O} \times \frac{2 \text{ mol H}}{1 \text{ mol } H_2O} \times \frac{1.008 \text{ g H}}{1 \text{ mol H}} = 0.056 \text{ g H}$$

Get g O by subtraction. (Since the analysis was performed by combustion, an unspecified amount of O_2 was a reactant, and thus not all the O in the CO_2 and H_2O produced came from vanillin.)

1.05 g vanillin - 0.663 g C - 0.056 g H = 0.331 g O

$0.663 \text{ g C} \times \frac{1 \text{ mol C}}{12.01 \text{ g C}} = 0.0552 \text{ mol C}; 0.0552 \div 0.0207 = 2.67$

$0.056 \text{ g H} \times \frac{1 \text{ mol H}}{1.008 \text{ g H}} = 0.0556 \text{ mol C}; 0.0556 \div 0.0207 = 2.68$

$0.331 \text{ g O} \times \frac{1 \text{ mol O}}{16.00 \text{ g O}} = 0.0207 \text{ mol O}; 0.0207 \div 0.0207 = 1.00$

Multiplying the numbers above by **3** to obtain an integer ratio of moles, the empirical formula of vanillin is $\mathbf{C_8H_8O_3}$.

3.83 The mass percentage is determined by the relative number of atoms of the element times the atomic weight, divided by the total formula mass. Thus, the mass percent of bromine in $KBrO_x$ is given by $0.5292 = \frac{79.91}{39.10 + 79.91 + x(16.00)}$. Solving for x, we obtain x = 2.00. Thus, the formula is $KBrO_2$.

3.84 Let x = g AgCl in the mixture, y = g AgBr in the mixture.

Assume 100 g. x + y = 100.

Also, g Ag from AgCl + g Ag from AgBr = 60.94

% Ag in AgCl (g AgCl) + % Ag in AgBr (g AgBr) = 60.94

$$\frac{107.9 \text{ g Ag}}{143.3 \text{ g AgCl}} (x) + \frac{107.9}{187.8} (y) = 60.94$$

0.7530 x + 0.5745 y = 60.94. Substituting y = 100 - x,

0.7530 x + 0.5745 (100 - x) = 60.94; 0.7530 x - 0.5745x = 60.94 - 57.45

0.1785 x = 3.49; x = 19.60 g AgCl, y = 80.40 g AgBr

$$\text{g Cl} = \frac{35.45 \text{ g Cl}}{143.3 \text{ g AgCl}} \times 19.60 \text{ g AgCl} = 4.85 \text{ g Cl in 100 g sample} = 4.85\% \text{ Cl}$$

$$\text{g Br} = \frac{79.90 \text{ g Br}}{187.8 \text{ g AgBr}} \times 80.40 \text{ g AgBr} = 34.21 \text{ g Br in 100 g sample} = 34.21\% \text{ Br}$$

Total = 60.94% Ag + 4.85% Cl + 34.21% Br = 100.00%

3.85 $1.00 \text{ g NaHCO}_3 \times \frac{1 \text{ mol NaHCO}_3}{84.01 \text{ g/mol}} = 1.19 \times 10^{-2} \text{ mol NaHCO}_3$

Abbreviate citric acid as H_3Cit

$$1.19 \times 10^{-2} \text{ mol NaHCO}_3 \times \frac{1 \text{ mol H}_3\text{Cit}}{3 \text{ mol NaHCO}_3} \times \frac{192.1 \text{ g H}_3\text{Cit}}{1 \text{ mol H}_3\text{Cit}} = 0.762 \text{ g citric acid}$$

3.86 $2C_{57}H_{110}O_6 + \mathbf{169}O_2 \rightarrow \mathbf{114}CO_2 + \mathbf{110}H_2O$

molar mass of fat = 57(12.01) + 110(1.008) + 6(16.00) = 891.5

$$2.0 \text{ kg fat} \times \frac{1000 \text{ g}}{1 \text{ kg}} \times \frac{1 \text{ mol fat}}{891.5 \text{ g fat}} \times \frac{110 \text{ mol H}_2\text{O}}{2 \text{ mol fat}} \times \frac{18.02 \text{ g H}_2\text{O}}{1 \text{ mol H}_2\text{O}} \times \frac{1 \text{ kg}}{1000 \text{ g}} = 2.2 \text{ kg H}_2\text{O}$$

3.87 (a) Strategy: calculate the kg of air in the room and then the mass of HCN required to produce a dose of 300 mg HCN/kg air.

12 ft × 15 ft × 8.0 ft = 1440 ft^3 of air in the room

$$1440 \text{ ft}^3 \text{ air} \times \frac{(12 \text{ in})^3}{1 \text{ ft}^3} \times \frac{(2.54 \text{ cm})^3}{1 \text{ in}^3} \times \frac{0.00118 \text{ gm air}}{1 \text{ cm}^3 \text{ air}} \times \frac{1 \text{ kg}}{1000 \text{ g}} = 48.1 \text{ kg air}$$

(2 sig. figs)

$$48.1 \text{ kg air} \times \frac{300 \text{ mg HCN}}{1 \text{ kg air}} \times \frac{1 \text{ g}}{1000 \text{ mg}} = 14 \text{ g HCN (14.4 g HCN to 3 sig. figs)}$$

(b) $\mathbf{2}NaCN(s) + H_2SO_4(aq) \rightarrow Na_2SO_4(aq) + \mathbf{2}HCN(g)$

The question can be restated as: What mass of NaCN is required to produce 14 g of HCN according to the above reaction?

$$14 \text{ g HCN} \times \frac{1 \text{ mol HCN}}{27.03 \text{ g HCN}} \times \frac{2 \text{ mol NaCN}}{2 \text{ mol HCN}} \times \frac{49.01 \text{ g NaCN}}{1 \text{ mol NaCN}} = 25 \text{ g NaCN}$$

(If the mass from part (a) is 14.4 g HCN, 26.2 g of NaCN are needed.)

(c) $12 \text{ ft} \times 15 \text{ ft} \times \frac{1 \text{ yd}^2}{9 \text{ ft}^2} \times \frac{30 \text{ oz}}{1 \text{ yd}^2} \times \frac{1 \text{ lb}}{16 \text{ oz}} \times \frac{454 \text{ g}}{1 \text{ lb}} = 17{,}000$ g acrilan in the room

50% of the carpet burns, so the starting amount of CH_2CHCN is 0.50(17,000) = 8,500 g

$8{,}500 \text{ g } CH_2CHCN \times \frac{50.9 \text{ g HCN}}{100 \text{ g } CH_2CHCN} = 4{,}300$ g HCN possible

If the actual yield of combustion is 20%,

Actual g HCN = 4,333(0.20) = 860 g HCN produced

From part (a), 14 g of HCN is a lethal dose. The fire produces much more than a lethal dose of HCN.

3.88 $N_2(g) + 3H_2(g) \rightarrow 2NH_3(g)$

Determine the moles of N_2 and H_2 required to form the 2.0 moles of NH_3 present after the reaction has stopped.

$2.0 \text{ mol } NH_3 \times \frac{3 \text{ mol } H_2}{2 \text{ mol } NH_3} = 3.0 \text{ mol } H_2$ reacted

$2.0 \text{ mol } NH_3 \times \frac{1 \text{ mol } N_2}{2 \text{ mol } NH_3} = 1 \text{ mol } N_2$ reacted

mol H_2 initial = 2.0 mol H_2 remain + 3.0 mol H_2 reacted = 5.0 mol H_2
mol N_2 initial = 2.0 mol N_2 remain + 1.0 mol N_2 reacted = 3.0 mol N_2

In tabular form:	$N_2(g)$	+	$3H_2(g)$	$\rightarrow$	$2NH_3(g)$
initial	3.0 mol		5.0 mol		0 mol
reaction	-1.0 mol		-3.0 mol		+2.0 mol
final	2.0 mol		2.0 mol		2.0 mol

(Tables like this will be extremely useful for solving chemical equilibrium problems in Chapter 15.)

3.89 $\frac{0.0129 \text{ g AgCl}}{1.03 \text{ g ointment}} \times \frac{1 \text{ mol AgCl}}{143.3 \text{ g AgCl}} \times \frac{1 \text{ mol } C_{11}H_{12}O_5N_2Cl_2}{2 \text{ mol AgCl}}$

$\times \frac{323.1 \text{ g } C_{11}H_{12}O_5N_2Cl_2}{1 \text{ mol } C_{11}H_{12}O_5N_2Cl_2} = \frac{0.0141 \text{ g chloromycetin}}{1 \text{ g ointment}} = 1.41\%$ by weight

3.90 All of the O_2 is produced from $KClO_3$; get g $KClO_3$ from g O_2. All of the H_2O is produced from $KHCO_3$; get g $KHCO_3$ from g H_2O. The g H_2O produced also reveals the g CO_2 from the decomposition of $NaHCO_3$. The remaining CO_2 (13.2 g CO_2 - g CO_2 from $NaHCO_3$) is due to K_2CO_3 and g K_2CO_3 can be derived from it.

$$4.00\text{ g } O_2 \times \frac{1\text{ mol } O_2}{32.00\text{ g } O_2} \times \frac{2\text{ mol } KClO_3}{3\text{ mol } O_2} \times \frac{122.6\text{ g } KClO_3}{1\text{ mol } KClO_3} = 10.2\text{ g } KClO_3$$

$$1.80\text{ g } H_2O \times \frac{1\text{ mol } H_2O}{18.02\text{ g } H_2O} \times \frac{2\text{ mol } KHCO_3}{1\text{ mol } H_2O} \times \frac{100.1\text{ g } KHCO_3}{1\text{ mol } KHCO_3} = 20.0\text{ g } KHCO_3$$

$$1.80\text{ g } H_2O \times \frac{1\text{ mol } H_2O}{18.02\text{ g } H_2O} \times \frac{2\text{ mol } CO_2}{1\text{ mol } H_2O} \times \frac{44.0\text{ g } CO_2}{1\text{ mol } CO_2} = 8.80\text{ g } CO_2 \text{ from } KHCO_3$$

13.20 g CO_2 total - 8.80 CO_2 from $KHCO_3$ = 4.40 g CO_2 from K_2CO_3

$$4.40\text{ g } CO_2 \times \frac{1\text{ mol } CO_2}{44.01\text{ g } CO_2} \times \frac{1\text{ mol } K_2CO_3}{1\text{ mol } CO_2} \times \frac{138.2\text{ g } K_2CO_3}{1\text{ mol } K_2CO_3} = 13.8\ K_2CO_3$$

100.0 g mixture - 10.2 $KClO_3$ - 20.0 g $KHCO_3$ - 13.8 g K_2CO_3 = 56.0 g KCl

3.91 $S(s) + O_2(g) \rightarrow SO_2(g)$; $SO_2(g) + CaO(s) \rightarrow CaSO_3(s)$

$$2000\text{ tons coal} \times \frac{2000\text{ lb}}{1\text{ ton}} \times \frac{1\text{ kg}}{2.20\text{ lb}} \times \frac{1000\text{ g}}{1\text{ kg}} \times \frac{0.028\text{ g S}}{1\text{ g coal}} \times \frac{1\text{ mol S}}{32.1\text{ g S}}$$

$$\times \frac{1\text{ mol } SO_2}{1\text{ mol S}} \times \frac{1\text{ mol } CaSO_3}{1\text{ mol } SO_2} \times \frac{120\text{ g } CaSO_3}{1\text{ mol } CaSO_3} \times \frac{1\text{ kg } CaSO_3}{1000\text{ g } CaSO_3} = 1.9 \times 10^5\text{ kg } CaSO_3$$

Incidentally, this corresponds to over 200 tons of $CaSO_3$ per day as a waste byproduct.

3.92 (a) $\mathbf{2}C_2H_2(g) + \mathbf{5}O_2(g) \rightarrow \mathbf{4}CO_2(g) + \mathbf{2}H_2O(g)$

(b) Following the approach in Sample Exercise 3.15,

$$10.0\text{ g } C_2H_2 \times \frac{1\text{ mol } C_2H_2}{26.04\text{ g } C_2H_2} \times \frac{5\text{ mol } O_2}{2\text{ mol } C_2H_2} \times \frac{32.00\text{ g } O_2}{1\text{ mol } O_2} = 30.7\text{ g } O_2 \text{ required}$$

Only 10.0 g O_2 are available, so O_2 limits.

(c) Since O_2 limits, 0.0 g O_2 remain.

Next, calculate the g C_2H_2 consumed and the amounts of CO_2 and H_2O produced by reaction of 10.0 g O_2.

$$10.0\text{ g } O_2 \times \frac{1\text{ mol } O_2}{32.0\text{ g } O_2} \times \frac{2\text{ mol } C_2H_2}{5\text{ mol } O_2} \times \frac{26.04\text{ g } C_2H_2}{1\text{ mol } C_2H_2} = 3.26\text{ g } C_2H_2 \text{ consumed}$$

10.0 g C_2H_2 initial - 3.26 g consumed = 6.74 g C_2H_2 remain

$$10.0\text{ g } O_2 \times \frac{1\text{ mol } O_2}{32.00\text{ g } O_2} \times \frac{4\text{ mol } CO_2}{5\text{ mol } O_2} \times \frac{44.01\text{ g } CO_2}{1\text{ mol } CO_2} = 11.0\text{ g } CO_2 \text{ produced}$$

$$10.0\text{ g } O_2 \times \frac{1\text{ mol } O_2}{32.00\text{ g } O_2} \times \frac{2\text{ mol } H_2O}{5\text{ mol } O_2} \times \frac{18.02\text{ g } H_2O}{1\text{ mol } H_2O} = 2.25\text{ g } H_2O \text{ produced}$$

3.93 (a) $1.5 \times 10^5\ \text{g } C_9H_8O_4 \times \frac{1\ \text{mol } C_9H_8O_4}{180.2\ \text{g } C_9H_8O_4} \times \frac{1\ \text{mol } C_7H_6O_3}{1\ \text{mol } C_9H_8O_4} \times \frac{138.1\ \text{g } C_7H_6O_3}{1\ \text{mol } C_7H_6O_3}$

$= 1.2 \times 10^2\ \text{kg } C_7H_6O_3$

(b) If only 80 percent of the acid reacts, then we need 1/0.80 = 1.25 times as much to obtain the same mass of product: $1.25 \times 1.2 \times 10^2$ kg = 1.5×10^2 kg.

(c) Calculate the number of moles of each reactant:

$$1.85 \times 10^5\ \text{g } C_7H_6O_3 \times \frac{1\ \text{mol } C_7H_6O_3}{138.1\ \text{g } C_7H_6O_3} = 1.34 \times 10^3\ \text{mol } C_7H_6O_3$$

$$1.25 \times 10^5\ \text{g } C_4H_6O_3 \times \frac{1\ \text{mol } C_4H_6O_3}{102.1\ \text{g } C_4H_6O_3} = 1.22 \times 10^3\ \text{mol } C_4H_6O_3$$

We see that $C_4H_6O_3$ is the limiting reagent, because equal numbers of moles of the two reactants are consumed in the reaction.

$$1.22 \times 10^3\ \text{mol } C_4H_6O_3 \times \frac{1\ \text{mol } C_9H_8O_4}{1\ \text{mol } C_7H_6O_3} \times \frac{180.2\ \text{g } C_9H_8O_4}{1\ \text{mol } C_9H_8O_4} = 2.20 \times 10^5\ \text{g } C_9H_8O_4$$

(d) $\text{percent yield} = \frac{1.82 \times 10^5\ \text{g}}{2.20 \times 10^5\ \text{g}} \times 100 = 82.4\%$

3.94 (a) 4 peaks; there are 4 possible combinations of isotopes.

(b)

$^1H^{35}Cl$ 1.0078 amu + 34.969 amu = 35.977 amu

$^2H^{35}Cl$ 2.0140 amu + 34.969 amu = 36.983 amu

$^1H^{37}Cl$ 1.0078 amu + 36.966 amu = 37.974 amu

$^2H^{37}Cl$ 2.0140 amu + 36.966 amu = 38.980 amu

(c) The intensities will be the products of the natural abundances of the twoisotopes.

$^1H^{35}Cl$ (99.985)(0.7553) = 75.52

$^2H^{35}Cl$ (0.015)(0.7553) = 0.011

$^1H^{37}Cl$ (99.985)(0.2447) = 24.47

$^2H^{37}Cl$ (0.015)(0.2447) = 0.0022

Dividing through by 0.0022, the relative intensities in order of increasing mass are 34,000 : 5 : 11,000 : 1. The two ^{2}HCl peaks are very small relative to the other peaks.

(d) Assuming the very small peaks due to ^{2}HX can be observed, element **X** has 4 naturally occurring isotopes.

CHAPTER 4

Aqueous Reactions and Solution Stoichiometry

Solution Composition; Molarity

4.1 Concentration is the **ratio** of the amount of solute present in a certain quantity of solvent or solution. This ratio remains constant regardless of how much solution is present. Thus, concentration is an intensive property. The absolute concentration does depend on the amount of solute present, but once this ratio is established, it doesn't vary with the volume of solution present.

4.2 The concentration of the remaining solution is unchanged, assuming the original solution was thoroughly mixed. Molar concentration is a **ratio** of moles solute to liters solution. Although there are fewer moles solute remaining in the flask, there is also less solution volume, so the ratio of moles solute/solution volume remains the same.

4.3 The second solution is 5 times as concentrated as the first. An equal volume of the more concentrated solution will contain 5 times as much solute (5 times the number of moles and also 5 times the mass) as the 0.50 *M* solution. Thus, the mass of solute in the 2.50 *M* solution is 5×3.5 g = 17.5 g.

Mathematically:

$$\frac{\dfrac{2.50 \text{ mol solute}}{1 \text{ L solution}}}{\dfrac{0.50 \text{ mol solute}}{1 \text{ L solution}}} = \frac{\text{x grams solute}}{3.5 \text{ g solute}}$$

$$\frac{2.5 \text{ mol solute}}{0.50 \text{ mol solute}} = \frac{\text{x g solute}}{3.5 \text{ g solute}}; \quad 5.0(3.5 \text{ g solute}) = 17.5 \text{ g solute}$$

CHAPTER 4 Aqueous Reactions and Solution Stoichiometry

Solutions to Exercises

4.4 The term **0.50 mol HCl** defines an amount (~18 g) of the pure substance HCl. The term 0.50 *M* HCl is a ratio; it indicates that there are 0.50 mol of HCl solute in 1.0 liters of solution. This same ratio of moles solute to solution volume is present regardless of the volume of solution under consideration.

4.5 (a) $M = \frac{\text{mol solute}}{\text{L solution}}; \frac{0.0575 \text{ mol } NH_4Cl}{400 \text{ mL}} \times \frac{1000 \text{ mL}}{1 \text{ L}} = 0.144 \ M \ NH_4Cl$

(b) $\text{mol} = M \times \text{L}; \frac{3.5 \text{ mol } HNO_3}{1 \text{ L}} \times 0.0350 \text{ L} = 0.123 \text{ mol } HNO_3$

(c) $\text{L} = \frac{\text{mol}}{M}; \frac{0.250 \text{ mol KOH}}{1.50 \text{ mol KOH/L}} = 0.167 \text{ L or } 167 \text{ mL } 1.50 \ M \text{ KOH}$

4.6 (a) $M = \frac{\text{mol solute}}{\text{L solution}}; \frac{0.0834 \text{ mol } Na_2SO_4}{0.650 \text{ L}} = 0.128 \ M \ Na_2SO_4$

(b) $\text{mol} = M \times \text{L}; \frac{0.168 \text{ mol } KMnO_4}{1 \text{ L}} \times 0.0750 \text{ L} = 0.0126 \text{ mol } KMnO_4$

(c) $\text{L} = \frac{\text{mol}}{M}; \frac{0.175 \text{ mol HCl}}{11.6 \text{ mol HCl/L}} = 0.0151 \text{ L or } 15.1 \text{ mL}$

4.7 $M \times \text{L} = \text{mol solute}$, $\text{mol solute} \times \text{molar mass} = \text{g solute}$

(a) $\frac{0.125 \text{ mol KBr}}{1 \text{ L}} \times 0.200 \text{ L} \times \frac{119.0 \text{ g KBr}}{1 \text{ mol KBr}} = 2.98 \text{ g KBr}$

(b) $\frac{0.150 \text{ mol } Na_2SO_4}{1 \text{ L}} \times 100 \text{ mL} \times \frac{1 \text{ L}}{1000 \text{ mL}} \times \frac{142.0 \text{ g } Na_2SO_4}{1 \text{ mol } Na_2SO_4} = 2.13 \text{ g } Na_2SO_4$

(c) $\frac{0.0500 \text{ mol } KBrO_3}{1 \text{ L}} \times 250\text{mL} \times \frac{1 \text{ L}}{1000 \text{ mL}} \times \frac{167.0 \text{ g } KBrO_3}{1 \text{ mol } KBrO_3} = 2.09 \text{ g } KBrO_3$

(d) $1.70 \text{ mol } C_6H_{12}O_6 \times 50.0 \text{ mL} \times \frac{1 \text{ L}}{1000 \text{ mL}} \times \frac{180.2 \text{ g } C_6H_{12}O_6}{1 \text{ mol } C_6H_{12}O_6} = 15.3 \text{ g } C_6H_{12}O_6$

4.8 $\text{mol} = \frac{\text{g}}{\text{molar mass}}; \ M = \frac{\text{mol}}{\text{L}}$

(a) $5.75 \text{ g } NaNO_3 \times \frac{1 \text{ mol } NaNO_3}{85.00 \text{ g } NaNO_3} \times \frac{1}{0.250 \text{ L}} = 0.271 \ M \ NaNO_3$

(b) $22.57 \text{ g } H_2SO_4 \times \frac{1 \text{ mol } H_2SO_4}{98.076 \text{ g } H_2SO_4} \times \frac{1}{0.100 \text{ L}} = 2.301 \ M \ H_2SO_4$

(c) $1.48 \text{ g } AgNO_3 \times \frac{1 \text{ mol } AgNO_3}{169.9 \text{ g } AgNO_3} \times \frac{1}{0.0500 \text{ L}} = 0.174 \ M \ AgNO_3$

(d) $138 \text{ g } NiCl_2 \cdot 6H_2O \times \frac{1 \text{ mol } NiCl_2 \cdot 6H_2O}{237.7 \text{ g } NiCl_2 \cdot 6H_2O} \times \frac{1}{2.00 \text{ L}} = 0.290 \ M \ NiCl_2 \cdot 6H_2O$

Aqueous Reactions and Solution Stoichiometry

4.9 The number of moles of sucrose needed is

$$\frac{0.150\ \text{mol}}{1\ \text{L}} \times 0.250\ \text{L} = 0.0375\ \text{mol}$$

Weigh out $0.0375\ \text{mol}\ C_{12}H_{22}O_{11} \times \frac{342.3\ \text{g}\ C_{12}H_{22}O_{11}}{1\ \text{mol}\ C_{12}H_{22}O_{11}} = 12.8\ \text{g}\ C_{12}H_{22}O_{11}$

Add this amount of solid to a 250 mL volumetric flask, dissolve in a small volume of water, and add water to the mark on the neck of the flask. Agitate throroughly to ensure total mixing.

4.10 The amount of $AgNO_3$ needed is: $0.0200\ M \times 0.1000\ \text{L} = 2.000 \times 10^{-3}\ \text{mol}\ AgNO_3$

$$2.000 \times 10^{-3}\ \text{mol}\ AgNO_3 \times \frac{169.88\ \text{g}\ AgNO_3}{1\ \text{mol}\ AgNO_3} = 0.3398\ \text{g}\ AgNO_3$$

Add this amount of solid to a 100 mL volumetric flask, dissolve in a small amount of water, bring the total volume to exactly 100 mL and agitate well.

4.11 Calculate the moles of solute present in the final 500 mL of 0.100 M $C_{12}H_{22}O_{11}$ solution:

$$\text{moles}\ C_{12}H_{22}O_{11} = M \times \text{L} = \frac{0.100\ \text{mol}\ C_{12}H_{22}O_{11}}{1\ \text{L}} \times 0.500\ \text{L} = 0.0500\ \text{mol}\ C_{12}H_{22}O_{11}$$

Calculate the volume of 1.50 M glucose solution that would contain 0.0500 mol $C_{12}H_{22}O_{11}$:

$$\text{L} = \text{moles}/M\ ;\ 0.0500\ \text{mol}\ C_{12}H_{22}O_{11} \times \frac{1\ \text{L}}{1.50\ \text{mol}\ C_{12}H_{22}O_{11}} = 0.0333\ \text{L}$$

$$0.0333\ \text{L} \times \frac{1000\ \text{mL}}{1\ \text{L}} = 33.3\ \text{mL}$$

Thoroughly rinse, clean and fill a 50 mL buret with the 1.50 M $C_{12}H_{22}O_{11}$. Dispense 33.3 mL of this solution into a 500 mL volumetric flask, add water to the mark and mix thoroughly. (33.3 mL is a difficult volume to measure with a pipette.)

4.12 Dilute the 6.0 M HNO_3 to prepare 200 mL of 1.0 M HNO_3. To determine the volume of 6.0 M HNO_3 needed, calculate the moles HNO_3 present in 200 mL of 1.0 M HNO_3 and then the volume of 6.0 M solution which contains this number of moles.

$0.200\ \text{L} \times 1.0\ \text{L} = 0.200\ \text{mol}\ HNO_3$ needed;

$$\text{L} = \frac{\text{mol}}{M}\ ;\ \text{L}\ 6.0\ M\ HNO_3 = \frac{0.200\ \text{mol needed}}{6.0\ M} = 0.033\ \text{L} = 33\ \text{mL}$$

Thoroughly clean, rinse and fill a 50 mL buret with the 6.0 M HNO_3, taking precautions appropriate for working with a relatively concentrated acid. Dispense 33 mL of the 6.0 M acid into a 200 mL volumetric flask, add water to the mark and mix thoroughly. (A pipette could be used to measure the 33 mL of concentrated acid, but a buret is more safe and convenient for this volume of acid.)

4.13 Calculate the mass of acetic acid, $HC_2H_3O_2$, present in 10.0 mL of the pure liquid.

$$10.0 \text{ mL acetic acid} \times \frac{1.049 \text{ g acetic acid}}{1 \text{ mL acetic acid}} = 10.49 \text{ g acetic acid}$$

$$10.49 \text{ g } HC_2H_3O_2 \times \frac{1 \text{ mol } HC_2H_3O_2}{60.05 \text{ g } HC_2H_3O_2} = 0.1747 \text{ mol } HC_2H_3O_2$$

$$M = \frac{\text{mol}}{\text{L}} = \frac{0.1747 \text{ mol } HC_2H_3O_2}{0.1000 \text{ L solution}} = 1.747\ M\ HC_2H_3O_2$$

4.14 $$40.000 \text{ mL glycerol} \times \frac{1.2656 \text{ g glycerol}}{1 \text{ mL glycerol}} = 50.624 \text{ g glycerol}$$

$$50.624 \text{ g } C_3H_8O_3 \times \frac{1 \text{ mol } C_3H_8O_3}{92.094 \text{ g } C_3H_8O_3} = 0.54970 \text{ mol } C_3H_8O_3$$

$$M = \frac{0.54970 \text{ mol } C_3H_8O_3}{0.25000 \text{ L solution}} = 2.1988\ M\ C_3H_8O_3$$

Electrolytes

4.15 Tap water contains enough dissolved electrolytes to conduct a significant amount of electricity. Thus, water can complete a circuit between an electrical appliance and our body, producing a shock.

4.16 The apparatus in Figure 4.4 can be used to determine if the solute is an electrolyte. If the unknown is an electrolyte, it dissociates into ions when dissolved in water. When the electrodes of the apparatus are submerged in the 0.50 *M* unknown solution, if there are ions in the solution, the circuit will be completed and the bulb will glow. An electrolyte solution lights the bulb; a nonelectrolyte solution, such as glucose, does not.

4.17 (a) HBrO -- weak (b) HNO_3 -- strong (c) KOH -- strong (d) $CoSO_4$ -- strong
(e) $C_{12}H_{22}O_{11}$ -- non (f) O_2 -- non

4.18 (a) HF -- weak (b) C_2H_5OH -- non (c) NH_3 -- weak
(d) $KClO_3$ -- strong (e) $Cu(NO_3)_2$ -- strong

4.19 (a) $0.14\ M\ Na^+ + 0.14\ M\ OH^- = 0.28\ M$ total
(b) $0.25\ M\ Ca^{2+} + 0.50\ M\ Br^- = 0.75\ M$ total
(c) 0.25 *M* (CH_3OH is a molecular solute)
(d) Calculating the molarity of each ion individually and rounding to two significant figures: $0.13\ M\ K^+ + 0.13\ M\ ClO_3^- + 0.13\ M\ Na^+ + 0.067\ M\ SO_4^{2-} = 0.46\ M$ total

Calculating the total molarity directly:

$(0.010 \text{ mol } K^+ + 0.010 \text{ mol } ClO_3^- + 0.010 \text{ mol } Na^+ + 0.0050 \text{ mol } SO_4^{2-}) / 0.075 \text{ L}$
$= 0.47\ M$ total

Note that "intermediate" rounding to obtain the individual molarities produced a slightly different result.

4.20 (a) $H^+ : \frac{0.100\ M \times 20.0\ \text{mL} + 0.220\ M \times 10.0\ \text{mL}}{30.0\ \text{mL}} = 0.14\ M\ H^+$

$Cl^- :$ concentration Cl^- = concentration H^+ = 0.14 M Cl^-

(b) $Na^+ : \frac{2(0.300\ M \times 15.0\ \text{mL}) + (0.100\ M \times 10.0\ \text{mL})}{25.0\ \text{mL}} = 0.400\ M$

$SO_4^{2-} : \frac{0.300\ M \times 15.0\ \text{mL}}{25.0\ \text{mL}} = 0.180\ M$; $Cl^- : \frac{0.100\ M \times 10.0\ \text{mL}}{25.0\ \text{mL}} = 0.0400\ M$

(c) $K: \frac{3.50\ \text{g KCl}}{0.060\ \text{L}} \times \frac{1\ \text{mol}}{74.55\ \text{g}} = 0.782\ M$; Ca^{2+} : 0.500 M

Cl^- : 0.782 M (from KCl(s)) + 1.000 M (from $CaCl_2$(aq)) = 1.782 M

4.21 $KCl \rightarrow K^+ + Cl^-$; 0.20 M KCl = 0.20 M K^+

$K_2Cr_2O_7 \rightarrow$ **2** $K^+ + Cr_2O_7^{2-}$; 0.15 M $K_2Cr_2O_7$ = 0.30 M K^+

$K_3PO_4 \rightarrow$ **3** $K^+ + PO_4^{3+}$; 0.080 M K_3PO_4 = 0.24 M K^+

0.15M $K_2Cr_2O_7$ has the highest K^+ concentration.

4.22 $NaCl \rightarrow Na^+ + Cl^-$; 0.35 $M \times$ 0.040 L = 0.014 mol Cl^-

$CaCl_2 \rightarrow Ca^{2+}$ + **2** Cl^-; 0.25 $M \times$ 0.025 L = 0.0063 mol $CaCl_2 \times 2$ = 0.013 mol Cl^-

The NaCl solution has slightly more moles Cl^-.

Acids, Bases and Salts

4.23 Since the solution does conduct some electricity, but less than an equimolar NaCl solution (a strong electrolyte) the unknown solute must be a weak electrolyte. The weak electrolytes in the list of choices are NH_3 and H_3PO_3; since the solution is acidic, the unknown must be $\mathbf{H_3PO_3}$.

4.24 (a) HF -- acid, mixture of ions and molecules (weak electrolyte)
(b) CH_3CN -- none of the above, entirely molecules (nonelectrolyte)
(c) $NaClO_4$ -- salt, entirely ions (strong electrolyte)
(d) $Ba(OH)_2$ -- base, entirely ions (strong electrolyte)

4.25 (a) A monoprotic acid has one ionizable (acidic) H and a diprotic acid has two.

(b) A strong acid is completely ionized in aqueous solution while only a fraction of weak acid molecules are ionized.

(c) An acid is an H^+ donor, a substance that increases the concentration of H^+ in aqueous solution. A base is an H^+ acceptor and thus increases the concentration of OH^- in aqueous solution.

4.26 (a) NH_3 produces OH^- in aqueous solution by reacting with H_2O (hydrolysis):
$NH_3(aq) + H_2O(l) \rightleftharpoons NH_4^+(aq) + OH^-(aq)$. The OH^- causes the solution to be basic.

(b) The term "weak" refers to the tendency of HF to dissociate into H^+ and F^- in aqueous solution, not its reactivity toward other compounds.

4.27 In aqueous solution, HNO_3 exists entirely as H^+ and NO_3^- ions; the single arrow denotes complete ionization. HCN is only dissociated to a small extent; the double arrow denotes a mixture of H^+ ions, CN^- ions and neutral undissociated HCN molecules in solution.

4.28 H_2SO_4 is a **diprotic** acid; it has two ionizable hydrogens. The first hydrogen completely dissociates to form H^+ and HSO_4^-, but HSO_4^- only **partially** ionizes into H^+ and SO_4^{2-} (HSO_4^- is a weak electrolyte). Thus, a 0.1 M solution of H_2SO_4 contains a mixture of H^+, HSO_4^- and SO_4^{2-}, with a concentration of H^+ greater than 0.1 M but less than 0.2 M.

4.29 (a) $Ba(OH)_2(s) + 2HNO_3(aq) \rightarrow Ba(NO_3)_2(aq) + 2H_2O(l)$

(b) $H_3PO_4(aq) + 3KOH(aq) \rightarrow K_3PO_4(aq) + 3H_2O(l)$

(c) $2Y(OH)_3(s) + 3H_2SO_4(aq) \rightarrow Y_2(SO_4)_3(aq) + 6H_2O(l)$

4.30 (a) $HC_2H_3O_2(aq) + NaOH \rightarrow NaC_2H_3O_2(aq) + H_2O(l)$

Salt: sodium acetate

(b) $2HCl(aq) + Ca(OH)_2(s) \rightarrow CaCl_2(aq) + 2H_2O(l)$

Salt: calcium chloride

(c) $HCl(aq) + NH_3(aq) \rightarrow NH_4Cl(aq)$

Salt: ammonium chloride

Ionic Equations; Metathesis Reactions

4.31 (a) $Pb^{2+}(aq) + SO_4^{2-}(aq) \rightarrow PbSO_4(s)$; spectators: Na^+, NO_3^-

(b) $Zn(s) + 2H^+(aq) \rightarrow Zn^{2+}(aq) + H_2(g)$; spectator: Cl^-

(c) $FeO(s) + 2H^+(aq) \rightarrow H_2O(l) + Fe^{2+}(aq)$; spectator: ClO_4^-

4.32 (a) $Cr(OH)_3(s) + 3HNO_3(aq) \rightarrow 3H_2O(l) + Cr(NO_3)_3(aq)$

$Cr(OH)_3(s) + 3H^+(aq) \rightarrow 3H_2O(l) + Cr^{3+}(aq)$

(b) $Na_2CO_3(aq) + 2HCl(aq) \rightarrow 2NaCl(aq) + H_2O(l) + CO_2(g)$

$CO_3^{2-}(aq) + 2H^+(aq) \rightarrow H_2O(l) + CO_2(g)$

(c) $CuBr_2(aq) + 2NaOH(aq) \rightarrow Cu(OH)_2(s) + 2NaBr(aq)$

$Cu^{2+}(aq) + 2OH^-(aq) \rightarrow Cu(OH)_2(s)$

4.33 The driving force in a metathesis reaction is the formation of a product which removes ions from solution. This means that at least some reactant species present as ions exist in a different form in the products, as indicated by the net ionic equation.

Driving forces in Exercise 4.32:
(a) $H_2O(l)$ -- nonelectrolyte
(b) $H_2O(l)$, $CO_2(g)$ -- nonelectrolyte and gas, respectively
(c) $Cu(OH)_2(s)$ -- precipitate

4.34 There are many examples for each of these driving forces.

(a) $AgNO_3(aq) + NaCl(aq) \rightarrow$ **$AgCl(s)$** $+ NaNO_3(aq)$
(b) $NaHCO_3(aq) + HCl(aq) \rightarrow NaCl(aq) + H_2O(l) +$ **$CO_2(g)$**
(c) $HNO_2 + KOH(aq) \rightarrow KNO_2(aq) +$ **$H_2O(l)$**
(or any neutralization reaction where $H_2O(l)$ is a product)

4.35 Follow the rules in Table 4.3.

(a) Ag_2**S** -- insoluble (b) Ca**Cl_2** -- soluble
(c) Cs**Br** -- soluble (d) Sr**CO_3** -- insoluble
(e) $(NH_4)_2SO_3$ -- although not specifically stated in Table 4.3, NH_4^+ salts of most anions are soluble. Note that NH_4^+ salts are exceptions to the insolubility of most sulfides, carbonates and phosphates.

4.36 According to Table 4.3:

(a) K_3PO_4 -- soluble
(b) $Pb(C_2H_3O_2)_2$ -- soluble
(c) $Ga(OH)_3$ -- insoluble
(d) NaCN -- soluble; cations in group IA (including Na^+) produce soluble salts of most anions
(e) $BaSO_4$ -- insoluble

4.37 Pb^{2+} is not present or an insoluble hydroxide would have formed. $BaSO_4$ is insoluble and $Ba(OH)_2$ is soluble, so the solution must contain Ba^{2+}. It could also contain K^+, but since we are dealing with a single salt, we will assume that only Ba^{2+} is present.

4.38 Br^- and NO_3^- can be ruled out because their Ba^{2+} salts are soluble. (Actually all NO_3^- salts are soluble.) SO_4^{2-} forms insoluble salts with the three cations given; it must be the anion in question.

4.39 (a) $ZnS(s) + 2H^+(aq) \rightarrow H_2S(g) + Zn^{2+}(aq)$

(b) $Ba^{2+}(aq) + CO_3^{2-}(aq) \rightarrow BaCO_3(s)$

(c) $3H^+(aq) + PO_4^{3-}(aq) \rightleftharpoons H_3PO_4(aq)$ (H_3PO_4 is a relatively weak acid.)

(d) $H^+(aq) + OH^-(aq) \rightarrow H_2O(l)$

(e) $Sr^{2+}(aq) + SO_4^{2-}(aq) \rightarrow SrSO_4(s)$

(f) $SO_3^{2-}(aq) + 2H^+(aq) \rightleftharpoons H_2SO_3(aq) \rightleftharpoons H_2O(l) + SO_2(g)$

(g) $Pb^{2+}(aq) + H_2S(aq) \rightarrow PbS(s) + 2H^+(aq)$

(h) $Fe(OH)_3(s) + 3H^+(aq) \rightarrow Fe^{3+}(aq) + 3H_2O(l)$

4.40 (a) $Ba^{2+}(aq) + HSO_4^-(aq) \rightarrow BaSO_4(s) + H^+(aq)$

(b) No reaction. Both components are soluble electrolytes, as are possible metathesis reaction products.

(c) $2Ag^+(aq) + CO_3^{2-}(aq) \rightarrow Ag_2CO_3(s)$

(d) $H^+(aq) + OH^-(aq) \rightarrow H_2O(l)$

(e) $HC_2H_3O_2(aq) + OH^-(aq) \rightarrow H_2O(l) + C_2H_3O_2^-(aq)$

(f) $K_2SO_3(s) + 2H^+(aq) \rightarrow K^+(aq) + H_2O(l) + SO_2(g)$

(g) $Pb^{2+}(aq) + SO_4^{2-}(aq) \rightarrow PbSO_4(s)$

CHAPTER 4
Aqueous Reactions and Solution Stoichiometry

Reactions of Metals

4.41 The most easily oxidized metals are near the bottom of groups on the left side of the chart, especially groups 1A and 2A. The least easily oxidized metals are on the lower right of the transition metals, particularly those near the bottom of groups 8B and 1B.

4.42 (a) Platinum and gold are called the noble metals because they are especially unreactive and difficult to oxidize.

(b) The alkali and alkaline earth metals are called active because they are very easily oxidized and chemically reactive.

4.43 (a) $2HCl(aq) + Ni(s) \rightarrow NiCl_2(aq) + H_2(g)$; $Ni(s) + 2H^+(aq) \rightarrow Ni^{2+}(aq) + H_2(g)$

(b) $H_2SO_4(aq) + Fe(s) \rightarrow FeSO_4(aq) + H_2(g)$; $Fe(s) + 2H^+(aq) \rightarrow Fe^{2+}(aq) + H_2(g)$

(c) $2HBr(aq) + Zn(s) \rightarrow ZnBr_2(aq) + H_2(g)$; $Zn(s) + 2H^+(aq) \rightarrow Zn^{2+}(aq) + H_2(g)$

(d) $2HC_2H_3O_2(aq) + Mg(s) \rightarrow Mg(C_2H_3O_2)_2(aq) + H_2(g)$;

$Mg(s) + 2HC_2H_3O_2(aq) \rightarrow Mg^{2+}(aq) + 2C_2H_3O_2^-(aq) + H_2(g)$

4.44 (a) $Mn(s) + H_2SO_4(aq) \rightarrow MnSO_4(aq) + H_2(g)$; $Mn(s) + 2H^+(aq) \rightarrow Mn^{2+}(aq) + H_2(g)$

(b) $2Cr(s) + 6HBr(aq) \rightarrow 2CrBr_3(aq) + 3H_2(g)$; $2Cr(s) + 6H^+(aq) \rightarrow 2Cr^{3+}(aq) + 3H_2(g)$

(c) $Sn(s) + 2HCl(aq) \rightarrow SnCl_2(aq) + H_2(g)$; $Sn(s) + 2H^+(aq) \rightarrow Sn^{2+}(aq) + H_2(g)$

(d) $2Al(s) + HCHO_2(aq) \rightarrow 2Al(CHO_2)_3(aq) + 3H_2(g)$;

$2Al(s) + 6HCHO_2(aq) \rightarrow 2Al^{3+}(aq) + 6CHO_2^-(aq) + 3H_2(g)$

4.45 (a) $2Al(s) + 3NiCl_2(aq) \rightarrow 2AlCl_3(aq) + 3Ni(s)$

(b) $Ag(s) + Pb(NO_3)_2(aq) \rightarrow$ no reaction

(c) $2Cr(s) + 3NiSO_4(aq) \rightarrow Cr_2(SO_4)_3(aq) + 3Ni(s)$

(d) $Mn(s) + 2HBr(aq) \rightarrow MnBr_2(aq) + H_2(g)$

(e) $H_2(g) + CuCl_2(aq) \rightarrow Cu(s) + 2HCl(aq)$

(f) $Ba(s) + 2H_2O(l) \rightarrow Ba(OH)_2(aq) + H_2(g)$

(The most active metals can displace H^+ from H_2O as well as from acids.)

4.46 (a) $Zn(s) + 2AgNO_3(aq) \rightarrow 2Ag(s) + Zn(NO_3)_2(aq)$

(b) $Fe(s) + Al_2(SO_4)_3(aq) \rightarrow$ no reaction

(c) $Co(s) + 2HCl(aq) \rightarrow CoCl_2(aq) + H_2(g)$

(d) $FeCl_2(aq) + H_2(g) \rightarrow$ no reaction

(e) $2Li(s) + 2H_2O(l) \rightarrow 2LiOH(aq) + H_2(g)$

(The most active metals can displace H^+ from H_2O as well as from acids.)

4.47 (a) i. $Zn(s) + Cd^{2+}(aq) \rightarrow Cd(s) + Zn^{2+}(aq)$

ii. $Cd(s) + Ni^{2+}(aq) \rightarrow Ni(s) + Cd^{2+}(aq)$

(b) According to Table 4.5, the most active metals are most easily oxidized, and Zn is more active than Ni. Observation (i) indicates that Cd is less active than Zn; (ii) indicates that Cd is more active than Ni. Cd is between Zn and Ni on the activity series.

(c) Place an iron strip in $CdCl_2(aq)$. If Cd(s) is deposited, Cd is less active than Fe; if there is no reaction, Cd is more active than Fe. Do the same test with Co if Cd is less active than Fe or with Cr if Cd is more active than Fe.

4.48 (a) $Br_2 + 2NaI \rightarrow 2NaBr + I_2$ indicates that Br_2 is more easily reduced than I_2. $Cl_2 + 2NaBr \rightarrow 2NaCl + Br_2$ shows that Cl_2 is more easily reduced than Br_2. The order for ease of reduction is $Cl_2 > Br_2 > I_2$. Conversely, the order for ease of oxidation is $I^- > Br^- > Cl^-$.

(b) Since the halogens are nonmetals, they tend to form anions when they react chemically. Nonmetallic character decreases going down a family and so does the tendency to gain electrons during a chemical reaction. Thus, the ease of reduction of the halogen, X_2, decreases going down the family and the ease of oxidation of the halide, X^-, increases going down the family.

(c) $Cl_2 + 2KI \rightarrow 2KCl + I_2$; $Br_2 + LiCl \rightarrow$ no reaction

4.49 Gold is less active than hydrogen, so it will not be oxidized by H^+ from HCl or HNO_3. Since Cl^- cannot be reduced any further, HCl is not an oxidizing acid and by itself cannot react with Au. HNO_3 is an oxidizing acid, but NO_3^- alone is not a strong enough oxidizing agent to drive the reaction to the right. However, when NO_3^- and Cl^- are both present, NO_3^- initiates the oxidation of Au to Au^{3+} to form a very stable $AuCl_4^-$ anion. The stability of the $AuCl_4^-$ anion is the "driving force" for the oxidation of Au in aqua regia (see the Aura of Gold in Section 4.6).

4.50 No. In equimolar solutions of the two acids, the H^+concentration is greater in the HCl(aq) solution because HCl is a strong acid and HF is not. Thus, HCl would be more effective at dissolving glass if the process were due to the action of H^+. The ability of HF(aq) to dissolve glass must be due to the properties of $F^-(aq)$ or undissociated HF(aq).

Solution Stoichiometry; Titrations

4.51 (a) Write the balanced equation for the reaction in question:
$HClO_4(aq) + NaOH(aq) \rightarrow NaClO_4(aq) + H_2O(l)$

Calculate the moles of the known substance, in this case NaOH.

$$\text{moles NaOH} = M \times L = \frac{0.0875 \text{ mol NaOH}}{1 \text{ L}} \times 0.0500 \text{ L} = 0.00438 \text{ mol NaOH}$$

Apply the mole ratio (moles unknown/moles known) from the chemical equation.

$$0.00438 \text{ mol NaOH} \times \frac{1 \text{ mol } HClO_4}{1 \text{ mol NaOH}} = 0.00438 \text{ mol } HClO_4$$

Calculate the desired quantity of unknown, in this case the volume of 0.105 M $HClO_4$ solution.

$$L = \frac{\text{mol}}{M};\ L = 0.00438 \text{ mol } HClO_4 \times \frac{1 \text{ L}}{0.105 \text{ mol } HClO_4} = 0.0417 \text{ L} = 41.7 \text{ mL}$$

(b) Following the procedure outlined in part (a):
$2HCl(aq) + Mg(OH)_2(s) \rightarrow MgCl_2(aq) + 2H_2O(l)$

$$2.87 \text{ g } Mg(OH)_2 \times \frac{1 \text{ mol } Mg(OH)_2}{58.33 \text{ g } Mg(OH)_2} = 0.0492 \text{ mol } Mg(OH)_2$$

$$0.0492 \text{ mol } Mg(OH)_2 \times \frac{2 \text{ mol HCl}}{1 \text{ mol } Mg(OH)_2} = 0.0984 \text{ mol HCl}$$

$$L = \frac{\text{mol}}{M} = 0.0984 \text{ mol HCl} \times \frac{1 \text{ L HCl}}{0.158 \text{ mol HCl}} = 0.623 \text{ L} = 623 \text{ mL}$$

(c) $AgNO_3(aq) + KCl(aq) \rightarrow AgCl(s) + KNO_3(aq)$

$$895 \text{ mg KCl} \times \frac{1 \times 10^{-3} \text{ g}}{1 \text{ mg}} \times \frac{1 \text{ mol KCl}}{74.55 \text{ g KCl}} \times \frac{1 \text{ mol } AgNO_3}{1 \text{ mol KCl}} = 0.0120 \text{ mol } AgNO_3$$

$$M = \frac{\text{mol}}{L} = \frac{0.0120 \text{ mol } AgNO_3}{0.0258 \text{ L}} = 0.465 \text{ M } AgNO_3$$

(d) $HCl(aq) + KOH(aq) \rightarrow KCl(aq) + H_2O(l)$

$$\frac{0.117 \text{ mol HCl}}{1 \text{ L}} \times 0.0358 \text{ L} \times \frac{1 \text{ mol KOH}}{1 \text{ mol HCl}} \times \frac{56.11 \text{ g KOH}}{1 \text{ mol KOH}} = 0.235 \text{ g KOH}$$

4.52 (a) $2HCl(aq) + Ba(OH)_2(aq) \rightarrow BaCl_2(aq) + 2H_2O(l)$

$$\frac{0.101 \text{ mol } Ba(OH)_2}{1 \text{ L } Ba(OH)_2} \times 0.0350 \text{ L } Ba(OH)_2 \times \frac{2 \text{ mol HCl}}{1 \text{ mol } Ba(OH)_2}$$

$$\times \frac{1 \text{ L HCl}}{0.210 \text{ mol HCl}} = 0.0337 \text{ L or } 33.7 \text{ mL HCl soln}$$

(b) $H_2SO_4(aq) + 2NaOH(aq) \rightarrow Na_2SO_4(aq) + 2H_2O(l)$

$$75.0 \text{ g NaOH} \times \frac{1 \text{ mol NaOH}}{40.00 \text{ g NaOH}} \times \frac{1 \text{ mol } H_2SO_4}{2 \text{ mol NaOH}} \times \frac{1 \text{ L } H_2SO_4}{3.50 \text{ mol } H_2SO_4}$$

$$= 0.268 \text{ L or } 268 \text{ mL } H_2SO_4 \text{ soln}$$

(c) $BaCl_2(aq) + Na_2SO_4(aq) \rightarrow BaSO_4(s) + 2NaCl(aq)$

$$544 \text{ mg} = 0.544 \text{ g } BaSO_4 \times \frac{1 \text{ mol } Na_2SO_4}{142.1 \text{ g } Na_2SO_4} \times \frac{1 \text{ mol } BaCl_2}{1 \text{ mol } BaSO_4} \times \frac{1}{0.0452 \text{ L}}$$

$$= 0.0847 \text{ M } BaCl_2$$

(d) $2HCl(aq) + Ca(OH)_2(aq) \rightarrow CaCl_2(aq) + 2H_2O(l)$

$$0.0427 \text{ L HCl} \times \frac{0.250 \text{ mol HCl}}{1 \text{ L HCl}} \times \frac{1 \text{ mol } Ca(OH)_2}{2 \text{ mol HCl}} \times \frac{74.10 \text{ g } Ca(OH)_2}{1 \text{ mol } Ca(OH)_2}$$

$$= 0.396 \text{ g } Ca(OH)_2$$

4.53 See Exercise 4.51(a) for a more detailed approach.

$$\frac{6.0 \text{ mol } H_2SO_4}{1 \text{ L}} \times 0.025 \text{ L} \times \frac{2 \text{ mol } NaHCO_3}{1 \text{ mol } H_2SO_4} \times \frac{84.01 \text{ g } NaHCO_3}{1 \text{ mol } NaHCO_3} = 25 \text{ g } NaHCO_3$$

4.54 See Exercise 4.51 (a) for a more detailed approach.

$$\frac{0.960 \text{ mol NaOH}}{1 \text{ L}} \times 0.0349 \text{ L} \times \frac{1 \text{ mol } HC_2H_3O_2}{1 \text{ mol NaOH}} \times \frac{60.05 \text{ g } HC_2H_3O_2}{1 \text{ mol } HC_2H_3O_2}$$

$$= 0.201 \text{ g } HC_2H_3O_2 \text{ in } 2.50 \text{ mL}$$

$$1.00 \text{ qt vinegar} \times \frac{1 \text{ L}}{1.057 \text{ qt}} \times \frac{1000 \text{ mL}}{1 \text{ L}} \times \frac{0.201 \text{ g } HC_2H_3O_2}{2.50 \text{ mL vinegar}} = 76.1 \text{ g } HC_2H_3O_2 / \text{qt}$$

4.55 The neutralization reaction here is:

$2HBr(aq) + Ca(OH)_2(aq) \rightarrow CaBr_2(aq) + 2H_2O(l)$

$$0.0488 \text{ L HBr soln} \times \frac{5.00 \times 10^{-2} \text{ mol HBr}}{1 \text{ L soln}} \times \frac{1 \text{ mol } Ca(OH)_2}{2 \text{ mol HBr}} \times \frac{1}{0.100 \text{ L of } Ca(OH)_2}$$

$$= 1.22 \times 10^{-2} \text{ M } Ca(OH)_2$$

From the molarity of the saturated solution, we can calculate the gram solubility of $Ca(OH)_2$ in 100 mL of H_2O.

$$0.100 \text{ L soln} \times \frac{1.22 \times 10^{-2} \text{ mol } Ca(OH)_2}{1 \text{ L soln}} \times \frac{74.10 \text{ g } Ca(OH)_2}{1 \text{ mol } Ca(OH)_2}$$

$$= 0.0904 \text{ g } Ca(OH)_2 \text{ in } 100 \text{ mL soln}$$

4.56 The balanced equation for the titration is:

$$Sr(NO_3)_2(aq) + Na_2CrO_4(aq) \rightarrow SrCrO_4(s) + 2NaNO_3(aq)$$

Beginning with a 0.100 L sample, we can do the following conversions:

volume soln $\rightarrow$ g $Sr(NO_3)_2$ $\rightarrow$ mol $Sr(NO_3)_2$ $\rightarrow$ mol Na_2CrO_4 $\rightarrow$ vol Na_2CrO_4 soln

$$0.100 \text{ L soln} \times \frac{6.67 \text{ g Sr(NO}_3)_2}{0.750 \text{ L soln}} \times \frac{1 \text{ mol Sr(NO}_3)_2}{211.6 \text{ g Sr(NO}_3)_2} \times \frac{1 \text{ mol Na}_2\text{CrO}_4}{1 \text{ mol Sr(NO}_3)_2}$$

$$\times \frac{1 \text{ L soln}}{0.0460 \text{ mol Na}_2\text{CrO}_4} = 0.0914 \text{ L Na}_2\text{CrO}_4 \text{ soln}$$

Additional Exercises

4.57 (a) $0.050 \text{ L soln} \times \frac{0.200 \text{ mol NaCl}}{1 \text{ L soln}} = 1.00 \times 10^{-2} \text{ mol NaCl}$

$0.100 \text{ L soln} \times \frac{0.100 \text{ mol NaCl}}{1 \text{ L soln}} = 1.00 \times 10^{-2} \text{ mol NaCl}$

Total moles NaCl = 2.00×10^{-2}, total volume = 0.050 L + 0.100 L = 0.150 L

$\text{Molarity} = \frac{2.00 \times 10^{-2} \text{ mol}}{0.150 \text{ L}} = 0.133 \text{ M}$

(b) $0.0245 \text{ L soln} \times \frac{1.50 \text{ mol NaOH}}{1 \text{ L soln}} = 0.0368 \text{ mol NaOH}$

$0.0205 \text{ L soln} \times \frac{0.850 \text{ mol NaOH}}{1 \text{ L soln}} = 0.0174 \text{ mol NaOH}$

Total moles NaOH = 0.0542, total volume = 0.0450 L

$\text{Molarity} = \frac{0.0542 \text{ mol NaOH}}{0.0450 \text{ L}} = 1.20 \text{ M}$

4.58 $\frac{6.00 \text{ mol HCl}}{1 \text{ L}} \times 0.0500 \text{ L} = 0.300 \text{ mol HCl}$

$\frac{1.00 \text{ mol HCl}}{1 \text{ L}} \times 0.1000 \text{ L} = 0.100 \text{ mol HCl}$

$M = \frac{\text{total mol HCl}}{0.250 \text{ L solution}} = \frac{0.400 \text{ mol HCl}}{0.250 \text{ L solution}} = 1.60\ M \text{ HCl}$

4.59 (a) $\frac{50 \text{ pg}}{1 \text{ mL}} \times \frac{1 \times 10^{-12} \text{ g}}{1 \text{ pg}} \times \frac{1 \times 10^{3} \text{ mL}}{\text{L}} \times \frac{1 \text{ mol Na}}{23.0 \text{ g Na}} = 2.2 \times 10^{-9}\ M$

(b) $\frac{2.2 \times 10^{-9} \text{ mol Na}}{1 \text{ L soln}} \times \frac{1 \text{ L}}{1 \times 10^{3} \text{ cm}^3} \times \frac{6.02 \times 10^{23} \text{ Na atom}}{1 \text{ mol Na}}$

$= 1.3 \times 10^{12}$ Na atoms or Na^+ ions/cm^3

4.60 (a,b) Expt. 1 No reaction.

Expt. 2 $2Ag^+(aq) + CrO_4^{2-}(aq) \rightarrow Ag_2CrO_4(s)$ red precipitate

Expt. 3 No reaction

Expt. 4 $2Ag^+(aq) + C_2O_4^{2-}(aq) \rightarrow AgC_2O_4(s)$ white precipitate

Expt. 5 $Ca^{2+}(aq) + C_2O_4^{2-}(aq) \rightarrow CaC_2O_4(s)$ white precipitate

Expt. 6 $Ag^+(aq) + Cl^-(aq) \rightarrow AgCl(s)$ white precipitate

(c) The silver salts of both ions are insoluble, but many silver salts are insoluble (Expt. 6). The calcium salt of CrO_4^{2-} is soluble (Expt. 3), while the calcium salt of $C_2O_4^{2-}(aq)$ is insoluble (Expt. 5). Thus, chromate salts appear more soluble than oxalate salts.

4.61 (a) $Al(OH)_3(s) + 3H^+(aq) \rightarrow Al^{3+}(aq) + 3H_2O(l)$

(b) $Mg(OH)_2(s) + 2H^+(aq) \rightarrow Mg^{2+}(aq) + 2H_2O(l)$

(c) $MgCO_3(s) + 2H^+(aq) \rightarrow Mg^{2+}(aq) + H_2O(l) + CO_2(g)$

(d) $NaAl(CO)_3(OH)_2(s) + 4H^+(aq) \rightarrow Na^+(aq) + Al^{3+}(aq) + 3H_2O(l) + CO_2(g)$

(e) $CaCO_3(s) + 2H^+(aq) \rightarrow Ca^{2+}(aq) + H_2O(l) + CO_2(g)$

[In (c), (d), and (e), one could also write the equation for formation of bicarbonate, e.g., $MgCO_3(s) + H^+(aq) \rightarrow Mg^{2+} + HCO_3^-(aq)$.]

4.62 The two precipitates formed are due to $AgCl(s)$ and $SrSO_4(s)$. Since no precipitate forms on addition of hydroxide ion to the remaining solution, the other two possibilities, Ni^{2+} and Mn^{2+}, are absent.

4.63 (a) $2\,Ti^{4+}(aq) + Zn(s) \rightarrow 2\,Ti^{3+}(aq) + Zn^{2+}(aq)$

The coefficients are required because the number of electrons lost by Zn ($2\,e^-$) must equal the number of electrons gained by Ti^{+4} ($2 \times 1\,e^-$).

(b) This reaction does not indicate the position of Ti on the activity series in Table 4.5. All reactions in Table 4.5 involve oxidation of an elemental metal to a metal ion or compound. This reaction does not produce Ti metal and thus cannot be compared to the other reactions.

4.64 A metal on Table 4.5 is able to displace the metal cations below it from their compounds. That is, zinc will reduce the cations below it to their metals.

(a) $Zn(s) + Na^+(aq) \rightarrow$ no reaction

(b) $Zn(s) + Pb^{2+}(aq) \rightarrow Zn^{2+}(aq) + Pb^{\circ}(s)$

(c) $Zn(s) + Mg^{2+}(aq) \rightarrow$ no reaction

(d) $Zn(s) + Fe^{2+}(aq) \rightarrow Zn^{2+}(aq) + Fe^{\circ}(s)$

(e) $Zn(s) + Cu^{2+}(aq) \rightarrow Zn^{2+}(aq) + Cu^{\circ}(s)$

(f) $Zn(s) + Al^{3+}(aq) \rightarrow$ no reaction

Aqueous Reactions and Solution Stoichiometry

4.65 (a) A - La_2O_3 Metals often react with the oxygen in air to produce metal oxides.

B - $La(OH)_3$ When metals react with water (HOH) to form H_2, OH^- remains.

C - $LaCl_3$ Most chlorides are soluble.

D - $La_2(SO_4)_3$ Sulfuric acid provides SO_4^{2-} ions.

(b) $4La(s) + 3\,O_2(g) \rightarrow 2La_2O_3(s)$

$2La(s) + 6HOH(l) \rightarrow 2La(OH)_3(s) + 3H_2(g)$

(There are no spectator ions in either of these reactions.)

molecular: $La_2O_3(s) + 6HCl(aq) \rightarrow 2LaCl_3(aq) + 3H_2O(l)$

net ionic: $La_2O_3(s) + 6H^+(aq) \rightarrow 2La^{3+}(aq) + 3H_2O(l)$

molecular: $La(OH)_3(s) + 3HCl(aq) \rightarrow LaCl_3(aq) + 3H_2O(l)$

net ionic: $La(OH)_3(s) + 3H^+(aq) \rightarrow La^{3+}(aq) + 3H_2O(l)$

molecular: $2LaCl_3(aq) + 3H_2SO_4(aq) \rightarrow La_2(SO_4)_3(s) + 6HCl(aq)$

net ionic: $2La^{3+}(aq) + 3SO_4^{2-}(aq) \rightarrow La_2(SO_4)_3(s)$

(c) La metal is oxidized by water to produce $H_2(g)$, so La is definitely above H on the activity series. [In fact, since an acid is not required to oxidize La, it is probably one of the more active metals. We know Al is not oxidized by water, or it could ot be used in cookware. La is somewhere above Al on the activity series.]

4.66 Since platinum metal is less active than hydrogen, it will not be oxidized by H^+ from either acid. Cl^- is not an oxidizing agent, and NO_3^- alone is not a strong enough oxidizing agent to effectively oxidize Pt metal to Pt^{4+}. In a mixture of the two acids, NO_3^- initiates the oxidation and Cl^- stabilzes the Pt^{4+} ion produced. The stabile $PtCl_6^{2-}$ effectively removes Pt^{4+} from solution and its formation is the driving force for the oxidation reaction (see Exercise 4.49).

4.67 $H_2C_4H_4O_6 + 2OH^-(aq) \rightarrow C_4H_4O_6^{2-}(aq) + 2H_2O(l)$

$$0.02262 \text{ L NaOH soln} = \frac{0.2000 \text{ mol NaOH}}{1 \text{ L}} \times \frac{1 \text{ mol } H_2C_4H_4O_6}{2 \text{ mol NaOH}} \times \frac{1}{0.04000 \text{ L } H_2C_4H_4O_6}$$

$$= 0.05655 \text{ M } H_2C_4H_4O_6 \text{ soln}$$

4.68 $0.0250 \text{ L soln} \times \frac{0.102 \text{ mol } Ag^+}{1 \text{ L soln}} \times \frac{1 \text{ mol } Ag_3AsO_4}{3 \text{ mol } Ag^+} \times \frac{1 \text{ mol As}}{1 \text{ mol } Ag_3AsO_4} \times \frac{74.92 \text{ g As}}{1 \text{ mol As}}$
$= 0.0637 \text{ g As}$

$\text{weight percent} = \frac{0.0637 \text{ g As}}{1.22 \text{ g sample}} \times 100 = 5.22 \text{ \% As}$

4.69 (a) mol HCl initial - mol NH_3 from air = mol HCl remaining

= mol NaOH required for titration

mol NaOH = 0.0588 M $\times$ 0.0131 L = 7.70×10^{-4} mol NaOH

= 7.70×10^{-4} mol HCl remain

mol HCl initial - mol HCl remaining = mol NH_3 from air

(0.0105 M HCl $\times$ 0.100 L) - 7.70×10^{-4} mol HCl

= 10.5×10^{-4} mol HCl - 7.70×10^{-4} mol HCl = 2.8×10^{-4} mol NH_3

$2.8 \times 10^{-4} \text{ mol } NH_3 \times \frac{17.03 \text{ g } NH_3}{1 \text{ mol } NH_3} = 4.8 \times 10^{-3} \text{ g } NH_3$

(b) ppm is defined as mL NH_3 / 1×10^6 mL air. Calculate "mL" NH_3 from g NH_3 and density.

$4.8 \times 10^{-3} \text{ g } NH_3 \times \frac{1 \text{ L } NH_3}{0.771 \text{ g } NH_3} \times \frac{1000 \text{ mL}}{1 \text{ L}} = 6.2 \text{ mL } NH_3$

Calculate total volume of air processed.

$\frac{10.0 \text{ L}}{1 \text{ min}} \times 10.0 \text{ min} \times \frac{1000 \text{ mL}}{1 \text{ L}} = 1.00 \times 10^5 \text{ mL air}$

$\frac{6.2 \text{ mL } NH_3}{1.00 \times 10^5 \text{ mL air}} \times \frac{10 \text{ mL}}{10 \text{ mL}} = \frac{62 \text{ mL } NH_3}{1.00 \times 10^6 \text{ mL air}} = 62 \text{ ppm}$

(c) 62 ppm > 50 ppm. The manufacturer is **not** in compliance.

CHAPTER 5

Energy Relationships in Chemistry: Thermochemistry

Nature of Energy

5.1 (a) **Work** is a force applied over a distance.

(b) The amount of work done is the magnitude of the force times the distance over which it is applied. $w = F \times d$.

5.2 (a) **Heat** is energy that is transferred due to a difference in temperature.

(b) Heat is transferred from one system to another until the two systems are at the same temperature.

5.3 (a) Electrostatic repulsion; no work is done because the particles are held stationary.

(b) Gravity; work is done because the force of gravity is opposed and the paper clip is lifted.

5.4 (a) Magnetic attraction; work is done because the magnets are moved apart.

(b) Mechanical force; work is done because the force of the coiled spring is opposed as the spring is stretched over a distance.

5.5 (a) Since 1 J = 1 kg - m^2/s^2, convert g $\rightarrow$ kg to obtain E_k in joules.

$$E_k = 1/2\ mv^2 = 1/2 \times 45\ g \times \frac{1\ kg}{1000\ g} \times \left(\frac{61\ m}{1\ s}\right)^2 = \frac{84\ kg \cdot m^2}{1\ s^2} = 84\ J$$

(b) $84\ J \times \frac{1\ Cal}{4.184\ J} = 20\ Cal$

(c) As the ball hits the tree, its speed (and hence its kinetic energy) drops to zero. Most of the kinetic energy is transferred to the potential energy of a slightly deformed golf ball and some is absorbed by the tree. As the ball bounces off the tree, the potential energy is reconverted to kinetic energy.

Energy Relationships in Chemistry: Thermochemistry

5.6 (a) $m = 4.8 \times 10^3 \text{ lb} \times \frac{1 \text{ kg}}{2.20 \text{ lb}} = 2.2 \times 10^3 \text{ kg}$, $v = \frac{45 \text{ mi}}{1 \text{ hr}} \times \frac{1 \text{ hr}}{3600 \text{ s}} = \frac{20 \text{ m}}{1 \text{ s}}$

$$E_k = \frac{1}{2} \times 2.2 \times 10^3 \text{ kg} \times \left(\frac{20 \text{ m}}{1 \text{ s}}\right)^2 = 4.4 \times 10^5 \text{ J} \times \frac{1 \text{ kJ}}{1000 \text{ J}} = 4.4 \times 10^2 \text{ kJ}$$

(b) $4.4 \times 10^5 \text{ J} \times \frac{1 \text{ cal}}{4.184 \text{ J}} = 1.1 \times 10^3 \text{ cal}$

(c) It is converted to heat (mainly in the tires and brake pads) and some work as the brake pads and tires are worn in stopping.

5.7 Given: 1 Btu = 1 lb H_2O/1°F
Know: 1 cal = 1 g H_2O/1°C; 1 cal • °C = 1 g
1 cal = 4.184 J
Strategy: Btu → cal → J (lb/°F → g/°C → J)

This strategy also requires changing °F to °C. Since this involves the magnitude of a degree on each scale, rather than a specific temperature, the 32 in the temperature relationship is not needed.

100°C = 180° F; 9°F = 5°C

$$\frac{1 \text{ lb}}{1°\text{F}} \times \frac{453.6 \text{ g}}{1 \text{ lb}} \times \frac{9°\text{F}}{5°\text{C}} \times \frac{1 \text{ cal} \cdot °\text{C}}{1 \text{ g}} \times \frac{4.184 \text{ J}}{1 \text{ cal}} = 3416 \text{ J}$$

1 Btu = 3416 J

5.8 Given: 1 kwh
1 watt = 1 J/s; 1 watt • s = 1 J

Strategy: kwh → wh → ws → J

$$1 \text{ kwh} \times \frac{1000 \text{ w}}{1 \text{ kw}} \times \frac{60 \text{ min}}{\text{h}} \times \frac{60 \text{ s}}{\text{min}} \times \frac{1 \text{ J}}{1 \text{ w} \cdot \text{s}} = 3.6 \times 10^6 \text{ J}$$

1 kwh = 3.6×10^6 J

First Law of Thermodynamics

5.9 (a) In any chemical or physical change, energy can be neither created nor destroyed, but it can be changed in form.

(b) The well-defined part of the universe whose energy changes are being described.

(c) The energy change of a system is equal in magnitude and opposite in sign to that of its surroundings.

5.10 (a) $\Delta E = q + w$

(b) *Surroundings* is everything that is not part of the system.
Surroundings = universe - system.

(c) $\Delta E_{sys} = -\Delta E_{surr}$. Energy is transferred between the system and the surroundings. The amount of energy lost by the system must equal the amount of energy gained by the surroundings and vice versa.

5.11 We need in each case to evaluate q and w in the expression $\Delta E = q + w$.

(a) q is positive because the system gains heat; w is negative because the system does work on the surroundings. ΔE = +145 J - 167 J = -22 J. The internal energy of the system has decreased.

(b) ΔE = -3.0 kJ - 0.750 kJ = -3.8 kJ. (The correct answer has 1 decimal place). The internal energy of the system has decreased.

(c) q is negative because the system loses heat; w is positive because work is done on the system. ΔE = -128 J + 143 J = +15 J. The internal energy of the system has increased.

5.12 We need in each case to evaluate q and w in the expression $\Delta E = q + w$. If ΔE is positive, the interal energy of the system increases; if ΔE is negative, it decreases.

(a) q = 0; work done on the surroundings is negative. Thus ΔE = 0 - 450 J = -450 J. The internal energy of the system decreases.

(b) q is positive, because heat has come into the system, which is the 200 g water. w is essentially zero. Thus, ΔE = +8360 J. The internal energy of the system increases.

(c) q is negative, but w is positive. Thus, ΔE = -146 J + 300 J = +154 J. In this case the system gained more internal energy via the work done on it than it lost as heat outflow.

5.13 (a) A "state function" is a property of a system that depends only on the physical state (pressure, temperature, etc.) of the system, not on the route used by the system to get to the current state.

(b) Internal energy and enthalpy are state functions; work is not a state function.

(c) Temperature is a state function; regardless of how hot or cold the sample has been, the temperature depends only on its present condition.

5.14 (a) Independent. Potential energy is a state function.

(b) Dependent. Some of the energy released could be employed in performing work, as is done in the body when sugar is metabolized; heat is not a state function.

(c) Dependent. The work accomplished depends on whether the gasoline is used in an engine, burned in an open flame, or in some other manner. Work is not a state function.

Enthalpy

5.15 (a) exothermic (b) $-\Delta H$ (c) $\Delta H = q$

5.16 (a) $+\Delta H$ (b) endothermic (c) constant pressure

Energy Relationships in Chemistry: Thermochemistry

5.17 (a) exothermic (b) endothermic (c) exothermic (d) exothermic (e) endothermic

5.18 (a) $+\Delta H$. Boiling is endothermic.

(b) $-\Delta H$. Combustion is usually exothermic.

(c) $-\Delta H$. The reaction generated the energy necessary to raise the temperature of the solution. (In this case, the substances involved in the reaction are the system and the solvent is part of the surroundings.)

(d) $+\Delta H$. The decomposition of water requires a large energy input. (The reverse of this reaction is described at the end of Section 5.1.)

5.19 q = -135 kJ (heat is given off by the system), w = -63 kJ (work is done by the system).

$\Delta E = q + w = -135 \text{ kJ} - 63 \text{ kJ} = -198 \text{ kJ}$. $\Delta H = q = -135 \text{ kJ}$ (at constant pressure).

5.20 The gas is the system. If 600 J of heat is added, q = + 600 J. Work done by the system decreases the overall energy of the system, so w = -140 J.

$\Delta E = q + w = +600 \text{ J} - 140 \text{ J} = +460 \text{ J}$. At constant pressure, $\Delta H = q$; $\Delta H = +600 \text{ J}$.

5.21 (a) $CH_3OH(l) + 2\,O_2(g) \rightarrow 2H_2O(l) + CO_2(g)$

(b)

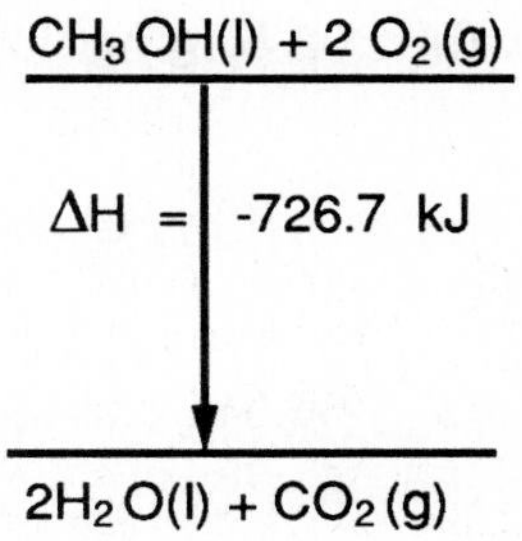

5.22 (a) $2NH_3(g) \rightarrow N_2(g) + 3H_2(g)$

$NH_3(g) \rightarrow 1/2\,N_2(g) + 3/2\,H_2(g)$

(b)

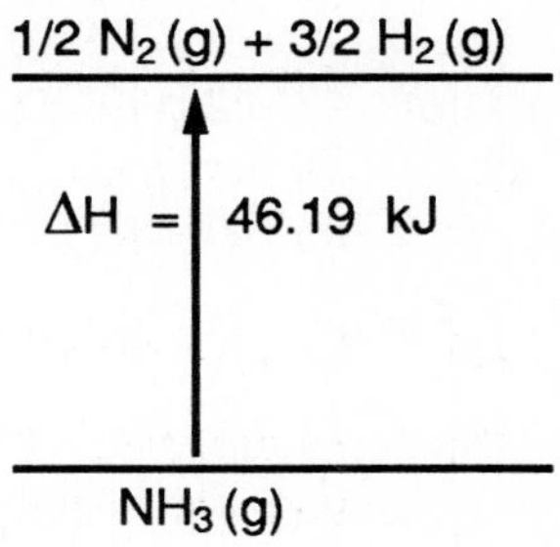

5.23 (a) Endothermic (ΔH is positive).

(b) $10.0 \text{ g } N_2O \times \dfrac{1 \text{ mol}}{44.02 \text{ g}} \times \dfrac{+163.2 \text{ kJ}}{2 \text{ mol } N_2O} = +18.5$ kJ of heat transferred (ΔH_{sys} increases)

(c) $1.00 \text{ kJ} \times \dfrac{2 \text{ mol } N_2}{163.2 \text{ kJ}} \times \dfrac{28.02 \text{ g } N_2}{1 \text{ mol } N_2} = 0.343 \text{ g } N_2$

(d) $15.0 \text{ g } N_2O \times \dfrac{1 \text{ mol}}{44.02 \text{ g}} \times \dfrac{-163.2 \text{ kJ}}{2 \text{ mol } N_2O} = 27.8$ kJ of heat produced (ΔH_{sys} decreases)

Energy Relationships in Chemistry: Thermochemistry

5.24 (a) Exothermic (ΔH is negative).

(b) $8.0 \text{ g Na} \times \frac{1 \text{ mol}}{22.99 \text{ g}} \times \frac{-821.8 \text{ kJ}}{2 \text{ mol Na}} = 1.4 \times 10^2 \text{ kJ}$

(143 kJ rounded to 2 significant figures)

(c) $10.0 \text{ kJ} \times \frac{2 \text{ mol NaCl}}{821.8 \text{ kJ}} \times \frac{58.44 \text{ g NaCl}}{1 \text{ mol NaCl}} = 1.42 \text{ g NaCl produced}$

(d) $2 \text{ NaCl(s)} \rightarrow 2\text{Na(s)} + Cl_2\text{(g)} \quad \Delta H = +821.8 \text{ kJ}$

This is the reverse of the reaction given above, so the sign of ΔH is reversed.

$25.0 \text{ g NaCl} \times \frac{1 \text{ mol NaCl}}{58.44 \text{ g NaCl}} \times \frac{821.8 \text{ kJ}}{2 \text{ mol NaCl}} = 176 \text{ kJ absorbed}$

5.25 (a) $0.200 \text{ mol AgCl} \times \frac{-65.5 \text{ kJ}}{\text{mol AgCl}} = -13.1 \text{ kJ}$

(b) $2.50 \text{ g AgCl} \times \frac{1 \text{ mol AgCl}}{143.3 \text{ g AgCl}} \times \frac{-65.5 \text{ kJ}}{\text{mol AgCl}} = -1.14 \text{ kJ}$

(c) $0.350 \text{ mol AgCl} \times \frac{-65.5 \text{ kJ}}{1 \text{ mol AgCl}} = +22.9 \text{ kJ}$ (sign of ΔH reversed)

5.26 (a) $0.345 \text{ mol } O_2 \times \frac{-89.4 \text{ kJ}}{3 \text{ mol } O_2} = -10.3 \text{ kJ}$

(b) $7.85 \text{ g KCl} \times \frac{1 \text{ mol KCl}}{74.55 \text{ g KCl}} \times \frac{-89.4 \text{ kJ}}{2 \text{ mol KCl}} = -4.71 \text{ kJ}$

(c) $9.22 \text{ g } KClO_3 \times \frac{1 \text{ mol } KClO_3}{122.6 \text{ g } KClO_3} \times \frac{-89.4 \text{ kJ}}{2 \text{ mol } KClO_3} = +3.36 \text{ kJ}$

5.27 Enthalpy of $H_2O(s) < H_2O(l) < H_2O(g)$. Heat must be added to convert s $\rightarrow$ l $\rightarrow$ g.

5.28 (a) O_3 has the higher enthalpy (ΔH is positive).

(b) Since the enthalpy of the product is higher than the enthalpy of the reactants, energy would have to be supplied for the reaction to occur. You would not expect it to proceed readily. (However, under certain circumstances (see Chapter 19), endothermic reactions can proceed readily.)

Energy Relationships in Chemistry: Thermochemistry

Calorimetry

The specific heat of water to four significant figures, **4.184 J/g • K**, will be used in many of the following exercises; temperature units of K and °C will be used interchangeably.

5.29 (a) $\frac{4.18\ J}{1\ g \bullet K}$ or $\frac{4.18\ J}{1\ g \bullet {}^\circ C}$ (b) $\frac{348\ g\ H_2O \times 4.18\ J}{1\ g \bullet {}^\circ C} = \frac{1.46 \times 10^3\ J}{{}^\circ C}$

(c) $2.06\ kg\ H_2O \times \frac{1000\ g}{1\ kg} \times \frac{4.184\ J}{1\ g \bullet {}^\circ C} \times \frac{1\ kJ}{1000\ J} \times 41.23\ {}^\circ C = 355\ kJ$

5.30 (a) $\frac{4.184\ J}{1\ g \bullet {}^\circ C} \times \frac{18.02\ g\ H_2O}{1\ mol\ H_2O} = \frac{75.40\ J}{mol \bullet {}^\circ C}$

(b) $6.35\ mol\ H_2O \times \frac{75.40\ J}{mol \bullet {}^\circ C} = \frac{479\ J}{{}^\circ C}$

(c) $165\ mol\ H_2O \times \frac{75.40\ J}{mol \bullet {}^\circ C} \times (47.32^\circ C - 10.55^\circ C) = 4.57 \times 10^5\ J = 457\ kJ$

5.31 $193\ g\ ethanol \times \frac{2.46\ J}{1\ g \bullet {}^\circ C} \times 16.00^\circ C = 7.60 \times 10^3\ J$

5.32 $382\ g\ Pb \times \frac{0.129\ J}{1\ g \bullet {}^\circ C} \times 14.7^\circ C = 724\ J$

5.33 Since the temperature of the water increases, the dissolving process is exothermic and the sign of ΔH is negative. The heat lost by the NaOH(s) dissolving equals the heat gained by the solution.

Calculate the heat gained by the solution. The temperature change is 37.8 - 21.6 = 16.2°C. The total mass of solution is (100.0 g H_2O + 6.50 g NaOH) = 106.5 g.

$106.5\ g\ solution \times 4.184\ \frac{J}{1g \bullet {}^\circ C} \times 16.2^\circ C \times \frac{1\ kJ}{1000\ J} = 7.22\ kJ$

This is the amount of heat lost when 6.50 g of NaOH dissolves.

The heat loss per mole NaOH is

$\frac{-7.22\ kJ}{6.50\ g\ NaOH} \times \frac{40.00\ g\ NaOH}{1\ mol\ NaOH} = -44.4\ kJ/mol \quad \Delta H = q_p = -44.4\ kJ/mol\ NaOH$

5.34 Following the logic in Exercise 5.33, the dissolving process is endothermic, ΔH is positive. The total mass of the solution is (60.0 g H_2O + 4.25 g NH_4NO_3) = 64.25 g. The temperature change of the solution is 22.0 - 16.9 = 5.10°C. The heat lost by the water is

$64.25\ g\ H_2O \times \frac{4.184\ J}{1\ g \bullet {}^\circ C} \times 5.10^\circ C \times \frac{1\ kJ}{1000\ J} = 1.37\ kJ$

Thus, 1.37 kJ is absorbed when 4.25 g NH_4NO_3(s) dissolves.

$\frac{1.37\ kJ}{4.25\ g\ NH_4NO_3} \times \frac{80.05\ g\ NH_4NO_3}{1\ mol\ NH_4NO_3} = \frac{+25.8\ kJ}{mol\ NH_4NO_3}$

5.35 $q_{bomb} = -q_{rxn}$; $\Delta T = 28.78°C - 21.36°C = 7.42°C$

$$q_{bomb} = \frac{11.66 \text{ kJ}}{1°C} \times 7.42°C = 86.5 \text{ kJ}$$

$$\Delta H_{rxn} = q_{rxn} = -q_{bomb} = \frac{-86.5 \text{ kJ}}{1.80 \text{ g } C_8H_{18}} = \frac{-48.1 \text{ kJ}}{1 \text{ g } C_8H_{18}}$$

$$\Delta H_{rxn} = \frac{-48.1 \text{ kJ}}{1 \text{ g } C_8H_{18}} \times \frac{114.2 \text{ g } C_8H_{18}}{1 \text{ mol } C_8H_{18}} = -5.49 \times 10^3 \frac{\text{kJ}}{1 \text{ mol } C_8H_{18}}$$

5.36 $q_{bomb} = -q_{rxn}$; $\Delta T = 30.57°C - 23.44°C = 7.13°C$

$$q_{bomb} = \frac{7.854 \text{ kJ}}{1°C} \times 7.13°C = 56.0 \text{ kJ}$$

$$\Delta H_{rxn} = q_{rxn} = -q_{bomb} = \frac{-56.0 \text{ kJ}}{2.20 \text{ g } C_6H_4O_2} = \frac{-25.4 \text{ kJ}}{1 \text{ g } C_6H_4O_2}$$

$$\Delta H_{rxn} = \frac{-25.4 \text{ kJ}}{1 \text{ g } C_6H_4O_2} \times \frac{108.1 \text{ g } C_6H_4O_2}{1 \text{ mol } C_6H_4O_2} = \frac{-2.75 \times 10^3 \text{ kJ}}{1 \text{ mol } C_6H_4O_2}$$

5.37 (a) $$C = 1.200 \text{ g } HC_7H_5O_2 \times \frac{26.38 \text{ kJ}}{1 \text{ g } HC_7H_5O_2} \times \frac{1}{3.65°C} = \frac{8.67 \text{ kJ}}{°C}$$

(b) $C_{total} = C_{H_2O} + C_{empty\ calorimeter}$

$$C_{H_2O} = 1.500 \text{ kg } H_2O \times \frac{4.184 \text{ kJ}}{1 \text{ kg} \cdot °C} = \frac{6.276 \text{ kJ}}{°C}$$

$$C_{empty\ calorimeter} = \frac{8.67 \text{ kJ}}{1°C} - \frac{6.287 \text{ kJ}}{1°C} = \frac{2.39 \text{ kJ}}{°C}$$

(c) $$q = 1.200 \text{ g } HC_7H_5O_2 \times \frac{26.38 \text{ kJ}}{1 \text{ g } HC_7H_5O_2} = 31.66 \text{ kJ produced}$$

$$C_{H_2O} = 1000 \text{ g } H_2O \times \frac{4.184 \text{ kJ}}{1 \text{ kg} \cdot °C} = \frac{4.184 \text{ kJ}}{°C} \; ; \; C_{empty\ calorimeter} = \frac{2.39 \text{ kJ}}{°C}$$

$$31.66 \text{ kJ} = \left(\frac{4.184 \text{ kJ}}{1°C} + \frac{2.39 \text{ kJ}}{1°C}\right) \times \Delta T; \; \Delta T = 4.82°C$$

5.38 (a) $$C_{total} = 2.500 \text{ g glucose} \times \frac{15.57 \text{ kJ}}{1 \text{ g glucose}} \times \frac{1}{2.70°C} = \frac{14.4 \text{ kJ}}{°C}$$

(b) $$C_{H_2O} = 2.700 \text{ kg } H_2O \times \frac{4.184 \text{ kJ}}{1 \text{ kg} \cdot °C} = \frac{11.30 \text{ kJ}}{°C}$$

$$C_{empty\ calorimeter} = \frac{14.4 \text{ kJ}}{1°C} - \frac{11.3 \text{ kJ}}{1°C} = \frac{3.1 \text{ kJ}}{°C}$$

(c) $q = 2.500 \text{ g glucose} \times \frac{15.57 \text{ kJ}}{1 \text{ g glucose}} = 38.93 \text{ kJ produced}$

$C_{H_2O} = 2.000 \text{ kg } H_2O \times \frac{4.184 \text{ kJ}}{1 \text{ kg} \cdot {}^\circ C} = \frac{8.368 \text{ kJ}}{{}^\circ C}$

$C_{total} = \frac{8.368 \text{ kJ}}{1^\circ C} + \frac{3.1 \text{ kJ}}{1^\circ C} = \frac{11.5 \text{ kJ}}{{}^\circ C}$

$38.98 \text{ kJ} = \frac{11.5 \text{ kJ}}{1^\circ C} \times \Delta T$; $\Delta T = 3.39^\circ C$

Hess's Law

5.39 If a reaction can be described as a series of steps, ΔH for the reaction is the sum of the enthalpy changes for each step. As long as we can describe a route where ΔH for each step is known, ΔH for any process can be calculated.

5.40 As a state function, ΔH is independent of path from reactants to products. Thus ΔH is the same whether the reaction occurs in one step or several steps. Hess's law says that ΔH for an overall process is the sum of the ΔH's for the individual steps.

5.41
$2S(s) + 3O_3(g) \rightarrow 2SO_3(g)$ $\Delta H = -790 \text{ kJ}$
$2SO_3(g) \rightarrow 2SO_2(g) + O_2(g)$ $\Delta H = +196 \text{ kJ}$

$2S(s) + 2O_2(g) \rightarrow 2SO_2(g)$ $\Delta H = +594 \text{ kJ}$
$S(s) + O_2(g) \rightarrow SO_2(g)$ $\Delta H = \frac{1}{2}(594 \text{ kJ}) = +297 \text{ kJ}$

5.42
$2H_2(g) + 2F_2(g) \rightarrow 4HF(g)$ $\Delta H = 2(-537 \text{ kJ})$
$2H_2O(l) \rightarrow O_2(g) + 2H_2(g)$ $\Delta H = -(-572 \text{ kJ})$

$2F_2(g) + 2H_2O(l) \rightarrow 4HF(g) + O_2(g)$ $\Delta H = -502 \text{ kJ}$

5.43
$C_2H_4(g) \rightarrow 2H_2(g) + 2C(s)$ $\Delta H = -52.3 \text{ kJ}$
$2C(s) + 4F_2(g) \rightarrow 2CF_4(g)$ $\Delta H = 2(-680) \text{ kJ}$
$2H_2(g) + 2F_2(g) \rightarrow 4HF(g)$ $\Delta H = 2(-537) \text{ kJ}$

$C_2H_4(g) + 6F_2(g) \rightarrow 2CF_4(g) + 4HF(g)$ $\Delta H = -2.49 \times 10^3 \text{ kJ}$

5.44
$2CO_2(g) + 3H_2O(l) \rightarrow C_2H_6(g) + \frac{7}{2} O_2(g)$ $\Delta H = \frac{1}{2}(+3120 \text{ kJ})$
$2C(s) + 2O_2(g) \rightarrow 2CO_{2(g)}$ $\Delta H = 2(-394)$
$3H_2(g) + \frac{3}{2} O_2(g) \rightarrow 3H_2O(l)$ $\Delta H = \frac{3}{2}(-572)$

$2C(s) + 3H_2(g) \rightarrow C_2H_6(g)$ $\Delta H = -86 \text{ kJ}$

Energy Relationships in Chemistry: Thermochemistry

Heats of Formation

5.45 (a) $1/2\ N_2(g) + 3/2\ H_2(g) \rightarrow NH_3(g)$ $\Delta H_f = -46.19$ kJ

(b) $1/2\ N_2(g) + 1/2\ O_2(g) + 1/2\ Cl_2(g) \rightarrow NOCl(g)$ $\Delta H_f = +52.6$ kJ

(c) $C(s) + 2H_2(g) + 1/2\ O_2(g) \rightarrow CH_3OH(l)$ $\Delta H_f = -238.6$ kJ

(d) $K(s) + 1/2\ Cl_2(g) + 3/2\ O_2(g) \rightarrow KClO_3(s)$ $\Delta H_f = -391.2$ kJ

5.46 (a) $2C(s) + 2H_2(g) + O_2(g) \rightarrow HC_2H_3O_2(l)$ $\Delta H_f = -487.0$ kJ

(b) $Na(s) + 1/2\ H_2(g) + C(s) + 3/2\ O_2(g) \rightarrow NaHCO_3(s)$ $\Delta H_f = -947.7$ kJ

(c) $N_2(g) + 2H_2(g) + 3/2\ O_2(g) \rightarrow NH_4NO_3(s)$ $\Delta H_f = -365.6$ kJ

(d) $Fe(s) + 3/2\ Cl_2(g) \rightarrow FeCl_3(s)$ $\Delta H_f = -400$ kJ

5.47 $\Delta H^\circ_{rxn} = \Delta H^\circ_f\ Al_2O_3(s) + 2\Delta H^\circ_f\ Fe(s) - \Delta H^\circ_f\ Fe_2O_3 - 2\Delta H^\circ_f\ Al(s)$

$\Delta H^\circ_{rxn} = (-1669.8\text{ kJ}) + 2(0) - (-822.16\text{ kJ}) - 2(0) = -847.6\text{ kJ}$

5.48 Use heats of formation to calculate ΔH for the combustion of butane.

$C_4H_{10}(l) + 13/2\ O_2(g) \rightarrow 4CO_2(g) + 5\ H_2O(l)$

$\Delta H^\circ_{rxn} = 4\Delta H^\circ_f CO_2(g) + 5\Delta H^\circ_f\ H_2O(l) - \Delta H^\circ_f\ C_4H_{10}(l)$

$\Delta H^\circ_{rxn} = 4(-393.5\text{ kJ}) + 5(-285.8\text{ kJ}) - (147.6\text{ kJ})$

$\Delta H^\circ_{rxn} = -3150.6\text{ kJ/mol } C_4H_{10}$

$$1.0\text{ g } C_4H_{10} \times \frac{1\text{ mol } C_4H_{10}}{58.123\text{ g } C_4H_{10}} \times \frac{-3150.6\text{ kJ}}{1\text{ mol } C_4H_{10}} = 54.206\text{ kJ/g}$$

5.49 (a) $\Delta H^\circ_{rxn} = \Delta H^\circ_f\ H_2O(g) + \Delta H^\circ_f\ NH_4CN(s) - 2\Delta H^\circ_f\ NH_3(g) - \Delta H^\circ_f\ CO(g)$

$= -241.8\text{ kJ} + 0.0 - 2(-46.19\text{ kJ}) - (-110.5\text{ kJ}) = -38.9\text{ kJ}$

(b) $\Delta H^\circ_{rxn} = \Delta H^\circ_f\ N_2O(g) + 2\Delta H^\circ_f\ H_2O(g) - \Delta H^\circ_f\ NH_4NO_3(s)$

$= 81.6\text{ kJ} + 2(-241.8\text{ kJ}) - (-365.6\text{ kJ}) = -36.4\text{ kJ}$

(c) $\Delta H^\circ_{rxn} = \Delta H^\circ_f\ P_4O_{10}(s) - \Delta H^\circ_f\ P_4O_6(s) - 2\Delta H^\circ_f\ O_2(g)$

$= -2940.1\text{ kJ} - (-1640.1\text{ kJ}) - 2(0) = -1300.0\text{ kJ}$

(d) $\Delta H^\circ_{rxn} = \Delta H^\circ_f\ POCl_3(l) + \Delta H^\circ_f\ KCl(s) - 3\Delta H^\circ_f\ PCl_3(l) - \Delta H^\circ_f\ KClO_3(s)$

$= 3(-597.0\text{ kJ}) + (-435.9\text{ kJ}) - 3(-319.6\text{ kJ}) - (-391.2\text{ kJ}) = -876.9\text{ kJ}$

5.50 (a) $\Delta H^\circ_{rxn} = 2\Delta H^\circ_f\ NO(g) + \Delta H^\circ_f\ Cl_2(g) - 2\Delta H^\circ_f\ NOCl(g)$

$= 2(+90.37\text{ kJ}) + 0 - 2(+52.6\text{ kJ}) = +75.5\text{ kJ}$

(b) $\Delta H^\circ_{rxn} = 2\Delta H^\circ_f NH_3(g) + \Delta H^\circ_f\ N_2(g) + 3\Delta H^\circ_f\ H_2(g)$

$= 2(-46.19\text{ kJ}) - 0 + 3(0) = -92.38\text{ kJ}$

(c) $\Delta H^\circ_{rxn} = \Delta H^\circ_f\ CH_4(g) + \Delta H^\circ_f\ CO_2(g) - \Delta H^\circ_f\ HC_2H_3O_2(l)$
$= -74.8\ kJ - 393.5\ kJ - (-487.0\ kJ) = +18.7\ kJ$

(d) $\Delta H^\circ_{rxn} = 4\Delta H^\circ_f\ NO(g) + 6\Delta H^\circ_f\ H_2O(g) - 4\Delta H^\circ_f\ NH_3(g) - 5\Delta H^\circ_f\ O_2(g)$
$= 4(+90.37\ KJ) + 6(-241.8\ kJ) - 4(-46.19\ kJ) - 5(0) = -904.6\ kJ$

5.51 $\Delta H^\circ_{rxn} = \Delta H^\circ_f\ Ca(OH)_2(s) + \Delta H^\circ_f\ C_2H_2(g) - 2\Delta H^\circ_f\ H_2O(l) - \Delta H^\circ_f\ CaC_2(s)$
$-127.2\ kJ = -986.2\ kJ + 226.7\ kJ - 2(-285.85\ kJ) - \Delta H^\circ_f\ CaC_2(s)$
ΔH°_f for $CaC_2(s) = -60.6\ kJ$

5.52 $\Delta H^\circ_{rxn} = 3\Delta H^\circ_f\ CO_2(g) + 3\Delta H^\circ_f\ H_2O(l) - \Delta H^\circ_f\ C_3H_6O(l)$
$-1790\ kJ = 3(-393.5\ kJ) + 3(-285.85\ kJ) - \Delta H^\circ_f\ C_3H_6O(l)$
$\Delta H^\circ_f\ C_3H_6O(l) = -248.0\ kJ$

5.53

$Mg(s) + 1/2\ O_2(g) \rightarrow MgO(s)$	$\Delta H^\circ = 1/2(-1203.6\ kJ)$
$MgO(s) + H_2O(l) \rightarrow Mg(OH)_2(s)$	$\Delta H^\circ = -(37.1\ kJ)$
$H_2(g) + 1/2\ O_2(g) \rightarrow H_2O(l)$	$\Delta H^\circ = 1/2(-571.7\ kJ)$
$Mg(s) + O_2(g) + H_2(g) \rightarrow Mg(OH)_2(s)$	ΔH°_f or $Mg(OH)_2(s) = -924.8\ kJ$

5.54

$2B(s) + 3/2\ O_2(g) \rightarrow B_2O_3(s)$	$\Delta H^\circ = 1/2(-2509.1\ kJ)$
$3H_2(g) + 3/2\ O_2(g) \rightarrow 3H_2O(l)$	$\Delta H^\circ = 3/2(-571.7\ kJ)$
$B_2O_3(s) + 3H_2O(l) \rightarrow B_2H_6(g) + 3\ O_2(g)$	$\Delta H^\circ = -(-2147.5\ kJ)$
$2B(s) + 3H_2(g) \rightarrow B_2H_6(g)$	$\Delta H^\circ_f = +35.4\ kJ$

Foods and Fuels

5.55 Calculate the Cal (kcal) due to each nutritional component of the potato chips, then sum.

$$15\ g\ carbohydrates \times \frac{17\ kJ}{1\ g\ carbohydrate} = 255 \text{ or } 2.6 \times 10^2\ kJ$$

$$1\ g\ protein \times \frac{17\ kJ}{1\ g\ protein} = 17 \text{ or } 0.2 \times 10^2\ kJ$$

$$10\ g\ fat \times \frac{38\ kJ}{1\ g\ fat} = 380 \text{ or } 3.8 \times 10^2\ kJ$$

total energy $= 2.6 \times 10^2\ kJ + 0.2 \times 10^2\ kJ + 3.8 \times 16^2\ kJ = 6.6 \times 10^2\ kJ$

$$6.6 \times 10^2\ kJ \times \frac{1\ kcal}{4.184\ kJ} \times \frac{1\ Cal}{1\ kcal} = 1.6 \times 10^2\ Cal$$

5.56 Calculate the mass % of carbohydrate, fat and protein in the peanut brittle.

total mass of brittle = 1.00 lb brittle $\times \frac{453.6 \text{ g}}{1 \text{ lb}}$ = 453.6 g brittle

$\frac{214 \text{ g carbohydrate}}{453.6 \text{ g brittle}} \times 100 = 47.2$ % carbohydrate

$\frac{146 \text{ g fat}}{453.6 \text{ g brittle}} \times 100 = 32.2$ % fat ; $\frac{79 \text{ g protein}}{453.6 \text{ g brittle}} \times 100 = 17.4$ % protein

Calculate the fuel value due to each component in the 50 g bar.

$$50 \text{ g brittle} \times \frac{0.472 \text{ g carbohydrate}}{1 \text{ g brittle}} \times \frac{17 \text{ kJ}}{1 \text{ g carbohydrate}} = 4.0 \times 10^2 \text{ kJ}$$

$$50 \text{ g brittle} \times \frac{0.322 \text{ g fat}}{1 \text{ g brittle}} \times \frac{38 \text{ kJ}}{1 \text{ g fat}} = 6.1 \times 10^2 \text{ kJ}$$

$$50 \text{ g brittle} \times \frac{0.174 \text{ g protein}}{1 \text{ g brittle}} \times \frac{17 \text{ kJ}}{1 \text{ g protein}} = 1.5 \times 10^2 \text{ kJ}$$

fuel value = 4.0×10^2 kJ + 6.1×10^2 kJ + 1.5×10^2 kJ = 11.6×10^2 kJ

$$11.6 \times 10^2 \text{ kJ} \times \frac{1 \text{ kcal}}{4.184 \text{ kJ}} \times \frac{1 \text{ Cal}}{1 \text{ kcal}} = 2.77 \times 10^2 = 277 \text{ Cal}$$

5.57 $16.0 \text{ g } C_6H_{12}O_6 \times \frac{1 \text{ mol } C_6H_{15}O_6}{180.2 \text{ g } C_6H_{12}O_6} \times \frac{2812 \text{ kJ}}{\text{mol } C_6H_{12}O_6} \times \frac{1 \text{ Cal}}{4.184 \text{ kJ}} = 59.7 \text{ Cal}$

5.58 $355 \text{ mL} \times \frac{1.0 \text{ g beer}}{1 \text{ mL}} \times \frac{0.037 \text{ g ethanol}}{1 \text{ g beer}} \times \frac{1 \text{ mol ethanol}}{46.1 \text{ g ethanol}} \times \frac{1371 \text{ kJ}}{1 \text{ mol ethanol}} \times \frac{1 \text{ Cal}}{4.184 \text{ kJ}}$

= 93 Cal

Notice that this is more than half the calorie content of ordinary beers (about 150 Cal/12 oz bottle), and nearly all the calorie content of so-called "light" beers.

5.59 (a) $C_3H_4(g) + 4O_2(g) \rightarrow 3CO_2(g) + 2H_2O(g)$

$\Delta H = 3(-393.5 \text{ kJ}) + 2(-241.8 \text{ kJ}) - (185.4 \text{ kJ}) = -1849.5 \text{ kJ}$

$$\frac{-1849.5 \text{ kJ}}{1 \text{ mol } C_3H_4} \times \frac{1 \text{ mol } C_3H_4}{40.065 \text{ g } C_3H_4} \times \frac{1000 \text{ g } C_3H_4}{1 \text{ kg } C_3H_4} = \frac{4.6162 \times 10^4 \text{ kJ}}{1 \text{ kg } C_3H_4}$$

(b) $C_3H_6(g) + \frac{9}{2} O_2(g) \rightarrow 3CO_2(g) + 3H_2O(g)$

$\Delta H = 3(-393.5 \text{ kJ}) + 3(-241.8 \text{ kJ}) - (20.4 \text{ kJ}) = -1926.3 \text{ kJ}$

$$\frac{-1926.3 \text{ kJ}}{1 \text{ mol } C_3H_6} \times \frac{1 \text{ mol } C_3H_6}{42.080 \text{ g } C_3H_6} \times \frac{1000 \text{ g } C_3H_6}{1 \text{ kg } C_3H_6} = \frac{4.5777 \times 10^4 \text{ kJ}}{1 \text{ kg } C_3H_6}$$

(c) $C_3H_8(g) + O_2(g) \rightarrow 3CO_2(g) + 4H_2O(g)$

$\Delta H = 3(-393.5 \text{ kJ}) + 4(-241.8 \text{ kJ}) - (-103.8 \text{ kJ}) = -2043.9 \text{ kJ}$

$$\frac{-2043.9 \text{ kJ}}{1 \text{ mol } C_3H_8} \times \frac{1 \text{ mol } C_3H_8}{44.096 \text{ g } C_3H_8} \times \frac{1000 \text{ g } C_3H_8}{1 \text{ kg } C_3H_8} = \frac{4.6351 \times 10^4 \text{ kJ}}{1 \text{ kg } C_3H_8}$$

These three substances yield nearly identical quantities of heat per unit mass, but propane is marginally higher than the other two.

5.60 For nitroethane:

$$\frac{1348 \text{ kJ}}{1 \text{ mol } C_2H_5NO_2} \times \frac{1 \text{ mol } C_2H_5NO_2}{75.072 \text{ g } C_2H_5NO_2} \times \frac{1.052 \text{ g } C_2H_5NO_2}{1 \text{ cm}^3} = 18.89 \text{ kJ/cm}^3$$

For ethanol:

$$\frac{1371 \text{ kJ}}{1 \text{ mol } C_2H_5OH} \times \frac{1 \text{ mol } C_2H_5OH}{46.069 \text{ g } C_2H_5OH} \times \frac{0.789 \text{ g } C_2H_5OH}{1 \text{ cm}^3} = 23.5 \text{ kJ/cm}^3$$

For diethylether:

$$\frac{2727 \text{ kJ}}{1 \text{ mol } (C_2H_5)_2O} \times \frac{1 \text{ mol } (C_2H_5)_2O}{74.124 \text{ g } (C_2H_5)O} \times \frac{0.714 \text{ g } (C_2H_5)_2O}{1 \text{ cm}^3} = 26.3 \text{ kJ/cm}^3$$

Thus, **diethyl ether** would provide the most energy per unit volume.

Additional Exercises

5.61 Heat must be removed from the system (the temperature decreases) when water freezes. If ΔH_{sys} is negative, the process is exothermic. Melting ice is the opposite of freezing; it must be endothermic. When ice melts, the temperature of the water increases, heat is gained by the system, the process is endothermic.

5.62 In the process described, one mole of solid CO_2 is converted to one mole of gaseous CO_2. The volume of the gas is much greater than the volume of the solid. Thus the system (that is, the mole of CO_2) must work against atmospheric pressure. To accomplish this work while maintaining a constant temperature requires the absorption of additional heat beyond that required to increase the internal energy of the CO_2.

5.63 $w = \Delta E - q$; $\Delta E = +1.32$ kJ, $q = +1.55$ kJ; $w = 1.32 \text{ kJ} - 1.55 \text{ kJ} = -0.23$ kJ

The negative sign for w indicates that work is done by the system on the surroundings.

5.64 $\Delta E = q + w = +38.95 \text{ kJ} - 2.47 \text{ kJ} = +36.48 \text{ kJ}$

$\Delta H = q_p = +38.95$ kJ

5.65 (a) Endothermic (ΔH is positive)

(b) The products have the higher enthalpy.

(c) $$5.00 \text{ g } C_2H_5OH \times \frac{1 \text{ mol } C_2H_5OH}{46.07 \text{ g } C_2H_5OH} \times \frac{-69.4 \text{ kJ}}{2 \text{ mol } C_2H_5OH} = 3.77 \text{ kJ}$$

(d) $$95.0 \text{ g } C_2H_5OH \times \frac{1 \text{ mol } C_2H_5OH}{46.07 \text{ g } C_2H_5OH} \times \frac{-69.4 \text{ kJ}}{2 \text{ mol } C_2H_5OH} = 71.6 \text{ kJ}$$

5.66 We need to find the heat capacity of 1000 gal H_2O.

$$C_{H_2O} = 1000 \text{ gal } H_2O \times \frac{4 \text{ qts}}{1 \text{ gal}} \times \frac{1 \text{ L}}{1.06 \text{ qt}} \times \frac{10^3 \text{ cm}^3}{1 \text{ L}} \times \frac{1 \text{ g}}{1 \text{ cm}^3} \times \frac{4.184 \text{ J}}{1 \text{ g} \cdot {}^\circ\text{C}}$$

$$= 1.58 \times 10^7 \text{ J/}^\circ\text{C} = 1.58 \times 10^4 \text{ kJ/}^\circ\text{C} \text{ ; then,}$$

$$\frac{1.58 \times 10^7 \text{ J}}{1^\circ\text{C}} \times \frac{1^\circ\text{C} \cdot \text{g}}{0.85 \text{ J}} \times \frac{1 \text{ kg}}{10^3 \text{ g}} \times \frac{1 \text{ brick}}{1.8 \text{ kg}} = 1.0 \times 10^4 \text{ bricks}$$

5.67 Ignoring the heat capacity of the cup, the heat lost by the Mo(s) is the heat gained by the water.

Calculate the heat gained by the water. $\Delta T_{H_2O} = 28.0^\circ C - 24.6^\circ C = 3.4^\circ C$

$$150 \text{ g } H_2O \times \frac{4.184 \text{ J}}{1 \text{ g} \cdot {}^\circ\text{C}} \times 3.4^\circ\text{C} = 2.1 \times 10^3 \text{ J} \quad \text{gained by } H_2O\text{, lost by Mo}$$

Calculate the specific heat of Mo(s). $\Delta T_{Mo} = 100^\circ C - 28.0^\circ C = 72^\circ C$

$$2.1 \times 10^3 \text{ J lost} \times \frac{1}{110 \text{ g Mo}} \times \frac{1}{72^\circ\text{C}} = \frac{0.27 \text{ J}}{\text{g} \cdot {}^\circ\text{C}}$$

5.68 $$q = -6.2^\circ\text{C} \times 100 \text{ g} \times \frac{4.184 \text{ J}}{1 \text{ g} \cdot {}^\circ\text{C}} = -2.6 \times 10^3 \text{ J} = -2.6 \text{ kJ}$$

The reaction as carried out involves only 0.050 mol of $CuSO_4$ and the stoichiometrically equivalent amount of KOH. On a molar basis,

$$\Delta H = \frac{-2.6 \text{ kJ}}{0.050 \text{ mol}} = -52 \text{ kJ} \quad \text{for the reaction as written}$$

5.69 (a) From the mass of benzoic acid that produces a certain temperature change, we can calculate the heat capacity of the calorimeter.

$$\frac{0.235 \text{ g benzoic acid}}{1.642^\circ\text{C change observed}} \times \frac{26.38 \text{ kJ}}{1 \text{ g benzoic acid}} = \frac{3.78 \text{ kJ}}{{}^\circ\text{C}}$$

Now we can use this experimentally determined heat capacity with the data for caffeine.

$$\frac{1.525^\circ\text{C rise}}{0.265 \text{ g caffeine}} \times \frac{3.78 \text{ kJ}}{1^\circ\text{C}} \times \frac{194.2 \text{ g caffeine}}{1 \text{ mol caffeine}} = 4.22 \times 10^3 \text{ kJ/mol caffeine}$$

(b) The overall uncertainty is approximately equal to the sum of the uncertainties due to each effect. The uncertainty in the mass measurement is 0.235/0.001 or 0.265/0.001, about 1 part in 235 or 1 part in 265. The uncertainty in the temperature measurements is 1.642/0.002 or 1.525/0.002, about 1 part in 820 or 1 part in 760. Thus the uncertainty in heat of combustion from each measurement is

$$\frac{4220}{235} = 18 \text{ kJ}; \ \frac{4220}{265} = 16 \text{ kJ}; \ \frac{4220}{820} = 5 \text{ kJ}; \ \frac{4220}{760} = 6 \text{ kJ}.$$

The sum of these uncertainties is 45 kJ. In fact, the overall uncertainty is less than this because independent errors in measurement do tend to partially cancel.

5.70 $$\frac{2.556°\text{C}}{3.556 \text{ g sample}} \times \frac{13.62 \text{ kJ}}{1°\text{C}} = \frac{9.790 \text{ kJ}}{1 \text{ g sample}}$$

Let us calculate the heat that would be evolved from a mass of impure sample equal to the molar mass of pure salicylic acid, 138.1 g/mol.

$$\frac{9.790 \text{ kJ}}{1 \text{ g}} \times \frac{138.1 \text{ g } C_7O_3H_6}{1 \text{ mol}} = \frac{1352 \text{ kJ}}{1 \text{ mol}}$$

This is less than the heat expected for pure acid, 3.00×10^3 kJ/mol. The percentage salicylic acid in the sample is then

$$\frac{1.35 \times 10^3}{3.00 \times 10^3} \times 100 = 45.0\%$$

The boric oxide contaminant is thus 55.0% by weight.

5.71 (a) $\Delta H° = 2\Delta H_f^\circ\ H_2O(l) + \Delta H_f^\circ\ O_2(g) - 2\Delta H_f^\circ\ H_2O_2(l)$

$= 2(-285.85 \text{ kJ}) + 0 - 2(-187.8 \text{ kJ})$

$\Delta H° = -196.1 \text{ kJ}$

(b) $\Delta H° = \Delta H_f^\circ\ CaCO_3(s) - \Delta H_f^\circ\ CaO(s) - \Delta H_f^\circ\ O_2(g)$

$= -1207.1 \text{ kJ} - (-635.5 \text{ kJ}) - (-393.5 \text{ kJ})$

$= -178.1 \text{ kJ}$

(c) $\Delta H° = \Delta H_f^\circ\ N_2(g) + 2\Delta H_f^\circ\ H_2O(g) - \Delta H_f^\circ\ N_2H_4(g) - \Delta H_f^\circ\ O_2(g)$

$= 0 + 2(-241.8 \text{ kJ}) - (95.40 \text{ kJ}) - 0$

$= -579.0 \text{ kJ}$

5.72 $\Delta H^\circ_{rxn} = \Delta H_f^\circ\ TiO_2(s) + 4\Delta H_f^\circ\ HCl(g) - \Delta H_f^\circ\ TiCl_4(l) - 2\Delta H_f^\circ\ H_2O(l)$

$\Delta H_f^\circ\ TiO_2(s) = \Delta H^\circ_{rxn} + \Delta H_f^\circ\ TiCl_4(l) + 2\Delta H_f^\circ\ H_2O(l) - 4\Delta H_f^\circ\ HCl(g)$

$= 67.0 \text{ kJ} + (-804.2 \text{ kJ}) + 2(-285.85 \text{ kJ}) - 4(-92.30 \text{ kJ})$

$= -939.7 \text{ kJ}$

5.73 (a) $C_6H_{12}O_6(s) + 6O_2(g) \rightarrow 6CO_2(g) + 6H_2O(l)$

$\Delta H^\circ_{rxn} = 6\Delta H^\circ_f\ CO_2(g) + 6\Delta H^\circ_f\ H_2O(l) - \Delta H^\circ_f\ C_6H_{12}O_6(s) - 6\Delta H^\circ_f\ O_2(g)$

$= 6(-393.5\text{ kJ}) + 6(-285.8\text{ kJ}) - (-1260\text{ kJ}) - 6(0)$

$= -2816\text{ kJ/mol } C_6H_{12}O_6$

$C_{12}H_{22}O_{11}(s) + 12O_2(g) \rightarrow 12CO_2(g) + 11H_2O(l)$

$\Delta H^\circ_{rxn} = 12\Delta H^\circ_f\ CO_2(g) + 11\Delta H^\circ_f\ H_2O(l) - \Delta H^\circ_f\ C_{12}H_{22}O_{11}(s) - 12\Delta H^\circ_f\ O_2(g)$

$= 12(-393.5\text{ kJ}) + 11(-285.8\text{ kJ}) - (-2221\text{ kJ}) - 12(0)$

$= -5645\text{ kJ/mol } C_{12}H_{22}O_{11}$

(b) $\frac{-2816\text{ kJ}}{1\text{ mol } C_6H_{12}O_6} \times \frac{1\text{ mol } C_6H_{12}O_6}{180.2\text{ g } C_6H_{12}O_6} = \frac{15.63\text{ kJ}}{1\text{ g } C_6H_{12}O_6} \rightarrow 16\text{ kJ/g}$

$\frac{-5645\text{ kJ}}{1\text{ mol } C_{12}H_{22}O_{11}} \times \frac{1\text{ mol } C_{12}H_{22}O_{11}}{342.3\text{ g } C_{12}H_{22}O_{11}} = \frac{16.49\text{ kJ}}{1\text{ g } C_{12}H_{22}O_{11}} \rightarrow 16\text{ kJ/g}$

(c) The average fuel value of carbohydrates (Section 5.8) is 17 kJ/g. These two carbohydrates have fuel values (16 kJ/g) slightly lower but in line with this average. (More complex carbohydrates supply more energy and raise the average value.)

5.74 (a) For comparison, balance the equations so that 1 mole of CH_4 is burned in each.

$CH_4(g) + O_2(g) \rightarrow C(s) + 2H_2O(l)$ $\quad \Delta H^\circ = -496.8\text{ kJ}$

$CH_4(g) + \frac{3}{2}O_2(g) \rightarrow CO(g) + 2H_2O(l)$ $\quad \Delta H^\circ = -607.2\text{ kJ}$

$CH_4(g) + 2O_2(g) \rightarrow CO_2(g) + 2H_2O(l)$ $\quad \Delta H^\circ = -890.2\text{ kJ}$

(b) $\Delta H^\circ_{rxn} = \Delta H^\circ_f C(s) + 2\Delta H^\circ_f\ H_2O(l) - \Delta H^\circ_f\ CH_4(g) - \Delta H^\circ_f\ O_2(g)$

$= 0 + 2(-285.85\text{ kJ}) - (-74.8) - 0 = -496.9\text{ kJ}$

$\Delta H^\circ_{rxn} = \Delta H^\circ_f\ CO(g) + 2\Delta H^\circ_f\ H_2O(l) - \Delta H^\circ_f\ CH_4(g) - \frac{3}{2}\Delta H^\circ_f\ O_2(g)$

$= (-110.5\text{ kJ}) + 2(-285.85\text{ kJ}) - (-74.8\text{ kJ}) - \frac{3}{2}(0) = -607.4\text{ kJ}$

$\Delta H^\circ_{rxn} = \Delta H^\circ_f\ CO_2(g) + 2\Delta H^\circ_f\ H_2O(l) - \Delta H^\circ_f\ CH_4(g) - \Delta H^\circ_f\ O_2(g)$

$= -393.5\text{ kJ} + 2(-285.85\text{ kJ}) - (-74.8\text{ kJ}) - 2(0) = -890.4\text{ kJ}$

(c) Assuming that $O_2(g)$ is present in excess, the reaction which produces $CO_2(g)$ represents the most negative $\Delta H°$ per mole of CH_4 burned. More of the potential energy of the reactants is released as heat during the reaction to give products of lower potential energy. The reaction which produces $CO_2(g)$ is the most "downhill" (Figure 5.5) in enthalpy.

5.75 (a) $3C_2H_2(g) \rightarrow C_6H_6(l)$

$$\Delta H°_{rxn} = \Delta H°_f C_6H_6(l) - 3\Delta H°_f C_2H_2(g) = 49.0 \text{ kJ} - 3(226.7 \text{ kJ}) = -631.1 \text{ kJ}$$

(b) Since the reaction is exothermic (ΔH is negative), the product, 1 mole of $C_6H_6(l)$, has less enthalpy than the reactants, 3 moles of $C_2H_2(g)$.

(c) The fuel value of a substance is the amount of heat (kJ) produced when 1 gram of the substance is burned. Calculate the molar heat of combustion (kJ/mol) and use this to find kJ/g of fuel.

$$C_2H_2(g) + \frac{5}{2} O_2(g) \rightarrow 2CO_2(g) + H_2O(l)$$

$$\Delta H°_{rxn} = 2\Delta H°_f\ CO_2(g) + \Delta H°_f\ H_2O(l) - \Delta H°_f\ C_2H_2(g) - \frac{5}{2}\Delta H°_f\ O_2(g)$$

$$= 2(-393.5 \text{ kJ}) + (-285.85 \text{ kJ}) - 226.7 \text{ kJ} - \frac{5}{2}(0) = -1299.6 \text{ kJ/mol } C_2H_2$$

$$\frac{-1299.6 \text{ kJ}}{1 \text{ mol } C_2H_2} \times \frac{1 \text{ mol } C_2H_2}{26.036 \text{ g } C_2H_2} = 49.916 \rightarrow 50 \text{ kJ/g}$$

$$C_6H_6(g) + \frac{15}{2} O_2(g) \rightarrow 6CO_2(g) + 3H_2O(l)$$

$$\Delta H°_{rxn} = 6\Delta H°_f\ CO_2(g) + 3\Delta H°_f\ H_2O(l) - \Delta H°_f C_6H_6(l) - \frac{15}{2}\Delta H°_f O_2(g)$$

$$= 6(-393.5 \text{ kJ}) + 3(-285.85 \text{ kJ}) - 49.0 \text{ kJ} - \frac{15}{2}(0) = -3267.6 \text{ kJ/mol } C_6H_6$$

$$\frac{-3267.6 \text{ kJ}}{1 \text{ mol } C_6H_6} \times \frac{1 \text{ mol } C_6H_6}{78.114 \text{ g } C_6H_6} = 41.831 \rightarrow 42 \text{ kJ/g}$$

5.76 The reaction for which we want ΔH is:

$$4NH_3(l) + 3O_2(g) \rightarrow 2N_2(g) + 6H_2O(g)$$

Before we can calculate ΔH for this, we must calculate ΔH_f for $NH_3(l)$.

We have that ΔH_f for $NH_3(g)$ is -46.2 kJ, and that for $NH_3(l) \rightarrow NH_3(g)$, $\Delta H = 4.6$ kJ

Thus, $\Delta H_{rxn} = \Delta H_f\ NH_3(g) - \Delta H_f\ NH_3(l)$.

$4.6\ kJ = -46.2\ kJ - \Delta H_f\ NH_3(l)\ ;\ \Delta H_f NH_3(l) = -50.8$ kJ/mol

Then for the overall reaction, the enthalpy change is:

$\Delta H_{rxn} = 6\Delta H_f\ H_2O(g) + 2\Delta H_f\ N_2 - 4\Delta H_f\ NH_3(l) - 3\Delta H_f\ O_2$

$= 6(-241.8\ kJ) + 0 - 4(-50.8\ kJ) - 0 = -1247.6$ kJ

$$\frac{-1247.6\ kJ}{4\ mol\ NH_3} \times \frac{1\ mol\ NH_3}{17.0\ g\ NH_3} \times \frac{0.81\ g\ NH_3}{1\ cm^3} \times \frac{1000\ cm^3}{1\ L} = \frac{1.5 \times 10^4\ kJ}{L\ NH_3}$$

(This result has 2 significant figures because the density is expressed to 2 figures.)

$2CH_3OH(l) + 3O_2(g) \rightarrow 2CO_2(g) + 4H_2O(g)$

$\Delta H = 2(-393.5\ kJ) + 4(-241.8\ kJ) - 2(-239\ kJ) = -1276$ kJ

$$\frac{-1276\ kJ}{2\ mol\ CH_3OH} \times \frac{1\ mol\ CH_3OH}{32.04\ g\ CH_3OH} \times \frac{0.792\ g\ CH_3OH}{1\ cm^3} \times \frac{1000\ cm^3}{1\ L} = \frac{1.58 \times 10^4\ kJ}{1\ L\ CH_3OH}$$

In terms of heat obtained per unit volume of fuel, methanol is a slightly better fuel than liquid ammonia.

5.77 **1,3-butadiene**, C_4H_6, $M = 54.092$ g/mol

(a) $C_4H_6(g) + \frac{11}{2}O_2(g) \rightarrow 4CO_2(g) + 3H_2O(l)$

$\Delta H^{o}_{rxn} = 4\Delta H^{o}_f\ CO_2(g) + 3\Delta H^{o}_f\ H_2O(l) - \Delta H^{o}_f\ C_4H_6(g) + \frac{11}{2}\Delta H^{o}_f\ O_2(g)$

$= 3(-393.5\ kJ) + 3(-285.85\ kJ) - 111.9\ kJ + \frac{11}{2}(0) = -2543.4$ kJ/mol C_4H_6

(b) $\frac{-2543.4\ kJ}{1\ mol\ C_4H_6} \times \frac{1\ mol\ C_4H_6}{54.092\ g} = 47.020 \rightarrow 47$ kJ/g

(c) $\%H = \frac{6(1.008)}{54.092} \times 100 = 11.18\%$ H

1-butene, C_4H_8, $M = 56.108$ g/mol

(a) $C_4H_8(g) + 6O_2(g) \rightarrow 4CO_2(g) + 4H_2O(l)$

$\Delta H^{o}_{rxn} = 4\Delta H^{o}_f\ CO_2(g) + 4\Delta H^{o}_f\ H_2O(l) - \Delta H^{o}_f\ C_4H_8(g) - 6\Delta H^{o}_f\ O_2(g)$

$= 4(-393.5\ kJ) + 4(-285.85\ kJ) - 1.2\ kJ - 6(0) = -2718.6$ kJ/mol C_4H_8

(b) $\dfrac{-2718.6}{1 \text{ mol } C_4H_8} \times \dfrac{1 \text{ mol } C_4H_8}{56.108 \text{ g } C_4H_8} = 48.453 \rightarrow 48 \text{ kJ/g}$

(c) $\% \text{ H} = \dfrac{8(1.008)}{56.108} \times 100 = 14.37\% \text{ H}$

n-butane, C_4H_{10}, $\mathcal{M} = 58.124$ g/mol

(a) $C_4H_{10}(g) + \frac{13}{2} O_2(g) \rightarrow 4CO_2(g) + 5H_2O(l)$

$$\Delta H^o_{rxn} = 4\Delta H^o_f\ CO_2(g) + 5\Delta H^o_f\ H_2O(l) - \Delta H^o_f\ C_4H_{10}(g) - \frac{13}{2}\ \Delta H^o_f\ O_2(g)$$

$$= 4(-393.5 \text{ kJ}) + 5(-285.85 \text{ kJ}) - (-124.7 \text{ kJ}) - \frac{13}{2}\ (0) \quad = -2878.6 \text{ kJ/mol } C_4H_{10}$$

(b) $\dfrac{-2878.6 \text{ kJ}}{1 \text{ mol } C_4H_{10}} \times \dfrac{1 \text{ mol } C_4H_{10}}{58.124 \text{ g } C_4H_{10}} = 49.525 \rightarrow 50 \text{ kJ/g}$

(c) $\%\text{H} = \dfrac{10(1.008)}{58.124} \times 100 = 17.34\% \text{ H}$

(d) It is certainly true that as the mass % H increases, the fuel value (kJ/g) of the hydrocarbon increases, given the same number of C atoms. A graph of the data in parts (b) and (c) (see below) suggests that mass % H and fuel value are directly proportionalgt when the number of C atoms is constant.

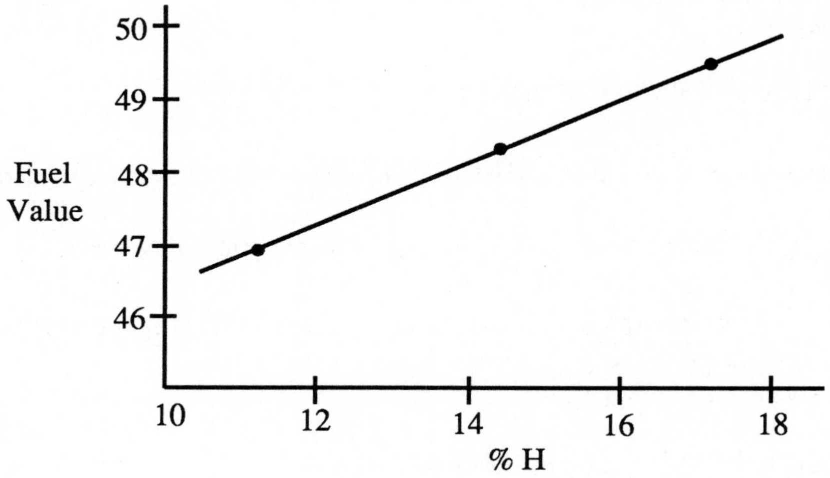

CHAPTER 6

Electronic Structures of Atoms

Radiant Energy

6.1 (a) meters (m) (b) 1/seconds (s^{-1}) (c) meters/second ($m \cdot s^{-1}$)

6.2 Wavelength (λ) and frequency (ν) are inversely proportional; the proportionality constant is the speed of light (c). $\nu = c/\lambda$.

The range of wavelengths in the visible portion of the electromagnetic spectrum is 400-700 nm.

6.3 Wavelength of (a) gamma rays < (d) yellow (visible) light < (e) red (visible) light < (b) 93.1 MHz FM (radio) waves < (c) 680 kHz or 0.680 MHz AM (radio) waves

6.4 Wavelength of (b) X-rays < (c) ultraviolet < (e) visible < (a) infrared < (d) microwave

6.5 (a) $\lambda = \frac{c}{\nu}$; $\frac{3.00 \times 10^8 \text{ m}}{1 \text{ s}} \times \frac{1 \text{ s}}{6.24 \times 10^{14}} = 4.81 \times 10^{-7} \text{ m} = 481 \text{ nm}$

(b) $\nu = \frac{c}{\lambda}$; $\frac{3.00 \times 10^8 \text{ m}}{1 \text{ s}} \times \frac{1}{3.55 \ \mu\text{m}} \times \frac{1 \ \mu\text{m}}{1 \times 10^{-6} \text{ m}} = 8.45 \times 10^{13} \text{ s}^{-1}$

(c) In order to "see" radiation, its wavelength must lie in the visible region of the electromagnetic spectrum, between 400 and 700 nm (Figure 6.3).
In part (a), λ = 481 nm; the radiation is visible.
In part (b), $\lambda = 3.55 \times 10^{-6} \text{ m} \times \frac{1 \text{ nm}}{1 \times 10^{-9} \text{ m}}$ = 3550 nm; the radiation is not visible.

(d) $\frac{3.00 \times 10^8 \text{ m}}{1 \text{ s}} \times \frac{60 \text{ s}}{1 \text{ min}} \times 2.50 \text{ min} = 4.50 \times 10^{10} \text{ m}$

6.6 (a) $\lambda = \frac{c}{\nu}$; $\frac{3.00 \times 10^8 \text{ m}}{1 \text{ s}} \times \frac{1 \text{ s}}{8.73 \times 10^8} = 0.344 \text{ m}$

(b) $\nu = \frac{c}{\lambda}$; $\frac{3.00 \times 10^8 \text{ m}}{1 \text{ s}} \times \frac{1}{650 \text{ nm}} \times \frac{1 \text{ nm}}{1 \times 10^{-9} \text{ m}} = 4.62 \times 10^{14} \text{ s}^{-1}$

(c) The radiation in (a) is in the microwave/TV region and the radiation in (b) is in the visible region. Neither can be detected with an infrared detector.

(d) $4.2 \text{ ns} \times \frac{1 \times 10^{-9} \text{ s}}{1 \text{ ns}} \times \frac{3 \times 10^8 \text{ m}}{1 \text{ s}} = 1.3 \text{ m}$

6.7 $\nu = \frac{c}{\lambda}$; $\frac{3.00 \times 10^8 \text{m}}{1 \text{ s}} \times \frac{1}{616 \text{ nm}} \times \frac{1 \text{ nm}}{1 \times 10^{-9} \text{ m}} = 4.87 \times 10^{14} \text{ s}^{-1}$

The color is orange.

6.8 $\lambda = \frac{c}{\nu}$; $\frac{3.00 \times 10^8 \text{m}}{1 \text{ s}} \times \frac{1}{6.59 \times 10^{14}} \times \frac{1 \text{ nm}}{1 \times 10^{-9} \text{ m}} = 455 \text{ nm}$

The color is dark blue or indigo.

Quantum Effects and Photons

6.9 Energy can only be absorbed or emitted in specific amounts or multiples of these amounts. This minimum amount of energy is called a **quantum** and is equal to a constant times the frequency of the radiation absorbed or emitted. $\mathbf{E = h\nu}$

6.10 In everyday activities, we deal with macroscopic objects such as our bodies or our cars, which gain and lose total amounts of energy much larger than a single quantum, hν. The gain or loss of the relatively minescule quantum of energy is unnoticed.

6.11 (a) $E = h\nu = \frac{hc}{\lambda} = 6.63 \times 10^{-34} \text{ J} \cdot \text{s} \times \frac{3.00 \times 10^8 \text{ m}}{1 \text{ s}} \times \frac{1}{405 \text{ nm}} \times \frac{1 \text{ nm}}{1 \times 10^{-9} \text{ m}}$

$= 4.91 \times 10^{-19} \text{ J}$

(b) $E = h\nu = 6.63 \times 10^{-34} \text{ J} \cdot \text{s} \times \frac{6.3 \times 10^{13}}{1 \text{ s}} = 4.2 \times 10^{-20} \text{ J}$

(c) $\lambda = \frac{hc}{E} = 6.63 \times 10^{-34} \text{ J} \cdot \text{s} \times \frac{3.00 \times 10^8 \text{ m}}{1 \text{ s}} \times \frac{1}{2.83 \times 10^{-19} \text{ J}} = 7.03 \times 10^{-7} \text{ m} = 703 \text{ nm}$

6.12 (a) $E = \frac{hc}{\lambda} = 6.63 \times 10^{-34} \text{ J} \cdot \text{s} \times \frac{3.00 \times 10^8 \text{ m}}{1 \text{ s}} \times \frac{1}{803 \text{ nm}} \times \frac{1 \text{ nm}}{1 \times 10^{-9} \text{ m}} = 2.48 \times 10^{-19} \text{ J}$

(b) $E = h\nu = 6.63 \times 10^{-34} \text{ J} \cdot \text{s} \times \frac{7.9 \times 10^{14}}{1 \text{ s}} = 5.2 \times 10^{-19} \text{ J}$

(c) $\nu = \frac{E}{h} = \frac{1.88 \times 10^{-18}\ J}{6.63 \times 10^{-34}\ J \cdot s} = 2.84 \times 10^{15}\ s^{-1}$

6.13 X-ray: $E = \frac{hc}{\lambda} = 6.63 \times 10^{-34}\ J \cdot s \times \frac{3.00 \times 10^{8}\ m}{1\ s} \times \frac{1}{1.54\ \text{Å}} \times \frac{1\ \text{Å}}{1 \times 10^{-10}\ m} = 1.29 \times 10^{-15}\ J$

microwave: $E = h\nu = 6.63 \times 10^{-34}\ J \cdot s \times \frac{5.87 \times 10^{10}}{1\ s} = 3.89 \times 10^{-23}\ J$

The X-ray photons have a much larger energy than the microwave photons. $E_{micro} < E_{X\text{-}ray}$

6.14 (a) $E = \frac{hc}{\lambda} = 6.63 \times 10^{-34}\ J \cdot s \times \frac{3.00 \times 10^{8}\ m}{1\ s} \times \frac{1}{8.7\ \mu m} \times \frac{1 \mu m}{1 \times 10^{-6}\ m} = 2.3 \times 10^{-20}\ J$

$E = \frac{hc}{\lambda} = 6.63 \times 10^{-34}\ J \cdot s \times \frac{3.00 \times 10^{8}\ m}{1\ s} \times \frac{1}{160\ nm} \times \frac{1\ nm}{1 \times 10^{-9}\ m} = 1.24 \times 10^{-18}\ J$

(b) The 8.7 μm photon is in the infrared and the 160 nm (1.60×10^{-7} m) photon is in the ultraviolet; the ultraviolet photon has the greater energy.

6.15 $E_{photon} = \frac{hc}{\lambda} = \frac{6.63 \times 10^{-34}\ J \cdot s}{351\ nm} \times \frac{3.00 \times 10^{8}\ m}{1\ s} \times \frac{1\ nm}{1 \times 10^{-9}\ m} = 5.67 \times 10^{-19}\ J$

$\frac{E_{total}}{E_{photon}}$ = # of photons; $8300\ J \times \frac{1\ \text{photon}}{5.67 \times 10^{-19}\ J} = 1.46 \times 10^{22}$ photons

6.16 The energy of each photon is hc/λ:

$6.63 \times 10^{-34}\ J \cdot s \times \frac{3.00 \times 10^{8}\ m}{1\ s} \times \frac{1}{550\ nm} \times \frac{1 nm}{1 \times 10^{-9}\ m} = 3.62 \times 10^{-19}$ J/photon

$\frac{1.45 \times 10^{-17}\ J\ \text{signal}}{3.62 \times 10^{-19}\ J/\text{photon}} = 40$ photons (The answer must be an integer; energy is quantized.)

6.17 $\frac{495 \times 10^{3}\ J}{mol\ O_2} \times \frac{1\ mol}{6.02 \times 10^{23}\ \text{photons}} = 8.22 \times 10^{-19}$ J/photon

$\lambda = \frac{hc}{E} = \frac{6.63 \times 10^{-34}\ J \cdot s}{8.22 \times 10^{-19}\ J} \times \frac{3.00 \times 10^{8}\ m}{1\ s} = 2.42 \times 10^{-7}\ m = 242\ nm$

According to Figure 6.3, this is ultraviolet radiation.

6.18 $\frac{80 \text{ kJ}}{1 \text{ mol}} \times \frac{1000 \text{ J}}{1 \text{ kJ}} \times \frac{1 \text{ mol}}{6.02 \times 10^{23} \text{ photons}} = 1.33 \times 10^{-19}$ J/photon

$$\lambda = \frac{hc}{E} = 6.63 \times 10^{-34} \text{ J} \cdot \text{s} \times \frac{3.00 \times 10^{8} \text{ m}}{1 \text{ s}} \times \frac{1}{1.33 \times 10^{-19} \text{ J}} = 1.50 \times 10^{-6} \text{ m}$$

The film could be used for infrared photography (Figure 6.3).

6.19 (a) $E = h\nu = 6.63 \times 10^{-34} \text{ J} \cdot \text{s} \times 5.57 \times 10^{14} \text{ s}^{-1} = 3.69 \times 10^{-19} \text{ J}$

(b) $\lambda = \frac{c}{\nu} = \frac{3.00 \times 10^{8} \text{ m}}{1 \text{ s}} \times \frac{1 \text{ s}}{5.57 \times 10^{14}} = 5.39 \times 10^{-7} \text{ m} = 539 \text{ nm}$

(c) According to Figure 6.3, this is **visible** radiation.

(d) $E_{510} = \frac{hc}{\lambda} = 6.63 \times 10^{-34} \text{ J} \cdot \text{s} \times \frac{3.00 \times 10^{8} \text{ m}}{1 \text{ s}} \times \frac{1}{510 \text{ nm}} \times \frac{1 \text{ nm}}{1 \times 10^{-9} \text{ m}}$

$= 3.90 \times 10^{-19}$ J

The excess energy of the 510 nm photon is converted into the kinetic energy of the emitted electron.

$E_{KE} = E_{510} - E_{min} = 3.90 \times 10^{-19} \text{ J} - 3.69 \times 10^{-19} \text{ J} = 0.21 \times 10^{-19}$ J/electron

6.20 (a) $\frac{277 \text{ kJ}}{\text{mol Ca}} \times \frac{1 \text{ mol}}{6.022 \times 10^{23} \text{ photons}} \times \frac{1000 \text{ J}}{1 \text{ kJ}} = 4.60 \times 10^{-19} \text{ J}$

(b) $\nu = \frac{E}{h} = \frac{4.60 \times 10^{-19} \text{ J}}{6.63 \times 10^{-34} \text{ J} \cdot \text{s}} = 6.94 \times 10^{14} \text{ s}^{-1}$

(c) $\lambda = \frac{hc}{E} = \frac{6.63 \times 10^{-34} \text{ J} \cdot \text{s} \times 3.00 \times 10^{8} \text{ m/s}}{4.60 \times 10^{-19} \text{ J}} = 4.32 \times 10^{-7} \text{ m} = 432 \text{ nm}$

(d) $E_{350 \text{ nm}} = \frac{hc}{\lambda} = \frac{6.63 \times 10^{-34} \text{ J} \cdot \text{s} \times 3.00 \times 10^{8} \text{ m/s}}{350 \text{ nm}} \times \frac{1 \text{ nm}}{1 \times 10^{-9} \text{ m}} = 5.68 \times 10^{-19} \text{ J}$

The excess energy of the 350 nm photon is converted into the kinetic energy of the emitted electron. The maximum possible kinetic energy/electron is the difference between the minimum energy per photon (4.60×10^{-19} J) and the energy of the 350 nm photon (5.68×10^{-19} J).

maximum KE/electron $= 5.68 \times 10^{-19} \text{ J} - 4.60 \times 10^{-19} \text{ J} = \frac{1.08 \times 10^{-19} \text{ J}}{\text{electron}}$

$$\frac{1.08 \times 10^{-19} \text{ J}}{\text{electron}} \times \frac{6.022 \times 10^{23} \text{ electrons}}{\text{mol}} \times \frac{1 \text{ kJ}}{1000 \text{ J}} = 65.2 \text{ kJ/mol}$$

Bohr's Model; Matter Waves

6.21 A continuous spectrum contains light of all wavelengths; a line spectrum contains only a few specific wavelengths. A rainbow is a continuous visible spectrum. The light given off by a neon sign or a mercury vapor lamp are examples of line spectra; only certain wavelengths (colors) are present. See Figures 6.9 and 6.12 in the text.

6.22 The Bohr theory states that only certain energy charges are allowed within the atom. These allowed energy charges correspond to the specific wavelengths in the line spectrum, $\Delta E = hc/\lambda$.

6.23 An isolated electron is assigned an energy of zero; the closer the electron comes to the nucleus, the more negative its energy. Thus, as an electron moves closer to the nucleus, the energy of the electron decreases and the excess energy is emitted. Conversely, as an electron moves further from the nucleus, the energy of the electron increases and energy must be absorbed.

(a) As the principle quantum number decreases, the electron moves closed to the nucleus and energy is **emitted**.

(b) An increase in the radius of the orbit means the electron moves further from the nucleus; energy is **absorbed**.

(c) Totally removing an electron from the atom requires that energy is **absorbed**.

6.24 (a) absorbed (b) emitted (c) absorbed

6.25 (a) $$\Delta E = R_H \left[\frac{1}{n_i^2} - \frac{1}{n_f^2} \right] = 2.18 \times 10^{-18} \text{ J } (1/1 - 1/9) = 1.94 \times 10^{-18} \text{ J}$$

$$\nu = \frac{E}{h} = \frac{1.94 \times 10^{-18} \text{ J}}{6.63 \times 10^{-34} \text{ J-s}} = 2.92 \times 10^{15} \text{ s}^{-1}$$

$$\lambda = \frac{c}{\nu} = \frac{3.00 \times 10^8 \text{ m}}{1 \text{ s}} \times \frac{1 \text{ s}}{2.92 \times 10^{15}} = 1.03 \times 10^{-7} \text{ m}$$

Since the sign of ΔE is positive, energy is absorbed.

(b) $\Delta E = 2.18 \times 10^{-18} \text{ J } (1/4 - 1/25) = 4.58 \times 10^{-19} \text{ J}$ absorbed

$$\nu = \frac{4.58 \times 10^{-19} \text{ J}}{6.63 \times 10^{-34} \text{ J-s}} = 6.91 \times 10^{14} \text{ s}^{-1} \; ; \; \lambda = \frac{3.00 \times 10^8 \text{ m/s}}{6.91 \times 10^{14}\text{/s}} = 4.34 \times 10^{-7} \text{ m}$$

(c) $\Delta E = 2.18 \times 10^{-18} \text{ J } (1/36 - 1/49) = 1.61 \times 10^{-20} \text{ J}$ absorbed

$$\nu = \frac{1.61 \times 10^{-20} \text{ J}}{6.63 \times 10^{-34} \text{ J-s}} = 2.43 \times 10^{13} \text{ s}^{-1} \; ; \; \lambda = \frac{3.00 \times 10^8 \text{ m/s}}{2.43 \times 10^{13}\text{/s}} = 1.24 \times 10^{-5} \text{ m}$$

6.26 (a) $\Delta E = R_H\left[\frac{1}{n_i^2} - \frac{1}{n_f^2}\right] = 2.18 \times 10^{-18}$ J (1/16 - 1/1) = -2.04×10^{-18} J

$$\nu = \frac{E}{h} = \frac{2.04 \times 10^{-18} \text{ J}}{6.63 \times 10^{-34} \text{ J-s}} = 3.08 \times 10^{15} \text{ s}^{-1};\ \lambda = \frac{c}{\nu} = \frac{3.00 \times 10^8 \text{ m/s}}{3.08 \times 10^{15}\text{/s}} = 9.74 \times 10^{-8} \text{ m}$$

Energy is emitted.

(b) $\Delta E = 2.18 \times 10^{-18}$ (1/25 - 1/9) = -1.55×10^{-19}

$$\nu = \frac{1.55 \times 10^{-19} \text{ J}}{6.63 \times 10^{-34} \text{ J-s}} = 2.34 \times 10^{14} \text{ s}^{-1};\ \lambda = \frac{3.00 \times 10^8 \text{ m/s}}{2.34 \times 10^{14}\text{/s}} = 1.28 \times 10^{-6} \text{ m}$$

Energy is emitted.

(c) $\Delta E = 2.18 \times 10^{-18}$ J (1/36 - 1/4) = -4.84×10^{-19}

$$\nu = \frac{4.84 \times 10^{-19} \text{ J}}{6.63 \times 10^{-34} \text{ J-s}} = 7.30 \times 10^{14} \text{ s}^{-1};\ \lambda = \frac{3.00 \times 10^8 \text{ m/s}}{7.30 \times 10^{14}\text{/s}} = 4.11 \times 10^{-7} \text{ m}$$

Energy is emitted.

6.27 $\Delta E = R_H \times \left[\frac{1}{n_i^2} - \frac{1}{n_f^2}\right] = 2.18 \times 10^{-18}$ J (1/25 - 1/4) = -4.58×10^{-19} J

$$\lambda = \frac{hc}{\Delta E} = \frac{6.63 \times 10^{-34} \text{ J-s}}{4.58 \times 10^{-19} \text{ J}} \times \frac{3.00 \times 10^8 \text{ m}}{1 \text{ s}} = 4.34 \times 10^{-7} \text{ m} = 434 \text{ nm}$$

Light of this wavelength lies in the blue region of the visible spectrum. There is an emission line for hydrogen at this wavelength.

6.28 $$\lambda = \frac{hc}{\Delta E} = \frac{6.63 \times 10^{-34} \text{ J} \bullet \text{s} \times 3.00 \times 10^8 \text{ m/s}}{2.18 \times 10^{-18} \text{ J (1/1 - 1/4)}} = 1.22 \times 10^{-7} \text{ m}$$

According to Figure 6.3, energy with this wavelength is in the ultraviolet region of the spectrum.

6.29 $\lambda = \frac{h}{mv}$; $1 \text{ J} = \frac{1 \text{ kg} \bullet \text{m}^2}{\text{s}^2}$; Change mass to kg and velocity to m/s in each case.

(a) $$\frac{60 \text{ km}}{1 \text{ hr}} \times \frac{1000 \text{ m}}{1 \text{ km}} \times \frac{1 \text{ hr}}{60 \text{ min}} \times \frac{1 \text{ min}}{60 \text{ s}} = \frac{16.67 \text{ m}}{1 \text{ s}}$$ (17 m/s to 2 sig. figs.)

$$\lambda = \frac{6.63 \times 10^{-34} \text{ kg} \bullet \text{m}^2 \bullet \text{s}}{1 \text{ s}^2} \times \frac{1}{85 \text{ kg}} \times \frac{1 \text{ s}}{16.67 \text{ m}} = 4.7 \times 10^{-37} \text{ m}$$

(b) $50\text{ g} \times \frac{1\text{ kg}}{1000\text{ g}} = 0.050\text{ kg}$

$$\lambda = \frac{6.63 \times 10^{-34}\text{ kg} \cdot \text{m}^2 \cdot \text{s}}{1\text{ s}^2} \times \frac{1}{0.050\text{ kg}} \times \frac{1\text{ s}}{400\text{ m}} = 3.3 \times 10^{-35}\text{ m}$$

(c) We need to calculate the mass of a single Li atom in kg.

$$\frac{6.94\text{ g Li}}{1\text{ mol Li}} \times \frac{1\text{ kg}}{1000\text{ g}} \times \frac{1\text{ mol}}{6.02 \times 10^{23}\text{ Li atoms}} = 1.15 \times 10^{-26}\text{ kg}$$

$$\lambda = \frac{6.63 \times 10^{-34}\text{ kg} \cdot \text{m}^2 \cdot \text{s}}{1\text{ s}^2} \times \frac{1}{1.15 \times 10^{-26}\text{ kg}} \times \frac{1\text{ s}}{6.5 \times 10^5\text{ m}} = 8.9 \times 10^{-14}\text{ m}$$

6.30 $\lambda = \frac{h}{mv}$; Change mass to kg and velocity to m/s.

(a) $3000\text{ lb} \times \frac{1\text{ kg}}{2.205\text{ lb}} = 1361\text{ kg}$;

$$\frac{55\text{ mi}}{1\text{ hr}} \times \frac{1\text{ hr}}{3600\text{ s}} \times \frac{1.6093\text{ km}}{1\text{ mi}} \times \frac{1000\text{ m}}{1\text{ km}} = \frac{24.59\text{ m}}{1\text{ s}}$$ (25 m/s to 2 sig. figs.)

$$\lambda = \frac{6.63 \times 10^{-34}\text{ kg} \cdot \text{m}^2 \cdot \text{s}}{1\text{ s}} \times \frac{1}{1361\text{ kg}} \times \frac{1\text{ s}}{24.59\text{ m}} = 2.0 \times 10^{-38}\text{ m}$$

(The answer has 2 sig. figs. because the speed was given to 2 sig. figs.)

(b) $5.0\text{ oz} \times \frac{28.3\text{ g}}{1\text{ oz}} \times \frac{1\text{ kg}}{1000\text{ g}} = 0.1415\text{ kg}$ (0.14 kg to 2 sig. figs.)

$$\frac{89\text{ mi}}{1\text{ hr}} \times \frac{1\text{ hr}}{3600\text{ s}} \times \frac{1.6093\text{ km}}{1\text{ mi}} \times \frac{1000\text{ m}}{1\text{ km}} = \frac{39.79\text{ m}}{1\text{ s}}$$ (40 m/s to 2 sig. figs.)

$$\frac{6.63 \times 10^{-34}\text{ kg} \cdot \text{m}^2 \cdot \text{s}}{1\text{ s}^2} \times \frac{1}{0.1415\text{ kg}} \times \frac{1\text{ s}}{39.79\text{ m}} = 1.2 \times 10^{-34}\text{ m}$$

(The answer has 2 sig. figs. because the mass and speed were given to 2 sig. figs.)

(c) $\frac{4.002\text{ g He}}{1\text{ mol He}} \times \frac{1\text{ kg}}{1000\text{ g}} \times \frac{1\text{ mol He}}{6.02 \times 10^{23}\text{ atoms}} = 6.64 \times 10^{-27}\text{ kg/He atom}$

$$\lambda = \frac{6.63 \times 10^{-34}\text{ kg} \cdot \text{m}^2 \cdot \text{s}}{1\text{ s}^2} \times \frac{1}{6.64 \times 10^{-27}\text{ kg}} \times \frac{1\text{ s}}{8.5 \times 10^5\text{ m}} = 1.2 \times 10^{-13}\text{ m}$$

6.31 $v = \frac{h}{m\lambda}$; $\lambda = 0.88\text{ Å} \times \frac{1 \times 10^{-10}\text{ m}}{1\text{ Å}} = 8.8 \times 10^{-11}\text{ m}$; $m = 1.67 \times 10^{-27}\text{ kg}$

$$v = \frac{6.63 \times 10^{-34}\text{ kg} \cdot \text{m}^2 \cdot \text{s}}{1\text{ s}^2} \times \frac{1}{1.67 \times 10^{-27}\text{ kg}} \times \frac{1}{8.8 \times 10^{-11}\text{ m}} = \frac{4.5 \times 10^3\text{ m}}{1\text{ s}}$$

6.32 $m_e = 9.11 \times 10^{-31}$ kg (back cover of text)

$$\lambda = \frac{6.63 \times 10^{-34}\ \text{kg} \cdot \text{m}^2 \cdot \text{s}}{1\ \text{s}^2} \times \frac{1}{9.11 \times 10^{-31}\ \text{kg}} \times \frac{1\ \text{s}}{5.93 \times 10^6\ \text{m}} = 1.23 \times 10^{-10}\ \text{m}$$

$$1.23 \times 10^{-10}\ \text{m} \times \frac{1\ \text{Å}}{1 \times 10^{-10}\ \text{m}} = 1.23\ \text{Å}$$

Since atomic radii and interatomic distances are on the order of 1-5 Å (Section 2.3), the wavelength of this electron is comparable to the size of atoms.

Quantum Mechanics and Atomic Orbitals

6.33 The square of the wave function has the physical significance of an amplitude, or probability. The quantity ψ^2 at a given point in space is the probability of locating the electron within a small volume element around that point at any given instant. The total probability, that is, the sum of ψ^2 over all the space around the nucleus, must equal 1.

6.34 In the Bohr model, the electron is treated as a small object that moves about the nucleus in circular orbits. A Bohr orbit specifies the exact path and energy of the electron. In the quantum-mechanical model, the wave properties of the electron are considered; any attempt to describe the exact path of an electron is inconsistent with the Heisenberg uncertainty principle. The quantum mechanical model is a statistical model which tells us the probability of finding an electron in certain regions around the nucleus. Thus, quantum mechanics would give the probability of finding the electron at 0.53 Å and this probability would always be less than 100%.

6.35 (a) $n = 5$, $l = 4, 3, 2, 1, 0$ (b) $l = 2$, $m_l = -2, -1, 0, 1, 2$

6.36 (a) $n = 4$, $l = 3, 2, 1, 0$ (b) $l = 3$, $m_l = -3, -2, -1, 0, 1, 2, 3$

6.37 (a) $n = 4$, $l = 2$, $m_l = -2$; $n = 4$, $l = 2$, $m_l = -1$; $n = 4$, $l = 2$, $m_l = 0$;

$n = 4$, $l = 2$, $m_l = 1$; $n = 4$, $l = 2$, $m_l = 2$

(b) $n = 3$, $l = 0$, $m_l = 0$; $n = 3$, $l = 1$, $m_l = -1, 0, 1$;

$n = 3$, $l = 2$, $m_l = -2, -1, 0, 1, 2$

6.38 (a) 4, 3, 3 ; 4, 3, 2 ; 4, 3, 1 ; 4, 3, 0 ; 4, 3, -1 ; 4, 3, -2 ; 4, 3, -3

(b) 2, 1, 1 ; 2, 1, 0 ; 2, 1, -1 ; 2, 0 , 0

6.39 (a) permissible, 2p (b) forbidden, for $l = 0$, m_l can only equal 0
(c) permissible, 4d (d) forbidden, for n = 3, the largest l value is 2

6.40 (a) forbidden, for $n = 1$, l can only equal 0 (b) allowed, 3s (c) allowed, 4p
(d) forbidden, for $l = 1$, the maximum value for m_l is 1

6.41 3d, 6p, 3s are permissible; 3f forbidden, for $n = 3$, the maximum l value is 2; 2d forbidden, for $n = 2$, the maximum l value is 1.

6.42 5f, 4d, 5s are permissible; 1p forbidden, for $n = 1$, $l = 0$ is the only permissible value; 2f forbidden, for $n = 2$, the maximum l value is 1.

6.43 (a)

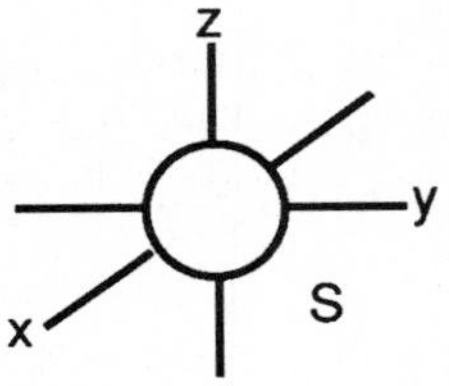

(b)

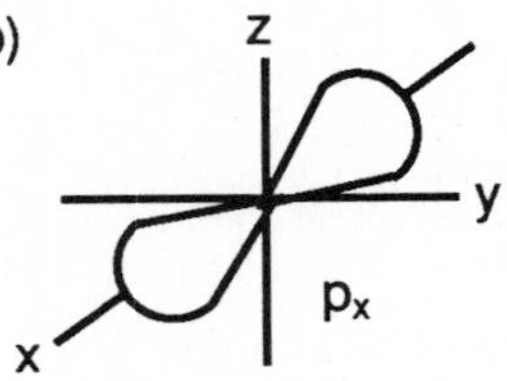

(c)

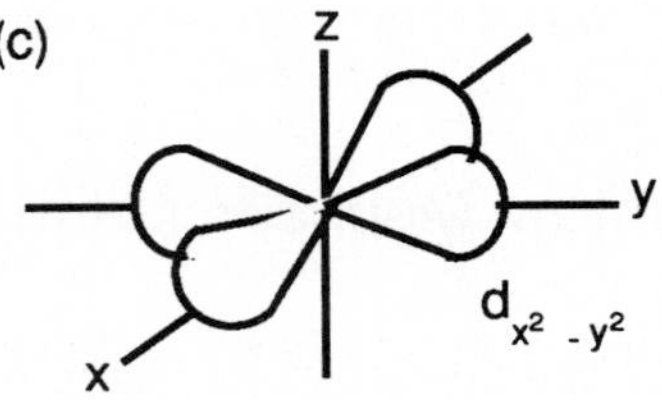

6.44 (a)

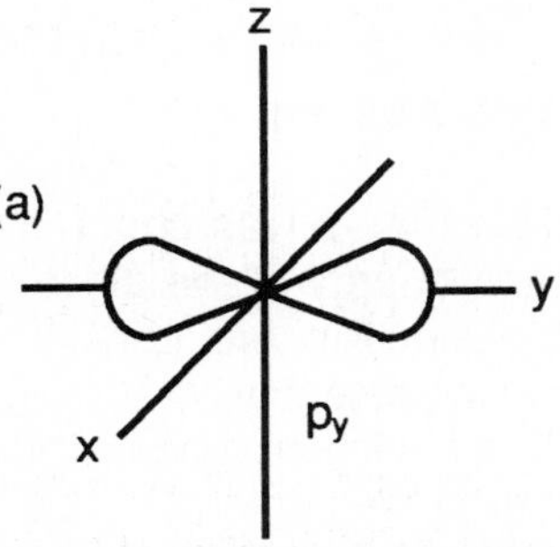

(b)

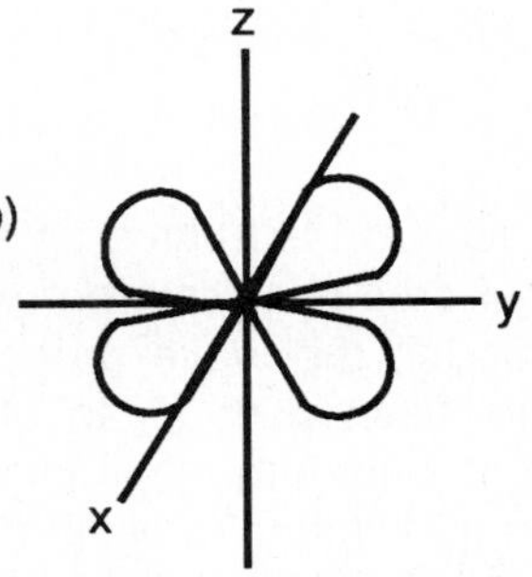

(c)

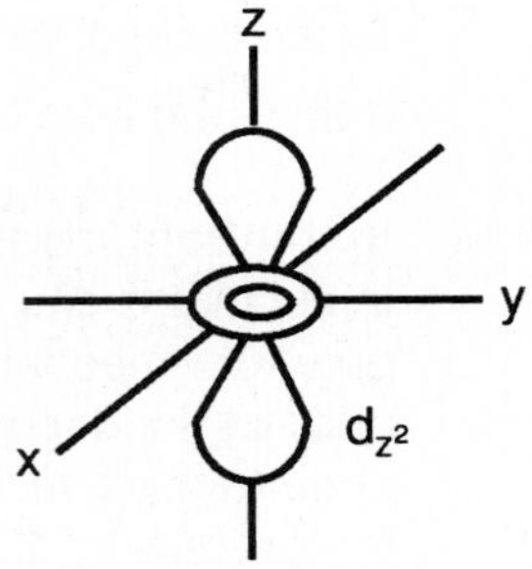

6.45 (a) The 2s and 3s orbitals have the same overall spherical shape, but the 3s orbital has a larger radial extension and one more node than the 2s orbital. Since the 3s orbital is "larger" than the 2s, there is a greater probability of finding an electron further from the nucleus in the 3s orbital.

(b) The shapes of the 2s and $2p_x$ orbitals are quite different (spherical vs dumbell), while the average distance from the nucleus of an electron occupying either orbital is similar.

(c) In the hydrogen atom, orbitals with the same *n* value are degenerate and energy increases with increasing *n* value. Thus, 2s and $2p_x$ have the same energy and 3s is at a higher energy.

6.46 The 2s orbitals in H and Li^{2+} are the same spherical shape. The Li^{2+} ion has a larger nuclear charge, which means that the energy associated with the 2s orbital will be more negative than that of the 2s orbital in H. In the one-electron model, a more negative energy means a smaller radial extension, so the 2s orbital in Li^{2+} is smaller than the 2s orbital in H.

Many Electron Atoms

6.47 (a) In the hydrogen atom, orbitals with the same principle quantum number, n, are degenerate.

(b) In a many-electron atom, orbitals with the same principle and azimuthal quantum numbers, n and *l*, are degenerate.

6.48 Within a given shell, the energies of the subshells increase in the order s < p < d < f. Within a subshell, the energies of the orbitals are the same.

6.49 A 3s electron has a greater probability of being close to the nucleus than a 3d electron, so it is less effectively shielded from the nuclear charge by the inner electrons (1s, 2s, 2p) than a 3d electron. Also, the 3d is shielded by the 3s and 3p electrons, while the 3s is not.

6.50 The average distance from the nucleus of a 2s electron is less than that of a 2p electron, because a 2s electron has a greater probability of being close to the nucleus and experiences a larger Z_{eff}. According to Figure 6.18, an electron in any s orbital has a finite probability of existing at the nucleus.

6.51 A 3s electron in magnesium experiences a nuclear charge (Z = 12) one higher than that in sodium (Z = 11). This added unit of nuclear charge is partially, but not completely, cancelled by the presence of a second 3s electron. The overall effect is an increase in the effective nuclear charge experienced by a 3s electron in Mg.

6.52 Although 4s electrons are shielded by 3d electrons and a 4s electron in Cu experiences slightly more shielding than a 4s electron in Ti, this does not completely counteract the larger value of Z for Cu (29 vs 22); Z_{eff} for a 4s electron in Cu is greater.

6.53 (a) 10 (b) 2 (c) 6 (d) 14

6.54 (a) 18 (b) 10 (c) 2 (d) 1

6.55 Li - $1s^22s^1$; 1s electrons: 1, 0, 0, 1/2; 1, 0, 0, - 1/2

2s electron: 2, 0, 0, 1/2 <u>or</u> 2, 0, 0, -1/2

6.56 2, 1, 1, $\frac{1}{2}$; 2, 1, 1 -$\frac{1}{2}$; 2, 1, 0, $\frac{1}{2}$; 2, 1, 0, -$\frac{1}{2}$; 2, 1, -1, $\frac{1}{2}$; 2, 1, -1, -$\frac{1}{2}$

Electron Configurations

6.57 (a) Ca - [Ar] $4s^2$ (b) Ge - [Ar] $4s^23d^{10}4p^2$ (c) Br - [Ar] $4s^23d^{10}4p^5$

(d) Co - [Ar] $4s^23d^7$ (e) Eu - [Xe] $6s^24f^7$ (f) Hf - [Xe] $6s^24f^{14}5d^2$

6.58 (a) Cs - [Xe] $6s^1$ (b) Mn - [Ar] $4s^23d^5$ (c) Ni - [Ar] $4s^23d^8$

(d) Sb - [Kr] $5s^24d^{10}5p^3$ (e) Lu - [Xe] $6s^24f^{14}5d^1$ (f) Pb - [Xe] $6s^24f^{14}5d^{10}6p^2$

6.59 We will consider the valence electrons to be those beyond the noble gas core.

(a) Si [↑↓] (3s) [↑][↑][ ] (3p) 2 unpaired electrons

(b) Te [↑↓] (5s) [↑↓][↑↓][↑↓][↑↓][↑↓] (4d) [↑↓][↑][↑] (5p) 2 unpaired electrons

(c) Fe [↑↓] (4s) [↑↓][↑][↑][↑][↑] (3d) 4 unpaired electrons

(d) Eu [↑↓] (6s) [↑][↑][↑][↑][↑][↑][↑] (4f) 7 unpaired

(e) Bi [↑↓] (6s) [↑↓][↑↓][↑↓][↑↓][↑↓][↑↓][↑↓] (4f) [↑↓][↑↓][↑↓][↑↓][↑↓] (5d) [↑][↑][↑] (6p) 3 unpaired electrons

6.60 (a) B [↑↓] (3s) [↑][ ][ ] (2p) 1 unpaired electron

(b) Ga [↑↓] (4s) [↑↓][↑↓][↑↓][↑↓][↑↓] (3d) [↑][ ][ ] (4p) 1 unpaired electron

(c) Sn [↑↓] (5s) [↑↓][↑↓][↑↓][↑↓][↑↓] (4d) [↑][↑][ ] (5p) 2 unpaired electrons

(d) Ru [↑↓] (5s) [↑↓][↑][↑][↑][↑] (4d) 4 unpaired electrons

(e) Ir [↑↓] (6s) [↑↓][↑↓][↑↓][↑↓][↑↓][↑↓][↑↓] (4f) [↑↓][↑↓][↑][↑][↑] (5d) 3 unpaired electrons

6.61 (a) O (b) Cl (c) K (d) Cr

6.62 (a) 1A (alkali metals) (b) 5A (pnicogens) (c) 4A (representative elements)
(d) 8B (transition metals)

Additional Exercises

6.63 (a) $\lambda_A = 6.0 \times 10^{-7}$ m, $\lambda_B = 12 \times 10^{-7}$ m

(b) $\nu = \frac{c}{\lambda}$; $\nu_A = \frac{3.00 \times 10^8 \text{ m}}{1 \text{ s}} \times \frac{1}{6.0 \times 10^{-7} \text{ m}} = 5.0 \times 10^{14} \text{ s}^{-1}$

$\nu_B = \frac{3.00 \times 10^8 \text{ m}}{1 \text{ s}} \times \frac{1}{12 \times 10^{-7} \text{ m}} = 2.5 \times 10^{14} \text{ s}^{-1}$

(c) A - visible, B - infrared

6.64 (a) Ag -- violet (this wavelength is technically in the UV region),
Au -- violet (UV), Ba -- dark blue, Ca -- dark blue,
Cu -- violet (UV), Fe -- violet (UV), K -- dark blue,
Mg -- violet(UV), Na -- yellow/orange, Ni -- violet(UV)

(b) Au -- shortest wavelength, highest energy;
Na -- longest wavelength, lowest energy

(c) $\lambda = \frac{c}{\nu} = \frac{3.00 \times 10^8 \text{ m/s}}{6.59 \times 10^{14}/\text{s}} \times \frac{1 \text{ nm}}{1 \times 10^{-9} \text{ m}} = 455$ nm, **Ba**

6.65 Light travels 3.00×10^8 m in 1 second. Change mi $\rightarrow$ m so that the distance units will be the same, and s $\rightarrow$ hr.

$$2.82 \times 10^9 \text{ mi} \times \frac{1.609 \text{ km}}{1 \text{ mi}} \times \frac{1000 \text{ m}}{1 \text{ km}} \times \frac{1 \text{ s}}{3.00 \times 10^8 \text{ m}} \times \frac{1 \text{ hr}}{3600 \text{ s}} = 4.20 \text{ hr}$$

6.66 (a) $\nu = \frac{c}{\lambda} = \frac{3.00 \times 10^8 \text{ m}}{1 \text{ s}} \times \frac{1}{7.80 \times 10^{-7} \text{ m}} = 3.85 \times 10^{14} \text{ s}^{-1}$

(b) $E = \frac{hc}{\lambda} = \frac{6.63 \times 10^{-34} \text{ J-s}}{7.80 \times 10^{-7} \text{ m}} \times \frac{3.00 \times 10^8 \text{ m}}{1 \text{ s}} = 2.55 \times 10^{-19}$ J

6.67 $\frac{\Delta E}{1 \text{ molecule}} = E_{absorbed} - E_{emitted} = \frac{hc}{460 \times 10^{-9} \text{ m}} - \frac{hc}{660 \times 10^{-9} \text{ m}}$

$$\frac{\Delta E}{1 \text{ molecule}} = 6.63 \times 10^{-34} \text{ J-s} \times \frac{3.00 \times 10^8 \text{ m}}{1 \text{ s}} \left(\frac{1}{460 \times 10^{-9} \text{ m}}\right) - \left(\frac{1}{660 \times 10^{-9} \text{ m}}\right)$$

$$\Delta E = \frac{1.31 \times 10^{-19} \text{ J}}{1 \text{ molecule}} \times \frac{6.02 \times 10^{23} \text{ molecules}}{1 \text{ mol}} = +78.9 \text{ kJ/mol}$$

Net energy is absorbed by the system.

6.68 (a) $E = \frac{hc}{\lambda} = 6.63 \times 10^{-34}\ J \cdot s \times \frac{3.00 \times 10^8\ m}{1\ s} \times \frac{1}{488\ nm} \times \frac{1\ nm}{1 \times 10^{-9}\ m}$

$= 4.08 \times 10^{-19}$ J/photon

(b) $10\ watts = \frac{10\ J}{s}$; $1.5\ \mu s \times \frac{1 \times 10^{-6}\ s}{1\ \mu s} = 1.5 \times 10^{-6}\ s$

$\frac{10\ J}{s} \times \frac{1\ photon}{4.08 \times 10^{-19}\ J} \times 1.5 \times 10^{-6}\ s = 3.7 \times 10^{13}$ photons

6.69 $\frac{8.6 \times 10^{-13}\ C}{1\ s} \times \frac{1e^-}{1.602 \times 10^{-19}\ C} \times \frac{1\ photon}{1\ e^-} = 5.4 \times 10^6$ photons/s

$\frac{E}{photon} = \frac{hc}{\lambda} = \frac{6.63 \times 10^{-34}\ J \cdot s}{550\ nm} \times \frac{3.00 \times 10^8\ m}{1\ s} \times \frac{1\ nm}{1 \times 10^{-9}\ m} \times \frac{5.4 \times 10^6\ photons}{s}$

$= 2.0 \times 10^{-12}$ J/s

6.70 $\Delta H^\circ_{rxn} = \Delta H^\circ_f\ O_2(g) + \Delta H^\circ_f\ O(g) - \Delta H^\circ_f\ O_3(g)$

$= 0 + 247.5\ kJ - 142.3\ kJ$

$\Delta H^\circ_{rxn} = +105.2\ kJ$

$\frac{105.2\ kJ}{mol\ O_3} \times \frac{1\ mol\ O_3}{6.022 \times 10^{23}\ atoms} \times \frac{1000\ J}{1\ kJ} = \frac{1.747 \times 10^{-19}\ J}{O_3\ atom}$

$\Delta E = \frac{hc}{\lambda}$; $\lambda = \frac{hc}{\Delta E} = \frac{6.626 \times 10^{-34}\ J \cdot s \times 2.998 \times 10^8\ m/s}{1.747 \times 10^{-19}\ J} = 1.137 \times 10^{-6}\ m$

Radiation with this wavelength is in the infrared portion of the spectrum. (Clearly, processes other than simple photodissociation cause O_3 to absorb ultraviolet radiation.)

6.71 We know the wavelength of microwave radiation, the volume of water to be heated and the desired temperature change. We need to calculate: (a) the total energy required to heat the water and (b) the energy of a single photon in order to find (c) the number of photons required.

(a) From Chapter 5, the heat capacity of liquid water is 4.184 J/g°C.

To find the mass of 100 mL of water at 20°C, use the density of water given in Appendix B.

$100\ mL \times \frac{0.997\ g}{1\ mL} = 99.7\ g\ H_2O$

$\frac{4.184\ J}{1\ g\ ^\circ C} \times 99.7\ g \times (100^\circ C - 20^\circ C) = 3.337 \times 10^4\ J = 33\ kJ$

(b) $E = \frac{hc}{\lambda} = 6.63 \times 10^{-34}\ J \cdot s \times \frac{3.00 \times 10^{8}\ m}{1\ s} \times \frac{1}{0.125\ m} = \frac{1.59 \times 10^{-24}\ J}{1\ photon}$

(c) $3.337 \times 10^{4}\ J \times \frac{1\ photon}{1.59 \times 10^{-24}\ J} = 2.1 \times 10^{28}$ photons

(The answer has 2 sig. figs. because the temperature change, 80°C, has 2 sig. figs.)

6.72 The emission lines correspond to electronic transitions within the neon atoms. These transitions occur between electronic energy states with very well defined energies. Thus only certain energy changes can occur with the neon atom. These specific changes in energy give rise to emission at specific wavelengths.

6.73 $\Delta E = R_H \left(\frac{1}{n_i^2} - \frac{1}{n_f^2}\right) = 2.18 \times 10^{-18}\ J \left(\frac{1}{1^2} - \frac{1}{\infty^2}\right) = \frac{2.18 \times 10^{-18}\ J}{1\ atom}$

$\frac{2.18 \times 10^{-18}\ J}{1\ atom} \times \frac{6.02 \times 10^{23}\ atoms}{1\ mol} \times \frac{1\ kJ}{1000\ J} = 1.31 \times 10^{3}\ kJ/mol$

6.74 $\Delta E = \frac{hc}{\lambda} = \frac{6.63 \times 10^{-34}\ J\text{-}s}{1875.6 \times 10^{-9}\ m} \times \frac{3.00 \times 10^{8}\ m}{1\ s} = 1.06 \times 10^{-19}\ J$

$\frac{\Delta E}{R_H} = \left(\frac{1}{n_i^2} - \frac{1}{n_f^2}\right)$; $\frac{-1.06 \times 10^{-19}\ J}{2.18 \times 10^{-18}\ J} = \left(\frac{1}{n_i^2} - \frac{1}{n_f^2}\right) = -0.0486$

Since this is an emission line, the sign of ΔE is negative and $n_f < n_i$. The line is in the infrared region and represents a **smaller ΔE** than lines in the visible region. Since lines in the visible region correspond to transitions from higher states to n = 2, the only way to achieve a smaller ΔE is $n_f \geq 3$.

Try $n_i = 4$, $n_f = 3$ $\left(\frac{1}{4^2} - \frac{1}{3^2}\right) = \frac{1}{16} - \frac{1}{9} = -0.0486$; $n_i = 4$, $n_f = 3$

6.75 (a) $\Delta x \geq \frac{h}{2\pi(\Delta p)}$; $\Delta x \geq \frac{h}{2\pi(m\Delta v)}$; $\Delta v = 0.010\ (3.0 \times 10^{6}\ m/s) = 3.0 \times 10^{4}\ m/s$

$\Delta x \geq \frac{6.63 \times 10^{-34}\ kg\text{-}m^2}{1\ s} \times \frac{1}{6.28} \times \frac{1}{9.11 \times 10^{-31}\ kg} \times \frac{1\ s}{3.0 \times 10^{4}\ m} = 3.9 \times 10^{-9}\ m$

(b) $\Delta v = 0.010\ (200\ \text{m/s}) = 2.0\ \text{m/s}$

$$\Delta x \geq \frac{6.63 \times 10^{-34}\ \text{kg - m}^2}{1\ \text{s}} \times \frac{1}{6.28} \times \frac{1}{0.012\ \text{kg}} \times \frac{1\ \text{s}}{2.0\ \text{m}} = 4.4 \times 10^{-33}\ \text{m}$$

Since Δx and Δp are inversely proportional, the electron, with the smaller Δp, has the larger Δx. The diameter of an atom is on the order of Ångstroms (10^{-10} m) and Δx for the electron is similar. The Δx of the bullet (4×10^{-33} m) is insignificant compared to the size of a bullet (on the order of millimeters).

6.76 (a) 2s (b) 4d (c) 5p (d) 3d (e) 4f

6.77 (a) 25 (b) 3 (c) 7 (d) 1 (e) 1 (f) 5

6.78 In a Cl atom, the 3p electrons experience the smallest effective nuclear charge because they are shielded by all the core electrons and to some extent by the 3s electrons. The 1s electrons experience the greatest effective nuclear charge because they are not shielded by inner electrons; only their own mutual repulsion mitigates the full charge of the nucleus.

6.79 (d) 3p < (a) 4s < (b) 3d = (c) 3d

6.80 The 2p subshell consists of three orbitals of equal energy, so it can accommodate more electrons than the single 2s orbital. Also, there are six sets of n, l, m_l, m_s values that can be written to describe electrons in the 2p subshell and only two sets for electrons in the 2s. Since no two electrons can have the same four quantum numbers (Pauli Principle), the 2p subshell must hold more electrons.

6.81 (a) Cd: $1s^2 2s^2 2p^6 3s^2 3p^6 4s^2 3d^{10} 4p^6 5s^2 4d^{10}$

(b) As: $1s^2 2s^2 2p^6 3s^2 3p^6 4s^2 3d^{10} 4p^3$

(c) La: $1s^2 2s^2 2p^6 3s^2 3p^6 4s^2 3d^{10} 4p^6 5s^2 4d^{10} 5p^6 6s^2 5d^1$

(d) Pd: $1s^2 2s^2 2p^6 3s^2 3p^6 4s^2 3d^{10} 4p^6 5s^2 4d^8$

(e) S: $1s^2 2s^2 2p^6 3s^2 3p^4$

6.82

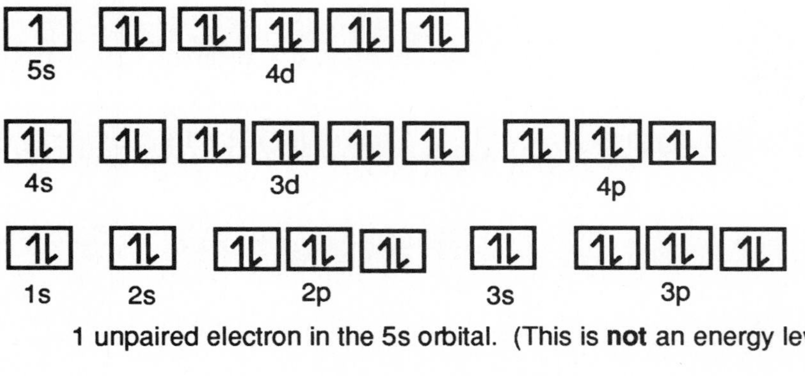

1 unpaired electron in the 5s orbital. (This is **not** an energy level diagram.)

6.83 (a) Ca, Zn, Kr (b) K, Sc, Cu (one unpaired 4s electron), Ga, Br (c) Cr, with the electron configuratrion $[Ar]4s^1\ 3d^5$, has **six** unpaired electrons; Mn, $[Ar]4s^2 3d^5$, has only five.

6.84 (a) O - excited (b) Br - ground (c) P - excited (d) In - ground

6.85 (a) Support: The emission spectrum of H is a **line** spectrum; only certain energy charges are allowed.

(b) Refute: According to the uncertainty principle, the exact position and momentum of an electron cannot be known simultaneously.

(c) Support: In the line spectra of multielectron atoms, apparently single lines are actually closely spaced pairs, corresponding to electrons in two different spin states. In 1921, Stern and Gerlock showed that an inhomogeneous magnetic field split a beam of Ag atoms in two, suggesting two different magnetic fields due to the unpaired electrons in the Ag atoms.

(d) Support: Z_{eff} is reduced by electron-electron repulsion. All 3p electrons in Cl^- are paired in orbitals, while there is one unpaired e^- in a neutral Cl atom. The average repulsions experienced by a 3p electron in Cl are less and Z_{eff} is greater. Physically, the radius of Cl^- is greater than the radius of Cl (see Chapter 7), indicating that the electrons are less tightly held.

(e) Refute: The Actinides have both 5f and 4f electrons, so there must be several elements with smaller atomic numbers than the Actinides which also have 4f electrons (the Lanthanides and the transition elements following them on the Periodic Table).

CHAPTER 7

Periodic Properties of the Elements

Periodic Table; Electron Shells

7.1 (a) Mendeleev noted that certain chemical and physical properties recur periodically when the elements are arranged by increasing atomic weight. He arranged the known elements by increasing atomic weight so that elements with similar properties were in the same family or verticle column.

(b) Shortly after Rutherford postulated the nuclear model of the atom, Mosely proposed that each element be assigned an integer or atomic number which corresponded to the charge on the nucleus of the atom. If elements are arranged by increasing atomic number, a few seeming contradictions in the Mendeleev table (the positions of Ar and K or Te and I) are eliminated.

7.2 Mendeleev insisted that elements with similar chemical and physical properties be placed within a family or column of the table. Since many elements were as yet undiscovered, Mendeleev left blanks. He predicted properties for the "blanks" based on properties of other elements in the family.

7.3 On a plot of the radial electron density of an atom, there are certain distances from the nucleus where there are high electron densities. The number of these maxima corresponds to the number of "electron shells" or principle quantum levels in the atom.

7.4

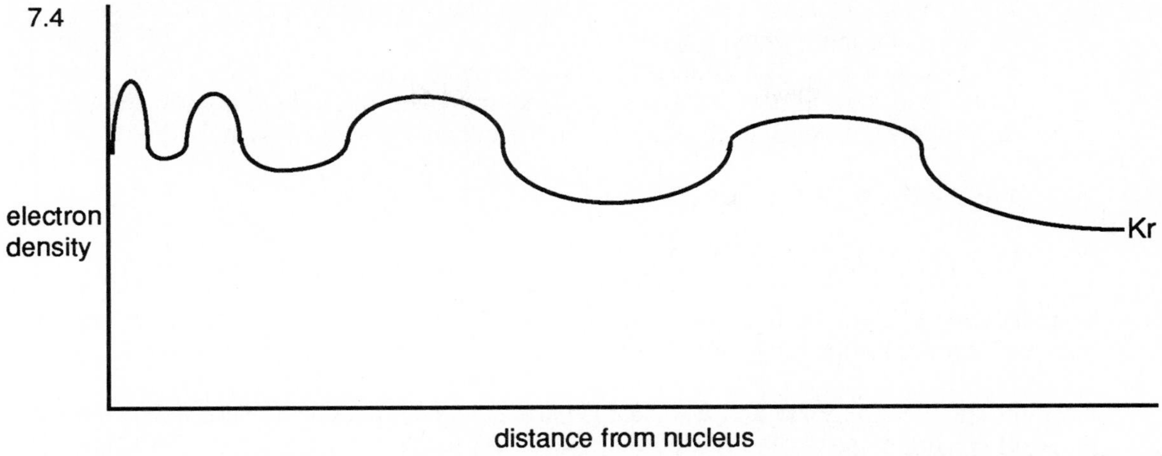

Kr has electrons in four shells, so there are four peaks in the radial electron density graph.

7.5 Rh < Ti < K < P < Mg

7.6 Neon has a larger nuclear charge (Z = 10) than carbon (Z = 6). The shielding of the n = 2 shells by the 1s electrons in the two atoms is approximately equal, so the n = 2 electrons in Ne experience a greater effective nuclear charge and are thus situated closer to the nucleus.

Sizes of Ions

7.7 Since the quantum mechanical description of the atom does not specify the exact location of electrons, there is no specific distance from the nucleus where the last electron can be found. Rather, the electron density decreases gradually as the distance from the nucleus increases. There is no quantum mechanical "edge" of an atom.

7.8 When assigning atomic radii, we assume that: 1) atoms are spherical objects that touch each other when they bond, 2) atomic radii are constant over time and 3) atomic radii are unchanged by bonding.

7.9 If the radius of a Cl atom is half the distance between the two atoms in a chlorine molecule, the radius of Cl is $\frac{1.99 \text{ Å}}{2}$ or 0.995 Å.

If the P-Cl distance is 2.04 Å and the Cl radius is 0.995 Å, the radius of P is 2.04 Å - 0.95 Å = 1.05 Å.

r_P = 1.05 Å, r_{Cl} = 0.995 Å. This is consistent with the trend in atomic radii across a row of the periodic chart.

7.10 $r_F = \frac{1.42 \text{ Å}}{2} = 0.710 \text{ Å}$ $\quad r_N = 1.37 \text{ Å} - 0.710 \text{ Å} = 0.66 \text{ Å}$

(According to the expected trend, the atomic radius of N should be greater than that of F. A more accurate result might be obtained by using the N-F bond lengths from several compounds and averaging their values.)

7.11 (a) Atomic radii **decrease** moving from left to right across a row and (b) **increase** from top to bottom within a group.

(c) B < Si < Al < Ge. The changes in radius moving across a row are usually smaller than those moving down a family, so although B and Si are both smaller than Al, B is probably smaller than Si, since it is in the second row and Si is in the third.

7.12 (a) F < O < P < Mg < Ca

(b) He < Ar < Ge < In < Rb

7.13 The electrons in a He atom experience a nuclear charge of 2 and are drawn closer to the nucleus than the electron in H, which feels a nuclear charge of only 1.

Even though Z = 10 for Ne and Z = 2 for He, the valence electrons in Ne are in n = 2 and those of He are closer to the nucleus in n = 1. Also, the 1 s electrons in Ne shield the valence electrons from the full nuclear charge. This shielding, coupled with the larger n value of the valence electrons in the Ne, means that its atomic radius is larger than that of He.

7.14 (a) Moving from left to right, the value of Z increases and the value of n remains unchanged. Since valence electrons do not effectively shield each other, those of elements on the right side of a row experience a greater nuclear charge than those on the left of a row. As effective nuclear charge increases, size decreases.

(b) Moving from top to bottom in a family, Z increases but so does n. As n increases, the valence electrons are effectively shielded by the core electrons. This has the effect of increasing the distance of the valence electrons from the nucleus while leaving the effective nuclear charge essentially constant. Thus, size increases going down a family.

Ionization Energies; Electron Affiities

7.15 $Sc^+ \rightarrow Sc^{2+} + 1e^-$; $Sc^{2+} \rightarrow Sc^{3+} + 1e^-$

7.16 $S^{3+} \rightarrow S^{4+} + 1e^-$ $\Delta H = +4565$ kJ

7.17 The electron configuration of Li^+ is $1s^2$ or [He] and that of Be^+ is $[He]2s^1$. Be^+ has one more valence electron to lose while Li^+ has the stable noble gas configuration of He. It requires much more energy to remove a 1s core electron close to the nucleus of Li^+ than a 2s valence electron further from the nucleus of Be^+.

7.18 Removing one electron from K produces K^+, which has the noble gas electron configuration of Ar. The second electron to be removed is then a core electron, and disrupting this stable electron configuration requires a substantial amount of energy.

7.19 (a) F (b) N (c) Hf (d) O (Ionization energy increases moving up and to the right on the periodic chart). (e) Ge

---

7.20 (a) Cl -- As the effective nuclear charge increases in moving from left to right in the third row, the energy required to remove an electron increases.

(b) Al (slightly) -- Al and Ga are rather similar; the buildup of nuclear charge due to the presence of 3d electrons in Ga offsets the larger value of the principle quantum number for its valence electrons.

(c) La - La and Cs compare as two elements in the same row, with the effects of increasing effective nuclear charge dominating their periodic properties.

(d) N - Valence electrons in N are closer to the nucleus (n = 2) and are shielded only by the [He] core, so they experience greater attraction for the nucleus and have a higher ionization energy.

7.21 First ionization energies increase slightly going from K to Ar and atomic sizes decrease. As valence electrons are drawn closer to the nucleus (atom size decreases), it requires more energy to completely remove them from the atom (first ionization energy increases). Each trend has a discontinuity at Ga, owing to the increased shielding of the 4p electrons by the filled 3d subshell.

7.22 Moving from He to Rn in group 8A, first ionization energies decrease and atomic size increases, because the n value and thus distance of the valence electrons from the nucleus increases going down a group.

7.23 The two ions are isoelectronic. Because they have the same electron configuration, $[Ne]3s^1$, electron repulsion and shielding effects should be similar. The nuclear charge is one greater in Si^{3+}, so the 3s electron of this ion experiences a larger effective nuclear charge. According to Table 7.2, I_4 for Si is greater than I_3 for Al; it requires more energy to remove a valence electron from Si^{3+} than from Al^{2+}.

7.24 To form AlO_2, the aluminum would need to lose four electrons. This would mean removing a 2p core electron from Al. The energy cost of doing this is given by I_4 in Table 7.2. Note that it is much greater than I_4 for Si, because the effective nuclear charge experienced by the 2p electron in the Al^{3+} ion is very high. The high energy cost of removing this electron rules out chemical behavior that corresponds to an Al^{4+} species.

7.25 When an electron is added to the 3p orbital of Cl, it experiences electron-electron repulsions, but it also experiences some electron-nuclear attraction, because the other five 3p electrons do not completely shield it from the nucleus. Also, a stable octet of electrons is formed. Cl^- is lower in energy than Cl, so the process is exothermic. In Ar the stable octet already exists. The added electron must go into a 4s orbital; in this orbital the electron is almost completely shielded from the attraction of the nucleus. This electronic arrangement is not lower in energy than that of Ar atoms, so the process is endothermic.

7.26 In Mg(g), the two outer shell electrons are paired in the 3s orbital. An added electron goes into a 3p orbital; it is effectively shielded from the nucleus, so electron-electron repulsions dominate over the electron-nuclear attraction and the process is endothermic. In Na(g) the added electron can pair with the **one** 3s electron already there. It "sees" some nuclear charge because it is not completely shielded by the other 3s electron, and the process is slightly exothermic.

Properties of Metals and Nonmetals

7.27 Metals are good conductors and nonmetals are poor ones. Metallic character increases going from right to left across a row and from top to bottom in a column. The order of increasing metallic character and electrical conductivity is S < Si < Ge < Ca.

7.28 The order of melting points should be F < Na < Mg < Al. Fluorine is a nonmetal and exists as a diatomic gas, so it has the lowest melting point. Na, Mg and Al are metals in the third period; their melting points should increase as metallic character decreases moving from left to right in the period up to a maximum at group 6A.

7.29 Metallic character increases moving down a family and to the left in a period.

(a) Li (b) Na (c) Sn (d) Al

7.30 (a) Metallic character increases going down a group: N < P < As < Sb < Bi

(b) Nonmetallic character increases going up and to the right in the chart.
Hg < In < Ge < S < F

7.31 Ionic: Na_2O, CaO, Fe_2O_3; molecular: N_2O, CO, P_2O_5, Cl_2O_7
Ionic compounds are formed by combining a metal and a nonmetal; covalent compounds are formed by two or more nonmetals.

7.32 Solids: Na_2S, BaO, $PbCl_2$, MgF_2; gases: NO, CO_2, OF_2. Ionic compounds are high melting solids, while covalent compounds exist in all three states. Covalent compounds with low molecular weight, such as NO, CO_2 and OF_2 are often gases.

7.33 When dissolved in water, an "acidic oxide" produces an acidic (pH < 7) solution. Oxides of nonmetals are acidic. Example: $SO_3(g)$. A "basic oxide" dissolved in water produces a basic (pH > 7) solution. Oxides of metals are basic. Example: CaO (quick lime).

7.34 The more **nonmetallic** the central atom, the more acidic the oxide. In order of increasing acidity: BaO < MgO < Al_2O_3 < SiO_2 < CO_2 < P_2O_5 < SO_3

7.35 (a) $Na_2O(s) + H_2O(l) \rightarrow 2NaOH(aq)$
(b) $CuO(s) + 2HNO_3(aq) \rightarrow Cu(NO_3)_2(aq) + H_2O(l)$
(c) $SO_3(g) + H_2O(l) \rightarrow H_2SO_4(aq)$
(d) $SeO_2(s) + 2NaOH(aq) \rightarrow Na_2SeO_3(aq) + H_2O(l)$

7.36 (a) $Cl_2O_7(g) + H_2O(l) \rightarrow 2HClO_4(aq)$
(b) $FeO(s) + 2HCl(aq) \rightarrow FeCl_2(aq) + H_2O(l)$
(c) $BaO(s) + H_2O(l) \rightarrow Ba(OH)_2(aq)$
(d) $CO_2(g) + 2KOH(aq) \rightarrow K_2CO_3(aq) + H_2O(l)$

Group Trends in Metals and Nonmetals

7.37

	Na	**Mg**
(a)	[Ne] $3s^1$	[Ne] $3s^2$
(b)	+1	+2
(c)	+496 kJ/mol	+738 kJ/mol
(d)	very reactive	reacts with steam, but not $H_2O(l)$
(e)	1.86 Å	1.60 Å

(b) When forming ions, both adopt the stable configuration of Ne, but Na loses one electron and Mg two electrons to achieve this configuration.

(c),(e) The nuclear charge of Mg (Z = 12) is greater than that of Na, so it requires more energy to remove a valence electron with the same n value from Mg than Na. It also means that the 2s electrons of Mg are held closer to the nucleus, so the atomic radius (e) is smaller than that of Na.

(d) Mg is less reactive because it has a filled subshell and it has a higher ionization energy.

7.38

	Rb	**Ag**
(a)	[Kr] $5s^1$	[Kr] $5s^1 4d^{10}$
(b)	+1	+1
(c)	extremely reactive	unreactive
(d)	2.2 Å	1.3 Å

(b) Each has 1 electron in the 5s subshell which it loses to form a +1 ion.

(c),(d) The 5s electron in Ag experiences a much greater Z_{eff} than the 5s electron in Rb because Z is much larger (47 vs 37) and the two electrons are approximately the same distance from the nucleus. (The extra shielding by the 4d electrons in Ag does not compensate for the higher Z value). Thus, Ag is less reactive (less likely to lose an electron) and has a smaller atomic radius.

7.39 (a) Sr is more reactive toward H_2O than Be is because its valence electrons are less tightly held (greater n value) than those of Be, and it is more easily oxidized.

(b) Na reacts with O_2 to form Na_2O_2, sodium **peroxide**. We expect metals to form metal oxides (O^{2-}) when reacted with O_2.

7.40 Z_{eff} experienced by the 4s electrons in Ca is greater than that experienced by the 4s electron in K, so the electrons are held more tightly and are more difficult to remove.

7.41 (a) $2K(s) + 2H_2O(l) \rightarrow 2KOH(aq) + H_2(g)$

(b) $Ba(s) + 2H_2O(l) \rightarrow Ba(OH)_2(aq) + H_2(g)$

(c) $6Li(s) + N_2(g) \rightarrow 2Li_3N(s)$

(d) $2Mg(s) + O_2(g) \rightarrow 2MgO(s)$

7.42 (a) $2Na(g) + Br_2(g) \rightarrow 2NaBr(s)$

(b) $2Na(l) + H_2(g) \rightarrow 2NaH(s)$

(c) $4Li(s) + O_2(g) \rightarrow 2Li_2O(s)$

(d) $SrO(s) + H_2O(l) \rightarrow Sr(OH)_2(aq)$

7.43 H - $1s^1$; Li - [He]$2s^1$; F - [He]$2s^22p^5$. Like Li, H has only one valence electron, and its most common oxidation number is +1, which both H and Li adopt after losing the single valence electron. Like F, H needs only one electron to adopt the stable electron configuration of the nearest noble gas. Both H and F can exist in the -1 oxidation state, when they have gained an electron to complete their valence shells.

7.44 (a) $2Na(s) + Cl_2(g) \rightarrow 2NaCl(s)$
$H_2(g) + Cl_2(g) \rightarrow 2HCl(g)$
Each product is diatomic and contains one atom from each of the reactants. NaCl is an ionic solid, while HCl is a (polar) covalent gas.

(b) $Ca(s) + F_2(s) \rightarrow CaF_2(s)$
$Ca(s) + H_2(g) \rightarrow CaH_2(s)$
Both products are ionic solids containing Ca^{2+} and the corresponding anion in a 1:2 ratio.

7.45

	O	F
(a)	[He] $2s^22p^4$	[He] $2s^22p^5$
(b)	-2	-1
(c)	+1314 kJ/mol	+1681 kJ/mol
(d)	unreactive	reacts exothermically to produce HF
(e)	reacts to form H_2O	reacts to form HF
(f)	0.73 Å	0.72 Å

(b) When forming ions, both adopt the stable electron configuration of Ne; O atoms gain two electrons and F atoms gain one electron.

(c) The ionization energy of F is greater because its valence electrons experience a greater Z_{eff} ($Z_F > Z_O$) and the average distance of the electrons from the nucleus is approximately the same.

(d),(e) The electron affinity of F is greater (more exothermic) than that of O, and F_2 can remove electrons from almost any substance, including H_2O and H_2.

(f) From trends in atomic size, one would expect F to be somewhat smaller than O. However, electron-electron repulsions are so great in the relatively cramped 2p orbitals of F that the larger Z_{eff} does not lead to a significantly smaller radius.

7.46

	F	Cl
(a)	[He] $2s^2 2p^5$	[Ne] $3s^2 3p^5$
(b)	-1	-1
(c)	1681 kJ/mol	1256 kJ/mol
(d)	reacts exothermically to form HF	unreactive
(e)	-332 kJ/mol	-349 kJ/mol
(f)	0.72 Å	0.99 Å

(b) F and Cl are in the same group, have the same valence electron configuration and common ionic charge.

(c),(f) The n = 2 valence electrons in F are closer to the nucleus and more tightly held than the n = 3 valence electrons in Cl. Therefore, the ionization energy of F is greater, and the atomic radius is smaller.

(d) In its reaction with H_2O, F is reduced; it gains an electron. Athough the electron affinity, a gas phase single atom property, of F is less negative than that of Cl, the tendency of F to hold its own electrons (high ionization energy) coupled with a relatively large exothermic electron affinity makes it extremely susceptible to reduction and chemical bond formation.

(e) Although F has a larger Z_{eff} than Cl, its small atomic radius gives rise to large repulsions when an extra electron is added, so the overall electron affinity of F is smaller (less exothermic) than that of Cl.

7.47 In 1962, N. Bartlett discovered that Xe, which has the lowest ionization energy of the nonradioactive Noble gases, would react with substances having a strong tendency to remove electrons, such as F_2. Thus, the term "inert" no longer described all the Group 8A elements. (Kr also reacts with F_2, but reactions of Ar, Ne and He are as yet unknown.)

7.48 Xe has a lower ionization energy than Ne. The valence electrons in Xe are much further from the nucleus than those of Ne (n = 5 vs n = 2) and much less tightly held by the nucleus; they are more "willing" to be shared than those in Ne. Also, Xe has empty 5d orbitals which can help to accommodate the bonding pairs of electrons, while Ne has all its valence orbitals filled.

7.49 (a) $S_8(s) + 16\,Li(s) \rightarrow 8Li_2S(s)$

(b) $2O_3(g) \rightarrow 3O_2(g)$

(c) $2KBr(aq) + 2H_2O(l) \rightarrow 2KOH(aq) + H_2(g) + Br_2(aq)$

(d) $Cl_2(g) + Ca(s) \rightarrow CaCl_2(s)$

7.50 (a) $I_2(s) + 2Li(s) \rightarrow 2LiI(s)$

(b) $S_8(s) + 8O_2(g) \rightarrow 8SO_2(g)$

(c) $Kr(g) + F_2(g) \rightarrow KrF_2(g)$

(d) $O_2(g) + K(s) \rightarrow KO_2(s)$ (potassium superoxide)

7.51 (a) Te has more metallic character and is a better electrical conductor.

(b) At room temperature, oxygen molecules are diatomic and exist in the gas phase. Sulfur molecules are 8-membered rings and exist in the solid state.

(c) Chlorine is generally more reactive than bromine because Cl atoms have a greater (more exothermic) electron affinity than Br atoms.

7.52 (a) Valence electrons in F atoms experience a much greater attraction for the nucleus than those in I atoms, because they are much closer to the nucleus. Thus, F atoms have a much greater tendency to gain electrons and are more reactive toward relatively inert substances such as Xe.

(b) O_2 is the allotrope routinely found in the atmosphere (21% of air is O_2). O_3 is produced only under special conditions.

(c) F_2 is extremely reactive because it can remove electrons from almost any substance, so a special apparatus which will not react with F_2 and will not leak F_2 must be used to carry out reactions.

Additional Exercises

7.53 Exceptions to Mendeleev's Periodic Law: Ar and K ; Co and Ni ; Te and I ; Th and Pa ; U and Np ; Pu and Am ; (possibly Cm and Bk, their approximately equal atomic weights are not known accurately). If placed in order of increasing atomic weight, these elements would not fall in the group (column) with which they have similar properties.

7.54 The nuclear charge of an atom increases as the atomic number increases. If the accompanying electron is added to the same shell as the element of the preceeding atomic number, there is an increase in the effective nuclear charge. As effective nuclear charge increases, the size of an atom decreases.

7.55 The neutral atom (on which I_1 is measured) always has more electrons than the 1^+ ion (on which I_2 is measured). With more electrons there is more electron-electron repulsion, making it easier to remove an electron. (This matter can also be viewed from the perspective of the screening effect. With more electrons there is greater shielding giving rise to a reduced effective nuclear charge. When the effective nuclear charge is smaller, it is easier to remove an electron.)

7.56 (a) Increasing ionization energy: Si < Se < C < O < F

(b) Increasing atomic radius: F < O < C < Si ≈ Se

F, O, C, and Si can be ordered according to the trends given in Figure 7.6. Se is difficult to place because it is both to the right and below Si; these two directions have conflicting trends. Also, it is in the fourth row and subject to the effects of the filling of the 3d subshell and accompanying increase in Z and Z_{eff}. Referring to the data in Figures 7.5 and 7.7, the ionization energy of Si is definitely less than that of Se, and the atomic radii are approximately equal.

7.57 (a) As electrons are removed, the electron-electron repulsion in the atom decreases, and it is harder to remove an electron. Also, the fewer total electrons, the less the outer electrons are shielded from the nucleus and the greater the effective nuclear charge.

(b) There is a huge increase from the third to the fourth ionization energy because the principle quantum number of the electron being removed changes from n = 2 to n= 1. This electron is closer to the nucleus and is not shielded by any other electrons; it experiences a much greater nuclear charge.

(c) The corresponding electrons being removed from B have smaller principle quantum numbers and are closer to the nucleus. Thus, they feel a greater attraction for the nucleus than those electrons in Al, and it requires more energy to remove them.

7.58 Both Na^+ and Mg^{2+} have the electron configuration of the noble gas Ne. In order for these ions to become Na^{2+} or Mg^{3+}, an electron would have to be removed from the completed second shell, a stable octet. This is energetically very unfavoralble and unlikely.

7.59 Zr - [Kr] $5s^2 4d^2$; Hf - [Xe] $6s^2 4f^{14} 5d^2$. The 6s electrons in Hf are incompletely shielded by the 4f electrons, so the much larger value of Z (72 for Hf, 40 for Zr) is not totally offset by shielding and the greater distance of the 6s electrons from the nucleus. That is, the 6s electrons in Hf experience a slightly greater **attraction** to the nucleus than those of Zr and the atomic radius of Hf is smaller.

7.60 (a) The atomic radius of Ca is greater than that of Zn because Zn has a significantly greater Z and Z_{eff}, which draws the 4s valence electrons closer to the nucleus.

(b) Ca^{+2} - [Ar] ; Zn^{+2} - [Ar] $3d^{10}$. In both Ca and Zn atoms, the outermost electrons are in the 4s subshell. The much larger Z_{eff} for Zn causes it to have a much smaller radius. In the +2 ions, both have lost their 4s electrons. The outermost electrons in Zn are in the 3d sublevel and are significantly shielded by the [Ar] core electrons. Thus, the radius of Zn^{2+} is closer to the radius of Ca^{2+} than the radii of the neutral ions.

7.61 The electron configurations and radii of the ions are listed below.

particle	electron configuration	radius
Li	$1s^2 2s^1$	135 pm
Li^+	$1s^2$	60 pm
Li^{2+}	$1s^1$	18 pm
Be^{2+}	$1s^2$	

The valence electrons in Li^+ are in a lower principle quantum level than the valence electron in neutral Li, and electrons in n = 1 are not shielded from the nuclear charge, so Li^+ is sigificantly smaller than Li.

There are two valence electrons in Li^+ and only one in Li^{2+}. Electron repulsion, which increases tne energy of the valence electrons and thus their distance from the nucleus, does not affect Li^{2+} , and it is sigificantly smaller than Li^+.

Be^{2+} has 2 valence electrons, so electron repulsion will cause them to be further from the nucleus than electrons in Li^{2+}. However, Z = 4 for Be and 3 for Li, so Be^{2+} will be smaller than Li^+. The radius of Be^{2+} will be somewhere between 60 pm (Li^+) and 18 pm (Li^{2+}).

7.62 Although the attraction of an electron for the nucleus decreases as the distance from the nucleus increases going down a group, the electron-electron repulsion also decreases because the valence orbitals are larger. That is, electrostatic attraction decreases but so does repulsion, so the overall trend toward smaller (less exothermic) electron affinity is not as pronounced as expected.

7.63 (a) F and O^- have the same electron configuration, [He] $2s^22p^5$; they are isoelectronic.

(b) In the first process, an electron is added to a neutral F atom, while in the second an electron is added to a negative O^- ion. Clearly, there is electrostatic repulsion to overcome in the second process. Also, since O^- and F are isoelectronic, but F has one more proton, an electron added to F experiences a greater effective nuclear charge. Both factors contribute to the $+\Delta H$ for the second process.

(c) Following the same trend, N^{2-} (g) will experience greater electrostatic repulsion and smaller Z_{eff} when gaining an electron, so the electron affinity will become smaller (less exothermic).

7.64 (a) P - [Ne] $3s^23p^3$; S - [Ne] $3s^23p^4$. In P, each 3p orbital contains a single electron, while in S one 3p orbital contains a pair of electrons. Removing an electron from S eliminates the need for electron pairing and reduces electrostatic repulsion, so the overall energy required to remove the electron is smaller than in P, even though Z is greater.

(b) C - [He] $2s^22p^2$; N - [He] $2s^22p^3$; O - [He] $2s^22p^4$. An electron added to a N atom must be paired in a relatively small 2p orbital, so the additional electron-electron repulsion more than compensates for the increase in Z and the electron affinity is smaller (less exothermic) than that of C. In an O atom, one 2p orbital already contains a pair of electrons, so the additional repulsion from an extra electron is offset by the increase in Z and the electron affinity is greater (more exothermic). Note from Figure 7.9 that the electron affinity of O is only slightly more exothermic than that of C, although the value of Z has increased by 2.

(c) O^+ - [Ne] $2s^22p^3$; O^{2+} - [Ne] $2s^22p^2$; F^+ - [Ne] $2s^22p^4$; F^{2+} - [Ne] $2s^22p^3$. The decrease in electron-electron repulsion going from F^+ to F^{2+} energetically favors ionization and causes it to be less endothermic than the corresponding process in O, where there is no significant decrease in repulsion.

(d) Mn^{2+} - [Ar] $3d^5$; Mn^{3+} - [Ar] $3d^4$; Cr^{2+} - [Ar] $3d^4$; Cr^{3+} - [Ar] $3d^3$; Fe^{2+} - [Ar] $3d^6$; Fe^{3+} - [Ar]$3d^5$. The third ionization energy of Mn is expected to be larger than that of Cr because of the larger Z value of Mn. The third ionization energy of Fe is **less** than that of Mn because going from $3d^6$ to $3d^5$ reduces electron repulsions, making the process less endothermic than predicted by nuclear charge arguments.

7.65 The electron affinity of Na^+ ($Na^+ + 1e^- \rightarrow Na$) is just the reverse of ionization energy of Na ($Na \rightarrow Na^+ + 1e^-$). EA of Na^+ = -IE of Na = -496 kJ/mol.

7.66

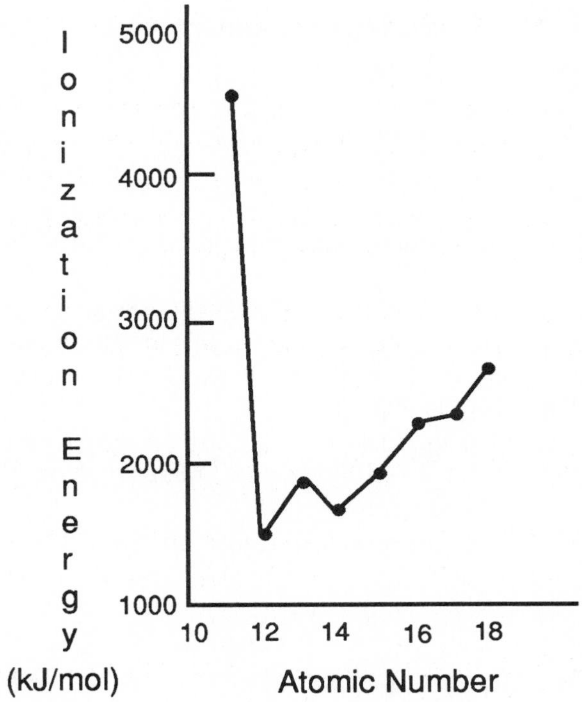

In the series Mg through Ar there is a general increase in I_2 because the effective nuclear charge seen by the electron increases. In Mg and Al a 3s electron is being removed. In Si a 3p electron is removed. There is a drop in going from Al to Si, because the 3p electron is better shielded from the nucleus than is the 3s. From Si to Ar the steady increase in I_2 comes from the fact that the 3p electrons do not completely shield one another from the nucleus. The value for Na is exceptionally large because I_2 for this element corresponds to removal of a 2p electron; that is, removal of a core electron which experiences a relatively high effective nuclear charge.

7.67 Moving one place to the right in a horizontal row of the table, for example, from Li to Be, there is an increase in ionization energy. Moving downward in a given family, for example from Be to Mg, there is usually a decrease in ionization energy. Similarly, atomic size decreases in moving one place to the right and increases in moving downward. Thus, two elements such as Li and Mg that are diagonally related tend to have similar ionization energies and atomic sizes. This in turn gives rise to some similarities in chemical behavior. Note, however, that the valences expected for the elements are not the same. That is, lithium still appears as Li^+, magnesium as Mg^{2+}.

7.68 Since Xe reacts with F_2 and O_2 has approximately the same ionization energy of Xe, O_2 will probably react with F_2. Possible products would be O_2F_2, analogous to XeF_2, or OF_2.

$$O_2(g) + F_2(g) \rightarrow O_2F_2(g)$$

$$O_2(g) + 2F_2(g) \rightarrow 2OF_2(g)$$

7.69 Fr (1A), Ra (2A) , Po (6A), At (7A), Rn (8A)

(a) most metallic character -- **Fr** (Metallic character decreases from left to right in a row.)

(b) most nonmetallic character -- **Rn**

(c) largest ionization energy -- **Rn** (Ionization energy increases from left to right in a row.)

(d) smallest ionization energy -- **Fr**

(e) greatest electron affinity -- **At** (Electron affinity becomes more exothermic from left to right in a row.)

(f) largest atomic radius -- **Fr** (Size decreases from left to right in a row.)

(g) appears least like the element above it -- **Fr** (According to the trends in melting point, Fr may be a gas at room temperature, while Cs is a liquid. The others should be in the same state as the element above them.)

(h) highest melting point -- **Ra** (The melting points of the group 2A metals are much higher than those of the other groups, even though the values decrease going down the group.)

(i) react most readily with H_2O -- **Fr** (It has the lowest ionization energy.)

7.70

Element	Density (g/cm^3)	Melting Point(°C)	Boiling Point(°C)	I_1	I_2	Formulas of Halides	Electron Configuration
Sb	6.684	630.5	1380	833.5	1592	SbX_3, SbX_5	$[Kr]5s^24d^{10}5p^3$
I	4.93	113.5	184.35	1009	1846	IX, IX_3,IX_5,IX_7	$[Kr]5s^24d^{10}5p^5$
Se	4.81	60 - 217	685	940.7	2074	Se_2X_2,SeX_4,SeX_6	$[Ar]4s^23d^{10}4p^4$
Te	6.25	452	1390	869.3	1795	TeX_2,TeX_4,TeX_6	$[Kr]5s^24d^{10}5p^4$

Because Te is a semimetal, its physical properties do not necessarily match the other elements in its group. In fact, with respect to density, boiling point and I_1, it is most similar to its left neighbor, Sb. (There are no obvious similarities in melting point, and the values for I_2 follow the predicted trend, increasing from left to right across a row.) Chemically, one could expect Te to resemble Se, since they have the same valence electron configuration, This is borne out by the observed formulas of the halides for the various elements.

CHAPTER **8**

Basic Concepts of Chemical Bonding

Lewis Symbols and Ionic Bonding

8.1 (a) Valence electrons are those that take part in chemical bonding, those in the outermost electron shell of the atom. This usually means the electrons beyond the core noble gas configuration of the atom.

(b) N - [He] $2s^2 2p^3$ N has 5 valence electrons.

valence electrons

8.2 (a) Atoms will gain, lose or share electrons to achieve the nearest noble gas electron configuration. Except for He, this corresponds to eight electrons in the valence shell, thus the term **octet** rule.

(b) S - [Ne] $3s^2 3p^6$ Sulfur atoms have six valence electrons, so they need to gain two electrons to achieve an octet.

8.3 (a) · Si · (b) Na · (c) :Se · (d) · Al ·

8.4 a) :Br · b) Mg · c) $[Ca]^{2+}$ or $[:Ca:]^{2+}$ d) $[:F:]^-$

8.5 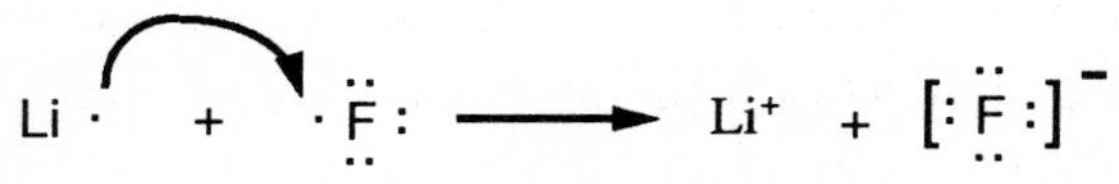

8.6 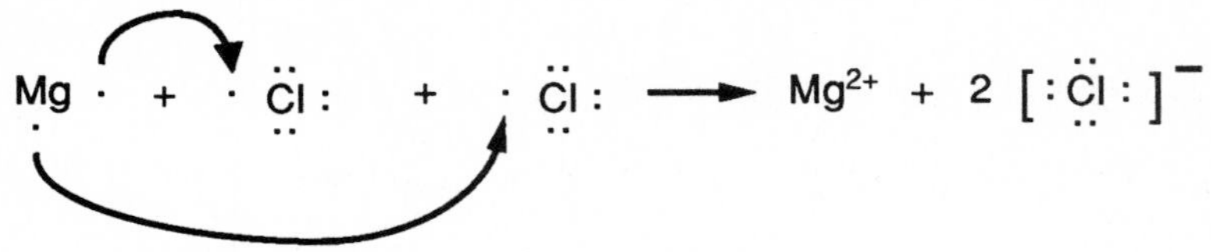

8.7 (a) AlF_3 (b) K_2S (c) Mg_3N (d) BaO

8.8 The issue for each formula is whether the ions have stable noble gas electron configurations.

(a) Stable. Rb^+ and O^{2-} both have octets.
(b) Not stable. F^- has an octet, but Ba^+ has a single valence electron.
(c) Not stable. O^{2-} has an octet, but Mg^+ has a single valence electron.
(d) Stable. Sc^{3+} and Br^- both have octets.
(e) Stable Na^+ and N^{3-} both have octets.

8.9 (a) I^- : $[Kr]5s^2\,4d^{10}5p^6$ = [Xe], noble gas configuration
(b) Se^{2-} : $[Ar]\,4s^23d^{10}4p^6$ = [Kr], noble gas configuration
(c) Sr^{2+}: [Kr] , noble gas configuration
(d) Ni^{2+}: $[Ar]\,3d^8$ (e) Pb^{2+}: $[Xe]\,6s^24f^{14}5d^{10}$

8.10 (a) As^{3-} : $[Ar]\,4s^23d^{10}4p^6$ = [Kr] , noble gas configuration
(b) Ag^+ : $[Kr]\,4d^{10}$ (c) Cs^+ : [Xe] , noble gas configuration
d) Co^{2+}: $[Ar]\,3d^7$ (e) Sc^{3+} : [Ar], noble gas configuration

8.11 (a) Lattice energy is the energy required to totally separate one mole of solid ionic compound into its gaseous ions.

(b) The magnitude of the lattice energy depends on the magnitudes of the charges of the two ions, their radii and the arrangement of ions in the lattice. The main factor is the charges, because the radii of ions do not vary over a wide range.

8.12 (a) According to Coulomb's Law, electrostatic attraction increases with increasing charges of the ions and decreases with increasing radius of the ions. Thus, lattice energy (a) **increases** as the charges of the ions increase and (b) **decreases** as the sizes of the ions decrease.

8.13 Coulomb's Law (Equation 8.3) predicts that as the oppositely charged ions approach each other, the energy of interaction will be large and negative. This more than compensates for the energy required to form Ca^{2+} and O^{2-} from the neutral atoms (see Figure 8.4 for the formation of NaCl).

8.14 $Ca(s) \rightarrow Ca(g)$; $Br_2(l) \rightarrow 2Br(g)$; $Ca(g) \rightarrow Ca^+(g) + 1e^-$;

$Ca^+(g) \rightarrow Ca^{2+}(g) + 1e^-$; $2Br(g) + 2e^- \rightarrow 2Br^-(g)$, exothermic;

$Ca^{2+}(g) + 2Br^-(g) \rightarrow CaBr_2(s)$, exothermic; $CaBr_2(s) \rightarrow Ca(s) + Br_2(l)$

8.15 The biggest factor is the magnitude of the charge of the cations, Li^+ vs. Mg^{2+}. Also, according to Table 8.3, Mg^{2+} is actually smaller than Li^+ (0.65 Å vs. 0.68 Å). Therefore, in MgH_2, twice the positive charge is manifested over a smaller distance, so the lattice energy is more than twice as large as that of LiH.

8.16 (a) O^{2-} is smaller than S^{2-} ; the closer approach of the oppositely charged ions in MgO leads to greater electrostatic attraction.

(b) The ions have 1+ and 1- charges in both compounds. However, the fact that the ionic radii are much smaller in LiF means that the ions can approach more closely, with a resultant increase in electrostatic attractive forces.

(c) The ions in CaO have 2+ and 2- charges, as compared with 1+ and 1- charges in KF.

8.17 $RCl(s) \rightarrow Rb^+(g) + Cl^-(g)$ ΔH (lattice energy) = ?

By analogy to NaCl, Figure 8.4, the lattice energy is:

$\Delta H_{latt} = -\Delta H_f^\circ\ RbCl(s) + \Delta H_f^\circ\ Rb(g) + \Delta H_f^\circ\ Cl(g) + I_1(Rb) + E(Cl)$

$= -(-430.5\text{ kJ}) + 85.8\text{ kJ} + 121.7\text{ kJ} + 403\text{ kJ} + (-349\text{ kJ}) = +692\text{ kJ}$

This value is smaller than that for NaCl (+788 kJ) because Rb^+ has a larger ionic radius than Na^+. This means that the value of d in the denominator of Equation 8.3 is larger for RbCl, and the potential energy of the electrostatic attraction is smaller.

8.18 By analogy to Figure 8.4:

$\Delta H_{latt} = -\Delta H_f^\circ\ CaF_2 + \Delta H_f^\circ\ Ca(g) + \Delta H_f^\circ\ F(g) + I_1(Ca) + I_2(Ca) + 2E(F)$

$= -(-1219.6\text{ kJ}) + 179.3\text{ kJ} + 2(80.0\text{ kJ}) + 590\text{ kJ} + 1145\text{ kJ} + 2(-332\text{ kJ}) = +2630\text{ kJ}$

From Table 8.2, the lattice energy of NaF, + 911 kJ/mol, is considerably less than that of CaF_2. The +2 charge of Ca^{2+} leads to much greater electrostatic attractions and a higher lattice energy.

Sizes of Ions

8.19 (a) As Z stays constant and the number of electrons increases, the electron-electron repulsions increase, the electrons spread apart and the ions become larger. Thus, I^- is larger than I, which is larger than I^+.

(b) Going down a family, the increased distance of the electrons from the nucleus and increased shielding by inner electrons outweights the buildup of Z and causes the size of particles with like charge to increase.

(c) Fe^{2+}: $[Ar]3d^6$, Fe^{3+}: $[Ar]3d^5$ Since there are five 3d orbitals, in Fe^{2+} at least one orbital must contain a pair of electrons. Removing one electron to form Fe^{3+} significantly reduces repulsion, increasing the nuclear charge experienced by each of the other d electrons and decreasing the size of the ion.

8.20 (a) The additional electrostatic repulsion produced by adding an electron to a neutral atom decreases the Z_{eff} of the valence electrons, causing them to be less tightly bound to the nucleus and the size of the anion to be larger.

(b) Electrostatic repulsions are reduced by removing an electron from a neutral atom, Z_{eff} increases, and the cation is smaller.

8.21 (a) In an isoelectronic series, the number of electrons and the electron configurations of the particles are the same, but the values of Z are different.

(b) Cl^-: **Ar**; Se^{2-}: **Kr**; Mg^{2+}: **Ne**

8.22 (a) K^+, Ca^{2+} (b) Ca^{2+}, Sc^{3+} (c) S^{2-}, Ar (d) Co^{3+}, Fe^{2+}

8.23 Since the electron configurations in an isoelectronic series are the same, repulsion and shielding effects do not vary for the different particles. As Z increases, Z_{eff} increases, the valence electrons are more strongly attracted to the nucleus and the size of the particle decreases.

8.24 (a) This is an isoelectronic series with the electron configuration of Ar; P^{3-} has the smallest value of Z and the largest radius.

(b) This is an isoelectronic series with the same electron configurtion as Kr; Br^- has the smallest value of Z and the largest radius.

8.25 (a) $Li^+ < K^+ < Rb^+$ (b) $Mg^{2+} < Na^+ < Br^-$ (c) $K^+ < Ar < Cl^- < S^{2-}$ (d) $Ar < Cl < Cl^-$

8.26 (a) $Se < Se^{2-} < Te^{2-}$ (b) $Co^{3+} < Fe^{3+} < Fe^{2+}$ (c) $Ti^{4+} < Sc^{3+} < Ca$ (d) $Be^{2+} < Na^+ < Ne$

Covalent Bonding, Electronegativity and Bond Polarity

8.27   H· + H· + ·S̈: ⟶ H—S̈:
                                    |
                                    H

8.28   :F̈· + :F̈· + :F̈· + ·N̈· ⟶ F—N̈—F
                                              |
                                              F

8.29   (a) Ö = C = Ö   (b) :N≡N: or :C≡O:

8.30   The greater the number of electron pairs shared between the two atoms, the shorter the distance between the atoms (that is, the shorter the chemical bond). Thus, the length of C-O is greater than the length of C=O, which is greater than the length of C≡O. Note that the identities of the two atoms must remain the same for the comparison to be valid.

8.31   (a)   Electronegativity is the ability of an atom in a molecule (a bonded atom) to attract electrons to itself.
        (b)   Fluorine, F, is the most electronegative element.

8.32   (a)   The electronegativity of the elements increases going from left to right across a row of the periodic chart.
        (b)   Electronegativity decreases going down a family of the periodic chart.
        (c)   Generally, the trends in electronegativity are the same as those in ionization energy and opposite those in electron affinity. That is, the more positive the ionization energy and the more negative the electron affinity (omitting a few exceptions), the greater the electronegativity of an element.

8.33   (a) P < S < O   (b) Mg < Al < Si   (c) S < Br < Cl   (d) Si < C < N

8.34   (a)   C   (b) Cl   (c) Si would be predicted from periodic trends, but Ge is actually more electronegative, due to incomplete shielding by the 3d electrons   (d) Se

8.35   All except (b) are polar to some extent, because the two elements involved differ in electronegativity. The more electronegative element in each polar bond is: (a) O, (c) Br, (d) N, (e) F.

8.36   In each case, assume that the greater the electronegativity difference (ΔEN) between the atoms, the more polar the bond.

	$\delta-$ $\delta+$	$\delta+$ $\delta-$
(a) N—Cl	N — Br	N—F
3.0 3.0	3.0 2.8	3.0 4.0

This is an order that cannot be predicted by trends alone. Since N has a relatively high electronegativity, the direction of the dipole changes within the series. The N-F bond has the largest electronegativity difference (ΔEN), N-Cl the smallest, and N-Br is intermediate, but the negative end of the dipole is no longer at the halogen.

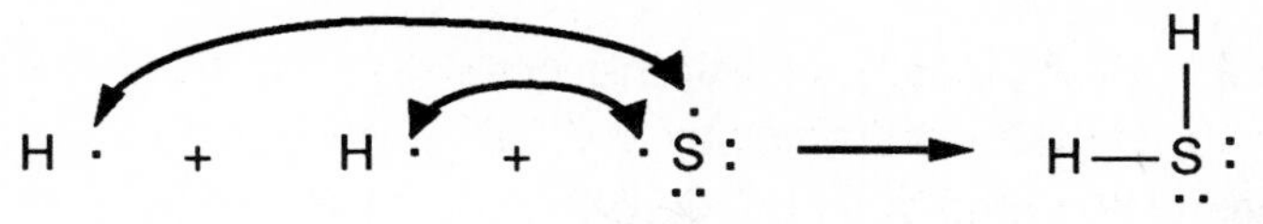

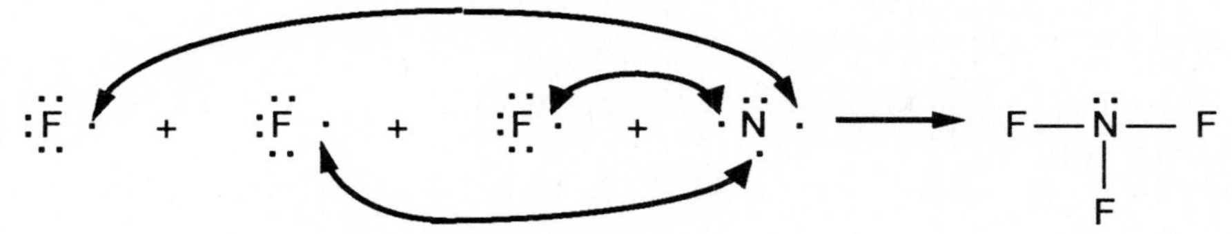

	δ+ δ-	δ+ δ-	δ+ δ-
(b)	O—F	H—F	B—F
	3.5 4.0	2.1 4.0	2.0 4.0

Again, this order cannot be predicted strictly on trends. H has a higher electronegativity value than expected by its position on the periodic chart. Thus, the B-F bond has a slightly greater ΔEN than the H-F bond.

		δ+ δ-	δ- δ+
(c)	N—N	N—O	N—P
		3.0 3.5	3.0 2.1

Again, the electronegativity values must be used to obtain a correct order. Trends correctly predict the direction of these dipoles (O is more electronegative than N and P is less electronegative), but the magnitudes of the differences can only be determined by using numerical values of electronegativity.

Lewis Structures; Resonance Structures

8.37 Counting the **correct number of valence electrons** is the foundation of every Lewis structure.

(a) Count valence electrons: 5 + (3 × 1) = 8 e^-, 4 e^- pairs. Follow the procedure in Sample Exercise 8.6.

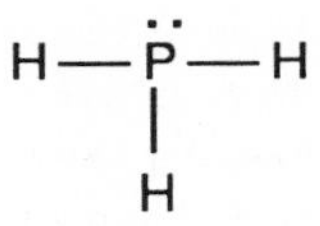

(b) Valence electrons: [7 + (3 × 6) + 1] = 26 e^-, 13 e^-pairs (The single electron is for the -1 charge on the ion.)

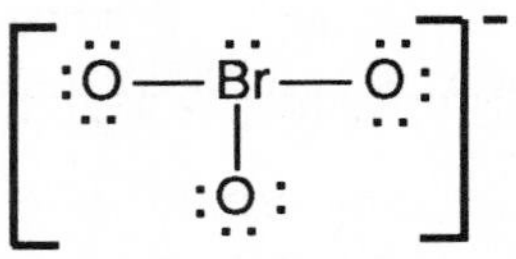

i. Place the Br atom in the middle and connect each O atom with a single bond; this requires 3 e^- pairs.

ii. Complete the octets of the O atoms with lone pairs of electrons; this requires an additional 9 e^- pairs.

iii. The remaining e^- pair completes the octet of the central Br atom.

(c) 10 valence e^-, 5 e^- pairs (d) 20 valence e^-, 10 e^- pairs (e) 20 valence e^-, 10 e^- pairs

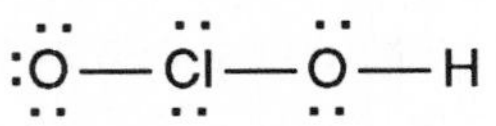

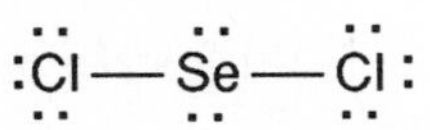

8.38 (a) 8 e$^-$, 4 e$^-$ pairs

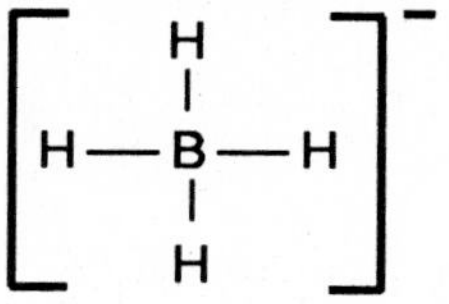

(b) 14 e$^-$, 7 e$^-$ pairs

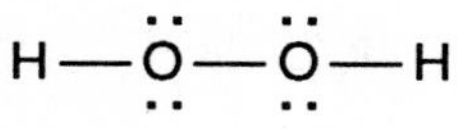

(c) 14 e$^-$, 7 e$^-$ pairs

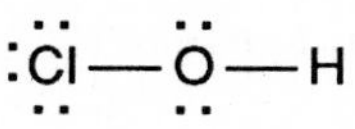

(d) 10 e$^-$, 5 e$^-$ pairs

There are too few e$^-$ pairs to complete the octets of the C atoms with lone pairs, so multiple bonds must be used.

H—C≡C—H

(e) 18 e$^-$, 9 e$^-$ pairs

In the second form, Cl has a formal charge of 1+ and O of 1-, so the first form, where all the formal charges are zero, is the major contributor.

8.39 (a) 24 e$^-$, 12 e$^-$ pairs

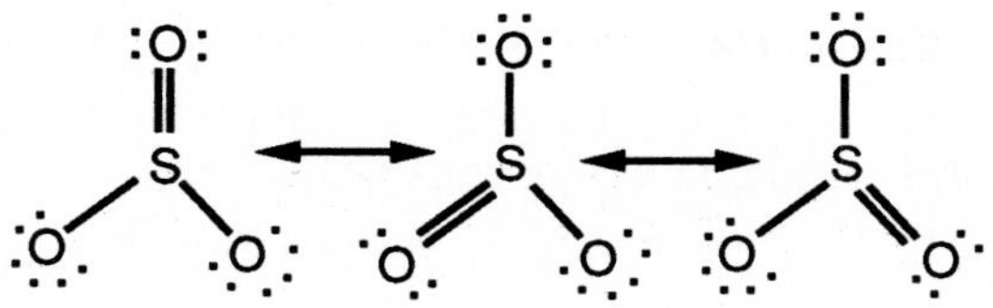

(b) 34 e$^-$, 17 e$^-$ pairs

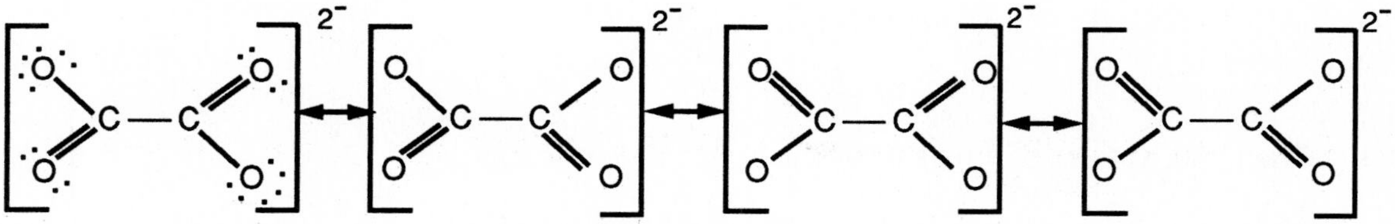

(nonbonded pairs on terminal O atoms have been omitted on all but the first form)

(c) 24 e$^-$, 12 e$^-$ pairs

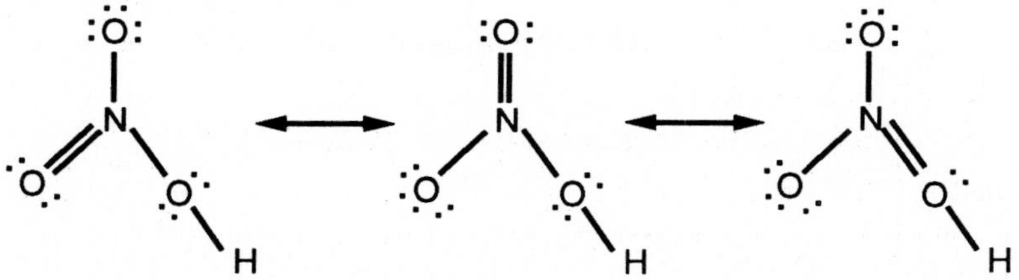

The third form is not a significant contributor because of the nonzero formal charges on all three O atoms.

(d)   25 e⁻; 12 e⁻ pairs, 1 odd e⁻

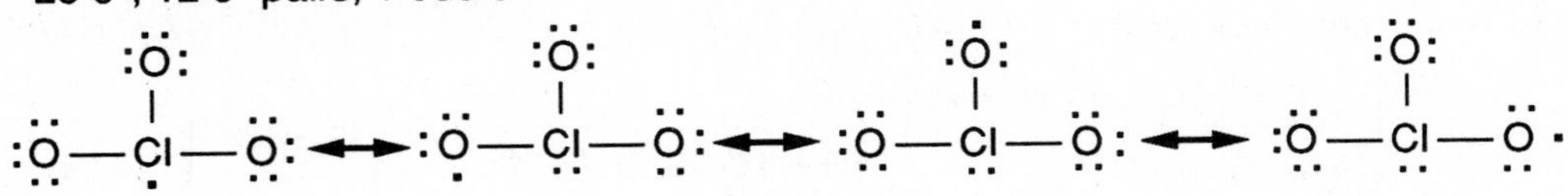

$$:\overset{..}{\underset{..}{O}}: \quad\overset{|}{\underset{..}{Cl}}\quad :\overset{..}{\underset{..}{O}}: \longleftrightarrow :\overset{..}{\underset{..}{O}}: \quad\overset{|}{\underset{..}{Cl}}\quad :\overset{..}{\underset{..}{O}}: \longleftrightarrow :\overset{..}{\underset{..}{O}}: \quad\overset{|}{\underset{..}{Cl}}\quad :\overset{..}{\underset{..}{O}}: \longleftrightarrow :\overset{..}{\underset{..}{O}}: \quad\overset{|}{\underset{..}{Cl}}\quad :\overset{..}{\underset{..}{O}}\cdot$$

8.40   (a)   18 e⁻, 9 e⁻ pairs

$$\left[\ \overset{..}{O}=\overset{..}{N}-\overset{..}{\underset{..}{O}}:\right]^{-} \longleftrightarrow \left[:\overset{..}{\underset{..}{O}}-\overset{..}{N}=\overset{..}{O}\ \right]^{-}$$

(b)   24 e⁻, 12 e⁻ pairs

$$\left[\begin{array}{c}:\overset{..}{O}:\\|\\\overset{..}{O}=C-\overset{..}{\underset{..}{O}}\end{array}\right]^{2-} \longleftrightarrow \left[\begin{array}{c}:\overset{..}{O}:\\||\\:\overset{..}{\underset{..}{O}}-C-\overset{..}{\underset{..}{O}}:\end{array}\right]^{2-} \longleftrightarrow \left[\begin{array}{c}:\overset{..}{\underset{..}{O}}:\\|\\:\overset{..}{\underset{..}{O}}-C=\overset{..}{\underset{..}{O}}\end{array}\right]^{2-}$$

(c)   16 e⁻, 8 e⁻ pairs

$$\left[\ \overset{..}{\underset{}{S}}=C=\overset{..}{N}\ \right]^{-} \longleftrightarrow \left[:\overset{..}{\underset{..}{S}}-C\equiv N:\right]^{-} \longleftrightarrow \left[:S\equiv C-\overset{..}{\underset{..}{N}}:\right]^{-}$$
$$\;\;0\quad\;\;0\;\;\;-1 \qquad\qquad\quad -1\;\;\;\;0\;\;\;0 \qquad\qquad\quad +1\;\;\;\;0\;\;\;-2$$

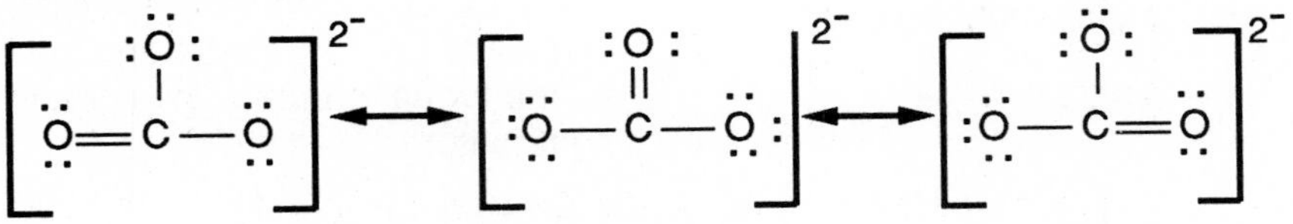

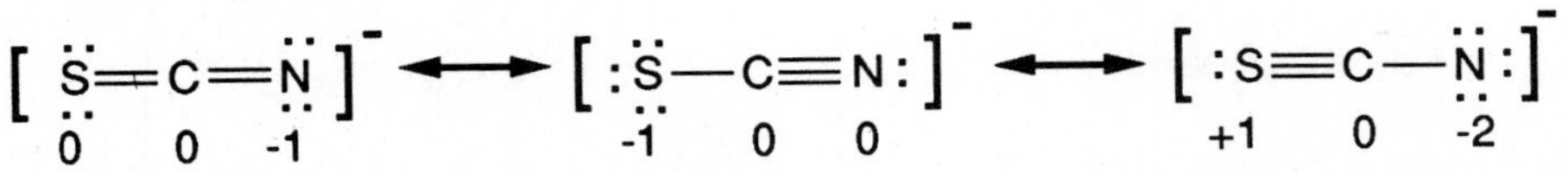

The numbers below the atoms are formal charges. The third form is not a significant contributor because of the additional formal charges.

(d)   18 e⁻, 9 e⁻ pairs

$$\left[\begin{array}{c}:O:\\||\\H-C-\overset{..}{\underset{..}{O}}:\end{array}\right]^{-} \longleftrightarrow \left[\begin{array}{c}:\overset{..}{\underset{..}{O}}:\\|\\H-C=\overset{..}{\underset{..}{O}}\end{array}\right]^{-}$$

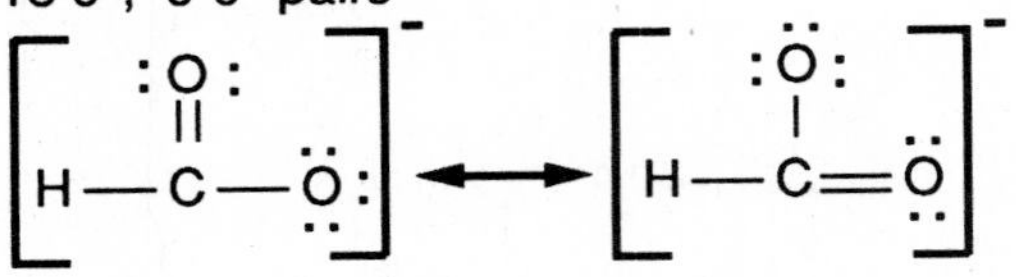

8.41   The Lewis structures are as follows:

CO $:C\equiv O:$ CO₂ $\overset{..}{O}=C=\overset{..}{O}$

   10 e⁻, 5 e⁻ pairs   16 e⁻, 8 e⁻ pairs

$$CO_3^{2-} \quad \left[\begin{array}{c}:\overset{..}{O}:\\|\\\overset{..}{O}=C-\overset{..}{\underset{..}{O}}\end{array}\right]^{2-} \longleftrightarrow \left[\begin{array}{c}:O:\\||\\:\overset{..}{\underset{..}{O}}-C-\overset{..}{\underset{..}{O}}:\end{array}\right]^{2-} \longleftrightarrow \left[\begin{array}{c}:\overset{..}{O}:\\|\\:\overset{..}{\underset{..}{O}}-C=\overset{..}{\underset{..}{O}}\end{array}\right]^{2-}$$

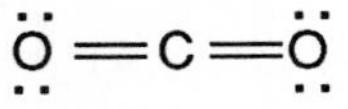

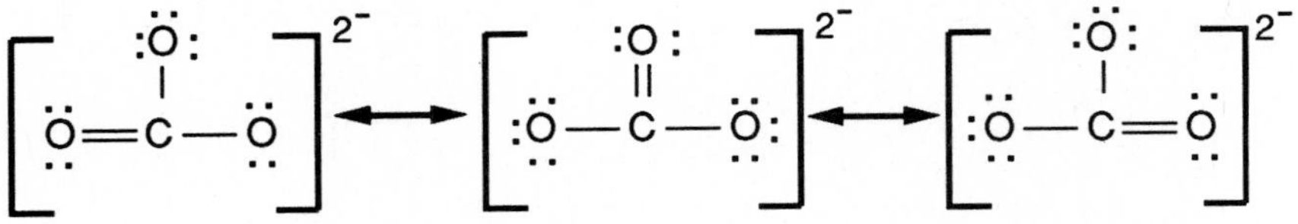

24 e⁻, 12 e⁻ pairs

According to Exercise 8.30, the more pairs of electrons shared by two atoms, the shorter the bond between the atoms. The average number of electron pairs shared by C and O in the three species is 3 for CO, 2 for CO_2 and 1.33 for CO_3^{2-}. This is also the order of increasing bond length: $CO < CO_2 < CO_3^{2-}$.

8.42 The Lewis structures are as follows:

5 e^- pairs 5 e^- pairs, 1 odd e^- 9 e^- pairs

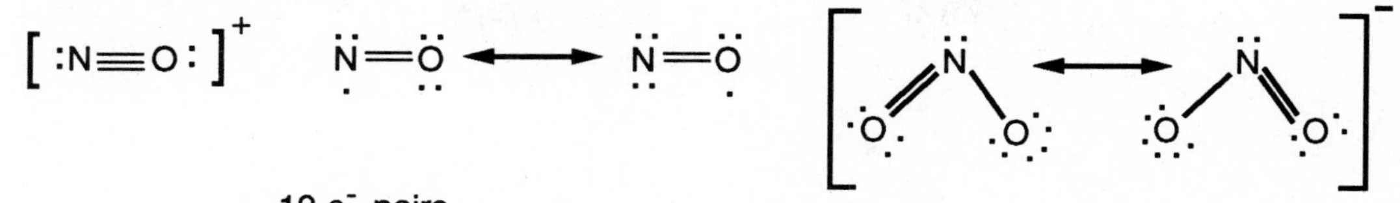

12 e^- pairs

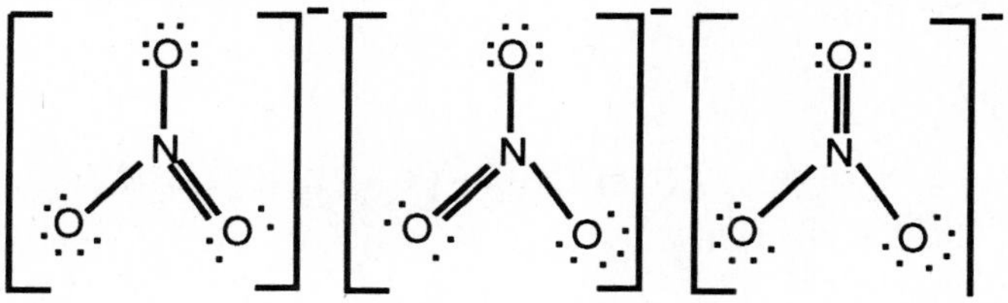

The average number of electron pairs in the N-O bond is 3.0 for NO^+ , 2.0 for NO , 1.5 for NO_2^- and 1.33 for NO_3^- . The more electron pairs shared between two atoms, the shorter the bond (see Exercise 8.30). Thus the N-O bond lengths vary in the order NO^+ < NO < NO_2^- < NO_3^- .

8.43 Formal charge (FC) = # valence e^- - (# nonbonding e^- + 1/2 # bonding e^-)

(a) 18 e^-, 9 e^- pairs

FC for the central O = 6 - [2 + 1/2 (6)] = +1

(b) 48 e^-, 24 e^- pairs

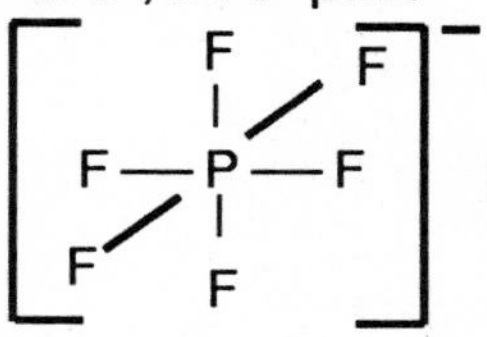

FC for P = 5 - [0 + 1/2 (12)] = -1

The three nonbonded pairs on each F have been omitted.

(c) 17 e^-; 8 e^- pairs, 1 odd e^-

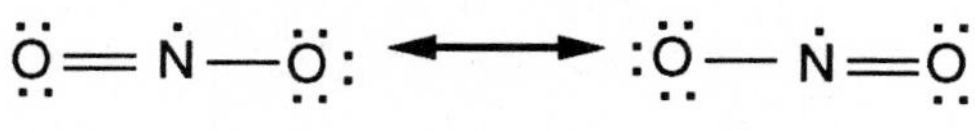

The odd electron is probably on N because it is less electronegative than O.

Assuming the odd electron is on N, FC for N = 5 - [1 + 1/2 (6)] = +1.

(d)   28 e⁻ ,   14 e⁻ pairs

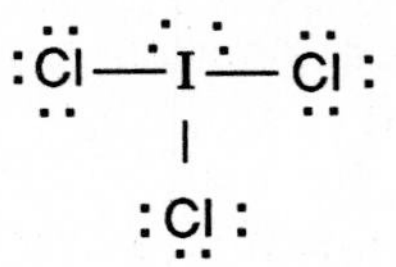

:Cl—I—Cl: FC for I = 7 - [4 + 1/2 (6)] = 0

:Cl:

(e)   32 e⁻ ,   16 e⁻ pairs

:O:

:O—Cl—O—H   FC for Cl = 7 - [0 + 1/2 (8)] = +3

:O:

8.44 (a)   16 e⁻ ,   8 e⁻ pairs

N=N=O   N=O=N   The first choice is preferable, because the
-1 +1   0   -1 +2 -1   formal charges are smaller.

(b)   10 e⁻ ,   5 e⁻ pairs

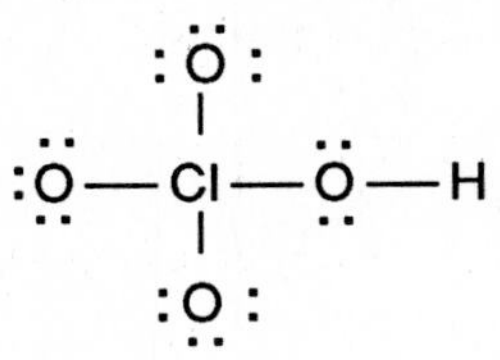

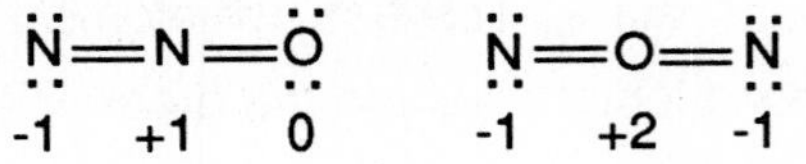

H—C≡N: H—N≡C: The HCN is preferred.
0   0   0   0   +1 -1

(c)   18 e⁻ ,   9 e⁻ pairs

N=O—Br: O=N—Br: The ONBr arrangement is preferred.
-1 +1   0   0   0   0

(d)   16 e⁻ ,   8 e⁻ pairs.   There are three resonance structures for each atomic
       arrangement.

[N=S=C]⁻   [S=N=C]⁻   [S=C=N]⁻
  -1 +2 -2   0 +1 -2   0   0   0

↕   ↕   ↕

[:N—S≡C:]⁻   [:S—N≡C:]⁻   [:S—C≡N:]⁻
  -2 +2 -1   -1 +1 -1   -1   0   0

↕   ↕   ↕

[:N≡S—C:]⁻   [:S≡N—C:]⁻   [:S≡C—N:]⁻
  0   +2 -3   +1 +1 -3   +2   0   -2

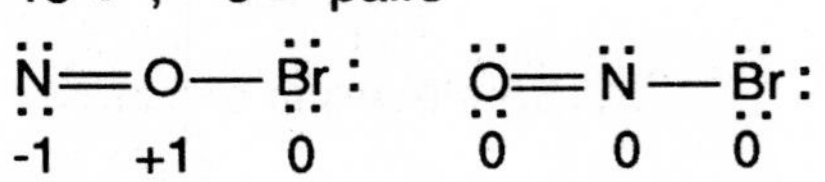

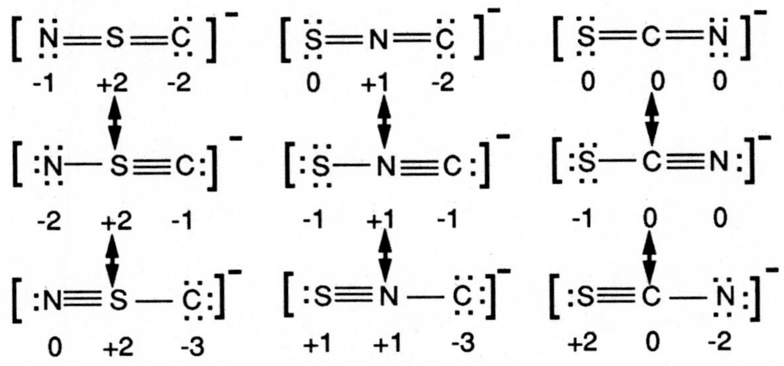

The SCN⁻ skeleton is clearly favored.

Exceptions to the Octet Rule

8.45 The most common exceptions to the octet rule are molecules with more than eight electrons around one or more atoms, usually the central atom.

8.46 In the third period, atoms have the space and available orbitals to accommodate extra electrons. Since atomic radius increases going down a family, elements in the third period and beyond are less subject to destabilization from additional electron-electron repulsions. Also, the third shell contains d orbitals that are relatively close in energy to 3s and 3p orbitals (the ones that accommodate the octet) and provide an allowed energy state for the extra electrons.

8.47 (a) 17 e^-, 8.5 e^- pairs ;
odd electron molecule

The odd electron is probably on N because it is less electronegative than O.

(b) 32 e^-, 16 e^- pairs

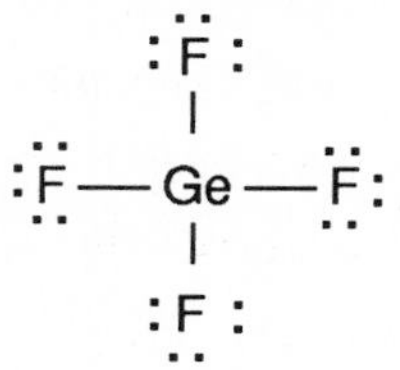

(c) 34 e^-, 17 e^- pairs

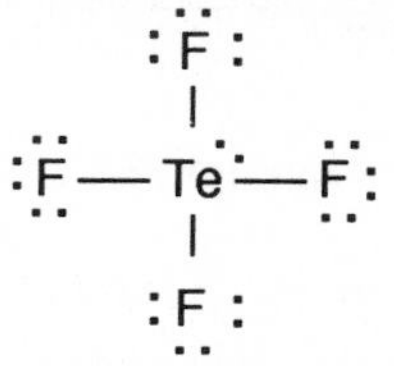

10 e^- around the central Te

(d) 24 e^-, 12 e^- pairs

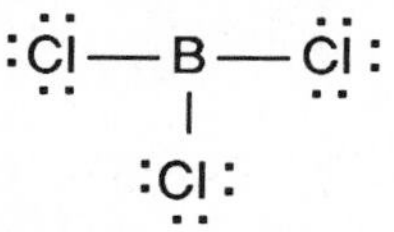

6 electrons around B

(e) 36 e^-, 18 e^- pairs

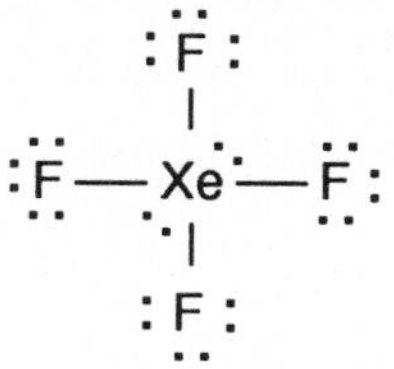

12 e^- around the central Xe

8.48 (a) 26 e^-, 13 e^- pairs

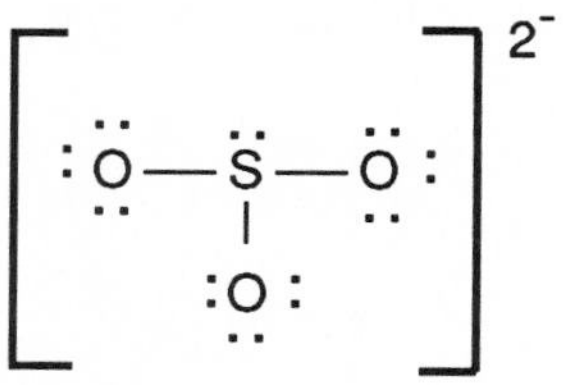

(b) 6 e^-, 3 e^- pairs, impossible to satisfy octet rule with only 6 valence electrons.

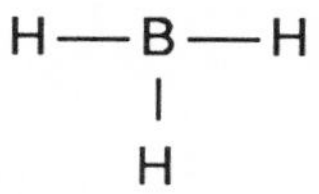

6 electrons around B

(c) 22 e^-, 11 e^- pairs

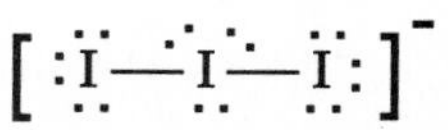

10 e^- around central I

(d) 48 e⁻, 24 e⁻ pairs

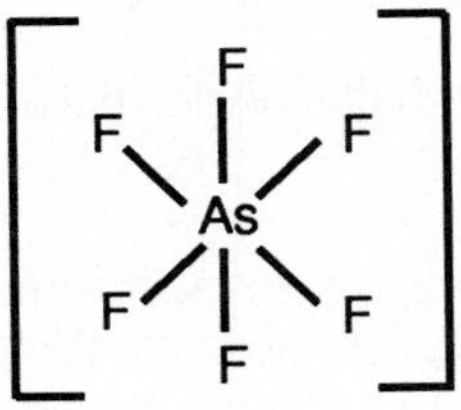

12 e⁻ around As; three nonbonded pairs on each F have been omitted

(e) 17 e⁻. 8.5 e⁻ pairs, odd electron molecule

8.49 (a) 16 e⁻, 8 e⁻ pairs

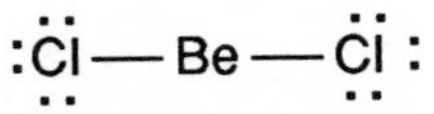

This structure violates the octet rule; Be has only 4 e⁻ around it.

(b)

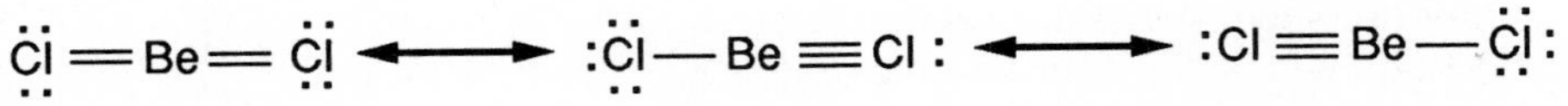

(c) The formal charges on each of the atoms in the four resonance structures are:

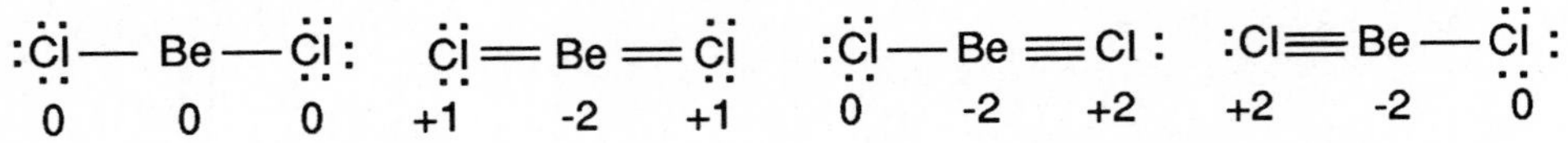

Since formal charges are minimized on the structure which violates the octet rule, this form is probably most important.

8.50 (a) 19 e⁻, 9.5 e⁻ pairs odd electron molecule

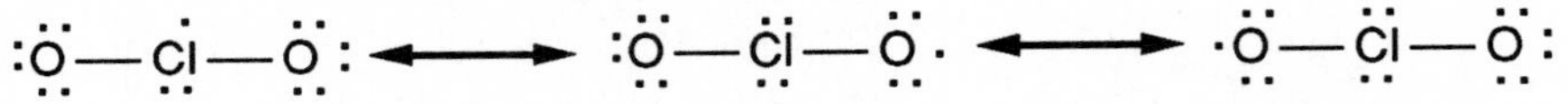

(b) None of the structures satisfies the octet rule. In each structure, one atom has only 7 e⁻ around it. If a molecule has an odd number of electrons in the valence shell, no Lewis structure can satisfy the octet rule.

(c)

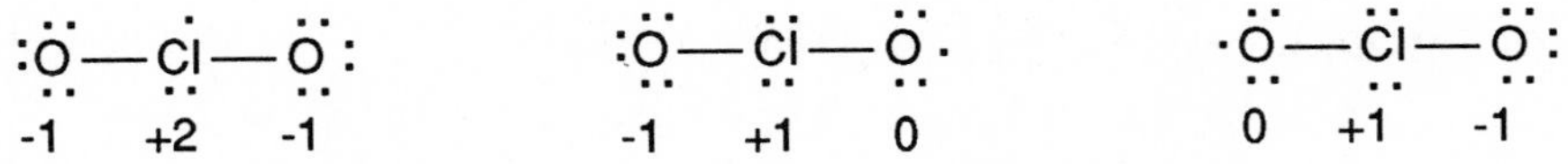

Formal charge arguments predict that the two resonance structures with the odd electron on O are most important. This contradicts electronegativity arguments, which would predict that the less electronegative atom, Cl, would be more likely to have fewer than 8 e⁻ around it.

CHAPTER 8 Basic Concepts of Chemical Bonding

Solutions to Exercises

Bond Energies

8.51 (a) $\Delta H = D(H\text{-}Br) + 3D(C\text{-}H) + D(C\text{-}O) + D(O\text{-}H) - 2D(O\text{-}H) - 3D(C\text{-}H) - D(C\text{-}Br)$
$= D(H\text{-}Br) + D(C\text{-}O) - D(O\text{-}H) - D(C\text{-}Br)$
$= 366 + 358 - 463 - 276 = -15$ kJ

(b) $\Delta H = 3D(C\text{-}Cl) + D(C\text{-}H) - 2D(C\text{-}Cl) - D(C{=}O) - D(H\text{-}Cl)$
$= D(C\text{-}Cl) + D(C\text{-}H) - D(C{=}O) - D(H\text{-}Cl)$
$= 328 + 413 - 799 - 431 = -489$ kJ

(c) $\Delta H = D(C{\equiv}O) + 2D(O\text{-}H) - D(H\text{-}H) - 2D(C{=}O)$
$= 1072 + 2(463) - 436 - 2(799) = -36$ kJ

8.52 (a) $\Delta H = 2D(O\text{-}H) + D(O\text{-}O) + 4D(C\text{-}H) + D(C{=}C)$
$- 2D(O\text{-}H) - 2D(O\text{-}C) - 4D(C\text{-}H) - D(C\text{-}C)$
$\Delta H = D(O\text{-}O) + D(C{=}C) - 2D(O\text{-}C) - D(C\text{-}C)$
$= 146 + 614 - 2(358) - 348 = -304$ kJ

(b) $\Delta H = 5D(C\text{-}H) + D(C{\equiv}N) + D(C{=}C) - 5D(C\text{-}H) - D(C{\equiv}N) - 2D(C\text{-}C)$
$= D(C{=}C) - 2D(C\text{-}C) = 614 - 2(348) = -82$ kJ

(c) $\Delta H = 6D(N\text{-}Cl) - 3D(Cl\text{-}Cl) - D(N{\equiv}N)$
$= 6(200) - 3(242) - 941 = -467$ kJ

8.53 First Lewis structures must be drawn so that bonds can be visualized:

(a) 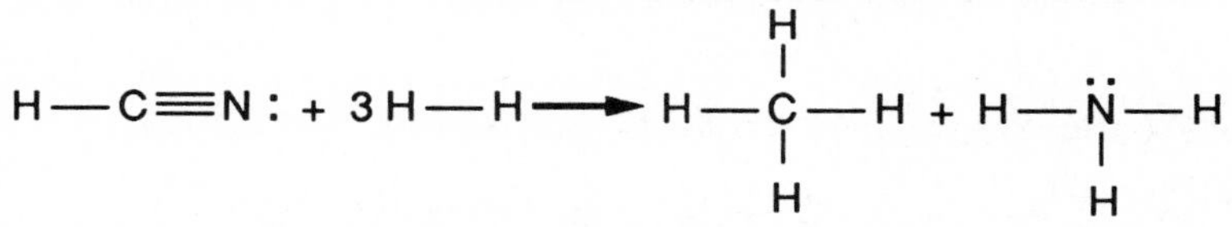

$\Delta H = D(C{\equiv}N) + D(C\text{-}H) + 3(H\text{-}H) - 4D(C\text{-}H) - 3D(N\text{-}H)$
$= 891 + 3(436) - 3(413) - 3(391) = -213$ kJ

(b) 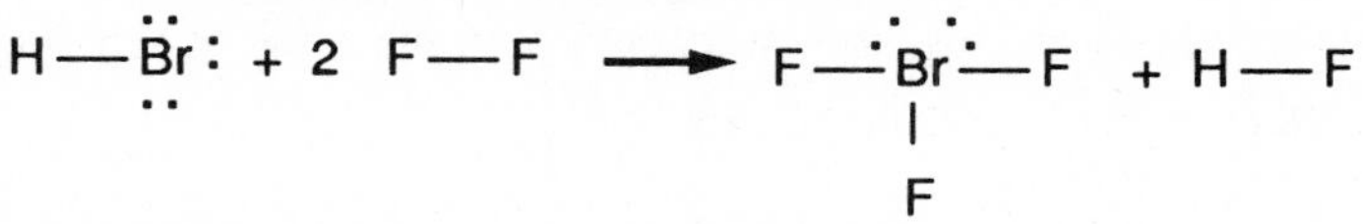

(Nonbonded pairs on F have been omitted.)

$\Delta H = D(H\text{-}Br) + 2D(F\text{-}F) - D(H\text{-}F) - 3D(Br\text{-}F)$
$= 366 + 2(155) - 567 - 3(237) = -602$ kJ

8.54 (a) 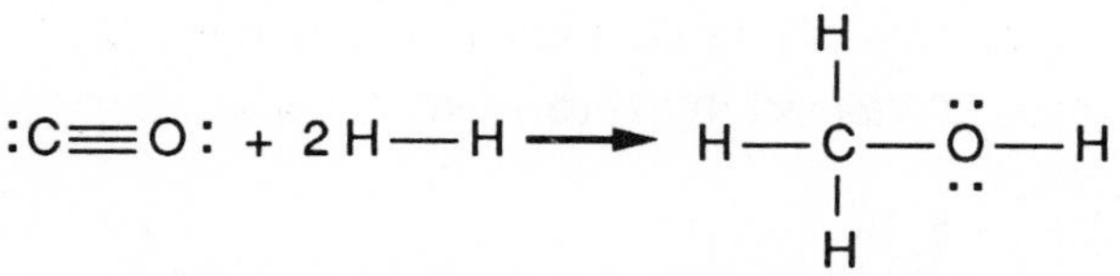

$\Delta H = D(C{\equiv}O) + 2D(H\text{-}H) - 3D(C\text{-}H) - D(O\text{-}H) - D(C\text{-}O)$
$= 1072 + 2(436) - 3(413) - 463 - 358 = -116$ kJ

(b) 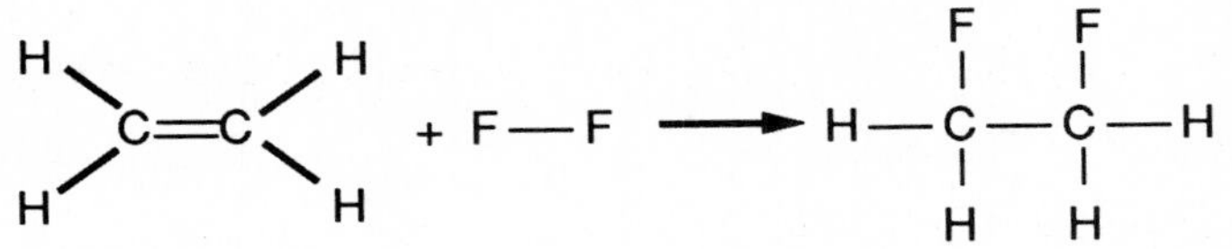

(Nonbonded pairs on F have been omitted.)

ΔH = 4D(C-H) + D(C=C) + D(F-F) - 4D(C-H) - 2D(C-F) - D(C-C)
= D(C=C) + D(F-F) - 2D(C-F) - D(C-C)
= 614 + 155 - 2(485) - 348 = -549 kJ

8.55 The average Ti-Cl bond dissociation energy is just the average of the four values listed, 430 kJ/mol.

8.56 Since bond energies are estimated for gas phase molecules, the calculated value for the C-C bond will be most appropriate if the reaction under consideration has all reactants and products in the gas phase. We can use Hess's Law to calculate ΔH for the formation of $C_2H_6(g)$ from gaseous atoms.

$2C(s) + 3H_2(g) \rightarrow C_2H_6(g)$	$\Delta H_f^\circ = -84.68$ kJ
$6H(g) \rightarrow 3H_2(g)$	$\Delta H = -6(217.94)$
$2C(g) \rightarrow 2C(s)$	$\Delta H = -2(718.4$ kJ)
$2C(g) + 6H(g) \rightarrow C_2H_6(g)$	$\Delta H = -2829$ kJ

ΔH = -6D(C-H) - D(C-C)

D(C-C) = -6D(C-H) - ΔH

= -6(413 kJ) - (-2829) + +351 kJ

This is very similar to the average C-C bond energy, 348 kJ, repeated in Table 8.4.

Oxidation Numbers

8.57 (a) P, 0 (b) As, +3 (c) C, +4 (d) P, +5 (e) N, +3 (f) Br, +3 (g) O, -1 (h) Pb, +2

8.58 (a) S, +4 (b) Hg, +1 (c) Co, +2 (d) Xe, +6 (e) Mn, +7 (f) N, +3 (g) Cl, +7 (h) U, +6

8.59 (a) N +5, -3 (b) Se +6, -2 (c) Ca +2, 0 (d) Mn +7, 0

8.60 (a) As +5, -3 (b) Br +7, -1 (c) Sc +3, 0 (d) Cr +6, 0

8.61 (a) FeF_3 (b) MoO_3 (c) $AsBr_5$ (d) vanadium(III) oxide
(e) cobalt(III) fluoride (f) manganese(II) sulfide

8.62 (a) MnO_2 (b) Ga_2S_3 (c) SeF_6 (d) Copper(I) oxide or cuprous oxide
(e) chlorine trifluoride
(f) tellurium trioxide or tellurium(VI) oxide

8.63 (a) Xenon is oxidized from 0 to +4; fluorine is reduced from 0 to -1
(b) copper is reduced from +2 to +1; iodine is oxidized from -1 to 0
(c) nitrogen is oxidized from -3 to +3; chlorine is reduced from 0 to -1
(d) iodine is reduced from 0 to -1; sulfur is oxidized from +4 to +6
(e) sulfur is oxidized from -2 to +4; oxygen is reduced from 0 to -2

8.64 (a) Iodine is oxidized from -1 to 0; fluorine is reduced from 0 to -1
(b) manganese is reduced from +4 to +2; chlorine is oxidized from -1 to 0
(c) oxygen is oxidized from -1 to 0 and reduced from -1 to -2
$2H_2O_2(aq) \rightarrow 2H_2O(l) + O_2(g)$ (disproportionation)
(d) carbon is oxidized from -4 to -2; O is reduced from 0 to -2
(e) iron is oxidized from +2 to +3; manganese is reduced from +7 to +2

Additional Exercises

8.65 (a) Group 4A or 4B (b) Group 2A (c) Group 5A or 5B

8.66 (a) Cu^+, $[Ar]3d^{10}$, does **not** have noble gas configuration
(b) Cd^{2+}, $[Kr]4d^{10}$, does **not** have noble gas configuration
(c) Ti^{4+}, [Ar], noble gas configuration
(d) Sc^{3+}, [Ar], noble gas configuration

8.67 (a) Ni^{2+}: $[Ar]\ 3d^8$ (b) Cr^{3+}: $[Ar]\ 3d^3$
(c) Bi^{3+}: $[Xe]\ 6s^2 4f^{14} 5d^{10}$ (d) Sc^{3+}: [Ar]

8.68 Each elemental transition metal has two ns electrons in its outer shell, which will be lost before (n - 1) d electrons upon ionization. There is a small energy difference between the ns and (n - 1) d subshells so the +2 oxidation state will be stable, regardless of further ionization involving the (n-1) d electrons.

8.69 The steady decrease in lattice energies is due to the increasing size of the cation. The potential energy of interaction varies as 1/d, where d is the distance between ionic centers. Thus, the larger the cation, the larger the d and the lower the lattice energy.

8.70 $E = k\,Q_1Q_2/r$; $k = 8.99 \times 10^9$ J-m/coul^2

(a) $$E = \frac{-8.99 \times 10^9 \text{ J-m}}{1 \text{ coul}^2} \times \frac{(2 \times 1.60 \times 10^{-19} \text{ coul})^2}{(0.65 + 1.45) \times 10^{-10} \text{ m}} = -4.38 \times 10^{-18} \text{ J}$$

The sign of E is negative because one of the interacting ions is an anion; this is an attractive interaction.

On a molar basis: -4.39×10^{-18} J $\times$ 6.02×10^{23} = -2.64×10^6 J

(b) $$E = \frac{-8.99 \times 10^9 \text{ J-m}}{1 \text{ coul}^2} \times \frac{(1.60 \times 10^{-19} \text{ coul})^2}{(0.98 + 1.96) \times 10^{-10} \text{ m}} = -7.83 \times 10^{-19} \text{ J}$$

On a molar basis: -4.71×10^5 J

8.71 $$E = \frac{-8.99 \times 10^9 \text{ J-m}}{1 \text{ coul}^2} \times \frac{(1.60 \times 10^{-19} \text{ coul})^2}{(0.98 + 1.81) \times 10^{-10} \text{ m}} = -8.25 \times 10^{-19} \text{ J}$$

On a molar basis: $(-8.25 \times 10^{-19}$ J$)(6.02 \times 10^{23})$ = -496 kJ

Note that its absolute value is less than the lattice energy, 785 kJ/mol. The difference represents the added energy of putting all the Na^+Cl^- ion pairs together in a three-dimensional array, as shown in Figure 8.3.

8.72 The pathway to the formation of SrO(s) can be written:

$Sr(s) \rightarrow Sr(g)$	ΔH_f Sr(g)
$1/2\ O_2(g) \rightarrow O(g)$	ΔH_f O(g)
$Sr(g) \rightarrow Sr^+(g) + 1\ e^-$	I_1(Sr)
$Sr^+(g) \rightarrow Sr^{2+}(g) + 1\ e^-$	I_2(Sr)
$O(g) + 1\ e^- \rightarrow O^-(g)$	E_1(O)
$O^-(g) + 1\ e^- \rightarrow O^{2-}(g)$	E_2(O)
$Sr^{2+}(g) + O^{2-}(g) \rightarrow SrO(s)$	$-\Delta H_{lat}$ SrO(s)
$Sr(s) + 1/2\ O_2(g) \rightarrow SrO(s)$	ΔH_f SrO(s)

ΔH_f° SrO(s) = ΔH_f° Sr(g) + ΔH_f° O(g) + I_1(Sr) + I_2(Sr) + E_1(O) + E_2(O) - ΔH_{lat} SrO(s)

E_2(O) = ΔH_f° SrO(s) + ΔH_{lat} SrO(s) - ΔH_f° Sr(g) - ΔH_f° O(g) - I_1(Sr) - I_2(Sr) - E_1(O)

= -592.0 kJ + 3369 kJ - 164.4 kJ - 247.5 kJ - 549 kJ - 1064 kJ - (-141 kJ)

= + 893 kJ

8.73 I_3^- has a Lewis structure with an expanded octet of electrons around the central I:

$$\left[\; :\ddot{\underset{..}{I}}\!-\!\ddot{\underset{..}{I}}\!-\!\ddot{\underset{..}{I}}:\; \right]^-$$

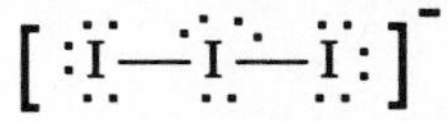

F cannot accommodate an expanded octet because it is too small and has no available d orbitals in its valence shell.

8.74 According to periodic trends in electronegativity, O is the most electronegative element in the group and P the least. Thus, the P-O bond is most polar because its atoms have the greatest difference in electronegativity. The P-P bond is least polar, because ΔEN is necessarily zero.

8.75 $:N\equiv N-\ddot{\underset{..}{O}}:$ ⟷ $:\ddot{N}-N\equiv O:$ ⟷ $:\ddot{N}=N=\ddot{O}:$

       0   1   -1   -2   1   +1   -1   1   0

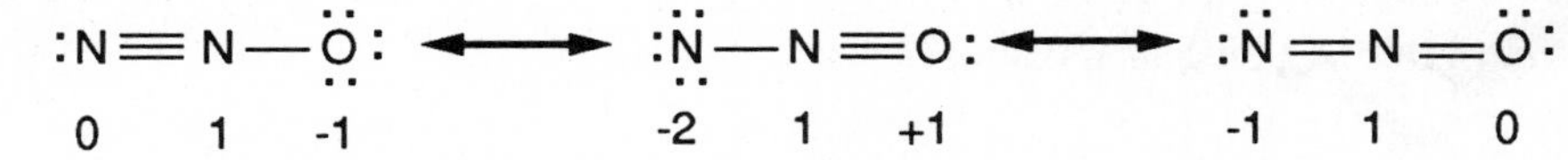

The formal charges are listed below the atoms in the three resonance structures. The structures on the left and right both minimize formal charge to the same extent, so they are likely to be more important to an accurate description of the molecule than the middle structure.

8.76   0 $:\ddot{\underset{..}{F}}-\underset{\underset{0}{\underset{\displaystyle :\ddot{F}:}{|}}}{\overset{0}{B}}-\ddot{\underset{..}{F}}:$ 0   0 $:\ddot{\underset{..}{F}}-\underset{\underset{0}{\underset{\displaystyle :\ddot{F}:}{|}}}{\overset{-1}{B}}=\ddot{F}:$ +1

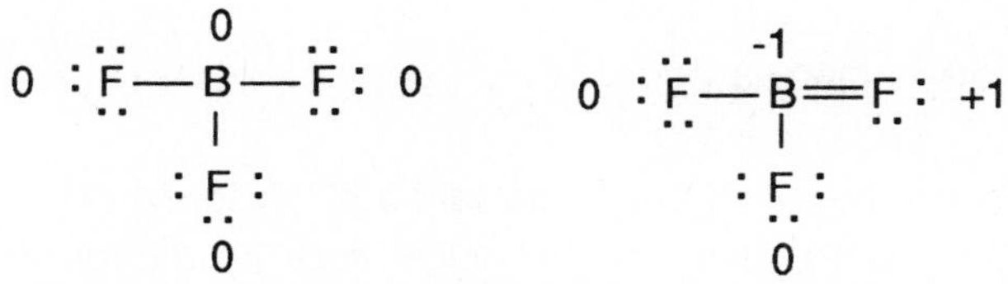

Clearly the structure with less than 8 electrons around B minimizes formal charges. Also, considering the electronegativities, a formal charge of +1 on F seems highly unlikely.

8.77 (a)   18 e$^-$, 9 e$^-$ pairs; SO_2 has the same number of valence electrons as O_3.

       $\ddot{\underset{..}{O}}=\ddot{S}-\ddot{\underset{..}{O}}:$ ⟷ $:\ddot{\underset{..}{O}}-\ddot{S}=\ddot{\underset{..}{O}}$ There are two resonance forms.

(b) $\ddot{\underset{..}{O}}=\ddot{S}=\ddot{\underset{..}{O}}$ There are no other resonance structures

               unless you consider $:O\equiv\ddot{S}\equiv O:$ etc.

(c) $\ddot{\underset{..}{O}}=\ddot{S}-\ddot{\underset{..}{O}}:$ ⟷ $:\ddot{\underset{..}{O}}-\ddot{S}=\ddot{\underset{..}{O}}$ ⟷ $\ddot{\underset{..}{O}}=\ddot{S}=\ddot{\underset{..}{O}}$

       0   +1   -1   -1   +1   0   0   0   0

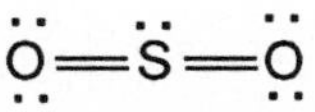

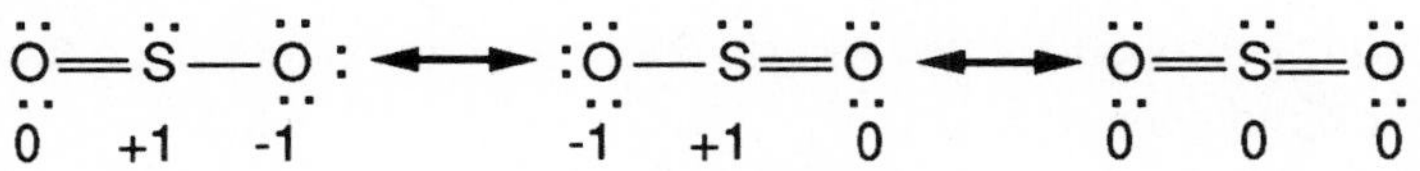

(d)   If the criterion of minimizing formal charges is used, the Lewis structure in (b) is a better representation of the bonding in SO_2.

8.78 (a) $:N{\equiv}N: + 3\,H{-}H \longrightarrow 2\ H{-}\ddot{N}{-}H$ (with a third H bonded below N)

$$\Delta H = D(N \equiv N) + 3\,D(H-H) - 6\,D(N-H)$$
$$= 941\text{ kJ} + 3\,(436\text{ kJ}) - 6\,(391\text{ kJ})$$
$$= -97\text{ kJ / 2 mol } NH_3$$; **exothermic**

(b) ΔH_f° $NH_3(g)$ = -46.19 kJ ; $\Delta H = 2(-46.19) = -92.4$ kJ

The ΔH calculated from bond energies is slightly more exothermic (more negative) than that obtained using ΔH_f° values.

8.79 $\Delta H = 8D(C\text{-}H) - D(C\text{-}C) - 6D(C\text{-}H) - D(H\text{-}H)$

$= 2D(C\text{-}H) - D(C\text{-}C) - D(H\text{-}H) = 2(413) - 348 - 436 = +42$ kJ

$\Delta H = 8D(C\text{-}H) + 1/2\ D(O{=}O) - D(C\text{-}C) - 6D(C\text{-}H) - 2D(O\text{-}H)$
$= 2D(C\text{-}H) + 1/2\ D(O{=}O) - D(C\text{-}C) - 2D(O\text{-}H)$
$= 2(413) + 1/2\ (495) - 348 - 2(463) = -200$ kJ

The fundamental difference in the two reactions is the formation of 1 mol of H-H bonds versus the formation of 2 mol of O-H bonds. The latter is much more exothermic, so the reaction involving oxygen is more exothermic.

8.80 $H_2(g) \rightarrow H(g) + H(g)$ $\Delta H = D(H\text{-}H) = +436$ kJ first reaction

$H_2(g) \rightarrow H(g) + H(g)$	D(H-H) = 436 kJ
$H(g) \rightarrow H^+(g) + 1e^-$	I(H) = 1312 kJ
$H(g) + 1e^- \rightarrow H^-(g)$	E(H) = -73 kJ
$H_2(g) \rightarrow H^+(g) + H^-(g)$	ΔH = 1675 kJ second reaction

The first reaction requires a much smaller energy input than the second, so it would be the preferred pathway for bond cleavage.

8.81

$NH(g) \rightarrow 1/2\ H_2(g) + 1/2\ N_2(g)$	$\Delta H_f^\circ = -360$ kJ
$1/2\ H_2(g) \rightarrow H(g)$	ΔH = 1/2 (436) kJ
$1/2\ N_2(g) \rightarrow N(g)$	ΔH = 1/2 (941) kJ
$NH(g) \rightarrow N(g) + H(g)$	ΔH = 328 kJ

8.82 (a) $\Delta H = 5D(C\text{-}H) + D(C\text{-}C) + D(C\text{-}O) + D(O\text{-}H) - [6D(C\text{-}H) + 2D(C\text{-}O)]$

$= D(C\text{-}C) + D(O\text{-}H) - D(C\text{-}H) - D(C\text{-}O)$

$= 348\text{ kJ} + 463\text{ kJ} - 413\text{ kJ} - 358\text{ kJ}$

$\Delta H = +40\text{ kJ}$; ethanol has the lower enthalpy

(b) $\Delta H = 4D(C\text{-}H) + D(C\text{-}C) + 2D(C\text{-}O) - [4D(C\text{-}H) + D(C\text{-}C) + D(C{=}O)]$

$= 2D(C\text{-}O) - D(C{=}O)$

$= 2(358\text{ kJ}) - 799\text{ kJ}$

$\Delta H = -83\text{ kJ}$; acetaldehyde has the lower enthalpy

(c) $\Delta H = 8D(C\text{-}H) + 4D(C\text{-}C) + D(C{=}C) - [8D(C\text{-}H) + 2D(C\text{-}C) + 2D(C{=}C)]$

$= 2D(C\text{-}C) - D(C{=}C)$

$= 2(348\text{ kJ}) - 614\text{ kJ}$

$\Delta H = +82\text{ kJ}$; cyclopentene has the lower enthalpy

(d) $\Delta H = 3D(C\text{-}H) + D(C\text{-}N) + D(C{\equiv}N) - [3D(C\text{-}H) + D(C\text{-}C) + D(C{\equiv}N)]$

$= D(C\text{-}N) - D(C\text{-}C)$

$= 293\text{ kJ} - 348\text{ kJ}$

$\Delta H = -55\text{ kJ}$; acetonitrile has the lower enthalpy

8.83 (a)

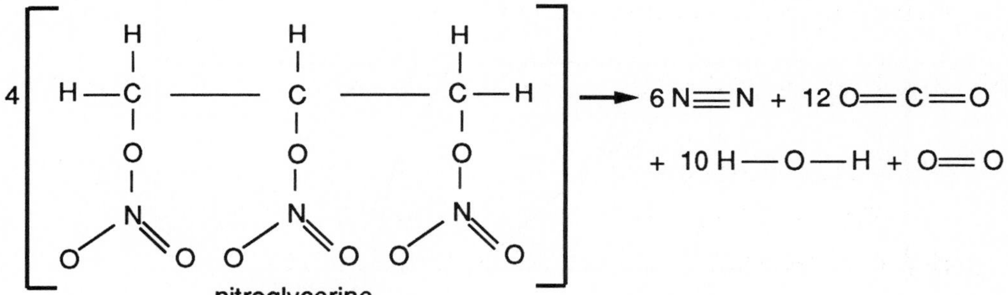

nitroglycerine

$\Delta H = 20D(C\text{-}H) + 8D(C\text{-}C) + 12D(C\text{-}O) + 24D(O\text{-}N) + 12D(N{=}O)$

$- [6D(N{\equiv}N) + 24D(C = O) + 20\,D(H\text{-}O) + D(O = O)]$

$= 20(413) + 8(348) + 12(358) + 24(201) + 12(607) - [6(941) + 24(799) + 20(463) + 495]$

$= -7129\text{ kJ}$

$$1.00\text{ g } C_3H_5N_3O_9 \times \frac{1\text{ mol } C_3H_5N_3O_9}{227.1\text{ g } C_3H_5N_3O_9} \times \frac{-7129\text{ kJ}}{4\text{ mol } C_3H_5N_3O_9} = 7.85\text{ kJ/g } C_3H_5N_3O_9$$

(b) $4C_7H_5N_3O_6(s) \rightarrow 6N_2(g) + 7CO_2(g) + 10H_2O(g) + 21C(s)$

8.84 $\underline{N}_2H_4$: -2 (b) $Na_2\underline{S}_2O_3$: +2 (c) $(\underline{N}H_4)_2SO_4$: -3 (d) $\underline{Cl}F_3$: +3 (e) $\underline{As}_2S_3$: +3

8.85 SeO_2 and N_2O_3 can be ruled out because both of them are low melting covalent compounds. SrO is not likely to react with NaOH, so it is most likely that the compound in question is GeO_2. Recall that oxides of metals in higher oxidation states are likely to be somewhat acidic in character. Thus, it is reasonable that GeO_2 should dissolve slightly in basic solution. Information that may be of interest: SrO, m.p. 2665°C, GeO_2, m.p. 1115°C.

8.86

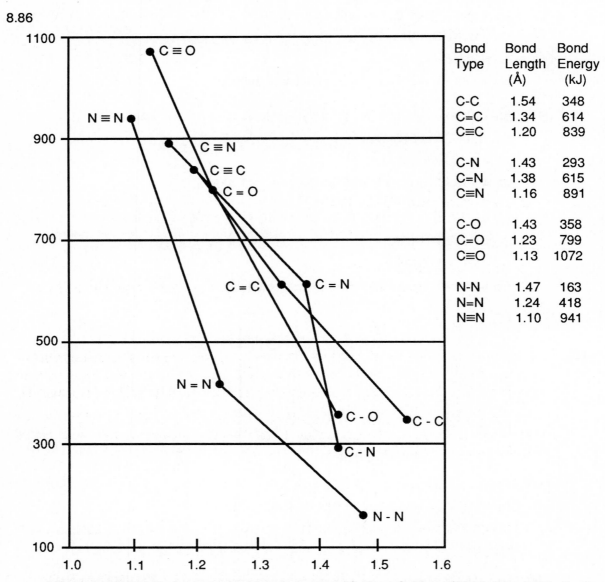

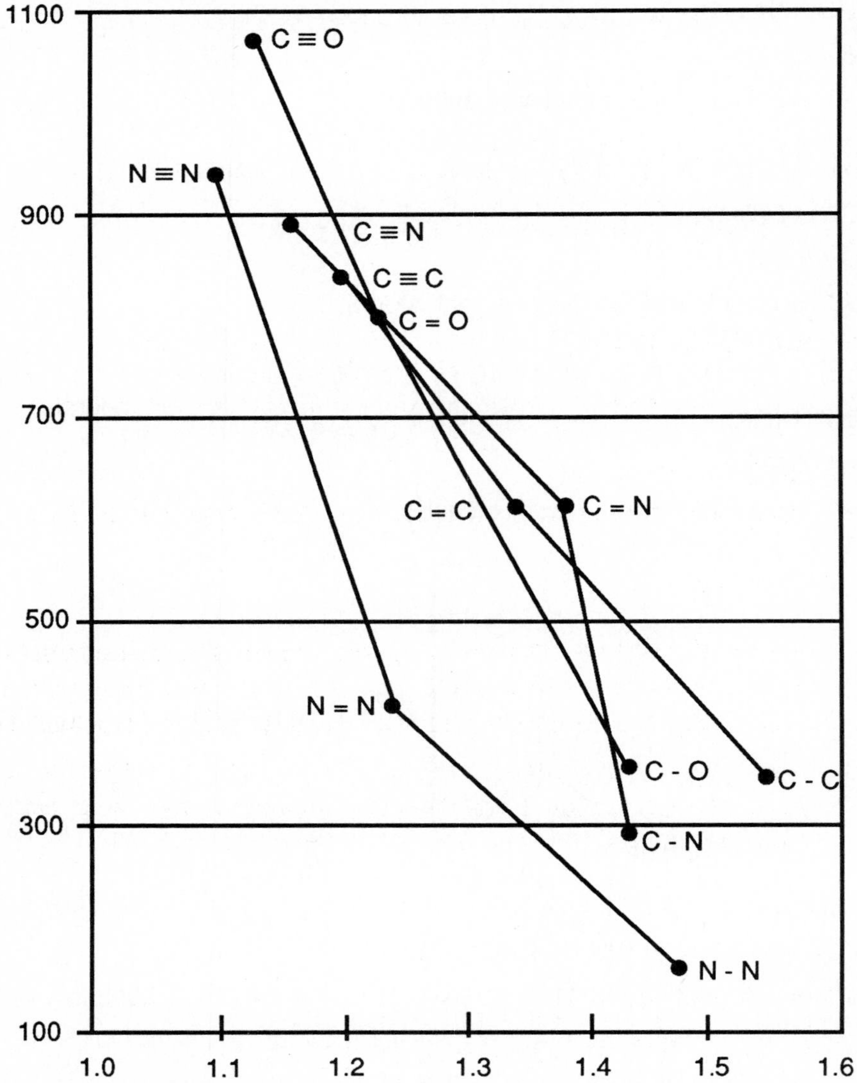

Bond Type	Bond Length (Å)	Bond Energy (kJ)
C-C	1.54	348
C=C	1.34	614
C≡C	1.20	839
C-N	1.43	293
C=N	1.38	615
C≡N	1.16	891
C-O	1.43	358
C=O	1.23	799
C≡O	1.13	1072
N-N	1.47	163
N=N	1.24	418
N≡N	1.10	941

When comparing the same pair of bonded atoms (C-N vs. C=N vs. C≡N), the shorter the bond the greater the bond energy, but the two quantities are not necessarily directly proportional. The plot clearly shows that there are no simple length/strength correlations for single bonds alone, double bonds alone, triple bonds alone, or among different pairs of bonded atoms (all C-C bonds vs. all C-N bonds, etc.).

CHAPTER 9

Molecular Geometry and Bonding Theories

Molecular Geometry; the VSEPR Model

9.1 (a) trigonal planar (b) tetrahedral (c) trigonal bipyramidal (d) octahedral

9.2 (a) 3 (or 5 if 90° angles are also present)
(b) 2 (5 if 120° angles are also present, 6 if more than one 180° angle is present)
(c) 4
(d) 6 (or 5 if 120° angles are also present)

9.3 The electron pair geometry indicated by VSEPR takes into account the total number of bonding (double and triple bonds count as one effective pair) and nonbonding pairs of electrons. The molecular geometry describes just the atomic positions. H_2O has the Lewis structure H—Ö—H (with two lone pairs on O); there are four pairs of electrons around oxygen so the electron pair geometry is tetrahedral, but the molecular geometry of the three atoms present is nonlinear, or bent.

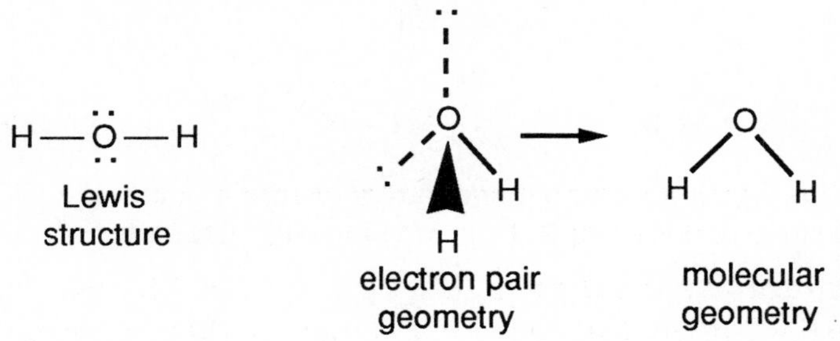

9.4

Lewis structure	electron pair geometry	molecular geometry
(a)	trigonal planar	**bent, 120° bond angle**
(b)	tetrahedral	**trigonal pyramid**
(c)	tetrahedral	**bent, 109° bond angle**
(d)	trigonal pyramid	**T-shaped**

9.5	Molecule or ion	Valence electrons	Lewis strucure		e⁻ pair geometry	Molecular geometry
(a)	Cl_2O	20	:Cl̈—Ö—Cl̈:		td	b
(b)	N_2O	16	:N≡N—Ö:	N—N—O	l	l

See Exercise 8.75 for other resonance forms and formal charge arguments. All resonance forms have the same e⁻ pair geometry and molecular geometry.

| (c) | SO_3 | 24 | :Ö=S—Ö: :Ö: (3 res forms) | | tr | tr |

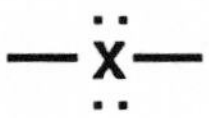

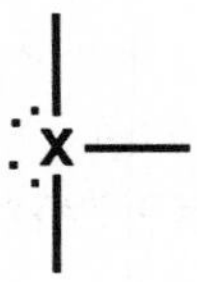

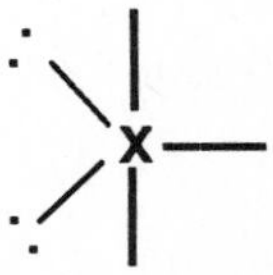

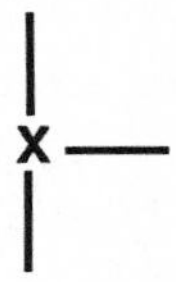

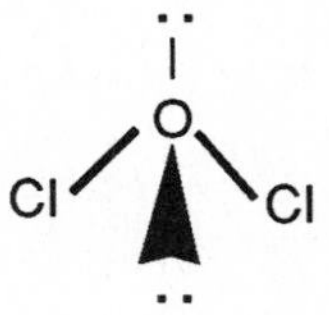

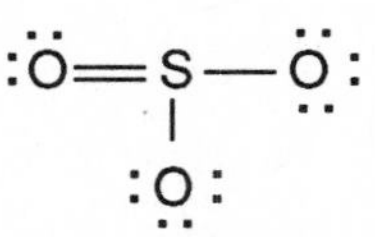

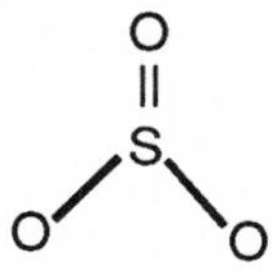

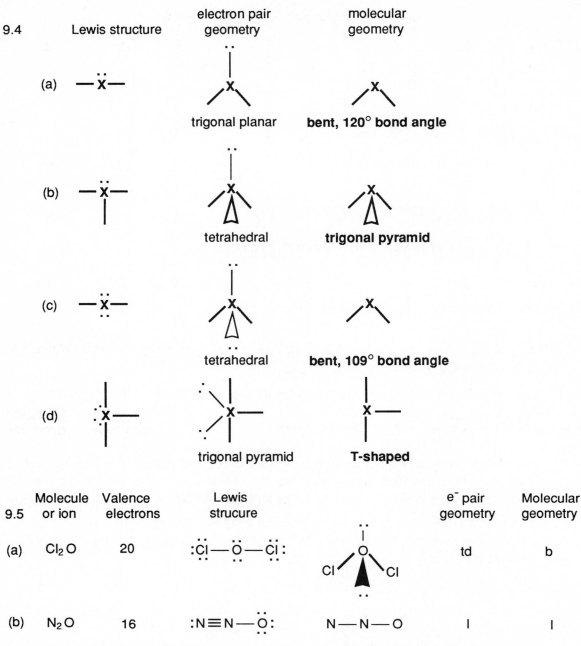

(d)	SO_3^{2-}	26	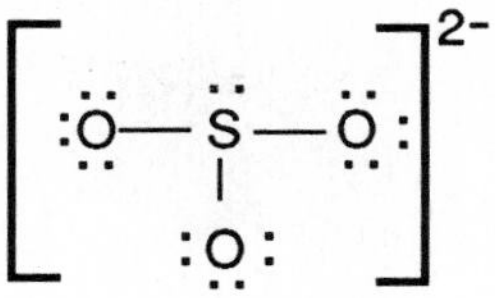	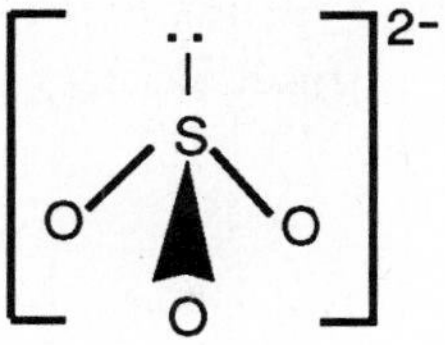	td	tp
(e)	PF_4^+	32	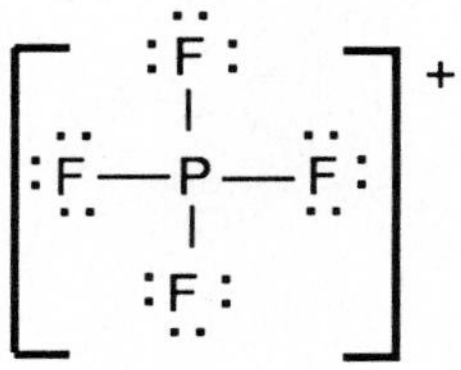	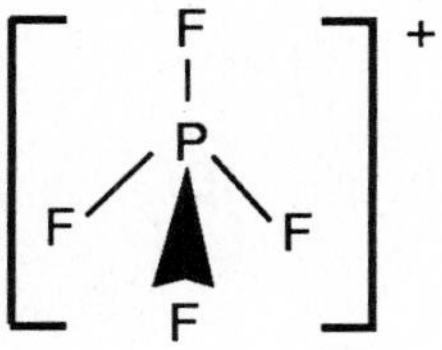	td	td
(f)	PF_4^-	34	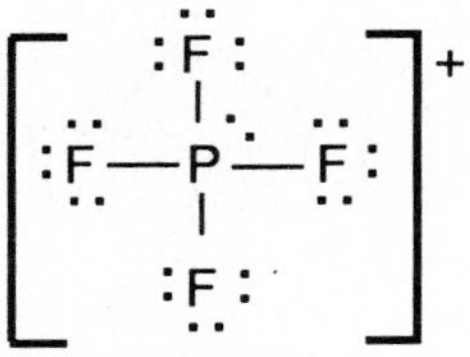	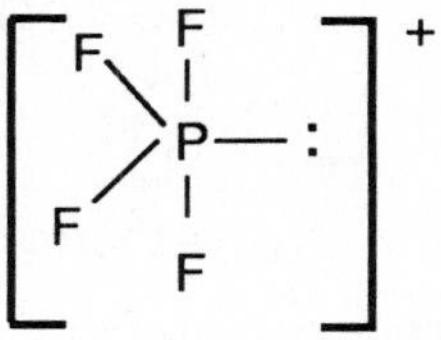	tbp	ss

bent (b), linear (l), seesaw (ss), tetrahedral (td), trigonal bipyramidal (tbp), trigonal planar (tr) trigonal pyramidal (tp), T-shaped (T)

9.6	Molecule or ion	Valence electrons	Lewis strucure		e^- pair geometry	Molecular geometry
(a)	SCl_2	20	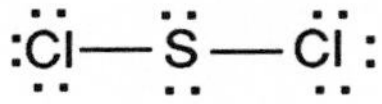	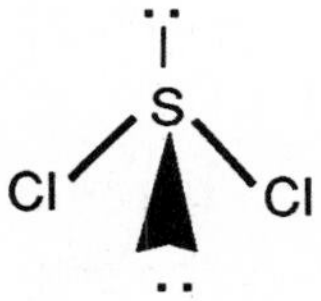	td	b
(b)	Cl_2SO	26	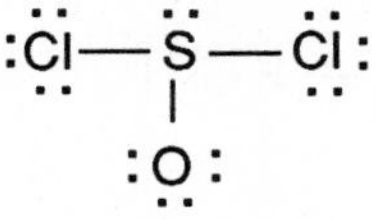	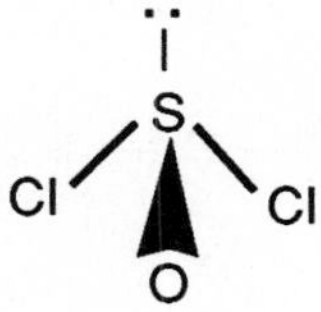	td	tp
(c)	ICl_2^-	22	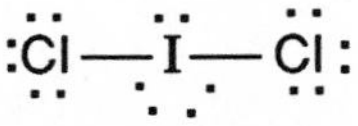	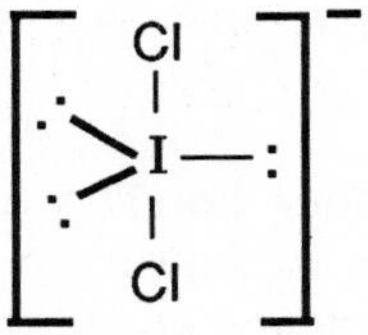	tbp	l

(d)	PCl_3	26	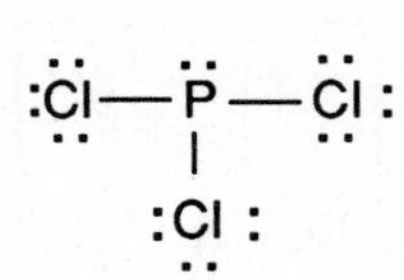	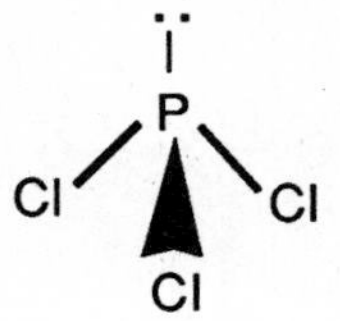	td	tp
(e)	ICl_3	28	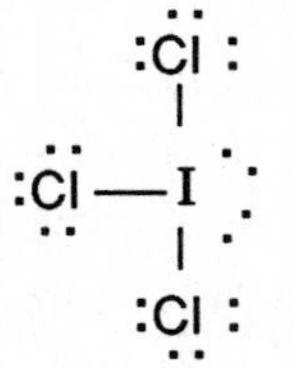	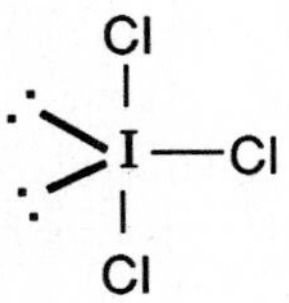	tbp	T
(f)	CO_3^{2-}	24	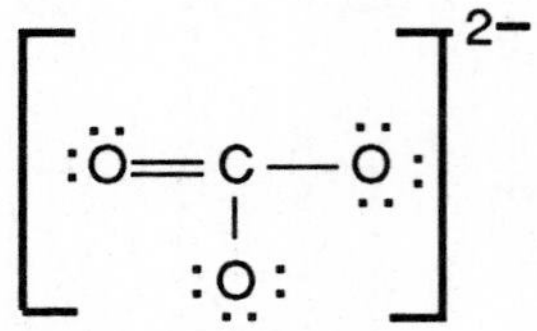 (3 res forms)	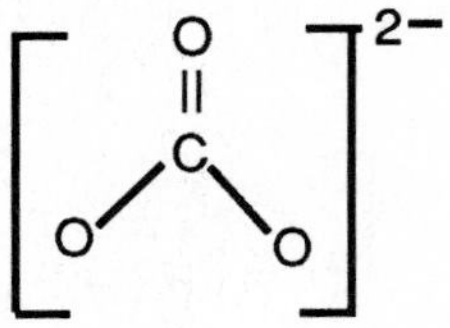	tr	tr

bent (b), linear (l), seesaw (ss), tetrahedral (td), trigonal bipyramidal (tbp), trigonal planar (tr) trigonal pyramidal (tp), T-shaped (T)

9.7 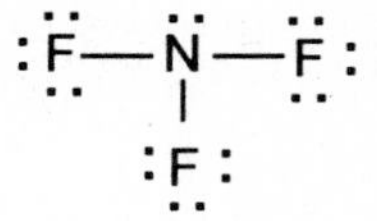 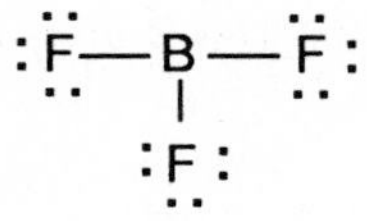 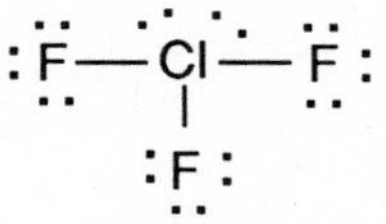

e^- pair geometry	td	tr	tpb
molecular geometry	tp	tr	T

The molecular geometries or shapes differ because there are a different number of nonbonding electron pairs around the central atom in each case.

9.8 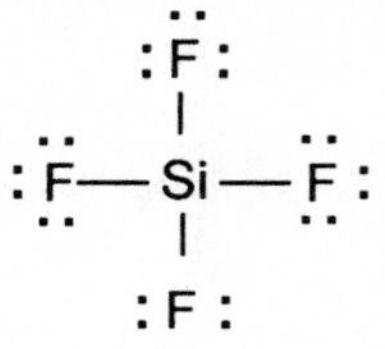 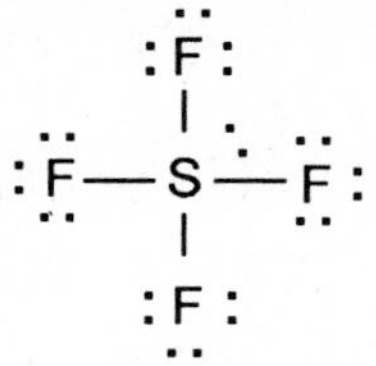 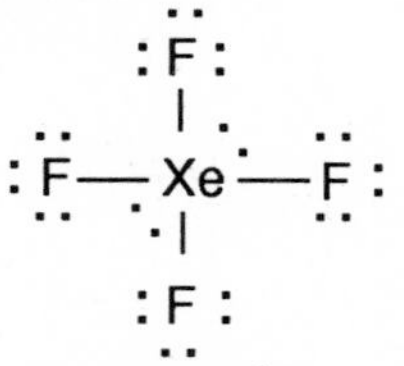

e^- pair geometry	td	tbp	octahedral (o)
molecular geometry	td	seesaw (ss)	square planar (s)

Although there are four bonding pairs of electrons in each molecule, the number of nonbonding electrons is different in each case and the molecular shape is influenced by the total number of valence electron pairs.

9.9 Write the Lewis structures:

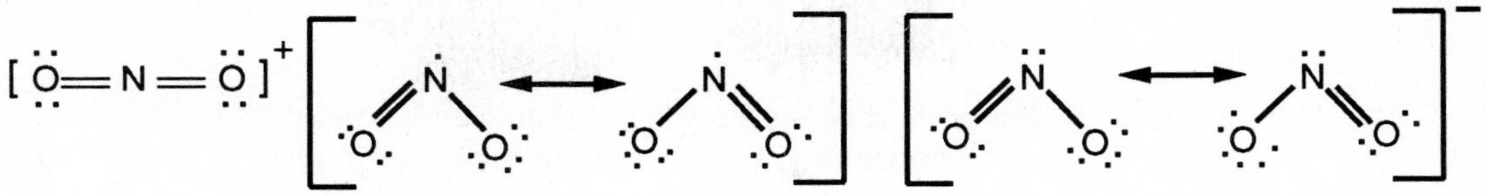

Note that in NO_2^+ (isoelectronic with CO_2) there are two multiple bonds, no unshared pairs, so we predict 180°. In NO_2 there are two bonds and an orbital with one electron. The repulsive effect of this single electron is less than that of an electron pair, so the O-N-O angle is larger than the 120° we would predict for three equivalent electron pairs. In NO_2^-, the repulsion from the unshared pair is greater than that from the shared pairs, so the O-N-O angle is less than the idealized 120°.

9.10 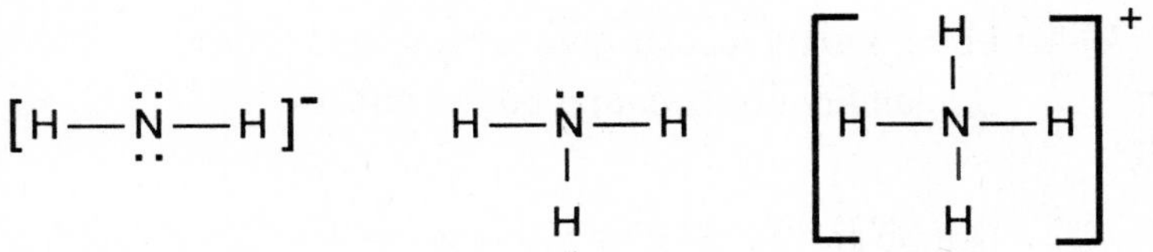

Each molecule has 4 pairs of electrons around the N atom, but the number of nonbonded pairs decreases from 2 to 0 going from NH_2^- to NH_4^+. Since lone pairs occupy more space than bonded pairs, the bond angles expand as the number of lone pairs decreases.

9.11 (a) 1 - 109°, 2 - 120° (b) 3 - 109°, 4 - 120°
(c) 5 - 109°, 6 - 109° (d) 7 - 180°, 8 - 109°

Molecular Polarity

9.12 (a) 1 - 109°, 2 - 109° (b) 3 - 109°, 4 - 109°
(c) 5 - 180° (d) 6 - 120°, 7 - 109°, 8 - 109°

9.13 (a) Br—Cl **Polar**, there is a small electronegativity difference.

(b) 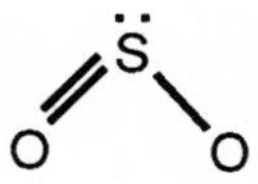 **Polar**, the molecule is bent so the bond dipoles do not cancel.

(c) SO_3 **Nonpolar**, in a symmetrical trigonal planar structure (See Exercise 9.5(c)), the bond dipoles cancel.

(d) 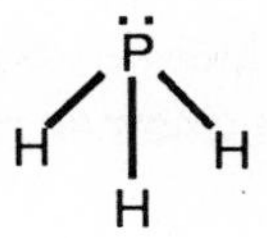 **Polar**, although the bond dipoles are essentially zero, there is an unequal charge distribution due to the nonbonded electron pair on P.

(e) $SiCl_4$ **Nonpolar**, in a symmetrical tetrahedron, the four Si-Cl bond dipoles cancel.

(f) SF_4 **Polar**, in a seesaw structure (see Exercise 9.8), the bond dipoles don't cancel and there is an unequal charge distribution due to the nonbonded electron pair on S.

9.14 (a) H—C≡N Polar, the molecule is linear, but the bond dipoles don't cancel.

(b) S = C = S Nonpolar, the molecule is linear and the bond dipoles cancel.

(c) BF_3 Nonpolar, in a symmetrical trigonal planar structure, (see Figure 9.11) the bond dipoles cancel.

(d) HCF_3 Polar, the molecular geometry is tetrahedral, but the bond dipoles are unequal and do not cancel.

(e)

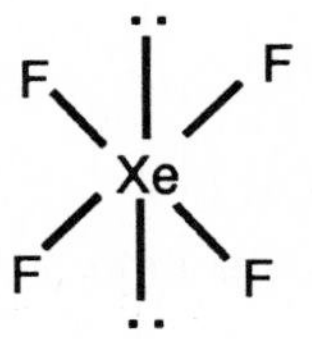

Nonpolar, the electron pair geometry is octahedral and the molecular geometry is square planar. The two nonbonded electron pairs are opposite each other and the four Xe-F bonds in the square plane also cancel.

(f)

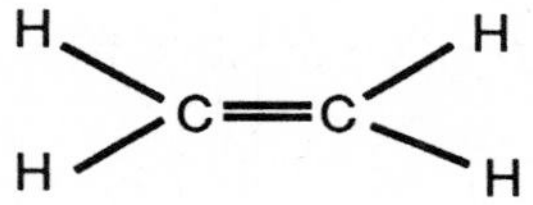

Nonpolar, the molecule is planar and the C-H bond dipoles cancel.

9.15 In $BeCl_2$ there are just two electron pairs about Be. The structure is linear, so the individual bond dipoles cancel. In SCl_2, there are four electron pairs about S. The two unshared pairs don't cancel with the two S-Cl bonding pairs.

9.16

μ = 1.03 D

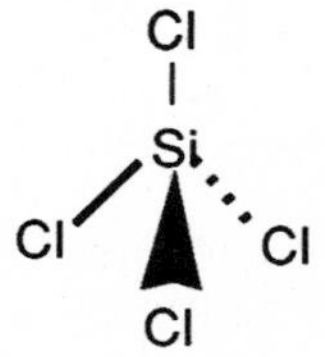

μ = O

$SiCl_4$ is a tetrahedral molecule with the central Si atom symmetrically surrounded by the four Cl atoms. Like CCl_4 (Figure 9.11) the individual Si-Cl bond dipoles cancel, and the molecule has a net dipole moment of zero. In PF_3, one of the positions in the tetrahedron is occupied by a nonbonding electron pair. The individual P-F bond dipoles do not cancel (nothing cancels a nonbonding electron pair except another nonbonding pair), and the molecule is polar.

9.17

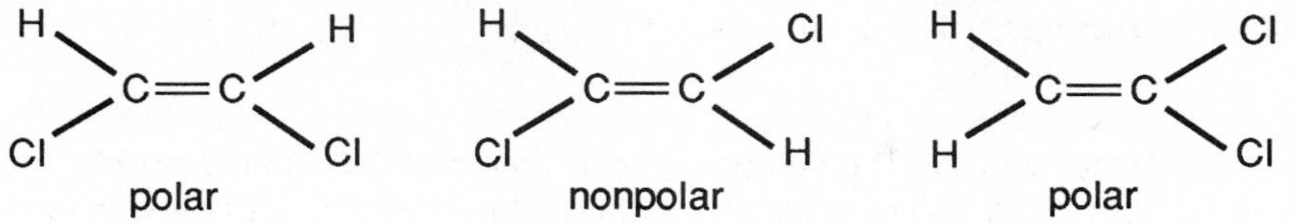

All three isomers are planar. The molecules on the left and right are polar because the C-Cl bond dipoles do not point in opposite directions. In the middle isomer, the C-Cl bonds and dipoles are pointing in opposite directions (as are the C-H bonds), the molecule is nonpolar and has a measured dipole moment of zero.

9.18 Each C-Cl bond is polar. The question is whether the vector sum of the C-Cl bond dipoles in each molecule will be nonzero. In the *ortho* and *meta* isomers, the C-Cl vectors are at 60° and 120° angles, respectively, and their resultant dipole moments are nonzero. In the *para* isomer, the C-Cl vectors are opposite, at an angle of 180°, with a resultant dipole moment of zero. The *ortho* and *meta* isomers are polar, the *para* isomer is nonpolar.

9.19 (a) Q is the charge at either end of the dipole.

$$0.82\ \text{D} \times \frac{3.33 \times 10^{-30}\ \text{C-m}}{1\ \text{D}} \times \frac{1}{1.41\ \text{Å}} \times \frac{1 \times 10^{10}\ \text{Å}}{1\ \text{m}} = 1.9 \times 10^{-20}\ \text{C}$$

A dipole moment of 0.82 D indicates a charge separation of 1.9×10^{-20} C .

(b) The percent of a full electronic charge (inside back cover of text) is

$$\frac{1.9 \times 10^{-20}\ \text{C}}{1.6 \times 10^{-19}\ \text{C}} \times 100 = 12\%$$

9.20 (a) Q is the charge at either end of the dipole.

$$Q = \frac{\mu}{\text{v}} = \frac{1.82\ \text{D}}{91.7\ \text{pm}} \times \frac{1\ \text{pm}}{1 \times 10^{-12}\ \text{m}} \times \frac{3.33 \times 10^{-30}\ \text{C-m}}{1\ \text{D}} = 6.61 \times 10^{-20}\ \text{C}$$

The calculated charge of H and F is 6.61×10^{-20} C . This can be thought of as the amount of charge "transferred" from H to F.

(b) The percent of a full electronic charge is

$$\frac{6.61 \times 10^{-20}\ \text{C}}{1.60 \times 10^{-19}\ \text{C}} \times 100 = 41.3\%$$

Comparing this with the result from Exercise 9.19, HF is a much more polar molecule than HBr.

Orbital Overlap; Hybrid Orbitals

9.21 (a) "Orbital overlap" occurs when a valence atomic orbital on one atom shares the same region of space with a valence atomic orbital on an adjacent atom.
(b) In valence bond theory, overlap of orbitals allows the two electrons in a chemical bond to mutually occupy the space between the bonded nuclei.
(c) Valence bond theory is a combination of the atomic orbital concept with the Lewis model of electron pair bonding.

9.22 (a)

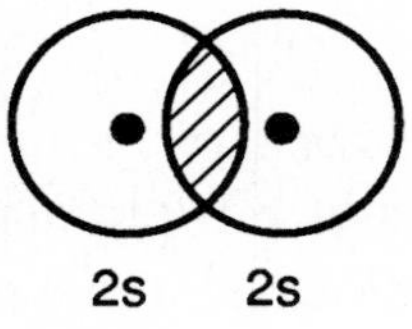

(b)

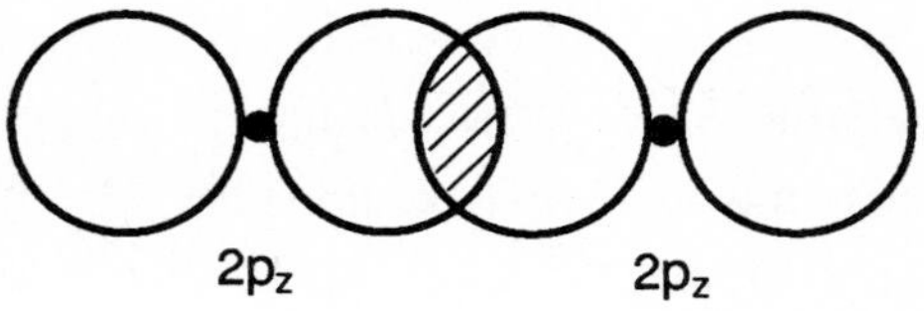

(c)

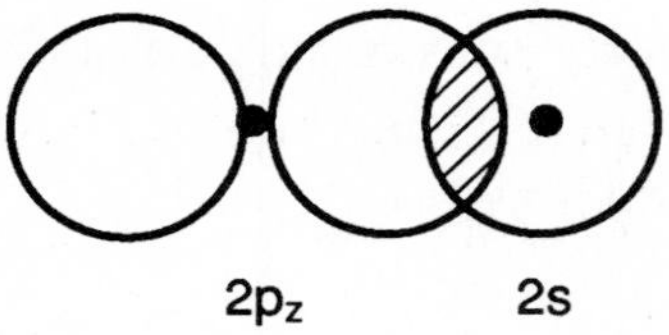

9.23 (a) sp^2 -- 120° angles in a plane
(b) sp^3d -- 90°, 120° and 180° bond angles (trigonal bipyramid)
(c) sp^3d^2 -- 90° and 180° bond angles (octahedron)

9.24 (a) sp -- 180° (b) sp^3 -- 109° (c) sp^2 -- 120°
(d) sp^3d^2 -- 90° and 180° (e) sp^3d -- 90°, 120° and 180°

9.25 26 e^-, 13 e^- pairs

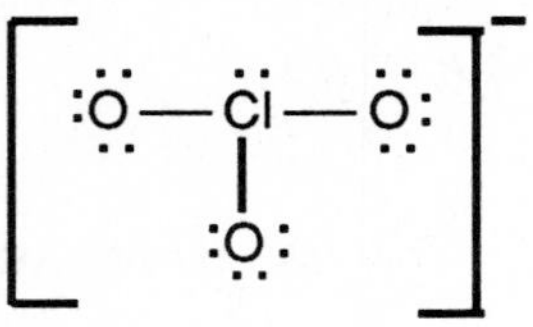

The electron pair geometry (four total pairs around Cl) is tetrahedral and molecular shape is trigonal pyramidal. Because there are four valence electron pairs on Cl, sp^3 hybrid orbitals are used for bonding. Each Cl-O σ bond is formed by the overlap of an sp^3 hybrid orbital on Cl with an atomic or hybrid orbital on O. (Because O is terminal, there is no angular data to indicate the type or orbital used in bonding.) The O-Cl-O bond angles should be approximately 109°.

9.26 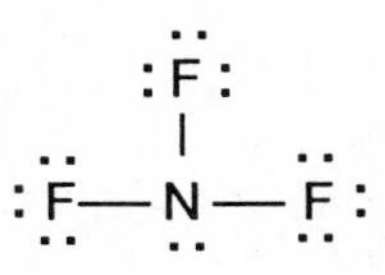

Electron pair geometry, tetrahedral
Molecular shape, trigonal pyramidal
Hybrid orbitals, sp^3
Bond angle, ~109°.

Each N-F σ bond is formed by the overlap of an sp^3 hybrid orbital on N with an orbital on F. (Because F is terminal, there is no angular data to indicate the type of orbital used in bonding.)

9.27 Draw the correct Lewis structure, determine the electron pair geometry and the appropriate set of hybrid orbitals.

(a) 24 e^-, 12 e^- pairs

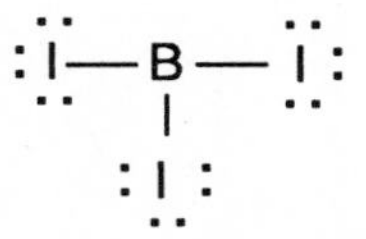

Note the B-I bond is probably too log for π overlap. However, resonance structures such as

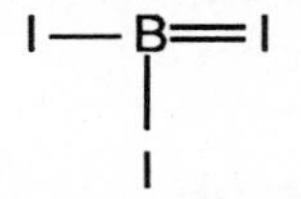

would use the same hybrid orbitals as the one given on the left.

3 e^- pairs around B
trigonal planar e^- pair geometry
sp^2 hybrid orbitals

(b) 26 e^-, 13 e^- pairs

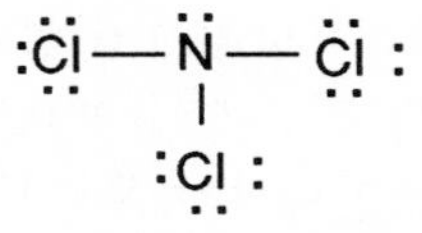

4 e^- pairs around N
tetrahedral e^- pair geometry
sp^3 hybrid orbitals

(c) 32 e^-, 16 e^- pairs

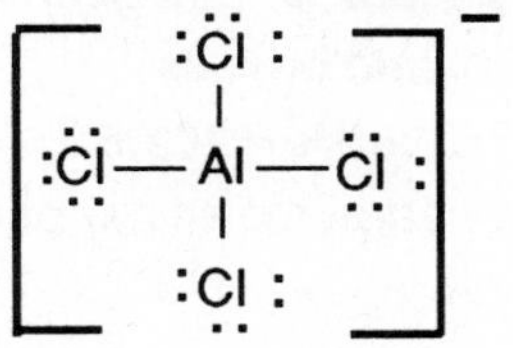

4 e^- pairs around Al
tetrahedral e^- pair geometry
sp^3 hybrid orbitals

(d) 40 e^-, 20 e^- pairs

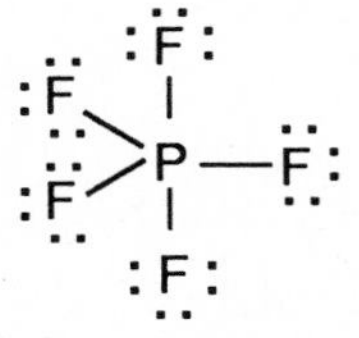

5 e^- pairs around P
trigonal bipyramidal e^- pair geometry
sp^3d hybrid orbitals

(e) 48 e^-, 24 e^- pairs

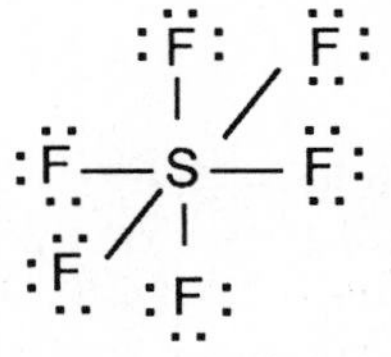

6 e^- pairs around S
octahedral e^- pair geometry
sp^3d^2 hybrid orbitals

9.28 (a) 20 e^- , 10 e^- pairs

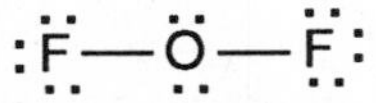

4 e^- pairs around O
tetrahedral e^- pair geometry
sp^3 hybrid orbitals
F-O-F angle < 109.5°
bent molecular geometry

(b) 26 e^- , 13 e^- pairs

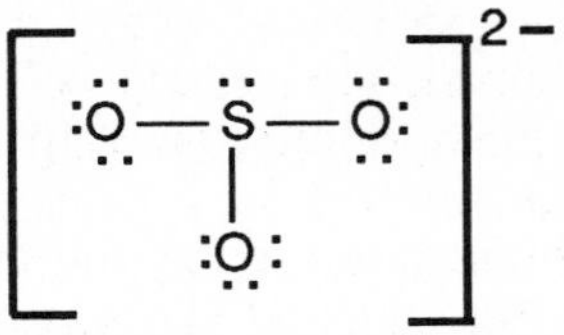

4 VSEPR pairs around S
tetrahedral e^- pair geometry
sp^3 hybrid orbitals
O-S-O angle ≈109.5°
trigonal pyramidal molecular geometry

(c) 30 e^- , 15 e^- pairs

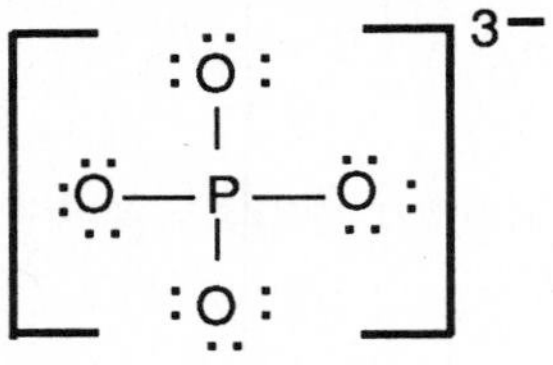

(resonance structures with double bonds can be drawn)
4 e^- pairs around P
tetrahedral e^- pair geometry
sp^3 hybrid orbitals
O-P-O angle ≈109.5°
tetrahedral molecular geometry

(d) 22 e^- , 11 e^- pairs

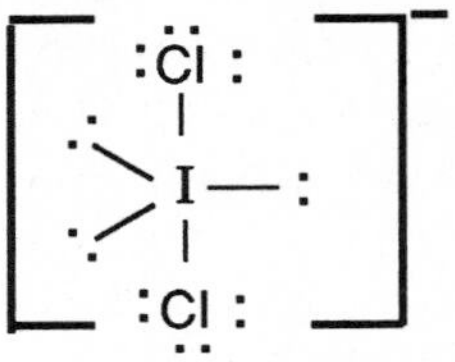

5 e^- pairs around Xe
trigonal bipyramid e^- pair geometry
sp^3d hybrid orbitals
Cl -I-Cl angle = 180°
linear molecular geometry
(In a trigonal bipyramid, placing nonbonding e^- pairs in the equatorial position minimizes e^- pair repulsion.)

(e) 36 e^- , 18 e^- pairs
(Xe has 8 valence e^-)

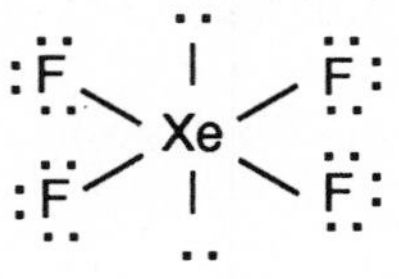

6 e^- pairs around Xe
octahedral e^- pair geometry
sp^3d^2 hybrid orbitals
F-Xe-F angles ≈ 90°, 180°
square planar molecular geometry
(In an octahedron, placing 2 lone pairs opposite each other minimizes repulsions.)

(f) 48 e^- , 24 e^- pairs

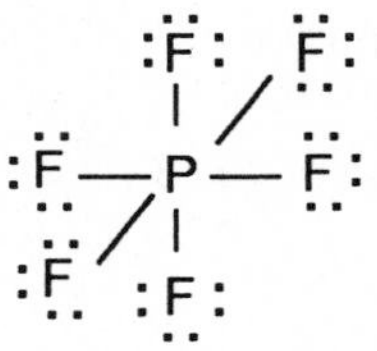

6 e^- pairs around P
octahedral e^- pair geometry
sp^3d^2 hybrid orbitals
F-P-F angles ≈ 90°, 180°
octahedral molecular geometry

Multiple Bonds

9.29 A sigma (σ) bond is formed by the end to end overlap of two orbitals, e.g., s + s, s + p, p + p, sp hydrid + p, etc., pointing along the internuclear axis. A σ bond has electron density symmetrically distributed along the internuclear axis of the two atoms forming the bond. A π bond is formed by the side to side overlap of two p orbitals oriented perpendicular to the internuclear axis. There are two regions of overlap, but no electron density along the internuclear axis. (Note that, due to the spherical nature of their electron distributions, s orbitals and hybrid orbitals can only form σ bonds; p orbitals can form σ or π bonds, depending on orientation.) Sigma (σ) bonds are generally stronger than π bonds, because the extent of orbital overlap is greater in a σ bond.

9.30 There can be at most one σ bond between any two atoms. Therefore:
a) a double bond contains one σ and one π bond;
b) a triple bond contains one σ and two π bonds.

9.31 The valence shell of an atom contains a single s and three p orbitals. If the s and two of the p orbitals form the hybrid set, a single valence p orbital remains on the atom. The atom can form one π bond.

9.32 (a) 2 unhybridized p orbitals (b) 0 unhybridized p orbitals
(c) 0 unhybridized p orbitals

9.33 (a) ~109° about the left most C, sp^3; ~120° about the right-hand C, sp^2
(b) the doubly bonded O can be viewed as sp^2, the other as sp^3; the nitrogen is sp^3 with approximately 109° bond angles
(c) nine σ bonds, one π bond

9.34 (a) 1 - 120°; 2 - 120°; 3- 109° (b) 1 - sp^2; 2- sp^2; 3 - sp^3
(c) 21 σ bonds

9.35 In a localized π bond, the electron density is concentrated strictly between the two atoms forming the bond. In a delocalized π bond, parallel p orbitals on more than two adjacent atoms overlap and the electron density is spread over all the atoms that contribute p orbitals to the network. There are still two regions of overlap, above and below the σ framework of the molecule.

9.36 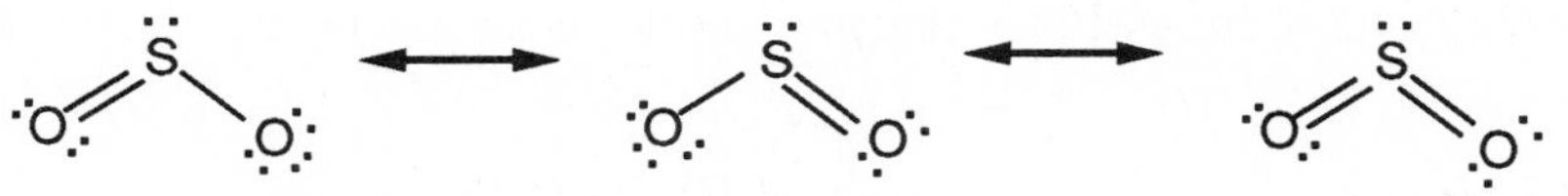

The existence of several resonance forms for SO_2 is a strong indication that the π bond is delocalized. From an orbital perspective, the electron pair geometry around S is trigonal planar, so the hybridization at S is sp^2. This leaves a p orbital on S and one on each O atom perpendicular to the trigonal plane of the molecule, in the correct orientation for delocalized π overlap. Physically, the two S-O bond lengths are equal, indicating that the two S-O bonds are equivalent, rather than one longer single bond and one shorter double bond.

9.37 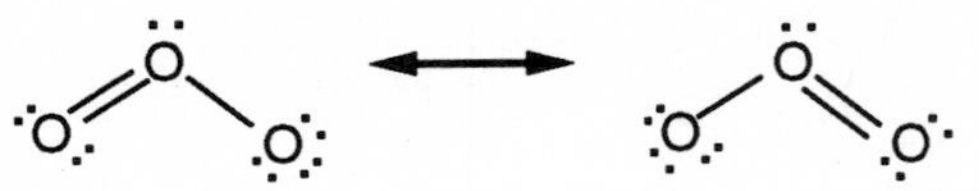

(a) There are two resonance forms for O_3 (a structure with two double bonds is not possible, since O cannot have an expanded octet). This indicates that the O-O bonds are neither single or double, but some intermediate between the two. This is consistent with equal O-O bond lengths that are longer than a double bond but shorter than a single bond.

(b) In both resonance structures, the central O atom in O_3 has three VSEPR pairs of electrons, indicating that sp^2 hybrid orbitals are used for sigma (σ) bonding. This leaves a single p atomic orbital on the central atom that can be aligned parallel to p atomic orbitals on the terminal O atoms (regardless of hybridization or absence of it on these atoms) to form a delocalized π bonding network. This delocalized π network is bent (the shape of the molecule predicted by VSEPR) and has electron density above and below the plane of the molecule. The electrons in the π network are distributed equally over the molecule. The O-O bond distance must be shorter than an O-O single bond to accommodate the geometric requirement for delocalized π overlap, and longer than a full O-O double bond because the π electron density is delocalized.

9.38 (a)

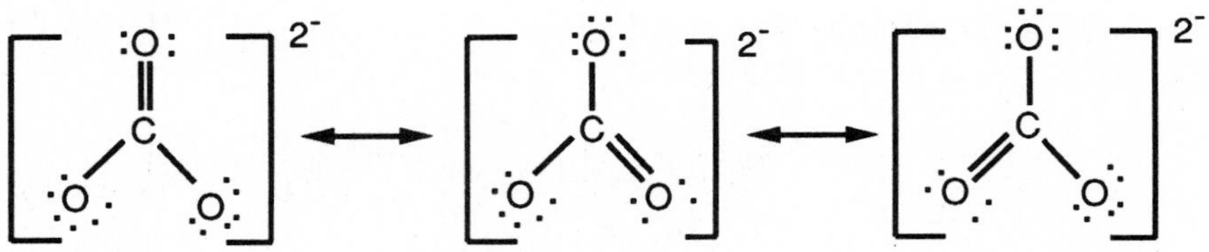

(b) 3 VSEPR e^- pairs around C in each resonance structure, therefore **trigonal planar** electron pair and molecular geometry.

(c)) 3 VSEPR e^- pairs, **sp^2 hybridization**

(d) The delocalized π network in CO_3^{2-} is formed by parallel overlap of one 2 p atomic orbital on each atom. The resulting π electron cloud is Y-shaped (the shape of the molecule) and has electron density above and below the plane of the molecule.

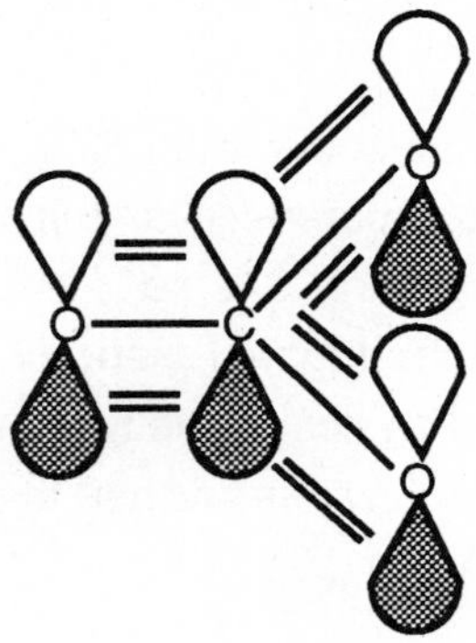

Molecular Orbitals

9.39 (a) Bonding MO's are lower in energy than the starting atomic orbitals, and antibonding MO's are higher.

(b) The electron density is concentrated between the nuclei in a bonding MO and away from the region between the nuclei in an antibonding MO.

9.40 In a σ molecular orbital, the electron density is spherically symmetric about the internuclear axis and is concentrated along this axis. In a π molecular orbital, the electron density is concentrated above and below the internuclear axis and zero along it.

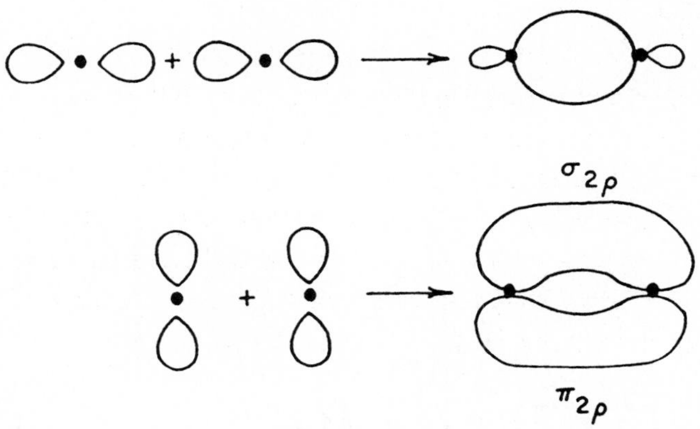

9.41 (a) Bond order is the net number of bonding electron pairs in a molecule [1/2 (bonding e^- - antibonding e^-)].

(b) Paramagnetism is the attraction of a substance into a magnetic field due to the presence of unpaired electrons.

(c) An energy level diagram shows the interacting atomic orbitals in the left and right columns and the resulting molecular orbitals and their relative energies in the center column.

9.42 As bond order increases, the bond length decreases, and the bond dissociation energy increases.

9.43 (a) CO

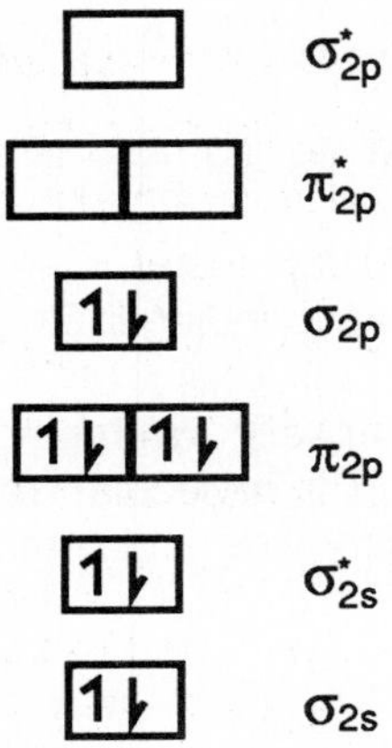

(b) NO^-; B.O. = (8-4) / 2 = 2, paramagnetic

(c) CN^-; B.O. = (8-2) / 2 = 3, diamagnetic

(d) OF ; B.O. = (8-5) / 2 = 1.5 paramagnetic

B.O. = (8-2) / 2 = 3.0, diamagnetic

9.44

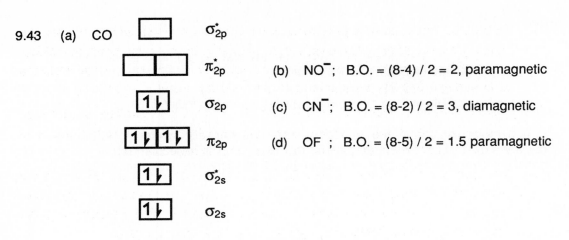

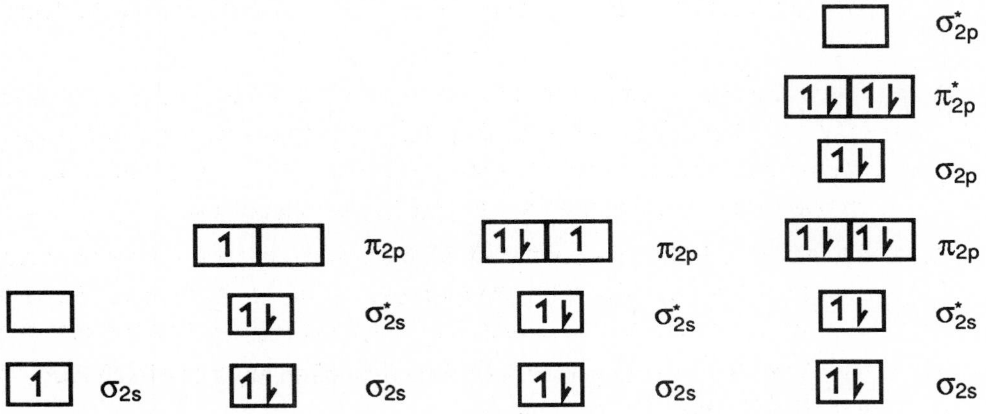

(a) Li_2^+ increases

(b) B_2^+ increases

(c) C_2^+ increases

(d) Ne_2^{2+} decreases

Addition of an electron increases stability if it occupies a bonding orbital and decreases stability if it occupies an antibonding orbital.

9.45 When an electron is added to O_2 it enters one of the π^*_{2p} orbitals. A second added electron enters the other π^*_{2p} orbital. Thus, each added electron reduces the O-O bond order. On the other hand, an electron removed from O_2 is taken from one of the π^*_{2p} orbitals. This has the effect of **increasing** the O-O bond order. Thus, in order of increasing bond length: O_2^+ (B.O = 2.5) < O_2 (B.O. = 2.0) < O_2^- (B.O. =1.5) < O_2^{2-} (B.O. = 1.0). Note that the higher the bond order, the shorter the bond length.

9.46 Bond order: $N_2 > N_2^+ \approx N_2^-$. Addition of an electron to an antibonding π orbtial **or** removal of an electron from a bonding π orbital will reduce the bond order of N_2 from 3 to 2.5. The relative lengths should be the reverse of bond order: $N_2 < N_2^+ \approx N_2^-$.

9.47 (a) Of the two molecular orbitals formed by the interaction of two atomic orbitals, the bonding molecular orbital is always lower in energy because the bonding molecular orbital concentrates electron density between the two bonding nuclei while the antibonding orbital channels electrons away from the nuclei.

(b) The σ^*_{1s} orbital is higher in energy than the H 1s atomic orbitals because an electron in this orbital is repelled from the area between the H nuclei and is actually farther from a single H nucleus than in the isolated H atom.

(c) The separation between related bonding and antibonding orbitals is directly related to the extent of overlap of the atomic orbitals. The Li 2s orbitals are larger than the 1s, the area of overlap is greater and the energy separation of the resulting σ_{2s} and σ^*_{2s} orbitals is larger than that between σ_{1s} and σ^*_{1s}.

9.48 (a) An electron in a σ_{2s} bonding molecular orbital is farther from the bonding nuclei than an electron in a σ^*_{1s} orbital. Thus, the energy of the σ_{2s} bonding orbital is higher.

(b) There are two 2p atomic orbitals on each B atom, $2p_x$ and $2p_y$, oriented correctly for π overlap. The two resulting π_{2p} bonding molecular orbitals are equal in energy but have different spatial orientations 90° apart, similar to the relative orientations of the degenerate $2p_x$ and $2p_y$ atomic orbitals.

(c) According to Rule 3 (Section 9.7), as the overlap of atomic orbitals increases, the energy of the resulting antibonding molecular orbital increases. Since the extent of head-to-head overlap between $2p_z$ orbitals is greater than the extent of side-to-side overlap of $2p_x$ or $2p_y$ orbitals, the σ^*_{2p} antibonding orbital is higher in energy than the π^*_{2p} orbitals.

Additional Exercises

9.49 (a) AsF_3; 26 valence e^-

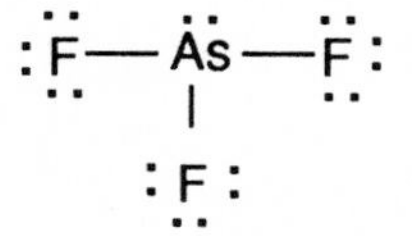

tetrahedral e^- pair geometry
trigonal pyramidal molecular geometry

(b) OCN^-; 16 valence e^-

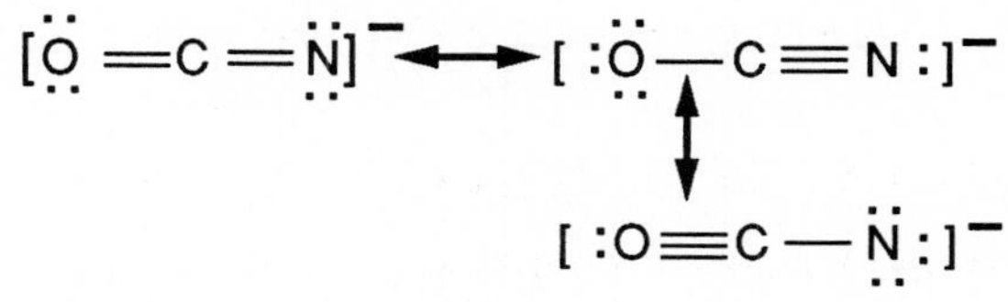

linear e^- pair geometry
linear "molecular" geometry

(c) H_2CO; 12 valence e^-

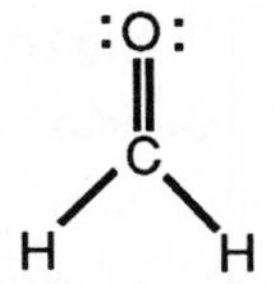

trigonal planar e^- pair geometry
trigonal planar molecular geometry

(d) I_3^-; 22 valence e^-

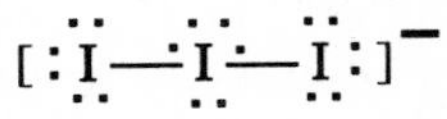

trigonal bipyramid e^- pair geometry
linear molecular geometry
(In a trigonal pyramid, nonbonded pairs lie in the trigonal plane.)

9.50 In PH_3 the unshared electron pair exerts a greater repulsive effect on the three bonding pairs, forcing them inward. In PH_4^+ all four electron pairs are equivalent, so the ion is strictly tetrahedral. PH_4^+ is formed by adding a proton, H^+, to the unshared electron pair of PH_3. This reduces the volume requirements of the unshared pair so the other three H-P-H angles can increase from 93° to 109.5°.

9.51 The σ bond electrons are localized in the region along the internuclear axes. The positions of the atoms and geometry of the molecule are thus closely tied to the locations of these electron pairs. Because the π bond electrons are distributed above and below the plane of the σ bond axis, these electron pairs do not, in effect, influence the geometry of the molecule.

9.52 (a) CO_2, 16 valence e^-

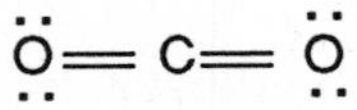

2 σ, 2 π

(b) NCS^-, 16 valence e^-

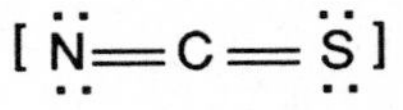

+ two other resonance structures

2 σ, 2 π (for any of the resonance structures)

(c) SO_4^{2-}, 32 valence e^-

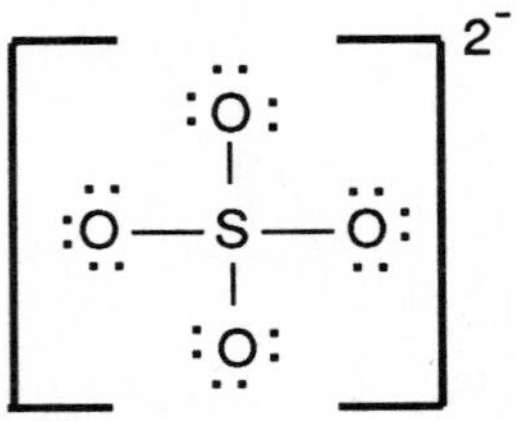

4 σ, 2 π

(d) HCO(OH), 18 valence e^-

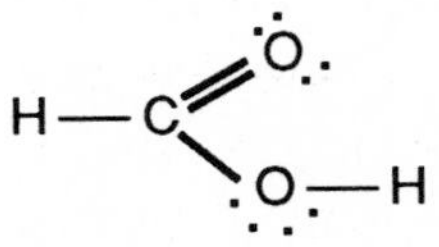

4 σ, 1 π

9.53 The compound on the right has a dipole moment. In the square planar *trans* structure on the left, all equivalent bond dipoles can be oriented opposite each other, for a net dipole moment of zero.

9.54

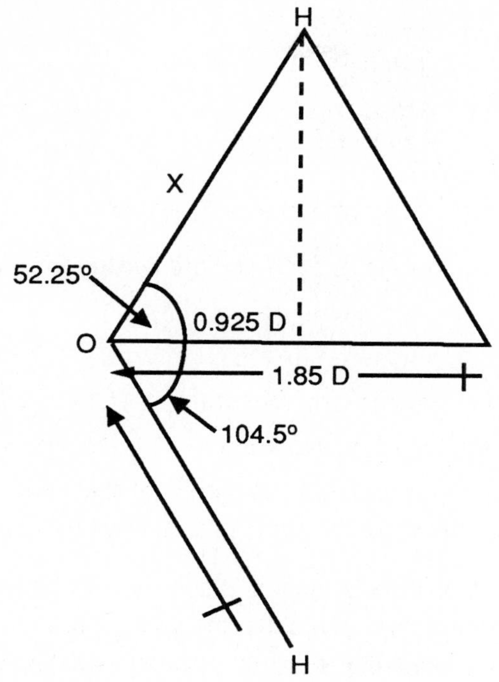

(a) The bond dipoles in H_2O lie along the O-H bonds with the positive end at H and the negative end at O. The dipole moment vector of the H_2O molecule is the resultant (vector sum) of the two bond dipoles. This vector bisects the H-O-H angle and has a magnitude of 1.85 D with the negative end pointing toward O.

(b) Since the dipole moment vector bisects the H-O-H bond angle, the angle between one H-O bond and the dipole moment vector is 1/2 the H-O-H bond angle, 52.25°. Dropping a perpendicular line from H to the dipole moment vector creates the right triangle pictured. If X = the magnitude of the O-H bond dipole, X cos (52.25) = 0.925 D. **X = 1.51 D.**

9.55 The BF_3 molecule has three B-F bonds of equal length arranged in a trigonal plane, 120° apart (see Exercise 9.26). These are labeled 1, 2 and 3 in the diagram. The resultant vector between 1 and 2, labelled 1 + 2, is equal and opposite to vector 3 and the net dipole moment is zero.

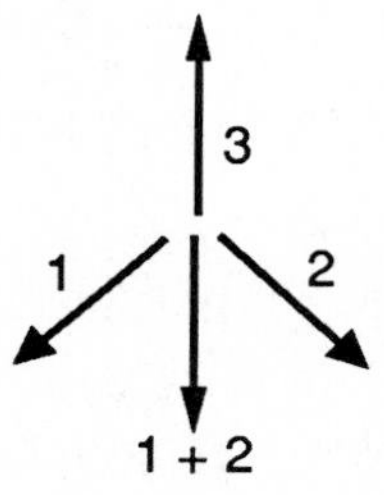

9.56

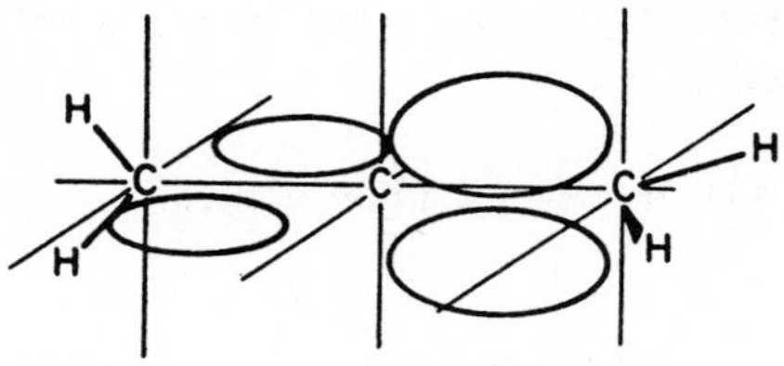

(a) The molecule is nonplanar. The CH_2 planes at each end are twisted 90° from one another.

(b) Allene has no dipole moment.

9.57 (a)

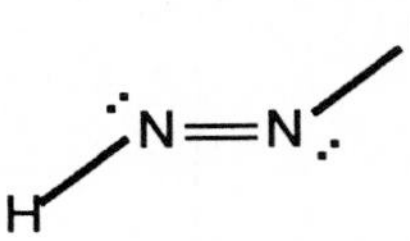

(b) The electron pair geometry around each N is trigonal planar with bond angles of ~120°. It is a geometric requirement of the π bond that the pure p orbitals on each N be parallel, and thus the trigonal planes of the sp^2 hybrids must also be aligned; the four atoms are then coplanar.

(c) Each N atom will use sp^2 hybrid orbitals for σ bonding and pure p orbitals for the π bond.

(d)

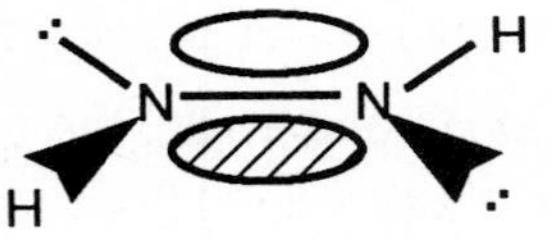

(e) The two N-H bonds must be on opposite sides of the molecule to produce a dipole moment of zero. If the two N-H bonds were on one side of the molecule and the two lone pairs on the other side, the molecule would have a net dipole moment.

9.58 (a) Assume 100 g of compound

$$2.1 \text{ g H} \times \frac{1 \text{ mol H}}{1.008 \text{ g H}} = 2.1 \text{ mol H};\ 2.1 \div 2.1 = 1$$

$$29.8 \text{ g N} \times \frac{1 \text{ mol N}}{14.01 \text{ g N}} = 2.13 \text{ mol N};\ 2.13 \div 2.1 \approx 1$$

$$68.1 \text{ g O} \times \frac{1 \text{ mol O}}{16.00 \text{ g O}} = 4.26 \text{ mol O};\ 4.26 \div 2.1 \approx 2$$

The empirical formula is HNO_2; formula weight = 47. Since the approximate molecular weight is 50, the **molecular formula is HNO_2.**

(b) Assume N is central, since it is unusual for O to be central, and part (d) indicates as much. HNO_2; 18 valence e$^-$

$$\ddot{O}=\ddot{N}-\ddot{O}-H \longleftrightarrow (\ :\ddot{\ddot{O}}-\ddot{N}=\ddot{O}-H\ )$$

$$-1 \qquad 0 \qquad +1$$

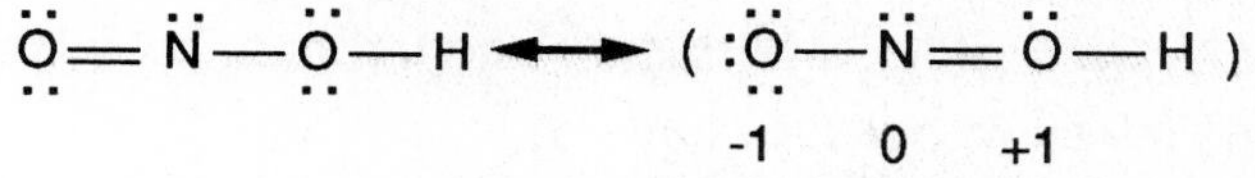

This resonance form is a minor
contributor due to unfavorable
formal charges.

(c) The electron pair geometry around N is trigonal planar; if the resonance structure on the right makes a significant contribution to the molecular structure, all 4 atoms would lie in a plane. If only the left structure contributes, the H could rotate in and out of the molecular plane. The contributions of the two resonance structures could be determined by measuring the O-N-O and N-O-H bond angles.

(d) 3 VSEPR e$^-$ pairs around N, sp^2 hybridization

(e) 3 σ, 1 π for both structures (or for H bound to N).

9.59 The carbons not in the ring all employ sp^3 hybrid orbitals. The ring carbons use an sp^2 hybrid set for the σ bond framework, with a p orbital involved in the delocalized π bonding in the ring. The two oxygen atoms of cumene hydroperoxide are approximately sp^3 hybridized (recall that there are two unshared pairs on each oxygen, not shown.)

9.60 The Lewis structures are as follows:

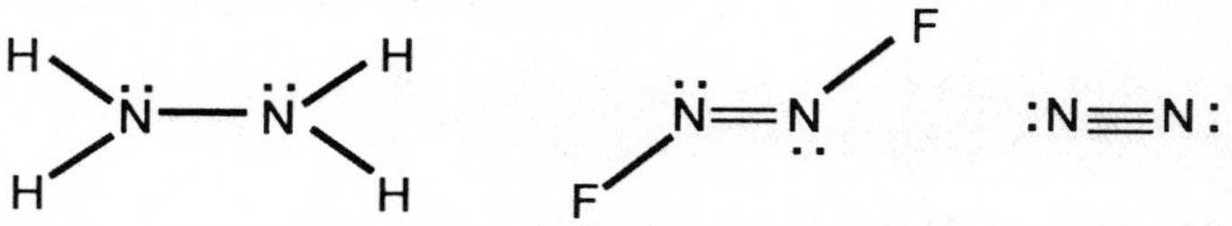

The number of electron pairs shared by the two N atoms increases in the series, and there is a corresponding decrease in the N-N distance.

9.61 (a) 16 e$^-$, 8 e$^-$ pairs

(b) $$[\ \ddot{N}=N=\ddot{N}\ ]^- \longleftrightarrow [:N\equiv N-\ddot{N}:]^- \longleftrightarrow [:\ddot{N}-N\equiv N:]^-$$

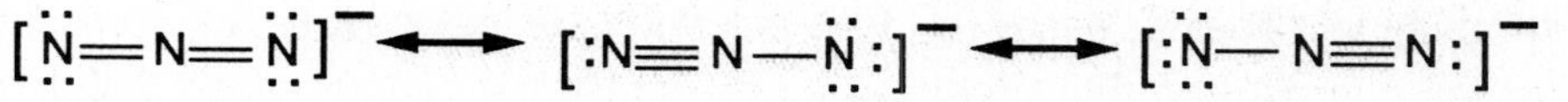

The observed bond length of 1.16 Å is intermediate between the values for N = N, 1.24 Å, and N ≡ N, 1.10 Å. This is consistent with the resonance structures, which indicate contribution from formally double and triple bonds to the true bonding picture in N_3^-.

(c) In each resonance structure, the central N has 2 VSEPR e$^-$ "pairs," so it must be sp hybridized. It is difficult to predict the hybridization of terminal atoms in molecules where there are resonance structures because there are a different number of VSEPR pairs around the terminal atoms in each structure. Since the "true" electronic arrangement is a combination of all resonance structures, we will assume that the terminal N-N bonds have some triple bond character and that the terminal N atoms are sp hybridized. (There is no experimental measure of hybridization at terminal atoms, since there are no bond angles to observe.)

(d) In each resonance structure, N-N σ bonds are formed by sp hybrids and π bonds are formed by unhybridized p orbitals. Nonbonding e$^-$ pairs can reside in sp hybrids or p atomic orbitals.

(e) Recall that electrons in 2s orbitals are on the average closer to the nucleus than electrons in 2p orbitals. Since sp hybrids have greater s orbital character, it is reasonable to expect the radial extension of sp orbitals to be smaller than that of sp^2 or sp^3 orbitals and σ bonds formed by sp orbitals to be slightly shorter than those formed by other hybrid orbitals, assuming the same bonded atoms.

There are no solitary σ bonds in N_3^-. That is, the two σ bonds in N_3^- are each accompanied by at least one π bond between the bonding pair of atoms. Sigma bonds which are part of a double or triple bond must be shorter so that the p orbitals can overlap enough for the π bond to form. Thus, the observation is not applicable to this molecule. (Comparison of C-H bond lengths in C_2H_2, C_2H_4, C_2H_6 and related molecules would confirm or deny the observation.)

9.62 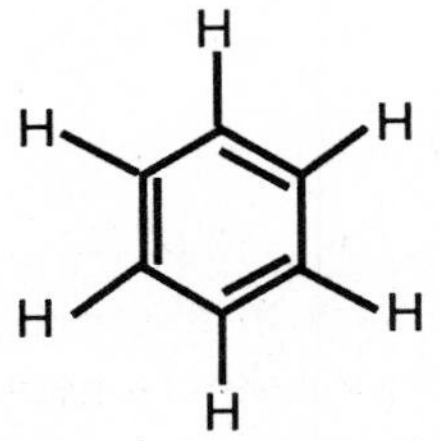 (g) ⟶ 6 C(g) + 6 H(g)

$\Delta H = 6D(C\text{-}H) + 3D(C\text{-}C) + 3D(C{=}C) - 0$ (The products are isolated atoms; there is no bond making.)
$= 6(413 \text{ kJ}) + 3(348 \text{ kJ}) + 3(614 \text{ kJ})$
$= 5364 \text{ kJ}$

According to Hess's Law:

$$\Delta H^\circ = 6\,\Delta H_f^\circ C(g) + 6\,\Delta H_f^\circ H(g) - \Delta H_f^\circ C_6H_6(g)$$
$$= 6(718.4 \text{ kJ}) + 6(217.94 \text{ kJ}) - (+82.9 \text{ kJ})$$
$$= 5535 \text{ kJ}$$

The difference in the two results, 171 kJ/mol C_6H_6, is due to the resonance stabilization in benzene. That is, because the π electrons are delocalized, the molecule has a lower overall energy than that predicted for the presence of 3 localized C-C and C=C bonds. Thus, the amount of energy actually required to decompose 1 mole of $C_6H_6(g)$, represented by the Hess's Law calculation, is greater than the sum of the localized bond energies (not taking resonance into account) from the first calculation above.

9.63 (a) XeF_6 50 e^-, 25 e^- pairs

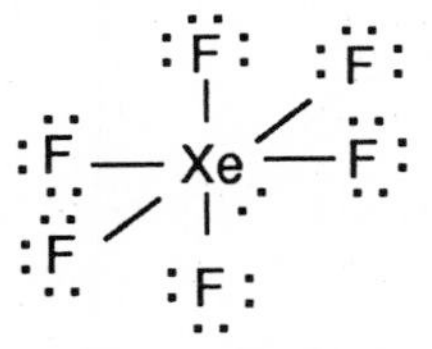

(b) There are 7 VSEPR e^- pairs around Xe, and the maximum number of e^- pairs in Table 9.3 is 6.

(c) Tie 7 balloons together and see what arrangement they adapt (seriously! see Figure 9.3). Alternately, go the chemical literature where VSEPR was first proposed and see if there is a preferred orientation for 7 e^- pairs.

(d) Since the hybrid orbitals for 5 VSEPR e^- pairs involved one d orbital and for 6 pairs, two d orbitals, a reasonable suggestion would be sp^3d^3.

(e) One of the 7 VSEPR pairs is a nonbonded pair. The question is whether it occupies an axial or equatorial position. The equatorial plane of a pentagonal bipyramid has F-Xe-F-angles of 72°. Placing the nonbonded pair in the equatorial plane would create severe repulsions between it and the adjacent bonded pairs. Thus, the nonbonded pair will reside in the axial position. The molecular structure is a pentagonal pyramid.

9.64 (a) 

To accommodate the π bonding by all 3 O atoms indicated in the resonance structures above, all O atoms are sp^2 hybridized.

(b) For the first resonance structure, both sigma bonds are formed by overlap of sp^2 hybrid orbitals, the π bond is formed by overlap of atomic p orbitals, one of the nonbonded pairs on the right terminal O atom is in a p atomic orbital, and the remaining 5 nonbonded pairs are in sp^2 hybrid orbitals.

(c) Only unhybridized p atomic orbitals can be used to form a delocalized π system.

(d) The unhybridized p orbital on each O atom is used to form the delocalized π system, and in both resonance structures one nonbonded electron pair resides in a p atomic orbital. The delocalized π system then contains 4 electrons, 2 from ther π bond and 2 from the nonbonded pair in the p orbital.

9.65 According to Table 8.5, the average C-C length is 1.54 Å, and the average C=C length is 1.34 Å. While the C=C bonds in butadiene appear "normal", the central C-C is significantly shorter than average. Examination of the bonding in butadiene reveals that each C atom is sp^2 hybridized and the π bonds are formed by the remaining unhybridized 2p orbital on each atom. Since the molecule is planar, these four 2p orbitals must be parallel and some delocalization of the π electrons occurs. This gives the central C-C single bond some double bond character, as indicated by the shorter, 1.48 Å vs 1.54 Å, bond length.

9.66 (a) Consider the bond order for each species.

	H_2	H_2^+	H_2^-	He_2^+	He_2
σ^*_{1s}			↑	↑	↑↓
σ_{1s}	↑↓	↑	↑↓	↑↓	↑↓
	BO = 1	BO = 0.5	BO = 0.5	BO = 0.5	BO = 0

Clearly He_2 has the weakest bond and H_2 the strongest. If antibonding orbitals are slightly more unfavorable than bonding orbitals are favorable (Section 9.6), then H_2^+ should have a slightly stronger bond than H_2^-. Since He_2^+ and H_2^- are isoelectronic, nuclear charge considerations prevail and He_2^+ should have a slightly stronger bond than H_2^-. Thus, the order of increasing bond strength is $He_2 < H_2^-$, He_2^+, $H_2^+ < H_2$, with small distinctions between the three ions with equal bond orders. There is no good way to compare H_2^+ and He_2^+ with the information given.

(b) For the three H species, the order of bond length is $H_2 < H_2^+ < H_2^-$. He_2^+ is probably shorter than H_2^-, due to greater nuclear charge. The trade off between increased nuclear charge for He_2^+ and absence of an electron in the σ^*_{1s} for H_2^+ makes it difficult to predict their relative bond lengths.

$H_2 < H_2^+ \approx He_2^+ < H_2^- < He_2$

9.67 Ne_2 16 e^- Ne_2^+ 15 e^-

	Ne_2	Ne_2^+
σ^*_{2p}	↑↓	↑
π^*_{2p}	↑↓ ↑↓	↑↓ ↑↓
σ_{2p}	↑↓	↑↓
π_{2p}	↑↓ ↑↓	↑↓ ↑↓
σ^*_{2s}	↑↓	↑↓
σ_{2s}	↑↓	↑↓

NO = 1/2 (8-8) = 0 BO = 1/2 (8-7) = 0.5

Ne_2 has a bond order of zero and is not energetically favored over isolated Ne atoms; it should not exist. Ne_2^+ has a bond order of 0.5 and is slightly lower in energy than isolated Ne atoms, so it will probably exist under special experimental conditions, but be unstable.

9.68 If the σ_{2p} molecular orbital were lower than the π_{2p} orbitals, the orbital energy level diagram for B_2 would be:

☐	σ^*_{2p}
☐☐	π^*_{2p}
☐☐	π_{2p}
⇅	σ_{2p}
⇅	σ^*_{2s}
⇅	σ_{2s}

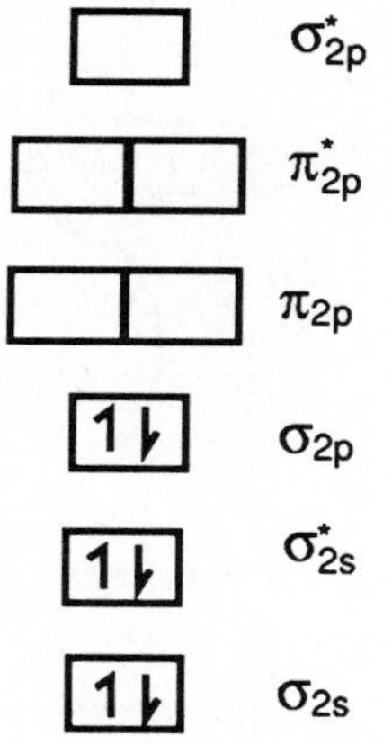

There would be no unpaired electrons, and the molecule would be diamagnetic. Switching the order of σ_{2p} and π_{2p} gives 1 unpaired electron in each degenerate π_{2p} orbital and the observed paramagnetism.

9.69 (a) 11 valence electrons, $\sigma^2_{2s}\ \sigma^{*2}_{2s}\ \pi^4_{2p}\ \sigma^2_{2p}\ \pi^{*1}_{2p}$ (b) paramagnetic

    (c) The bond order of NO is 1/2 (8 - 3) = 2.5. The electron that is lost is in an antibonding molecular orbital, so the bond order in NO^+ is 3.0. The increase in bond order is the driving force for the formation of NO^+.

    (d) To form NO^-, an electron is added to an antibonding orbital, and the new bond order is 1/2 (8 - 4) = 2. The order of increasing bond order and bond strength is $NO^- < NO < NO^+$.

    (e) NO^+ is isoelectronic with N_2, and NO^- is isoelectronic with O_2

9.70 (a) CH_3^+ 6 e^-,   3 e^- pairs

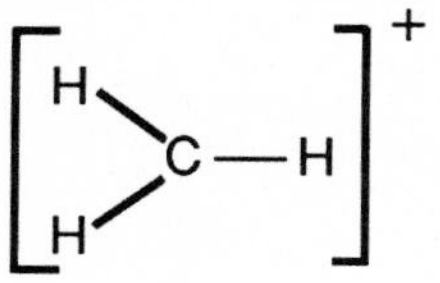

3 VSEPR e^- pairs around C
trigonal planar e^- pair geometry
trigonal planar molecular geometry

    (b) No. CH_3^- has 4 VSEPR e^- pairs around C. The e^- pair geometry is tetrahedral, and the molecular geometry is trigonal pyramidal.

(c)

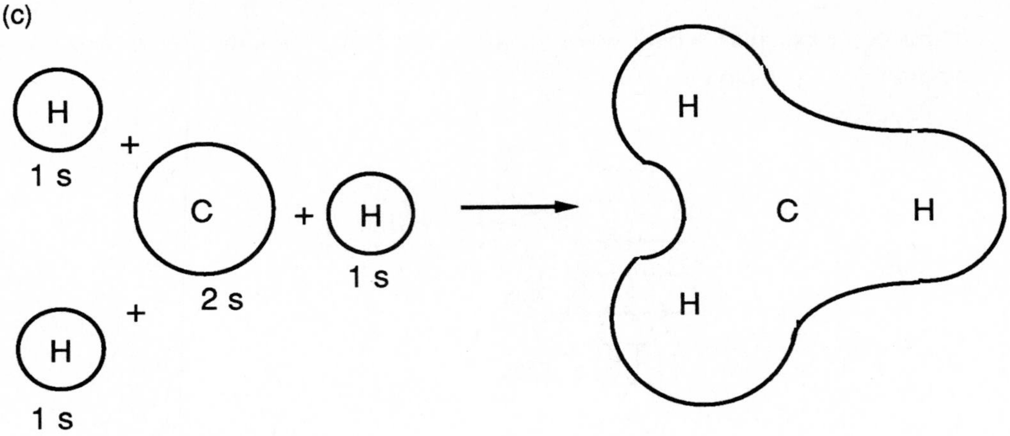

9.71 (a) $3d_{z^2}$

(b) Ignoring the donut of the d_{z^2} orbital

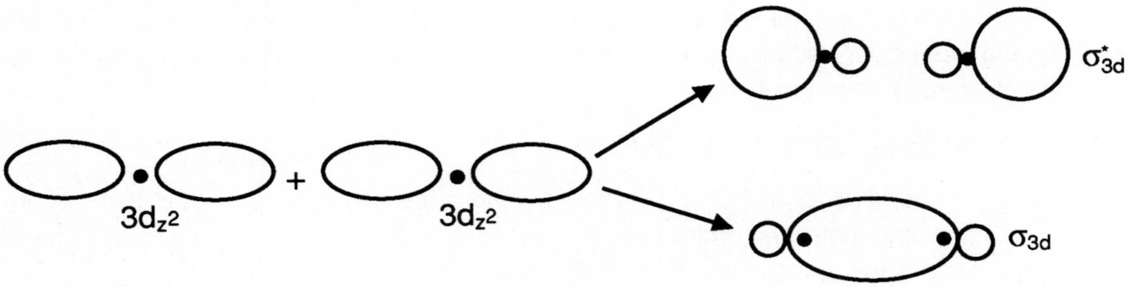

(c) Sc - [Ar] $4s^2 3d^1$ Omitting the core electrons, there are 6 e^- in the energy level diagram.

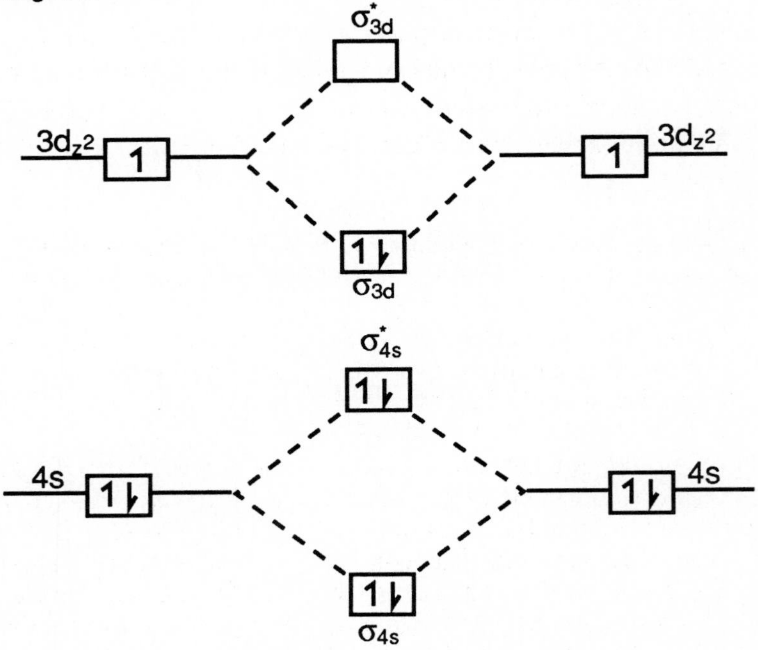

(d) The bond order in SC_2 is 1/2(4-2) = 1.0.

CHAPTER 10

Gases

Gas Characteristics; Pressure

10.1 The key to the difference in densities is that Cl_2 and Br_2 are in different states at 1 atmosphere pressure and room temperature. Most of the volume occupied by a gas is empty space; the molecules are very far apart. In a liquid, the molecules are touching. For $Br_2(l)$, 3.19 g of Br_2 molecules occupy almost all of the 1.0 cm^3 volume. On the other hand, most of the 1.0 L volume that contains 2.90 g of Cl_2 molecules is empty space.

10.2 In the liquid state, the molecules are touching. In order for $H_2O(l)$ and $CCl_4(l)$ to mix, one kind of molecule must be able to displace the other. For molecules as dissimilar as H_2O and CCl_4, this is not possible. In the vapor (or gas) phase, the molecules are so far apart that they do not interact with each other to any appreciable extent. Thus, there is no barrier to mixing, regardless of the identity of the molecule. All mixtures of gases are homogeneous.

10.3 (a) $F = m \times a$. Since both people have the same mass and both experience the acceleration of gravity, the forces they exert on the floor are exactly equal.

(b) $P = F / A$. The two forces are equal, but the person standing on one foot exerts this force over a smaller area. Thus, the person standing on one foot exerts a greater pressure on the floor.

10.4 The height of the mercury column in a barometer is a measure of atmospheric pressure. The Hg column is steady when the pressure exerted by the atmosphere is equal to the downward pressure due to the mass of the Hg column. Pressure is defined as force **per unit area**. Thus, the larger the diameter of the Hg column, the greater the mass of Hg and the greater the **total force** exerted by the Hg column. However, this force is distributed over a larger cross-sectional area, so the ratio of force/area, the pressure, is the same as that of a smaller diameter column. The same length Hg column is required to balance atmospheric pressure, regardless of the diameter of the column.

Mathematically: $P = \frac{F}{A}$; $F = m \times a$; $m = d \times V$ (density)(volume) ; $V = A \times$ height

Substituting $P = \frac{d \times V \times a}{A} = \frac{d \times \cancel{A} \times \text{height} \times a}{\cancel{A}}$

$P = d \times \text{height} \times a$

The area (A) terms cancel, and pressure is independent of cross-sectional area of the tube.

10.5 (a) $P_{Hg} = P_{H_2O}$; Using the relationship derived in 10.4: $(d \times h \times a)_{H_2O} = (d \times h \times a)_{Hg}$

Since a, the acceleration due to gravity, is equal in both liquids,

$(d \times h)_{H_2O} = (d \times h)_{Hg}$

$1.00 \text{ g/mL} \times h_{H_2O} = 13.6 \text{ g/mL} \times 760 \text{ mm}$

$$h_{H_2O} = \frac{13.6 \text{ g/mL} \times 760 \text{ mm}}{1.00 \text{ g/mL}} = 1.03 \times 10^4 \text{ mm} = 10.3 \text{ m}$$

(b) Pressure due to H_2O:

$1 \text{ atm} = 1.03 \times 10^4 \text{ mm } H_2O$ (from part (a))

$$25 \text{ ft } H_2O \times \frac{12 \text{ in}}{1 \text{ ft}} \times \frac{2.54 \text{ cm}}{1 \text{ in}} \times \frac{10 \text{ mm}}{1 \text{ cm}} \times \frac{1 \text{ atm}}{1.03 \text{ x } 10^4 \text{ mm}} = 0.74 \text{ atm}$$

$P_{total} = P_{atm} + P_{H_2O} = 1.00 \text{ atm} \times 0.74 \text{ atm} = 1.74 \text{ atm}$

10.6 Using the relationship derived in Exercise 10.5 for two liquids under the influence of gravity, $(d \times h)_{dbp} = (d \times h)_{Hg}$

$\frac{1.05 \text{ g}}{1 \text{ mL}} \times h_{dbp} = \frac{13.6 \text{ g}}{1 \text{ mL}} \times 760 \text{ mm}$; At 760 torr, the height of a Hg barometer is 760 mm.

$$h_{dbp} = \frac{13.6 \text{ g/mL} \times 760 \text{ mm}}{1.05 \text{ g/mL}} = 9.84 \times 10^3 \text{ mm} = 9.84 \text{ m}$$

10.7 (a) $0.660 \text{ atm} \times \frac{101.325 \text{ kPa}}{1 \text{ atm}} = 66.9 \text{ kPa}$

(b) $347 \text{ torr} \times \frac{1 \text{ atm}}{760 \text{ torr}} = 0.457 \text{ atm}$

(c) $825 \text{ mm Hg} \times \frac{1 \text{ atm}}{760 \text{ mm Hg}} = 1.09 \text{ atm}$

(d) $0.917 \text{ atm} \times \frac{760 \text{ torr}}{1 \text{ atm}} = 697 \text{ torr}$

10.8 (a) $98{,}200 \text{ Pa} \times \frac{1 \text{ atm}}{101{,}325 \text{ Pa}} = 0.969 \text{ atm (or } 0.96916 \text{ atm)}$

(98,200 could have 3 or 5 sig. figs.)

(b) $1.43 \text{ atm} \times \frac{760 \text{ mm Hg}}{1 \text{ atm}} = 1.09 \times 10^3 \text{ mm Hg}$

(c) $659 \text{ mm Hg} \times \frac{1 \text{ torr}}{1 \text{ mm Hg}} = 659 \text{ mm Hg}$

(d) $1.00 \times 10^3 \text{ torr} \times \frac{1 \text{ atm}}{760 \text{ torr}} = 1.32 \text{ atm}$

10.9 (a) $30.45 \text{ in Hg} \times \frac{25.4 \text{ mm}}{1 \text{ in}} \times \frac{1 \text{ torr}}{1 \text{ mm Hg}} = 773.4 \text{ torr}$

[The result has 4 sig. figs. because 25.4 mm/in is considered to be an exact number. (section 1.5)]

(b) The pressure in Chicago is greater than **standard atmospheric pressure**, 760 torr, so it makes sense to classify this weather system as a "high pressure system."

10.10 (a) $\frac{1.6 \text{ Earth atm}}{1 \text{ Titan atm}} \times \frac{101.3 \text{ kPa}}{1 \text{ Earth atm}} = \frac{1.6 \times 10^2 \text{ kPa}}{1 \text{ Titan atm}}$

(b) $\frac{90 \text{ Earth atm}}{1 \text{ Venus atm}} \times \frac{101.3 \text{ kPa}}{1 \text{ Earth atm}} = \frac{9.1 \times 10^3 \text{ kPa}}{1 \text{ Venus atm}}$

10.11 $P = \frac{m \times a}{A} = \frac{130 \text{ lb}}{0.50 \text{ in}^2} \times \frac{9.81 \text{ m}}{1 \text{ s}^2} \times \frac{0.454 \text{ kg}}{1 \text{ lb}} \times \frac{39.4^2 \text{ in}^2}{1 \text{ m}^2} = 1.8 \times 10^3 \text{ kPa}$

$1.8 \times 10^3 \text{ kPa} \times \frac{1 \text{ atm}}{101.3 \text{ kPa}} = 18 \text{ atm}$

10.12 $1 \text{ Pa} = \frac{1 \text{ N}}{\text{m}^2} = \frac{1 \text{ kg} \cdot \text{m}}{\text{s}^2} \times \frac{1}{\text{m}^2} = \frac{1 \text{ kg}}{\text{m} \cdot \text{s}^2}$ Change mass to kg and area to m^2.

mass: $\frac{2.70 \text{ g}}{\text{cm}^3} \times (20.0 \text{ cm})^3 \times \frac{1 \text{ kg}}{1000 \text{ g}} = 21.6 \text{ kg}$

area: $(20.0 \text{ cm})^2 \times \frac{1 \text{ m}^2}{(100 \text{ cm})^2} = 0.0400 \text{ m}^2$

$P = \frac{F}{A} = \frac{m \times a}{A} = \frac{21.6 \text{ kg}}{0.0400 \text{ m}^2} \times \frac{9.81 \text{ m}}{1 \text{ s}^2} = 5.30 \times 10^3 \text{ Pa}$

10.13 (a) It **is not** necessary to know atmospheric pressure to use a closed-end manometer. The gas on one side of the Hg is working against a vacuum on the other side of the Hg. The pressure of the gas is just the difference in the heights of the two arms.

(b) It **is** necessary to know atmospheric pressure to use an open-end manometer. In an open-end manometer the pressure of the sample is balanced against atmospheric pressure. If the Hg level is higher in the open end than in the end exposed to the sample, the pressure of the sample is atmospheric pressure plus the difference in heights of the two sides. If the Hg level is lower in the open end than in the other end, the pressure of the sample is atmospheric pressure minus the difference in heights of the two sides.

10.14 (a) The gas is exerting 27 mm Hg (torr) more pressure than the atmosphere.

$$P_{gas} = P_{atm} + 27 \text{ torr} = 0.981 \text{ atm} \times \frac{760 \text{ torr}}{1 \text{ atm}} + 27 \text{ torr} = 773 \text{ torr}$$

(b) The atmosphere is exerting 56 mm Hg (torr) more pressure than the gas.

$$P_{gas} = P_{atm} - 56 \text{ torr} = 1.03 \text{ atm} \times \frac{760 \text{ torr}}{1 \text{ atm}} - 56 \text{ torr} = 727 \text{ torr}$$

The Gas Laws; The Ideal Gas Equation

10.15 (a) If two quantities, X and Y, are **directly proportional**, a change in X causes a proportional change in Y in the **same** direction. Mathematically, if X and Y are directly proportional and c is a constant, Y = cX or Y/X = c.

If X and Y are **inversely proportional**, a change in X causes a change in Y in the **opposite** direction. Mathematically X • Y = c or Y = c/X.

(b) Volume (V) is inversely proportional to pressure (P); volume is directly proportional to absolute temperature (T); volume is directly proportional to quantity of gas (n).

10.16 In the following equations, C indicates a constant, which can have a different value in each relationship.

(a) $PV = C$ or $P \propto 1/V$ when n and T are constant

(b) $V \propto n$ or $V = Cn$ when T and P are constant

(c) $P \propto T$ or $P = CT$ when n and V are constant

(d) $PV \propto T$ or $PV = CT$ when n is constant

10.17 $P_1V_1 = P_2V_2$; The proportionality holds true for any pressure or volume units.

$P_1 = 715$ torr, $V_1 = 10.7$ L

(a) $$P_2 = 1.40 \text{ atm} \times \frac{760 \text{ torr}}{1 \text{ atm}} = 1.06 \times 10^3 \text{ torr}$$

$$V_2 = \frac{P_1V_1}{P_2} = \frac{715 \text{ torr} \times 10.7 \text{ L}}{1.06 \times 10^3 \text{ torr}} = 7.22 \text{ L}$$ (7.19 L if P_2 is not rounded to 1.06×10^3.)

(b) $V_2 = 15.5$ L; $P_2 = \frac{P_1 V_1}{V_2} = \frac{715 \text{ torr} \times 10.7 \text{ L}}{15.5 \text{ L}} = 494$ torr

10.18 $\frac{V_1}{T_1} = \frac{V_2}{T_2}$; T must be in Kelvins for the relationship to be true.

$V_1 = 6.75$ L, $T_1 = 31.0°C = 304$ K

(a) $T_2 = 125°C = 398$ K; $V_2 = \frac{V_1 T_2}{T_1} = \frac{6.75 \text{ L} \times 398 \text{ K}}{304 \text{ K}} = 8.84$ L

(b) $V_2 = 5.00$ L; $T_2 = \frac{T_1 V_2}{V_1} = \frac{304 \text{ K} \times 5.00 \text{ L}}{6.75 \text{ L}} = 225 \text{ K} = -48° \text{ C}$

10.19 (a) Avogadro's hypothesis states that equal volumes of gases at the same temperature and pressure contain equal numbers of molecules. Since molecules react in the ratios of small whole numbers, it follows that the volumes of reacting gases (at the same temperature and pressure) are in the ratios of small whole numbers.

(b) Since the two gases are at the same temperature and pressure, the ratio of the numbers of atoms is the same as the ratio of volumes. There are 1.5 times as many Xe atoms as Ne atoms.

10.20 According to Avogadro's hypothesis, the mole ratios in the chemical equation will be volume ratios for the gases if they are at the same temperature and pressure.

$$N_2(g) + 3H_2(g) \rightarrow 2NH_3(g)$$

The volumes of H_2 and N_2 are in a stoichiometric $\frac{2.1 \text{ L}}{0.70 \text{ L}}$ or $\frac{3 \text{ vol } H_2}{1 \text{ vol } N_2}$ ratio, so either can be used to determine the volume of $NH_3(g)$ produced.

$0.70 \text{ L } N_2 \times \frac{2 \text{ mol } NH_3}{1 \text{ mol } N_2} = 1.4 \text{ L } NH_3(g)$ produced.

10.21 (a) An ideal gas exhibits pressure, volume and temperature relationships which are described by the equation PV = nRT. (An ideal gas obeys the Ideal Gas Law.)

(b) PV = nRT; P in atmospheres
V in liters
n in moles
T in kelvins

10.22 (a) STP stands for standard temperature, 0°C (or 273 K), and standard pressure, 1 atm.

(b) $V = \frac{nRT}{P}$; $V = 1 \text{ mol} \times \frac{0.0821 \text{ L} \cdot \text{atm}}{\text{K} \cdot \text{mol}} \times \frac{273 \text{ K}}{1 \text{ atm}}$

V = 22.4 L for 1 mole of gas at STP

(c) $25°C = 273 = 298\ K$

$$V = \frac{nRT}{P};\ V = 1\ mol \times \frac{0.0821\ L \cdot atm}{K \cdot mol} \times \frac{298\ K}{1\ atm}$$

$V = 24.5\ L$ for 1 mol of gas at 1 atm and 25°C

10.23 (a) $n = 2.59 \times 10^{-2}\ mol,\ V = 121\ mL = 0.121\ L,\ T = -15°C = 258\ K,\ P = ?$

$$P = \frac{nRT}{V} = 0.0259\ mol \times \frac{0.0821\ L \cdot atm}{K \cdot mol} \times \frac{258\ K}{0.121\ L} = 4.53\ atm$$

(b) $V = 4.65\ L,\ T = 35°C = 308\ K,\ P = 1150\ torr \times \frac{1\ atm}{760\ torr} = 1.51\ atm,\ n = ?$

$$n = \frac{PV}{RT} = 1.51\ atm \times \frac{K \cdot mol}{0.0821\ L \cdot atm} \times \frac{4.65\ L}{308\ K} = 0.278\ mol$$

(c) $n = 1.76\ mol,\ P = 0.76\ atm,\ T = 47°C = 320\ K,\ V = ?$

$$V = \frac{nRT}{P} = 1.76\ mol \times \frac{0.0821\ L \cdot atm}{K \cdot mol} \times \frac{320\ K}{0.76\ atm} = 61\ L\ (60.8\ L)$$

(d) $n = 9.87 \times 10^{-2}\ mol,\ V = 164\ mL = 0.164\ L$

$P = 722\ torr \times \frac{1 atm}{760\ torr} = 0.950\ atm,\ T = ?$

$$T = \frac{PV}{nR} = 0.950\ atm \times \frac{0.164\ L}{0.0987\ mol} \times \frac{1\ K \cdot mol}{0.0821\ L \cdot atm} = 19.2\ K$$

10.24 (a) $n = 1.04\ mol,\ V = 21.8\ L,\ T = 25°C = 298\ K,\ P = ?$

$$P = \frac{nRT}{V} = 1.04\ mol \times \frac{0.0821\ L \cdot atm}{K \cdot atm} \times \frac{298\ K}{21.8\ L} = 1.17\ atm$$

(b) $n = 6.72 \times 10^{-3}\ mol,\ T = 265°C = 538\ K,\ P = 23.0\ torr \times \frac{1\ atm}{760\ torr} = 0.0303\ atm,\ V = ?$

$$V = \frac{nRT}{P} = 6.72 \times 10^{-3} mol \times \frac{0.0821\ L \cdot atm}{K \cdot mol} \times \frac{538\ K}{0.0303\ atm} = 9.80\ L$$

(c) $V = 1.50\ L,\ T = 37°C = 310\ K,\ P = 725\ torr \times \frac{1\ atm}{760\ torr} = 0.954\ atm,\ n = ?$

$$n = \frac{PV}{RT} = 0.954\ atm \times \frac{K \cdot mol}{0.0821\ L \cdot atm} \times \frac{1.50\ L}{310\ K} = 5.62 \times 10^{-2}\ mol$$

(d) $n = 0.270\ mol,\ V = 15.0\ L,\ P = 2.54\ atm,\ T = ?$

$$T = \frac{PV}{nR} = \frac{2.54\ atm \times 15.0\ L}{0.270\ mol} \times \frac{K \cdot mol}{0.0821\ L \cdot atm} = 1.72 \times 10^{3}\ K$$

10.25 Air is a mixture of N_2 and O_2, but for the purpose of calculating pressure, only the total number of gas molecules is important, not the identity of these molecules.

$$V = 1.05 \text{ L}, \ T = 37°\text{C} = 310 \text{ K}, \ P = 740 \text{ mm Hg} \times \frac{1 \text{ atm}}{760 \text{ mm}} = 0.974 \text{ atm}$$

$$n = \frac{PV}{RT} = 0.974 \text{ atm} \times \frac{\text{K} \cdot \text{mol}}{0.0821 \text{ L} \cdot \text{atm}} \times \frac{1.05 \text{ L}}{310 \text{ K}} = 0.0402 \text{ mol of gas}$$

$$0.0402 \text{ mol} \times \frac{6.022 \times 10^{23} \text{ molecules}}{1 \text{ mol}} = 2.42 \times 10^{22} \text{ gas molecules}$$

10.26 Find the volume of the tube in cm^3; $1 \text{ cm}^3 = 1 \text{ mL}$.

$$r = d/2 = 2.0 \text{ cm}/2 = 1.0 \text{ cm}; \ h = 4.0 \text{ m} = 4.0 \times 10^2 \text{ cm}$$

$$V = \pi r^2 h = 3.14159 \times (1.0 \text{ cm})^2 \times (4.0 \times 10^2 \text{ cm}) = 1.257 \times 10^3 \text{ cm}^3 = 1.257 \text{ L}$$

$$PV = \frac{g}{\mathcal{M}} RT; \ g = \frac{\mathcal{M} \ PV}{RT}; \ P = 1.5 \text{ mm Hg} \times \frac{1 \text{ atm}}{760 \text{ mm Hg}} = 1.97 \times 10^{-3} \text{ atm}$$

$$g = \frac{20.18 \text{ g Ne}}{1 \text{ mol Ne}} \times \frac{\text{K} \cdot \text{mol}}{0.0821 \text{ L} \cdot \text{atm}} \times \frac{1.97 \times 10^{-3} \text{ atm} \times 1.257 \text{ L}}{308 \text{ K}} = 2.0 \times 10^{-3} \text{ g Ne}$$

10.27 (a) $$V_2 = \frac{P_1 V_1 T_2}{P_2 T_1} = \frac{1.00 \text{ atm} \times 0.600 \text{ L} \times 273 \text{ K}}{0.205 \text{ atm} \times 309 \text{ K}} = 2.59 \text{ L}$$

(b) $$V_2 = \frac{1.00 \text{ atm} \times 0.600 \text{ L} \times 273 \text{ K}}{1.00 \text{ atm} \times 309 \text{ K}} = 0.530 \text{ L}$$

10.28 (a) $$V_2 = \frac{P_1 V_1 T_2}{P_2 T_1} = \frac{740 \text{ torr} \times 6.18 \text{ L} \times 380 \text{ K}}{680 \text{ torr} \times 306 \text{ K}} = 8.35 \text{ L}$$

(b) $$V_2 = \frac{740 \text{ torr} \times 6.18 \text{ L} \times 273 \text{ K}}{760 \text{ torr} \times 306 \text{ K}} = 5.37 \text{ L}$$

(c) $$T_2 = \frac{P_2 V_2 T_1}{P_1 V_1} = \frac{800 \text{ m Hg} \times 3.00 \text{ L} \times 306 \text{ K}}{740 \text{ mm Hg} \times 6.18 \text{ L}} = 161 \text{ K}$$

(d) $$P_2 = \frac{P_1 V_1 T_2}{V_2 T_1} = \frac{740 \text{ torr} \times 6.18 \text{ L} \times 340 \text{ K}}{5.00 \text{ L} \times 306 \text{ K}} = 1.02 \times 10^3 \text{ torr}$$

10.29 (a) $$g = \frac{\mathcal{M} \ PV}{RT} = \frac{32.0 \text{ g } O_2}{1 \text{ mol } O_2} \times \frac{\text{K} \cdot \text{mol}}{0.0821 \text{ L} \cdot \text{atm}} \times 1.8 \times 10^4 \text{ kPa} \times \frac{1 \text{ atm}}{101.3 \text{ kPa}} \times \frac{42.0 \text{ L}}{296 \text{ K}}$$

$$= 9.83 \times 10^3 \text{ g } O_2$$

(b) $$V_2 = \frac{P_1 V_1 T_2}{T_1 P_2} = \frac{18{,}000 \text{ kPa} \times 42.0 \text{ L} \times 273 \text{ K}}{296 \text{ k} \times 101.3 \text{ kPa}} = 6.88 \times 10^3 \text{ L}$$

10.30 (a) $g = \frac{\mathcal{M}\ PV}{RT}$; $P = 160\ lb/in^2 \times \frac{1\ atm}{14.7\ lb/in^2} = 10.9\ atm$

$$g = \frac{38.0\ g\ F_2}{1\ mol\ F_2} \times \frac{K \cdot mol}{0.0821\ L \cdot atm} \times \frac{10.9\ atm \times 30.0\ L}{299\ K} = 506\ g\ F_2$$

(b) $V_2 = \frac{P_1 V_1 T_2}{T_1 P_2} = \frac{160\ lb/in^2 \times 30.0\ L \times 273\ K}{299\ K \times 14.7\ lb/in^2} = 298\ L$

10.31 (a) $5.2\ g \times 1\ h \times \frac{0.8\ mL\ O_2}{1\ g \cdot hr} = 4\ mL\ O_2$ consumed

$$n = \frac{PV}{RT} = 1\ atm \times \frac{K \cdot mol}{0.0821\ L \cdot atm} \times \frac{0.004\ L}{297\ K} = 2 \times 10^{-4}\ mol$$

(b) $1\ qt\ air \times \frac{0.946\ L}{1\ qt} \times 0.21\%\ O_2$ in air $= 0.199\ L\ O_2$ available

$$n = 1\ atm \times \frac{K \cdot mol}{0.0821\ L \cdot atm} \times \frac{0.199\ L}{297\ K} = 8 \times 10^{-3}\ mol\ O_2 \text{ available}$$

roach uses $\frac{2 \times 10^{-4}\ mol}{1\ hr} \times 48\ hr = 8 \times 10^{-3}\ mol\ O_2$ consumed

Not only does the roach use 20% of the available O_2, it needs all the O_2 in the jar.

10.32 (a) $P = \frac{gRT}{\mathcal{M}V}$; mass $= 1800 \times 10^{-9}\ g = 1.8 \times 10^{-6}\ g$; $V = 1\ m^3 = 1 \times 10^3\ L$

$$P = \frac{1.8 \times 10^{-6}\ g\ Hg \times 1\ mol\ Hg}{200.6\ g\ Hg} \times \frac{0.0821\ L \cdot atm}{K \cdot mol} \times \frac{283\ K}{1 \times 10^3\ L} = 2.1 \times 10^{-10}\ atm$$

(b) $\frac{1.8 \times 10^{-6}\ g\ Hg}{1\ m^3} \times \frac{1\ mol\ Hg}{200.6\ g\ Hg} \times \frac{6.02 \times 10^{23}\ Hg\ atoms}{1\ mol\ Hg} = \frac{5.4 \times 10^{15}\ Hg\ atoms}{m^3}$

(c) $1600\ km^3 \times \frac{1000^3\ m^3}{1\ km^3} \times \frac{1.8 \times 10^{-6}\ g\ Hg}{1\ m^3} = \frac{2.9 \times 10^6\ g\ Hg}{day}$

10.33 (a) Density of a gas $= \frac{g}{L}$; $PV = \frac{g\ RT}{\mathcal{M}}$; $\frac{g}{V} = \frac{\mathcal{M}\ P}{RT} = d$

$P = 1.00\ atm$, $T = 35°C = 308\ K$, $\mathcal{M}$ of $NO_2 = 46.01\ g/mol$

$$d = \frac{46.01\ g}{mol} \times \frac{K \cdot mol}{0.0821\ L \cdot atm} \times \frac{1.00\ atm}{308\ K} = 1.82\ g/L$$

(b) $\mathcal{M} = \frac{gRT}{VP} = \frac{4.40\ g}{3.50\ L} \times \frac{0.0821\ L \cdot atm}{K \cdot mol} \times \frac{314\ K}{560\ torr} \times \frac{760\ torr}{1\ atm} = \frac{44.0\ g}{mol}$

10.34 (a) $d = \frac{\mathcal{M}P}{RT}$; $P = \frac{600 \text{ torr} \times 1 \text{ atm}}{760 \text{ torr}} = 0.789$ atm, T =100°C = 373 K,

$\mathcal{M}$ of $SF_4 = \frac{108.07 \text{ g}}{1 \text{ mol}}$

$$d = \frac{108.07 \text{ g}}{1 \text{ mol}} \times \frac{\text{K} \cdot \text{mol}}{0.0821 \text{ L} \cdot \text{atm}} \times \frac{0.789 \text{ atm}}{373 \text{ K}} = 2.78 \text{ g/L}$$

(b) $\mathcal{M} = \frac{dRT}{P} = \frac{3.67 \text{ g}}{1 \text{ L}} \times \frac{0.0821 \text{ L} \cdot \text{atm}}{\text{K} \cdot \text{mol}} \times \frac{288 \text{ K}}{825 \text{ torr}} \times \frac{760 \text{ torr}}{1 \text{ atm}} = \frac{79.9 \text{ g}}{\text{mol}}$

10.35 $\mathcal{M} = \frac{gRT}{VP} = \frac{1.05 \text{ g}}{0.500 \text{ L}} \times \frac{0.0821 \text{ L} \cdot \text{atm}}{\text{K} \cdot \text{mol}} \times \frac{298 \text{ K}}{750 \text{ mm Hg}} \times \frac{760 \text{ mm Hg}}{1 \text{ atm}} = \frac{52.1 \text{ g}}{\text{mol}}$

$$0.462 \text{ g C} \times \frac{1 \text{ mol C}}{12.01 \text{ g C}} = 0.0385 \text{ mol C}$$

$$0.538 \text{ g N} \times \frac{1 \text{ mol}}{14.01 \text{ g N}} = 0.0384 \text{ mol N}$$

The mole ratio is 1C:1N and the empirical formula is CN; the formula weight of CN = 12 + 14 = 26 g. Since molar mass of 52.1 g is twice the empirical formula weight, the molecular formula is C_2N_2.

10.36 $\mathcal{M} = \frac{gRT}{VP} = \frac{1.56 \text{ g}}{1.00 \text{ L}} \times \frac{0.0821 \text{ L} \cdot \text{atm}}{\text{K} \cdot \text{mol}} \times \frac{323 \text{ K}}{0.984 \text{ atm}} = 42.0$ g/mol

Assume 100 g cyclopropane

$$100 \text{ g} \times 0.857 \text{ \%C} = 85.7 \text{ g C} \times \frac{1 \text{ mol C}}{12.01 \text{ g}} = \frac{7.136 \text{ mol C}}{7.136} = 1 \text{ mol C}$$

$$100 \text{ g} \times 0.143 \text{ \% H} = 14.3 \text{ g H} \times \frac{1 \text{ mol H}}{1.008 \text{ g}} = \frac{14.19 \text{ mol H}}{7.136} = 2 \text{ mol H}$$

The empirical formula of cyclopropane is CH_2 and the empirical formula weight is 12 + 2 = 14 g. The ratio of molar mass to empirical formula weight, 42.0 g ÷ 14 g, is 3; therefore, there are three empirical formula units in one cyclopropane molecule. The molecular formula is $3 \times (CH_2) = C_3H_6$.

10.37 $\mathcal{M} = \frac{gRT}{VP} = \frac{1.012 \text{ g}}{0.354 \text{ L}} \times \frac{0.0821 \text{ L} \cdot \text{atm}}{\text{K} \cdot \text{atm}} \times \frac{372 \text{ K}}{742 \text{ mm Hg}} \times \frac{760 \text{ mm Hg}}{1 \text{ atm}} = \frac{89.4 \text{ g}}{\text{mol}}$

10.38 $\mathcal{M} = \frac{gRT}{VP} = \frac{0.846 \text{ g}}{0.354 \text{ L}} \times \frac{0.0821 \text{ L} \cdot \text{atm}}{\text{K} \cdot \text{atm}} \times \frac{373 \text{ K}}{752 \text{ mm Hg}} \times \frac{760 \text{ mm Hg}}{1 \text{ atm}} = \frac{74.0 \text{ g}}{\text{mol}}$

Partial Pressures

10.39 (a) The total pressure of a mixture of gases is equal to the sum of the pressures the individual gases would exert if present in the same container alone.

(b) Partial pressure is the pressure exerted by a single component of a gaseous mixture at the same temperature and volume as the mixture.

10.40 No. Only the total pressure of a gaseous mixture can be measured by a manometer, not the partial pressures of the components. The partial pressure of a gaseous component is directly related to the moles of that gas in the mixture. Thus, the moles of a component are measured stoichiometrically and partial pressure is then calculated from moles.

10.41 (a) $P_{He} = \frac{nRT}{V} = 0.432 \text{ mol} \times \frac{0.0821 \text{ L} \cdot \text{atm}}{\text{K} \cdot \text{atm}} \times \frac{298 \text{ K}}{7.00 \text{ L}} = 1.51 \text{ atm}$

$P_{Ne} = \frac{nRT}{V} = 0.357 \text{ mol} \times \frac{0.0821 \text{ L} \cdot \text{atm}}{\text{K} \cdot \text{atm}} \times \frac{298 \text{ K}}{7.00 \text{ L}} = 1.25 \text{ atm}$

$P_{Ar} = \frac{nRT}{V} = 0.103 \text{ mol} \times \frac{0.0821 \text{ L} \cdot \text{atm}}{\text{K} \cdot \text{atm}} \times \frac{298 \text{ K}}{7.00 \text{ L}} = 0.360 \text{ atm}$

(b) $P_t = 1.51 \text{ atm} + 1.25 \text{ atm} + 0.360 \text{ atm} = 3.12 \text{ atm}$

10.42 (a) $5.00\text{g } CH_4 \times \frac{1 \text{ mol } CH_4}{16.04 \text{ g } CH_4} = 0.312 \text{ mol } CH_4$

$P_{CH_4} = \frac{nRT}{V} = 0.312 \text{ mol} \times \frac{0.0821 \text{ L} \cdot \text{atm}}{\text{K} \cdot \text{mol}} \times \frac{273 \text{ K}}{1.50 \text{ L}} = 4.66 \text{ atm}$

$5.00 \text{ g } C_2H_4 \times \frac{1 \text{ mol } C_2H_4}{28.05 \text{ g } C_2H_4} = 0.178 \text{ mol } C_2H_4$

$P_{C_2H_4} = \frac{nRT}{V} = 0.178 \text{ mol } C_2H_4 \times \frac{0.0821 \text{ L} \cdot \text{atm}}{\text{K} \cdot \text{mol}} \times \frac{273 \text{ K}}{1.50 \text{ L}} = 2.66 \text{ atm}$

$5.00 \text{ g } C_4H_{10} \times \frac{1 \text{ mol } C_4H_{10}}{58.12 \text{ g } C_4H_{10}} = 0.0860 \text{ mol } C_4H_{10}$

$P_{C_4H_{10}} = \frac{nRT}{V} = 0.0860 \text{ mol} \times \frac{0.0821 \text{ L} \cdot \text{atm}}{\text{K} \cdot \text{mol}} \times \frac{273 \text{ K}}{1.50 \text{ L}} = 1.29 \text{ atm}$

(b) $P_t = 4.66 \text{ atm} + 2.66 \text{ atm} + 1.29 \text{ atm} = 8.61 \text{ atm}$

10.43 The partial pressure of each component is equal to the mole fraction of that gas times the total pressure of the mixture. Find the mole fraction of each component and then its partial pressure.

$n_t = 0.45 \text{ mol } N_2 + 0.25 \text{ mol } O_2 + 0.10 \text{ mol } CO_2 = 0.80 \text{ mol}$

$\chi_{N_2} = \frac{0.45}{0.80} = 0.56; \quad P_{N_2} = 0.56 \times 1.32 \text{ atm} = 0.74 \text{ atm}$

$\chi_{O_2} = \frac{0.25}{0.80} = 0.31$; $P_{O_2} = 0.31 \times 1.32$ atm $= 0.41$ atm

$\chi_{CO_2} = \frac{0.10}{0.80} = 0.12$; $P_{CO_2} = 0.12 \times 1.32$ atm $= 0.16$ atm

10.44 $n_{N_2} = 3.50 \text{ g } N_2 \times \frac{1 \text{ mol}}{28.02 \text{ g}} = 0.125$ mol; $n_{H_2} = 1.30 \text{ g } H_2 \times \frac{1 \text{ mol}}{2.016 \text{ g}} = 0.645$ mol

$n_{H_3} = 5.27 \text{ g } NH_3 \times \frac{1 \text{ mol}}{17.03 \text{ g}} = 0.309$ mol; $n_t = 0.125 + 0.645 + 0.309 = 1.079$ mol

$P_{N_2} = \frac{n_{N_2}}{n_t} \times P_t = \frac{0.125}{1.079} \times 2.50 \text{ atm} = 0.290$ atm

$P_{H_2} = \frac{0.645}{1.079} \times 2.50 \text{ atm} = 1.49$ atm ; $P_{NH_3} = \frac{0.309}{1.079} \times 2.50 \text{ atm} = 0.716$ atm

10.45 (a) The partial pressure of gas A is **not affected** by the addition of gas C. The partial pressure of A depends only on moles of A, the volume of the container and conditions; none of these factors change when gas C is added.

(b) The total pressure in the vessel **increases** when gas C is added, because the total number of moles of gas increases.

(c) The mole fraction of gas B **decreases** when gas C is added. The moles of gas B stay the same, but the total moles increase, so the mole fraction of B (n_B/n_t) decreases.

10.46 (a) $6.00 \text{ g } SO_2 \times \frac{1 \text{ mol } SO_2}{64.06 \text{ g } SO_2} = 0.0937 \text{ mol } SO_2$; $\Delta T = 333 \text{ K} - 291 \text{ K} = 42 \text{ K}$

$\Delta P_{SO_2} = \frac{nR\Delta T}{V} = 0.0937 \text{ mol} \times \frac{0.0821 \text{ L} \cdot \text{atm}}{\text{K} \cdot \text{mol}} \times \frac{42 \text{ K}}{5.00 \text{ L}} = +0.065 \text{ atm}$

As the flask is heated, the partial pressure of $SO_2(g)$ increases by 0.065 atm.

(b) $7.50 \text{ g } SO_3 \times \frac{1 \text{ mol } SO_3}{80.06 \text{ g } SO_3} = 0.0937 \text{ mol } SO_3$; $\Delta T = 42 \text{ K}$

$n_T = 0.0937 \text{ mol } SO_2 + 0.0937 \text{ mol } SO_3 = 0.1874$ mol gas

$\Delta P_T = \frac{n_T R\Delta T}{V} = 0.1874 \text{ mol} \times \frac{0.0821 \text{ L} \cdot \text{atm}}{\text{K} \cdot \text{mol}} \times \frac{42 \text{ K}}{5.00 \text{ L}} = +0.13 \text{ atm}$

The total pressure in the flask increases by 0.13 atm when the flask is heated.

(c) Neither the moles of each gas nor the total moles of gas are affected by heating, so the mole fraction of $SO_3(g)$ is unchanged. $\chi_{SO_3} = \frac{0.0937 \text{ mol}}{0.1874 \text{ mol}} = 0.500$.

10.47 $P_{N_2} = \frac{P_1 V_1 T_2}{V_2 T_1} = \frac{4.60 \text{ atm} \times 1.00 \text{ L} \times 293 \text{ K}}{10.0 \text{ L} \times 299 \text{ K}} = 0.451 \text{ atm}$

$P_{O_2} = \frac{P_1 V_1 T_2}{V_2 T_1} = \frac{3.50 \text{ atm} \times 5.00 \text{ L} \times 293 \text{ K}}{10.0 \text{ L} \times 299 \text{ K}} = 1.71 \text{ atm}$

$P_T = 0.451 \text{ atm} + 1.71 \text{ atm} = 2.16 \text{ atm}$

10.48 It is simplest to calculate the partial pressure of each gas as it expands into the total volume, then sum the partial pressures.

$P_2 = P_1V_1/V_2$

$P_{N_2} = 635 \text{ mm Hg } (1.0 \text{ L}/2.5 \text{ L}) = 254 \text{ mm Hg}$

$P_{Ne} = 212 \text{ mm Hg } (1.0 \text{ L}/2.5 \text{ L}) = 85 \text{ mm Hg}$

$P_{H_2} = 418 \text{ mm Hg } (0.5 \text{ L}/2.5 \text{ L}) = \underline{84 \text{ mm Hg}}$

Total Pressure = 423 mm Hg = 4.2×10^2 mm Hg (two significant figures)

Quantities of Gases in Chemical Reactions

10.49 $n_{H_2} = \frac{P_{H_2}V}{RT}$; $P = 740 \text{ mm Hg} \times \frac{1 \text{ atm}}{760 \text{ mm Hg}} = 0.974 \text{ atm}$

$n_{H_2} = 0.974 \text{ atm} \times \frac{\text{K} \cdot \text{mol}}{0.0821 \text{ L} \cdot \text{atm}} \times \frac{10.0 \text{ L}}{296 \text{ K}} = 0.401 \text{ mol } H_2$

$0.401 \text{ mol } H_2 \times \frac{1 \text{ mol } CaH_2}{2 \text{ mol } H_2} \times \frac{42.10 \text{ g } CaH_2}{1 \text{ mol } CaH_2} = 8.44 \text{ g } CaH_2$

10.50 $\text{mol } O_2 = \frac{PV}{RT} = 3.5 \times 10^{-6} \text{ mm Hg} \times \frac{1 \text{ atm}}{760 \text{ mm Hg}} \times \frac{\text{K} \cdot \text{mol}}{0.0821 \text{ L} \cdot \text{atm}} \times \frac{0.382 \text{ L}}{300 \text{ K}}$

$= 7.1 \times 10^{-11} \text{ mol } O_2$

$7.1 \times 10^{-11} \text{ mol } O_2 \times \frac{2 \text{ mol Mg}}{1 \text{ mol } O_2} \times \frac{24.3 \text{ g Mg}}{1 \text{ mol Mg}} = 3.4 \times 10^{-9} \text{ g Mg}$

10.51 g glucose $\rightarrow$ mol glucose $\rightarrow$ mol CO_2 $\rightarrow$ V CO_2

$5.00 \text{ g} \times \frac{1 \text{ mol glucose}}{180.1 \text{ g}} = 0.0278 \text{ mol glucose} \times \frac{6 \text{ mol } CO_2}{1 \text{ mol glucose}} = 0.167 \text{ mol } CO_2$

$V = \frac{nRT}{P} = 0.167 \text{ mol} \times \frac{0.0821 \text{ L} \cdot \text{atm}}{\text{K} \cdot \text{mol}} \times \frac{310 \text{ K}}{1.00 \text{ atm}} = 4.25 \text{ L } CO_2$

10.52 kg $H_2SO_4 \rightarrow$ g $H_2SO_4 \rightarrow$ mol $H_2SO_4 \rightarrow$ mol $NH_3 \rightarrow$ V NH_3

$$150 \text{ kg} \times \frac{1000 \text{ g}}{1 \text{ Kg}} = 1.50 \times 10^5 \text{ g } H_2SO_4 \times \frac{1 \text{ mol}}{98.08 \text{ g}} = 1.53 \times 10^3 \text{ mol } H_2SO_4$$

$$1.53 \times 10^3 \text{ mol } H_2SO_4 \times \frac{2 \text{ mol } NH_3}{1 \text{ mol } H_2SO_4} = 3.06 \times 10^3 \times \text{mol } NH_3$$

$$V_{NH_3} = \frac{nRT}{P} = 3.06 \times 10^3 \text{ mol} \times \frac{0.0821 \text{ L} \cdot \text{atm}}{\text{K} \cdot \text{mol}} \times \frac{293 \text{ K}}{25.0 \text{ atm}} = 2.94 \times 10^3 \text{ L } NH_3$$

10.53 The gas sample is a mixture of $H_2(g)$ and $H_2O(g)$. Find the partial pressure of $H_2(g)$ and then the moles of $H_2(g)$ and Zn(s).

$P_t = 725$ mm Hg $= P_{H_2} + P_{H_2O}$

From appendix C, the vapor pressure of H_2O at 24°C = 22.4 mm Hg

$$P_{H_2} = 725 \text{ mm Hg} - 22.4 \text{ mm Hg} = 703 \text{ mm Hg} \times \frac{1 \text{ atm}}{760 \text{ mm Hg}} = 0.925 \text{ atm}$$

$$n_{H_2} = \frac{P_{H_2}V}{RT} = 0.925 \text{ atm} \times \frac{\text{K} \cdot \text{mol}}{0.0821 \text{ L} \cdot \text{atm}} \times \frac{0.124 \text{ L}}{297 \text{ K}} = 0.00470 \text{ mol } H_2$$

$$0.00470 \text{ mol } H_2 \times \frac{1 \text{ mol Zn}}{1 \text{ mol } H_2} \times \frac{65.38 \text{ g Zn}}{1 \text{ mol Zn}} = 0.307 \text{ g Zn}$$

10.54 $V = \frac{nRT}{P}$; find mol O_2 by stoichiometry; find P_{O_2} by partial pressures.

P_{H_2O} at 23°C = 21.1 mm Hg

$$P_{O_2} = 742 \text{ mm Hg} - 21.1 \text{ mm Hg} = 721 \text{ mm Hg} \times \frac{1 \text{ atm}}{760 \text{ mm Hg}} = 0.949 \text{ atm}$$

$$0.2890 \text{ g } KClO_3 \times \frac{1 \text{ mol } KClO_3}{122.55 \text{ g}} \times \frac{3 \text{ mol } O_2}{2 \text{ mol } KClO_3} = 0.003537 \text{ mol } O_2$$

$$V = 0.003537 \text{ mol} \times \frac{0.0821 \text{ L} \cdot \text{atm}}{\text{K} \cdot \text{mol}} \times \frac{296 \text{ K}}{0.949 \text{ atm}} = 0.0906 \text{ L} = 90.6 \text{ mL}$$

Kinetic - Molecular Theory; Graham's Law

10.55 (a) They have the same number of molecules (equal volumes of gases at the same temperature and pressure contain equal numbers of molecules).

(b) SF_6 is more dense because it has the larger molar mass. Since the volumes of the samples and the number of molecules are equal, the gas with the larger molar mass will have the greater density.

(c) The average kinetic energies are equal (statement 5, section 10.8).

(d) N_2 will effuse faster. The lighter the gas molecules, the faster they will effuse (Graham's Law).

10.56 (a) vessel A (b) vessel B (c) vessel B (d) vessel A

10.57 (a) False. The average kinetic energy per molecule in a collection of gas molecules is the same for all gases at the same temperature. (b) True. (c) False. The molecules in a gas sample at a given temperature exhibit a distribution of kinetic energies. (d) True.

10.58 (a) increase in temperature at constant volume, decrease in volume, increase in pressure (b) decrease in temperature (c) increase in volume (d) increase in temperature

10.59 (a) The larger the molecular weight, the slower the average speed (at constant temperature). In order of increasing speed (and decreasing molar mass): $SF_6 < HI < Cl_2 < H_2S < CO$

(b) $$u = \sqrt{\frac{3RT}{M}} = \left(\frac{3 \times 8.314 \text{ kg} \cdot \text{m}^2/\text{s}^2 \cdot \text{K} \cdot \text{mol} \times 298 \text{ K}}{28.0 \times 10^{-3} \text{ kg/mol}}\right)^{\frac{1}{2}} = 515 \text{ m/s}$$

10.60 (a) In order of increasing speed (and decreasing molar mass):

$CO_2 \approx N_2O < F_2 < HF < H_2$

(b) $$u_{H_2} = \left(\frac{3 \times 8.314 \text{ Kg} \cdot \text{m}^2/\text{s}^2 \cdot \text{K} \cdot \text{mol} \times 300 \text{ K}}{2.02 \times 10^{-3} \text{ Kg/mol}}\right)^{\frac{1}{2}} = 1.92 \times 10^3 \text{ m/s}$$

$$u_{CO_2} = \left(\frac{3 \times 8.314 \text{ Kg} \cdot \text{m}^2/\text{s}^2 \cdot \text{K} \cdot \text{mol} \times 300 \text{ K}}{44.0 \times 10^{-3} \text{ Kg/mol}}\right)^{\frac{1}{2}} = 4.12 \times 10^2 \text{ m/s}$$

As expected, the lighter molecule moves at the greater speed.

10.61 The heavier the molecule, the slower the rate of effusion. Thus, the order for increasing rate of effusion is in order of decreasing mass.

rate $^{2}H^{37}Cl$ < rate $^{1}H^{37}Cl$ < rate $^{2}H^{35}Cl$ < rate $^{1}H^{35}Cl$

10.62 $\dfrac{\text{rate } ^{235}\text{U}}{\text{rate } ^{238}\text{U}} = \sqrt{\dfrac{238.05}{235.04}} = \sqrt{1.0128} = 1.0064$

There is a slightly greater rate enhancement for ^{235}U (g) atoms (1.0064) than $^{235}UF_6$ (g) molecules (1.0043), because ^{235}U is a greater percentage (100%) of the mass of the diffusing particles than in $^{235}UF_6$ molecules. The masses of the isotopes were taken from *The Handbook of Chemistry and Physics.*

10.63 The time required is proportional to the reciprocal of the effusion rate.

$$\frac{\text{rate (X)}}{\text{rate } (O_2)} = \frac{28 \text{ s}}{72 \text{ s}} = \left[\frac{32 \text{ g } O_2}{\mathcal{M}_x}\right]^{\frac{1}{2}} ; \ \mathcal{M}_x = 32 \text{ g O} \times \left[\frac{72}{28}\right]^{\frac{1}{2}} = 210 \text{ g/mol}$$

(two significant figures)

10.64 $\dfrac{\text{rate (sulfide)}}{\text{rate (Ar)}} = \left[\dfrac{39.9}{\mathcal{M}\text{ (sulfide)}}\right]^{\frac{1}{2}} = 0.28$

$\mathcal{M}$ (sulfide) = $39.9 \div 0.28^2$ = 510 g/mol (two significant figures)

The unit formula of arsenic(III) sulfide is As_2S_3 which has a formula mass of 246.1. Twice this is 490 g/mol, close to the value estimated from the effusion experiment. Thus, the formula of the vapor phase molecule is As_4S_6.

Nonideal-Gas Behavior

10.65 (a) Nonideal gas behavior is observed at very high pressures and/or low temperatures.

(b) The real volumes of gas molecules and attractive intermolecular forces between molecules cause gases to behave nonideally.

10.66 Ideal gas behavior is most likely to occur at high temperature and low pressure, so the atmosphere on Mercury is more likely to obey the ideal gas law. The higher temperature on Mercury means that the kinetic energies of the molecules will be larger relative to intermolecular attractive forces. Further, the gravitational attractive forces on Mercury are lower because the planet has a much smaller mass. This means that for the same column mass of gas (Figure 10.1), atmospheric pressure on Mercury will be lower.

10.67 The ratio PV/RT is equal to the number of moles of molecules in an ideal gas sample; this number should be a constant for all pressure, volume and temperature conditions. If the value of this ratio changes with increasing pressure, the gas sample is not behaving ideally (according to the ideal gas equation).

10.68 (a) Intermolecular forces cause the molecules to "stick" together and behave as if there are fewer net particles in the sample.

(b) The real volume of gas molecules causes the amount of free space in the gas sample to be less than the container volume. Using the container volume to calculate PV/RT gives a value larger than that for an ideal gas (which assumes that the total container volume is free space).

(c) As the temperature of a gas increases, the average kinetic energy of the particles increases. The increased kinetic energy overcomes the attractive forces between molecules and keeps them separate.

10.69 The constants *a* and *b* are part of the correction terms in the van der Waals equation. The smaller the values of *a* and *b*, the smaller the corrections and the more ideal the gas. Ar (a = 1.34, b = 0.0322) will behave more like an ideal gas than CO_2 (a = 3.59, b = 0.0427) at high pressures.

10.70 The constant *a* is a measure of the strength of the intermolecular attractions among gas molecules; *b* is a measure of molecular volume. Both increase with increasing molecular mass and structural complexity.

10.71 (a) $$P = 1.00 \text{ mol} \times \frac{0.0821 \text{ L} \cdot \text{atm}}{\text{K} \cdot \text{mol}} \times \frac{313 \text{ K}}{28.0 \text{ L}} = 0.918 \text{ atm}$$

$$P = \frac{nRT}{V \cdot nb} - \frac{an^2}{V^2} = \frac{1.00 \times 0.0821 \times 313}{28.0 - (1.00 \times 0.1383)} - \frac{20.4(1.00)^2}{(28.0)^2} = 0.896 \text{ atm}$$

10.72 (a) At STP, the molar volume $= 1 \text{ mol} \times \frac{0.0821 \text{ L} \cdot \text{atm}}{\text{K} \cdot \text{mol}} \times \frac{273 \text{ K}}{1 \text{ atm}} \times = 22.4 \text{ L}$

Dividing the value for *b*, 0.0322 L/mol, by 4, we obtain 0.0080 L. Thus, the volume of the Ar atoms is (0.0080/22.4)100 = 0.036% of the total volume.

(b) At 100 atm pressure the molar volume is 0.244 L, and the volume of the Ar atoms is 3.6% of the total volume.

Additional Exercises

10.73 A mercury barometer with water trapped in its tip would not read the correct pressure. The standard relationship between height of an Hg column and atmospheric pressure (760 mm Hg = 1 atm) assumes that there is a vacuum in the closed end of the barometer and that gravity is the only downward force on the Hg column. Water at the top of the Hg column would establish a vapor pressure which would exert additional downward pressure and partially counterbalance the pressure of the atmosphere. The Hg column would read lower than the actual atmospheric pressure.

10.74 $PV = \frac{g\,RT}{\mathcal{M}}$; $g = \frac{\mathcal{M}\,PV}{RT}$; Change m^3 to L, then calculate grams (or kg).

$$2.0 \times 10^5\ m^3 \times \frac{10^3\ dm^3}{1\ m^3} \times \frac{1\ L}{1\ dm^3} = 2.0 \times 10^8\ L\ H_2$$

$$g = \frac{2.02\ g\ H_2}{1\ mol\ H_2} \times \frac{K \bullet mol}{0.0821\ L \bullet atm} \times \frac{1\ atm \times 2.0 \times 10^8\ L}{300\ K} = 1.6 \times 10^7\ g = 1.6 \times 10^4\ kg\ H_2$$

10.75 (a) $n = \frac{PV}{RT} = 3.00\ atm \times \frac{K \bullet atm}{0.0821\ L \bullet atm} \times \frac{1.00\ L}{300\ K} = 0.122\ mol\ C_3H_8\,(g)$

(b) $\frac{0.590\ g\ C_3H_8\ (l)}{1\ mL} \times 1.00 \times 10^3\ mL \times \frac{1\ mol\ C_3H_8}{44.094\ g} = 13.4\ mol\ C_3H_8\,(l)$

(c) A 1.00 L container holds many more moles (molecules) of $C_3H_8(l)$ because in the liquid phase the molecules are touching. In the gas phase, the molecules are far apart (statement 2, section 10.8) and many fewer molecules will fit in the 1.00 L container.

10.76 $P = \frac{nRT}{V}$; $n = 1 \times 10^{-5}$ mol , V = 0.200 L, T = 23° C = 296 K

$$P = 1 \times 10^{-5}\ mol \times \frac{0.0821\ L \bullet atm}{K \bullet mol} \times \frac{296\ K}{0.200\ L} \times \frac{760\ mm\ Hg}{1\ atm} = 0.9\ mm\ Hg$$

10.77 $\frac{P_1}{T_1} = \frac{P_2}{T_2}$; $P_1 = 2.2$ atm, $T_1 = 297$ K, $P_2 = 3.0$ atm, $T_2 = ?$
(T must be in kelvins for the P, T relationship to hold true.)

$$T_2 = \frac{P_2\,T_1}{P_1} = \frac{3.0\ atm \times 297\ K}{2.2\ atm} = 405\ K\ \text{or}\ 132\ °C$$

10.78 (a) $n = \frac{PV}{RT} = 0.980\ atm \times \frac{K \bullet mol}{0.0821\ L \bullet atm} \times \frac{0.600\ L}{353\ K} = 0.203\ mol\ air$

$$mol\ O_2 = 0.203\ mol\ air \times \frac{0.2095\ mol\ O2}{1\ mol\ air} = 0.0425\ mol\ O_2$$

(b) $C_8H_{18}(l) + 25/2\ O_2(g) \rightarrow 8CO_2(g) + 9H_2O(g)$

$$0.0425\ mol\ O_2 \times \frac{1\ mol\ C_8H_{18}}{12.5\ mol\ O_2} \times \frac{114.2\ g\ C_8H_{18}}{1\ mol\ C_8H_{18}} = 0.388\ g\ C_8H_{18}$$

10.79 Calculate mol O_3/L, then molecules/L.

$$PV = nRT\ ;\ \frac{n}{V} = \frac{P}{RT} = \frac{3.0 \times 10^{-3}\ atm}{250\ K} \times \frac{K \bullet mol}{0.0821\ L \bullet atm} = \frac{1.46 \times 10^{-4}\ mol\ O_3}{L}$$

$$\frac{1.46 \times 10^{-4}\ mol\ O_3}{L} \times \frac{6.022 \times 10^{23}\ molecules}{mol\ O_3} = \frac{8.8 \times 10^{19}\ O_3\ molecules}{L}$$

10.80 Volume of laboratory = $110\ m^2 \times 2.7\ m \times \frac{1000\ L}{1\ m^3} = 2.97 \times 10^5$ L

Calculate the **total** moles of gas in the labortory at the conditions given.

$$n_T = \frac{PV}{RT} = 1.00\ \text{atm} \times \frac{\text{K} \cdot \text{mol}}{0.0821\ \text{L} \cdot \text{atm}} \times \frac{2.97 \times 10^5\ \text{L}}{297\ \text{K}} = 1.22 \times 10^4\ \text{mol g}$$

An $Ni(CO)_4$ concentration of 1 part in 10^9 means 1 mol $Ni(CO)_4$ in 1×10^9 total moles of gas.

$$\frac{\text{X mol Ni(CO)}_4}{1.22 \times 10^4\ \text{mol gas}} = \frac{1}{10^9} = 1.22 \times 10^{-5}\ \text{mol Ni(CO)}_4$$

$$1.22 \times 10^{-5}\ \text{mol Ni(CO)}_4 \times \frac{170.74\ \text{g Ni(CO)}_4}{1\ \text{mol Ni(CO)}_4} = 2.1 \times 10^{-3}\ \text{g Ni(CO)}_4$$

10.81 (a) $5.00\ \text{g HCl} \times \frac{1\ \text{mol HCl}}{36.46\ \text{g HCl}} = 0.137\ \text{mol HCl}$; $5.00\ \text{g NH}_3 \times \frac{1\ \text{mol NH}_3}{17.03\ \text{g NH}_3} = 0.294\ \text{mol}$

The gases react in a 1:1 mole ratio. HCl is the limiting reactant and is completely consumed. (0.294 mol - 0.137 mol =) 0.157 mol NH_3 remain in the system.

$NH_3(g)$ is the only gas remaining after reaction. $V_T = 4.00$ L

(b) $P = \frac{nRT}{V} = 0.157\ \text{mol} \times \frac{0.0821\ \text{L} \cdot \text{atm}}{\text{K} \cdot \text{mol}} \times \frac{298\ \text{K}}{4.00\ \text{L}} = 0.960\ \text{atm}$

10.82 The number of moles of CO_2 is equal to the total number of moles of carbon atoms in the CH_4/C_2H_2 mixture. Let x = moles CH_4, y = moles C_2H_4. Then, total moles gas in the mixture = x + y. Total moles CO_2 = x + 2y.

$$\frac{x + y}{x + 2y} = \frac{70.5\ \text{mm Hg}}{96.4\ \text{mm Hg}}\ ;\ \frac{x}{y} = 1.72\ ;\ x = 1.72\ y$$

Let x + y = 1; y = 0.37 = the fraction that is acetylene, C_2H_2.

Notice that the only property of gases needed here is that equal numbers of moles of gas exert equal pressures under the same conditions of volume and temperature.

10.83 $\mathcal{M}_{avg} = \frac{dRT}{P} = \frac{1.104\ \text{g}}{1\ \text{L}} \times \frac{0.0821\ \text{L} \cdot \text{atm}}{\text{K} \cdot \text{mol}} \times \frac{300\ \text{K}}{435\ \text{mm Hg}} \times \frac{760\ \text{mm Hg}}{1\ \text{atm}} = \frac{47.5\ \text{g}}{\text{mol}}$

x = mole fraction O_2; 1 - x = mole fraction Kr

47.5 g = x(32.00) + (1 - x)(83.80)

36.3 = 51.8 x; x = 0.701; 70.1% O_2

10.84 After reaction, the flask contains $IF_5(g)$ and whichever reactant is in excess. Determine the limiting reactant, which regulates the moles of IF_5 produced and moles of excess reactant.

$$I_2(s) + 5F_2(g) \rightarrow 2IF_5(g)$$

$$10.0 \text{ g } I_2 \times \frac{1 \text{ mol } I_2}{253.8 \text{ g } I_2} \times \frac{5 \text{ mol } F_2}{1 \text{ mol } I_2} = 0.197 \text{ mol } F_2 \text{ needed}$$

$$10.0 \text{ g } F_2 \times \frac{1 \text{ mol } F_2}{38.00 \text{ g } F_2} = 0.263 \text{ mol } F_2 \text{ available}$$

I_2 is the limiting reactant; F_2 is in excess.

0.263 mol F_2 available - 0.197 mol F_2 reacted = 0.066 mol F_2 remain.

$$10.0 \text{ g } I_2 \times \frac{1 \text{ mol } I_2}{253.8 \text{ g } I_2} \times \frac{2 \text{ mol } IF_5}{1 \text{ mol } I_2} = 0.0788 \text{ mol } IF_5 \text{ produced}$$

(a) $$P_{IF_5} = \frac{nRT}{V} = 0.0788 \text{ mol} \times \frac{0.0821 \text{ L} \cdot \text{atm}}{\text{K} \cdot \text{mol}} \times \frac{398 \text{ K}}{5.00 \text{ L}} = 0.515 \text{ atm}$$

(b) $$\chi_{IF_5} = \frac{\text{mol } IF_5}{\text{mol } IF_5 + \text{mol } F_2} = \frac{0.0788}{0.0788 + 0.066} = 0.544$$

10.85 Calculate the number of moles of Ar in the vessel:

n = (339.712 - 337.428)/39.948 = 0.05717 mol

The total number of moles of the mixed gas is the same (Avogadro's Law). Thus, the average atomic weight is (339.218 - 337.428)/0.05717 = 31.31. Let the mole fraction of Ne be x. Then,

x (20.183) + (1 - x) (39.948) = 31.31 ; 8.638 = 19.765x ; x = 0.437

Neon is thus 43.7 mole percent of the mixture.

10.86 The balloon will expand; H_2 ($\mathcal{M}$ = 2 g/mol) will effuse in through the walls of the balloon faster than He ($\mathcal{M}$ = 4 g/mol) will effuse out, because the gas with the smaller molar mass effuses more rapidly.

10.87 (a) The quantity $\frac{d}{P} = \frac{\mathcal{M}}{RT}$ should be a constant at all pressures for an ideal gas. It is not, however, because of the nonideal behavior. If we graph d/P vs P, the ratio should approach ideal behavior a low P. At P = 0, d/P = 2.2525. Using this value in the formula $\mathcal{M} = \frac{d}{P} \times RT$, $\mathcal{M}$ = 50.49 g/mol.

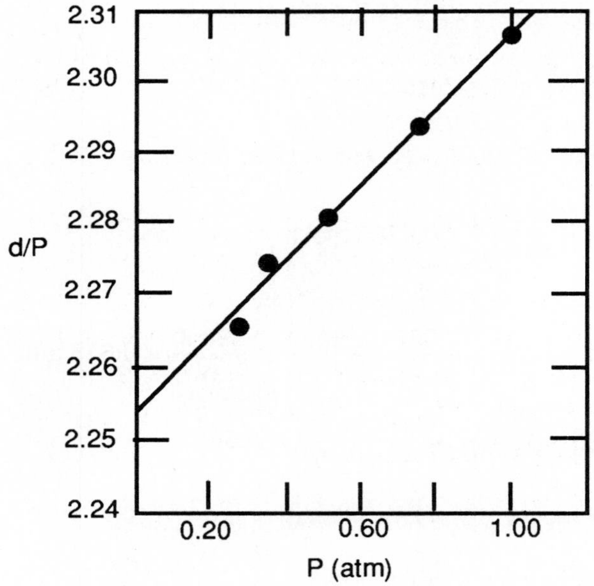

P	d/P
1.00	2.307
0.666	2.292
0.500	2.280
0.333	2.274
0.250	2.264

(b) The ratio d/P varies with pressure because of the finite volumes of gas molecules and attractive intermolecular forces.

10.88 In the expanded container volume, there are fewer collisions with the walls and with other gas particles. Attractive and repulsive forces which might change the speed of the particles become less important, and the particles maintain essentially the same average speed, kinetic energy and temperature after the expansion.

10.89

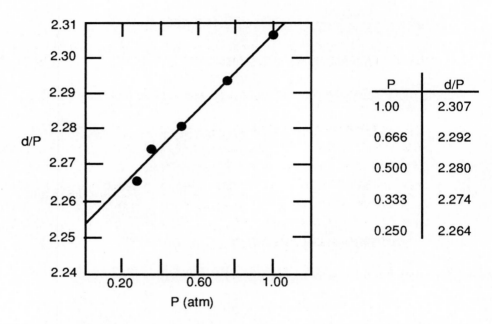

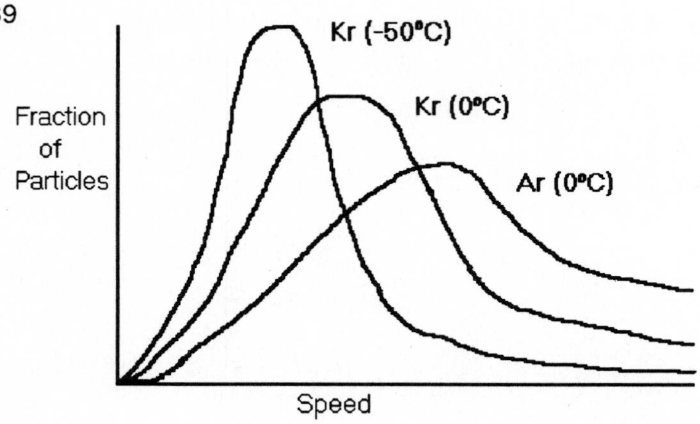

10.90 (a) $80.00 \text{ kg } N_2(g) \times \frac{1000 \text{ g}}{1 \text{ kg}} \times \frac{1 \text{ mol } N_2}{28.02 \text{ g N}} = 2855 \text{ mol } N_2$

$$P = \frac{nRT}{V} = 2855 \text{ mol} \times \frac{0.08206 \text{ L} \cdot \text{atm}}{\text{K} \cdot \text{mol}} \times \frac{573 \text{ K}}{1000.0 \text{ L}} = 134.2 \text{ atm}$$

(b) According to Sample Exercise 10.15,

$$P = \frac{nRT}{V - nb} - \frac{n^2 a}{V^2}$$

$$P = \frac{(2855 \text{ mol})(0.08206 \text{ L} \cdot \text{atm/K} \cdot \text{mol})(573 \text{ K})}{1000.0 \text{ L} - (2855 \text{ mol})(0.0391 \text{ L/mol})} - \frac{(2855 \text{ mol})^2 (1.39 \text{ L}^2 \cdot \text{atm/mol}^2)}{(1000.0 \text{ L})^2}$$

$$P = \frac{134{,}243 \text{ L} \cdot \text{atm}}{1000.0 \text{ L} - 111.6 \text{ L}} - 11.3 \text{ atm} = 151.1 \text{ atm} - 11.3 \text{ atm} = 139.8 \text{ atm}$$

(c) The pressure corrected for the real volume of the N_2 molecules is 151.1 atm, 16.9 atm higher than the ideal pressure of 134.2 atm. The 11.3 atm correction for intermolecular forces reduces the calculated pressure somewhat, but the "real" pressure is still higher than the ideal pressure. The correction for the real volume of molecules dominates. Even though the value of b is small, the number of moles of N_2 is large enough so that the molecular volume correction is larger than the attractive forces correction.

CHAPTER ***11***

Intermolecular Forces, Liquids and Solids

Kinetic-Molecular Theory

11.1 The solid state is characterized by a high degree of order. The units of the solid, ions or molecules, are arranged in a regular array. The packing pattern is one that achieves the lowest energy. In a liquid, molecules are still in close contact, but their orientations with respect to near neighbors are no longer regular. The molecules possess sufficient kinetic energy to move with respect to one another, so there is no long-range order. The difference between liquids and solids lies in the fact that liquids possess sufficient energy to overcome the forces that keep each molecule in a specific position in relation to its neighbors. Liquids are fluid and take the shape of their container. Solids are rigid, they have their own volume and do not flow.

11.2 The degree of order is most similar in a liquid and a solid; in both these states, the molecules are touching. In the gas phase, particles are far apart and their positions are changing rapidly and continuously.

11.3 Gases are highly compressible because the average distances between gas molecules are large compared with the sizes of the molecules themselves. In a solid or liquid the molecules are in close contact. Solids and liquids are thus not very compressible, because we are working against the repulsive forces between molecules.

11.4 Density is the ratio of the mass of a substance to the volume it occupies. For the same substance in different states, mass will be the same. In the liquid and solid states, the particles are touching and there is very little empty space, so the volumes occupied by a unit mass are very similar and the densities are similar. In the gas phase, the molecules are far apart, so a unit mass occupies a much greater volume than the liquid or solid, and the density of the gas phase is much less.

11.5 As the temperature of a substance is increased, the average kinetic energy of the particles increases. In a collection of particles (molecules), the state is determined by the strength of interparticle forces relative to the average kinetic energy of the particles. As the average kinetic energy increases more particles are able to overcome intermolecular attractive forces and move to a less ordered state, from solid to liquid to gas.

11.6 At constant temperature, the average kinetic energy of a collection of particles is constant. Compression brings particles closer together and increases the number of particle-particle collisions. With more collisions, the likelihood of intermolecular attractions causing the particles to coalesce (liquefy) is greater.

Intermolecular Forces

11.7 (a) Dipole-dipole forces are the electrostatic attractions between the positive end of a neutral polar molecule and the negative end of another (Figure 11.3)
Example: acetone

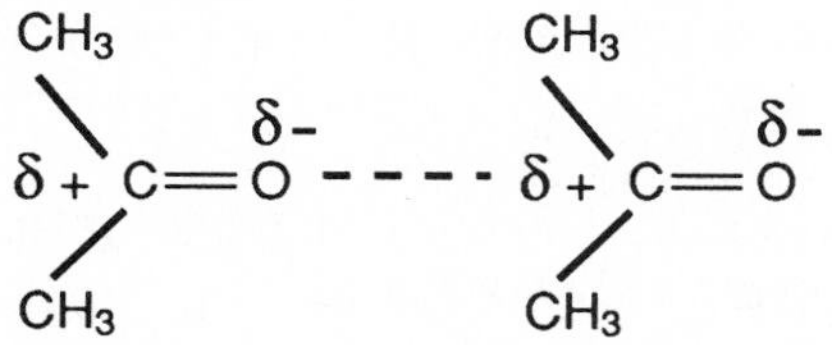

(b) Ion-dipole forces are the attractive forces between one end of the dipole in a polar covalent molecule and an oppositely charged ion. They occur when an ionic solute dissolves in a polar covalent solvent. Example: H_2O, $CuCl_2$

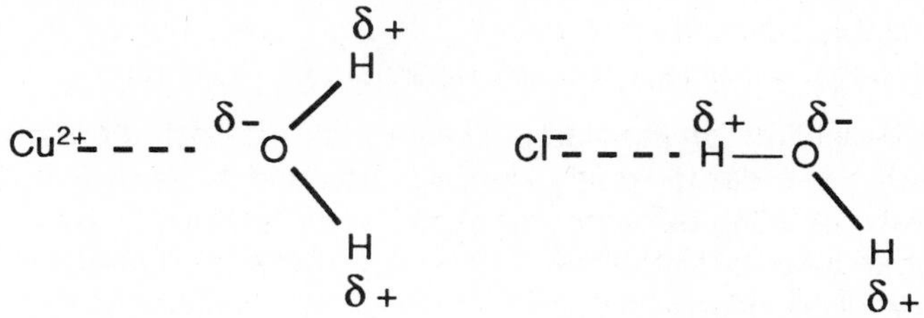

(c) London dispersion forces are attractive forces between nonpolar molecules or neutral atoms produced by **transient dipoles** created by the motion of electrons in the atom or molecule. Example: forces among I_2 molecules which hold them in the solid state at room temperature.

(d) Van der Waals forces is a collective term for dipole-dipole and London dispersion forces. Examples: forces between slightly polar CH_2Cl_2 molecules; (a) and (c) above.

(e) A hydrogen bond is a relatively strong intermolecular force between the H atom in a polar bond and the lone pair on a nearby electronegative atom. The strongest of these occur between F-H, O-H or N-H donors and F, O or N acceptors. Examples: HF, H_2O, NH_3.

11.8 (a) London dispersion forces (b) dipole-dipole forces (c) hydrogen bonding

11.9 (a) Ion-ion forces (ionic bonds) (b) London dispersion forces (c) ion-dipole forces
(d) hydrogen bonds

11.10 (a) Hydrogen bonds (b) ionic bonds (c) London dispersion forces
(d) dipole-dipole forces

11.11 (a) HCl has stronger dipole-dipole forces because it is a more polar molecule. (Strictly speaking, it does not have hydrogen bonding.)

(b) HI has stronger London dispersion forces because it is a larger molecule with a more polarizable electron cloud.

11.12 CO is a polar covalent molecule ($\Delta EN = 1.0$) and N_2 is nonpolar. Dipole-dipole forces between CO molecules are stronger than London dispersion forces between N_2 molecules. A higher temperature and greater average kinetic energy is required to overcome the dipole-dipole forces in CO and separate (vaporize) the molecules.

11.13 (a) Polarizability is the ease with which the charge distribution (electron cloud) in a molecule can be distorted to produce a transient dipole.

(b) Te is most polarizable because its valence electrons are farthest from the nucleus and least tightly held.

(c) Polarizability increases as molecular size (and thus molecular weight) increases. In order of increasing polarizability: $CH_4 < SiH_4 < SiCl_4 < GeCl_4 < GeBr_4$

11.14 (a) A more polarizable molecule can develop a larger transient dipole, increasing the strength of electrostatic attractions among polarized molecules.

(b) Helium is a collection of individual isolated atoms with no tendency to form chemical bonds. In order to liquefy He(g), there must be attractive forces between He atoms strong enough to overcome their kinetic energies. Since He(g) can be liquefied, some kind of attractive interaction between particles with normally uniform charge distributions (dispersion forces) must exist.

(c) The strengths of dispersion forces increase with increasing molecular size, because the relatively diffuse electron clouds are less tightly held by distant nuclei.

11.15 (a) O_2 -- Both substances experience London dispersion forces; the higher the molecular weight, the stronger the forces, the higher the boiling point.

(b) SiH_4 -- same reason as in part (a)

(c) NaCl -- Ionic bonding in NaCl is much stronger than dipole-dipole forces in CH_3Cl.

(d) C_2H_5OH -- hydrogen bonding in C_2H_5OH is stronger than dipole-dipole forces in C_2H_5Cl; the compounds are similar in size so molecular weight differences are not a factor.

11.16 (a) HF has the higher boiling point because hydrogen bonding is stronger than dipole-dipole forces.

(b) $CHBr_3$ has the higher boiling point because it has the higher molecular weight which leads to greater polarizability and stronger dispersion forces.

(c) ICl has the higher boiling point because it is a polar molecule. For molecules with similar structures and molecular weights, dipole-dipole forces are stronger than dispersion forces.

(d) These are nonpolar covalent molecules with equal molecular weights. The rod-like shape of $CH_3CH_2CH_2CH_3$ allows more interaction between molecules, leading to stronger dispersion forces and a higher boiling point.

11.17 Surface tension (Section 11.3), high boiling point (relative to H_2S, H_2Se, H_2Te, Figure 11.6), high heat capacity per gram, high enthalpy of vaporization; the solid is less dense than the liquid; it is a liquid at room temperature despite its low molecular weight.

11.18 (a) Water expands when it freezes to maximize the number of hydrogen bonding interactions in the structure. In the solid state, each H atom is involved in one hydrogen bond and each O atom participates in two hydrogen bonds. The H_2O molecules must be far enough apart to allow for the steric requirements of these four interactions.

(b) In order to maximize the number of H-bonding interactions, water molecules adopt a hexagonal arrangement in ice. This pattern at the molecular level leads to the hexagonal macroscopic shape of snowflakes.

(c) Since ice, $H_2O(s)$, floats on water, it insulates the $H_2O(l)$ from extreme cold, allowing aquatic life to exist. If $H_2O(s)$ sank to the bottom of natural waters, temperatures of the $H_2O(l)$ (unfrozen water) would be too low to support living organisms.

Viscosity and Surface Tension

11.19 Viscosities and surface tensions of liquids both increase as intermolecular forces become stronger.

11.20 As temperature increases, the average kinetic energy of the molecules increases and intermolecular attractions are more easily overcome. Surface tensions and viscosities decrease.

11.21 (a) Ethanol molecules experience hydrogen bonding, a stronger intermolecular force than the weak dipole-dipole forces between ether molecules. The stronger forces between ethanol molecules make the liquid more resistant to flow and thus more viscous.

(b) The shape of a meniscus depends on the strength of the cohesive forces within a liqiuid relative to the adhesive forces between the walls of the capillary and the liquid. Because polar water molecules have strong cohesive intermolecular interactions relative to the weak adhesive forces between the polar water molecules and nonpolar polyethylene, the meniscus is concave-downward.

11.22 (a) $CHBr_3$ has a higher molecular weight, is more polarizable and has stronger dispersion forces, so the surface tension is greater (see Exercise 11.16(b)).

(b) As temperature increases, the viscosity of the oil decreases because the average kinetic energies of the molecules increase (Exercise 11.20).

(c) Adhesive forces between polar water and nonpolar car wax are weak, so the large surface tension of water draws the liquid into the shape with the smallest surface area, a sphere.

Changes of State

11.23 Endothermic: melting (s $\rightarrow$ l), vaporization (l $\rightarrow$ g), sublimation (s $\rightarrow$ g)
Exothermic: condensation (g $\rightarrow$ l), freezing (l $\rightarrow$ s), deposition (g $\rightarrow$ s)

11.24 (a) Ice, $H_2O(s)$, sublimes to water vapor, $H_2O(g)$.

(b) The heat energy required to increase the kinetic energy of molecules enough to melt the solid does not produce a large separation of molecules. The specific order is disrupted, but the molecules remain close together. On the other hand, when a liquid is vaporized, the intermolecular forces which maintain close molecular contacts must be overcome. Because molecules are being separated, the energy requirement is higher than for melting.

(c) Liquid ethyl chloride at room temperature is far above its boiling point. When the liquid boils, heat is required to overcome the intermolecular forces between molecules. This heat is extracted from the surface that contacts the liquid.

(d) At 279°C, CS_2 is above its critical temperature. The molecules have too much kinetic energy to be liquified, regardless of pressure.

11.25 Evaporation of 10 g of water requires:

$$10.0 \text{ g } H_2O \times \frac{2.4 \text{ kJ}}{1 \text{ g } H_2O} = 24.0 \text{ kJ or } 2.4 \times 10^4 \text{ J}$$

Cooling a certain amount of water by 13°C:

$$2.4 \times 10^4 \text{ J} \times \frac{1 \text{ g} \cdot \text{K}}{4.18 \text{ J}} \times \frac{1}{13° \text{C}} = 4.4 \times 10^2 \text{ g } H_2O$$

11.26 Energy released when 100 g of H_2O is cooled from 18°C to 0°C:

$$\frac{4.184 \text{ J}}{\text{g} \cdot \text{K}} \times 100 \text{ g } H_2O \times 18° \text{C} = 7.5 \times 10^3 \text{ J} = 7.5 \text{ kJ}$$

Energy released when 100 g of H_2O is frozen (there is no change in temperature during a change of state):

$$\frac{334 \text{ J}}{\text{g}} \times 100 \text{ g } H_2O = 3.34 \times 10^4 \text{ J} = 33.4 \text{ kJ}$$

Total energy released = 7.5 kJ + 33.4 kJ = 40.9 kJ

Mass of freon which will absorb 40.9 kJ when vaporized:

$$40.9 \times \frac{1 \times 10^3\ \text{J}}{1\ \text{kJ}} \times \frac{1\ \text{g}\ CCl_2F_2}{289\ \text{J}} = 142\ \text{g}\ CCl_2F_2$$

11.27 Consider the process in steps, using the appropriate thermochemical constant.

Heat the liquid from -50°C to 23.8°C (223 K to 296.8 K), using the specific heat of the liquid:

$$10.0\ \text{g}\ CCl_3F \times \frac{0.87\ \text{J}}{\text{g} \cdot \text{K}} \times 73.8\ \text{K} \times \frac{1\ \text{kJ}}{1000\ \text{J}} = 0.64\ \text{kJ}$$

Boil the liquid at 23.8°C (296.8 K), using the enthalpy of vaporization.

$$10.0\ \text{g}\ CCl_3F \times \frac{1\ \text{mol}\ CCl_3F}{137.4\ \text{g}\ CCl_3F} \times \frac{24.75\ \text{kJ}}{\text{mol}} = 1.80\ \text{kJ}$$

Heat the gas from 23.8°C to 50°C (296.8 K to 323 K), using the specific heat of the gas.

$$10.0\ \text{g}\ CCl_3F \times \frac{0.59\ \text{J}}{\text{g} \cdot \text{K}} \times 26.2\ \text{K} \times \frac{1\ \text{kJ}}{1000\ \text{J}} = 0.15\ \text{kJ}$$

The total energy required is 0.64 kJ + 1.80 kJ + 0.15 kJ = 2.59 kJ.

11.28 Heat the solid from -150°C to -114°C (123 K to159 K), using the specific heat of the solid.

$$50.0\ \text{g}\ C_2H_5OH \times \frac{0.97\ \text{J}}{\text{g} \cdot \text{K}} \times 36\ \text{K} \times \frac{1\ \text{kJ}}{1000\ \text{J}} = 1.7\ \text{kJ}$$

At -114°C (159 K), melt the solid, using its enthalpy of fusion.

$$50\ \text{g}\ C_2H_5OH \times \frac{1\ \text{mol}\ C_2H_5OH}{46.07\ \text{g}\ C_2H_5OH} \times \frac{5.02\ \text{kJ}}{1\ \text{mol}} = 5.45\ \text{kJ}$$

Heat the liquid from -114°C to 78°C (159 K to 351 K), using the specific heat of the liquid.

$$50.0\ \text{g}\ C_2H_5OH \times \frac{2.3\ \text{J}}{\text{g} \cdot \text{K}} \times 192\ \text{K} \times \frac{1\ \text{kJ}}{1000\ \text{J}} = 22\ \text{kJ}$$

At 78°C (351 K), vaporize the liquid, using its enthalpy of vaporization.

$$50.0\ \text{g}\ C_2H_5OH \times \frac{1\ \text{mol}\ C_2H_5OH}{46.07\ \text{g}\ C_2H_5OH} \times \frac{38.56\ \text{kJ}}{1\ \text{mol}} = 41.85\ \text{kJ}$$

The total energy required is 1.7 kJ + 5.45 kJ + 22 kJ + 41.85 kJ = 71 kJ.

11.29 (a) The critical temperature is the highest temperature at which a gas can be liquefied, regardless of pressure.

(b) The critical pressure is the pressure required to cause liquefaction at the critical temperature.

11.30 (a) Those gases whose critical temperatures are greater than 298 K can be liquefied at room temperature. The gases in Table 11.4 that meet this criterion are: NH_3, CO_2, H_2O.

(b) The temperature of $N_2(l)$ is 77 K. All of the gases in Table 11.4 have critical temperatures higher than 77 K, so all of them can be liquified at this temperature, given sufficient pressure.

Vapor Pressure and Boiling Point

11.31 The boiling point is the temperature at which the vapor pressure of a liquid equals the external pressure acting on the surface of the liquid. If the external pressure is changed, then the temperature required to produce that vapor pressure will also change.

Melting of a solid occurs when the vapor pressures of the solid and liquid phases are equal. Both of these vapor pressures will vary to a degree with the external pressure acting on the solid or liquid, but the variation is not great, and both solid and liquid vapor pressures tend to change in the same way with applied pressure.

11.32 (a) No effect.

(b) Vapor pressure increases with increasing temperature because average kinetic energies of molecules increase.

(c) Vapor pressure would decrease with increasing intermolecular attractive forces because fewer molecules would have sufficient kinetic energy to overcome the attractive forces and escape to the vapor phase.

(d) No effect.

(e) No effect. The pressure of the air above the liquid determines how quickly liquid-vapor equilibrium is reached but not the equilibrium vapor pressure.

11.33 $CBr_4 < CHBr_3 < CH_2Br_2 < CH_2Cl_2 < CH_3Cl < CH_4$

The weaker the intermolecular forces, the higher the vapor pressure, the more volatile the compound. The order of increasing volatility is the order of decreasing strength of intermolecular forces. By analogy to the boiling points of HCl and HBr (Section 11.2), the trend will be dominated by dispersion forces, even though four of the molecules ($CHBr_3$, CH_2Br_2, CH_2Cl_2 and CH_3Cl) are polar. Thus, the order of increasing volatility is the order of decreasing molar mass and decreasing strength of dispersion forces.

11.34 (a) The less volatile compound, $AsCl_3$, has the stronger intermolecular forces. In the more volatile one, PCl_3, more molecules have sufficient kinetic energy to overcome intermolecular attractive forces and escape to the gas phase. Since both compounds have the same distribution of kinetic energies at 25°C, the attractive forces must be stronger and harder to overcome in $AsCl_3$. This makes sense, because $AsCl_3$ is heavier and has stronger dispersion forces than PCl_3. $AsCl_3$ is also slightly more polar, but dispersion forces account for most of the volatility difference between these two compounds.

(b) PCl_3; the more volatile compound has the higher vapor pressure.

(c) $AsCl_3$; the less volatile compound requires a greater increase in temperature (kinetic energy) to promote enough molecules to the gas phase so that vapor pressure is equal to atmospheric pressure.

11.35 The water in the two pans is at the same temperature, the boiling point of water at the atmospheric pressure of the room. During a phase change, the temperature of a system is constant. All energy gained from the surroundings is used to accomplish the transition, in this case to vaporize the liquid water. The pan of water that is boiling vigorously is gaining more energy and the liquid is being vaporized more quickly than in the other pan, but the temperature of the phase change is the same.

11.36 (a) On a humid day, there are more gaseous water molecules in the air and more are recaptured by the surface of the liquid, making evaporation slower.

(b) At high altitude, atmospheric pressure is lower and water boils at a lower temperature. The eggs must be cooked longer at the lower temperature.

11.37 The boiling point is the temperature at which the vapor pressure of a liquid equals atmospheric pressure.

(a) The boiling point of ethanol at 200 torr is ~48°C, or, at 48°C, the vapor pressure of ethanol is 200 torr.

(b) At a pressure of 12 torr, water would boil at 14°C, or, the vapor pressure of water at 14°C is 12 torr.

11.38 (a) A pressure of 1.3 atm corresponds to $1.3 \text{ atm} \times \frac{760 \text{ torr}}{1 \text{ atm}} = 988 \text{ torr}$. According to Appendix B, the vapor pressure of water reaches 988 torr somewhere between 106-108°C. By linear interpolation, the boiling point should be near $106°C + \left[\frac{(988 - 938) \text{ torr}}{(1004 - 938) \text{ torr}} \times 2°C\right] = 107.5°C$.

(b) The vapor pressure of diethyl ether at 20°C is approximately 450 torr. Thus, at 20°C diethyl ether would boil at an external pressure of 450 torr.

11.39 The boiling point of a liquid is the temperature at which its vapor pressure equals atmospheric pressure. According to Appendix B, the vapor pressure of water is 680 torr at approximately 97°C.

11.40 From Appendix B, the temperature at which the vapor pressure of water is 350 torr is approximately 80°C.

Phase Diagrams

11.41 The liquid/gas line of a phase diagram ends at the critical point, the temperature and pressure beyond which the gas and liquid phases are indistinguishable. At temperatures higher than the critical temperature, a gas cannot be liquefied, regardless of pressure.

1.42 The triple point on a phase diagram represents the temperature at which the gas, liquid and solid phases are in equilibrium.

11.43 (a) The water vapor would condense to form a solid at a pressure of around 4 torr. At higher pressure, perhaps 5 atm or so, the solid would melt to form liquid water. This occurs because the melting point of ice, which is 0°C at 1 atm, decreases with increasing pressure.

(b) In thinking about this exercise, keep in mind that **total** pressure is being maintained constant at 0.3 atm. That pressure is made up of water vapor pressure and some other pressure, which could come from an inert gas. The water at -1.0°C and 0.30 atm is in the solid form. Upon heating, it melts to form liquid water slightly above 0°C. The liquid converts to a vapor when the temperature reaches the point at which the vapor pressure of water reaches 0.3 atm (228 torr Hg). From Appendix B we see that this occurs just below 70°C.

11.44 (a) Solid CO_2 sublimes to form $CO_2(g)$ at a temperature of about -60°C.

(b) Solid CO_2 melts to form $CO_2(l)$ at a temperature of about -50°C. The $CO_2(l)$ boils when the temperature reaches approximately -40°C.

11.45 (a)

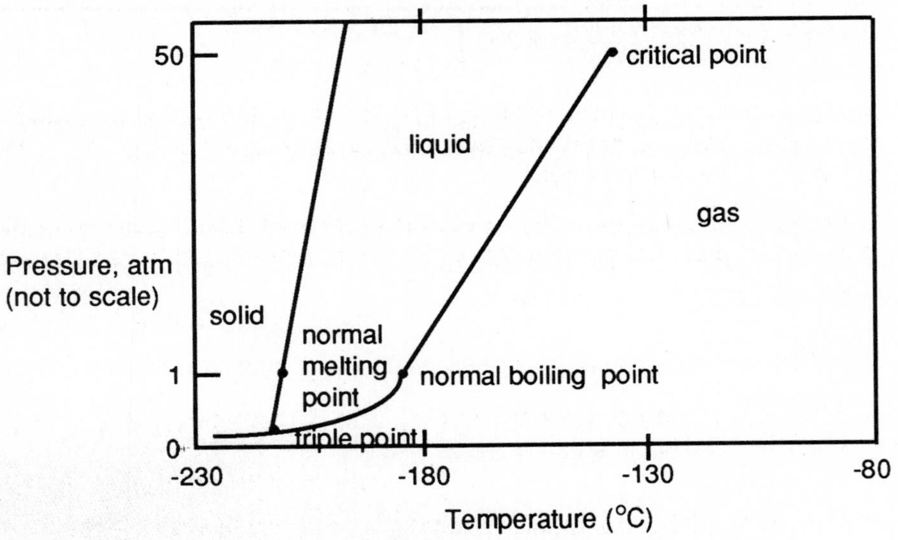

(b) $O_2(s)$ is denser than $O_2(l)$ because the solid-liquid line on the phase diagram is normal. That is, as pressure increases, the melting temperature increases. [Note that the solid-liquid line for O_2 is nearly vertical, indicating a small difference in the densities of $O_2(s)$ and $O_2(l)$].

(c) O_2 will melt when heated at a pressure of 1 atm, since this is a much greater pressure than the pressure at the triple point.

11.46 (a)

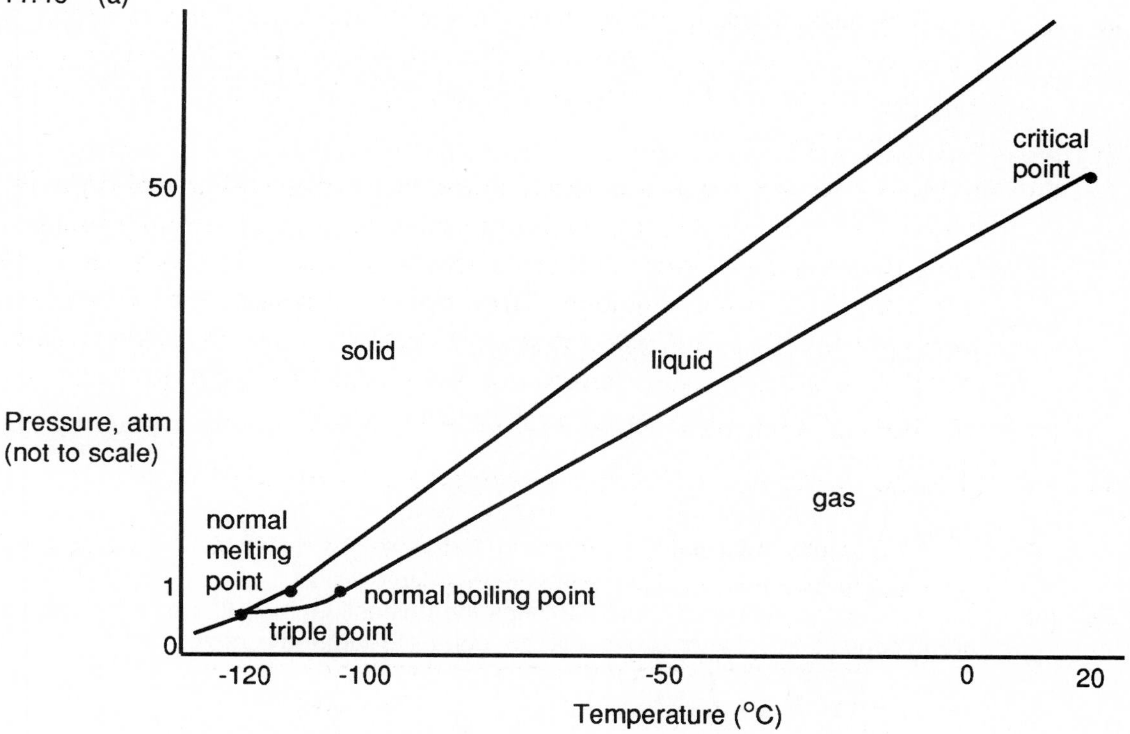

(b) Xe(s) will not float in Xe(l). The solid-liquid line on the phase diagram is normal, the melting point of Xe(s) increases with increasing temperature, which means that Xe(s) is denser than Xe(l).

(c) Cooling Xe(g) at 100 torr will cause deposition of the solid. A pressure of 100 torr is below the pressure of the triple point, so the gas will change directly to the solid upon cooling.

Structure of Solids

11.47 In a crystalline solid, the component particles (ions or molecules) are arranged in an ordered repeating pattern. In an amorphous solid, there is no orderly structure.

11.48 In amorphous silica (SiO_2) the regular structure of quartz is disrupted; the loose, disordered structure, Figure 11.26(b), has many vacant "pockets" throughout. There are fewer SiO_2 groups per volume in the amorphous solid; the packing is less efficient and less dense.

11.49 The unit cell is the building block of the crystal lattice. When repeated in three dimensions, it produces the crystalline solid. It is a parallelepiped with the characteristic distances and angles, a, b, c, α, β and γ. Unit cells can be primitive (lattice points only at the corners of the parallelepiped) or centered (lattice points at the corners and at the middle of faces or the middle of the parallelepiped).

11.50 (a) 8 corners × 1/8 sphere/corner = 1 sphere
(b) 8 corners × 1/8 sphere/corner + 1 center × 1 sphere/center = 2 spheres
(c) 8 corners × 1/8 sphere/corner + 6 faces × 1/2 sphere/face = 4 spheres

11.51 (a) 4 [See Exercise 11.50(c).]

(b) Each sphere is in contact with 12 nearest neighbors; its coordination number is thus 12.

(c) The length of the face diagonal of a face-centered cubic unit cell is four times the radius of the metal and $\sqrt{2}$ times the unit cell dimension (usually designated a for cubic cells).

$$4 \times 1.44 \text{ Å} = \sqrt{2} \times a$$

$$a = \frac{4 \times 1.44 \text{ Å}}{1.414} = 4.07 \text{ Å or } 4.07 \times 10^{-8} \text{ cm}$$

(d) The density of the metal is the mass of the unit cell contents divided by the volume of the unit cell.

$$\text{density} = \frac{4 \text{ Ag atoms}}{(4.07 \times 10^{-8} \text{ cm})^3} \times \frac{107.868 \text{ g Ag}}{6.022 \times 10^{23} \text{ Ag atoms}} = \frac{10.6 \text{ g}}{\text{cm}^3}$$

11.52 (a) The face diagonal length is just $\sqrt{2}$ times the unit cell length. Because the atoms are in contact, this diagonal has a length equal to four atomic radii.

$$r = (\sqrt{2} \times 4.95 \text{ Å})/4 = 1.75 \text{ Å}$$

(b) The density of the metal is the mass of the unit cell contents divided by the volume of the unit cell. There are four Pb atoms in a face-centered cubic unit cell.

$$\text{density} = \frac{4 \text{ Pb atoms}}{(4.95 \text{ Å})^3} \times \frac{207.2 \text{ g Pb}}{6.022 \times 10^{23} \text{ Pb atoms}} \times \frac{(1 \times 10^8 \text{ Å})^3}{1 \text{ cm}^3} = \frac{11.3 \text{ g}}{\text{cm}^3}$$

11.53 (a) Each sphere is in contact with 12 nearest neighbors; its coordination number is thus 12.
(b) Each sphere has a coordination number of six.
(c) Each sphere has a coordination number of eight.

11.54 (a) Na^+: 6 (b) Zn^{2+}: 4 (c) Ca^{2+}: 8

11.55 In the face-centered cubic structure, there are four NiO units in the unit cell. Density is the mass of the unit cell contents divided by the unit cell volume (a^3).

$$\text{density} = \frac{4 \text{ NiO units}}{(4.18 \text{ Å})^3} \times \frac{74.7 \text{ g NiO}}{6.022 \times 10^{23} \text{ NiO units}} \times \left(\frac{1 \text{ Å}}{1 \times 10^{-8} \text{ cm}}\right)^3 = \frac{6.79 \text{ g}}{\text{cm}^3}$$

11.56 There are four PbSe units in the unit cell. The unit cell edge is designated *a*.

$$8.27 \text{ g/cm}^3 = \frac{4 \text{ PbSe units}}{a^3} \times \frac{286.2 \text{ g}}{6.022 \times 10^{23} \text{ PbSe units}} \times \left[\frac{1 \text{ Å}}{10^{-8} \text{ cm}}\right]^3$$

$a^3 = 229.87 \text{ Å}^3$, $a = 6.13$ Å

11.57 The volume of the unit cell is $(2.86 \times 10^{-8} \text{ cm})^3$. The mass of the unit cell is:

$$\frac{7.92 \text{ g}}{\text{cm}^3} \times \frac{2.34 \times 10^{-23} \text{ cm}^3}{\text{unit cell}} = \frac{1.85 \times 10^{-22} \text{ g}}{\text{unit cell}}$$

There are two atoms of the element present in the body-centered cubic unit cell. Thus the atomic weight is:

$$\frac{1.85 \times 10^{-22} \text{ g}}{\text{unit cell}} \times \frac{1 \text{ unit cell}}{2 \text{ atoms}} \times \frac{6.02 \times 10^{23} \text{ atoms}}{1 \text{ mol}} = \frac{55.7 \text{ g}}{\text{mol}}$$

11.58 Avogadro's number is the number of KCl formula units in 74.55 g of KCl.

$$74.55 \text{ g KCl} \times \frac{1 \text{ cm}^3}{1.984 \text{ g}} \times \frac{(1 \times 10^{10} \text{ pm})^3}{1 \text{ cm}^3} \times \frac{4 \text{ KCl units}}{628^3 \text{ pm}^3} = 6.07 \times 10^{23} \text{ KCl formula units}$$

Bonding in Solids

11.59 (a) Hydrogen bonding, dipole-dipole forces, London dispersion forces
(b) covalent chemical bonds (mainly)
(c) ionic bonds (mainly)
(d) metallic bonds

11.60 (a) Molecular (b) molecular (c) metallic (d) ionic
(e) network covalent (f) molecular

11.61 Carbon (graphite and diamond) and SiO_2 (quartz)

11.62 In molecular solids, relatively weak intermolecular forces (hydrogen bonding, dipole-dipole, dispersion) bind the molecules in the lattice, so relatively little energy is required to disrupt these forces. In network-covalent solids, covalent bonds join atoms into an extended network. Melting or deforming a network-covalent solid means breaking these covalent bonds, which requires a large amount of energy.

11.63 (a) KBr -- strong ionic versus weak dispersion forces
(b) SiO_2 -- covalent bonds establish the network structure of the lattice versus weak dispersion forces holding CO_2 molecules together
(c) Se -- network covalent lattice versus weak dispersion forces
(d) MgF_2 -- due to higher charge on Mg^{2+} than Na^+

11.64 (a) C_6Cl_6 -- both are influenced by dispersion forces, C_6Cl_6 has the higher molecular weight
(b) HF -- it has hydrogen bonding and HCl does not
(c) SiO_2 -- both have regular lattices, but the covalent bonds in the SiO_2 lattice require more energy to break than the ionic bonds in KO_2
(d) Xe -- greater atomic weight, stronger dispersion forces

11.65 According to Table 11.6, the solid could be either ionic with low water solubility or network covalent. Due to the extremely high sublimation temperature, it is probably network covalent.

11.66 Because of its relatively high melting point and properties as a conducting solution, the solid must be ionic.

11.67 (a) According to Figure 11.44(b), there are 4 AgI units in the "zinc blende" unit cell [4 complete Ag^+ spheres, 6(1/2) + 8(1/8) I^- sphere's.]

$$5.69\frac{g}{cm^3} = \frac{4\text{AgI units}}{a^3} \times \frac{234.8\text{ g}}{6.022 \times 10^{23}\text{ AgI units}} \times \left(\frac{1\text{ Å}}{1 \times 10^{-8}\text{ cm}}\right)^3$$

$a^3 = 274.10\text{ Å}^3$, $a = 6.50$ Å

(b) In an orthonormal coordinate system, the distance between two points (x_1, y_1, z_1) and (x_2, y_2, z_2) is $\sqrt{(x_1 - x_2)^2 + (y_1 - y_2)^2 + (z_1 - z_2)^2}$.

On Figure 11.44(b), select a right-handed coordinate system and select a bonded pair of ions. One possibility is shown below.

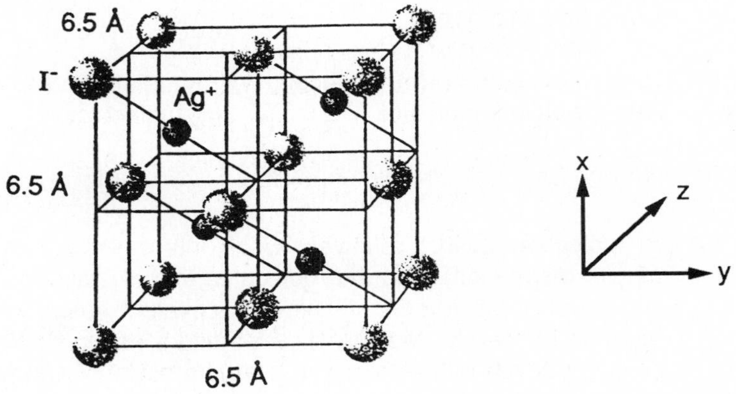

In a cubic unit cell, the lengths of all three cell edges are the same, in this case 6.50 Å.

For Ag^+: x = 0.75 (6.50 Å), y = 0.25(6.50 Å), z = 0.25(6.50 Å)
x = 4.875, y = 1.625, z = 1.625
I^-: x = 6.50, y = 0, z = 0

The Ag-I distance is then $\sqrt{(4.875 - 6.50)^2 + (1.625 - 0)^2 + (1.625 - 0)^2}$.

$$\text{Ag - I} = \sqrt{(1.625)^2 + (1.625)^2 + (1.625)^2} = 2.81 \text{ Å}$$

11.68 (a) The U atoms in UO_2 are represented by the smaller spheres in figure 11.44(c). The chemical formula requires twice as many O^{2-} ions as U^{4+} ions. There are eight complete large spheres and four total (8 × 1/8 + 6 × 1/2) small spheres, so the small ones must represent U^{4+}. (It is probably true that O^{2-} has a physically larger radius than U^{4+}, but the elements' large separation on the periodic chart makes the relative radii difficult to estimate from trends.)

(b) According to Figure 11.44(c), there are four UO_2 units in the "fluorite" unit cell.

$$\frac{4 \text{ UO}_2 \text{ units}}{(5.468 \text{ Å})^3} \times \frac{270.03 \text{ g}}{6.022 \times 10^{23} \text{ UO}_2 \text{ units}} \times \left(\frac{1 \text{ Å}}{1 \times 10^{-8} \text{ cm}}\right)^3 = 10.97 \text{ g/cm}^3$$

Additional Exercises

11.69 $Br_2(l) \rightarrow Br_2(g)$ $\Delta H = 30.71$ kJ $Br_2(g) \rightarrow 2Br(g)$ $\Delta H = 192.9$ kJ

$I_2(s) \rightarrow I_2(g)$ $\Delta H = 62.25$ kJ k $I_2(g) \rightarrow 2I(g)$ $\Delta SH = 150.95$ kJ

(a) Breaking one mole of Br-Br covalent bonds requires 192.9 kJ, roughly six times the energy required to overcome the intermolecular dispersion forces in one mole of $Br_2(l)$, 30.71 kJ.

(b) The enthalpy of sublimation for $I_2(s)$ is twice as large as the enthalpy of vaporization for $Br_2(l)$. Since I_2 is a larger molecule with a higher molecular weight, it should experience stronger dispersion forces and require more energy to separate the molecules into the gas phase. On the other hand, the bond energy of the I-I bond is less than that of the Br-Br bond. Although bond strength is not in general related to bond length, it is true for diatomic molecules composed of halogen atoms (homo- or heteronuclear) that the shorter the bond length, the greater the bond energy (see Table 8.4).

11.70 (a) Dipole-dipole attractions (polar covalent molecules): SO_2, IF, HBr
(b) hydrogen bonding (O-H, N-H or F-H bonds): CH_3NH_2, HCOOH

11.71 (a) The *cis* isomer has stronger dipole-dipole forces; the trans isomer is nonpolar. The higher boiling point of the *cis* isomer supports this conclusion.

(b) While boiling points are primarily a measure of strength of intermolecular forces, melting points are influenced by crystal packing efficiency as well as intermolecular forces. Since the nonpolar *trans* isomer with the weaker intermolecular forces has the higher melting point, it must pack more efficiently.

11.72 In dibromethane, CH_2Br_2, the dispersion force contribution will be larger than for CH_2Cl_2, because bromine is more polarizable than the lighter element chlorine. At the same time, the dipole-dipole contribution for CH_2Cl_2 is greater than for CH_2Br_2 because CH_2Cl_2 has a larger dipole moment. Just the opposite comparisons apply to CH_2F_2, which is less polarizable and has a higher dipole moment than CH_2Cl_2.

11.73 (a) Viscosity (b) boiling point (c) surface tension (d) enthalpy of vaporization, ΔH_{vap}

11.74 In Figure 11.13, the less viscous liquid on the right has filled its container to a greater height, assuming equal starting masses, flow times and cross-sectional area on the containers. In units of kg/m•s, if mass (kg) and time (s) are equal for two samples, the less viscous sample will travel a greater distance (m), leading to a smaller calculated viscosity value.

11.75 Propylamine experiences hydrogen bonding interactions while trimethylamine, with no N-H bonds, does not. Also, the rod-like shape of propylamine (see Exercise 11.16) leads to stronger dispersion forces than in trimethylamine. The stronger intermolecular forces in propylamine lead to the lower vapor pressure.

11.76 The two O-H groups in ethylene glycol are involved in many hydrogen bond interactions, leading to its high boiling point and viscosity, relative to pentane, which experiences only dispersion forces.

11.77 $$n = \frac{PV}{RT} = 735 \text{ torr} \times \frac{1 \text{ atm}}{760 \text{ torr}} \times \frac{\text{K} \cdot \text{mol}}{0.0821 \text{ L} \cdot \text{atm}} \times \frac{3.00 \text{ L}}{290 \text{ K}} = 0.122 \text{ mol } C_4H_{10}$$

$$0.122 \text{ mol } C_4H_{10} \times \frac{21.3 \text{ kJ}}{1 \text{ mol } C_4H_{10}} = 2.60 \text{ kJ}$$

11.78

$$X(s) \rightarrow X(l) \qquad \Delta H_{fus}$$

$$X(l) \rightarrow X(g) \qquad \Delta H_{vap}$$

$$X(s) \rightarrow X(g) \qquad \Delta H_{fus} + \Delta H_{vap} = \Delta H_{sub}$$

Sublimation, the process of a substance changing directly from the solid to the gas phase, can be thought of as stepwise fusion followed by vaporization. Hess's Law states that if a process can be written as a series of steps, the enthalpy change for the overall process is the sum of the enthalpy changes for the individual steps. Whether sublimation actually occurs by this mechanism, that is, whether the substance ever exists in the liquid state, is irrelevant. If we can write a stepwise path for which the energies of the steps are known, the overall enthalpy change can be calculated.

11.79 $PV = \frac{g}{\mathcal{M}} \times RT$; $g = \frac{PV\mathcal{M}}{RT}$; T = 313 K; $\mathcal{M}$ = 18.02 g/mol

P (the vapor pressure of H_2O at 40°C) = $55.3 \text{ torr} \times \frac{1 \text{ atm}}{760 \text{ torr}} = 0.0728 \text{ atm}$

$$V = 4.0 \text{ m} \times 4.0 \text{ m} \times 3.0 \text{ m} \times \frac{10^3 \text{ dm}^3}{1 \text{ m}^3} \times \frac{1 \text{ L}}{1 \text{ dm}^3} = 4.8 \times 10^4 \text{ L}$$

$$g = \frac{0.0728 \text{ atm} \times 4.8 \times 10^4 \text{ L} \times 18.02 \text{ g}}{313 \text{ K} \times \text{mol}} \times \frac{\text{K} \cdot \text{mol}}{0.0821 \text{ L} \cdot \text{atm}} = 2.4 \times 10^3 \text{ g} = 2.4 \text{ kg } H_2O$$

11.80 (a) If the Clausius-Clapeyron equation is obeyed, a graph of ln VP vs 1/T(K) should be linear. Here are the data in form for graphing.

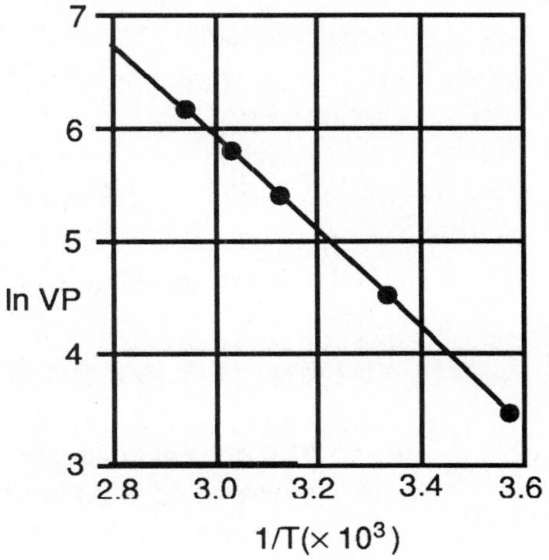

T(K)	1/T	VP(torr)	ln(VP)
280.0	3.571×10^{-3}	32.42	3.479
300.0	3.333×10^{-3}	92.47	4.527
320.0	3.125×10^{-3}	225.1	5.417
330.0	3.030×10^{-3}	334.4	5.812
340.0	2.941×10^{-3}	482.9	6.180

According to the graph, the Clausius-Clapeyron equation is obeyed, to a first approximation.

$\Delta H_{vap} = -\text{slope} \times R$

$$\text{slope} = \frac{3.479 - 6.180}{(3.571 - 2.941) \times 10^{-3}} = -\frac{2.701}{0.630 \times 10^{-3}}$$

$$= -4.29 \times 10^3$$

$$\Delta H_{vap} = -(-4.29 \times 10^3) \times 8.314 \text{ J/K} \cdot \text{mol}$$

$$= 35.7 \text{ kJ/mol}$$

(b) The normal boiling point is the temperature at which the vapor pressure of the liquid equals atmospheric pressure, 760 torr. From the graph,

ln 760 = 6.63, 1/T for this VP = 2.828×10^{-3}; T = 353.6 K

11.81 (a) The Clausius-Clapeyron equation is $\ln P = \frac{-\Delta H_{vap}}{RT} + C$.

For two vapor pressures, P_1 and P_2, measured at corresponding temperatures T_1 and T_2, the relationship is

$$\ln P_1 - \ln P_2 = \frac{-\Delta H_{vap}}{RT_1} + C - \left(\frac{-\Delta H_{vap}}{RT_2} + C\right)$$

$$\ln P_1 - \ln P_2 = \frac{-\Delta H_{vap}}{R}\left(\frac{1}{T_1} - \frac{1}{T_2}\right) + C - C$$

$$\ln\frac{P_1}{P_2} = \frac{-\Delta H_{vap}}{R}\left(\frac{1}{T_1} - \frac{1}{T_2}\right)$$

(b) $P_1 = 10.00$ torr, $T_1 = 716$ K; $P_2 = 400.0$ torr, $T_2 = 981$ K

$$\ln\frac{10.00}{400.0} = \frac{-\Delta H_{vap}}{8.314 \text{ J/K} \cdot \text{mol}}\left(\frac{1}{716} - \frac{1}{981}\right)$$

$$-3.6889 \text{ (8.314 J/K} \cdot \text{mol)} = -\Delta H_{vap} \text{ } (3.773 \times 10^{-4}\text{/K})$$

$$\Delta H_{vap} = 8.13 \times 10^4 \text{ J/mol} = 81.3 \text{ kJ/mol}$$

(c) The normal boiling point of a liquid is the temperature at which the vapor pressure of the liquid is 760 torr.

$P_1 = 400.0$ torr, $T_1 = 981$ K; $P_2 = 760$ torr, T_2 = b.p. of potassium

$$\ln\left(\frac{400.0}{760.0}\right) = \frac{-8.13 \times 10^4 \text{ J/mol}}{8.314 \text{ J/K} \cdot \text{mol}}\left(\frac{1}{981 \text{ K}} - \frac{1}{T_2}\right)$$

$$\frac{-0.6419}{-9.779 \times 10^3} = 1.0194 \times 10^{-3} - \frac{1}{T_2}; \quad \frac{1}{T_2} = 1.0194 \times 10^{-3} - 6.564 \times 10^{-5}$$

$$\frac{1}{T_2} = 9.537 \times 10^{-4}; \quad T_2 = 1048 \text{ K } (775°\text{C})$$

(d) P_1 = VP of K(l) at 100°C, $T_1 = 373$ K; $P_2 = 10.00$ torr, $T_2 = 716$ K

$$\ln\frac{P_1}{10.00 \text{ torr}} = \frac{-8.13 \times 10^4 \text{ J/mol}}{8.314 \text{ J/K} \cdot \text{mol}}\left(\frac{1}{373} - \frac{1}{716}\right)$$

$$\ln\frac{P_1}{10.00 \text{ torr}} = \frac{-8.13 \times 10^4 \text{ J/mol}}{8.314 \text{ J/K} \cdot \text{mol}} \times 1.284 \times 10^{-3} = -12.559$$

$$\frac{P_1}{10.00 \text{ torr}} = e^{-12.559} = 3.51 \times 10^{-6}; \quad P_1 = 3.51 \times 10^{-5} \text{ torr}$$

11.82 Physical data for the two compounds from the *Handbook of Chemistry and Physics*:

	$\mathcal{M}$	dipole moment	boiling point
CH_2Cl_2	85 g/mol	1.60 D	40.0°C
CH_3I	142 g/mol	1.62 D	42.4°C

(a) The two substances have very similar molecular structures; each is an unsymmetrical tetrahedron with a single central carbon atom and no hydrogen bonding. Since the structures are very similar, the magnitudes of the dipole-dipole forces should be similar. This is verified by their very similar dipole moments. The heavier compound, CH_3I, will have slightly stronger London dispersion forces. Since the nature and magnitude of the intermolecular forces in the two compounds is nearly the same, it is very difficult to predict which will be more volatile (or which will have the higher boiling point as in part (b)).

(b) Given the structural similarities discussed in part (a), one would expect the boiling points to be very similar, and they are. Based on its larger molecular weight (and dipole-dipole forces being essentially equal) one might predict that CH_3I would have a slightly higher boiling point; this is verified by the known boiling points.

(c) According to Equation 11.1, $\ln P = \frac{-\Delta H_{vap}}{RT} + C$
A plot of ln P vs. 1/T for each compound is linear. Since the order of volatility changes with temperature for the two compounds, the two lines must cross at some temperature; the slopes of the two lines, ΔH_{vap} for the two compounds, and the y-intercepts, C, must be different.

(d)

	CH_2Cl_2		CH_3I	
lnVP	T(K)	1/T	T(K)	1/T
2.303	229.9	4.350×10^{-3}	227.4	4.398×10^{-3}
3.689	250.9	3.986×10^{-3}	249.0	4.016×10^{-3}
4.605	266.9	3.747×10^{-3}	266.2	3.757×10^{-3}
5.991	297.3	3.364×10^{-3}	298.5	3.350×10^{-3}

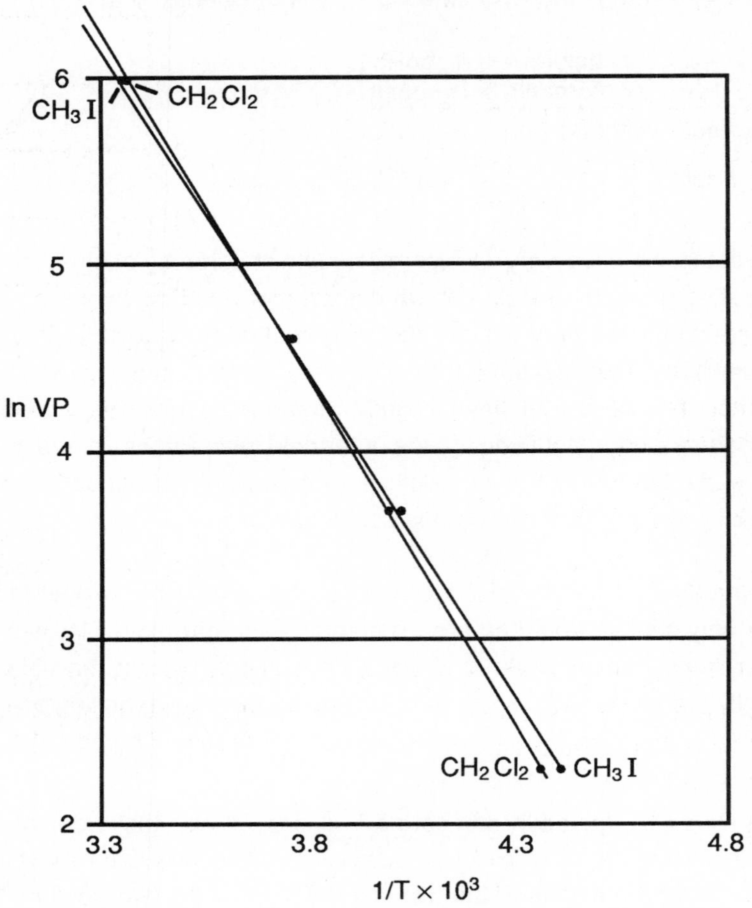

For CH_2Cl_2, $-\Delta H_{vap}/R$ = slope =

$$\frac{(5.991 - 2.303)}{(3.364 \times 10^{-3} - 4.350 \times 10^{-3})} = -\frac{3.688}{0.987 \times 10^{-3}} = -3.74 \times 10^3 = -\Delta H_{vap}/R$$

$\Delta H_{vap} = 8.314\ (3.74 \times 10^3) = 3.107 \times 10^4$ J/mol = 31.07 kJ/mol

For CH_3I, $-\Delta H_{vap}/R$ = slope =

$$\frac{(5.991 - 2.303)}{(3.350 \times 10^{-3} - 4.398 \times 10^{-3})} = -\frac{3.688}{1.048 \times 10^{-3}} = -3.519 \times 10^3 = -\Delta H_{vap}/R$$

$\Delta H_{vap} = 8.314\ (3.519 \times 10^3) = 2.926 \times 10^4$ J/mol = 29.26 kJ/mol

11.83 Graph ln (VP) vs 1/T, where T is absolute temperature.

T(K)	1/T	VP(torr)	ln(VP)
253.2	3.99×10^{-3}	0.640	-0.446
257.2	3.89×10^{-3}	1.132	0.1240
261.2	3.83×10^{-3}	1.632	0.4898
265.2	3.77×10^{-3}	2.326	0.8442
269.2	3.71×10^{-3}	3.280	1.188
273.2	3.66×10^{-3}	4.579	1.521

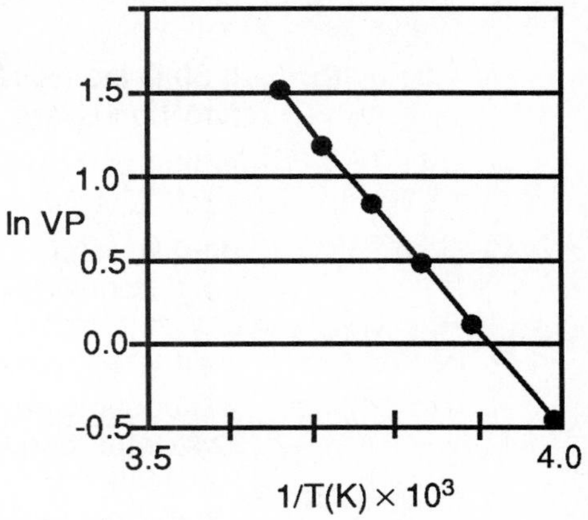

The graph is shown on the right. The slope can be estimated by reading ln(VP) at two well-separated values of 1/T:

$$\text{slope} = \frac{0.1240 - 1.188}{3.89 \times 10^{-3} - 3.71 \times 10^{-3}} = -5.91 \times 10^3$$

This slope equals $-\Delta H_s / R$. Thus, $-\Delta H_s = 8.314(-5.91 \times 10^3)$; $\Delta H_s = 49.1$ kJ/mol

11.84 The length of the body diagonal in a body-centered cubic cell is 4r, where r is the radius of a Cr atom. To deduce the value of r, look at the triangle formed by the cube side (length = *a*), a face diagonal (length = $\sqrt{2}\ a$) and the body diagonal (length = 4r).

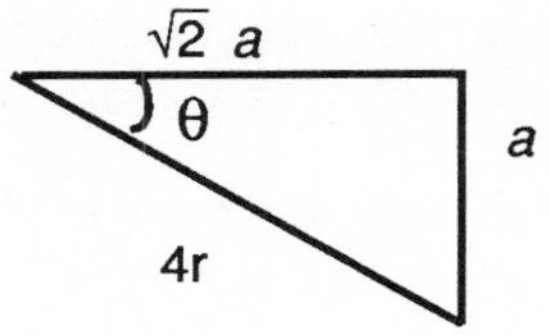

The angle θ is arctan $1/\sqrt{2} = 35.26°$.
$\sin\theta = a/4r$; $r = 2.884$ Å / 4 sin 35.26° = 1.248 Å.

11.85 The most effective diffraction of light by a grating occurs when the wavelength of light and the separation of the slits in the grating are similar. When X-rays are diffracted by a crystal, layers of atoms serve as the "slits." The most effective diffraction occurs when the distances between layers of atoms are similar to the wavelength of the X-rays. Typical interlayer distances in crystals range from 2 Å to 20 Å. Visible light, 400-700 nm or 4,000 to 7,000 Å, is too long to be diffracted effectively by crystals. Molybdenum X-rays of 0.71 Å are on the same order of magnitude as interlayer distances in crystals and are diffracted.

11.86 (a) There are 90 bonds in the C_{60} molecule. [Hint: Use a soccer ball to count.]

(b) If 60 C atoms each share four pairs of electrons with atoms other than C, a total of 240 bonds are required. If C is bound only to C, as is the case in C_{60}, two C atoms participate in each single bond, so 120 single bonds are required to satisfy each C atom. Since there are only 90 bonds in C_{60}, some of these must be double bonds. In a double bond, two C atoms each share two pairs of electrons; each double bond takes care of four of the required bonds. Thus,

60 single bonds × 2 C atoms
+ 30 double bonds × 2 C atoms × 2 pairs of electrons
240 total "bonds"

Of the 90 bonds in C_{60}, 30 are formally double bonds. There is probably an extensive delocalized π bonding network in C_{60}.

11.87 (a) Both diamond (d = 3.5 g/cm^3) and graphite (d = 2.3 g/cm^3) are network covalent solids with efficient packing arrangements in the solid state; there is relatively little empty space in their respective crystal lattices. Diamond, with bonded C-C distances of 1.54 Å in all directions, is more dense than graphite, with shorter C-C distances within carbon sheets but longer 3.41 Å separations between sheets (Figure 11.41). Buckminsterfullerene has much more empty space, both inside each C_{60} "ball" and between balls, than either diamond or graphite, so its density will be considerably less than 2.3 g/cm^3.

(b) In a face-centered-cubic unit cell, there are 4 complete C_{60} units.

$$\frac{4\ C_{60}\ \text{units}}{(14.2\ \text{Å})^3} \times \frac{729.66\ \text{g}}{6.022 \times 10^{23}\ C_{60}\ \text{units}} \times \left(\frac{1\ \text{Å}}{1 \times 10^{-8}\ \text{cm}}\right)^3 = 1.67\ \text{g/cm}^3$$

(1.67 g/cm^3 is the smallest density of the three allotropes, diamond, graphite and buckminsterfullerene.)

CHAPTER 12

Modern Materials

Liquid Crystals

12.1 A liquid is clear and fluid while a liquid crystal is cloudy and often viscous. There is greater molecular order in a liquid crystal than in a liquid.

A crystal is rigid and has its own well-defined volume while a liquid crystal is fluid (even if it is viscous). There is greater molecular order in a solid than in a liquid crystal.

12.2 The molecules in a liquid crystal are ordered in one or two directions (Figure 12.4), but not in all three directions like molecules in the solid state. This partial ordering gives liquid crystals physical properties that are intermediate between those of liquids and solids.

12.3 Reinitzer observed that cholesteryl benzoate has a phase that exhibits properties intermediate between those of the solid and liquid phases. This "liquid-crystalline" phase, formed by melting at 145°C, is opaque, changes color from red to blue as the temperature is increased, and becomes clear at 179°C.

12.4 An isotropic (non-liquid crystalline) liquid is clear. A liquid crystalline material has a milky appearance and undergoes a transition to clear liquid at some temperature above its melting point. It is easily affected by electric or magnetic fields and often is considerably more viscous than an isotropic liquid.

12.5 Long, somewhat rigid, rod-like molecules (Figure 12.6) are the most likely candidates for liquid-crystalline properties. The rigid rods can have flexible tails, as in cholesteric liquid crystals (Figure 12.8).

12.6 Benzene rings, C=N and N=N bonds contribute to the rigidity of the molecule. The extensive π electron density increases the polorizability of the molecules, increasing the strength of dispersion forces and encouraging a more ordered molecular arrangement.

12.7   The three major classes of liquid-crystalline materials are nematic, smectic and cholesteric. Nematic and smectic phases are formed by rod-like molecules aligned with their long axes parallel. Smectic phases have some additional ordering, which often involves alignment of the ends of the molecular rods to form sheets. In cholesteric phases, the molecules are stacked in layers and are often twisted with respect to molecules in adjacent layers, (Figure 12.7).

12.8   In the smetic A phase, the long axis of the molecule is perpendicular to the layer; in the smetic C phase, the rod axis of the molecule is tilted relative to the layer (Figure 12.4).

12.9   The presence of a second substance creates new solute-liquid crystal intermolecular forces which disrupt the orienting forces present in the pure liquid crystalline phase. For the liquid crystalline phase to be stable, the ordering of the molecules must be quite long-range. That is, it must extend over many molecules. A single solute-liquid crystal interaction reduces the long range order, and very few of these disruptions can be tolerated before the stability of the liquid-crystalline phase is destroyed.

12.10   The presence of polar groups or nonbonded pairs of electrons leads to relatively strong dipole-dipole interactions between molecules. These are a significant part of the orientating forces necessary for liquid crystal formation.

Polymers

12.11   **Polymers** are substances with high molecular mass formed by joining together many **monomers**, small molecules with lower molecular mass. Monomers are the repeating units of a polymer.

12.12   All monomers that undergo addition polymerization contain multiple bonds. The two electrons from the π bond are used to form two new single bonds with two other monomers.

12.13   By analogy to polyisoprene, Table 12.1,

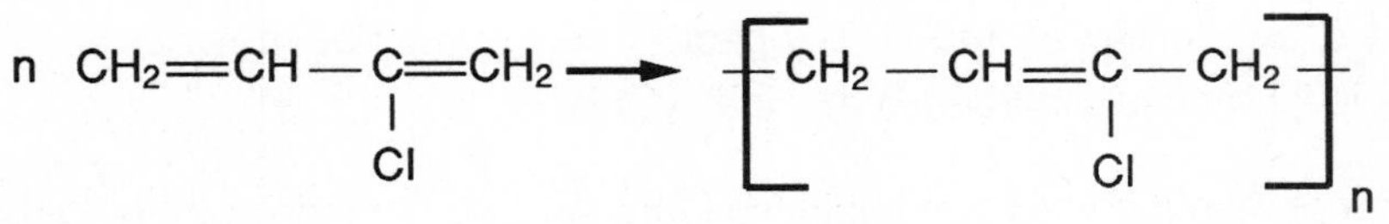

$$n \; CH_2{=}CH - C{=}CH_2 \longrightarrow \left[CH_2 - CH{=}C - CH_2 \right]_n$$
$$\hspace{4.5cm} | \hspace{5.5cm} |$$
$$\hspace{4.5cm} Cl \hspace{5.5cm} Cl$$

12.14 $n \; CH_2{=}CH \longrightarrow CH_3 - CH\left[CH_2 - CH \right]_n CH_2 - CH_2$
$$\hspace{2cm} | \hspace{3cm} | \hspace{2cm} | \hspace{2.5cm} |$$
$$\hspace{2cm} CN \hspace{2.5cm} CN \hspace{2cm} CN \hspace{2.5cm} CN$$

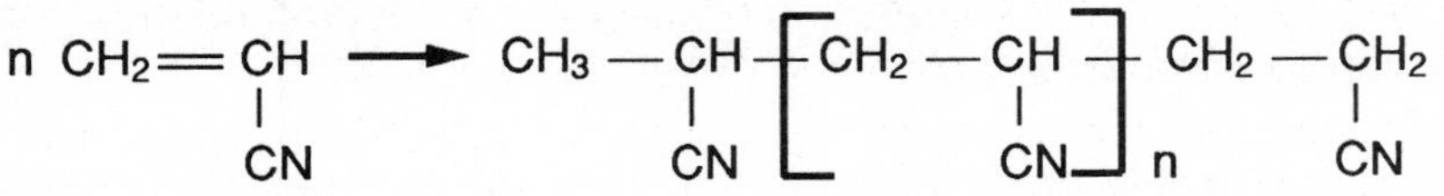

12.15   In a **condensation** reaction two molecules are combined to form a larger molecule by elimination of a small molecule such as H_2O. Polymers are formed when condensation occurs at both ends of the reacting molecules (monomers).

12.16   Condensation reaction to form an ester:

$$CH_3 - \overset{\overset{O}{\|}}{C} - \boxed{OH \; + \; H}O - CH_2 - CH_3 \;\rightarrow\; CH_3 - \overset{\overset{O}{\|}}{C} - O - CH_2CH_3 \; + H_2O$$

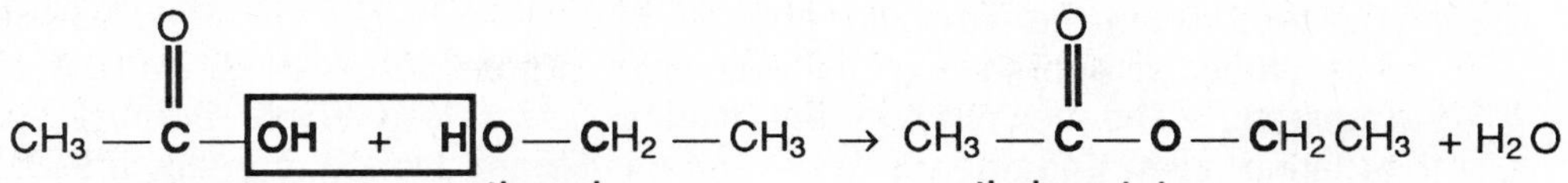

acetic acid   ethanol   ethyl acetate

If a dicarboxylic acid (two —COOH groups, usually at opposite ends of the molecule) and a dialcohol (two —OH groups, usually at opposite ends of the molecule) are combined, there is the potential for propagation of the polymer chain at both ends of both monomers. Polyethylene terephthalate (Table 12.1) is an example of a polyester formed from the monomers ethylene glycol and terephthalic acid.

HO — CH_2CH_2 — OH   HOOC— ◯ —COOH

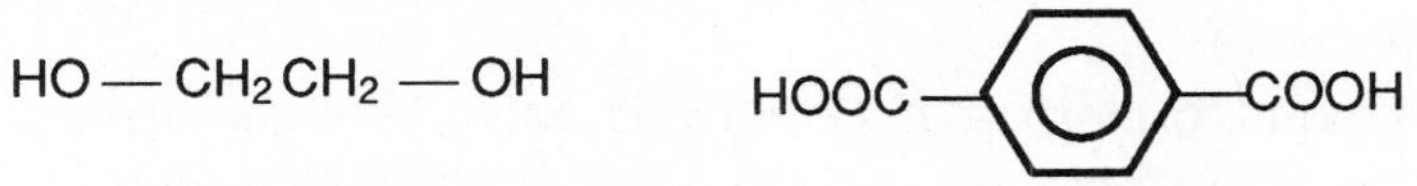

ethylene glycol   terephthalic acid

12.17

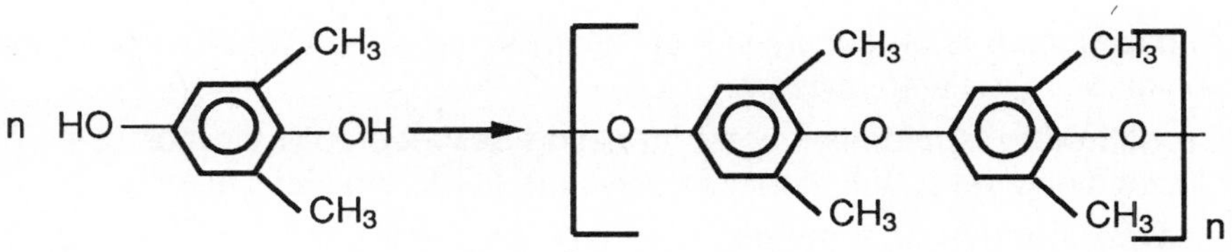

12.18   When nylon polymers are made, H_2O is produced as the C-N bonds are formed. Reversing this process (adding H_2O across the C-N bond), we see that the monomers used to produce Nomex are:

HOOC ◯ COOH   and H_2N ◯ NH_2

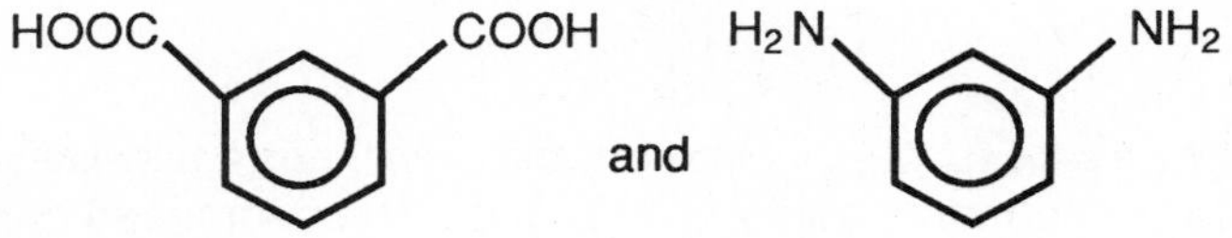

12.19   (a)   Increased crystallinity means increased ordering of polymer chains, which produces a harder material.

(b)   Crosslinking is the formation of chemical bonds between polymer chains, reducing flexibility of the molecular backbone and increasing hardness of the material.

(c)   Plasticizers reduce interactions between polymer chains, increasing the flexibility of the molecular backbone and decreasing hardness of the material.

(d)   Increasing strength of intermolecular interactions between chains has an effect similar to crosslinking; the hardness of the material increases.

(e)   Chain branching decreases interactions between polymer chains and decreases the hardness of the material.

12.20   (a)   Most of a polymer backbone is composed of σ bonds. The geometry around individual atoms is tetrahedral with bond angles of 109°, so the polymer is not flat, and there is relatively free rotation around the σ bonds. This flexibility of the molecular chains causes flexibility of the bulk material. Flexibility is enhanced by molecular features that inhibit order, such as branching, and diminished by features that encourage order, such as crosslinking or delocalized π electron density.

(b) Most polymers are amorphous; they lack long range order, because of the size and flexibility of the molecular chains. In order to hold a very long polymer chain in a particular orientation, relatively strong interactions throughout the chain are required. The flexibility in the chains themselves works against systematic ordering; even if the heads of several chains are aligned, the tails are likely to be moving in another direction.

12.21 (a) An **elastomer** is a polymer material that recovers its shape when released from a distorting force. A typical elastomeric polymer can be stretched to at least twice its original length and return to its original dimensions upon release.

(b) A **thermoplastic** material can be shaped and reshaped by application of heat and/or pressure.

(c) A **thermosetting** plastic can be shaped once, through chemical reaction in the shape-forming process, but cannot easily be reshaped, due to the presence of chemical bonds that crosslink the polymer chains.

12.22 (a) A **plasticizer** is a substance of relatively low molecular weight added to a polymeric material to soften it.

(b) **Crosslinking** is chemical bond formation between polymer molecules at various places along the chain. It makes polymers harder and less elastic.

(c) In a **condensation reaction**, two molecules are linked by formation of a new covalent bond and a small molecule, usually water, is "split out" as the new bond forms. Many polymers are formed by condensation reactions between appropriate monomers.

Ceramics

12.23 Engineering ceramics are resistant to heat, corrosion and wear, not easily deformed by stress and significantly less dense than traditional materials. The decreased density can reduce the total mass of the engineered product by 50 percent or more.

12.24 Structurally, polymers are formed from organic monomers held together by covalent bonds, whereas ceramics are formed from inorganic materials linked by ionic or highly polar covalent bonds. Ceramics are often stabilized by a three-dimensional bonding network, whereas in polymers, covalent bonds link atoms into a long, chain-like molecule with only weak interactions between molecules. Polymers are nearly always amorphous, and though ceramics can be amorphous, they are often crystalline.

In terms of physical properties, ceramics are generally much harder, more heat-stable and higher melting than polymers. (These properties are true in general for network solids relative to molecular solids, as described in Table 11.6.)

12.25 By analogy to the ZnS structure, the C atoms form a face-centered cubic array with Si atoms occupying alternate tetrahedral holes in the lattice. This means that the coordination numbers of both Si and C are 4; each Si is bound to 4 C atoms in a tetrahedral arrangement, and each C is bound to 4 Si atoms in a tetrahedral arrangement, producing an extended three-dimensional network. ZnS, an ionic solid, sublimes at 1185° and 1 atm pressure and melts at 1850° and 150 atm pressure. The considerably higher melting point of SiC, 2800° at 1 atm, indicates that SiC is probably not a purely ionic solid and that the Si-C bonding network has significant covalent character. This is reasonable, since the electronegativities of Si and C are similar (Figure 8.7). SiC is high-melting because a great deal of chemical energy is stored in the covalent Si-C bonds, and it is hard because the three-dimensional lattice resists any change that would weaken the Si-C bonding network.

12.26 (a) In zirconia, ZrO_2, the bonding is largely ionic. The strong interionic forces produce a rigid three-dimensional lattice that remains stable under the extreme temperature conditions of an engine or gas turbine.

(b) In $Mg_3Al_2(SiO_4)_3$, the ionic attractions are very strong because of the high charges on the metal ions (Mg^{2+} = 2+, Al^{3+} = 3+). This produces a rigid three-dimensional lattice and a very hard substance which is appropriate for use as an abrasive.

12.27 Ceramic materials typically have rigid three-dimensional lattices; movements of atoms with respect to one another require the breaking of chemical bonds. For this reason, most ceramics are not very flexible, especially compared to an organic polymer solid. As bonds in a ceramic are broken under great stress, the break tends to propagate itself by applying stress to adjacent atoms.

12.28 To increase the resistance of a ceramic material to mechanical failure, very pure, small (less than 1 micron in diameter) particles are produced (often by a sol-gel process), and then sintered to form the desired object. Also, ceramic fibers (the same molecular composition as the host or a different one) can be imbedded in the host ceramic lattice to toughen the material.

12.29 When obtained by adding water to a metal alkoxide (Equation 12.6), the metal hydroxide is produced in the form of a sol, a uniform suspension of very small particles. Beginning with solid metal hydroxide and adding the appropriate solvents offers little control over the particle size of the solid or the uniformity of the sol. The characteristics of the sol are very important in determining the ultimate purity, particle size and fracture resistance of the ceramic produced.

12.30 Sintering produces a solid composed of extremely small spheres and having uniform chemical composition throughout. The spheres are connected by covalent bonds formed by condensation reactions occurring at the surface of the spheres. Although it would be impossible to achieve perfect uniformity in three-dimensions over a large volume of solid, sintering improves uniformity by producing many bonding connections between each tiny sphere and its neighbors.

12.31 A composite is a material in which ceramic fibers or whiskers have been imbedded in a host ceramic matrix; the host and fibers may have the same or different compositions. The fibers absorb much of the microscopic stress that would otherwise initiate a fracture. Thus, a composite is much more crack resistant than the pure host ceramic.

12.32 The ceramic fibers of a composite have great strength along the fiber direction. By adsorbing most of an applied stress, the fiber reduces the load on the surrounding ceramic matrix, decreasing the chance that a crack will be initiated. Since fibers are distributed in all orientations in the composite, they reduce stress applied in any direction.

12.33 (a) A thermistor temperature sensor has a limited electrical conductivity that varies reproducibly with temperature, so that a particular temperature corresponds to a well defined electrical current. Temperature can be measured by measuring the current flow through the thermistor.

(b) Machine cutting tools take advantage of the hardness and fracture resistance offered by SiC reinforced Al_2O_3 to cut hard metals and alloys.

(c) SiO_2 is a structurally stable electrical insulator. It serves as an inert base for deposition of conducting or semi-conducting materials.

(d) SiC has the highest melting point of any ceramic material listed in Table 12.4. Its high melting point, resistance to corrosion and low density (producing a lighter part than a heat resistant metal alloy) make it a particularly appropriate material for an aircraft engine component.

12.34 The three most important characteristics are: (a) high thermal stability and high melting point; (b) low thermal conductivity and (c) low density. The tiles are specifically formulated to possess a porous structure with especially low density.

12.35 A superconducting material exhibits no resistance to the flow of an electrical current. A related property, known as the Meissner effect, is that all magnetic field lines are excluded from the volume of the material.

12.36 $YBa_2Cu_3O_7$ is a ceramic; nearly all superconducting materials known before Bednorz and Mueller's 1986 discovery were metals or intermetallic compounds. $YBa_2Cu_3O_7$ has a much higher superconducting transition temperature, T_c = 95 K, than previously known superconducting materials, for which the highest observed T_c was 23 K.

Thin Films

12.37 A "thin film" is a very thin, 0.1 to 300 μm coating of a material deposited on a substrate.

12.38 In general, a thin film should:
(a) be chemically stable in its working environment
(b) adhere to its substrate
(c) have a uniform thickness
(d) have an easily controllable composition
(e) be nearly free of imperfections.

12.39 A tantalum carbide thin film on a cutting tool should be heat, fracture and corrosion resistant, adhere well to the metal substrate of the tool, have a uniform thickness, be of precisely known composition and be free of imperfections or dislocations (which reduce mechanical strength).

12.40 A thin film of SnO_2 applied to a glass bottle increases the ease with which bottles slide by each other and decreases scratching and abrasion to the bottle.

12.41 There are three major methods of producing thin films.

(a) In **vacuum deposition**, a substance is vaporized or evaporated by heating under vacuum and then deposited on the desired substrate.

(b) In **sputtering**, ions accelerated to high energies by applying a high voltage are allowed to strike the target material, knocking atoms from its surface. These target material atoms are further accelerated toward the substrate, forming a thin film.

(c) In **chemical vapor deposition**, two gas phase substances react at the substrate surface to form a stable product which is deposited as a thin film.

12.42 (a) MgO, vacuum deposition or sputtering; (b) Teflon (an organic polymer with high thermal stability), sputtering; (c) Ti, vacuum deposition, sputtering or chemical vapor deposition (CVD).

Additional Exercises

12.43 In order to show liquid-crystalline behavior, a substance must be ordered in one or more directions. Molecules with strictly C-C backbones have so much flexibility that long range ordering is unlikely, even in one direction. Substances that exhibit liquid crystalline behavior have some structural feature, such as C=C, C=N, a benzene ring or a cholesterol ring system, to add rigidity to the molecule.

12.44

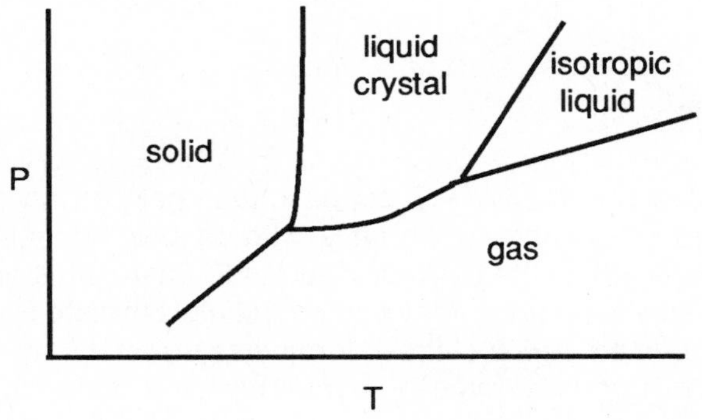

12.45 Application of an electric field in the direction perpendicular to the planes of the plates would cause the molecules to orient along the direction of the applied field, vertical in the diagram below.

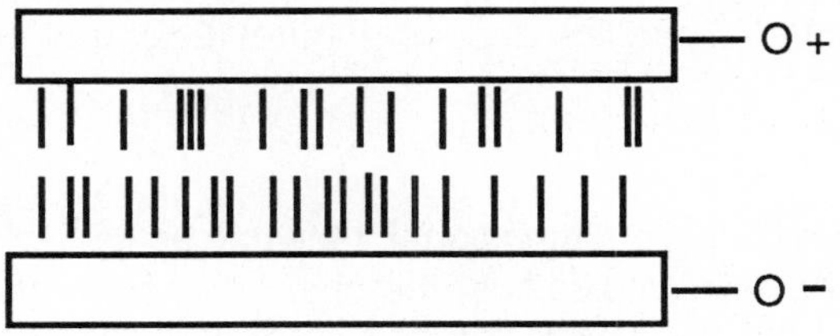

Because the molecules have thermal energy which causes them to move randomly, alignment parallel to the applied field is not complete, but some net reorientation does take place. This happens because the molecules in a nematic substance generally have a dipole moment (partial charge separation) oriented roughly parallel to the long direction of the molecule. Also, the molecules are more polarizable in their long direction, so London dispersion forces between molecules are stronger when the molecules are aligned in the direction of the applied field.

12.46 Hydrogen bonding is a much stronger intermolecular interaction than London dispersion forces (Section 11.2). The stronger the interactions between polymer chains, the more rigid the backbone and the harder the material. Thus, nylon is a harder material than polyethylene.

12.47

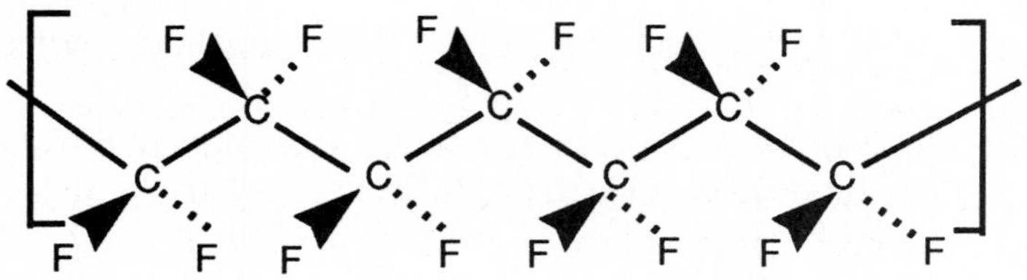

Teflon is formed by addition polymerization.

12.48 Polyethylene and polypropylene both have the same 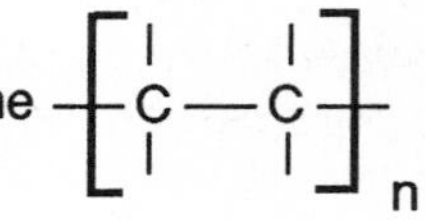 backbone,

but alternating C atoms in the polypropylene chain have methyl, $-CH_3$, substituents.

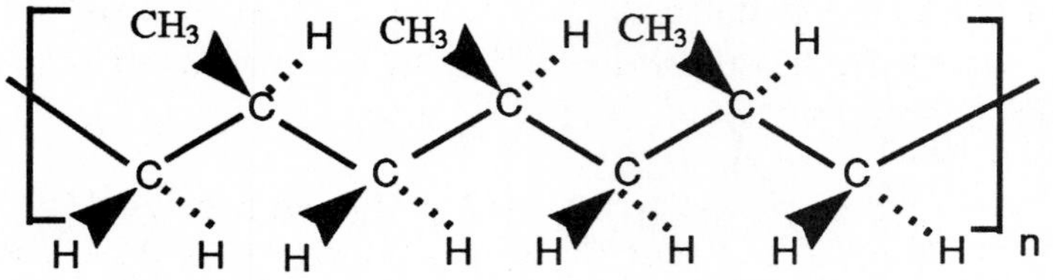

Because of the possibility of rotation about C — C bonds in the chain as the polymer is formed, the methyl groups can be oriented differently with respect to one another, depending on the conditions under which the polymer is formed. The methyl groups can be all on one side of the chain, as shown above; they can regularly alternate from front to back or they can be randomly oriented. Each of these forms has distinct properties. The regular structure shown has the highest degree of crystallinity and generally the best mechanical properties.

12.49 (a) Polymer (b) ceramic (c) ceramic (d) polymer (e) liquid crystal (an organic molecule with a characteristic long axis and the kinds of functional groups often found in compounds with liquid-crystalline phases (Figure 12.5); not enough repeating units to be a polymer)

12.50 Ceramics are usually three-dimensional network solids, whereas plastics most often consist of large, chain-like molecules (the chain may be branched) held loosely together by relatively weak van der Waals forces. Ceramics are rigid precisely because of the many strong bonding interactions intrinsic to the network. Once a crack forms, atoms near the defect are subject to great stress, and the crack is propagated. They are stable to high temperatures because tremendous kinetic energy (temperature) is required for an atom to break free from the bonding network. On the other hand, plastics are flexible because the molecules themselves are flexible (free rotation around the sigma bonds in the polymer chain), and it is easy for the molecules to move relative to one another (weak intermolecular forces). (However, recall that rigidity of the plastic increases as crosslinking of the polymer chains increases. The melamine-formaldehyde polymer in Figure 12.20 is a very rigid, brittle polymer.) Plastics are not thermally stable because their largely organic molecules are subject to oxidation and/or bond breaking at high temperatures.

12.51 For composites to demonstrate significantly greater structural strength and crack resistance than the pure host ceramic, the imbedded fibers must have a strong interaction with the matrix. If the fibers expand by a different amount upon heating, their interaction with the matrix will decrease (the host bonding network could be disrupted) and the ability of the fiber to withstand stress will not be transferred to the bulk composite material.

12.52 In ceramic material with a three-dimensional network bonding structure the surface atoms necessarily possess unfilled valence orbitals, or at least are bonded to interior atoms via higher energy bonds than normal. Often the otherwise unused orbitals on the surface are occupied by weak bonding to adsorbed molecules such as H_2O, O_2 or N_2.

In a composite, the interface between the fiber and the host matrix involves some form of bonding between the two sets of surface layer atoms. It could consist of covalent bond formation, for example between SiC fibers and SiC matrix, or possibly extensive hydrogen bonding involving surface OH groups, as in Al_2O_3 or SiO_2. In ionic materials, the two ionic lattices could intermingle at the surface to form ionic bonds.

When the composite is processed at high temperatures, the weaker bonding arrangements on the surfaces of the fibers and host ceramic are disrupted and new bonds are formed between the two surfaces. Thus, adsorbed water could be driven out or converted to -OH bonds by reaction on an oxide surface or surface bonding in each component could be rearranged to create possibilities for bonding between the two surfaces. Small and uniform particle size of the host ceramic material maximizes surface area and sites available for bonding between the two components.

12.53 (a)

$$\left[\begin{array}{c} CH_3 \\ | \\ -Si- \\ | \\ CH_3 \end{array} \right]_n \xrightarrow{400^\circ C} \left[\begin{array}{c} H \\ | \\ -Si-CH_2- \\ | \\ CH_3 \end{array} \right]_n$$

$$\left[\begin{array}{c} H \\ | \\ -Si-CH_2- \\ | \\ CH_3 \end{array} \right]_n \xrightarrow{1200^\circ C} [SiC]_n + CH_4(g) + H_2(g)$$

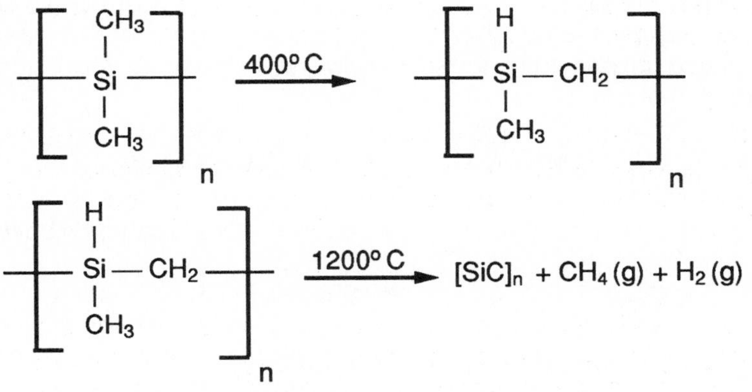

(b) $2NbBr_5(g) + 5H_2(g) \longrightarrow 2Nb(s) + 10HBr(g)$

(c) $SiCl_4(l) + C_2H_5OH(l) \longrightarrow Si(OC_2H_5)_4(s) + 4HCl(g)$

(d) 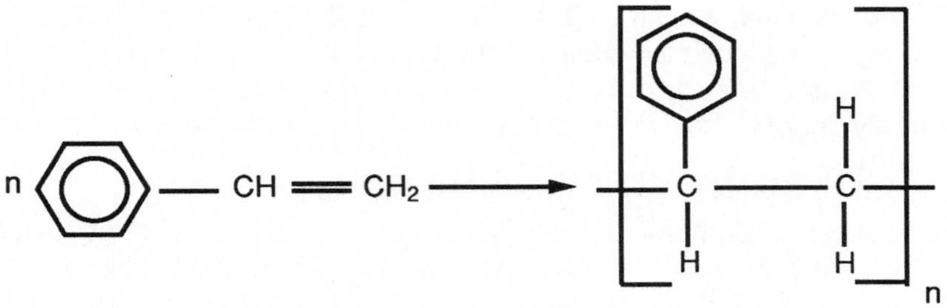

12.54 If the expected oxidation states on Y and Ba are +3 and +2, respectively, the average oxidation state of Cu is +2 1/3. That is, two Cu ions are in the +2 state and one is in the +3 state. Y^{3+} and Ba^{2+} have the stable electron configurations of their nearest nobles gases, while Cu has an incomplete d orbital set. Cu(II) is d^9, Cu(III) is d^8 and both have unpaired electrons. Although the mechanism by which the copper 3d electrons interact through the bridging oxygen atoms to form a superconducting state is still not clear, it is evident that the electronic structure of the copper ions is essential to the observed superconductivity.

12.55 (a) In liquid crystal displays, the glass plates which hold the liquid-crystalline material are coated with a thin film of electrically conducting material. Application of electrical current to certain areas of the display causes the liquid crystal molecules in these areas to change orientations and a character to appear on the display. Also, thin films of electrically conducting materials are used in photovoltaic devices such as solar cells.

(b) Metal oxide thin films are applied to optical lenses to reduce the amount of light reflected by the lense and as protection against wear.

(c) Very tough, hard ceramics such as tantalum carbide are coated on cutting tools.

12.56 These compounds cannot be vaporized without destroying their chemical identities. Anions with names ending in "ite" or "ate" are oxyanions with nonmetallic elements as central atoms. Under the conditions of vacuum deposition, high temperatures and low pressures, these anions tend to chemically decompose to form gaseous nonmetal oxides. For example:

$$MnSO_4(s) \longrightarrow MnO(s) + SO_3(g)$$

12.57 The formula of the compound deposited as a thin film is indicated by boldface type.

(a) $SiBr_4(g) + 2H_2(g) + CO_2(g) \longrightarrow$ **$SiO_2(s)$** $+ 4HBr(g) + 2CO(g)$

(b) $TiCl_4(g) + 2H_2O(g) \longrightarrow$ **$TiO_2(s)$** $+ 4HCl(g)$

(c) $GeH_4(g) \longrightarrow$ **Ge(s)** $+ 2H_2(g)$

The H_2 carrier gas dilutes the $GeH_4(g)$ so the reaction occurs more evenly and at a controlled rate; it does not participate in the reaction.

CHAPTER 13

Solutions

The Solution Process

13.1 In the solution process, solute-solute and solvent-solvent interactions are overcome and solute-solvent interactions are established. The solubility of a solute in a particular solvent thus depends on the strength of the "new" solute-solvent interactions relative to the strengths of "old" solute-solute and solvent-solvent interactions.

13.2 The overall energy change associated with dissolution depends on the relative magnitudes of the solute-solute, solvent-solvent, and solute-solvent interactions. If disrupting the "old" solute-solute and solvent-solvent interactions requires more energy than is released by the "new" solute-solvent interactions, the process is endothermic. When $NH_4NO_3(s)$ dissolves in water, disrupting the ion-ion interactions among NH_4^+ and NO_3^- ions of the solute and the hydrogen bonding among H_2O molecules of the solvent requires more energy than is released by subsequent ion-dipole and hydrogen bonding interactions among NH_4^+ and H_2O and NO_3^- and H_2O.

13.3 (a) ion-dipole forces (b) London dispersion forces
(c) hydrogen bonding (d) dipole-dipole forces

13.4 (a) hydrogen bonding (b) London dispersion forces
(c) ion-dipole forces (d) London dispersion forces

13.5 A solute is more likely to dissolve in a solvent if the strengths of the solute-solute, solvent-solvent and solute-solvent interactions are similar. If solvent-solute interactions are significantly weaker than solute-solute or solvent-solvent interactions, mixing will not occur.

13.6 (a) CH_3OH is more soluble in H_2O because both have intermolecular hydrogen bonding interactions. CH_3CH_3 is nonpolar and does not have strong intermolecular interactions with H_2O.

(b) C_2H_5OH is more soluble in H_2O because both have intermolecular hydrogen bonding interactions. CH_3OCH_3 is much less polar than C_2H_5OH and does not experience hydrogen bonding.

(c) Ionic $CaCl_2$ is more soluble because ion-dipole solute-solvent interactions are more similar to ionic solute-solute and hydrogen-bonding solvent-solvent interactions than the weak dispersion forces between CCl_4 and H_2O.

13.7 (a) There are no solute-solvent interactions between ionic KCl and nonpolar covalent C_6H_6 that are strong enough to compete with the ionic lattice forces in KCl.

(b) $Br_2(l)$, $I_2(s)$ and $CCl_4(l)$ are all nonpolar covalent molecules that experience London dispersion forces. The strength of dispersion forces increases with molecular weight, so $Br_2(l)$ (MW = 159.8 amu) experiences forces more similar to those in CCl_4 (MW = 153.8 amu) than I_2 (MW = 253.8 amu) does.

13.8 (a) Dispersion forces between C_6H_{12} and CH_3OH cannot compete with hydrogen bonding interactions among CH_3OH molecules; solute-solvent forces are not strong enough to overcome solute-solute interactions.

(b) Although the polar ends of ethanol molecules undergo hydrogen bonding, the essentially nonpolar ethyl ($-C_2H_5$) group is compatible with nonpolar or slightly polar solutes such as $CHCl_3$. The extensive hydrogen bonding (solvent-solvent interactions) in water makes it a poor solvent for solute molecules where van der Waals forces are predominant.

13.9 (a) ΔH_{soln} is determined by the relative magnitudes of the "old" solute-solute (ΔH_1) and solvent-solvent (ΔH_2) interactions and the new solute-solvent interactions (ΔH_3); $\Delta H_{soln} = \Delta H_1 + \Delta H_2 + \Delta H_3$. Since the solute and solvent in this case experience very similar London dispersion forces, the energy required to separate them individually and the energy released when they are mixed are approximately equal. $\Delta H_1 + \Delta H_2 \approx -\Delta H_3$. Thus, ΔH_{soln} is nearly zero.

(b) Mixing hexane and heptane produces a homogeneous solution from two pure substances, and the randomness of the system increases. Since no strong intermolecular forces prevent the molecules from mixing, they do so spontaneously due to the increase in disorder.

13.10 KBr is quite soluble in water because of the sizeable increase in disorder of the system (ordered KBr lattice $\rightarrow$ freely moving hydrated ions) associated with the dissolving process. An increase in disorder or randomness in a process tends to make that process spontaneous.

Concentrations of Solutions

13.11 (a) $$\text{mass \%} = \frac{\text{mass solute}}{\text{total mass solution}} \times 100 = \frac{10.5\text{ g Na}_2\text{SO}_4}{10.5\text{ g Na}_2\text{SO}_4 + 175\text{ g H}_2\text{O}} \times 100$$

$$= 5.66\%$$

(b) $$\text{ppm} = \frac{\text{mass solute}}{\text{total mass solution}} \times 10^6 \;;\; \frac{5.0\text{ g gold}}{1\text{ ton ore}} \times \frac{1\text{ ton}}{2000\text{ lb}} \times \frac{1\text{ lb}}{454\text{ g}} \times 10^6 = 5.5\text{ ppm}$$

13.12 (a) $$\text{mass \%} = \frac{\text{mass solute}}{\text{total mass solution}} \times 100$$

$$\text{mass solute} = 0.050\text{ mol I}_2 \times \frac{253.8\text{ g I}_2}{1\text{ mol I}_2} = 13\text{ g I}_2$$

$$\text{mass \% I}_2 = \frac{13\text{ g I}_2}{13\text{ g I}_2 + 50.0\text{ g CCl}_4} \times 100 = 21\%\text{ I}_2$$

(b) $$\text{ppm} = \frac{\text{mass solute}}{\text{total mass solution}} \times 10^6 = \frac{0.412\text{ g Ca}^{2+}}{1 \times 10^3\text{ gH}_2\text{O}} \times 10^6 = 412\text{ ppm Ca}^{2+}$$

13.13 (a) $$6.00\text{ g CH}_3\text{OH} \times \frac{1\text{ mol CH}_3\text{OH}}{32.0\text{ g CH}_3\text{OH}} = 0.188\text{ mol CH}_3\text{OH}$$

$$480\text{ g H}_2\text{O} \times \frac{1\text{ mol H}_2\text{O}}{18.0\text{ g H}_2\text{O}} = 26.7\text{ mol H}_2\text{O}$$

$$\chi_{\text{CH}_3\text{OH}} = \frac{0.188}{0.188 + 26.7} = 6.98 \times 10^{-3}$$

(b) $$\frac{4.13\text{ g CH}_3\text{OH}}{32.0\text{ g/mol}} = 0.129\text{ mol CH}_3\text{OH} \;;\; \frac{48.6\text{ g CCl}_4}{154\text{ g/mol}} = 0.316\text{ mol CCl}_4$$

$$\chi_{\text{CH}_3\text{OH}} = \frac{0.129}{0.129 + 0.316} = 0.290$$

13.14 (a) $$\frac{120\text{ g C}_2\text{H}_6\text{O}_2}{62.1\text{ g/mol}} = 1.93\text{ mol C}_2\text{H}_6\text{O}_2 \;;\; \frac{120\text{ g H}_2\text{O}}{18.0\text{ g/mol}} = 6.66\text{ mol H}_2\text{O}$$

$$\chi_{\text{C}_2\text{H}_6\text{O}_2} = \frac{1.93}{1.93 + 6.66} = 0.225$$

(b) $$1.93\text{ mol C}_2\text{H}_6\text{O}_2 \;;\; \frac{1.20 \times 10^3\text{ g C}_3\text{H}_6\text{O}}{58.1\text{ g/mol}} = 20.7\text{ mol C}_3\text{H}_6\text{O}$$

$$\chi_{\text{C}_2\text{H}_6\text{O}_2} = \frac{1.93}{1.93 + 20.7} = 0.0853$$

13.15 (a) $$M = \frac{\text{mol solute}}{\text{L soln}} \;;\; \frac{10.5\text{ g NaOH}}{0.350\text{ L soln}} \times \frac{1\text{ mol NaOH}}{40.00\text{ g NaOH}} = 0.750\ M$$

(b) $$\frac{40.7\text{ g LiClO}_4 \cdot 3\text{H}_2\text{O}}{0.125\text{ L soln}} \times \frac{1\text{ mol LiClO}_4 \cdot 3\text{H}_2\text{O}}{160.4\text{ g LiClO}_4 \cdot 3\text{H}_2\text{O}} = 2.03\ M$$

(c) $M_c \times L_c = M_d \times L_d$; $1.50\ M\ HNO_3 \times 0.0400\ L = ?M\ HNO_3 \times 0.500\ L$

500 mL of 0.120 M HNO_3

13.16 (a) $M = \frac{\text{mol solute}}{\text{L soln}}$; $\frac{25.0\ g\ MgI_2}{0.200\ L\ soln} \times \frac{1\ mol\ MgI_2}{278.1\ g\ MgI_2} = 0.0449\ M$

(b) $\frac{3.40\ g\ Cr(NO_3)_3 \cdot 9H_2O}{0.0750\ L\ soln} \times \frac{1\ mol\ Cr(NO_3)_3 \cdot 9H_2O}{400.2\ g\ Cr(NO_3)_3 \cdot 9H_2O} = 0.113\ M$

(c) $M_c \times L_c = M_d \times L_d$; $6.00\ M\ H_2SO_4 \times 0.0750\ L = ?\ M\ H_2SO_4 \times 1.00\ L$
1.00 L of 0.450 M H_2SO_4

13.17 (a) $m = \frac{\text{moles solute}}{\text{kg solvent}}$; $\frac{13.0\ g\ C_6H_6}{17.0\ g\ CCl_4} \times \frac{1\ mol\ C_6H_6}{78.11\ g\ C_6H_3} \times \frac{1000\ g\ CCl_4}{1\ kg\ CCl_4} = 9.79\ m\ C_6H_6$

(b) $\frac{5.85\ g\ NaCl}{0.250\ L\ H_2O} \times \frac{1\ mol\ NaCl}{58.44\ g\ NaCl} \times \frac{1\ L\ H_2O}{0.997\ kg\ H_2O} = 0.402\ M\ NaCl$

13.18 (a) $m = \frac{\text{moles solute}}{\text{kg solvent}}$; $\frac{2.1\ g\ S_8}{0.095\ kg\ C_{10}H_8} \times \frac{1\ mol\ S_8}{256.5\ g\ S_8} = 0.086\ m\ S_8$

(b) $\frac{1.50\ mol\ NaCl}{15.0\ mol\ H_2O} \times \frac{1\ mol\ H_2O}{18.0\ g\ H_2O} \times \frac{1000\ g}{1\ kg} = 5.56\ m\ NaCl$

13.19 Given: 100.0 mL of $CH_3CN(l)$, 0.786 g/mL; 20.0 mL CH_3OH, 0.791 g/mL

(a) $mol\ CH_3CN = \frac{0.786\ g}{1\ mL} \times 100.0\ mL \times \frac{1\ mol\ CH_3CN}{41.05\ g\ CH_3CN} = 1.91\ mol$

$mol\ CH_3OH = \frac{0.791\ g}{1\ mL} \times 20.0\ mL \times \frac{1\ mol\ CH_3OH}{32.04\ g\ CH_3OH} = 0.494\ mol$

$\chi_{CH_3OH} = \frac{0.494\ mol\ CH_3OH}{1.91\ mol\ CH_3CN + 0.494\ mol\ CH_3OH} = 0.205$

(b) Assuming CH_3OH is the solute and CH_3CN is the solvent,

$100.0\ mL\ CH_3CN \times \frac{0.786\ g}{1\ mL} \times \frac{1\ kg}{1000\ g} = 0.0786\ kg\ CH_3CN$

$m_{CH_3OH} = \frac{0.494\ mol\ CH_3OH}{0.0786\ kg\ CH_3CN} = 6.28\ m\ CH_3OH$

(c) The total volume of the solution is 120.0 mL, assuming volumes are additive.

$M = \frac{0.494\ mol\ CH_3OH}{0.120\ L\ solution} = 4.12\ M\ CH_3OH$

13.20 Given: 15.0 g C_4H_4S, 1.065 g/mL; 250.0 mL C_7H_8, 0.867 g/mL

(a) $\text{mol } C_4H_4S = 15.0 \text{ g } C_4H_4S \times \frac{1 \text{ mol } C_4H_4S}{84.15 \text{ g } C_4H_4S} = 0.178 \text{ mol } C_4H_4S$

$\text{mol } C_7H_8 = \frac{0.867 \text{ g}}{1 \text{ mL}} \times 250.0 \text{ mL} \times \frac{1 \text{ mol } C_7H_8}{92.14 \text{ g } C_7H_8} = 2.35 \text{ mol}$

$\chi_{C_4H_4S} = \frac{0.178 \text{ mol } C_4H_4S}{0.178 \text{ mol } C_4H_4S + 2.35 \text{ mol } C_7H_8} = 0.0704$

(b) $m_{C_4H_4S} = \frac{\text{mol } C_4H_4S}{\text{kg } C_7H_8}$; $250.0 \text{ mL} \times \frac{0.867 \text{ g}}{1 \text{ mL}} \times \frac{1 \text{ kg}}{1000 \text{ g}} = 0.217 \text{ kg } C_7H_8$

$m_{C_4H_4S} = \frac{0.178 \text{ mol } C_4H_4S}{0.217 \text{ kg } C_7H_8} = 0.820\ m\ C_4H_4S$

(c) $15.0 \text{ g } C_4H_4S \times \frac{1 \text{ mL}}{1.065 \text{ g}} = 14.1 \text{ mL } C_4H_4S$;

$V_{soln} = 14.1 \text{ mL } C_4H_4S + 250.0 \text{ mL } C_7H_8 = 264.1 \text{ mL}$

$M_{C_4H_4S} = \frac{0.178 \text{ mol } C_4H_4S}{0.2641 \text{ L soln}} = 0.674\ M\ C_4H_4S$

13.21 (a) $\text{weight \% } H_2O = \frac{46.0 \text{ g } H_2O}{46.0 \text{ g } H_2O + 66.0 \text{ g } C_3H_6O} \times 100 = 41.1\%$

(b) $\frac{46.0 \text{ g } H_2O}{18.02 \text{ g/mol}} = 2.55 \text{ mol } H_2O$; $\frac{66.0 \text{ g } C_3H_6O}{58.08 \text{ g/mol}} = 1.14 \text{ mol } C_3H_6O$

$\chi_{H_2O} = \frac{2.55}{2.55 + 1.14} = 0.691$

(c) $m \text{ of } H_2O = \frac{2.55 \text{ mol } H_2O}{0.0660 \text{ kg } C_3H_6O} = 38.6\ m\ H_2O$

(d) $M \text{ of } H_2O = \frac{2.55 \text{ mol } H_2O}{(46.0 + 66.0) \text{ g soln}} \times \frac{0.926 \text{ g soln}}{1 \text{ mL soln}} \times \frac{1000 \text{ mL}}{1 \text{ L}} = 21.1\ M$

13.22 (a) $\frac{571.6 \text{ g } H_2SO_4}{1 \text{ L soln}} \times \frac{1 \text{ L soln}}{1329 \text{ g soln}} = \frac{0.430 \text{ g } H_2SO_4}{1 \text{ g soln}}$

Weight percentage is thus $0.430 \times 100 = 43.0\%$.

(b) In a liter of solution there are 1329 - 572 = 757 g H_2O.

$$\frac{571.6 \text{ g } H_2SO_4}{98.1 \text{ g } 1 \text{ mol}} = 5.83 \text{ mol } H_2SO_4 \;;\; \frac{757 \text{ g } H_2O}{18.0 \text{ g/mol}} = 42.0 \text{ mol } H_2O$$

$$\chi_{H_2SO_4} = \frac{5.83}{42.0 + 5.83} = 0.122$$

$$\text{molality} = \frac{5.83 \text{ mol } H_2SO_4}{0.757 \text{ kg } H_2O} = 7.60\ m$$

$$\text{molarity} = \frac{5.83 \text{ mol } H_2SO_4}{1 \text{ L soln}} = 5.83\ M$$

13.23 Assume 100 g of solution

100 g soln × 69% HNO_3 = 69 g HNO_3 + 31 g H_2O

$$M = \frac{69 \text{ g } HNO_3}{100 \text{ g soln}} \times \frac{1.42 \text{ g soln}}{1 \text{ mL soln}} \times \frac{1 \text{ mol } HNO_3}{63.0 \text{ g } HNO_3} \times \frac{1000 \text{ mL}}{1 \text{ L}} = 16\ M$$

(2 significant figures)

13.24 100 g solution contains 28 g NH_3 + 72 g H_2O

$$m = \frac{28 \text{ g } NH_3}{100 \text{ g soln}} \times \frac{0.90 \text{ g}}{1 \text{ cm}^3} \times \frac{1000 \text{ cm}^3}{1 \text{ L}} \times \frac{1 \text{ mol } NH_3}{17.0 \text{ g } NH_3} = 15\ m$$

13.25 (a) $\frac{0.122 \text{ mol } MgCl_2}{1 \text{ L soln}} \times 0.214 \text{ L} = 2.61 \times 10^{-2} \text{ mol } MgCl_2$

(b) $\frac{2.00 \text{ mol HCl}}{1 \text{ L soln}} \times 3.50\ \mu\text{L} \times \frac{1 \times 10^{-6} \text{ L}}{1\ \mu\text{L}} = 7.00 \times 10^{-6} \text{ mol HCl}$

(c) $\frac{0.0455 \text{ mol } (NH_4)_2CrO_4}{1 \text{ L soln}} \times 0.156 \text{ L} = 7.10 \times 10^{-3} \text{ mol } (NH_4)_2CrO_4$

13.26 (a) $60.0 \text{ g solution} \times \frac{0.0125 \text{ g KI}}{1 \text{ g soln}} \times \frac{1 \text{ mol KI}}{166 \text{ g KI}} = 4.52 \times 10^{-3} \text{ mol KI}$

(b) $250 \text{ g solution} \times \frac{0.00460 \text{ g NaCl}}{1 \text{ g soln}} \times \frac{1 \text{ mol NaCl}}{58.4 \text{ g NaCl}} = 1.97 \times 10^{-2} \text{ mol NaCl}$

(c) $\frac{1.25 \text{ mol } H_2SO_4}{1 \text{ L}} \times 0.600 \text{ L} = 0.750 \text{ mol } H_2SO_4$

13.27 (a) $\frac{1.50 \times 10^{-2} \text{ mol KBr}}{1 \text{ L soln}} \times 1.40 \text{ L} \times \frac{119 \text{ g KBr}}{1 \text{ mol KBr}} = 2.50 \text{ g KBr}$

Weigh out this much KBr, dissolve in water, dilute with stirring to 1.40 L.

(b) Determine the mass percent of KBr:

$$\frac{0.400 \text{ mol KBr}}{1000 \text{ g } H_2O} \times \frac{119 \text{ g KBr}}{1 \text{ mol KBr}} = \frac{47.6 \text{ g KBr}}{1000 \text{ g } H_2O}$$

Thus, mass fraction $= \frac{47.6 \text{ g KBr}}{1000 + 47.6} = 0.0454$

In 250 g of the 0.400 *m* solution, there are

$$(250 \text{ g soln}) \times \frac{0.0454 \text{ g KBr}}{1 \text{ g soln}} = 11.4 \text{ g KBr}$$

Weigh out 11.4 g KBr, dissolve it in 250 - 11.4 = 238.6 g H_2O to make exactly 250 g of 0.420 *m* solution.

(c) Calculate the total mass of 1.50 L of solution, and from the weight % of KBr, the mass of KBr required.

$$1.50 \text{ L soln} \times \frac{1000 \text{ mL}}{1 \text{ L}} \times \frac{1.10 \text{ g soln}}{1 \text{ mL}} = 1650 \text{ g soln}$$

0.120(1650 g soln) = 198 g KBr

Weigh 198 g KBr and dissolve in (1650 - 198) = 1452 g H_2O.

Strictly speaking, the mass of H_2O should be given to 3 significant figures, so

1.45×10^3 g H_2O or $1.45 \times 10^3 \text{ g } H_2O \times \frac{1.00 \text{ mL}}{0.997 \text{ g } H_2O} = 1.45 \times 10^3 \text{ mL } H_2O$

(d) Calculate moles KBr needed to precipitate 21.0 g AgBr.

$$21.0 \text{ g AgBr} \times \frac{1 \text{ mol AgBr}}{188 \text{ g AgBr}} \times \frac{1 \text{ mol KBr}}{1 \text{ mol AgBr}} = 0.112 \text{ mol KBr}$$

$$0.112 \text{ mol KBr} \times \frac{1 \text{ L soln}}{0.200 \text{ mol KBr}} = 0.560 \text{ L soln}$$

Weight out 0.112 mol KBr or 13.3 g KBr, dissolve it in a small amount of water and dilute to 0.560 L.

13.28 (a) $\frac{0.200 \text{ mol } Na_2CO_3}{1 \text{ L soln}} \times 0.500 \text{ L} \times \frac{106.0 \text{ g } Na_2CO_3}{1 \text{ mol } Na_2CO_3} = 10.6 \text{ g } Na_2CO_3$

Weigh 10.8 g Na_2CO_3, place it in a 500 mL volumetric flask, dissolve in a small amount of water, continue adding water with thorough mixing up to the mark on the neck of the flask.

(b) Determine the mass % of $(NH_4)_2SO_4$ in the solution:

$$\frac{1.00\text{ mol }(NH_4)_2SO_4}{1000\text{ g }H_2O} \times \frac{132\text{ g }(NH_4)_2SO_4}{1\text{ mol }(NH_4)_2SO_4} = \frac{132\text{ g }(NH_4)_2SO_4}{1000\text{ g }H_2O}$$

$$\text{mass \%} = \frac{132\text{ g }(NH_4)_2SO_4}{1000\text{ g }H_2O + 132\text{ g }(NH_4)_2SO_4} \times 100 = 11.7\%$$

In 150 g of solution, there are 0.117(150) = 17.6 g $(NH_4)_2SO_4$.

Weigh out 17.6 g $(NH_4)_2SO_4$ and dissolve it in 150 - 17.6 = 132.4 g H_2O to make exactly 150 g of solution.

(132.4 g $H_2O \times$ 0.997 g H_2O/mL @ 25°C = 132.0 mL H_2O)

(c) $1.50\text{ L} \times \frac{1000\text{ mL}}{1\text{ L}} \times \frac{1.20\text{ g}}{1\text{ mL}} = 1800\text{ g solution}$ 0.200(1800 g soln) = 360 g $Pb(NO_3)_2$

Weigh 360 g $Pb(NO_3)$ and add (1800 - 360) = 1440 g H_2O to make exactly 1800 g or 1.50 L of solution.

$$\left(1440\text{ g }H_2O \times \frac{1\text{ mL }H_2O}{0.997\text{ g }H_2O} = 1444\text{ mL }H_2O\right)$$

(d) Calculate the mol HCl necessary to neutralize 5.0 g $Ba(OH)_2$.

$Ba(OH)_2(g) + 2HCl(aq) \rightarrow BaCl_2(aq) + 2H_2O(l)$

$$5.0\text{ g }Ba(OH)_2 \times \frac{1\text{ mol }Ba(OH)_2}{171\text{ g }Ba(OH)_2} \times \frac{2\text{ mol HCl}}{1\text{ mol }Ba(OH)_2} = 0.0585\text{ mol HCl}$$

$$M = \frac{\text{mol}}{\text{L}}\ ;\ \text{L} = \frac{\text{mol}}{M} = \frac{0.0585\text{ mol HCl}}{0.50\ M\text{ HCL}} = 0.120\text{ L} = 120\text{ mL}$$

120 mL of 0.50 M HCl are needed.

$M_c \times L_c = M_d \times L_d$; $6.0\ M \times L_c = 0.50\ M \times 0.12$ L ; $L_c = 0.010$ L = 10 mL

Using a pipette, measure exactly 10 mL of 6.0 M HCl and dilute with water to a total volume of 120 mL.

13.29 (a) In the process $Zn^{2+}(aq) + 2e^- \rightarrow Zn(s)$, two moles of electrons are transferred for each mole of Zn^{2+}; one mole of Zn^{2+} = 2 equivalents of Zn^{2+}.

$$4.50\text{ g }ZnI_2 \times \frac{1\text{ mol }ZnI_2}{319.19\text{ g }ZnI_2} \times \frac{2\text{ equivalents }ZnI_2}{1\text{ mol }ZnI_2} = 0.0282\text{ equivalents }ZnI_2$$

$$N = \frac{\text{equivalents }ZnI_2}{\text{L soln}} = \frac{0.0282\text{ equivalents }ZnI_2}{0.100\text{ L}} = 0.282\ N$$

(b) If 3 H's react, 1 mol H_3PO_4 = 3 equivalents H_3PO_4,

$$1\,M = \frac{1\text{ mol}}{1\text{ L}} = \frac{3\text{ equiv.}}{1\text{ L}};\ 1M = 3N \text{ or } M = \frac{N}{3}$$

$$M = \frac{0.135\,N}{3} = 0.0450\,M\ H_3PO_4$$

13.30 (a) $Sn^{2+} \rightarrow Sn^{4+} + 2e^-$; one mole of Sn^{2+} yields two moles of electrons; one mole of Sn^{2+} = two equivalents of Sn^{2+}

mol $SnBr_2$ = $M \times$ L = 0.0500 $M \times$ 0.0100 L = 5.00×10^{-4} mol Sn^{2+}

$$5.00 \times 10^{-4}\text{ mol } Sn^{2+} \times \frac{2\text{ equiv. } Sn^{2+}}{1\text{ mol } Sn^{2+}} = 1.00 \times 10^{-3}\text{ equivalents } Sn^{2+}$$

(b) $MnO_4^- + 5\,e^- \rightarrow Mn^{2+}$; 1 mol MnO_4^- = 5 equiv. MnO_4^-

$$0.500\text{ g } KMnO_4 \times \frac{1\text{ mol } KMnO_4}{158.0\text{ g } KMnO_4} \times \frac{5\text{ equivalents } KMnO_4}{1\text{ mol } KMnO_4} \times \frac{1}{0.100\text{ L}} = 0.158\,N\ KMnO_4$$

Saturated Solutions; Factors Affecting Solubility

13.31 (a) A saturated solution is one that is in equilibrium with undissolved solute. If additional solute is added to a saturated solution, it will not dissolve.

(b) A supersaturated solution contains more dissolved solute than is needed to form a saturated solution.

(c) Supersaturated solutions can be prepared by saturating a solution at high temperature and then carefully cooling the solution. Supersaturated solutions exist because not enough solute molecules are aligned properly for crystallization to occur. Addition of a seed crystal usually initiates crystallization and returns the solution to a saturated state.

13.32 (a) Yes (b) no

(c) Upon slow cooling, the solution has become supersaturated; not enough Na^+ and $C_2H_3O_2^-$ ions aligned themselves so that $NaC_2H_3O_2(s)$ formed when the solution reached the temperature of saturation.

13.33 (a) Not saturated (b) saturated (c) saturated (d) not saturated

13.34 (a) Not saturated (b) saturated (c) not saturated (d) saturated

13.35 Most ionic compounds have endothermic heats of solution, so their solubility increases with increasing temperature. The solubility of gases **decreases** with increasing temperature. (As the temperature increases, the average kinetic energy of the gas molecules increases, and they are more likely to escape from solution.)

13.36 The solubilities of liquids and solids are not greatly affected by external pressure. For gases, Henry's Law states that the concentration of a gas in solution is directly proportional to the partial pressure of that gas above the solution. That is, the greater the pressure of the gas above the solution, the greater the solubility of the gas.

13.37 $C_{He} = 3.7 \times 10^{-4}\ M/\text{atm} \times 2.5\ \text{atm} = 9.2 \times 10^{-4}\ M$

$C_{N_2} = 6.0 \times 10^{-4}\ M/\text{atm} \times 2.5\ \text{atm} = 1.5 \times 10^{-3}\ M$

13.38 $\frac{1.38 \times 10^{-3}\ M}{1\ \text{atm}\ O_2\ \text{pressure}} \times 0.21\ \text{atm}\ O_2\ \text{pressure} = 2.9 \times 10^{-4}\ M$

Colligative Properties

13.39 The total number of particles in a solution, or concentration, determines its colligative properties.

13.40 Colligative properties: (b) vapor-pressure lowering (c) osmotic pressure (e) boiling point elevation

13.41 For a solution with a nonvolatile solute, the vapor pressure of the solution is equal to the mole fraction of the *solvent* times the vapor pressure of the pure solvent.

13.42 (a) An ideal solution is one that obeys Raoult's Law. Solutions are most ideal when the solute and solvent have similar structures and intermolecular interactions.

(b) Comparing the structures of the solute and solvent gives a good indication of the ideality of the solution. Quantitatively, the vapor pressure of the solution could be measured and compared to the vapor pressure calculated from Raoult's law. The ratio P/P° should be constant for any value of $\chi_{solvent}$.

13.43 For this problem, it will be convenient to express Raoult's law in terms of the lowering of the vapor pressure of the solvent, ΔP_A.

$$\Delta P_A = P_A^\circ - \chi_A P_A^\circ = P_A^\circ (1 - \chi_A)$$

$1 - \chi_A = \chi_B$, the mole fraction of *solute* particles

$\Delta P_A^\circ = \chi_B P_A^\circ$; the vapor presure of the solvent (A) is lowered according to the mole fraction of solute (B) particles present.

(a) Calculate χ_B by vapor pressure lowering; $\chi_B = \Delta P_A/P_A^\circ$. Given moles solvent, calculate moles solute from the definition of mole fraction.

$$\chi_{C_2H_6O_2} = \frac{8.1\ \text{torr}}{100\ \text{torr}} = 0.081$$

$$\frac{1.00 \times 10^3\ \text{g}\ C_2H_5OH}{46.1\ \text{g/mol}} = 21.7\ \text{mol}\ C_2H_5OH; \quad \text{let } y = \text{mol}\ C_2H_6O_2$$

$$\chi_{C_2H_6O_2} = \frac{y \text{ mol } C_2H_6O_2}{y \text{ mol } C_2H_6O_2 + 21.7 \text{ mol } C_2H_5OH} = 0.081 = \frac{y}{y + 21.7}$$

$0.081\,y + 1.758 = y;\ \ 0.919\,y = 1.758;\ \ y = 1.91 \text{ mol } C_2H_6O_2$

$$1.91 \text{ mol } C_2H_6O_2 \times \frac{62.1 \text{ g}}{1 \text{ mol}} = 119 \text{ g } C_2H_6O_2$$

(b) $\Delta P_A = \chi_B P^o_A$; the vapor pressure of solvent is lowered according to the mole fraction of solute particles present.

P_{H_2O} at 50° C = 92.5 torr; $\frac{500 \text{ g } H_2O}{18.0 \text{ g/mol}} = 27.8 \text{ mol } H_2O$

$$\chi_{ions} = \frac{6.50 \text{ torr}}{92.5 \text{ torr}} = \frac{y \text{ mol ions}}{y \text{ mol ions} + 27.8 \text{ mol } H_2O}$$

$0.0703 = \frac{y}{y + 27.8}$; $0\ .0703\,y + 1.95 = y$; $0.930\,y = 1.95$, $y = 2.10$ mol of ions

$$2.10 \text{ mol ions} \times \frac{1 \text{ mol KBr}}{2 \text{ mol ions}} \times \frac{119 \text{ g KBr}}{\text{mol KBr}} = 125 \text{ g KBr}$$

13.44 H_2O vapor pressure will be determined by the mole fraction of H_2O in the solution. The vapor pressure of pure H_2O at 338 K = 187.5 mm Hg.

(a) $\frac{10.00 \text{ g } C_{12}H_{22}O_{11}}{342.3 \text{ g/mol}} = 0.02921 \text{ mol};\ \ \frac{82.0 \text{ g } H_2O}{18.0 \text{ g/mol}} = 4.55 \text{ mol}$

$$P_{H_2O} = \chi_{H_2O} P^o_{H_2O} = \frac{4.55 \text{ mol } H_2O}{4.55 + 0.02921} \times 187.5 \text{ torr} = 186 \text{ torr}$$

(b) $\frac{5.00 \text{ g } Na_2SO_4}{142 \text{ g } Na_2SO_4\text{/mol}} \times \frac{3 \text{ mol ions}}{1 \text{ mol } Na_2SO_4} = 0.106 \text{ mol ions} \times \frac{115 \text{ g } H_2O}{18.0 \text{ g/mol}} = 6.39 \text{ mol } H_2O$

$$P_{H_2O} = \frac{6.39 \text{ mol } H_2O}{6.39 + 0.106} \times 187.5 \text{ torr} = 184 \text{ torr}$$

(c) $\frac{32.5 \text{ g } C_3H_8O_3}{92.10 \text{ g/mol}} = 0.353 \text{ mol};\ \ \frac{120 \text{ g } H_2O}{18.0 \text{ g/mol}} = 6.67 \text{ mol}$

$$P_{H_2O} = \frac{6.67 \text{ mol } H_2O}{6.67 + 0.353} \times 187.5 \text{ torr} = 178 \text{ torr}$$

13.45 At 63.5°C, $P^o_{H_2O} = 175$ torr, $P^o_{Eth} = 400$ torr.

Let G = the mass of H_2O and C_2H_5OH combined.

(a) $$\chi_{Eth} = \frac{\dfrac{G}{46.07 \text{ g } C_2H_5OH}}{\dfrac{G}{46.07 \text{ g } C_2H_5OH} + \dfrac{G}{18.02 \text{ g } H_2O}}$$

Multiplying top and bottom of the right side of the equation by 1/G gives:

$$\chi_{Eth} = \frac{1/46.07}{1/46.07 + 1/18.02} = \frac{0.02171}{0.02171 + 0.05549} = 0.2812$$

(b) $P_T = P_{Eth} + P_{H_2O}$; $P_{Eth} = \chi_{Eth}P^\circ_{Eth}$; $P_{H_2O} = \chi_{H_2O}P^\circ_{H_2O}$

$\chi_{Eth} = 0.2812$, $\chi_{H_2O} = 1 - 0.2812 = 0.7188$

$P_{Eth} = 0.2812\ (400\ \text{torr}) = 112\ \text{torr}$; $P_{H_2O} = 0.7188(175\ \text{torr}) = 126\ \text{torr}$

$P_T = 112\ \text{torr} + 126\ \text{torr} = 238\ \text{torr}$

(c) χ_{Eth} in vapor $= \frac{P_{Eth}}{P_{total}} = \frac{112\ \text{torr}}{238\ \text{torr}} = 0.471$

13.46 (a) Since C_6H_6 and C_7H_8 form an ideal solution, we can use Raoult's Law. Since both components are volatile, both contribute to the total vapor pressure of 40 torr.

$P_T = P_{C_6H_6} + P_{C_7H_8}$; $P_{C_6H_6} = \chi_{C_6H_6}P^\circ_{C_6H_6}$; $P_{C_7H_8} = \chi_{C_7H_8}P^\circ_{C_7H_8}$

$\chi_{C_7H_8} = 1 - \chi_{C_6H_6}$; $P_T = \chi_{C_6H_6}P^\circ_{C_6H_6} + (1 - \chi_{C_6H_6})\ P^\circ_{C_7H_8}$

$40\ \text{torr} = \chi_{C_6H_6}(75\ \text{torr}) + (1 - \chi_{C_6H_6})22\ \text{torr}$

$18\ \text{torr} = 53\ \text{torr}\ (\chi_{C_6H_6})$; $\chi_{C_6H_6} = \frac{18\ \text{torr}}{53\ \text{torr}} = 0.34$; $\chi_{C_7H_8} = 0.66$

(b) $P_{C_6H_6} = 0.34(75\ \text{torr}) = 25.5\ \text{torr}$; $P_{C_7H_8} = 0.66(22\ \text{torr}) = 14.5\ \text{torr}$

In the vapor, $\chi_{C_6H_6} = \frac{P_{C_6H_6}}{P_T} = \frac{25.5\ \text{torr}}{40\ \text{torr}} = 0.64$; $\chi_{C_7H_8} = 0.36$

13.47 $\Delta T = K(m)$ first calculate the **molality** of the solute particles.

(a) 0.45 *m* (b) $\frac{6.7\ \text{g}\ C_{10}H_8}{0.150\ \text{kg}\ CHCl_3} \times \frac{1\ \text{mol}\ C_{10}H_8}{128\ \text{g}\ C_{10}H_8} = 0.35\ m$

(c) $\frac{5.0\ \text{g CsI}}{0.0500\ \text{kg}\ H_2O} \times \frac{1\ \text{mol CsI}}{260\ \text{g CsI}} \times \frac{2\ \text{mol ions}}{1\ \text{mol CsI}} = 0.77\ m$

Then, f.p. = $T_f - K_f(m)$; b.p. = $T_b + K_b(m)$; T in °C

	m	T_f	$-K_f(m)$		f.p.	T_b	$+K_b(m)$	b.p.
(a)	0.45	-114.6	-1.99(0.45) =	-0.89	-115.5	78.4	1.22(0.45) = 0.55	79.0
(b)	0.35	-63.5	-4.68(0.35) =	-1.64	-65.1	61.2	3.63(0.35) = 1.27	62.5
(c)	0.77	0.0	-1.86(0.77) =	-1.43	-1.4	100.0	0.52(0.77) = 0.40	100.4

13.48 $\Delta T = K(m)$ first calculate the **molality** of the solute particles.

(a) 0.15 *m* (b) $\dfrac{20.0\text{ g } C_{12}H_{26}}{0.350\text{ kg } CCl_4} \times \dfrac{1\text{ mol } C_{12}H_{26}}{170.3\text{ g } C_{12}H_{26}} = 0.336\ m$

(c) $\dfrac{2.75\text{ g } MgCl_2}{0.0800\text{ kg } H_2O} \times \dfrac{1\text{ mol } MgCl_2}{95.21\text{ g } MgCl_2} \times \dfrac{3\text{ mol ions}}{1\text{ mol } MgCl_2} = 1.08\ m$

Then, f.p. = $T_f - K_f(m)$; b.p. = $T_b + K_b(m)$; T in °C

	m	T_f	$-K_f(m)$	f.p.	T_b	$+K_b(m)$	b.p.
(a)	0.15	-114.6	1.99(0.15) = -0.30	-114.9	78.4	1.22(0.15) = 0.18	78.6
(b)	0.336	-22.3	29.8(0.336) = -10.0	-32.3	76.8	5.02(0.336) = 1.69	78.5
(c)	1.08	0.0	1.86(1.08) = -2.01	-2.0	100.0	0.52(1.08) = 0.56	100.0

13.49 0.03 *m* glycerine < 0.03 *m* $HC_7H_5O_2$ < 0.02 *m* KBr. $HC_7H_5O_2$ is partially ionized, and so produces more dissolved particles than a glycerine solution of the same molality; the solution of KBr is 0.040 *m* in ionic solute particles.

13.50 The more particles in solution, the lower the freezing point. Since LiBr and $Zn(NO_3)_2$ are electrolytes, the particle concentrations in these two solutions are 0.12 *m* and 0.09 *m*, respectively (although ion-ion attractive forces may decrease the "real" concentrations somewhat). Thus, the order of **decreasing** freezing points is 0.080 *m* glucose > 0.030 *m* $Zn(NO_3)_2$ > 0.06 *m* LiBr.

13.51 The solvent is water. $\Delta T_f = 0.00°C - 8.95°C = -8.95°C$

$$\Delta T_f = -K_f(m);\quad m = -\frac{\Delta T_f}{K_f}$$

$$\Delta T_b = K_b(m) = K_b\left(-\frac{\Delta T_f}{K_f}\right) = 0.52\left(-\frac{8.95°C}{1.86}\right) = +2.5°C$$

b.p. = $T_b + \Delta T_b$ = 100.0°C = 2.5°C = 102.5°C

13.52 From Exercise 13.51, $\Delta T_b = K_b\left(-\dfrac{\Delta T_f}{K_f}\right)$; $\Delta T_f = -\dfrac{K_f}{K_b} \times \Delta T_b$

The solvent is CCl_4; $K_b = 5.02$, $K_f = 29.8$, $T_b = 76.8$, $\Delta T_b = 80.9 - 76.8 = +4.1$.

$$\Delta T_f = -\frac{29.8}{5.02} \times 4.1°C = -24°C;\ \text{f.p.} = T_f + \Delta T_f = -22.3°C - 24°C = -46.3°C$$

13.53 $\pi = MRT$; T = 20°C + 273 = 293 K

$$M = \frac{\text{mol urea}}{\text{L soln}} = \frac{1.10\text{ g } (NH_2)_2CO}{0.1\text{ L soln}} \times \frac{1\text{ mol urea}}{60.1\text{ g } (NH_2)_2CO} = 0.183\ M$$

$$\pi = \frac{0.183\text{ mol}}{\text{L}} \times \frac{0.0821\text{ L-atm}}{\text{K-mol}} \times 293\text{ K} = 4.40\text{ atm}$$

13.54 $\pi = MRT$; $T = 20^\circ C + 273 = 293\ K$

$$M\text{ (of ions)} = \frac{\text{mol NaCl} \times 2}{\text{L soln}} = \frac{3.4\text{ g NaCl}}{1\text{ L soln}} \times \frac{1\text{ mol NaCl}}{58.4\text{ g NaCl}} \times \frac{2\text{ mol ions}}{1\text{ mol NaCl}} = 0.116\ M$$

$$\pi = \frac{0.116\text{ mol}}{\text{L}} \times \frac{0.0821\text{ L-atm}}{\text{K-mol}} \times 293\text{ K} = 2.8\text{ atm}$$

13.55 $\Delta T_b = K_b m$; $m = \frac{\Delta T_b}{K_b} = \frac{+0.49}{5.02} = 0.098\ m$ adrenaline

$$m = \frac{\text{mol adrenaline}}{\text{kg } CCl_4} = \frac{\text{g adrenaline}}{\mathcal{M}\text{ adrenaline} \times \text{kg } CCl_4}$$

$$\mathcal{M}\text{ adrenaline} = \frac{\text{g adrenaline}}{m \times \text{kg } CCl_4} = \frac{0.64\text{ g adrenaline}}{0.098\ m \times 0.036\text{ kg } CCl_4} = 1.8 \times 10^2\text{ g/mol adrenaline}$$

(2 significant figures)

13.56 $\Delta T_f = 5.5 - 4.1 = 1.4$; $m = \frac{\Delta T_f}{K_f} = \frac{1.4}{5.12} = 0.273\ m$

As in Exercise 13.55, $\mathcal{M}$ lauryl alcohol $= \frac{\text{g l. alcohol}}{m \times \text{kg } C_6H_6} = \frac{5.00\text{ g l. alcohol}}{0.273 \times 0.100\text{ kg } C_6H_6}$

$= 1.8 \times 10^2$ g/mol lauryl alcohol (2 significant figures)

13.57 $\pi = MRT$; $M = \frac{\pi}{RT}$; $T = 25^\circ C + 273 = 298\ K$

$$M = 0.953\text{ torr} \times \frac{1\text{ atm}}{760\text{ torr}} \times \frac{\text{K} \cdot \text{mol}}{0.0821\text{ L} \cdot \text{atm}} \times \frac{1}{298\text{ K}} = 5.125 \times 10^{-5}\ M$$

$\text{mol} = M \times \text{L} = 5.125 \times 10^{-5} \times 0.210\text{ L} = 1.076 \times 10^{-5}$ mol lysozyme

$$\mathcal{M} = \frac{\text{g}}{\text{mol}} = \frac{0.150\text{ g}}{1.076 \times 10^{-5}\text{ mol}} = 1.39 \times 10^4\text{ g/mol lysozyme}$$

13.58 $\pi = MRT$; $M = \pi/RT = 2.86\text{ atm} \times \frac{\text{K} \cdot \text{mol}}{0.0821\text{ L} \cdot \text{atm}} \times \frac{1}{298\text{ K}} = 0.117\ M$

We have $\frac{16.0\text{ g sugar}}{0.200\text{ L } H_2O} = 80$ g sugar/L soln

Thus, 80 g sugar = 0.117 mol sugar; $\mathcal{M} = 684$ g/mol

13.59 (a) $i = \frac{\pi\text{ (measured)}}{\pi\text{ (calculated for a nonelectrolyte)}}$

Calculate π for a 0.010 M solution of nonelectrolyte.

$$\pi = MRT = \frac{0.010\text{ mol}}{1\text{ L}} \times 298\text{ K} \times \frac{0.0821\text{ L} \cdot \text{atm}}{\text{K} \cdot \text{mol}} = 0.245\text{ atm}$$

$i = \frac{0.674\text{ atm (actual)}}{0.245\text{ atm (calculated)}} = 2.75$ (The ideal value for $CaCl_2$ is 3.)

(b) As the solution becomes more concentrated, there is more aggregation and fewer effective particles in solution, so the measured value of π will decrease and the value of *i* will decrease.

13.60 (a) ΔT_f (measured) = $i \times \Delta T_f$ (calculated)

ΔT_f (calculated) = $K_f m = 1.86 \times 0.100\ M = 0.186^\circ C$

NaCl: ΔT_f (measured) = 1.87 (0.186°C) = 0.348°C; T_f = 0.0 - 0.348 = -0.348°C

$MgSO_4$: ΔT_f (measured) = 1.21 (0.186°C) = 0.225°C; T_f = 0.0 - 0.225 = -0.225°C

(b) The greater charges on Mg^{2+} and SO_4^{2-} lead to greater ionic aggregation in solution, a smaller value of *i* and a smaller ΔT_f.

Colloids

13.61 In a solution, the dispersed (solute) particles are much smaller, 0 - 10 Å diameter, than the dispersed particles of a colloid, 10 - 2000 Å. The larger dispersed particles in a colloid scatter light as it passes through. Thus, a light beam can be seen passing through a collloid, but is invisible passing through a solution. Most colloids are cloudy, whereas solutions are transparent. White vinegar is a solution, milk is a colloid.

13.62 The outline of a light beam passing through a colloid is visible, whereas light passing through a true solution is invisible unless collected on a screen. The suspended particles in a colloid, much larger than molecules, have dimensions comparable to the wavelength of visible light (400 - 700 nm). Because of this, the photons interact with colloid particles and are scattered, producing the outline of the light beam.

1363 (a) Hydrophobic (b) hydrophilic (c) hydrophobic

13.64 (a) Colloid particles must have low mass. When particle mass becomes large enough so that gravitational and interparticle forces are greater than the kinetic energies of the particles, settling and aggregation can occur.

(b) Hydrophobic colloids do not attract a sheath of water molecules around them and thus tend to aggregate from aqueous solution. They can be stabilized as colloids by adsorbing charges on their surfaces. The charged particles interact with solvent water, stabilizing the colloid.

(c) Charges on colloid particles can stabilize them against aggregation. Particles carrying like charges repel one another and are thus prevented from aggregating and settling out.

13.65 Colloid particles are stabilized by attractive intermolecular forces with the dispersing medium (solvent) and do not coalesce because of electrostatic repulsions between groups at the surface of the dispersed particles. Colloids can be coagulated by heating (more collisions, greater chance that particles will coalesce); hydrophilic colloids can be coagulated by adding electrolytes, which neutralize surface charges allowing the colloid particles to collide more freely.

13.66 (a) The nonpolar hydrophobic tails of soap particles (the hydrocarbon chain of stearate ions) establish attractive intermolecular dispersion forces with the nonpolar oil molecules, while the charged hydrophilic head of the soap particles interacts with H_2O to keep the oil molecules suspended. (This is the mechanism by which laundry detergents remove greasy dirt from clothes.)

(b) The electrolytes from the acid neutralize the surface charges of the suspended particles in milk, causing the colloid to coagulate.

Additional Exercises

13.67 For ionic solids, the exothermic part of the solution process is step (3), surrounding the separated ions by solvent molecules. The released energy comes from the attractive interaction of the solvent with the separated ions. In the hydrates, one water molecule is already associated with the cation, reducing the total energy released during solvation.

13.68 The outer periphery of the BHT molecule is mostly hydrocarbon-like groups, such as $-CH_3$. The one OH group is rather buried inside, and probably does little to enhance solubility in water. Thus, BHT accumulates in the essentially nonpolar fats, and has a fairly long residence time in the body.

13.69 Assume 100 g of solution.

100 g soln × 10.0% NaCl = 10.0 g NaCl + 90.0 g H_2O

(a) $m = \frac{\text{mol NaCl}}{\text{kg } H_2O}, \frac{10.0 \text{ g NaCl}}{0.0900 \text{ kg } H_2O} \times \frac{1 \text{ mol NaCl}}{58.44 \text{ g NaCl}} = 1.90\ m$

(b) $M = \frac{\text{mol NaCl}}{\text{L soln}} = \frac{10.0 \text{ g NaCl}}{100 \text{ g soln}} \times \frac{1.071 \text{ g soln}}{1 \text{ mL soln}} \times \frac{1 \text{ mol NaCl}}{58.44 \text{ g NaCl}} \times \frac{1000 \text{ mL}}{1 \text{ L}} = 1.83\ M$

13.70 (a) $\frac{1.80 \text{ mol } CH_3CN}{1 \text{ L soln}} \times \frac{86.8 \text{ g LiBr}}{1 \text{ mol LiBr}} = \frac{156 \text{ g LiBr}}{1 \text{ L soln}}$

1 L soln = 826 g soln; g CH_3CN = 826 - 156 = 670 g

$m \text{ LiBr} = \frac{1.80 \text{ mol LiBr}}{0.670 \text{ kg } CH_3CN} = 2.69\ m$

(b) $\frac{670 \text{ g } CH_3CN}{41.0 \text{ g/mol}} = 16.3 \text{ mol } CH_3CN;\ \chi_{LiBr} = \frac{1.80}{1.80 + 16.3} = 0.0994$

(c) $\text{mass \%} = \frac{670 \text{ g LiBr}}{826 \text{ g soln}} \times 100 = 81.1\% \text{ LiBr}$

13.71 $8.00\text{ L soln} \times \frac{1.208\text{ g}}{1\text{ cm}^3} \times \frac{1000\text{ cm}^3}{1\text{ L}} \times \frac{0.180\text{ g CuSO}_4}{1\text{ g soln}} = 1.74 \times 10^3\text{ g CuSO}_4$

$$1.74 \times 10^3\text{ g CuSO}_4 \times \frac{249.6\text{ g CuSO}_4 \cdot 5H_2O}{159.6\text{ g CuSO}_4} = 2.72 \times 10^3\text{ g CuSO}_4 \cdot 5H_2O$$

$$\frac{2.72 \times 10^3\text{ g CuSO}_4 \cdot 5H_2O}{8.00\text{ L}} \times \frac{1\text{ mol CuSO}_4 \cdot 5H_2O}{249.6\text{ g CuSO}_4 \cdot 5H_2O} = 1.36\ M$$

13.72 $\frac{2.16\text{ g } C_7H_6O_2}{122\text{ g/mol}} = 0.0177\text{ mol } C_7H_6O_2$

$$180\text{ mL CCl}_4 \times \frac{1.59\text{ g}}{1\text{ mL}} = 286\text{ g CCl}_4\ ;\quad \frac{286\text{ g CCl}_4}{154\text{ g/mol}} = 1.86\text{ mol CCl}_4$$

$$180\text{ mL } C_2H_5OH \times \frac{0.782\text{ g}}{1\text{ mL}} = 141\text{ g } C_2H_5OH;\quad \frac{141\text{ g } C_2H_5OH}{46.1\text{ g/mol}} = 3.05\text{ mol } C_2H_5OH$$

	$\mathbf{CCl_4}$	$\mathbf{C_2H_5OH}$
(a)	$\chi = \frac{0.0177}{0.0177 + 1.86} = = 9.43 \times 10^{-3}$	$\chi = \frac{0.0177}{0.0177 + 3.05} = 5.77 \times 10^{-3}$
	$m = \frac{0.0177\text{ mol } C_7H_6O_2}{0.0286\text{ kg CCl}_4} = 0.0619\ m$	$m = \frac{0.0177\text{ mol } C_7H_6O_2}{0.141\text{ kg } C_2H_5OH} = 0.126\text{ m}$

(b) CCl_4: $2.16\text{ g } C_7H_6O_2 + 286\text{ g CCl}_4 = 288\text{ g soln} \times \frac{1\text{ mL}}{1.59\text{ g}} = 181\text{ mL soln}$

$$M = \frac{0.0177\text{ mol } C_7H_6O_2}{0.181\text{ L soln}} = 0.0978\ M$$

C_2H_5OH: $2.16\text{ g } C_7H_6O_2 + 141\text{ g } C_2H_5OH = 143\text{ g soln} \times \frac{1\text{ mL}}{0.782\text{ g}} = 183\text{ mL soln}$

$$M = \frac{0.0177\text{ mol } C_7H_6O_2}{0.183\text{ L soln}} = 0.0967\ M$$

(c) The molarities vary slightly because of the assumption that density of the solution equals density of the solvent; nonetheless, the molarities are close. The molalities differ by a factor of about two, because the densities of the two solvents are not the same. When equal volumes of the solvents are used, the masses of solvent are very different and the molalities of the solutions are different.

13.73 (a) $m = \frac{\text{mol Na(s)}}{\text{kg Hg(l)}}$; $1.0\text{ cm}^3\text{ Na(s)} \times \frac{0.97\text{ g}}{1\text{ cm}^3} \times \frac{1\text{ mol}}{23.0\text{ g Na}} = 0.042\text{ mol Na}$

$$20.0\text{ cm}^3\text{ Hg(l)} \times \frac{13.6\text{ g}}{1\text{ cm}^3} \times \frac{1\text{ kg}}{1000\text{ g}} = 0.272\text{ kg Hg(l)};\quad m = \frac{0.042\text{ mol Na}}{0.272\text{ Kg Hg(l)}} = 0.15\ m$$

(b) $M = \frac{\text{mol Na(s)}}{\text{L soln}} = \frac{0.042\text{ mol Na}}{0.021\text{ mL soln}} = 2.0\ M$

(c) Clearly, molality and molarity are not the same for this amalgum. Only in the instance that kg solvent and liters solution are nearly equal do the two concentration units have similar values. Although the density of the solution (13.0 g/cm^3) and the density of the solvent (Hg(l), 13.6 g/cm^3) are similar, the density of the solvent is much different than 1 g/mL, so kilograms solvent and liters solution are different and so are the two ways to express concentration.

13.74 In this equilibrium system, molecules move from the surface of the solid into solution, while molecules in solution are deposited on the surface of the solid. As molecules leave the surface of the small particles of powder, the reverse process preferentially deposits other molecules on the surface of a single crystal. Eventually, all molecules that were present in the 50 g of powder are deposited on the surface of a 50 g crystal; this can only happen if the dissolution and deposition processes are ongoing.

13.75 (a) The solution is saturated.

$$\frac{92.4 \text{ g}}{100 \text{ mL}} \times 500 \text{ mL} = 462 \text{ g } SmCl_3 \text{ in solution}$$

500 g $SmCl_3$ - 462 g $SmCl_3$ in solution = 38 g $SmCl_3$(s) undissolved

(b) $$m = \frac{\text{mol } SmCl_3 \text{ dissolved}}{\text{kg } H_2O} = \frac{462 \text{ g } SmCl_3}{0.500 \text{ kg } H_2O} \times \frac{1 \text{ mol } SmCl_3}{256.8 \text{ g } SmCl_3} = 3.60\ m\ SmCl_3$$

13.76 Since these are very dilute solutions, assume that the density of the solution ≈ the density of H_2O ≈ 1.0 g/mL at 25°C. Then, 100 g solution = 100 g H_2O = 0.100 kg H_2O.

(a) $$CF_4: \frac{0.0015 \text{ g } CF_4}{0.100 \text{ kg } H_2O} \times \frac{1 \text{ mol } CF_4}{88.00 \text{ g } CF_4} = 1.7 \times 10^{-4}\ m$$

$$CClF_3: \frac{0.009 \text{ g } CClF_3}{0.100 \text{ kg } H_2O} \times \frac{1 \text{ mol } CClF_3}{104.46 \text{ g } CClF_3} = 8.6 \times 10^{-4}\ m$$

$$CCl_2F_2: \frac{0.028 \text{ g } CCl_2F_2}{0.100 \text{ kg } H_2O} \times \frac{1 \text{ mol } CCl_2F_2}{120.9 \text{ g } CCl_2F_2} = 2.3 \times 10^{-3}\ m$$

$$CHClF_2: \frac{0.30 \text{ g } CHClF_2}{0.100 \text{ kg } H_2O} \times \frac{1 \text{ mol } CHClF_2}{86.47 \text{ g } CHClF_2} = 3.5 \times 10^{-2}\ m$$

(b) $$m = \frac{\text{mol solute}}{\text{kg solvent}};\ M = \frac{\text{mol solute}}{\text{L solution}}$$

Molality and molarity are numerically similar when kilograms solvent and liters solution are nearly equal. This is true when solutions are dilute, so that the density of the solution is essentially the density of the solvent, and when the density of the solvent is nearly 1 g/mL. That is, for dilute aqueous solutions such as the ones in this problem, $M \approx m$.

(c) Water is a polar solvent; the solubility of solutes increases as their polarity increases. All the fluorocarbons listed have tetrahedral molecular structures. CF_4, a symmetrical tetrahedron, is nonpolar and has the lowest solubility. As more different atoms are bound to the central carbon, the electron density distribution in the molecule becomes less symmetrical and the molecular polarity increases. The most polar fluorocarbon, $CHClF_2$, has the greatest solubility in H_2O.

(d) $C_g = k P_g$. Assume $M = m$ for $CHClF_2$. P_g = 1 atm

$$k = \frac{C_g}{P_g} = \frac{M}{P_g};\ k = \frac{3.5 \times 10^{-2}\ M}{1.0\ \text{atm}} = 3.5 \times 10^{-2}\ \text{mol/L} \cdot \text{atm}$$

This value is greater than the Henry's law constant for $N_2(g)$, because $N_2(g)$ is nonpolar and less soluble in water than $CHClF_2$. In fact, the Henry's law constant for nonpolar CF_4, 1.7×10^{-4} mol/L • atm is similar to the value for N_2, 6.8×10^{-4} mol/L • atm.

13.77 $P_{Rn} = \chi_{Rn} P_{total}$; $P_{Rn} = 3.5 \times 10^{-6}(36\ \text{atm}) = 1.26 \times 10^{-5}$ atm

$$C_{Rn} = k P_{Rn};\ C_{Rn} = \frac{7.27 \times 10^{-3}\ M}{1\ \text{atm}} \times 1.26 \times 10^{-5}\ \text{atm} = 9.2 \times 10^{-8}\ M$$

13.78 At the high external pressures (~ 10 atm) experienced by divers, the gases they breathe are much more soluble in body fluids than at atmospheric pressure. When the divers return to the surface, the previously dissolved gases form bubbles in the bloodstream and elsewhere. These bubbles inhibit nerve functions and lead to decompression sickness, "the bends." He(g) is much less soluble in body fluids than $N_2(g)$ so it produces fewer bubbles and less severe decompression effects when the diver returns to atmospheric pressure.

13.79 Mole fraction ethanol, $\chi_{C_2H_5OH} = \dfrac{P_{C_2H_5OH}}{P^o_{C_2H_5OH}} = \dfrac{8\ \text{torr}}{100\ \text{torr}} = 0.08$

$$\frac{620 \times 10^3\ \text{g}\ C_{24}H_{50}}{338.6\ \text{g/mol}} = 1.83 \times 10^3\ \text{mol}\ C_{24}H_{50};\ \text{let y = mol}\ C_2H_5OH$$

$$\chi_{C_2H_5OH} = 0.08 = \frac{y}{y + 1.83 \times 10^3};\ 0.92\,y = 146.4;\ y = 1.6 \times 10^2\ \text{mol}\ C_2H_5OH$$

$$1.6 \times 10^2\ \text{mol}\ C_2H_5OH \times \frac{46\ \text{g}\ C_2H_5OH}{1\ \text{mol}} = 7.4 \times 10^3\ \text{g or 7.4 kg}\ C_2H_5OH$$

13.80 Assume the radiator solution is prepared by mixing 1.00 L of each of the liquids. To calculate the freezing and boiling points, we need the molality of this solution. Assume H_2O is the solvent.

$m = \dfrac{\text{mol } C_2H_6O_2}{\text{kg } H_2O}$; kg H_2O = 1.00 L × $\dfrac{1000 \text{ mL}}{1 \text{ L}} \times \dfrac{1.00 \text{ g}}{\text{mL}}$ = 1.00 kg

mol $C_2H_6O_2$ = 1.00 L × $\dfrac{1000 \text{ mL}}{1 \text{ L}} \times \dfrac{1.12 \text{ g}}{1 \text{ mL}} \times \dfrac{1 \text{ mol } C_2H_6O_2}{62.07 \text{ g } C_2H_6O_2}$ = 18.04 mol C_2H_6O

$m = \dfrac{18.04 \text{ mol } C_2H_6O}{1.00 \text{ kg } H_2O} = 18.04\ m$

$\Delta T_f = K_f m$ = -1.86(18.04) = -33.6°C; f.p. = 0.0 - 33.6 = -33.6°C

$\Delta T_b = K_b m$ = 0.52(18.04) = 9.4°C ; b.p. = 100.0 + 9.4 = +109.4°C

13.81 (a) 0.100 *m* K_2SO_4 is 0.300 *m* in particles. H_2O is the solvent.

$\Delta T_f = K_f m$ = -1.86(0.300) = -0.558 ; T_f = 0.0 - 0.558 = -0.558°C

(b) ΔT_f(nonelectrolyte) = -1.86(0.100) = -0.186 ; T_f = 0.0 - 0.186 = -0.186°C

T_f(measured) = *i* × T_f (nonelectrolyte)

From Table 13.6, *i* for 0.100 *m* K_2SO_4 = 2.32

T_f(measured) = 2.32(-0.186°C) = -0.432

13.82 The compound with the larger *i* value is the stronger electrolyte.

$i = \dfrac{\Delta T_f \text{ (measured)}}{\Delta T_f \text{ (calculated)}}$ The idealized value is 3 for both salts.

$Hg(NO_3)_2$: $m = \dfrac{10.0 \text{ g } Hg(NO_3)_2}{1.00 \text{ kg } H_2O} \times \dfrac{1 \text{ mol } Hg(NO_3)_2}{324.6 \text{ g } Hg(NO_3)_2} = 0.0308\ m$

ΔT_f (nonelectrolyte) = -1.86 (0.0308) = -0.0573°C

$i = \dfrac{-0.172°C}{-0.0573°C} = 3.00$

$HgCl_2$: $m = \dfrac{10.0 \text{ g } HgCl_2}{1.00 \text{ kg } H_2O} \times \dfrac{1 \text{ mol } HgCl_2}{271.5 \text{ g } HgCl_2} = 0.0368\ m$

ΔT_f (nonelectrolyte) = -1.86(0.0368) = -0.0685°C

$i = \dfrac{-0.0685}{-0.0685} = 1.00$

With an *i* value of 3.00, $Hg(NO_3)_2$ is almost completely dissociated into ions; with an *i* value of 1.00, the $HgCl_2$ behaves essentially like a nonelectrolyte. Clearly, $Hg(NO_3)_2$ is the stronger electrolyte.

13.83 $\Delta T_f = K_f m$; $m = \frac{\Delta T_f}{K_f} = \frac{-0.043°C}{1.86} = 0.0231\ m$ particles (assuming no aggregation)

0.0200 mol HF were dissolved in 1 kg of water, producing 0.0231 mol of particles (H^+, F^-, HF).

Let x = mol HF dissociated = mol H^+ = mol F^-; mol HF = 0.0200 - x.

0.0231 mol particles = mol HF + mol H^+ + mol F^- = (0.0200 - x) + x + x
0.0200 + x = 0.0231; x = 0.0031 mol HF dissociated.

$$\%\ \text{ionization} = \frac{0.0031\ \text{mol dissociated}}{0.0200\ \text{mol HF initial}} \times 100 = 16\%\ \text{(2 significant figures)}$$

18.84 (a) $K_b = \frac{\Delta T_b}{m}$; $\Delta T_b = 47.46°C - 46.30°C = 1.16°C$

$$m = \frac{\text{mol solute}}{\text{kg } CS_2} = \frac{0.25\ \text{mol}}{400\ \text{mL } CS_2} \times \frac{1\ \text{mL } CS_2}{1.261\ \text{g } CS_2} \times \frac{1000\ \text{g}}{1\ \text{kg}} = 0.50\ m$$

$$K_b = \frac{1.16°C}{0.50\ m} = 2.3°C/m$$

(b) $$m = \frac{\Delta T_b}{K_b} = \frac{(47.08 - 46.30)°C}{2.3°C/m} = 0.34\ m$$

$$m = \frac{\text{mol unknown}}{\text{kg } CS_2};\ m \times \text{kg } CS_2 = \frac{\text{g unknown}}{\mathcal{M}\ \text{unknown}};\ \mathcal{M} = \frac{\text{g unknown}}{m \times \text{kg } CS_2}$$

$$50.0\ \text{mL } CS_2 \times \frac{1.261\ \text{g } CS_2}{1\ \text{mL}} \times \frac{1\ \text{kg}}{1000\ \text{g}} = 0.0631\ \text{kg } CS_2$$

$$\mathcal{M} = \frac{5.39\ \text{g unknown}}{0.34\ m \times 0.0631\ \text{kg } CS_2} = 2.5 \times 10^2\ \text{or}\ 250\ \text{g/mol}$$

13.85 $\chi_{CHCl_3} = \chi_{C_3H_6O} = 0.500$

(a) For an ideal solution, Raoult's Law is obeyed.

$P_T = P_{CHCl_3} + P_{C_3H_6O}$; $P_{CHCl_3} = 0.5(300\ \text{torr}) = 150$ torr

$P_{C_3H_6O} = 0.5(360\ \text{torr}) = 180$ torr; $P_T = 150$ torr + 180 torr = 330 torr

(b) The real solution has a lower vapor presure, 250 torr, than an ideal solution of the same composition, 330 torr. Thus, fewer molecules escape to the vapor phase from the liquid. This means that fewer molecules have sufficient kinetic energy to overcome intermolecular attractions. Clearly, even weak hydrogen bonds such as this one are stronger attractive forces than dipole-dipole or dispersion forces. These hydrogen bonds prevent molecules from escaping to the vapor phase and result in a lower than ideal vapor pressure for the solution.

(c) According to Coulomb's law, electrostatic attractive forces lead to an overall lowering of the energy of the system. Thus, when the two liquids mix and hydrogen bonds are formed, the energy of the system is decreased and $\Delta H_{soln} < O$; the solution process is exothermic.

13.86 $M = \frac{\pi}{RT} = \frac{6.41 \text{ atm}}{298 \text{ K}} \times \frac{\text{K} \cdot \text{mol}}{0.0821 \text{ L} \cdot \text{atm}} = 0.262\ M$

There are 0.262 mol of **particles** per liter of solution and 0.131 mol particles in 500 mL. Each mole of sucrose provides 1 mol of particles and each mole of NaCl provides 2.

Let x = g $C_{12}H_{22}O_{11}$, 15.0 - x = g NaCl

$$\frac{x \text{ g } C_{12}H_{22}O_{11}}{342.3 \text{ g } C_{12}H_{22}O_{11}} + \frac{2(15.0 - x)}{59.44 \text{ g NaCl}} = 0.131$$

58.44x + 2(342.3)(15.0 - x) = 0.131(58.44)(342.3)

58.44x - 684.6x + 10,269 = 2,621

-626.2x = -7648; x = 12.2 g $C_{12}H_{22}O_{11}$

$$\frac{12.2 \text{ g } C_{12}H_{22}O_{11}}{15.0 \text{ g mixture}} \times 100 = 81.3\% \text{ sucrose, } 18.7\% \text{ NaCl}$$

13.87 $M = \frac{\pi}{RT} = \frac{57.1 \text{ torr}}{298 \text{ K}} \times \frac{1 \text{ atm}}{760 \text{ torr}} \times \frac{\text{K} \cdot \text{mol}}{0.0821 \text{ L} \cdot \text{atm}} = 3.07 \times 10^{-3}\ M$

$$\frac{0.036 \text{ g solute}}{100 \text{ g } H_2O} \times \frac{1000 \text{ g } H_2O}{1 \text{ kg } H_2O} = 0.36 \text{ g solute/kg } H_2O$$

Assuming molarity and molality are the same in this dilute solution, we can then say 0.36 g solute = 3.07×10^{-3} mol; $\mathcal{M}$ = 117 g/mol. Because the salt is completely ionized, the formula weight of the lithium salt is **twice** this calculated value, or **234 g/mol**. The organic portion, $C_nH_{2n-1}O_2^-$, has a formula weight of 234 - 7 = 227 g. Subtracting 32 for the oxygens, and adding 1 to make the formula C_nH_{2n}, we have C_nH_{2n}, $\mathcal{M}$ = 196 g. Since each CH_2 unit has a mass of 14, n = 196/14 = 14. Thus the formula for our salt is $LiC_{14}H_{27}O_2$.

13.88 The solvent vapor pressure over each solution is determined by the total particle concentrations present in the solutions. When the particle concentrations are equal, the vapor pressures will be equal and equilibrium established. The particle concentration of the nonelectrolyte is just 0.050 *M*; the ion concentration of the NaCl is $2 \times 0.03\ M$ = 0.060 M. Solvent will diffuse from the less concentrated nonelectrolyte solution. Let x = volume of solvent transferred.

$$\frac{0.050\ M \times 20.0 \text{ mL}}{(20.0 - x) \text{ mL}} = \frac{0.060\ M \times 20.0 \text{ mL}}{(20.0 + x) \text{ mL}}; \quad 1.0(20.0 + x) = 1.2(20.0 - x)$$

2.2x = 4.0, x = 1.8 mL transferred

The nonelectrolyte beaker contains 20.0 - 1.8 = 18.2 mL solution; the NaCl beaker contains 20.0 + 1.8 = 21.8 mL solution.

CHAPTER 14

Chemical Kinetics

Reaction Rates

14.1 (a) $-\Delta[H_2O]/2\Delta t = \Delta[H_2]/2\Delta t = \Delta[O_2]\Delta t$

(b) $-\Delta[CO]/\Delta t = -\Delta[H_2]/2\Delta t = \Delta[CH_3OH]/\Delta t$

(c) $-\Delta[S_2O_8^{2-}]\Delta t = -\Delta[I^-]/2\Delta t = \Delta[SO_4^{2-}]/2\Delta t = \Delta[I_2]/\Delta t$

14.2 (a) $-\Delta[N_2]/\Delta t = -\Delta[H_2]/3\Delta t = \Delta[NH_3]_2/2\Delta t$

(b) $-\Delta[NO]/\Delta t = -\Delta[Cl_2]/\Delta t = \Delta[NOCl]/2\Delta$

(c) $-\Delta[B_2H_6]/\Delta t = -\Delta[O_2]/3\Delta t = \Delta[B_2O_3]/\Delta t = [H_2O]/3\Delta t$

14.3 (a) $$\frac{\Delta[H_2O]}{2\Delta t} = \frac{-\Delta[H_2]}{2\Delta t} = \frac{-\Delta[O_2]}{\Delta t}$$

H_2 is burning, $\frac{-\Delta[H_2]}{\Delta t} = 4.6$ mol/s

O_2 is consumed, $\frac{-\Delta[O_2]}{\Delta t} = \frac{-\Delta[H_2]}{2\Delta t} = \frac{4.6 \text{ mol/s}}{2} = 2.3$ mol/s

H_2O is produced, $\frac{+\Delta[H_2O]}{\Delta t} = \frac{-\Delta[H_2]}{\Delta t} = 4.6$ mol/s

(b) The change in total pressure is the sum of the changes of each partial pressure. NO and Cl_2 are disappearing and NOCl is appearing.

$-\Delta P_{NO}/\Delta t = -30$ torr/min
$-\Delta P_{Cl_2}/\Delta t = \Delta P_{NO}/2\Delta t = -15$ torr/min
$+\Delta P_{NOCl}/\Delta t = -\Delta P_{NO}/\Delta t = +30$ torr/min
$\Delta P_T/\Delta t = -30$ torr/min $- 15$ torr/min $+ 30$ torr/min
$= -15$ torr/min

14.4 (a) $\frac{-\Delta[CH_4]}{\Delta t} = \frac{+[CO_2]}{\Delta t} = \frac{\Delta[H_2O]}{2\Delta t}$ If CH_4 is burning, $-\Delta[CH_4]/\Delta t$, at 0.40 *M*/s, CO_2 is being produced, $+\Delta[CO_2]/\Delta t$, at 0.40 *M*/s and H_2O is being produced, $+\Delta[H_2O]/\Delta t$, at 0.80 *M*/s.

(b) $\frac{+\Delta[NH_3]}{\Delta t} = \frac{-2\Delta[N_2]}{\Delta t} = \frac{2\Delta[H_2]}{3\Delta t}$; $\frac{+\Delta[NH_3]}{\Delta t} = 100$ torr/hr

$\frac{-\Delta[N_2]}{\Delta t} = 50$ torr/hr ; $\frac{-\Delta[H_2]}{\Delta t} = 150$ torr/hr

ΔP_{total} = (+100 torr - 50 torr - 150 torr) = -100 torr/hr

14.5

Time (min)	Time Interval (min)	Concentration (*M*)	Conc. Change	Rate (*M*/min)
0		1.85		
79	79	1.67	0.18	2.3×10^{-3}
158	79	1.52	0.15	1.9×10^{-3}
316	158	1.30	0.22	1.4×10^{-3}
632	316	1.00	0.30	0.95×10^{-3}

14.6

Time (min)	Time Interval (min)	Concentration (*M*)	Conc. Change	Rate (*M*/min)
0		0.0165		
2,000	2,000	0.0110	0.0055	28×10^{-7}
5,000	3,000	0.00591	0.0051	17×10^{-7}
8,000	3,000	0.00314	0.0028	9.2×10^{-7}
12,000	4,000	0.00137	0.0018	4.4×10^{-7}
15,000	3,000	0.00074	0.00063	2.1×10^{-7}

14.7 From the slopes of the lines in the figure at right, the rates are 1.9×10^{-3} *M*/min at t = 100, 9.4×10^{-4} *M*/min at t = 500.

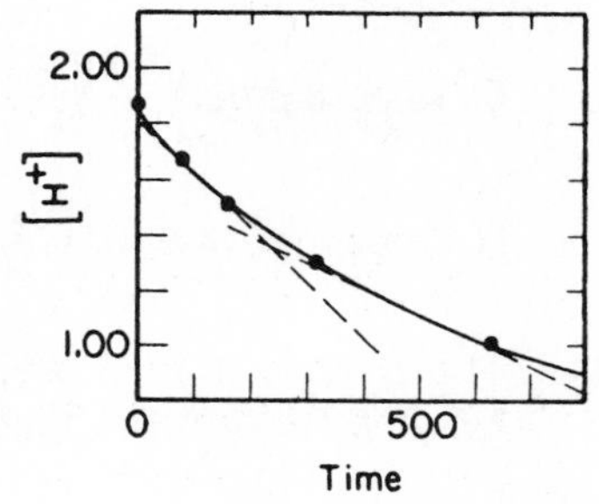

14.8 From the slopes of the lines in the figure at right, the rates are 1.7×10^{-6} *M*/min at 3500 min and 2.1×10^{-7} *M*/min at 13,500 min.

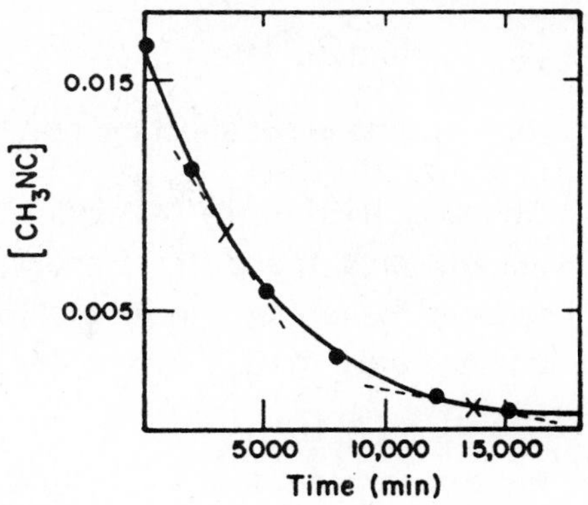

Rate Laws

14.9 (a) rate = $k[N_2O_5] = 6.08 \times 10^{-4}\ s^{-1}\ [N_2O_5]$

(b) rate = $6.08 \times 10^{-4} s^{-1}$ (0.100 *M*) = 6.08×10^{-5} *M*/s

(c) rate = $6.08 \times 10^{-4} s^{-1}$ (0.200 *M*) = 12.16×10^{-5} = 1.22×10^{-4} *M*/s

When the concentration of N_2O_5 doubles, the rate of the reaction doubles.

14.10 (a) $\frac{-[NO]}{\Delta t} = \frac{-\Delta[H_2]}{\Delta t} = k[H_2][NO]^2$

(b) rate = $(6.0 \times 10^4\ M^{-2}s^{-1})(0.050\ M)^2(0.010\ M) = 1.5\ M/s$

(c) rate = $(6.0 \times 10^4\ M^{-2}s^{-1})(0.10\ M)^2(0.010\ M) = 6.0\ M/s$

(Note that doubling [NO] causes a quadrupling in rate.)

14.11 (a,b) rate = $k[CH_3Br][OH^-]$; $k = \frac{\text{rate}}{[CH_3Br][OH^-]}$

at 298 K, $k = \frac{0.28\ M/s}{(0.010\ M)(0.10\ M)} = 2.8 \times 10^2\ M^{-1}s^{-1}$

(c) Since the rate law is first order in $[OH^-]$, if $[OH^-]$ is tripled, the rate triples.

14.12 (a,b) rate = $k[NO]^2[O_2]$; $k = \text{rate}/[NO_2]^2[O_2]$

$$k = \frac{7.2 \times 10^{-5}\ M/s}{(0.030\ M)^2(0.040\ M)} = 2.0\ M^{-2}s^{-1}$$

(c) Since the reaction is second order in NO, if the [NO] is decreased by a factor of 2, the rate would decrease by a factor of 2^2, or 4.

14.13 (a) Since the rate is directly proportional to [A], it is first order in [A]. A change in [B] does not change the rate, so [B] does not appear in the rate law. rate = k[A].

(b) The rate is first order in [A] and second order in [B] (an n-fold increase in [B] causes an n^2 increase in the rate). rate = $[A][B]^2$

14.14 (a) rate = $[A]^2[B]$

(b) [X] does not affect the rate; the rate is directly proportional to [Y]. rate = k[Y]

14.15 Doubling [NO] while holding $[O_2]$ constant increases the rate by a factor of 4 (experiments 1 and 3). Reducing $[O_2]$ by a factor of 2 while holding [NO] constant reduces the rate by a factor of 2 (experiments 2 and 3). The rate is second order in [NO] and first order in $[O_2]$. rate = $k[NO]^2[O_2]$

From experiment 1: $k = \frac{1.41 \times 10^{-2}\ M/s}{(0.0126\ M)^2 (0.0125\ M)} = 7.11 \times 10^3\ M^{-2}s^{-1}$

14.16 (a) Doubling $[NH_3]$ while holding $[BF_3]$ constant doubles the rate (experiments 1 and 2). Doubling $[BF_3]$ while holding $[NH_3]$ constant doubles the rate (experiments 4 and 5). Thus, the reaction is first order in both BF_3 and NH_3; rate = $k[BF_3]\,[NH_3]$.

(b) The reaction is second order overall.

(c) At t = 0, $k = \frac{0.2130\ M/s}{(0.250\ M)\ (0.250\ M)} = 3.41\ M^{-1}\,s^{-1}$

(Any of the five sets of initial concentrations and rates could be used to calculate the rate constant k.)

14.17 (a) The rate increases four-fold on doubling [NO], it doubles on doubling $[Br_2]$. Thus, rate = $k[NO]^2[Br_2]$.

(b) $k = \frac{12\ M/s}{[0.10\ M]^2 [0.10\ M]} = 1.2 \times 10^4\ M^{-2}\,s^{-1}$

(c) From the balanced equation, $\Delta[NOBr]/2\Delta t = -\Delta[Br_2]/\Delta t$. That is, NOBr concentration increases at twice the rate the Br_2 concentration decreases.

(d) $\frac{\Delta[NOBr]}{\Delta t} = \frac{-\Delta[NO]}{\Delta t} = (1.2 \times 10^4\ M^{-2}s^{-1})\ (0.15\ M)^2 (0.25\ M) = 68\ M/s$

(e) $\frac{-\Delta[Br_2]}{\Delta t} = \frac{-\Delta[NO]}{2\Delta t} = 1/2\ (1.2 \times 10^4\ M^{-2}s^{-1})\ (0.075\ M)^2\ (0.185\ M) = 6.2\ M/s$

14.18 (a) A doubling of $[S_2O_8^{2-}]$ causes a doubling in rate; halving $[I^-]$ causes a halving in rate. Therefore, rate = $k[S_2O_8^{2-}]\,[I^-]$.

(b) $k = \frac{(1.4 \times 10^{-5}\ M/s)}{(0.038\ M)\ (0.060\ M)} = 6.1 \times 10^{-3}\ M^{-1}\,s^{-1}$

(c) rate = $(6.1 \times 10^{-3}\ M^{-1}s^{-1})(0.025\ M)(0.100\ M) = 1.5 \times 10^{-5}\ M/s$

(d) $\text{rate} = \dfrac{-\Delta[S_2O_8^{2-}]}{\Delta t} = \dfrac{\Delta[SO_4^{2-}]}{2\Delta t}$; $\dfrac{\Delta[SO_4^{2-}]}{\Delta t} = \dfrac{-2\Delta[S_2O_8^{2-}]}{\Delta t}$

$$\frac{\Delta[SO_4^{2-}]}{\Delta t} = 2(6.1 \times 10^{-3}\ M^{-1}\,s^{-1})\,(0.025\ M)\,(0.050\ M) = 1.5 \times 10^{-5}\ M/s$$

Change of Concentration with Time

14.19 Half-life, $t_{1/2}$, is the time required to reduce the concentration of a reactant to one-half of its initial value. For first order reactions, $t_{1/2} = 0.693/k$; the half-life depends only on the rate constant, k, and is independent of reactant concentration.

14.20 For a first order reaction, $t_{1/2}$ depends only on k. Calculate k from Equation 14.9 and then solve for $t_{1/2}$.

$\ln\dfrac{[A]_t}{[A]_0} = -kt$; $[A]_t = 0.25[A]_0$, $t = 36$ min

$\ln\dfrac{0.25[A]_0}{[A]_0} = -k(36\ \text{min})$; $\ln 0.25 = -36\ k$

$-1.386 = -36\ k$; $k = 0.039\ \text{min}^{-1}$

$$t_{1/2} = \frac{0.693}{k} = \frac{0.693}{0.039\ \text{min}} = 18\ \text{min}$$

14.21 For a first order reaction, $t_{1/2} = 0.693/k$, $k = 5.1 \times 10^{-4}\ s^{-1}$

$$t_{1/2} = \frac{0.693}{5.1 \times 10^{-4}\ s^{-1}} = 1.4 \times 10^{3}\ s\ \ (23\ \text{min})$$

14.22 $t_{1/2} = 0.91$ s; $t_{1/2} = 0.693/k$, $k = 0.693/t_{1/2}$

$$k = \frac{0.693}{0.91\ s} = 0.76\ s^{-1}$$

14.23 (a) Rearranging Equation [14.9] for a first order reaction:

$\ln[A]_t = -kt + \ln[A]_o$

1.5 min = 90 s ; $[N_2O_5]_o = (0.300\ \text{mol}/0.500\ \text{L}) = 0.600\ M$

$\ln[N_2O_5]_{90} = -(6.82 \times 10^{-3}\,s^{-1})\,(90\ s) + \ln(0.600)$

$\ln[N_2O_5]_{90} = -0.614 + (-0.511) = -1.125$

$[N_2O_5]_{90} = 0.325\ M$; mol $N_2O_5 = 0.325\ M \times 0.500\ \text{L} = 0.162$ mol

(b) $[N_2O_5]_t = 0.030 \text{ mol}/0.500 \text{ L} = 0.060\ M$; $[N_2O_5]_o = 0.600\ M$

$\ln(0.060) = -(6.82 \times 10^{-3} s^{-1})(t) + \ln(0.600)$

$$t = \frac{-[\ln(0.060) - \ln(0.600)]}{(6.82 \times 10^{-3} s^{-1})} = 338 \text{ s} \times \frac{1 \text{ min}}{60 \text{ s}} = 5.63 \text{ min}$$

(c) $t_{1/2} = 0.693/k = 0.693/6.82 \times 10^{-3} s^{-1} = 102$ s or 1.69 min

14.24 (a) For a first order reaction:

$\ln[A]_t = -kt + \ln[A]_o$; $k = 1.65 \text{ yr}^{-1}$, $t = 3$ months $= 0.250$ yr

$\ln[A]_{0.25} = -1.65 \text{ yr}^{-1} \times 0.250 \text{ yr} + \ln(6.0 \times 10^{-3})$

$\ln[A]_{0.25} = -5.53$; $[A]_{0.25} = 3.97 \times 10^{-3}\ M$

At = 1.0 yr, $\ln[A]_{1.0} = (-1.65 \text{ yr}^{-1} \times 1.0 \text{ yr}) + \ln 6.0 \times 10^{-3}$

$\ln[A]_{1.0} = -6.77$; $[A]_{1.0} = 1.15 \times 10^{-3}\ M$

(b) $\ln[A]_t - \ln[A]_o = -kt$; $t = \frac{2.30}{k}(\ln[A]_o - \ln[A]_t)$

$$t = \frac{[\ln(6.0 \times 10^{-3}) - \ln(1.0 \times 10^{-3})]}{1.65 \text{ yr}^{-1}} = \frac{1.792}{1.65 \text{ yr}^{-1}} = 1.1 \text{ yr}$$

(c) $t_{1/2} = 0.693/k = 0.693/1.65 \text{ yr}^{-1} = 0.420$ yr $= 5.04$ mo $= 153$ days

14.25

t(s)	$P_{SO_2Cl_2}$	$\ln P_{SO_2Cl_2}$
0	1.000	0
2500	0.947	-0.0545
5000	0.895	-0.111
7500	0.848	-0.165
10000	0.803	-0.219

Graph $\ln P_{SO_2Cl_2}$ vs. time. (Pressure is a satisfactory concentration unit for a gas, since the concentration in moles/liter is proportional to P.) The graph is linear with slope $-2.19 \times 10^{-5}\ s^{-1}$ as shown on the figure. The rate constant k = - slope = $2.19 \times 10^{-5}\ s^{-1}$.

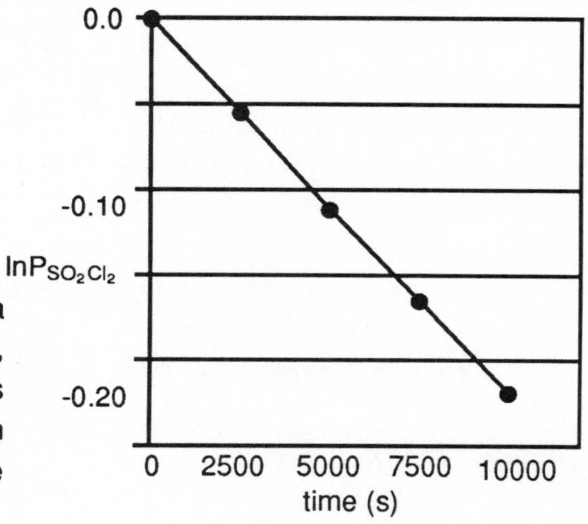

14.26

t(s)	P_{CH_3NC}	$\ln P_{CH_3NC}$
0	502	6.219
2000	335	5.814
5000	180	5.193
8000	95.5	4.559
12000	41.7	3.731
15000	22.4	3.109

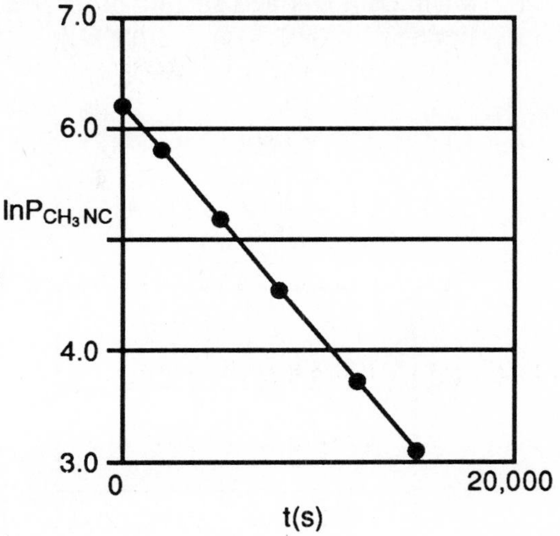

A graph of ln P vs. t is linear, with a slope of $-2.07 \times 10^{-4}\ s^{-1}$. The rate constant $k = -\text{slope} = 2.07 \times 10^{-4}\ s^{-1}$.

14.27 Make both first- and second-order plots to see which is linear.

time(s)	$N_2O_5]$ (M)	$\ln[N_2O_5]$	$1/[N_2O_5]$
0	0.90	-0.105	1.1
5	0.75	-0.288	1.3
10	0.63	-0.462	1.6
20	0.44	-0.821	2.3
50	0.15	-1.90	6.7

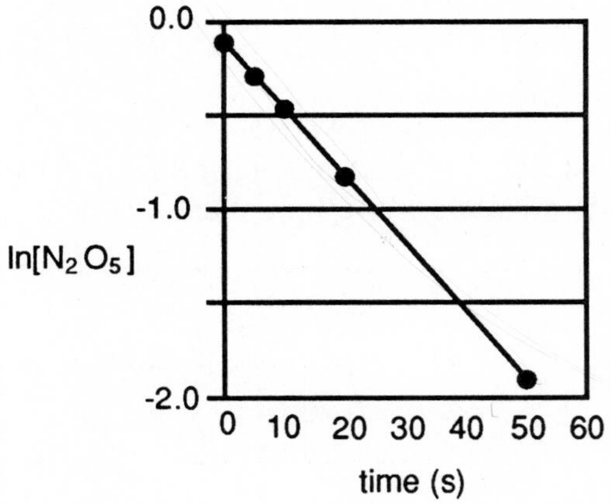

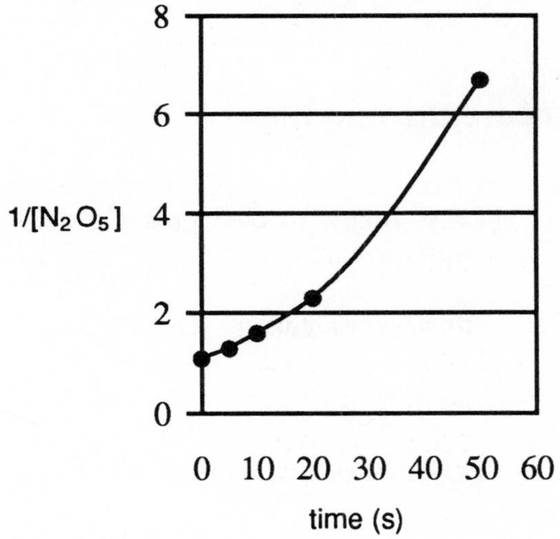

(a) The plot of $\ln[N_2O_5]$ vs. time is linear, so the reaction is first order in N_2O_5.

(b) $k = -\text{slope} = -(-1.61/45\ \text{min}) = 0.036\ \text{min}^{-1} = 6.0 \times 10^{-4} s^{-1}$

14.28 Make both first and second order plots to see which is linear.

time(s)	$[NO_2](M)$	$\ln[NO_2]$	$1/[NO_2]$
0.0	0.100	-2.303	10.0
5.0	0.017	-4.08	59
10.0	0.0090	-4.71	110
15.0	0.0062	-5.08	160
20.0	0.0047	-5.36	210

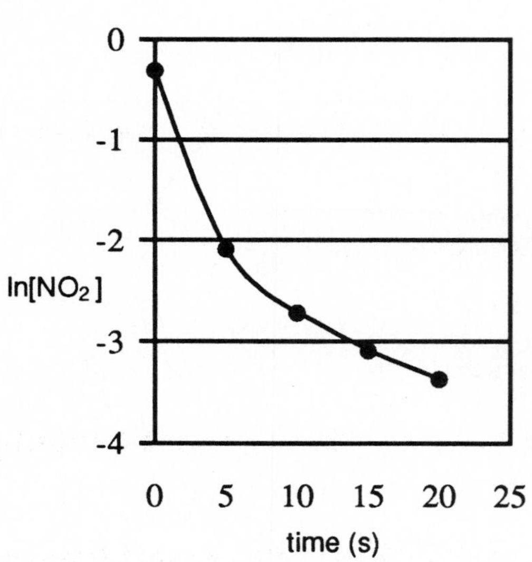

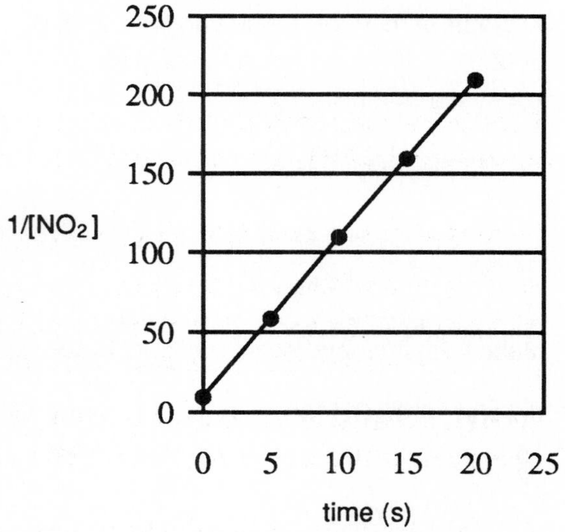

(a) The plot of $1/[NO_2]$ vs. time is linear, so the reaction is second order in NO_2.

(b) The slope of the line is $\frac{(210\text{-}10)\ M}{20\ \text{s}} = 10\ M^{-1}\ \text{s}^{-1} = \text{k}$.

Temperature and Rate

14.29 In order for a collision to lead to a chemical reaction, the reactants must have sufficient kinetic energy to overcome the activation energy barrier, and they must collide in an orientation that will lead to the desired products.

14.30 Effective orientation

```
O—O →← N—O
|
O
```

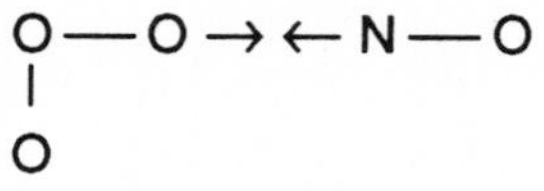

Ineffective orientation

```
O—O →← O—N
|
O
```

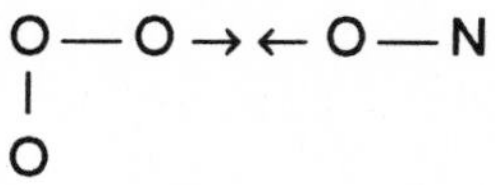

14.31 (a)

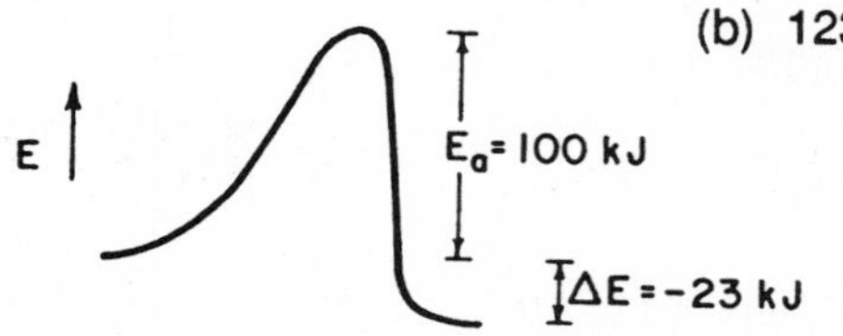

(b) 123 kJ/mol

14.32 (a)

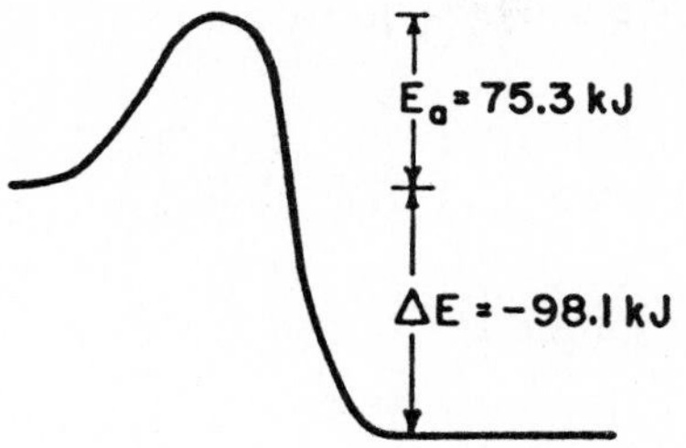

(b) 173.4 kJ/mol

14.33 Reaction rate depends only on E_a; it is independent of ΔE. Based on the magnitude of E_a, reaction (a) is fastest and reaction (c) is slowest.

14.34 In the reverse direction, $E_a(r) = E_a(f) - \Delta E(f)$. (See Figure 14.8)

(a) $E_a(r)$ = 30 kJ/mol - (-10 kJ/mol) = 40 kJ/mol
(b) $E_a(r)$ = 45 kJ/mol - 15 kJ/mol =30 kJ/mol
(c) $E_a(r)$ = 60 kJ/mol - 10 kJ/mol = 50 kJ/mol

Thus, in the reverse direction, reaction (b) is fastest and reaction (c) is slowest, since E_a is smallest for reaction (b) and largest for reaction (c).

14.35 No. The value of A, related to frequency and effectiveness of collisions, is different for each reaction and k is proportional to A.

14.36 From Equation [14.18], reactions with different variations of k with respect to temperature have different activation energies, E_a. That k for the two reactions is the same at a certain temperature is accidental. The reaction with the higher rate at 35° has the larger activation energy, because it was able to use the increase in energy more effectively.

14.37 According to Equation 14.17, $\ln\left(\frac{k_1}{k_2}\right) = \frac{E_a}{R}\left[\frac{1}{T_2} - \frac{1}{T_1}\right]$

(a) T_2 = 700°C = 973 K, T_1 = 600°C = 873 K

$$\ln\left(\frac{k_{873}}{k_{973}}\right) = \frac{182 \times 10^3 \text{ J/mol}}{8.314 \text{ J/mol}} \times \left[\frac{1}{973} - \frac{1}{873}\right] = -2.58\ ;\ \frac{k_{873}}{k_{973}} = 0.0760$$

$k_{873} = 0.0760\ (1.57 \times 10^{-3}\ M^{-1}s^{-1}) = 1.19 \times 10^{-4}\ M^{-1}s^{-1}$

(b) T_2 = 700°C = 973 K, T_1 = 800°C = 1073 K

$$\ln\left(\frac{k_{1073}}{k_{973}}\right) = \frac{182 \times 10^3 \text{ J/mol}}{8.314 \text{ J/mol}} \left[\frac{1}{973} - \frac{1}{1073}\right] = 2.10\ ;\ \frac{k_{1073}}{k_{973}} = 8.14$$

$k_{1073} = 8.14\ (1.57 \times 10^{-3}\ M^{-1}s^{-1}) = 1.28 \times 10^{-2}\ M^{-1}s^{-1}$

14.38 (a) $T_1 = 75°C + 273 = 348\ K$; $T_2 = 273\ K$; $k_2 = 3.0 \times 10^{-2} s^{-1}$

$$\ln\left(\frac{k_1}{k_2}\right) = \frac{E_a}{R}\left[\frac{1}{273} - \frac{1}{348}\right] = \frac{47.8 \times 10^3\ J/mol}{8.314\ J/mol}\left[\frac{1}{273} - \frac{1}{348}\right]$$

$$\ln\left(\frac{k_1}{k_2}\right) = 5.749 \times 10^3\ (7.894 \times 10^{-4}) = 4.5385\ ;\quad \frac{k_1}{k_2} = 93.5$$

$$k_1 = 93.5\ (3.0 \times 10^{-2} s^{-1}) = 2.8\ s^{-1}$$

(b) $$\ln\left(\frac{k_1}{k_2}\right) = \frac{125 \times 10^3\ J/mol}{8.314\ J/mol}\left[\frac{1}{273} - \frac{1}{348}\right] = 1.504 \times 10^4\ (7.894 \times 10^{-4}) = 11.873$$

$$\frac{k_1}{k_2} = 1.434 \times 10^5\ ;\quad k_1 = 1.434 \times 10^5\ (3.0 \times 10^{-2} s^{-1}) = 4.30 \times 10^3 s^{-1}$$

14.39

k	ln k	T(K)	1/T(x 10^3)
0.0521	-2.995	288	3.47
0.101	-2.293	298	3.36
0.184	-1.693	308	3.25
0.332	-1.103	318	3.14

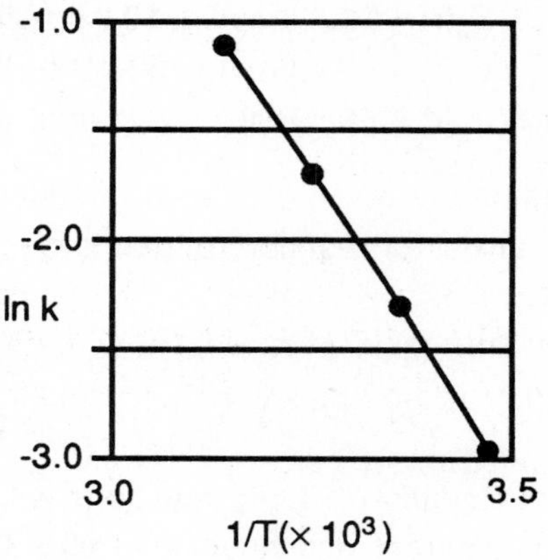

The slope, -5.6×10^3, equals $-E_a/R$. Thus, $E_a = 5.6 \times 10^3 \times 8.314$ J/mol = 47 kJ/mol.

14.40

k	ln k	T(K)	1/T(× 10^3)
0.028	-3.58	600	1.670.22
	-1.51	650	1.54
1.3	0.26	700	1.43
6.0	1.79	750	1.33
23	3.14	800	1.25

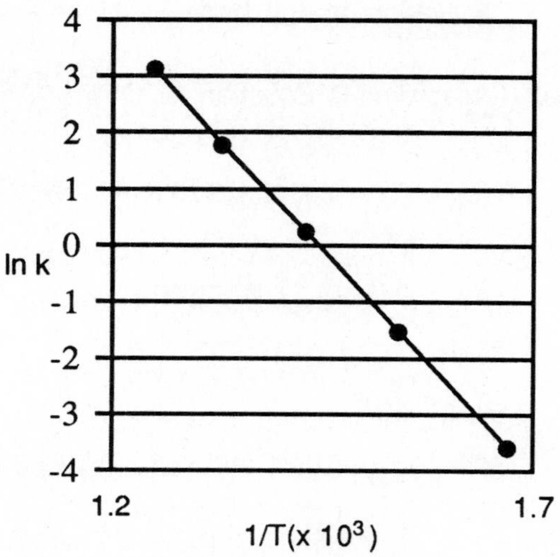

Using the relationship ln k = ln A - E_a/RT, the slope, -16×10^3, equals $-E_a/R$. Thus, $E_a = (16 \times 10^3 \times 8.314\ J/mol) = 1.3 \times 10^2$ kJ/mol. To calculate A, we will use the rate data at 700 K. From the equation given above, $0.262 = \ln A - 16.0 \times 10^3/700$; $\ln A = 0.262 + 22.857$. $A = 1.1 \times 10^{10}$.

14.41 $T_1 = 27°C = 300\ K$; $T_2 = 37°C = 310\ K$; $k_2 = 2k_1$, $k_1/k_2 = 1/2$.

$$\ln\left(\frac{k_1}{k_2}\right) = \frac{E_a}{R}\left[\frac{1}{T_2} - \frac{1}{T_1}\right] ; \quad \ln(0.5) = \frac{E_a}{8.314\ J/mol}\left[\frac{1}{310} - \frac{1}{300}\right]$$

$$E_a = \frac{\ln(0.5)\ (8.314\ J/mol)}{(-1.075 \times 10^{-4})} = \frac{+5.36 \times 10^4\ J}{mol} = \frac{+53.6\ kJ}{mol}$$

14.42 $T_1 = 40°C + 273 = 313\ K$; $T_2 = 0°C + 273 = 273\ K$

$$\ln\left(\frac{k_1}{k_2}\right) = \frac{E_a}{R}\left[\frac{1}{T_2} - \frac{1}{T_1}\right] = \frac{38.2\ kJ/mol}{8.314\ J/mol} \times \frac{1000\ J}{1\ kJ}\left[\frac{1}{273} - \frac{1}{313}\right]$$

$$\ln\left(\frac{k_1}{k_2}\right) = 4.595 \times 10^3\ (4.681 \times 10^{-4}) = 2.151 ; \quad \frac{k_1}{k_2} = 8.59$$

The reaction will occur 8.59 times faster at 40°C, assuming equal initial concentrations.

Reaction Mechanisms

14.43 The molecularity of a process indicates the number of molecules that participate as reactants in the process. A unimolecular process has one reactant molecule, a bimolecular process has two reactant molecules and a termolecular process has three reactant molecules. Termolecular processes are rare because it is highly unlikely that three molecules will simultaneously collide with the correct energy and orientation to form an activated complex.

14.44 For the reaction $A + B \rightarrow C$ (or any reaction), the rate law cannot be predicted based on the balanced chemical equation. The rate law depends on the number of elementary steps in the process and their relative speeds. That is, it depends on the reaction mechanism.

If the overall reaction is classified as an elementary step, then it occurs in a single step, and the rate law depends on the molecularity of the step. The elementary step $A + B \rightarrow C$ is bimolecular (two molecules, A and B, are involved in the activated complex), and the associated rate law is: rate = k[A] [B].

14.45 (a) bimolecular, rate = $k[N_2O][Cl]$
(b) unimolecular, rate = $k[Cl_2]$
(c) bimolecular, rate = $k[NO][Cl_2]$

14.46 (a) bimolecular, rate = $k[NO][O_3]$
(b) bimolecular, rate = $k[CO][Cl_2]$
(c) unimolecular, rate = $k[O_3]$

14.47 (a)

$$NO(g) + NO(g) \rightarrow N_2O_2(g)$$
$$N_2O_2(g) + H_2(g) \rightarrow N_2O(g) + H_2O(g)$$

$$2NO(g) + N_2O_2(g) + H_2(g) \rightarrow N_2O_2(g) + N_2O(g) + H_2O(g)$$
$$2NO(g) + H_2(g) \rightarrow N_2O(g) + H_2O(g)$$

(b) First step: $-\Delta[NO]/\Delta t = k[NO]\,[NO] = k[NO]^2$
Second step: $-\Delta[H_2]/\Delta t = k[H_2][N_2O_2]$

(c) N_2O_2 is the intermediate; it is produced in the first step and consumed in the second.

(d) Since $[H_2]$ appears in the rate law, the second step must be slow relative to the first.

14.48 (a)

i. $HBr + O_2 \rightarrow HOOBr$
ii. $HOOBr + HBr \rightarrow 2HOBr$
iii. $2HOBr + 2HBr \rightarrow 2H_2O + 2Br_2$

$$4HBr + O_2 \rightarrow 2H_2O + 2Br_2$$

(b) i. $\dfrac{-\Delta[HBr]}{\Delta t} = k[HBr]\,[O_2]$ ii. $-\Delta[HOOBr] = k[HOOBr]\,[HBr]$

iii. $\dfrac{-\Delta[HOBr]}{\Delta t} = k[HOBr]\,[HBr]$

(c) HOOBr and HOBr are both intermediates; HOOBr is produced in i and consumed in ii and HOBr is produced in ii and consumed in iii.

(d) The first order dependence of the reaction on HBr means that the second and third steps cannot be slow or rate-determining. It must be that the first step is rate-determining; thus neither HOOBr or HOBr accumulates enough to be determined.

14.49 Both (b) and (d) are consistent with the observed rate law. For a multistep reaction, the rate law is determined by the slow step and any preceeding steps. In both (b) and (d), the first step is the slow step and the rate law is $k[H_2]\,[ICl]$. The subsequent fast steps do not influence the rate law. Mechanisms (a) and (c) would lead to a rate law which is second order in ICl.

14.50 (a) rate = $k[NO][Cl_2]$

(b) Since the observed rate law is secondorder in [NO], the second step must be slow relative to the first step; the second step is rate determining.

Catalysis

14.51 A catalyst increases the rate of reaction by decreasing the activation energy, E_a, or increasing the frequency factor A. Lowering the activation energy is more common and more dramatic.

14.52 Catalysts generally <u>do</u> take part in the reaction, but there is no net consumption of the catalyst. A good example is the role of Br_2 and Br^- in catalyzing the decomposition of H_2O_2, Equations [14.27] and [14.28].

14.53 A homogeneous catalyst is in the same physical state as the reactants; a heterogeneous catalyst is in a different state and is usually a solid.

14.54 The activity of a heterogeneous catalyst depends on the total surface area and number of active sites per unit amount of catalyst, which are influenced by method of preparation and prior treatment. The total surface area depends on particle size and processing techniques and the number of active sites depends on the presence or absence of impurities and possible exposure to contaminants.

14.55 (a)

$$2[NO(g) + N_2O(g) \rightarrow N_2(g) + NO_2(g)]$$
$$2NO_2(g) \rightarrow 2NO(g) + O_2(g)$$
$$\overline{2N_2O(g) \rightarrow 2N_2(g) + O_2(g)}$$

(b) An intermediate is produced and then consumed during the course of the reaction. A catalyst is consumed and then reproduced. In other words, the catalyst is present when the reaction sequence begins and after the last step is completed. In this reaction, NO is the catalyst and NO_2 is an intermediate.

(c) Since NO is in the same state as the reactant, N_2O, the catalysis in this reaction is homogeneous.

(d) rate = $k[N_2O][NO]$ (Yes, even though NO is a catalyst, it appears in the rate law because it participates in the rate determining step.)

14.56 (a)

$$2[NO_2(g) + SO_2(g) \rightarrow NO(g) + SO_3(g)]$$
$$2NO(g) + O_2(g) \rightarrow 2NO_2(g)$$
$$\overline{2SO_2(g) + O_2(g) \rightarrow 2SO_3(g)}$$

(b) $NO_2(g)$ is a catalyst because it is consumed and then reproduced in the reaction sequence. (NO(g) is an intermediate; see Exercise 14.55(b)).

(c) Since NO_2 is in the same state as the other reactants, this is homogeneous catalysis.

14.57 Iron is a heterogeneous catalyst in this reaction because it is in a different state (s) than the reactants and products (aq). Thus, its activity depends on surface area and the number of active sites available. Finely ground iron filings would have significant advantages in both these areas.

14.58 Use of chemically stable supports such as alumina and silica makes it possible to obtain very large surface areas per unit mass of the precious metal catalyst. This is so because the metal can be deposited in a very thin, even monomolecular, layer on the surface of the support.

14.59 As illustrated in Figure 14.18, the two C-H bonds that exist on each carbon of the ethylene molecule before adsorption are retained in the process in which a D atom is added to each C (assuming we use D_2 rather than H_2). To put two deuteriums on a single carbon, it is necessary that one of the already existing C-H bonds in ethylene be broken while the molecule is adsorbed, so the H atom moves off as an adsorbed atom, and is replaced by a D. This requires a larger activation energy than simply adsorbing C_2H_4 and adding one D atom to each carbon.

14.60 Replacement of hydrogen by $-CH_3$ or other organic groups would reduce the tendency to undergo hydrogenation. It would be difficult for the C=C bond to get close enough to the metal so that the π electrons of the alkene could interact with the metal orbitals to form a stable binding arrangement. With a $-CH_3$ or other bulky group blocking the way, it would also be difficult for an adsorbed H to find its way into the vicinity of the C=C.

14.61 An enzyme is a very large biological molecule that acts as a catalyst for a specific reaction or reactions. The active site is the specific location on the large enzyme molecue where the reacting substance or substrate is bound. Binding to the active site can activate the substrate by distorting its molecular structure or electron density distribution, and enables the substrate to react very quickly.

14.62 An enzyme inhibitor prevents the substrate from binding to the active site, by either occupying the active site itself or attaching to some other area of the protein so that the shape of the active site is changed significantly, and the substrate no longer fits the "lock."

14.63 The blood or body fluids at the surface of the wound contain some enzyme, probably catalase, which catalyzes the decomposition of hydrogen peroxide.

14.64 Enzymes and proteins are biopolymers, with much of the same structural flexibility as synthetic polymers (Chapter 12). The three dimensional shape of the protein is determined by many relatively weak intermolecular interactions and is sensitive to changes in local environment. Changes in temperature change the kinetic energy of the various groups on the enzyme and their tendency to form intermolecular associations or break free from them. Thus, changing the temperature changes the overall shape of the protein and specifically the shape of the active site. At body temperature, the competition between kinetic energy driving groups apart and intermolecular attraction pulling them together forms an active site which is optimum for a specific substrate. At other temperatures, a different structural equilibrium is reached, the shape of the active site is slightly different and the enzyme is less active.

Additional Exercises

14.65 The rate of a reaction is determined by the number of "effective" collisions (correct energy and orientation) of the reacting molecules. Concentration determines the number of molecules available for collision. Temperature regulates the speed of the molecules and thus the total number of collisions. It also influences the kinetic energy with which the molecules collide. Catalysts bring molecules together in the correct orientation for reaction, increasing the number of "effective" collisions.

14.66 $\text{rate} = \frac{-\Delta[H_2S]}{\Delta t} = \frac{\Delta[Cl^-]}{2\Delta t} = k[H_2S][Cl_2]$

$$\frac{-\Delta[H_2S]}{\Delta t} = (3.5 \times 10^{-2}\ M^{-1}s^{-1})(1.6 \times 10^{-4}\ M)(0.070\ M) = 3.9 \times 10^{-7}\ M/s$$

$$\frac{\Delta[Cl^-]}{\Delta t} = \frac{2\Delta[H_2S]}{\Delta t} = 2(3.9 \times 10^{-1}\ M/s) = 7.8 \times 10^{-7}\ M/s$$

14.67 (a) Doubling the $[H_2O]$ while holding $[CH_3Cl]$ constant (experiments 1 and 4), increases the rate by a factor of 4. Increasing the $[CH_3Cl]$ by a factor of 1.5 and holding $[H_2O]$ constant (experiments 1 and 2) increases the rate by a factor of 1.5 The reaction is second order in H_2O and first order in CH_3Cl. The rate law is: rate = $k[CH_3Cl][H_2O]^2$.

(b) The reaction is third order overall.

(c) At t = 0, $k = \frac{\text{rate}}{[CH_3Cl][H_2O]^2} = \frac{22.7\ M/s}{(0.500\ M)^2(0.500\ M)} = 182\ M^{-2}s^{-1}$

14.68 (a) Note that the rate increases by a factor of four when $[C_2O_4^{2-}]$ doubles (compare Expt 3 with 4, or 2 with 1). The rate doubles when $[HgCl_2]$ doubles (compare Expt 1 with 4, or 2 with 3). The rate law is apparently: rate = $k[HgCl_2][C_2O_4^{2-}]^2$

(b) $k = \frac{\text{rate}}{[HgCl_2][C_2O_4^{2-}]^2}$ Using the data for Experiment 3,

$$k = \frac{(3.5 \times 10^{-5}\ M/s)}{[0.052\ M][0.30\ M]^2} = 7.5 \times 10^{-3}\ M^{-2}s^{-1}$$

(c) rate = $(7.5 \times 10^{-3}\ M^{-2}s^{-1})(0.080\ M)(0.10\ M)^2 = 6.0 \times 10^{-6}\ M/s$

14.69 (a) $k = (8.56 \times 10^{-5}\ M/s)/(0.200\ M) = 4.28 \times 10^{-4}\ s^{-1}$

(b) $\ln[\text{urea}] = -(4.28 \times 10^{-4}\ s^{-1} \times 5.00 \times 10^3\ s) + \ln(0.500)$

$= -2.14^\circ - 0.693 = -2.833$; $[\text{urea}] = 0.0588\ M$

(c) $t_{1/2} = 0.693/k = 0.693/4.28 \times 10^{-4}\ s^{-1} = 1.62 \times 10^3\ s$

14.70 (a) Because a plot of $\ln[SO_2Cl_2]$ vs. t is linear, the reaction is first order. The rate law is rate = $k[SO_2Cl_2]$. For a first order reaction, Equation [14.9] is appropriate.

$$\ln \frac{[A]_t}{[A]_o} = -kt \; ; \quad k = \frac{-\ln [A]_t / [A]_o}{t}$$

$$k = \frac{-\ln(0.280 / 0.400)}{240 \text{ s}} = \frac{-(-0.357)}{240 \text{ s}} = 1.49 \times 10^{-3} \text{ s}^{-1}$$

(b) For a first order reaction, $t_{1/2} = 0.693 / k$ (Equation [14.12]).

$t_{1/2} = 0.693 / 1.49 \times 10^{-3} \text{s}^{-1} = 465 \text{ s} = 7.75 \text{ min}$

14.71

Time (s)	$[C_5H_6]$ (*M*)	$\ln[C_5H_6]$	$1/[C_5H_6]$
0	0.0400	-3.22	25.0
50	0.0300	-3.51	33.3
100	0.0240	-3.73	41.7
150	0.0200	-3.91	50.0
200	0.0174	-4.05	57.5

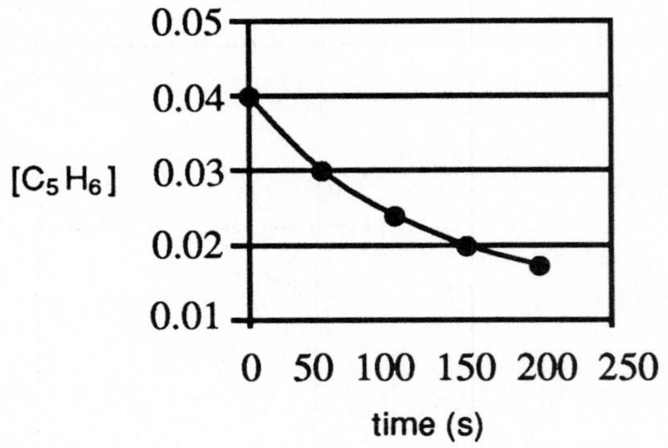

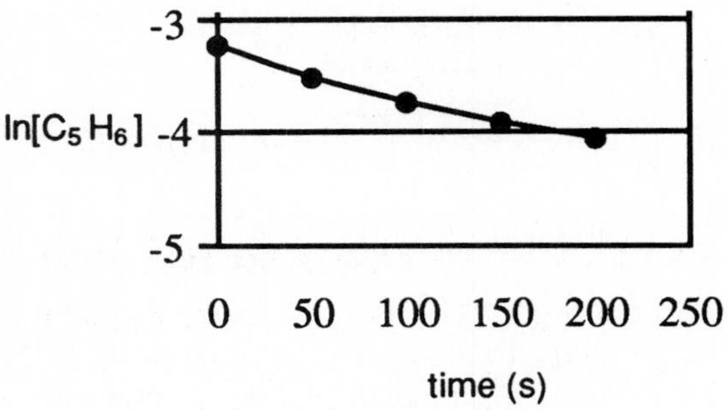

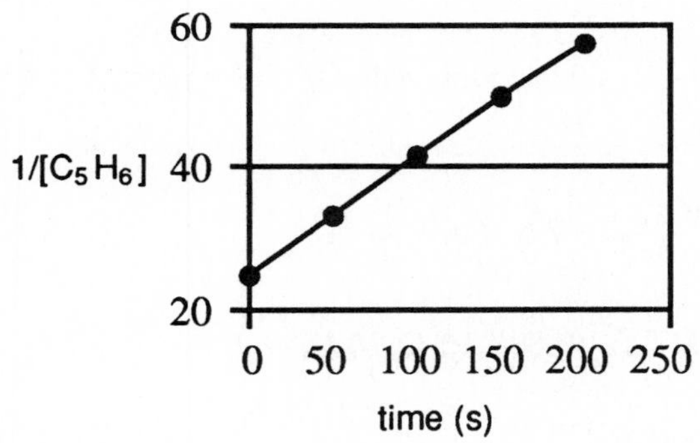

The plot of $1/[C_5H_6]$ vs. t is linear and the reaction is second order. The slope of this line is k.

$$k = \text{slope} = \frac{(50.0 - 25.0)\ M^{-1}}{(150 - 0)\ \text{s}} = 0.167\ M^{-1}\ \text{s}^{-1}$$

14.72

% uno	ln (% uno)	time (days)
79	4.37	1
63	4.14	2
50	3.91	3
40	3.69	4
32	3.47	5
25	3.22	6
20	3.00	7
10	2.30	10
1	0.00	20

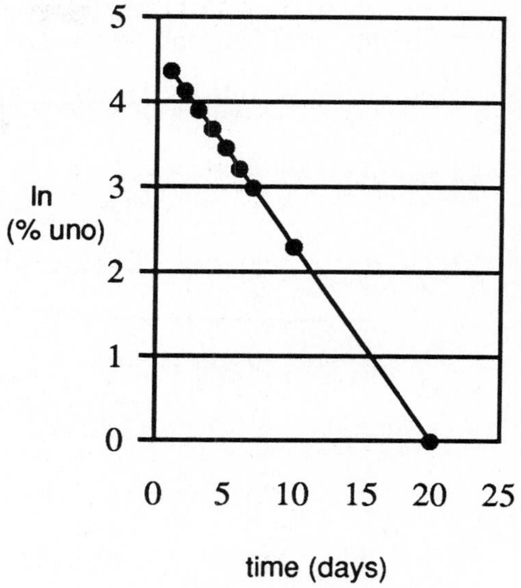

The data can be tested for first-order reaction behavior by graphing ln [percentage unoxidized organic matter] vs. time. That is, ln(100 - % oxidized organic matter) vs. time. The graph is linear with a slope = 0.10.

k = -slope. In this case, k = 0.23/day.

14.73

ln k	**1/T**
-11.51	3.33×10^{-3}
-9.90	3.13×10^{-3}
-8.52	2.94×10^{-3}
-7.60	$2.82 \times ^{-3}$

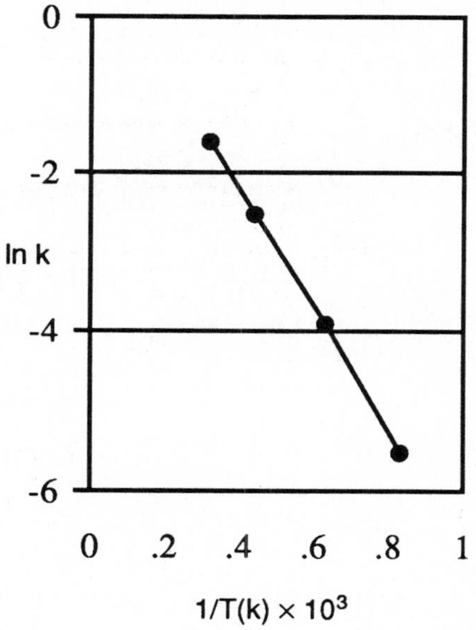

The activation energy E_a, equals (-slope) (8.314 J/mol). Thus, $E_a = 7.8 \times 10^3(8.314) = 6.5 \times 10^4$ J/mol = 65 kJ/mol

14.74 Convert temperatures to K, then use Equation [14.18].

$$\ln (k_{773}/k_{873}) = \frac{E_a}{R}\left[\frac{1}{873} - \frac{1}{773}\right]$$

$$\ln k_{773} - \ln(0.75) = \frac{-114 \times 10^3 \text{ J/mol}}{8.314 \text{ J/mol}} \times 1.48 \times 10^{-4}$$

$$\ln k_{773} = -2.029 - 0.288; \quad k_{773} = 9.86 \times 10^{-2}\ M^{-1}s^{-1}$$

14.75 (a) rate = $k[H_2O_2][I^-]$

(b) $2H_2O_2(aq) \rightarrow 2H_2O(l) + O_2(g)$

(c) $IO^-(aq)$ is the intermediate.

14.76 (a)

$$Cl_2(g) \rightleftharpoons 2Cl(g)$$
$$Cl(g) + CHCl_3(g) \rightarrow HCl(g) + CCl_3(g)$$
$$Cl(g) + CCl_3(g) \rightarrow CCl_4(g)$$

$$Cl_2(g) + 2Cl(g) + CHCl_3(g) + CCl_3(g) \rightarrow 2Cl(g) + HCl(g) + CCl_3(g) + CCl_4(g)$$
$$Cl_2(g) + CHCl_3(g) \rightarrow HCl(g) + CCl_4(g)$$

(b) $Cl(g)$, $CCl_3(g)$

(c) Step 1 - unimolecular, Step 2 - bimolecular, Step 3 - bimolecular

(d) Step 2, the slow step, is rate determining.

(e) If Step 2 is rate determining, rate = $k_2[CHCl_3][Cl]$. Cl is an intermediate formed in Step 1, an equilibrium. By definition, the rates of the forward and reverse processes are equal: $k_1[Cl_2] = k_{-1}[Cl]^2$. Solving for [Cl] in terms of $[Cl_2]$,

$$[Cl]^2 = \frac{k_1}{k_{-1}}[Cl_2]\ ;\ [Cl] = \left(\frac{k_1}{k_{-1}}[Cl_2]\right)^{1/2}$$

Substituting into the overall rate law

$$\text{rate} = k_2\left(\frac{k_1}{k_{-1}}\right)^{1/2}[CHCl_3][Cl_2]^{1/2} = k[CHCl_3][Cl_2]^{1/2}$$

(The overall order is 3/2.)

14.77 (a) $(CH_3)_3AuPH_3 \rightarrow C_2H_6 + (CH_3)AuPH_3$

(b) $(CH_3)_3Au$, $(CH_3)Au$ and PH_3 are intermediates.

(c) Step 1 is unimolecular, Step 2 is unimolecular, Step 3 is bimolecular.

(d) Step 2, the slow step, is rate determining.

(e) If Step 2 is rate determining, rate = $k_2[(CH_3)_3Au]$.

$(CH_3)_3Au$ is an intermediate formed in Step 1, an equilibrium. By definition, the rates of the forward and reverse processes in Step 1 are equal:

$k_1[(CH_3)_3AuPH_3] = k_{-1}[(CH_3)_3Au][PH_3]$ solving for $[(CH_3)_3Au]$,

$$[(CH_3)_3Au] = \frac{k_1[(CH_3)_3AuPH_3]}{k_{-1}[PH_3]}$$

Substituting into the rate law

$$\text{rate} = \left(\frac{k_2 k_1}{k_{-1}}\right)\frac{[(CH_3)_3AuPH_3]}{[PH_3]} = \frac{k[(CH_3)_3AuPH_3]}{[PH_3]}$$

(f) The rate is inversely proportional to $[PH_3]$, so adding PH_3 to the $(CH_3)_3AuPH_3$ solution would decrease the rate of the reaction.

14.78 Let k = the rate constant for the uncatalyzed reaction. According to Equation [14.17]:

$$\ln k = \frac{-E_a}{RT} + \ln A$$

Assuming that T and A are unchanged in the presence of a catalyst, let k_c = the catalyzed rate constant and E_{ac} the catalyzed activation energy.

Subtracting ln k from ln k_c

$$\ln k_c - \ln k = \left[-\frac{E_{ac}}{RT} + \ln A\right] - \left[-\frac{E_a}{RT} + \ln A\right].$$

$$\ln\left(\frac{k_c}{k}\right) = \frac{[E_a - E_{ac}]}{RT}\ ; \quad [E_a - E_{ac}] = RT\ln\left(\frac{k_c}{k}\right)$$

$$E_{ac} = E_a - RT\ln\left(\frac{k_c}{k}\right)\ ; \quad E_a = \frac{45.0\text{ kJ}}{\text{mol}} = \frac{4.50\times10^4\text{ J}}{\text{mol}}\ ; \quad T = 310\text{ K}$$

$$E_{ac} = \frac{4.50\times10^4\text{ J}}{\text{mol}} - \frac{8.314\text{ J}}{\text{K-mol}}\times310\text{ K}\times\ln\left(\frac{2.0\times10^{-2}\text{ s}^{-1}}{5.0\times10^{-4}\text{ s}^{-1}}\right)$$

$$E_{ac} = \frac{4.50\times10^4\text{ J}}{\text{mol}} - \frac{8.314\text{ J}}{\text{K-mol}}\times310\text{ K}\times3.69 = \frac{4.50\times10^4\text{ J}}{\text{mol}} - \frac{0.951\times10^4\text{ J}}{\text{mol}}$$

$$E_{ac} = \frac{3.55\times10^4\text{ J}}{\text{mol}} = \frac{35.5\text{ kJ}}{\text{mol}}$$

The enzyme must lower the activation energy by approximately 10 kJ/mol (9.5 kJ/mol) for the reaction to be viable in the body.

14.79 (a) A zero-order dependence on the concentration of a reactant means that the rate is independent of the concentration of that reactant, so long as there is any present.

(b) When there are many vacant active sites on the catalyst surfaces, increasing the ethylene pressure increases the likelihood that a site will be occupied by an ethylene molecule, so that reaction can occur faster. When these conditons hold, the rate is first-order in ethylene pressure. However, as ethylene pressure increases, an increasing fraction of the sites are occupied. When all (or nearly all) the active sites are occupied, increasing the ethylene pressure further does not increase the number of occupied active sites or the rate. In this pressure region the reaction is zero-order in ethylene pressure.

$$\ln k_c = \frac{-E_{ac}}{RT} + \ln A$$

14.80 Enzyme--carbonic anhydrase; substrate--carbonic acid (H_2CO_3); turnover number--1×10^7 molecules/s.

14.81 The fact that the rate doubles with a doubling of the concentration of sugar tells us that the fraction of enzyme tied up in the form of an enzyme-substrate complex is small. A doubling of the substrate concentration leads to a doubling of the concentration of enzyme-substrate complex, because most of the enzyme molecules are available to bind substrates. The behavior of innositol suggests that it acts as a competitor with sucrose for binding at the active sites of the enzyme system. Such a competition results in a lower effective concentration of active sites for binding of sucrose, and thus results in a lower reaction rate.

CHAPTER 15

Chemical Equilibrium

The Concept of Equilibrium; Equilibrium Expressions

15.1 (a) At equilibrium the forward and reverse reactions proceed at equal rates. Reactants are continually transformed into products, but products are also transformed into reactants at the same rate, so the net concentrations of reactants and products are constant at equilibrium.

(b) At equilibrium, the net concentrations of reactants and products are **constant** (see part (a)), but not necessarily equal. The relative concentrations of reactants and products at equilibrium are determined by their initial concentrations and the value of the equilibrium constant.

15.2 In this equilibrium system, ions move from the surface of the solid into solution, while ions in solution are deposited on the surface of the solid. As ions leave the surface of the small particles of powder, the reverse process preferentially deposits ions on the surface of a single crystal. Eventually, all ions that were present in the 10.0 g of powder are deposited on the surface of a 10.0 g crystal; this can only happen if the dissolution and deposition processes are ongoing.

15.3 $K = \frac{k_f}{k_r}$, Equation [15.3] ; $K = \frac{6.2 \times 10^6 \text{ s}^{-1}}{4.8 \times 10^3 \text{ s}^{-1}} = 1.3 \times 10^3$

15.4 $\text{rate}_f = k_f [A] [B]$; $\text{rate}_r = k_r [C] [D]$

At equilibrium, $\text{rate}_f = \text{rate}_r$: $k_f[A] [B] = k_r[C] [D]$; $\frac{k_f}{k_r} = \frac{[C] [D]}{[A] [B]} = K$

15.5 (a) $K_c = \frac{[HCl]^2}{[H_2][Cl_2]}$ (b) $K_c = \frac{[NO_2]^2 [O_2]}{[N_2O_5]^2}$ (c) $K_c = \frac{[CS_2][H_2]^4}{[CH_4][H_2S]^2}$

(d) $K_c = \frac{[CO]^2}{[CO_2]}$ (e) $K_c = \frac{[H_2O]}{[H_2]}$

Homogeneous: (a), (b), (c); heterogeneous: (d), (e)

15.6 (a) $K_c = \frac{[CO]^2[O_2]}{[CO_2]^2}$, $K_p = \frac{P_{CO}^2 P_{O_2}}{P_{CO_2}^2}$ (b) $K_c = \frac{[H_2O]^2[SO_2]^2}{[H_2S]^2[O_2]^3}$, $K_p = \frac{P_{H_2O}^2 P_{SO_2}^2}{P_{H_2S}^2 P_{O_2}^3}$

(c) $K_c = \frac{[H_2]}{[HCl]^2}$, $K_p = \frac{P_{H_2}}{P_{HCl}^2}$ (d) $K_c = \frac{[N_2]^2}{[NH_3]^4[O_2]^3}$, $K_p = \frac{P_{N_2}^2}{P_{NH_3}^4 P_{O_2}^3}$

(e) $K_c = [N_2]$, $K_p = P_{N_2}$

Homogeneous: (a), (b); heterogeneous: (c), (d), (e)

15.7 (a) $2SO_2(g) + O_2(g) \rightleftharpoons 2SO_3(g)$ is the reverse of the reaction given.

$K_c' = (K_c)^{-1} = 1/2.4 \times 10^{-3} = 4.2 \times 10^2$

(b) Since $K_c < 1$ (when SO_3 is the reactant) and $K_c' > 1$ (when SO_3 is the product), the equilibrium favors SO_3 at this temperature.

15.8 (a) $K_c' = (K_c)^{-1} = 1/2.4 \times 10^3 = 4.2 \times 10^{-4}$

(b) Since $K_c > 1$ when N_2 and O_2 are products and $K_c' < 1$ when N_2 and O_2 are reactants, the equilibrium favors N_2 and O_2 at this temperature.

Calculating Equilibrium Constants

15.9 $K_c = \frac{[H_2][I_2]}{[HI]^2} = \frac{(4.79 \times 10^{-4})(4.79 \times 10^{-4})}{(3.53 \times 10^{-3})^2} = 1.84 \times 10^{-2}$

15.10 $K_c = \frac{[H_2][CO]}{[H_2O]} = \frac{(4.0 \times 10^{-2})(4.0 \times 10^{-2})}{1.0 \times 10^{-2}} = 0.16$

$K_p = K_c(RT)^{\Delta n} = 0.16(0.0821 \times 1073) = 14$

15.11 (a) Since the reaction is carried out in a 1.00 L vessel, the moles of each component are equal to the molarity.

	$2NO$	+ $2H_2$	$\rightleftharpoons$ N_2	+ $2H_2O$
initial	0.100 *M*	0.050 *M*	0 *M*	0.100 *M*
change	-0.038 *M*	-0.038 *M*	+0.019 *M*	+0.038 *M*
equil.	0.062 *M*	0.012 *M*	0.019 *M*	0.138 *M*

First calculate the change in [NO], 0.100 - 0.062 = 0.038 *M*. From the stoichiometry of the reaction, calculate the change in the other concentrations. Finally, calculate the equilibrium concentrations.

(b) $K_c = \frac{[N_2][H_2O]^2}{[NO]^2[H_2]^2} = \frac{(0.019)(0.138)^2}{(0.062)^2(0.012)^2} = 6.5 \times 10^2$

15.12 (a) Calculate the initial concentrations of $H_2(g)$ and $Br_2(g)$ and the equilibrium concentration of H_2.

$$\frac{1.374\ H_2}{2.00\ L} \times \frac{1\ mol\ H_2}{2.016\ g\ H_2} = 0.341\ M\ H_2\ ;\quad \frac{70.31\ g\ Br_2}{2.00\ L} \times \frac{1\ mol\ Br_2}{159.8\ g\ Br_2} = 0.220\ M\ Br_2$$

$$\frac{0.566\ g\ H_2}{2.00\ L} \times \frac{1\ mol\ H_2}{2.016\ g\ H_2} = 0.140\ M\ H_2$$

	$H_2(g)$	+	$Br_2(g)$	$\rightleftharpoons$	$2HBr(g)$
initial	0.341 M		0.220 M		0
change	-0.201 M		-0.201 M		+2(0.201)
equil.	0.140 M		0.019 M		0.402 M

The change in $[H_2]$ is (0.341 M - 0.140 M = -0.201 M). The changes in $[Br_2]$ and [HBr] are set by stoichiometry, resulting in the equilibrium concentrations shown in the table.

(b) $$K_c = \frac{[HBr]^2}{[H_2][Br_2]} = \frac{(0.402)^2}{(0.140)(0.019)} = 61$$

15.13 (a) $K_p = K_c(RT)^{\Delta n}$; $\Delta n = +1$; T = 1285 + 273 = 1558 K

$K_p = 1.04 \times 10^{-3}\ [(0.0821)(1558)]^1 = 0.133$

(b) $K_p = K_c(RT)^{\Delta n}$; $K_c = K_p/(RT)^{\Delta n}$; $\Delta n = 0$

$K_c = K_p/(RT)^0 = K_p/1 = K_p = 0.11$

It is true in general that if $\Delta n = 0$, $K_p = K_c$.

15.14 (a) $K_p = K_c(RT)^{\Delta n}$; $\Delta n = -1$; T + 700 + 273 = 973 K

$$K_p = \frac{416}{(0.0821)(973)} = 5.21$$

(b) $K_p = K_c(RT)^{\Delta n}$; $K_c = K_p/(RT)^{\Delta n}$; $\Delta n = +2$, T = 480 + 273 = 753 K

$$K_c = \frac{0.0752}{(0.0821 \times 753)^2} = 1.97 \times 10^{-5}$$

15.15 (a) $$K_p = \frac{P_{PCl_5}}{P_{PCl_3} \times P_{Cl_2}} = \frac{1.30\ atm}{0.124\ atm \times 0.157\ atm} = 66.8$$

(b) Since $K_p > 1$, products (the numerator of the K_p expression) are favored over reactants (the denominator of the K_p expression).

15.16 $2NO(g) + Cl_2(g) \rightleftharpoons 2NOCl$

$$K_p = \frac{P^2_{NOCl}}{P^2_{NO} \times P_{Cl_2}} = 52.0\ ;\ P^2_{NOCl} = 52.0\,(P^2_{P_{NO}} \times P_{Cl_2})$$

$$P_{NOCl} = [52.0(P^2_{NO} \times P_{Cl_2})]^{1/2} = [52.0\,((0.095)^2 \times 0.171)]^{1/2}$$

$$P_{NOCl} = 0.28 \text{ atm}$$

15.17 (a)

	$N_2O_4(g)$ $\rightleftharpoons$	$2NO_2(g)$
initial	0.50 atm	0.50 atm
change	+0.10 atm	-0.20 atm
equil.	0.60 atm	0.30 atm

The change in $P_{N_2O_4}$ is (0.60 atm - 0.50 atm) = +0.10 atm, so the change in P_{NO_2} is - (2 × 0.10) = -0.20 atm and the equilibrium P_{NO_2} is (0.50 atm - 0.20 atm) = 0.30 atm.

(b) $$K_p = \frac{(P_{NO_2})^2}{P_{N_2O_4}} = \frac{(0.30)^2}{(0.60)} = 0.15$$

15.18 First, calculate the number of moles of each component present.

$$\frac{3.22 \text{ g NOBr}}{110 \text{ g/mol}} = 0.0293 \text{ mol NOBr};\quad \frac{3.08 \text{ g NO}}{30.0 \text{ g/mol}} = 0.103 \text{ mol NO}$$

$$\frac{4.19 \text{ g } Br_2}{160 \text{ g/mol}} = 0.0262 \text{ mol } Br_2$$

(a) In calculating K_c, divide each number of moles by 5.00 L to convert to moles/L, then insert into the expression for K_c to obtain:

$$K_c = \frac{[Br_2]\,[NO]^2}{[NOBr]^2} = \frac{(5.24 \times 10^{-3})\,(2.06 \times 10^{-2})^2}{(5.86 \times 10^{-3})^2} = 6.42 \times 10^{-2}$$

(b) $K_p = (6.42 \times 10^{-2})\,(0.0821 \times 373) = 1.98$

(c) The total moles of gas present is 0.0293 + 0.103 + 0.0262 = 0.158

$$P = (0.158 \text{ mol}) \times \frac{0.0821 \text{ L} \cdot \text{atm}}{1 \text{ mol} \cdot \text{K}} \times \frac{373 \text{ K}}{5.00 \text{ L}} = 0.968 \text{ atm}$$

Applications of Equilibrium Constants

15.19 $K_c = \dfrac{[CO][Cl_2]}{[COCl_2]} = 2.19 \times 10^{-10}$ at 100°C

(a) $Q = \dfrac{(3.31 \times 10^{-6})(3.31 \times 10^{-6})}{(5.00 \times 10^{-2})} = 2.19 \times 10^{-10}$; $Q = K_c$

The mixture is at equilibrium.

(b) $Q = \dfrac{(1.11 \times 10^{-5})(3.25 \times 10^{-6})}{(3.50 \times 10^{-3})} = 1.03 \times 10^{-8}$; $Q > K_c$.

The reaction will proceed to the left to attain equilibrium.

(c) $Q = \dfrac{(1.56 \times 10^{-6})(1.56 \times 10^{-6})}{(1.45)} = 1.68 \times 10^{-12}$, $Q < K_c$.

The reaction will proceed to the right to attain equilibrium.

15.20 Calculate the reaction quotient in each case, compare with

$K_p = \dfrac{P^2_{NH_3}}{P_{N_2} \times P^3_{H_2}} = 4.51 \times 10^{-5}$

(a) $Q = \dfrac{(105)^2}{(35)(495)^3} = 2.6 \times 10^{-6}$

Since $Q < K_p$, reaction will shift to the right to attain equilibrium.

(b) $Q = \dfrac{(35)^2}{(0)(595)^3} = \infty$

Since $Q > K_p$, reaction must shift to the left to attain equilibrium. There must be some N_2 present to attain equilibrium. In this example, the only source of N_2 is the decomposition of NH_3.

(c) $Q = \dfrac{(26)^2}{(42)^3(202)} = 4.52 \times 10^{-5}$; $Q = K_p$ Reaction is at equilibrium.

(d) $Q = \dfrac{(105)^2}{(5.0)(55)^3} = 1.3 \times 10^{-2}$; $Q > K_p$

Reaction will proceed to the left to attain equilibrium.

15.21 $N_2(g) + O_2(g) \rightleftharpoons 2NO(g) \quad K_c = \frac{[NO]^2}{[N_2][O_2]} = 4.1 \times 10^{-4}$

	$N_2(g)$	$O_2(g)$	$2NO(g)$
initial	0.20 *M*	0.10 *M*	0
change	-x	-x	+2x
equil.	(0.20-x) *M*	(0.10-x) *M*	+2x *M*

$4.1 \times 10^{-4} = \frac{(2x)^2}{(0.20\text{-}x)(0.10\text{-}x)} \approx \frac{(2x)^2}{(0.20)(0.10)}$ (assuming x is small)

$8.2 \times 10^{-6} = 4x^2$, $x = 1.4 \times 10^{-3}$ (x is small compared to 0.10)

$[NO] = 2x = 2.8 \times 10^{-3}$ *M*

15.22 $Br_2(g) + Cl_2(g) \rightleftharpoons 2BrCl(g) \quad K_c = \frac{[BrCl]^2}{[Br_2][Cl_2]} = 7.0$

	$Br_2(g)$	$Cl_2(g)$	$2BrCl(g)$
initial	0.50 *M*	0.50 *M*	0
change	-x	-x	+2x
equil.	(0.50-x) *M*	(0.50-x) *M*	+2x

$7.0 = \frac{(2x)^2}{(0.50\text{-}x)^2}$ (Assuming x is small leads to x = 0.66 *M*, [BrCl] = 1.3 *M*. Clearly x is not small compared to 0.50 *M*.)

$(7.0)^{1/2} = \frac{2x}{0.50\text{-}x}$, $2.6(0.50\text{-}x) = 2x$, $1.3 = 4.6x$, $x = 0.28$

[BrCl] = 2x = 0.56 *M* (To 3 sig figs, x = 0.285 *M*, [BrCl] = 0.569 *M*)

15.23 $K_c = 1.04 \times 10^{-3} = \frac{[Br]^2}{[Br_2]}$; $[Br_2] = \frac{0.245 \text{ g } Br_2}{159.8 \text{ g } Br_2/\text{mol} \times 0.200 \text{ L}} = 7.67 \times 10^{-3}$ *M*

$[Br] = (1.04 \times 10^{-3}[Br_2])^{1/2} = (1.04 \times 10^{-3}(7.67 \times 10^{-3}))^{1/2} = 2.82 \times 10^{-3}$ *M*

$\text{g Br} = \frac{2.82 \times 10^{-3} \text{ mol Br}}{1 \text{ L}} \times \frac{79.9 \text{ g Br}}{1 \text{ mol Br}} \times 0.200 \text{ L} = 0.0451 \text{ g Br}$

$[Br_2] = 7.67 \times 10^{-3}$ *M*; $[Br] = 2.82 \times 10^{-3}$ *M*; 0.0451 g Br

15.24 $PV = nRT$; $P = \frac{gRT}{\mathcal{M}V}$

$P_{H_2} = \frac{0.056 \text{ g } H_2}{2.016 \text{ g/mol}} \times \frac{0.0821 \text{ L} \cdot \text{atm}}{\text{K} \cdot \text{mol}} \times \frac{700 \text{ K}}{2.000 \text{ L}} = 0.80 \text{ atm}$

$P_{I_2} = \frac{4.36 \text{ g } I_2}{253.8 \text{ g/mol}} \times \frac{0.0821 \text{ L} \cdot \text{atm}}{\text{K} \cdot \text{mol}} \times \frac{700 \text{ K}}{2.000 \text{ L}} = 0.494 \text{ atm}$

$K_p = 55.3 = \frac{P_{HI}^2}{P_{H_2} \times P_{I_2}}$; $P_{HI} = [55.3(P_{H_2})(P_{I_2})]^{1/2} = [55.3(0.80)(0.494)]^{1/2} = 4.67 \text{ atm}$

$g_{HI} = \frac{\mathcal{M}_{HI} P_{HI} V}{RT} = \frac{128.0 \text{ g HI}}{\text{mol HI}} \times \frac{\text{K} \cdot \text{mol}}{0.0821 \text{ L} \cdot \text{atm}} \times \frac{4.67 \text{ atm} \times 2.000 \text{ L}}{700 \text{ K}} = 20.8 \text{ g HI}$

Rounding to two significant figures, 21 g HI are present.

15.25 $K_p = \frac{P^2_{CO}}{P_{CO_2}} = 167.5$. When $P_{CO} = 0.500$ atm

$P_{CO_2} = (0.500)^2/167.5 = 1.49 \times 10^{-3}$ atm

15.26 $K_p = \frac{P^2_{NO}\, P_{Br_2}}{P^2_{NOBr}}$

When $P_{NOBr} = P_{NO}$, these terms cancel and $P_{Br_2} = K_p = 0.416$ atm. This is true for all cases where $P_{NOBr} = P_{NO}$.

15.27 $K_c = [H_2S][NH_3] = 1.2 \times 10^{-4}$. The concentrations of H_2S and NH_3 will be the same; call this quantity y. Then, $y^2 = 1.2 \times 10^{-4}$, $y = 1.1 \times 10^{-2}$ *M*.

15.28 (a) Beginning with only $PH_3BCl_3(s)$, the equation requires that at equilibrium the partial pressures of PH_3 and BCl_3 are equal. Call this pressure y.

$K_p = P_{PH_3} \times P_{BCl_3}$; $1.57 = y^2$; $y = 1.25$ atm PH_3 and BCl_3

(b) Calculate the moles of PH_3 or BCl_3 that would produce this pressure.

$$n = \frac{PV}{RT} = \frac{1.25 \text{ atm} \times 0.500 \text{ L}}{353 \text{ K}} \times \frac{\text{K} \cdot \text{mol}}{0.0821 \text{ L} \cdot \text{atm}} = 0.0216 \text{ mol } PH_3$$

At least 0.0216 mol PH_3BCl_3 would be required.

$$0.0216 \text{ mol } PH_3BCl_3 \times \frac{151 \text{ g } PH_3BCl_3}{1 \text{ mol}} = 3.26 \text{ g } PH_3BCl_3$$

15.29 $K_c = 280 = \frac{[IBr]^2}{[I_2][Br_2]}$; $[IBr]$ initial $= \frac{0.500 \text{ mol}}{1.000 \text{ L}} = 0.500$ M

	I_2	+ Br_2	$\rightleftharpoons$ $2IBr$
initial	0 *M*	0 *M*	0.500 *M*
change	+ x	+ x	-2x
equil.	x	x	0.500-2x

Since no I_2 or Br_2 were present initially, the amounts present at equilibrium are produced by the reverse reaction and stoichiometrically equal. Let these amounts equal x. The amount of HBr that reacts is then 2x. Substitute the equilibrium concentrations (in terms of x) into the equilibrium expression and solve for x.

$K_c = 280 = \frac{(0.500 - 2x)^2}{x^2}$; taking the square root of both sides

$16.733 = \frac{0.500 - 2x}{x}$; $16.733x + 2x = 0.500$; $18.733x = 0.500$

$x = 0.0267$ *M*; $[I_2] = 0.0267$ *M*, $[Br_2] = 0.0267$ *M*

$[IBr] = 0.500 - 2x = 0.500 - 0.0534 = 0.447$ *M*

15.30 $K_c = \frac{[PCl_3][Cl_2]}{[PCl_5]}$; $[PCl_5]$ initial $= \frac{0.100 \text{ mol}}{5.00 \text{ L}} = 0.0200\ M$

	PCl_5	$\rightleftharpoons$ PCl_3	+ Cl_2
initial	0.0200 *M*	0	0
change	-x	+x	+x
equil.	0.0200-x	x	x

$K_c = 1.80 = \frac{x^2}{0.0200\text{-}x}$ Assume x is small compared to 0.0200.

$1.80 \approx \frac{x^2}{0.0200}$; $x^2 = 0.0360$; $x = 0.190\ M$

Clearly, our assumption is not true. Solve the quadratic formula to obtain the value of x.

$x^2 = 1.80(0.0200\text{-}x)$; $x^2 + 1.80x - 0.0360 = 0$

$$x = \frac{-b + \sqrt{b^2 - 4ac}}{2} = \frac{-1.80 + \sqrt{(1.80)^2 + 4(0.0360)}}{2}$$

$$x = \frac{-1.80 + \sqrt{3.384}}{2} = \frac{0.03957}{2} = 0.0198\ M$$

$[PCl_3] = [Cl_2] = 0.0198\ M$; $[PCl_5] = 0.0200\ M - 0.0198\ M = 2 \times 10^{-4}\ M$

(For this initial condition, the reaction essentially goes to completion.)

LeChatelier's Principle

15.31 (a) Increase (b) increase (c) decrease (d) no effect (e) no effect (f) no effect

15.32 (a) Shift equilibrium to the right; more CO will be formed.

(b) No effect on equilibrium; so long as there is any C(s) present the amount is not involved in the equilibrium.

(c) Equilibrium is shifted to the right, in the direction in which the reaction is endothermic.

(d) An increase in pressure will cause a shift to the left; that is, some of the CO will be converted to CO_2(g) and C(s).

(e) No effect on equilibrium.

(f) Removal of CO will cause a shift to the right; that is, more CO_2(g) will react with C(s), forming CO(g).

15.33 (a) No effect (b) no effect (c) increase equilibrium constant (d) no effect

15.34 $\Delta H^{\circ} = \Delta H_f^{\circ}\ CH_3OH(l) - \Delta H_f^{\circ}\ CO(g) - 2\Delta H_f^{\circ}\ H_2(g)$
$= -238.6\ kJ - (-110.5\ kJ) - 0\ kJ$
$= -128.1\ kJ$

(a) The reaction is exothermic; an increase in temperature would decrease the value of K_c and decrease the yield. A *low temperature* is needed to maximize yield.

(b) K_c decreases with increasing temperature.

(c) Assuming equal pressures of CO and H_2, increasing total pressure would increase the concentration of each gas, shifting the equilibrium toward products. The extent of conversion to CH_3OH increases as the total pressure increases.

Additional Exercises

15.35 (a) Since both the forward and reverse processes are elementary steps, we can write the rate laws directly from the chemical equation.

$$rate_f = k_f\ [CO]\ [Cl_2] = rate_r = k_r\ [COCl]\ [Cl]$$

$$\frac{k_f}{k_r} = \frac{[COCl]\ [Cl]}{[CO]\ [Cl_2]} = K$$

$$K = \frac{k_f}{k_r} = \frac{1.4 \times 10^{-28}\ M^{-1}\ s^{-1}}{9.3 \times 10^{10}\ M^{-1}\ s^{-1}} = 1.5 \times 10^{-39}$$

(b) Since the K is quite small, reactants are much more plentiful than products at equilibriuim.

15.36 $$[CO] = \frac{8.62\ g\ CO}{5.00\ L} \times \frac{1\ mol\ CO}{28.0\ g\ CO} = 0.0616\ M$$

$$[H_2] = \frac{2.60\ g\ H_2}{5.00\ L} \times \frac{1\ mol\ H_2}{2.02\ g\ H_2} = 0.257\ M$$

$$[CH_4] = \frac{43.0\ g\ CH_4}{5.00\ L} \times \frac{1\ mol\ CH_4}{16.0\ g\ CH_4} = 0.538\ M$$

$$[H_2O] = \frac{48.4\ g\ H_2O}{5.00\ L} \times \frac{1\ mol\ H_2O}{18.0\ g\ H_2O} = 0.538\ M$$

$$K_c = \frac{[CO]\ [H_2]^3}{[CH_4]\ [H_2O]} = \frac{(0.0616)\ (0.257)^3}{(0.538)\ (0.538)} = 3.61 \times 10^{-3}$$

15.37 (a) $H_2(g) + S(s) \rightleftharpoons H_2S(g)$

$$K_c = \frac{[H_2S]}{[H_2]}$$

(b) Calculate the molarity of H_2S and H_2. M = mol/L

$$[H_2S] = 0.46 \text{ g } H_2S \times \frac{1 \text{ mol } H_2S}{34.1 \text{ g } H_2S} \times \frac{1}{1.00 \text{ L}} = 0.0135\ M$$

$$[H_2] = 0.40 \text{ g } H_2 \times \frac{1 \text{ mol } H_2}{2.02 \text{ g } H_2} \times \frac{1}{1.00 \text{ L}} = 0.198\ M$$

$$K_c = \frac{(0.0135)}{(0.198)} = 0.068$$

(Intermediate rounding of $[H_2S]$ = 0.014 M and $[H_2]$ = 0.20 M, yields K_c = 0.070)

(c) Since S is a pure solid, its concentration doesn't change during the reaction, and [S] does not appear in the equilibrium expression.

15.38 If 9.0% of the NOCl reacts, this corresponds to a change of -0.090 mol in the amount of NOCl. Recall that the stoichiometry of the reaction is shown on the change line of the table.

$$2NOCl(g) \rightleftharpoons 2NO(g) + Cl_2(g)$$

	2NOCl(g)	2NO(g)	Cl_2(g)
initial	1.00 mol	0 mol	0 mol
change	-0.090 mol	+0.090 mol	+0.045 mol
equil.	0.91 mol	0.090 mol	0.045 mol

$$K_c = \frac{[NO]^2[Cl_2]}{[NOCl]^2}; \quad M = \frac{\text{mol}}{\text{L}};$$

$$[NOCl] = \frac{0.91 \text{ mol}}{1.00 \text{ L}} = 0.91\ M$$

$$[NO] = \frac{0.090 \text{ mol}}{1.00 \text{ L}} = 0.090\ M; \quad [Cl_2] = \frac{0.045 \text{ mol}}{1.00 \text{ L}} = 0.045\ M$$

$$K_c = \frac{(0.090)^2(0.045)}{(0.91)^2} = 4.4 \times 10^{-4}$$

15.39 (a) From the change of 2.0 mol SO_2, deduce the changes in the other amounts according to the reaction stoichiometry.

$$SO_2 + NO_2 \rightleftharpoons SO_3 + NO$$

	SO_2	NO_2	SO_3	NO
initial	3.0 mol	4.0 mol	1.0 mol	4.0 mol
change	-2.0 mol	-2.0 mol	+2.0 mol	+2.0 mol
equil.	1.0 mol	2.0 mol	3.0 mol	6.0 mol
[equil.]	0.50 M	1.0 M	1.5 M	3.0 M

If the volume is 2.00 L, the equilibrium concentrations are as shown.

(b) $$K_c = \frac{[SO_3][NO]}{[SO_2][NO_2]} = \frac{(1.5)(3.0)}{(0.50)(1.0)} = 9.0$$

15.40 (a)

	$A(g)$	$\rightleftharpoons$ $2B(g)$
initial	0.75 atm	0 atm
change	-0.25 atm	+0.50 atm
equil.	0.50 atm	0.50 atm

$P_T = P_A + P_B = 0.50 \text{ atm} + 0.50 \text{ atm} = 1.00 \text{ atm}$

(b) $K_p = \frac{(P_B)^2}{P_A} = \frac{(0.50)^2}{0.50} = 0.50$

(c) $K_c = \frac{K_p}{(RT)^{\Delta n}}$; $\Delta n = +1$, $T = 0°C + 273 = 273 \text{ K}$

$$K_c = \frac{0.50}{(0.0821 \times 273)^{+1}} = 0.022$$

15.41 (a) $K_p = \frac{P^2_{NH_3}}{P_{N_2} \times P^3_{H_2}} = 4.34 \times 10^{-3}$

$$P_{NH_3} = \frac{gRT}{\mathcal{M}V} = \frac{0.753 \text{ g}}{17.03 \text{ g/mol}} \times \frac{0.0821 \text{ L} \cdot \text{atm}}{\text{K} \cdot \text{mol}} \times \frac{573 \text{ K}}{1.000 \text{ L}} = 2.08 \text{ atm}$$

	$N_2(g)$	+ $3H_2(g)$	$\rightleftharpoons$ $2NH_3(g)$
initial	0 atm	0 atm	?
change	x	3x	-2x
equil.	x atm	3x atm	2.08 atm

(Remember, only the change line reflects the stoichiometry of the reaction.)

$K_p = \frac{(2.08)^2}{(x)(3x)^3} = 4.34 \times 10^{-3}$; $27x^4 = \frac{(2.08)^2}{4.34 \times 10^{-3}}$; $x^4 = 36.9$

$x = 2.46 \text{ atm} = P_{N_2}$; $P_{H_2} = 3x = 7.39 \text{ atm}$

$$g_{N_2} = \frac{\mathcal{M}PV}{RT} = \frac{28.02 \text{ g } N_2}{\text{mol } N_2} \times \frac{\text{K} \cdot \text{mol}}{0.0821 \text{ L} \cdot \text{atm}} \times \frac{2.46 \text{ atm} \times 1.000 \text{ L}}{573 \text{ K}} = 1.46 \text{ g } N_2$$

$$g_{H_2} = \frac{2.016 \text{ g } H_2}{\text{mol } H_2} \times \frac{\text{K} \cdot \text{mol}}{0.0821 \text{ L} \cdot \text{atm}} \times \frac{7.39 \text{ atm} \times 1.00 \text{ L}}{573 \text{ K}} = 0.317 \text{ g } H_2$$

(b) The initial $P_{NH_3} = 2.08 \text{ atm} + 2(2.46 \text{ atm}) = 7.00 \text{ atm}$

$$g_{NH_3} = \frac{\mathcal{M}PV}{RT} = \frac{17.03 \text{ g } NH_3}{\text{mol } NH_3} \times \frac{\text{K} \cdot \text{mol}}{0.0821 \text{ L} \cdot \text{atm}} \times \frac{7.00 \text{ atm} \times 1.000 \text{ L}}{573 \text{ K}} = 2.53 \text{ g } NH_3$$

(c) $P_t = P_{N_2} + P_{H_2} + P_{NH_3} = 2.46 \text{ atm} + 7.39 \text{ atm} + 2.08 \text{ atm} = 11.93 \text{ atm}$

15.42 $K_c = \frac{[Br]^2}{[Br_2]}$; $[Br_2] = \frac{4.53 \times 10^{-2} \text{ mol}}{0.200 \text{ L}} = 0.227\ M$; $1.04 \times 10^{-3} = \frac{[Br]^2}{0.227}$

$[Br] = 1.53 \times 10^{-2}\ M$; mol Br = $1.53 \times 10^{-2}\ M \times 0.200 \text{ L} = 3.07 \times 10^{-3}$ mol

15.43 $K_c = \frac{[I_2][Br_2]}{[IBr]^2}$; initial [IBr] = $\frac{0.040 \text{ mol}}{1.00 \text{ L}} = 0.040\ M$

	$2IBr$	$\rightleftharpoons$	I_2	+	Br_2
initial	0.040 *M*		0		0
change	-2x		x		x
equil.	0.040-2x		x		x

$K_c = 8.5 \times 10^{-3} = \frac{x^2}{(0.040\text{-}2x)^2}$; Taking the square root of both sides

$\frac{x}{0.040\text{-}2x} = \sqrt{8.5 \times 10^{-3}} = 0.092$; $x = 0.092(0.040\text{-}2x)$

$x + 0.18\,x = 0.0037$; $1.18\,x = 0.0037$, $x = 0.0031$

$[IBr] = 0.040 - 2(0.0031) = 0.034\ M$

15.44 (a) $K_p = K_c(RT)^{\Delta n}$; $K_c = \frac{K_p}{(RT)^{\Delta n}} = \frac{0.052}{(0.0821 \times 333)^2} = 7.0 \times 10^{-5}$

(b) $[BCl_3] = 0.0216 \text{ mol}/0.500 \text{ L} = 0.0432\ M$
PH_3BCl_3 is a solid and its concentration is taken as a constant, C.

	PH_3BCl_3	$\rightleftharpoons$	PH_3	+	BCl_3
initial	C		0 *M*		0.0432 *M*
change			+x *M*		+x *M*
equil.	C		x *M*		0.0432+x *M*

$K_c = [PH_3][BCl_3]$; $7.0 \times 10^{-5} = x(0.0432 + x)$

$x^2 + 0.0432x - 7.0 \times 10^{-5} = 0$

$x = \frac{-0.432 \pm [(0.0432)^2 - 4(-7.0 \times 10^{-5})]^{1/2}}{2} = 1.6 \times 10^{-3}\ M = [PH_3]$

Check: $(1.6 \times 10^{-3} + 0.0432)(1.6 \times 10^{-3}) = 7.2 \times 10^{-5}$; the solution is correct to two significant figures.

15.45 $K_c = \frac{[CO]^2}{[CO_2]}$

To calculate K_c, find concentrations in units of moles/L. From the ideal gas law, n/V = P/RT. Since the total pressure is 1 atm in all cases, n/V for CO_2 and CO at each temperature can be calculated. For example, at 850° (1123 K):

$$[CO_2] = \frac{0.0623\text{ atm}}{\left[\frac{0.0821\text{ L}\cdot\text{atm}}{\text{mol}\cdot\text{K}}\right](1123\text{ K})} = 6.75 \times 10^{-4}\ M$$

$$[CO] = \frac{0.9377\text{ atm}}{\left[\frac{0.0821\text{ L}\cdot\text{atm}}{\text{mol}\cdot\text{K}}\right](1123\text{ K})} = 1.02 \times 10^{-2}\ M$$

Temp (K)	$[CO_2]$	[CO]	K_c
1123	6.75×10^{-4}	1.02×10^{-2}	0.154
1223	1.31×10^{-4}	9.83×10^{-3}	0.735
1323	3.41×10^{-5}	9.18×10^{-3}	2.47
1473	5×10^{-6}	8.26×10^{-3}	14

Because K_c grows larger with increasing temperature, the reaction must be endothermic in the forward direction.

15.46 $K_p = P_{NH_3} \times P_{H_2S}$; $P_t = 0.614$ atm

If the equilibrium amounts of NH_3 and H_2S are due solely to the decomposition of $NH_4HS(s)$, the equilibrium pressures of the two gases are equal, and each is 1/2 of the total pressure.

$P_{NH_3} = P_{H_2S} = 0.614\text{ atm}/2 = 0.307\text{ atm}$

$K_p = (0.307)^2 = 0.0943$

15.47 First find the initial moles of SO_3 and then the total moles of gas at equilibrium.

$$\frac{0.831\text{ g } SO_3}{80.1\text{ g/mol}} = 0.0104\text{ mol } SO_3\ ;\ n_T = \frac{1.30\text{ atm} \times 1.00\text{ L}}{1100\text{ K}} \times \frac{\text{K}\cdot\text{mol}}{0.0821\text{ L}\cdot\text{atm}} = 0.0144\text{ mol}$$

	$2SO_3$ $\rightleftharpoons$	$2SO_2$	+ O_2
initial	0.0104	0	0
change	-2x	+2x	+x
equil.	0.0104-2x	2x	x
[equil.]	0.0024 *M*	0.0080 *M*	0.0040 *M*

$n_t = 0.0144 = 0.0104\text{-}2x + 2x + x;\ x = 0.0040\text{ mol } O_2$

Since the volume is 1 L, the equilibrium molar concentrations are equal to the moles of each component.

$$K_c = \frac{[SO_2]^2[O_2]}{[SO_3]^2} = \frac{(0.0080)^2(0.0040)}{(0.0024)^2} = 4.4 \times 10^{-2}$$

$$K_p = K_c(RT)^{\Delta n} = 4.4 \times 10^{-2}\ (0.0821 \times 1100)^1 = 4.0$$

15.48 In general, the reaction quotient is of the form $Q = \frac{[NOCl]^2}{[NO]^2[Cl_2]}$.

(a) $Q = \frac{(0.11)^2}{(0.15)^2(0.31)} = 1.7$

$Q > K_p$. Therefore, the reaction will shift toward reactants; i.e., to the left, in moving toward equilibrium.

(b) $Q = \frac{(0.050)^2}{(0.12)^2(0.10)} = 1.7$

$Q > K_p$. Therefore, the reaction will shift toward reactants, i.e., to the left, in moving toward equilibrium.

(c) $Q = \frac{(5.10 \times 10^{-3})^2}{(0.15)^2(0.20)} = 5.8 \times 10^{-3}$

$Q < K_p$. Therefore, the reaction mixture will shift in the direction of more product, i.e., to the right, in moving toward equilibrium.

15.49 $K_c = [CO_2] = 0.0108$

(a) $[CO_2] = 15.0\ g\ CO_2 \times \frac{1\ mol\ CO_2}{44.0\ g\ CO_2} = \frac{0.341\ mol}{10.0\ L} = 0.0341\ M$

$Q = 0.0341 > K_c$. The reaction proceeds to the left to achieve equilibrium and the amount of $CaCO_3(s)$ increases.

(b) $[CO_2] = 4.75\ g\ CO_2 \times \frac{1\ mol\ CO_2}{44.0\ g\ CO_2} \times \frac{1}{10.0\ L} = 0.0108\ M$

$Q = 0.0108 = K_c$. The mixture is at equilibrium and the amount of $CaCO_3(s)$ remains constant.

(c) $[CO_2] = 2.50\ g\ CO_2 \times \frac{1\ mol\ CO_2}{44.0\ g\ CO_2} \times \frac{1}{10.0\ L} = 0.00568\ M$

$Q = 0.00568 < K_c$. The reaction proceeds to the right to achieve equilibrium and the amount of $CaCO_3(s)$ decreases.

15.50 (a) To the left, to partially reduce the CO_2 concentration

(b) To the right, to form more CO_2

(c) To the left; a volume increase will produce a lower pressure, and the system will respond by a shift in equilibrium toward the side that produces more moles of gas.

(d) Increase in pressure will cause a shift to the right, in the direction of fewer moles of gas.

(e) Increase in temperature will cause a shift in equilibrium to the left, the direction in which the reaction is endothermic.

15.51 $K_c = K_p = \frac{P_{CO_2}}{P_{CO}} = 600$

If P_{CO} is 150 torr, P_{CO_2} can never exceed 760 - 150 = 610 torr. Then Q = 610/150 = 4.1. Since this is far less than K, the reaction will shift in the direction of more product. Reduction will therefore occur.

15.52 (a) $K_c = \frac{[Ni(CO)_4]}{[CO]^4}$

(b) Increasing the temperature to 200° C favors the reverse process (decomposition of $Ni(CO)_4(g)$) and thus the value of K is smaller at the higher temperature. This is the behavior expected from an exothermic reaction (heat is a product).

(c) At the temperature of the exhaust pipe, the $Ni(CO)_4$ product is a gas and is carried into the atmosphere with other exhaust gases. Thus, equilibrium is never established (we do not have a closed system) and the reaction proceeds to the right as $Ni(CO)_4$ product is removed.

15.53 (a) $K_c = \frac{K_p}{(RT)^{\Delta n}}$; $\Delta n = +1$, T = 700 K

$$K_c = \frac{0.76}{(0.0821)\,(700)^{+1}} = 0.013$$

(b)

	$CCl_4(g)$ $\rightleftharpoons$	$C(s)$ +	$2Cl_2(g)$
initial	2.00 atm		0 atm
change	-x atm		+2x atm
equil.	(2.00-x) atm		2x atm

$$K_p = 0.76 = \frac{P^2_{Cl_2}}{P_{CCl_4}} = \frac{(2x)^2}{(2.00\text{-}x)}$$

$1.52 - 0.76\,x = 4x^2$; $4x^2 + 0.76\,x - 1.52 = 0$

Using the quadratic formula, a = 4, b = 0.76, c = -1.52

$$x = \frac{-0.76 \pm \sqrt{(0.76)^2 - 4(4)(-1.52)}}{2(4)} = \frac{-0.76 + 4.99}{8} = 0.53 \text{ atm}$$

$$\text{Fraction } CCl_4 \text{ reacted} = \frac{x \text{ atm}}{2.00 \text{ atm}} = \frac{0.53}{2.00} = 0.26 = 26\%$$

(c) $P_{Cl_2} = 2x = 2(0.53) = 1.06$ atm

$P_{CCl_4} = 2.00 - x = 2.00 - 0.53 = 1.47$ atm

15.54 (a) $K_p = P_{NH_3} \times P_{H_2S}$. Before solid is added, $Q = P_{NH_3} \times P_{H_2S} = 0.100 \text{ atm} \times 0 = 0$. Q < K and the reaction will proceed to the right. However, no $NH_4HS(s)$ is present to produce $H_2S(g)$, so the reaction cannot proceed.

(b)

	$NH_4HS(s)$ ⇌	$NH_3(g)$	+	$H_2S(g)$
initial		0.100 atm		0 atm
change		+x atm		+x atm
equil.		0.100 + x atm		x atm

Since Q < K initially (part (a)), P_{NH_3} must increase along with P_{H_2S} until equilibrium is established.

$K_p = P_{NH_3} \times P_{H_2S}$; $0.120 = (0.100 + x)(x)$; $0 = x^2 + 0.100\,x - 0.120$

$0 = (x + 0.4)(x - 0.3)$; $x = 0.300$ (The negative result, x = -0.4 is meaningless.)

$P_t = P_{NH_3} + P_{H_2S} = (0.100 + 0.300) \text{ atm} + 0.300 \text{ atm} = 0.700$ atm

(c) The minimum amount of $NH_4HS(g)$ required is equal to the number of moles H_2S present at equilibrium (no unreacted solid would remain). We can calculate the mol H_2S present at equilibrium using the Ideal Gas Equation.

$$n_{H_2S} = \frac{P_{H_2S}V}{RT} = 0.300 \text{ atm} \times \frac{K \cdot \text{mol}}{0.0821 \text{ L} \cdot \text{atm}} \times \frac{10.00 \text{ L}}{298 \text{ K}} = 0.123 \text{ mol } H_2S$$

$$0.123 \text{ mol } H_2S \times \frac{1 \text{ mol } NH_4HS}{1 \text{ mol } H_2S} \times \frac{51.11 \text{ g } NH_4HS}{1 \text{ mol } NH_4HS} = 6.27 \text{ g } NH_4HS$$

15.55 (a) $$Q = \frac{P_{PCl_5}}{P_{PCl_3} \times P_{Cl_2}} = \frac{0.10}{(0.30)(0.60)} = 0.56$$

0.56 (Q) > 0.0870 (K), the reaction proceeds to the left.

(b)

	$PCl_3(g)$	+	$Cl_2(g)$	⇌	$PCl_5(g)$
initial	0.30 atm		0.60 atm		0.10 atm
change	+x atm		+x atm		-x atm
equil.	(0.30 + x) atm		(0.60 + x) atm		(0.10 - x) atm

(Since the reaction proceeds to the left, P_{PCl_5} must decrease and P_{PCl_3} and P_{Cl_2} must increase.)

$$K_p = 0.0870 = \frac{(0.10 - x)}{(0.30 + x)(0.60 + x)}\ ; \quad 0.0870 = \frac{(0.10 - x)}{(0.18 + 0.90\,x + x^2)}$$

$$0.0870\,x^2 + 0.0783\,x + 0.0157 = 0.10 - x\ ; \quad 0.0870\,x^2 + 1.0783\,x - 0.0843 = 0$$

$$x = \frac{-1.0783 \pm \sqrt{(1.0783)^2 - 4(0.0870)(-0.0843)}}{2(0.0870)} = \frac{-1.0783 + 1.0918}{0.174} = 0.078$$

P_{PCl_3} = (0.30 + 0.078) atm = 0.378 $\qquad$ P_{Cl_2} = (0.60 + 0.078) atm = 0.678 atm

P_{PCl_5} = (0.10 - 0.078) atm = 0.022 atm

To 2 decimal places, the pressures are 0.38, 0.68 and 0.02 atm, respectively. When substituting into the K_p expression, pressures to 3 decimal places yield a result much closer to 0.0870.

(c) Increasing the volume of the container favors the process where more moles of gas are produced, so the reverse reaction is favored and the equilibrium shifts to the left; the mole fraction of PCl_5 decreases.

(d) For an exothermic reaction, increasing the temperature decreases the value of K; more reactants and fewer products are present at equilibrium and the partial pressure of PCl_5 decreases.

15.56 First calculate K_c for the equilibrium

$$H_2 + I_2 \rightleftharpoons 2HI$$

$$K_c = \frac{[HI]^2}{[H_2][I_2]} = \frac{(0.155)^2}{(2.24 \times 10^{-2})\,(2.24 \times 10^{-2})} = 47.9$$

The added HI represents a concentration of $\frac{0.100 \text{ mol}}{5.00 \text{ L}}$ = 0.020 *M*.

	H_2	+	I_2	$\rightleftharpoons$	2HI
initial	2.24×10^{-2} *M*		2.24×10^{-2} *M*		0.155 +0.020 *M*
change	+ x *M*		+ x *M*		-2x *M*
equil.	2.24×10^{-2}+ x *M*		2.24×10^{-2}+ x *M*		0.175-2x *M*

$$\frac{(0.175 - 2x)^2}{(2.24 \times 10^{-2} + x)^2} = 47.9.$$ Take the square root of both sides:

$$\frac{0.175 - 2x}{2.24 \times 10^{-2} + x} = (47.9)^{1/2} = 6.92$$

$0.175 - 2x = 0.155 + 6.92x; \quad x = 2.24 \times 10^{-3}$

$[I_2] = [H_2] = 2.24 \times 10^{-2} + 2.24 \times 10^{-3} = 2.46 \times 10^{-2}\ M$

$[HI] = 0.175 - 2x = 0.170\ M$

15.57 (a) Since the volume of the vessel = 1.00 L, mol = *M*. The reaction will proceed to the left to establish equilibrium.

	A(g) +	2B(g) ⇌	2C(g)
initial	0 *M*	0 *M*	1.00 *M*
change	+x *M*	+2x *M*	-2x *M*
equil.	x *M*	2x *M*	(1.00 - 2x) *M*

At equilibrium, [A] = x *M*, [B] = 2x *M*.

(b) x must be less than 0.50 *M* (so that [C], 1.00 - 2x, is not less than zero).

(c) $K_c = \frac{[C]^2}{[A][B]^2}; \quad \frac{(1.00 - 2x)^2}{(x)(2x)^2} = 0.25$

$1.00 - 4x + 4x^2 = 0.25(4x^3); \quad x^3 - 4x^2 + 4x - 1 = 0$

(d)

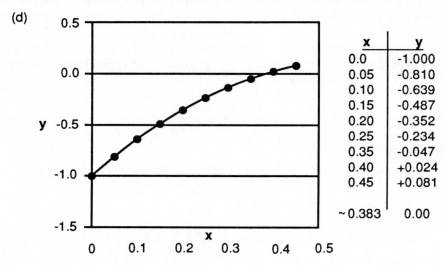

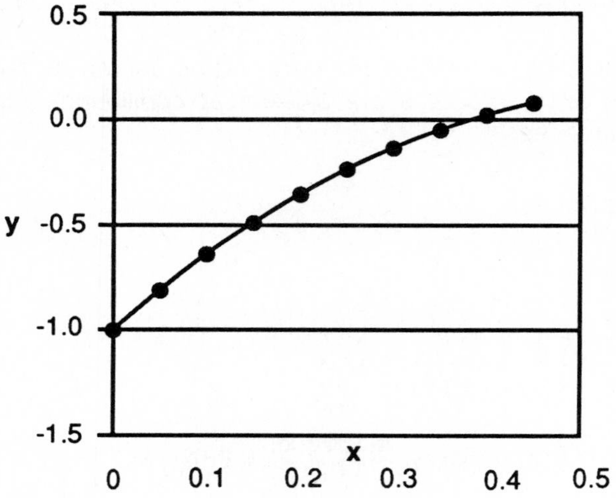

x	y
0.0	-1.000
0.05	-0.810
0.10	-0.639
0.15	-0.487
0.20	-0.352
0.25	-0.234
0.35	-0.047
0.40	+0.024
0.45	+0.081
~0.383	0.00

(e) From the plot, x ≈ 0.383 *M*

[A] = x = 0.383 *M*; [B] = 2x = 0.766 *M*

[C] = 1.00 - 2x = 0.234 *M*

Using the K_c expression as a check:

$K_c = 0.25; \quad \frac{(0.234)^2}{(0.383)(0.766)^2} = 0.24$; the estimated values are reasonable.

15.58 $K_p = \dfrac{P_{O_2} \times P^2_{CO}}{P^2_{CO_2}}$; P_{O_2} = (0.03) (1 atm) = 0.03 atm

P_{CO} = (0.002)(1 atm) = 0.002 atm; P_{CO_2} = (0.12)(1 atm) = 0.12 atm

$$Q = \frac{(0.03)(0.002)^2}{(0.12)^2} = 8.3 \times 10^{-6}$$

Since Q > K_p, the system will shift to the left to attain equilibrium. Thus a catalyst that promoted the attainment of equilibrium would result in a lower CO content in the exhaust.

15.59 The patent claim is false. A catalyst does not alter the position of equilibrium in a system, only the rate of approach to the equilibrium condition.

15.60 (a) At equilibrium, the forward and reverse reactions occur at equal rates.
(b) One expects the reactants to be favored at equilibrium since they are lower in energy.
(c) A catalyst lowers the activation energy for both the forward and reverse reactions; the "hill" would be lower.
(d) Since the activation energy is lowered for both processes, new rates are equal and the ratio of the rate constants, k_f/k_r, remains unchanged.
(e) Since the reaction is endothermic (the energy of the reactants is lower than that of the products, ΔE is positive) the value of K should increase with increasing temperature.

15.61 Consider the energy profile for an exothermic reaction.

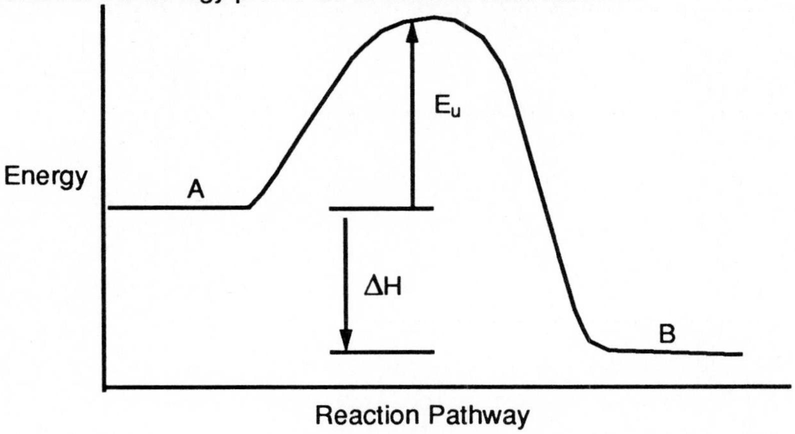

The activation energy in the forward direction, E_{af}, equals E_u, and the activation energy in the reverse reaction, E_{ar}, equals E_u - ΔH. (The same is true for an endothermic reaction because the sign of ΔH is the positive and $E_{ar} < E_{af}$). For the reaction in question,

$$K = \frac{k_f}{k_r} = \frac{A_f\, e^{-E_{af}/RT}}{A_r\, e^{-E_{ar}/RT}}$$

Since the ln form of the Ahrrenius equation is easier to manipulate, we will consider ln K.

$$\ln K = \ln\left(\frac{k_f}{k_r}\right) = \ln k_f - \ln k_r = \frac{-E_{af}}{RT} = \ln A_f - \left[\frac{-E_{ar}}{RT} + \ln A_r\right]$$

Substituting E_u for E_{af} and $(E_u - \Delta H)$ for E_{ar} ,

$$\ln K = \frac{-E_u}{RT} + \ln A_f - \left[\frac{-(E_u - \Delta H)}{RT} + \ln Ar\right]; \ \ln K = \frac{-E_u + (E_u - \Delta H)}{RT} + \ln A_f - \ln A_r$$

$$\ln K = \frac{-\Delta H}{RT} + \ln\frac{A_f}{A_r}$$

For the catalyzed reaction, $E_c < E_u$ and $E_{af} = E_c$, $E_{ar} = E_c - \Delta H$. The catalyst does not change the value of ΔH.

$$\ln K_c = \frac{-E_c + (E_c - \Delta H)}{RT} + \ln A_f - \ln A_r$$

$$\ln K_c = \frac{-\Delta H}{RT} + \ln\frac{A_f}{A_r}$$

Thus, assuming A_f and A_r are not changed by the catalyst, $\ln K = \ln K_c$ and $K = K_c$.

CHAPTER 16

Acid-Base Equilibria

Introduction: Dissociation of Water

16.1 Solutions of HCl and H_2SO_4 taste sour, turn litmus paper red (are acidic), neutralize solutions of bases, react with active metals to form $H_2(g)$ and conduct electricity. The two solutions have these properties in common because both solutes are strong acids. That is, they both dissociate completely in H_2O to form $H^+(aq)$ and an anion. (The first dissociation step for H_2SO_4 is complete, but HSO_4^- is not completely dissociated.) The presence of ions enables the solutions to conduct electricity; the presence of $H^+(aq)$ in excess of 1×10^{-7} *M* accounts for all the other properties listed.

16.2 When NaOH dissolves in water, it completely dissociates to form $Na^+(aq)$ and $OH^-(aq)$. CaO is the oxide of a metal; it dissolves in water according to the following process: $CaO(s) + H_2O(l) \rightarrow Ca^{2+}(aq) + 2OH^-(aq)$. Thus, the properties of both solutions are dominated by the presence of $OH^-(aq)$. Both solutions taste bitter, turn litmus paper blue (are basic), neutralize solutions of acids and conduct electricity.

16.3 (a) *Autoionization* is the dissociation of a neutral molecule (in the absence of any other reactant) into an anion and a cation. The equilibrium expression for the autoionization of water is $H_2O(l) \rightleftharpoons H^+(aq) + OH^-(aq)$.

(b) Pure water is a poor conductor of electricity because it contains very few ions. Ions, mobile charged particles, are required for the conduction of electricity in liquids.

(c) If a solution is *acidic*, it contains more H^+ than OH^- ($[H^+] > [OH^-]$).

16.4 (a) $H_2O(l) \rightleftharpoons H^+(aq) + OH^-(aq)$

(b) $K_w = [H^+][OH^-]$. The $[H_2O(l)]$ is omitted because water is a pure liquid. The molarity (mol/L) of pure solids or liquids does not change as equilibrium is established, so it is usually omitted from equilibrium expressions.

(c) If a solution is *basic*, it contains more OH^- than H^+ ($[OH^-] > [H^+]$).

16.5 In pure water, $[H^+] = [OH^-] = 1.0 \times 10^{-7}$ *M*. If $[H^+] > 1.0 \times 10^{-7}$ *M*, the solution is acidic; if $[OH^-] > 1.0 \times 10^{-7}$ *M*, the solution is basic.

(a) acidic (b) basic (c) If $[H^+] < 1 \times 10^{-7}$, then $[OH^-]$ must be $> 1 \times 10^{-7}$, because $[H^+][OH^-] = 1.0 \times 10^{-14}$; basic

(d) neutral (e) basic (barely)

16.6 In pure water, $[H^+] = [OH^-] = 1.0 \times 10^{-7}$ *M*. If $[H^+] > 1.0 \times 10^{-7}$ *M*, the solution is acidic; if $[OH^-] > 1.0 \times 10^{-7}$ *M*, the solution is basic.

(a) neutral (b) basic (c) acidic (barely)

(d) $[H^+] = 5.0 \times 10^{-10}$ *M* $< 1.0 \times 10^{-7}$; basic (e) $[OH^-] = 1.0 \times 10^{-7}$; neutral

16.7 $K_w = 1.0 \times 10^{-14} = [H^+][OH^-]$

(a) $(0.0050\ M)\,[OH^-] = 1.0 \times 10^{-14}$; $[OH^-] = 2.0 \times 10^{-12}\ M$

(b) $(1.3 \times 10^{-9}\ M)\,[OH^-] = 1.0 \times 10^{-14}$; $[OH^-] = 7.7 \times 10^{-6}\ M$

(c) $[OH^-] = 100\,[H^+]$; $[H^+] \cdot 100\,[H^+] = 1.0 \times 10^{-14}$; $[H^+]^2 = 1.0 \times 10^{-16}\ M$

$[H^+] = 1.0 \times 10^{-8}\ M$; $[OH^-] = 100\,(1.0 \times 10^{-8}\ M) = 1.0 \times 10^{-6}\ M$

16.8 $K_w = 1.0 \times 10^{-14} = [H^+][OH^-]$

(a) $[H^+](4.5 \times 10^{-11}\ M) = 1.0 \times 10^{-14}$; $[H^+] = 2.2 \times 10^{-4}\ M$

(b) $[H^+](1.50\ M) = 1.0 \times 10^{-14}$; $[H^+] = 6.7 \times 10^{-15}\ M$

(c) $[H^+] = 5[OH^-]$; $5[OH^-] \cdot [OH^-] = 1.0 \times 10^{-14}$; $[OH^-]^2 = 2.0 \times 10^{-15}$

$[OH^-] = 4.5 \times 10^{-8}\ M$; $[H^+] = 5(4.5 \times 10^{-8}\ M) = 2.3 \times 10^{-7}\ M$

16.9 (a) According to the Arrhenius definition, an acid when dissolved in water increases $[H^+]$. According to the Bronsted-Lowry definition, an acid is capable of donating H^+, regardless of physical state. The Arrhenius definition of an acid is confined to an aqueous solution; the Bronsted-Lowry definition applies to any physical state.

(b) $HCl(g) + NH_3(g) \rightarrow NH_4^+\ Cl^-(s)$ HCl is the B-L (Bronsted-Lowry) acid; it donates an H^+ to NH_3 to form NH_4^+. NH_3 is the B-L base; it accepts the H^+ from HCl.

16.10 (a) According to the Arrhenius definition, a base when dissolved in water increases $[OH^-]$. According to the Bronsted-Lowry theory, a base is an H^+ acceptor regardless of physical state. A Bronsted-Lowry base is not limited to aqueous solution and need not contain OH^- or produce it in aqueous solution.

(b) $NH_3(g) + H_2O(l) \rightleftharpoons NH_4^+(aq) + OH^-(aq)$ When NH_3 dissolves in water, it accepts H^+ from H_2O (B-L definition). In doing so, OH^- is produced (Arrhenius definition). Note that the OH^- produced was originally part of the H_2O molecule, not part of the NH_3 molecule.

CHAPTER 16
Acid-Base Equilibria

Solutions to Problems

16.11 A conjugate base has one less H^+ than its conjugate acid.

(a) HS^- (b) NO_3^- (c) ClO^- (d) CO_3^{2-} (e) NH_3

16.12 A conjugate acid has one more H^+ than its conjugate base.

(a) NH_4^+ (b) $HClO_3$ (c) HPO_4^{2-} (d) H_2CO_3 (e) H_3S^+

16.13

	acid	base	conjugate acid	conjugate base
(a)	H_2O	NH_2^-	NH_3	OH^-
(b)	$H_2C_2O_4$	H_2O	H_3O^+	$HC_2O_4^-$
(c)	$H^+(aq)$	HPO_4^{2-}	$H_2PO_4^-$	H_2O (understood)

16.14

	acid	base	conjugate acid	conjugate base
(a)	$HC_2O_4^-$	CO_3^{2-}	HCO_3^-	$C_2O_4^{2-}$
(b)	PH_4^+	H_2O	H_3O^+	PH_3
(c)	NH_4^+	CN^-	HCN	NH_3

16.15 (a) H_2O; moving from left to right across a row of the periodic chart, the acidity of the binary hydrides increases as the polarity of the X-H bond increases.

(b) $HClO_4$; $HClO_4$ is one of the six acids which completely dissociate in water; HF is a weak acid ($K_a < 1$).

16.16 (a) CN^- ; Cl^- is the conjugate base of the strong acid, HCl, which is completely dissociated in water. Cl^- has no tendency to accept $H^+(aq)$.

(b) CO_3^{2-} ; same reason as part (a).

16.17 (a) A substance is *amphoteric* if it can act as either an acid or a base. That is, the substance can either donate or accept an H^+.

(b) $H_2O(l) + H_2O(l) \rightleftharpoons H_3O^+(aq) + OH^-(aq)$

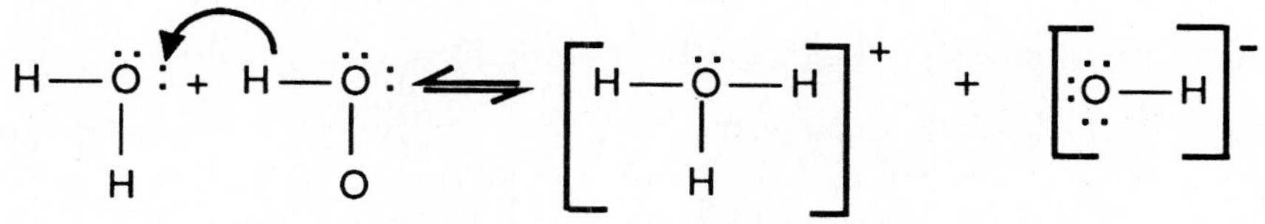

In the autoionization of water, one H_2O molecule acts as an H^+ donor (B-L acid) and the other acts as an H^+ acceptor (B-L base).

16.18 (a) $H_2PO_4^-(aq) + H_2O(l) \rightleftharpoons H_3PO_4(aq) + OH^-(aq)$

(b) $H_2PO_4^-(aq) + H_2O(l) \rightleftharpoons HPO_4^{2-}(aq) + H_3O^+(aq)$

(c) H_3PO_4 is the conjugate acid of $H_2PO_4^-$
HPO_4^{2-} is the conjugate base of $H_2PO_4^-$

(Note that conjugate acid/conjugate base pairs differ by a single H^+; H_3PO_4/HPO_4^- is not a conjugate acid/conjugate base pair.)

The pH Scale

16.19 A solution is acidic if pH < 7.00, neutral if pH = 7.00 and basic if pH > 7.00; pH + pOH = 14.00.

(a) acidic (b) basic (c) neutral (pH = 14.00 - pOH = 7.00)
(d) basic (pH = 14.00 - pOH = 11.06)

16.20 A solution is acidic if pH < 7.00, neutral if pH = 7.00 and basic if pH > 7.00; pH + pOH = 14.00.

(a) neutral (b) acidic (Since pH = -log $[H^+]$, a negative pH value indicates $[H^+] > 1$.)
(c) acidic (pH = 14.00 - pOH = 1.69) (d) basic (pH = 14.00 - pOH = 7.50)

16.21 When calculating logs, the number of significant figures in the number becomes the number of decimal places in the log. The digits to the left of the decimal in a log indicate the power of ten of the original number.

(a) $pH = -\log [H^+] = -\log (3.6 \times 10^{-3}) = -(-2.44) = 2.44$

(b) $pH = -\log [H^+] = -\log(4.7 \times 10^{-2}) = -(-1.33) = 1.33$

(c) $pH = 14.00 - pOH = 14.00 - 5.33 = 8.67$

(d) $pH = 14.00 - pOH$; $pOH = -\log [OH^-] = -\log(6.7 \times 10^{-2}) = 1.17$
$pH = 14.00 - 1.17 = 12.83$

16.22 (a) $pH = -\log[H^+] = -\log(4.2 \times 10^{-5}) = -(-4.38) = 4.38$

(b) $pH = -\log[H^+] = -\log(1.20) = -(0.079) = -0.079$ (Yes, pH can be negative.)

(c) $pH = 14.00 - pOH = 14.00 - 11.21 = 2.79$

(d) $pH = 14.00 - pOH$; $pOH = -\log[OH^-] = -\log(0.92) = -0.036$

$pH = 14.00 - (-0.036) = 14.04$

(Yes, pH can in principle be greater than 14.)

16.23 $[H^+]$ = antilog (-pH) or 10^{-pH}

(a) $[H^+] = 10^{-4.78} = 1.66 \times 10^{-5}\ M\ H^+$ (b) $[H^+] = 10^{-12.32} = 4.79 \times 10^{-13}\ M\ H^+$

(c) $[H^+] = 10^{-(-0.67)} = 4.68\ M\ H^+$ (d) $[H^+] = 10^{-7.1} = 7.94 \times 10^{-8}\ M\ H^+$

16.24 (a) pH = 6.21 pOH = 14.00 - pH = 7.79
$[H^+] = 10^{-pH} = 10^{-6.21} = 6.17 \times 10^{-7}\ M$ $[OH^-] = 10^{-pOH} = 10^{-7.79} = 1.62 \times 10^{-8}\ M$

(b) pH = 14.00 - pOH = 13.00 pOH = 1.00
$[H^+] = 10^{-pH} = 10^{-13.00} = 1.0 \times 10^{-13} M$ $[OH^-] = 10^{-pOH} = 10^{-1.00} = 1.0 \times 10^{-1}\ M$

(c) pH = 9.88 pOH = 14.00 - pH = 4.12
$[H^+] = 10^{-9.88} = 1.3 \times 10^{-10}\ M$ $[OH^-] = 10^{-4.12} = 7.6 \times 10^{-5}\ M$

(d) pH = 14.00 - pOH = 6.5 pOH = 7.5
$[H^+] = 10^{-6.5} = 3 \times 10^{-7}\ M$ $[OH^-] = 10^{-7.5} = 3 \times 10^{-8}\ M$

16.25 At 37°C, $K_w = 2.4 \times 10^{-14} = [H^+][OH^-]$.
In pure water, $[H^+] = [OH^-]$; $2.4 \times 10^{-14} = [H^+]^2$; $[H^+] = (2.4 \times 10^{-14})^{1/2}$
$[H^+] = [OH^-] = 1.5 \times 10^{-7}\ M$; pH = pOH = 6.81.

16.26 $K_w = [D^+][OD^-]$; for pure D_2O, $[D^+] = [OD^-]$
$8.9 \times 10^{-16} = [D^+]^2$; $[D^+] = [OD^-] = 3.0 \times 10^{-8}\ M$
pD = -log $[D^+]$ = -log (3.0×10^{-8}) = 7.52 ; pD = pOD = 7.52

16.27 A change of one pH unit (in either direction) is:

$$\Delta pH = pH_2 - pH_1 = -(\log [H^+]_2 - \log [H^+]_1) = -\log \frac{[H^+]_2}{[H^+]_1} = \pm 1$$

The antilog of +1 is 10; the antilog of -1 is 1×10^{-1}. Thus, a ΔpH of one unit represents an increase or decrease in $[H^+]$ by a factor of 10.

(a) $\Delta pH = \pm 2.00$ is a change of $10^{2.00}$; $[H^+]$ changes by a factor of 100.
(b) $\Delta pH = \pm 0.5$ is a change of $10^{0.50}$; $[H^+]$ changes by a factor of 3.2.

16.28 $[H^+]_A = 500\ [H^+]_B$ From Exercise 16.27, $\Delta pH = -\log \frac{[H^+]_B}{[H^+]_A}$

$$\Delta pH = -\log \frac{[H^+]_B}{500\ [H^+]_b} = -\log \left(\frac{1}{500}\right) = 2.70$$

The pH of solution A is 2.70 pH units lower than the pH of solution B, because $[H^+]_A$ is 500 times greater than $[H^+]_B$. The greater $[H^+]$, the lower the pH of the solution.

Strong Acids and Bases

16.29 (a) A *strong* acid is completely dissociated into ions (H^+ and some anion) in aqueous solution; a strong acid is a strong electrolyte.

(b) For a strong acid such as HCl, $[H^+]$ = initial acid concentration. $[H^+] = 0.500\ M$

(c) HCl, HBr, HI

16.30 (a) A *strong* base is completely dissociated into ions (OH^- and some cation) in aqueous solution; a strong base is a strong electrolyte.

(b) $Sr(OH)_2$ is a soluble strong base.

$Sr(OH)_2(aq) \rightarrow Sr^{2+}(aq) + \mathbf{2}OH^-(aq)$

$0.125\ M\ Sr(OH)_2(aq) = 0.250\ M\ OH^-$

(c) Base strength should not be confused with solubility. Base strength describes the tendency of a dissolved molecule (formula unit for ionic compounds such as $Mg(OH)_2$) to dissociate into cations and hydroxide ions. $Mg(OH)_2$ is a strong base because each $Mg(OH)_2$ unit that dissolves is dissociated into $Mg^{2+}(aq)$ and OH^- (aq). $Mg(OH)_2$ is not very soluble, so relatively few $Mg(OH)_2$ units dissolve when the solid compound is added to water.

16.31 For a strong acid, which is completely dissociated into ions, $[H^+]$ = the initial acid concentration.

(a) $0.025\ M\ HNO_3 = 0.025\ M\ H^+$; pH = -log (0.025) = 1.60

(b) $$\frac{0.824\text{ g } HClO_4}{0.500\text{ L soln}} \times \frac{1\text{ mol } HClO_4}{100.5\text{ g } HClO_4} = \frac{0.0164\text{ mol } HClO_4}{1\text{ L}} = 0.0164\ M\ HClO_4$$

$[H^+] = 0.0164\ M$; pH = -log (0.0164) = 1.785

(c) $M_c \times V_c = M_d \times V_d$

$1.5\ M$ HCl × 5.00 mL HCl = M_d HCl × 100 mL HCl

$$M_d\text{ HCl} = \frac{1.5\text{ M} \times 5.00\text{ mL}}{100\text{ mL}} = 0.075\ M\text{ HCl} = 0.075\ M\ H^+$$

pH = -log (0.075) = 1.12

(d) $$[H^+]_{total} = \frac{\text{mol } H^+ \text{ from HCl} + \text{mol } H^+ \text{ from HI}}{\text{total L solution}}\ ; \quad \text{mol} = M \times \text{L}$$

$$[H^+]_{total} = \frac{(0.020\ M\text{ HCl} \times 0.010\text{ L}) + (0.010\ M\text{ HI} \times 0.030\text{ L})}{0.040\text{ L}}$$

$$[H^+]_{total} = \frac{2.0 \times 10^{-4}\text{ mol } H^+ + 3.0 \times 10^{-4}\text{ mol } H^+}{0.040\text{ L}} = 0.0125\ M$$

pH = -log (0.0125) = 1.90

16.32 For a strong acid, $[H^+]$ = initial acid concentration.

(a) $1.8 \times 10^{-4}\ M\ HBr = 1.8 \times 10^{-4}\ M\ H^+$; $pH = -\log(1.8 \times 10^{-4}) = 3.74$

(b) $$\frac{1.02\ g\ HNO_3}{0.250\ L\ soln} \times \frac{1\ mol\ HNO_3}{63.02\ g\ HNO_3} = 0.0647\ M\ HNO_3$$

$[H^+] = 0.0647\ M$; $pH = -\log(0.0647) = 1.189$

(c) $M_c \times L_c = M_d \times L_d$; $0.500\ M \times 0.00200\ L = ?\ M \times 0.0500\ L$

$$M_d = \frac{0.500\ M \times 0.00200}{0.0500} = 0.0200\ M\ HCl$$

$[H^+] = 0.0200\ M$; $pH = -\log(0.075) = 1.699$

(d) $$[H^+]_{total} = \frac{mol\ H^+\ from\ HBr + mol\ H^+\ from\ HCl}{total\ L\ solution}$$

$$[H^+]_{total} = \frac{(0.0100\ M\ HBr \times 0.0100\ L) + (2.50 \times 10^{-3}\ M \times 0.0200\ L)}{0.\ 0300\ L}$$

$$[H^+]_{total} = \frac{1.00 \times 10^{-4}\ mol\ H^+ + 0.500 \times 10^{-4}\ mol\ H^+}{0.0300\ L} = 5.00 \times 10^{-3}\ M$$

$pH = -\log(5.00 \times 10^{-3}\ M) = 2.301$

16.33 For a strong base, which is completely dissociated into ions, $[OH^-]$ = the initial base concentration. Then, $pOH = -\log[OH^-]$ and $pH = 14 - pOH$.

(a) $0.050\ M\ KOH = 0.050\ M\ OH^-$; $pOH = -\log(0.050) = 1.30$; $pH = 14 - 1.30 = 12.70$

(b) $$\frac{2.33\ g\ NaOH}{0.500\ L} \times \frac{1\ mol\ NaOH}{40.00\ g\ NaOH} = \frac{0.116\ mol\ NaOH}{1\ L} = 0.116\ M\ NaOH$$

$[OH^-] = 0.116\ M$; $pOH = -\log(0.116) = 0.9337$

$pH = 14 - 0.9337 = 13.066$

(c) $M_c \times V_c = M_d \times L_d$

$0.150\ M\ Ca(OH)_2 \times 10.0\ mL = M_d Ca(OH)_2 \times 500\ mL$

$$M_d\ Ca(OH)_2 = \frac{0.150\ M\ Ca(OH)_2 \times 10.0\ mL}{500\ mL} = 0.00300\ M\ Ca(OH)_2$$

$Ca(OH)_2(aq) \rightarrow Ca^{2+}(aq) + \underline{2}OH^-(aq)$

$[OH^-] = 2[Ca(OH)_2] = 2(3.00 \times 10^{-3}\ M) = 6.00 \times 10^{-3}\ M$

$pOH = -\log(6.00 \times 10^{-3}) = 2.222$; $pH = 14 - 2.222 = 11.778$

(d) $[OH^-]_{total} = \dfrac{\text{mol } OH^- \text{ from } Ba(OH)_2 + \text{mol } OH^- \text{ from NaOH}}{\text{total L solution}}$

mol OH^- from NaOH = $M \times L = 6.8 \times 10^{-3}$ M NaOH $\times$ 0.0300 L
$= 2.0 \times 10^{-4}$ mol OH^-

mol OH^- from $Ba(OH)_2$ = 2 $\times$ mol $Ba(OH)_2$ (see part (c) above)
$= 2(M \times L) = 2(0.015\ M\ Ba(OH)_2 \times 0.0100\ L)$
$= 2(1.5 \times 10^{-4}$ mol $Ba(OH)_2) = 3.0 \times 10^{-4}$ mol OH^-

$[OH^-]_{total} = \dfrac{3.0 \times 10^{-4} \text{ mol } OH^- + 2.0 \times 10^{-4} \text{ mol OH}}{0.040\ L} = 0.0125\ M\ OH^-$

pOH = -log (0.0125) = 1.90 ; pH = 14 - pOH = 12.10

16.34 (a) $[OH^-] = 2[Sr(OH)_2] = 2(3.5 \times 10^{-4}\ M) = 7.0 \times 10^{-4}\ M\ OH^-$ (see Exercise 16.33 (c))

pOH = -log (7.0×10^{-4}) = 3.15; pH = 14 - pOH = 10.85

(b) $\dfrac{1.50 \text{ g LiOH}}{0.250 \text{ L soln}} \times \dfrac{1 \text{ mol LiOH}}{23.95 \text{ g LiOH}} = 0.250\ M \text{ LiOH} = [OH^-]$

pOH = -log (0.250) = 0.601 ; pH = 14 - pOH = 13.399

(c) $M_c \times L_c = M_d \times L_d$; $0.095\ M \times 0.00100\ L = ?\ M \times 2.00\ L$

$M_d = \dfrac{0.095\ M \times 0.00100\ L}{2.00\ L} = 4.75 \times 10^{-5}\ M \text{ NaOH} = [OH^-]$

pOH = -log (4.75×10^{-5}) = 4.32; pH = 14 - pOH = 9.68

(d) $[OH^-]_{total} = \dfrac{\text{mol } OH^- \text{ from KOH} + \text{mol } OH^- \text{ from } Ca(OH)_2}{\text{total L soln}}$

$[OH^-]_{total} = \dfrac{(0.0105\ M \times 0.0050\ L) + 2(3.5 \times 10^{-3} \times 0.0150\ L)}{0.0200\ L}$

$[OH^-]_{total} = \dfrac{5.250 \times 10^{-5} \text{ mol } OH^- + 10.50 \times 10^{-5} \text{ mol } OH^-}{0.0200\ L} = 7.88 \times 10^{-3}\ M$

pOH = -log (7.88×10^{-3}) = 2.103; pH = 14- pOH = 11.896

16.35 Upon dissolving, Li_2O dissociates to form Li^+ and O^{2-}. According to Equation 16.17, O^{2-} is completely protonated in aqueous solution.

Thus, initial $[Li_2O] = [O_2^-]$; $[OH^-] = 2[O^{2-}] = 2[Li_2O]$

$$[Li_2O] = \frac{\text{mol } Li_2O}{\text{L solution}} = 2.00 \text{ g } Li_2O \times \frac{1 \text{ mol } Li_2O}{29.88 \text{ g } Li_2O} \times \frac{1}{0.600 \text{ L}} = 0.112\ M$$

$[OH^-] = 0.223\ M$; pOH = 0.652 pH = 14.00 - pOH = 13.348

16.36 $NaH(aq) \rightarrow Na^+(aq) + H^-(aq)$
$H^-(aq) + H_2O(l) \rightarrow H_2(g) + OH^-(aq)$

Thus, initial $[NaH] = [OH^-]$

$$[NaH] = \frac{\text{mol NaH}}{\text{L solution}} = 5.00 \text{ g NaH} \times \frac{1 \text{ mol NaH}}{24.00 \text{ g NaH}} \times \frac{1}{0.900 \text{ L}} = 0.232\ M$$

$[OH^-] = 0.232\ M$; pOH = 0.636, pH = 14 - pOH = 13.364

16.37 Strongest: $HClO_4$ (one of the six strong acids); weakest: HClO (smallest K_a value of the other acids listed).

16.38 HBr is strongest, HCN weakest.

16.39 $HC_3H_5O_3(aq) \rightleftharpoons H^+(aq) + C_3H_5O_3^-(aq)$; $K_a = \frac{[H^+][C_3H_5O_3^-]}{[HC_3H_5O_3]}$

$[H^+] = [C_3H_5O_3^-] = 10^{-2.44} = 3.6 \times 10^{-3}\ M$

$[HC_3H_5O_3] = 0.10 - 3.6 \times 10^{-3} = 0.096\ M$

$$K_a = \frac{(3.6 \times 10^{-3})^2}{(0.096)} = 1.4 \times 10^{-4}$$

16.40 $HCrO_4^-(aq) \rightleftharpoons H^+(aq) + CrO_4^-(aq)$; $K_a = \frac{[H^+][CrO_4^{2-}]}{[HCrO_4^-]}$

$[H^+] = [CrO_4^{2-}]$ = antilog (-3.80) = $1.6 \times 10^{-4}\ M$

$[HCrO_4^-] = 0.050 - 1.6 \times 10^{-4} = 0.050\ M$

(That is, the portion that ionizes is negligible.)

$K_a = (1.6 \times 10^{-4})^2/(0.050) = 5.1 \times 10^{-7}$

16.41 (a) $HN_3(aq) \rightleftharpoons H^+(aq) + N_3^-(aq)$

initial	0.175 *M*	0	0
equil.	(0.175 - x)*M*	x *M*	x *M*

$$K_a = \frac{[H^+][N_3^-]}{[HN_3]} = \frac{(x)(x)}{(0.175 - x)} \approx \frac{x^2}{0.175} = 1.9 \times 10^{-5}$$

$x^2 = 3.3 \times 10^{-6}$; $x = [H^+] - 1.8 \times 10^{-3}$ *M*, pH = 2.74

$$\frac{[H^+]}{[HN_3]_i} \times 100 = \frac{1.8 \times 10^{-3}\ M}{0.175\ M} \times 100 = 1.0\%;$$ the assumption is valid

(b) $$K_a = \frac{[H^+][C_3H_5O_2^-]}{[HC_3H_5O_2]} = \frac{(x)(x)}{(0.040 - x)} \approx \frac{x^2}{0.040} = 1.3 \times 10^{-5}$$

$x^2 = 5.2 \times 10^{-7}$; $x = [H^+] = 7.2 \times 10^{-4}$ *M* ; pH = 3.14

$$\frac{7.2 \times 10^{-4}\ M}{0.040} \times 100 = 1.8\%;$$ the assumption is valid

(c) $C_2H_5NH_2(aq) + H_2O(l) \rightleftharpoons C_2H_5NH_3^+(aq) + OH^-(aq)$

initial	0.050 *M*	0	0
equil.	(0.050 - x) *M*	x *M*	x *M*

$$K_b = \frac{[C_2H_5NH_3^+][OH^-]}{[C_2H_5NH_2]} = \frac{(x)(x)}{(0.050 - x)} \approx \frac{x^2}{0.050} = 6.4 \times 10^{-4}$$

$x^2 = 3.2 \times 10^{-5}$; $x = [OH^-] = 5.7 \times 10^{-3}$ *M* ; pH = 11.76

$$\frac{5.7 \times 10^{-3}\ M}{0.050\ M} \times 100 = 11.3\%;$$ the assumption is <u>not</u> valid

To obtain a more precise result, the K_b expression is rewritten in standard quadratic form and solved via the quadratic formula.

$$\frac{x^2}{0.050 - x} = 6.4 \times 10^{-4}; \quad x^2 + 6.4 \times 10^{-4}x - 3.2 \times 10^{-5} = 0$$

$$x = \frac{-b \pm \sqrt{b^2 - 4ac}}{2a} = \frac{-6.4 \times 10^{-4} \pm \sqrt{(6.4 \times 10^{-4})^2 - 4(1)(-3.2 \times 10^{-5})}}{2}$$

$x = 5.3 \times 10^{-3}$ *M* OH^- ; pOH = 2.27, pH = 14.00 - pOH = 11.73

Note that the pH values obtained using the two algebraic techniques are very similar.

16.42 (a)

	$HOCl(aq)$	⇌	$H^+(aq)$ +	$OCl^-(aq)$
initial	0.045 *M*		0	0
equil.	(0.045-x) *M*		x*M*	x*M*

$$K_a = \frac{[H^+][OCl^-]}{[HOCl]} = \frac{(x)(x)}{(0.045\text{-}x)} \approx \frac{x^2}{0.045} = 3.0 \times 10^{-8}$$

$x^2 = 1.4 \times 10^{-9}$; $x = [H^+] = 3.7 \times 10^{-5}$ *M*, pH = 4.43

$\frac{3.7 \times 10^{-5}\ M}{0.045\ M} \times 100 = 0.083\%$ dissociation; the assumption is valid

(b) $$K_a = \frac{[H^+][C_6H_5O^-]}{[C_6H_5OH]} = \frac{(x)(x)}{(0.0068\text{-}x)} \approx \frac{x^2}{0.0068} = 1.3 \times 10^{-10}$$

$x^2 = 8.8 \times 10^{-13}$; $x = [H^+] = 9.4 \times 10^{-7}$ *M*, pH = 6.03

Because $[H^+]$ from the dissociation of phenol, 9.7×10^{-7} *M*, is on the order of magnitude of the $[H^+]$ expected from the dissociation of water, $[H^+]$ from the solvent should be considered.

$1.0 \times 10^{-14} = [H^+][OH^-]$; $[OH^-] = y$, $[H^+] = 9.4 \times 10^{-7} + y$
$(y + 9.4 \times 10^{-7})\,y = 1.0 \times 10^{-14}$

From the quadratic formula, $y = 1.1 \times 10^{-8}$ *M*. Since this is less than 5% of the $[H^+]$ due to phenol, it can be ignored.

(c)

	$HONH_2(aq) + H_2O(l)$	⇌	$HONH_3^+(aq)$ +	$OH^-(aq)$
initial	0.080 *M*		0	0
equil.	(0.080-x) *M*		x*M*	x*M*

$$K_b = \frac{[HONH_3^+][OH^-]}{[HONH_2]} = \frac{(x)(x)}{(0.080\text{-}x)} \approx \frac{x^2}{0.080} = 1.1 \times 10^{-8}$$

$x^2 = 8.8 \times 10^{-10}$; $x = [OH^-] = 3.0 \times 10^{-5}$ *M*, pH = 9.47

$\frac{3.0 \times 10^{-5}\ M}{0.080\ M} \times 100 = 0.038\%$ dissociation; the assumption is valid

16.43 $HIO_3(aq) \rightleftharpoons H^+(aq) + IO_3^-(aq)$

$$K_a = 1.7 \times 10^{-1} = \frac{[H^+][IO_3^-]}{[HIO_3]} = \frac{x^2}{0.020 - x}$$

Assuming that the % dissociation is small, $x = [H^+] = 0.058\ M$. This is an obviously incorrect result. The expression must be solved as a quadratic equation. Although the % dissociation depends on both K_a and initial acid concentration, when K_a is as large as 1.7×10^{-1}, it is a good bet that the quadratic equation will be required.

$x^2 = 0.17(0.020 - x)$; $x^2 + 0.17x - 0.0034 = 0$

$$x = \frac{-0.17 \pm \sqrt{(0.17)^2 - 4(1)(-0.0034)}}{2} = 0.018\ M\ H^+,\ pH = 1.74$$

Note that the % dissociation in this solution is large.

$$\frac{0.018\ M\ H^+}{0.020\ M\ HIO_3} \times 100 = 90\%$$

16.44 $(CH_3)_2NH(aq) + H_2O(l) \rightleftharpoons (CH_3)_2NH_2^+(aq) + OH^-(aq)$

$$K_b = 5.4 \times 10^{-4} = \frac{[(CH_3)_2NH_2^+][OH^-]}{[(CH_3)_2NH]} = \frac{x^2}{2.00 \times 10^{-3} - x}$$

Assuming % dissociation is small, $x = [OH^-] = 1.04 \times 10^{-3}\ M$; this is more than 50% dissociation. An accurate solution requires the quadratic formula. (Rule of thumb: if $[HA]_i/K_a > 400$, the assumption is valid. In this case, $[B]_i/K_b = 3.70$, so the quadratic is required.)

$x^2 = (5.4 \times 10^{-4})(2.00 \times 10^{-3} - x)$; $x^2 + 5.4 \times 10^{-4} - 1.08 \times 10^{-6} = 0$

$$x = \frac{-5.4 \times 10^{-4} + \sqrt{(5.4 \times 10^{-4})^2 - 4(1)(-1.08 \times 10^{-6})}}{2} = 8.05 \times 10^{-4}\ M\ OH^-$$

pOH = 3.09, pH = 10.91

$$\%\ \text{dissociation} = \frac{8.05 \times 10^{-4}\ M}{2.00 \times 10^{-3}\ M} \times 100 = 40.3\%$$

16.45 (a) $HN_3(aq) \rightleftharpoons H^+(aq) + N_3^-(aq)$

	$HN_3(aq)$	$H^+(aq)$	$N_3^-(aq)$
initial	0.400 M	0	0
equil.	(0.400-x) M	x M	x M

$$K_a = \frac{[H^+][N_3^-]}{[HN_3]} = 1.9 \times 10^{-5};\quad \frac{x^2}{(0.400-x)} \approx \frac{x^2}{0.400} = 1.9 \times 10^{-5}$$

$$x = 2.8 \times 10^{-3}\ M = [H^+];\quad \%\ \text{ionization} = \frac{2.8 \times 10^{-3}}{0.400} \times 100 = 0.70\%$$

(b) $1.9 \times 10^{-5} \approx \frac{x^2}{0.100}$; $x = 1.4 \times 10^{-3}\ M\ H^+$

$$\underline{\%\ \text{ionization}} = \frac{1.4 \times 10^{-3}\ M\ H^+}{0.100\ M\ HN_3} \times 100 = 1.4\%$$

(c) $1.9 \times 10^{-5} \approx \frac{x^2}{0.040}$; $x = 8.7 \times 10^{-4}\ M\ H^+$

$$\underline{\%\ \text{ionization}} = \frac{8.7 \times 10^{-4}\ M\ H^+}{0.040\ M\ HN_3} \times 100 = 2.2\%$$

Notice that a tenfold dilution [part (a) versus part (c)] leads to a slightly more than threefold increase in percent ionization.

16.46 (a) $HCrO_4^-(aq) \rightleftharpoons H^+(aq) + CrO_4^{2-}(aq)$

$$K_a = 3.0 \times 10^{-7} = \frac{[H^+][CrO_4^{2-}]}{[HCrO_4^-]} = \frac{x^2}{0.250\text{-}x}$$

$x^2 = 7.5 \times 10^{-8}$; $x = 2.7 \times 10^{-4}\ M\ H^+$

$$\underline{\%\ \text{ionization}} = \frac{2.7 \times 10^{-4}\ M\ H^+}{0.250\ M\ HCrO_4^-} \times 100 = 0.11\%$$

(b) $\frac{x^2}{0.0800} \approx 3.0 \times 10^{-7}$; $x = 1.6 \times 10^{-4}\ M\ H^+$

$$\underline{\%\ \text{ionization}} = \frac{1.6 \times 10^{-4}\ M\ H^+}{0.0800\ M\ HCrO_4^-} \times 100 = 0.19\%$$

(c) $\frac{x^2}{0.0200} \approx 3.0 \times 10^{-7}$; $x = 7.7 \times 10^{-5}\ M\ H^+$

$$\underline{\%\ \text{ionization}} = \frac{7.7 \times 10^{-5}\ M\ H^+}{0.0200\ M\ HCrO_4^-} \times 100 = 0.39\%$$

16.47 $[H^+] = 0.094 \times [HX]_{initial} = 0.019\ M$

	$HX(aq)$	$\rightleftharpoons$	$H^+(aq)$	+	$X^-(aq)$
initial	0.200 M		0		0
equil.	(0.200 - 0.019) M		0.019 M		0.019 M

$$K_a = \frac{[H^+][X^-]}{[HX]} = \frac{(0.019)^2}{0.181} = 2.0 \times 10^{-3}$$

16.48 $[H^+] = 0.132 \times [CH_2BrCOOH]_{initial} = 0.0132\ M$

	$CH_2BrCOOH(aq)$	$\rightleftharpoons$	$H^+(aq)$	+	$CH_2BrCOO^-(aq)$
initial	0.100 *M*		0		0
equil.	0.087 *M*		0.0132 *M*		0.0132 *M*

$$K_a = \frac{[H^+][CH_2BrCOO^-]}{[CH_2BrCOOH]} = \frac{(0.0132)^2}{0.087} = 2.0 \times 10^{-3}$$

16.49 Let the weak acid be HX. $HX(aq) \rightleftharpoons H^+(aq) + X^-(aq)$

$$K_a = \frac{[H^+][X^-]}{[HX]}\ ;\ [H^+] = [X^-] = y;\ K_a = \frac{y^2}{[HX] - y}$$; assume that % ionization is small

$$K_a = \frac{y^2}{[HX]}\ ;\ y = K_a^{1/2}\,[HX]^{1/2}$$

$$\%\ \text{ionization} = \frac{y}{[HX]} \times 100 = \frac{K_a^{1/2}[HX]^{1/2}}{[HX]} \times 100 = \frac{K_a^{1/2}}{[HX]^{1/2}} \times 100$$

That is, percent ionization varies inversely as the square root of concentration HX.

16.50 $HX(aq) \rightleftharpoons H^+(aq) + X^-(aq)$; $K_a = \frac{[H^+][X^-]}{[HX]}$

$[H^+] = [X^-]$; assume the % ionization is small; $K_a = \frac{[H^+]^2}{[HX]}$; $[H^+] = K_a^{1/2}\,[HX]^{1/2}$

$pH = -\log K_a^{1/2}[HX]^{1/2} = -\log K_a^{1/2} - \log[HX]^{1/2}$; $pH = -1/2 \log K_a - 1/2 \log [HX]$

This is the equation of a straight line; where the intercept is $-1/2 \log K_a$, the slope is $-1/2$, and the independent variable is log[HX].

16.51 $H_3C_6H_5O_7(aq) \rightleftharpoons H^+(aq) + H_2C_6H_5O_7^-(aq)$

$H_2C_6H_5O_7^-(aq) \rightleftharpoons H^+(aq) + HC_6H_5O_7^{2-}(aq)$

$HC_6H_5O_7^{2-}(aq) \rightleftharpoons H^+(aq) + C_6H_5O_7^{3-}(aq)$

To calculate the pH of a 0.050 *M* solution, assume initially that only the first ionization is important:

	$H_3C_6H_5O_7(aq)$	$\rightleftharpoons$	$H^+(aq)$	+	$H_2C_6H_5O_7^-(aq)$
initial	0.050 *M*		0		0
equil.	(0.050-x) *M*		x *M*		x *M*

$$K_{a1} = \frac{[H^+][H_2C_6H_5O_7^-]}{[H_3C_6H_5O_7]} = \frac{x^2}{(0.050-x)} = 7.4 \times 10^{-4}$$

$x^2 = (0.050-x)(7.4 \times 10^{-4})$; $x^2 \approx (0.050)(7.4 \times 10^{-4})$; $x = 6.1 \times 10^{-3}\ M$

Since this value for x is rather large in relation to 0.050, a better approximation for x can be obtained by substituting this first estimate into the expression for x^2, then solving again for x:

$$x^2 = (0.050\text{-}x)\ (7.4 \times 10^{-4}) = (0.050 - 6.1 \times 10^{-3})\ (7.4 \times 10^{-4})$$
$$x^2 = 3.2 \times 10^{-5}\ ;\quad x = 5.7 \times 10^{-3}\ M$$

The correction to the value of x, though not large, is significant. (This is the same result obtained from the quadratic formula.) Does the second ionization produce a significant additional concentration of H^+?

	$H_2C_6H_5O_7^-(aq)$ ⇌	$H^+(aq)$ +	$HC_6H_5O_7^{2-}(aq)$
initial	$5.7 \times 10^{-3}\ M$	$5.7 \times 10^{-3}\ M$	0
equil.	$(5.7 \times 10^{-3} - y)$	$(5.7 \times 10^{-3} + y)$	y

$$K_{a2} = \frac{[H^+][HC_6H_5O_7^{2-}]}{[H_2C_6H_5O_7^-]} = 1.7 \times 10^{-5}\ ;\quad \frac{(5.7 \times 10^{-3} + y)\ (y)}{(5.7 \times 10^{-3} - y)} = 1.7 \times 10^{-5}$$

Assume that y is small relative to 5.7×10^{-3}; that is, that additional ionization of $H_2C_6H_5O_7^-$ is small, then

$$\frac{(5.7 \times 10^{-3})y}{(5.7 \times 10^{-3})} = 1.7 \times 10^{-5}\ ;\quad y = 1.7 \times 10^{-5}\ M$$

This value is indeed small compared to 5.7×10^{-3} *M*. This indicates that the second ionization can be neglected. pH is therefore $-\log[5.7 \times 10^{-3}] = 2.24$.

16.52 $H_6C_4O_6(aq) \rightleftharpoons H^+(aq) + H_5C_4O_6^-(aq)$ $K_{a1} = 1.0 \times 10^{-3}$
$H_5C_4O_6^-(aq) \rightleftharpoons H^+(aq) + H_4C_4O_6^{2-}(aq)$ $K_{a2} = 4.6 \times 10^{-5}$

Begin by calculating the $[H^+]$ from the first ionization. The equilibrium concentrations are $[H^+] = [H_5C_4O_6^-] = x$, $[H_6C_4O_6] = 0.025 - x$.

$$K_{a1} = \frac{[H^+][H_5C_4O_6^-]}{[H_6C_4O_6]} = \frac{x^2}{0.025 - x}\ ;\quad x^2 + 1.0 \times 10^{-3}\,x - 2.5 \times 10^{-5} = 0$$

Using the quadratic formula, $x = 4.5 \times 10^{-3}$ *M* H^+ from the first ionization. Next calculate the H^+ contribution from the second ionization.

	$H_5C_4O_6^-(aq)$ ⇌	$H^+(aq)$ +	$H_4C_4O_6^{2-}(aq)$
initial	4.5×10^{-3}	4.5×10^{-3}	0
equil.	$(4.5 \times 10^{-3} - y)$	$(4.5 \times 10^{-3} + y)$	y

$$K_{a2} = \frac{(4.5 \times 10^{-3} + y)(y)}{(4.5 \times 10^{-3} - y)} = 4.6 \times 10^{-5}$$

If y is small relative to 4.5×10^{-3}, $4.6 \times 10^{-5} = \frac{(4.5 \times 10^{-3})(y)}{(4.5 \times 10^{-3})}$; $y = 4.6 \times 10^{-5}$ *M*

Since y, 4.6×10^{-5} *M* H^+, is only 1% of the $[H^+]$ from the first ionization, y can be safely ignored when calculating total $[H^+]$. Thus, pH = $-\log(4.5 \times 10^{-3}) = 2.35$.

16.53 (a) $C_3H_7NH_2(aq) + H_2O(l) \rightleftharpoons C_3H_7NH_3^+(aq) + OH^-(aq)$; $K_b = \frac{[C_3H_7NH_3^+][OH^-]}{[C_3H_7NH_2]}$

(b) $CN^-(aq) + H_2O(l) \rightleftharpoons HCN(aq) + OH^-(aq)$; $K_b = \frac{[HCN][OH^-]}{[CN^-]}$

(c) $CHO_2^-(aq) + H_2O(l) \rightleftharpoons HCHO_2(aq) + OH^-(aq)$; $K_b = \frac{[HCHO_2][OH^-]}{[CHO_2^-]}$

16.54 (a) $H_2NNH_2(aq) + H_2O(l) \rightleftharpoons H_2NNH_3^+(aq) + OH^-(aq)$; $K_b = \frac{[H_2NNH_3^+][OH^-]}{[H_2NNH_2]}$

(b) $CH_3NH_2(aq) + H_2O(l) \rightleftharpoons CH_3NH_3^+(aq) + OH^-(aq)$; $K_b = \frac{[CH_3NH_3^+][OH^-]}{[CH_3NH_2]}$

(c) $C_6H_5CO_2^-(aq) + H_2O(l) \rightleftharpoons HC_6H_5CO_2(aq) + OH^-(aq)$; $K_b = \frac{[HC_6H_5CO_2][OH^-]}{[C_6H_5CO_2^-]}$

K_a - K_b Relationship; Acid-Base Properties of Salts

16.55 (a)

$HCN(aq) \rightleftharpoons H^+(aq) + CN^-(aq)$ ionization of HCN

$CN^-(aq) + H_2O(l) \rightleftharpoons HCN(aq) + OH^-(aq)$ hydrolysis of CN^-

$HCN(aq) + CN^-(aq) + H_2O(l) \rightleftharpoons H^+(aq) + OH^-(aq) + HCN(aq) + CN^-(aq)$

After cancelling, terms that appear on both sides of the equation:

$H_2O(l) \rightleftharpoons H^+(aq) + OH^-(aq)$ autoionization of H_2O

(b) K_a for HCN = $\frac{[H^+][CN^-]}{[HCN]}$; K_b for CN^- = $\frac{[HCN][OH^-]}{[CN^-]}$

$$K_a \times K_b = \frac{[H^+][\cancel{CN^-}]}{[\cancel{HCN}]} \times \frac{[OH^-][\cancel{HCN}]}{[\cancel{CN^-}]} = [H^+][OH^-] = K_w$$

16.56 (a) $C_6H_5O^-(aq) + H_2O(l) \rightleftharpoons C_5H_5OH(aq) + OH^-(aq)$

Since $K_a \times K_b = K_w$, K_b for $C_6H_5O^-$ can be calculated from K_a for the conjugate acid C_6H_5OH and K_w.

$$K_b = \frac{K_w}{K_a} = \frac{1.0 \times 10^{-14}}{1.3 \times 10^{-10}} = 7.7 \times 10^{-5}$$

16.57 $K_b = K_w/K_a$ for the conjugate acid of the weak base in question.

(a) $K_b = K_w/K_a\ (HNO_2) = 1.0 \times 10^{-14}/4.5 \times 10^{-4} = 2.2 \times 10^{-11}$

(b) $K_b = K_w/K_a\ (HN_3) = 1.0 \times 10^{-14}/1.9 \times 10^{-5} = 5.3 \times 10^{-10}$

(c) $K_b = K_w/K_a\ (H_2PO_4^-) = 1.0 \times 10^{-14}/6.2 \times 10^{-8} = 1.6 \times 10^{-7}$

(d) $K_b = K_w/K_a\ (HCHO_2) = 1.0 \times 10^{-14}/1.8 \times 10^{-4} = 5.6 \times 10^{-11}$

16.58 $K_a = K_w/K_b$ for the conjugate base of the weak acid in question.

(a) $K_a = K_w/K_b(\ (CH_3)_2NH_2) = 1.0 \times 10^{-14}/5.4 \times 10^{-4} = 1.8 \times 10^{-11}$

(b) $K_a = K_w/K_b\ (H_2NNH_2) = 1.0 \times 10^{-14}/1.3 \times 10^{-6} = 7.7 \times 10^{-9}$

(c) $K_a = K_w/K_b\ (C_5H_5N) = 1.0 \times 10^{-14}/1.7 \times 10^{-9} = 5.9 \times 10^{-6}$

(d) $K_a = K_w/K_b\ (HONH_2) = 1.0 \times 10^{-14}/1.1 \times 10^{-8} = 9.1 \times 10^{-7}$

16.59 When the solute in an aqueous solution is a salt, evaluate the acid/base properties of the component ions.

(a) NaCN is a soluble salt and thus a strong electrolyte. When it is dissolved in H_2O, it dissociates completely into Na^+ and CN^-. $[NaCN] = [Na^+] = [CN^-] = 0.10$ M. Na^+ is the conjugate acid of the strong base NaOH and thus does not influence the pH of the solution. CN^-, on the other hand, is the conjugate base of the weak acid HCN and <u>does</u> influence the pH of the solution. Like any other weak base, it hydrolyzes water to produce $OH^-(aq)$. Solving the equilibrium problem to determine $[OH^-]$:

	$CN^-(aq) + H_2O(l)$	$\rightleftharpoons$	$HCN(aq)$	$+ OH^-(aq)$
initial	0.10 *M*		0	0
equil.	(0.10-x) *M*		x *M*	x *M*

$$K_b \text{ for } CN^- = \frac{[HCN][OH^-]}{[CN^-]} = \frac{K_w}{K_a \text{ for HCN}} = \frac{1 \times 10^{-14}}{4.90 \times 10^{-10}} = 2.0 \times 10^{-5}$$

$2.0 \times 10^{-5} = \frac{(x)(x)}{(0.10-x)}$; assume the percent of CN^- that hydrolyzes (x) is small

$x^2 = 0.10\ (2.0 \times 10^{-5})$; $x = [OH^-] = 1.4 \times 10^{-3}$ *M*

pOH = 2.85; pH = 14 - 2.85 = 11.15

(b) $Na_2CO_3(aq) \rightarrow 2Na^+(aq) + CO_3^{-2}(aq)$

CO_3^{2-} is the conjugate base of HCO_3^- and its hydrolysis reaction will determine the $[OH^-]$ and pH of the solution (see similar explanation for NaCN in part (a)). We will assume the process $HCO_3^-(aq) + H_2O(l) \rightleftharpoons H_2CO_3(aq) + OH^-$ will not add significantly to the $[OH^-]$ in solution because $[HCO_3^-(aq)]$ is so small. Solving the equilibrium problem for $[OH^-]$:

	$CO_3^{2-}(aq) + H_2O(l) \rightleftharpoons$	$HCO_3^-(aq)$	+ $OH^-(aq)$
initial	0.080 *M*	0	0
equil.	(0.080-x) *M*	x	x

$$K_b = \frac{[HCO_3^-][OH^-]}{[CO_3^{2-}]} = \frac{K_w}{K_a \text{ for } HCO_3^-} = \frac{1.0 \times 10^{-14}}{5.6 \times 10^{-11}} = 1.8 \times 10^{-4}$$

$$1.8 \times 10^{-4} = \frac{x^2}{(0.080\text{-}x)}\ ;\ x^2 = 0.080\ (1.8 \times 10^{-4})\ ;\ x = 3.8 \times 10^{-3}\ M\ OH^-$$

(Assume x is small compared to 0.080) ; pOH = 2.42; pH = 14 - 2.42 = 11.58

$$\%\ \text{hydrolysis} = \frac{3.8 \times 10^{-3}\ M\ OH^-}{0.080\ M\ CO_3^{2-}} \times 100 = 4.75\%$$; the assumption is valid.

(c) For the two salts present, Na^+ and Ca^{2+} are negligible acids. NO_2^- is the conjugate base of HNO_2 and will determine the pH of the solution.

Calculate total $[NO_2^-]$ present initially.

$[NO_2^-]_{total} = [NO_2^-]$ from $NaNO_2$ + $[NO_2^-]$ from $Ca(NO_2)_2$

$[NO_2^-]_{total} = 0.10\ M + 2(0.20\ M) = 0.50\ M$

The hydrolysis equilibrium is:

	$NO_2^-(aq) + H_2O(l) \rightleftharpoons$	HNO_2	+ OH^-
initial	0.50 *M*	0	0
equil.	(0.50-x) *M*	x *M*	x *M*

$$K_b = \frac{[HNO_2][OH^-]}{[NO_2^-]} = \frac{K_w}{K_a \text{ for } HNO_2} = \frac{1.0 \times 10^{-14}}{4.5 \times 10^{-4}} = 2.2 \times 10^{-11}$$

$$2.2 \times 10^{-11} = \frac{x^2}{(0.50\text{-}x)} \approx \frac{x^2}{0.50}\ ;\ x^2 = 0.50\ (2.2 \times 10^{-11})$$

$x = 3.3 \times 10^{-6}\ M\ OH^-$; pOH = 5.48 ; pH = 14 - 5.48 = 8.52

16.60 (a) Proceeding as in 16.59(a):

$$F^-(aq) + H_2O(l) \rightleftharpoons HF(aq) + OH^-(aq)$$

	F^-		HF	OH^-
initial	0.10 *M*		0 *M*	0 *M*
equil.	(0.10-x) *M*		x *M*	x *M*

$$K_b \text{ for } F^- = \frac{[HF][OH^-]}{[F^-]} = \frac{K_w}{K_a \text{ for HF}} = \frac{1.0 \times 10^{-14}}{6.8 \times 10^{-4}} = 1.5 \times 10^{-11}$$

$1.5 \times 10^{-11} = \frac{(x)(x)}{(0.10-x)}$; assume the amount of F^- that hydrolyzes is small

$x^2 = 0.10\,(1.5 \times 10^{-11})$; $x = [OH^-] = 1.2 \times 10^{-6}\ M$

pOH = 5.92 ; pH = 14 - 5.92 = 8.08

(b) $Na_2S(aq) \rightarrow S^{2-}(aq) + 2Na^+(aq)$

$$S^{2-}(aq) + H_2O(l) \rightleftharpoons HS^-(aq) + OH^-(aq)$$

Assume the further hydrolysis of HS^- does not significantly contribute to $[OH^-]$. As in part (a) above, $[OH^-] = [HS^-] = x$; $[S^{2-}] = 0.10\ M$

$$K_b = \frac{[HS^-][OH^-]}{[S^{2-}]} = \frac{K_w}{K_a \text{ for } HS^-} = \frac{1.0 \times 10^{-14}}{1.3 \times 10^{-13}} = 7.7 \times 10^{-2}$$

$7.7 \times 10^{-2} = \frac{x^2}{0.10-x}$; $x^2 + 7.7 \times 10^{-2}\,x - 7.7 \times 10^{-3} = 0$

Since K_b is relatively large, we will not assume x is small compared to 0.10 *M* S^{2-}.

$$x = \frac{-0.077 \pm \sqrt{(0.077)^2 - 4(1)(-7.7 \times 10^{-3})}}{2(1)} = \frac{-0.077 \pm \sqrt{0.0367}}{2}$$

$$x = \frac{-0.077 + 0.192}{2} = 0.057\ M\ OH^-$$

pH = 14 + log (0.057) = 12.76 (Note that S^{2-} is a relatively strong base.)

(c) As in exercise 16.59 (c), calculate total $[C_2H_3O_2^-]$.

$[C_2H_3O_2^-]_t = [C_2H_3O_2^-]$ from $NaC_2H_3O_2$ + $[C_2H_3O_2^-]$ from $Ba(C_2H_3O_2)_2$

$[C_2H_3O_2^-]_t = 0.085\ M + 2(0.055\ M) = 0.195\ M$

The hydrolysis equilibrium is $C_2H_3O_2^-(aq) + H_2O(l) \rightleftharpoons HC_2H_3O_2(aq) + OH^-(aq)$

$$K_b = \frac{[HC_2H_3O_2][OH^-]}{[C_2H_3O_2^-]} = \frac{K_w}{K_a \text{ for } HC_2H_3O_2} = \frac{1.0 \times 10^{-14}}{1.8 \times 10^{-5}} = 5.56 \times 10^{-10}$$

$[OH^-] = [HC_2H_3O_2] = x$, $[C_2H_3O_2^-] = 0.195\text{-}x$

$K_b = 5.56 \times 10^{-10} = \frac{x^2}{(0.195-x)}$; assume x is small compared to 0.195 *M*

$x^2 = 0.195\,(5.56 \times 10^{-10})$; $x = [OH^-] = 1.0 \times 10^{-5}$

$pH = 14 + \log(1.0 \times 10^{-5}) = 9.02$

16.61 (a) Basic because of the hydrolysis of $C_2H_3O_2^-$; K^+ is negligible.

(b) Basic because K_b for the base hydrolysis of HCO_3^- (2.3×10^{-8}) is greater than K_a for its acid dissociation (5.6×10^{-11}).

(c) Acidic because of the acid dissociation of $CH_3NH_3^+$; Br^- is negligible.

(d) Basic because of the hydrolysis of NO_2^- ; K^+ is negligible.

16.62 (a) Acidic, because K_a (6.4×10^{-5}) is greater than K_b (1.7×10^{-13}) for $HC_2O_4^-$. Note the K_a for $HC_2O_4^-$ is K_{a2} for $H_2C_2O_4$ and K_b for $HC_2O_4^-$ is K_w/K_{a1}.

(b) Neutral, neither Cs^+ nor I^- have acid or base properties.

(c) Acidic, because of hydrolysis of $Al^{3+}(aq)$.

(d) Basic, CN^- is a stronger base ($K_b = 2.0 \times 10^{-5}$) than NH_4^+ is an acid ($K_a = 5.6 \times 10^{-10}$).

16.63 The solution will be basic because of the hydrolysis of the sorbate anion, $C_6H_7O_2^-$. Calculate the initial molarity of $C_6H_7O_2^-$.

$$\frac{4.93 \text{ g } NaC_6H_7O_2}{0.500 \text{ L}} \times \frac{1 \text{ mol } NaC_6H_7O_2}{134.1 \text{ g } NaC_6H_7O_2} = 0.0735\ M\ NaC_6H_7O_2$$

$[C_6H_7O_2^-] = [NaC_6H_7O_2] = 0.0735\ M$

	$C_6H_7O_2^-(aq) + H_2O(l) \rightleftharpoons$	$HC_6H_7O_2(aq)$	$+\ OH^-(aq)$
initial	0.0735 *M*	0	0
equil.	(0.0735-x) *M*	x *M*	x *M*

$$K_b = \frac{[HC_6H_7O_2][OH^-]}{[C_6H_7O_2^-]} = \frac{K_w}{K_a \text{ for } HC_6H_7O_2} = \frac{1.0 \times 10^{-14}}{1.7 \times 10^{-5}} = 5.9 \times 10^{-10}$$

$5.9 \times 10^{-10} = \frac{x^2}{0.0735-x} \approx \frac{x^2}{0.0735}$; $x^2 = 0.0735\,(5.9 \times 10^{-10})$

$x = [OH^-] = 6.6 \times 10^{-6}\ M$; $pOH = -\log(6.6 \times 10^{-6}) = 5.18$; $pH = 14 - 5.18 = 8.82$

16.64 The solution is basic because of the hydrolysis of PO_4^{3-}. The molarity of PO_4^{3-} is

$$\frac{50.0\text{ g }Na_3PO_4}{1.00\text{ L soln}} \times \frac{1\text{ mol }Na_3PO_4}{163.9\text{ g }Na_3PO_4} = 0.305\ M\ PO_4^{3-}$$

$$PO_4^{3-}(aq) + H_2O(l) \rightleftharpoons HPO_4^{2-}(aq) + OH^-(aq)$$

$$K_b = \frac{[HPO_4^{2-}][OH^-]}{[PO_4^{3-}]} = \frac{K_w}{K_a \text{ for } HPO_4^{2-}} = \frac{1.0 \times 10^{-14}}{4.2 \times 10^{-13}} = 2.4 \times 10^{-2}$$

Ignoring the further hydrolysis of HPO_4^{2-}.

$[OH^-] = [HPO_4^{2-}] = x$, $[PO_4^{3-}] = 0.305\text{-}x$

$$2.4 \times 10^{-2} = \frac{x^2}{(0.305\text{-}x)}\ ;\ x^2 + 2.4 \times 10^{-2}x - 0.0073 = 0$$

Since K_b is relatively large, we will not assume x is small compared to 0.305.

$$x = \frac{-0.024 \pm \sqrt{(0.024)^2 - 4(1)(-0.0073)}}{2(1)} = \frac{-0.024 \pm \sqrt{0.0299}}{2(1)}$$

$x = 0.074\ M\ OH^-$; pH = 14 + log (0.074) = 12.87

Acid-Base Character and Chemical Structure

16.65 (a) As the electronegativity of the central atom (X) increases, more electron density is withdrawn from the X-O and O-H bonds, respectively. In water, the O-H bond is dissociated to a greater extent and the strength of the oxyacid increases.

(b) As the number of nonprotonated oxygen atoms in the molecule increases, they withdraw electron density from the other bonds in the molecule and the strength of the oxyacid increases.

16.66 (a) Acid strength increases as the polarity of the X-H bond increases and decreases as the strength of the X-H bond increases.

(b) Assuming the element, X, is more electronegative than H, as the electronegativity of X increases, the X-H bond becomes more polar and the strength of the acid increases. This trend holds true as electronegativity increases across a row of the periodic chart. However, as electronegativity decreases going down a family, acid strength increases because the strength of the H-X bond decreases, even though the H-X bond becomes less polar.

16.67 (a) H_2SO_3 (more electronegative central atom)
(b) H_3PO_4 (more nonprotonated O atoms)
(c) H_2SO_3 (more electronegative central atom)

16.68 (a) BrO^- ($HBrO_2$ has more nonprotonated O atoms and is the stronger acid, so BrO^- is the stronger base.)

(b) BrO^- (HClO is the stronger acid due to a more electronegative central atom, so BrO^- is the stronger base.)

(c) HPO_4^{2-} (larger negative charge, greater attraction for H^+)

16.69 (a) True

(b) False. In a series of acids that have the same central atom, acid strength increases with the number of nonprotonated oxygen atoms bonded to the central atom.

(c) False. H_2Te is a stronger acid than H_2S because the H-Te bond is longer, weaker and more easily dissociated than the H-S bond.

16.70 (a) True.

(b) False. For oxyacids with the same structure but different central atom, the acid strength increases as the electronegativity of the central atom increases.

(c) False. HF is a weak acid, weaker than the other hydrogen halides, primarily because the H-F bond energy is exceptionally high.

Lewis Acids and Bases

16.71

Theory	Acid	Base
Arrhenius	forms H^+ ions in water	produces OH^- in water
Bronsted-Lowry	proton (H^+) donor	proton acceptor
Lewis	electron pair acceptor	electron pair donor

The Bronsted-Lowry theory is more general than Arrhenius's definition, because it is based on a unified model for the processes responsible for acidic or basic character, and it shows the relationships between these processes. The Lewis theory is more general still because it does not restrict the acidic species to compounds having ionizable hydrogen. Any substance that can be viewed as an electron pair acceptor is defined as a Lewis acid.

16.72 (a) HBr acts as a proton donor, a Bronsted-Lowry acid, toward the water molecule, which acts as a proton acceptor, a Bronsted-Lowry base. The conjugate acid H_3O^+ ($H^+(aq)$) and conjugate base Br^- are formed in the reaction.

(b) Hydride ion, H^- acts like a Bronsted-Lowry base, accepting a proton from a water molecule to form $H_2(g)$ and the conjugate base OH^-. The balanced equation is

$$NaH(s) + H_2O(l) \rightarrow H_2(g) + Na^+(aq) + OH^-(aq)$$

(c) Pyridine acts like a Lewis base, donating an electron pair to SO_2, which acts as a Lewis acid. The SO_2 is able to act as an acceptor by shifting a pair of electrons from S to O:

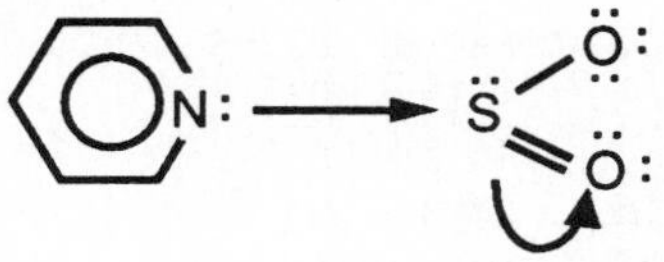

(d) In terms of Lewis acid-base theory, the reaction of SO_2 with water is very similar to the reaction of SO_2 with pyridine. SO_2 acts as an electron pair acceptor (Lewis acid), and water acts as the electron pair donor (Lewis base).

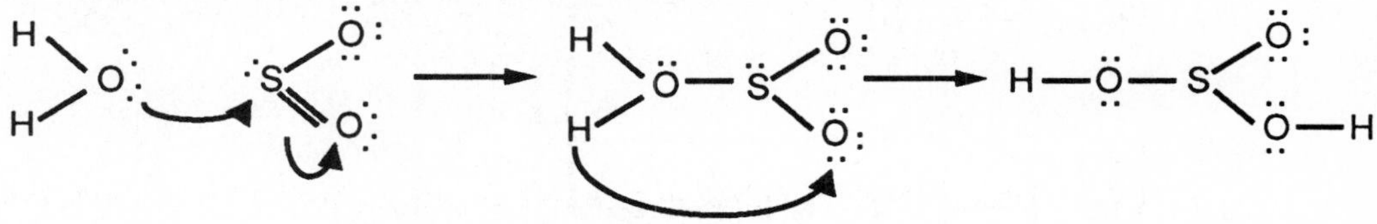

16.73

	Lewis Acid	**Lewis Base**
(a)	$Fe(ClO_4)_3$ or Fe^{3+}	H_2O
(b)	H_2O	CN^-
(c)	BF_3	$(CH_3)_3N$
(d)	HIO	NH_2^-

16.74

	Lewis Acid	**Lewis Base**
(a)	HNO_2 (or H^+)	OH^-
(b)	$FeBr_3(Fe^{3+})$	Br^-
(c)	Zn^{2+}	NH_3
(d)	SO_2	H_2O

16.75 (a) CdI_2 -- higher cation charge (b) $Fe(NO_3)_3$ -- higher cation charge
(c) $CrCl_3$ -- higher cation charge

16.76 (a) BH_3 -- has a vacant valence shell or orbital
(b) SO_3 -- electronegative oxygen atoms will draw electron density from S and make it an electron acceptor
(c) HBr -- H^+ accepts electrons much more readily than H^-
(d) Mo^{4+} -- higher cation charge

Additional Exercises

16.77 (a) $[H^+]$ = antilog (-6.1) or $10^{-6.1} = 8 \times 10^{-7}$ *M*; acidic
(b) $10^{-2.1} = 8 \times 10^{-3}$ *M*; acidic
(c) $10^{-2.4} = 4 \times 10^{-2}$ *M*; acidic
(d) $10^{-11.9} = 1 \times 10^{-12}$ *M*; basic

16.78 The strongest base has the largest K_b value.

NH_3: $K_b = 1.8 \times 10^{-5}$ (Appendix D)

H_2O: $K_b = K_a = 1.0 \times 10^{-14}$

OH^-: $K_b >> 1$ (OH^- is the strongest base possible in an aqueous solution.)

$$C_2H_3O_2^- : K_b = \frac{K_w}{K_a \text{ for } HC_2H_3O_2} = \frac{1.0 \times 10^{-14}}{1.8 \times 10^{-5}} = 5.6 \times 10^{-10}$$

Br^- : $K_b < 1.0 \times 10^{-14}$ (it cannot hydrolyze H_2O)

In order of decreasing base strength: $OH^- > NH_3 > C_2H_3O_2^- > H_2O > Br^-$.

16.79 (a)

conj. base — conj. acid

$$NH_3(l) + NH_3(l) \rightleftharpoons NH_4^+(lq) + NH_2^-(lq)$$

conj. acid — conj. base

(b) $K_c = [NH_4^+][NH_2^-]$

(c) The molar concentration of a pure liquid such as $NH_3(l)$ does not change as a reaction proceeds, so it is not included in the equilibrium expression.

(d) $[NH_4] = [NH_2^-]$; at -50°C, $K = 1 \times 10^{-30} = [NH_4^+]^2$

$[NH_4^+] = 100^{-15}$ *M*

(e) By analogy with the autoionization of water, dissolving a compound such as HCl produces an "acidic" solution because it adds the conjugate acid of water to the solution. NH_4^+ is the conjugate acid of $NH_3(l)$, so dissolving NH_4Cl will produce an "acidic" solution.

16.80 The equilibrium shifts in the forward direction with reaction between OH^- and H^+. The solution will be blue, corresponding to an excess of the conjugate base form, Bb^-, in solution.

16.81 A pH of 3.25 corresponds to $[H^+] = 5.6 \times 10^{-4}$ *M*.

(a) Because HCl is a strong acid, $[HCl] = 5.6 \times 10^{-4}$ *M*

$(0.200\text{ L})(5.6 \times 10^{-4}\text{ mol/L}) = 1.1 \times 10^{-4}$ mol HCl

(b) K_a for $HC_7H_5O_2$ is 6.5×10^{-5} (benzoic acid, Appendix D)

$$\frac{(5.6 \times 10^{-4})^2}{x} = 6.5 \times 10^{-5} \;;\; x = 4.8 \times 10^{-3}\ M$$

$$0.200\text{ L} \times \frac{4.8 \times 10^{-3}\text{ mol } HC_7H_5O_2}{1\text{ L}} = 9.6 \times 10^{-4}\text{ mol}$$

(c) HF has $K_a = 6.8 \times 10^{-4}$

$$\frac{(5.6 \times 10^{-4})^2}{x} = 6.8 \times 10^{-4} \;;\; x = 4.61 \times 10^{-4}\ M$$

This is the amount that remains in solution after ionization to form H^+ and F^-. The total HF is the sum of the ionized and nonionized portions:
$4.6 \times 10^{-4}\ M + 5.6 \times 10^{-4}\ M = 1.0 \times 10^{-3}\ M$

$$0.200\text{ L} \times \frac{1.0 \times 10^{-3}\text{ mol HF}}{1\text{ L}} = 2.0 \times 10^{-4}\text{ mol HF}$$

16.82 (a) A high O_2 concentration displaces protons from Hb, and thus produces a more acidic solution, with lower pH.

(b) $[H^+]$ = antilog (-7.4) = 4.0×10^{-8} *M*. The blood is very slightly basic.

(c) The equilibrium indicates that a high $[H^+]$ shifts the equilibrium toward the proton-bound from, HbH^+, which means a lower concentration of HbO_2 in the blood. Thus the ability of hemoglobin to transport oxygen is impeded.

16.83 Let $[H^+] = [NC_7H_4SO_3^-] = z$. K_a = -antilog (11.68) = 2.1×10^{-12}

$$\frac{z^2}{0.10 - z} \approx \frac{z^2}{0.10} = 2.1 \times 10^{-12} \qquad z = [H^+] = 4.6 \times 10^{-7}\ M;\ \text{pH} = 6.34$$

Notice that the pH is not much lower than it is in pure water. A correction for the contribution from ionization of water itself to the concentration of $[H^+]$ can be made, but it turns out that this contribution can be neglected even in this case where the source of acidity is not very great.

16.84 Calculate the initial concentration of $HC_9H_7O_4$.

$$\frac{325\text{ mg}}{1\text{ tablet}} \times 2\text{ tablets} \times \frac{1\text{ g}}{1000\text{ mg}} \times \frac{1\text{ mol } HC_9H_7O_4}{180.2\text{ g } HC_9H_7O_4} = 0.00361\text{ mol } HC_9H_7O_4$$

$$\frac{0.00361\text{ mol } HC_9H_7O_4}{0.250\text{ L}} = 0.0144\ M\ HC_9H_7O_4$$

	$HC_9H_7O_4(aq)$ $\rightleftharpoons$	$C_9H_7O_4^-(aq)$ +	$H^+(aq)$
initial	0.0144 *M*	0 *M*	0 *M*
equil.	(0.0144-x) *M*	x *M*	x *M*

$$K_a = 3.3 \times 10^{-4} = \frac{[H^+][C_7H_9O_4^-]}{[HC_7H_9O_4]} = \frac{x^2}{(0.0144\text{-}x)}$$

Assuming x is small compared to 0.0144,

$x^2 = 0.0144(3.3 \times 10^{-4})$; $x = [H^+] = 2.2 \times 10^{-3}\ M$

$$\%\ \text{dissociation} = \frac{2.2 \times 10^{-3}\ M}{0.0144\ M} \times 100 = 15\%$$; the assumption is not valid

Using the quadratic formula, $x^2 + 3.3 \times 10^{-4}\,x - 4.75 \times 10^{-6} = 0$

$$x = \frac{-3.3 \times 10^{-4} \pm \sqrt{(3.3 \times 10^{-4})^2 - 4(1)(-4.75 \times 10^{-6})}}{2(1)} = \frac{-3.3 \times 10^{-4} \pm \sqrt{1.91 \times 10^{-5}}}{2}$$

$x = 2.0 \times 10^{-3}\ M\ H^+$; $pH = -\log(2.0 \times 10^{-3}) = 2.70$

16.85 $1.0 \times 10^{-8}\ M\ HNO_3$ produces $1.0 \times 10^{-8}\ M\ H^+$

This is less than $[H^+]$ from the autoionization of water. We need to solve the K_w equilibrium. Let $y = [OH^-]$.

$K_w = 1.0 \times 10^{-14} = [H^+][OH^-]$; $1.0 \times 10^{-14} = (1.0 \times 10^{-8} + y)(y)$

$y^2 + 1.0 \times 10^{-8}\,y - 1.0 \times 10^{-14} = 0$

$$y = \frac{-1.0 \times 10^{-8} + \sqrt{(1.0 \times 10^{-8})^2 - 4(1)(-1.0 \times 10^{-14})}}{2} = 9.5 \times 10^{-8}\ M\ OH^-$$

$[H^+] = 1.0 \times 10^{-8}\ M\ (HNO_3) + 9.5 \times 10^{-8}(H_2O) = 1.05 \times 10^{-7}\ M\ H^+$; pH = 6.978

16.86 (a) $H_2X \rightarrow H^+ + HX^-$

Assuming HX^- does not dissociate, $[H^+] = 0.050\ M$, pH = 1.30

(b) $H_2X \rightarrow 2H^+ + X^-$; $0.050\ M\ H_2X = 0.10\ M\ H^+$; pH = 1.00

(c) The observed pH of a 0.050 *M* solution of H_2X is only slightly less than 1.30, the pH assuming no dissociation of HX^-. HX^- is not completely dissociated; H_2X, which is completely dissociated, is a stronger acid than HX^-.

(d) Since H_2X is a strong acid, HX^- has no tendency to act like a base. HX^- does act like a weak acid, so a solution of NaHX would be acidic.

16.87 Considering the stepwise dissociation of H_3PO_4:

	$H_3PO_4(aq)$	$\rightleftharpoons$	$H^+(aq)$ +	$H_2PO_4^-(aq)$
initial	0.050 *M*		0	0
equil.	(0.050 - x) *M*		x	x

$$K_{a1} = \frac{[H^+][H_2PO_4^-]}{[H_3PO_4]} = \frac{x^2}{(0.100\text{-}x)} = 7.5 \times 10^{-3}\ ;\ x^2 + 7.5 \times 10^{-3}\,x - 7.5 \times 10^{-4} = 0$$

$$x = \frac{-7.5 \times 10^{-3} \pm \sqrt{(7.5 \times 10^{-3})^2 - 4(1)(-7.5 \times 10^{-4})}}{2(1)} = \frac{-7.5 \times 10^{-3} \pm \sqrt{0.00306}}{2}$$

x = 0.024 *M* H^+, 0.024 *M* $H_2PO_4^-$ available for further dissociation

	$H_2PO_4^{2-}(aq)$	$\rightleftharpoons$	$H^+(aq)$	+	$HPO_4^-(aq)$
initial	0.024 *M*		0.024 *M*		0 *M*
equil.	(0.024 - y) *M*		(0.024 + y) *M*		y *M*

$$K_{a2} = \frac{[H^+][HPO_4^-]}{[H_2PO_4^{2-}]} = \frac{(y)(0.024 + y)}{(0.024 - y)} = 6.2 \times 10^{-8}$$

Since K_{a2} is very small, assume x is small compared to 0.024 *M*.

$$6.2 \times 10^{-8} = \frac{0.024\,y}{0.024}\ ;\ y = [HPO_4^-] = 6.2 \times 10^{-8}\ M$$

$[H_2PO_4^-] = (2.4 \times 10^{-2}\ M + 6.2 \times 10^{-8}\ M) \approx 2.4 \times 10^{-2}\ M$

$[H^+] = (2.4 \times 10^{-2}\ M + 6.2 \times 10^{-8}\ M) \approx 2.4 \times 10^{-2}\ M$

$[HPO_4^-]$ available for hydrolysis = $6.2 \times 10^{-8} M$

	HPO_4^-	$\rightleftharpoons$	$H^+(aq)$	+	$PO_4^{3-}(aq)$
initial	6.2×10^{-8} *M*		2.4×10^{-2} *M*		0 *M*
equil.	$(6.2 \times 10^{-8} - z)$ *M*		$(2.4 \times 10^{-2} + z)$ *M*		z *M*

$$K_{a3} = \frac{[H^+][PO_4^{3-}]}{[HPO_4^-]} = \frac{(2.4 \times 10^{-2} + z)(z)}{(6.2 \times 10^{-8} - z)} = 4.2 \times 10^{-13}$$

Assuming z is small compared to 6.2×10^{-8} (and 2.4×10^{-2}),

$$\frac{(2.4 \times 10^{-2})(z)}{(6.2 \times 10^{-8})} = 4.2 \times 10^{-13}\ ;\ z = 1.1 \times 10^{-18}\ M\ PO_4^{3-}$$

The contribution of z to $[HPO_4^-]$ and $[H^+]$ is negligible. In summary, after all dissociation steps have reached equilibrium:

$[H^+] = 2.4 \times 10^{-2}$ *M*, $[H_2PO_4^-] = 2.4 \times 10^{-2}$ M, $[HPO_4^{2-}] = 6.2 \times 10^{-8}$ *M*, $[PO_4^{3-}] = 1.1 \times 10^{-18}$ *M*

Note that the first dissociation step is the major source of H^+ and the others are important as sources of HPO_4^- and PO_4^{3-}. The $[PO_4^{3-}]$ is <u>very</u> small at equilibrium.

16.88 $C_{10}H_{15}ON(aq) + H_2O(l) \rightleftharpoons C_{10}H_{15}ONH^+(aq) + OH^-(aq)$

initial	0.035 *M*	0	0
equil.	(0.035 - y) *M*	y *M*	y *M*

$$K_b = \frac{[C_{10}H_{15}ONH^+][OH^-]}{[C_{10}H_{15}ON]} = \frac{y^2}{(0.035 - y)} = 1.4 \times 10^{-4}$$

$\frac{y^2}{0.035} \approx 1.4 \times 10^{-4}$; $y = 2.2 \times 10^{-3}$ $M = [OH^-]$; pOH = 2.66, pH = 11.34

This $[OH^-]$ represents approximately 6% ionization, but use of the quadratic formula gives $[OH^-] = 2.1 \times 10^{-3}$ and pH = 11.33, values not significantly different from those obtained using the approximation.

K_a for $C_{10}H_{15}ONH^+$ is $\frac{1.0 \times 10^{-14}}{1.4 \times 10^{-4}} = 7.1 \times 10^{-11}$; $pK_a = 10.15$

16.89 $C_{18}H_{21}NO_3(aq) + H_2O(l) \rightleftharpoons C_{18}H_{21}NHO_3^+ + OH^-(aq)$

initial	0.0050 *M*	0	0
equil.	(0.0050 - x) *M*	x *M*	x *M*

$$K_b = \frac{[C_{18}H_{21}HNO_3^+][OH^-]}{[C_{18}H_{21}NO_3]} = \frac{x^2}{0.0050 - x}$$

pOH = 14.00 - 9.95 = 4.05 ; $[OH^-] = 10^{-4.05} = 8.9 \times 10^{-5}$ *M*

$$K_b = \frac{(8.9 \times 10^{-5})^2}{(0.0050 - 8.9 \times 10^{-5})} = 1.6 \times 10^{-6}$$

16.90 (a) (i)

$HCO_3^-(aq) \rightleftharpoons H^+(aq) + CO_3^{2-}(aq)$ $K_1 = K_{a2}$ for $H_2CO_3 = 5.6 \times 10^{-11}$

$H^+(aq) + OH^-(aq) \rightleftharpoons H_2O(l)$ $K_2 = 1/K_w = 1 \times 10^{14}$

$HCO_3^-(aq) + OH^-(aq) \rightleftharpoons CO_3^{2-}(aq) + H_2O(l)$ $K = K_1 \times K_2 = 5.6 \times 10^3$

(ii) $NH_4^+(aq) \rightleftharpoons H^+(aq) + NH_3(aq)$ $K_1 = K_a$ for $NH_4^+ = 5.6 \times 10^{-10}$

$CO_3^{2-}(aq) + H^+(aq) \rightleftharpoons HCO_3^-(aq)$ $K_2 = 1/K_{a2}$ for $H_2CO_3 = 1.8 \times 10^{10}$

$NH_4^+(aq) + CO_3^{2-}(aq) \rightleftharpoons HCO_3^-(aq) + NH_3(aq)$ $K = K_1 \times K_2 = 10$

(b) Both (i) and (ii) have K > 1, although K = 10 is not much greater than 1. Both could be written with a single arrow. (This is true in general when a strong acid or strong base, $H^+(aq)$ or $OH^-(aq)$, is a reactant.)

16.91 Call each compound in the neutral form Q.
Then, $Q(aq) + H_2O(l) \rightleftharpoons QH^+(aq) + OH^-$. $K_b = [QH^+][OH^-]/[Q]$

The ratio in question is $[QH^+]/[Q]$, which equals $K_b/[OH^-]$. At pH = 2.5, pOH = 11.5, $[OH^-]$ = antilog(-11.5) = 3.1×10^{-12} M. Now calculate $K_b/[OH^-]$ for each compound:

Nicotine $\frac{[QH^+]}{[Q]} = 7 \times 10^{-7}/3.1 \times 10^{-12} = 2 \times 10^5$

Caffeine $\frac{[QH^+]}{[Q]} = 4 \times 10^{-14}/3.1 \times 10^{-12} = 1 \times 10^{-2}$

Strychnine $\frac{[QH^+]}{[Q]} = 1 \times 10^{-6}/3.1 \times 10^{-12} = 3 \times 10^5$

Quinine $\frac{[QH^+]}{[Q]} = 1.1 \times 10^{-6}/3.1 \times 10^{-12} = 3.5 \times 10^5$

For all the compounds except caffeine the protonated form is in much higher concentration than the neutral form. However, for caffeine, a very weak base, the neutral form dominates.

16.92 (a) The apparent molality of the solution is given by

$$1.90°C \times \frac{1\ m}{1.86°\ C} = 1.02\ m$$

The ionization equilibrium, and molalities of the various species are:

	$HF(aq)$	$\rightleftharpoons$	$H^+(aq)$	+	$F^-(aq)$
equil.	1.00 - x m		x m		x m

The total molality is thus (1.00 - x + x + x) = 1.00 + x = 1.02. Thus, x = 0.02. For such a dilute aqueous solution, molality and molarity are essentially identical (Section 13.2). Thus, we can write

$$\%\ \text{dissociation} = \frac{[H^+]}{[HF]} \times 100 = \frac{0.02}{1.00} \times 100 = 2\%$$

(b) Assuming $M \approx m$, $K_a = \frac{[H^+][F^-]}{[HF]} = \frac{(0.02)^2}{0.98} = 4 \times 10^{-4}$.

The precision of the K_a is not very good; to improve upon it, we would need to know the magnitude of the freezing point depression more precisely.

16.93 (a) Consider the formation of the zwitterion as a series of steps (Hess's Law).

$$NH_2 - CH_2 - COOH + H_2O \rightleftharpoons NH_2 - CH_2 - COO^- + H_3O^+ \quad K_a$$
$$NH_2 - CH_2 - COOH + H_2O \rightleftharpoons {}^+NH_3 - CH_2 - COOH + OH^- \quad K_b$$
$$H_3O^+ + OH^- \rightleftharpoons 2H_2O \quad 1/K_w$$

$$NH_2 - CH_2 - COOH \rightleftharpoons {}^+NH_3 - CH_2 - COO^- \quad \frac{K_a \times K_b}{K_w}$$

$$K = \frac{K_a \times K_b}{K_w} = \frac{(4.3 \times 10^{-3})(6.0 \times 10^{-5})}{1.0 \times 10^{-14}} = 2.6 \times 10^7$$

The large value of K indicates that formation of the zwitterion is favorable. The assumption is that the same $NH_2 - CH_2 - COOH$ molecule is acting like an acid ($-COOH \rightarrow H^+ + -COO^-$) and a base ($-NH_2 + H^+ \rightarrow NH_3^+$), simultaneously. Glycine is both a stronger acid and a stronger base than water, so the H^+ transfer should be intramolecular, as long as there are no other acids or bases in the solution.

(b) Since glycine exists as the zwitterion in aqueous solution, the pH is determined by the dissociation equilibrium below.

$${}^+NH_3 - CH_2 - COO^- + H_20 \rightleftharpoons NH_2 - CH_2 - COO^- + H_3O^+$$

$$K_a = \frac{[NH_2 - CH_2 - COO^-][H_3O^+]}{[{}^+NH_3 - CH_2 - COO^-]} = \frac{K_w}{K_b} = \frac{1.0 \times 10^{-14}}{6.0 \times 10^{-5}} = 1.7 \times 10^{-10}$$

$$x = [H_3O^+] = [NH_2 - CH_2 - COO^-]; \quad K_a = 1.7 \times 10^{-10} = \frac{(x)(x)}{(0.050-x)} \approx \frac{x^2}{0.050}$$

$x = [H_3O^+] = 2.9 \times 10^{-6}$ *M*, pH = 5.54

(c) In a strongly acidic solution the CO_2^- function would be protonated, so glycine would exist as ${}^+H_3NCH_2COOH$. In strongly basic solution the $-NH_3^+$ group could be deprotonated, so glycine would be in the form $H_2NCH_2CO_2^-$.

16.94

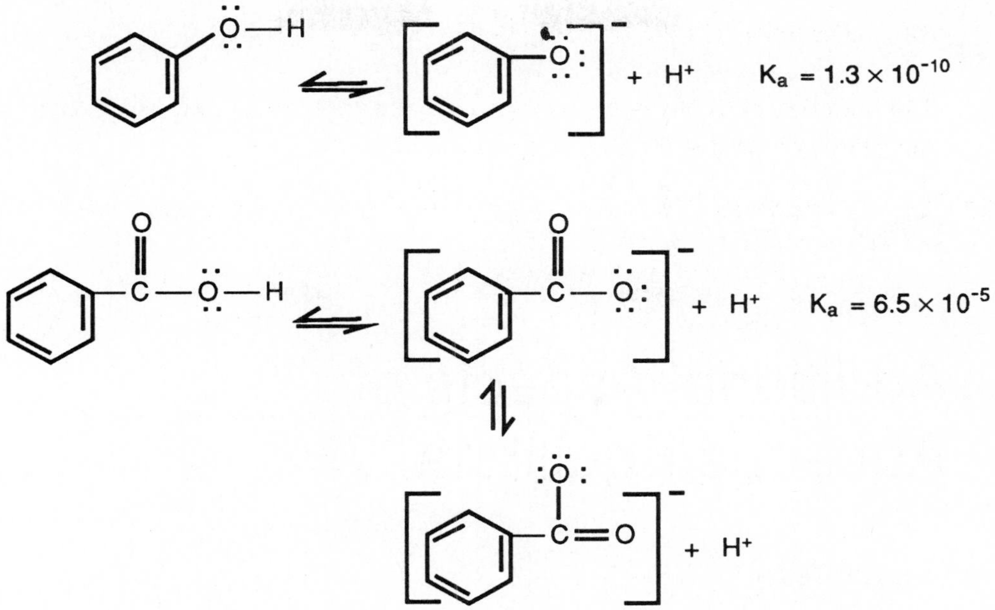

Benzoic acid is a carboxylic acid. Loss of H^+ to form benzoate anion allows an extension of the delocalized π bonding network. The same phenomenon is not present in phenolate anion, the conjugate base of phenol. Benzoic acid has a greater tendency to donate H^+ because of resonance stabilization in the resulting conjugate base, benzoate ion.

16.95 The general Lewis structure for these acids is

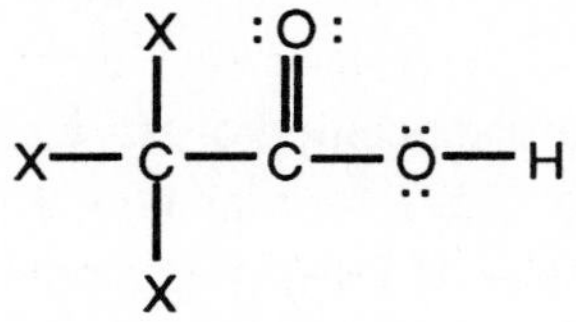

Replacement of H by the more electronegative chlorine atoms causes the central carbon to become more positively charged, thus withdrawing more electrons from the attached COOH group, in turn causing the O-H bond to be more polar, so that H^+ is more readily ionized.

To calculate pH proceed as usual, except that the full quadratic form must be used for all but acetic acid.

Acid	pH
acetic	2.87
chloroacetic	1.95
dichloroacetic	1.36
trichloroacetic	1.14

CHAPTER 17

Additional Aspects of Aqueous Equilibria

Common-Ion Effect

17.1 (a) The extent of dissociation of a weak electrolyte is decreased when a strong electrolyte containing an ion in common with the weak electrolyte is added to it.

(b) NaOCl

(c) For a generic weak base B, $K_b = \frac{[BH^+][OH^-]}{[B]}$. If an external source of BH^+ such as BH^+Cl^- is added to a solution of B(aq), $[BH^+]$ increases, decreasing $[OH^-]$ and increasing [B], effectively suppressing the dissociation of B.

17.2 (a) The presence of A^- from a source other than the dissociation of HA stresses the equilibrium $HA(aq) \rightleftharpoons H^+(aq) + A^-(aq)$. In order to maintain the ratio $[H^+][A^-]/[HA]$ as a constant, $[H^+]$ must decrease and [HA] must increase; the equilibrium "shifts" to the left.

(b) NH_4Cl

(c) The conjugate base of a weak acid HA is A^-. The presence of A^- from a source other than the dissociation of HA causes the equilibrium to shift to the left, as explained in part (a). The result is that the concentration of undissociated acid HA(aq) increases.

17.3 (a) Increase; formate ion is a moderately strong base (b) decrease; NH_4^+ is a weak acid, ClO_4^- is a <u>very</u> weak base (c) no change (d) decrease

17.4 (a) Increase, addition of a base (b) possible decrease or no change; depending on the concentration of the strong acid HCL, addition of the weak acid NH_4^+ will increase the $[H^+]$ slightly or have a negligible effect (c) increase, CN^- is a moderately strong base (d) decrease, $C_5H_5NH^+$ is an acid

---

17.5 (a)

	$HC_3H_5O_2(aq)$	$\rightleftharpoons$	$H^+(aq)$	+	$C_3H_5O_2^-(aq)$
i	0.16 *M*				0.080 *M*
c	-x		+ x		+x
e	(0.16 - x) *M*		+ x *M*		(0.080 + x) *M*

$$K_a = 1.3 \times 10^{-5} = \frac{[H^+][C_3H_5O_2^-]}{[HC_3H_5O_2]} = \frac{(x)(0.080 + x)}{(0.16 - x)}$$

Assume x is small compared to 0.080 and 0.16.

$$1.3 \times 10^{-5} = \frac{0.080\ x}{0.16};\ x = 2.6 \times 10^{-5} = [H^+],\ pH = 4.59$$

(b)

	$(CH_3)_3N(aq) + H_2O(l)$	$\rightleftharpoons$	$(CH_3)_3NH^+(aq)$	+	$OH^-(aq)$
i	0.15 *M*		0.12 *M*		
c	- x		+ x		+ x
e	(0.15 - x) *M*		(0.12 + x) *M*		+ x *M*

$$K_b = 6.4 \times 10^{-5} = \frac{[OH^-][(CH_3)_3NH^+]}{[(CH_3)_3N]} = \frac{x(0.12 + x)}{(0.15 - x)} \approx \frac{0.12\ x}{0.15}$$

$x = 8.0 \times 10^{-5} = [OH^-]$, pOH = 4.10, pH = 14.00 - 4.10 = 9.90

17.6 (a)

	$HCHO_2(aq)$	$\rightleftharpoons$	$H^+(aq)$	+	$CHO_2^-(aq)$
i	0.180 *M*				0.100 *M*
c	-x		+x		+x
e	(0.180-x) *M*		+x *M*		(0.100 + x) *M*

$$K_a = 1.8 \times 10^{-4} = \frac{[H^+][CHO_2^-]}{[HCHO_2]} = \frac{(x)(0.100 + x)}{(0.180 - x)} \approx \frac{0.100\ x}{0.180}$$

$x = 3.2 \times 10^{-4}\ M = [H^+]$, pH = 3.49

Since the extent of dissociation of a weak acid or base is suppressed by the presence of a conjugate salt, the 5% rule usually holds true in buffer solutions.

(b)

	$C_5H_5N(aq) + H_2O(l)$	$\rightleftharpoons$	$C_5H_5NH^+(aq)$	+	$OH^-(aq)$
i	0.0750 *M*		0.0500 *M*		
c	-x		+x		+x
e	(0.0750-x) *M*		(0.0500 + x) *M*		+x *M*

$$K_b = 1.7 \times 10^{-9} = \frac{[C_5H_5NH^+][OH^-]}{[C_5H_5N]} = \frac{(0.0500 + x)(x)}{(0.0750 - x)} \approx \frac{0.0500\ x}{0.0750}$$

$x = 2.6 \times 10^{-9}\ M = [OH^-]$, pOH = 8.59, pH = 14.00 - 8.59 = 5.41

Acid-Base Titrations

17.7 (a) $40.0 \text{ mL } HNO_3 \times \frac{0.0350 \text{ mol } HNO_3}{1000 \text{ mL soln}} \times \frac{1 \text{ mol NaOH}}{1 \text{ mol } HNO_3} \times \frac{1000 \text{ mL soln}}{0.0350 \text{ mol NaOH}}$

$= 40.0 \text{ mL NaOH soln}$

(b) $65.0 \text{ mL HBr} \times \frac{0.0620 \text{ } M \text{ HBr}}{1000 \text{ mL soln}} \times \frac{1 \text{ mol NaOH}}{1 \text{ mol HBr}} \times \frac{1000 \text{ mL soln}}{0.0350 \text{ mol NaOH}}$

$= 115 \text{ mL NaOH soln}$

(c) $\frac{1.65 \text{ g HCl}}{1 \text{ L soln}} \times \frac{1 \text{ L}}{1000 \text{ mL}} \times \frac{1 \text{ mol HCl}}{36.46 \text{ g HCl}} = 0.0453 \text{ } M \text{ HCl}$

$80.0 \text{ mL HCl} \times \frac{0.0453 \text{ mol HCl}}{1000 \text{ mL}} \times \frac{1 \text{ mol NaOH}}{1 \text{ mol HCl}} \times \frac{1000 \text{ mL soln}}{0.0350 \text{ mol NaOH}}$

$= 104 \text{ mL NaOH soln}$

17.8 (a) $50.0 \text{ mL NaOH} \times \frac{1.50 \text{ mol NaOH}}{1000 \text{ mL soln}} \times \frac{1 \text{ mol HCl}}{1 \text{ mol NaOH}} \times \frac{1000 \text{ mL soln}}{0.0750 \text{ mol HCl}}$

$= 1.00 \times 10^3 \text{ mL HCl soln}$

(b) $43.5 \text{ mL KOH} \times \frac{0.0320 \text{ mol KOH}}{1000 \text{ mL soln}} \times \frac{1 \text{ mol HCl}}{1 \text{ mol KOH}} \times \frac{1000 \text{ mL soln}}{0.0750 \text{ mol HCl}}$

$= 18.6 \text{ mL HCl soln}$

(c) $27.0 \text{ mL } Ca(OH)_2 \times \frac{0.0600 \text{ mol } Ca(OH)_2}{1000 \text{ mL soln}} \times \frac{2 \text{ mol HCl}}{1 \text{ mol } Ca(OH)_2} \times \frac{1000 \text{ mL soln}}{0.0750 \text{ mol HCl}}$

$= 43.2 \text{ mL HCl soln}$

17.9 Since the combining ratio is 1 mol NaOH/1 mol HBr,

$M_A \times V_A = M_B \times V_b \text{ ; } M_{HBr} = \frac{M_{NaOH} \times V_{NaOH}}{V_{HBr}}$

(a) $\frac{0.100 \text{ } M \times 27.5 \text{ mL NaOH}}{20.0 \text{ mL HBr}} = 0.138 \text{ } M \text{ HBr}$

(b) $\frac{0.100 \text{ } M \times 21.8 \text{ mL NaOH}}{20.0 \text{ mL HBr}} = 0.109 \text{ } M \text{ HBr}$

(c) $\frac{0.100 \text{ } M \times 48.9 \text{ mL NaOH}}{20.0 \text{ mL HBr}} = 0.245 \text{ } M \text{ HBr}$

17.10 Since the combining ratio is 1 mol HNO_3/1 mol KOH,

$M_A \times V_A = M_B \times V_B \text{; } M_{KOH} = \frac{M_{HNO_3} \times V_{HNO_3}}{V_{KOH}}$

(a) $\frac{0.0750 \text{ } M \times 39.1 \text{ mL } HNO_3}{40.0 \text{ mL KOH}} = 0.0733 \text{ } M \text{ KOH, pOH} = 1.135\text{, pH} = 12.865$

(b) $\dfrac{0.750\ M \times 20.7\text{ mL HNO}_3}{40.0\text{ mL}} = 0.0388\ M$ KOH, pOH = 1.411, pH = 12.589

(c) $\dfrac{0.0750\ M \times 48.5\text{ mL HNO}_3}{40.0\text{ mL}} = 0.0909\ M$ KOH, pOH = 1.041, pH = 12.959

17.11 Construct a table similar to Table 17.1.

moles $H^+ = M_{HBr} \times L_{HBr} = 0.200\ M \times 0.0200\text{ L} = 4.00 \times 10^{-3}$ mol

moles $OH^- = M_{NaOH} \times L_{NaOH} = 0.200\ M \times L_{NaOH}$

	mL_{HBr}	mL_{NaOH}	Total Volume	Moles H^+	Moles OH^-	Molarity Excess Ion	pH
(a)	20.0	15.0	35.0	4.00×10^{-3}	3.00×10^{-3}	$0.0286(H^+)$	1.544
(b)	20.0	19.9	39.9	4.00×10^{-3}	3.98×10^{-3}	$5 \times 10^{-4}(H^+)$	3.3
(c)	20.0	20.0	40.0	4.00×10^{-3}	4.00×10^{-3}	$1 \times 10^{-7}(H^+)$	7.0
(d)	20.0	20.1	40.1	4.00×10^{-3}	4.02×10^{-3}	$5 \times 10^{-4}(OH^-)$	10.7
(e)	20.0	35.0	55.0	4.00×10^{-3}	7.00×10^{-3}	$0.0546(OH^-)$	12.737

molarity of excess ion = $\dfrac{\text{moles ion}}{\text{total vol in L}}$

(a) $\dfrac{4.00 \times 10^{-3}\text{ mol H}^+ - 3.00 \times 10^{-3}\text{ mol OH}^-}{0.0350\text{ L}} = 0.0286\ M\ H^+$

(b) $\dfrac{4.00 \times 10^{-3}\text{ mol H}^+ - 3.98 \times 10^{-3}\text{ mol OH}^-}{0.0399\text{ L}} = 5 \times 10^{-4}\ M\ H^+$

(c) equivalence point, mol H^+ = mol OH^-
NaBr does not hydrolyze, so $[H^+] = [OH^-] = 1 \times 10^{-7}\ M$

(d) $\dfrac{4.02 \times 10^{-3}\text{ mol OH}^- - 4.00 \times 10^{-3}\text{ mol H}^+}{0.041\text{ L}} = 5 \times 10^{-4}\ M\ OH^-$

(e) $\dfrac{7.00 \times 10^{-3}\text{ mol OH}^- - 4.00 \times 10^{-3}\text{ mol H}^+}{0.0550\text{ L}} = 0.0546\ M\ OH^-$

17.12 Construct a table similar to Table 17.1

moles $OH^- = M_{KOH} \times L_{KOH} = 0.200\ M \times 0.0300\text{ L} = 6.00 \times 10^{-3}$ mol

moles $H^+ = M_{HClO_4} \times L_{HClO_4} = 0.150\ M \times L_{HClO_4}$

	mL_{KOH}	mL_{HClO_4}	Total Volume	Moles OH^-	Moles H^+	Molarity Excess Ion	pH
(a)	30.0	30.0	60.0	6.00×10^{-3}	4.50×10^{-3}	$0.0250(OH^-)$	12.398
(b)	30.0	39.5	69.5	6.00×10^{-3}	5.93×10^{-3}	$1 \times 10^{-3}(OH^-)$	11.0
(c)	30.0	39.9	69.9	6.00×10^{-3}	5.99×10^{-3}	$1 \times 10^{-4}(OH^-)$	10.2
(d)	30.0	40.0	70.0	6.00×10^{-3}	6.00×10^{-3}	$1 \times 10^{-7}(H^+)$	7.0
(e)	30.0	40.1	70.1	6.00×10^{-3}	6.02×10^{-3}	$3 \times 10^{-4}(H^+)$	3.5

molarity of excess ion = $\dfrac{\text{moles ion}}{\text{total vol in L}}$

(a) $\frac{6.00 \times 10^{-3} \text{ mol OH}^- - 4.50 \times 10^{-3} \text{ mol H}^+}{0.0600 \text{ L}} = 0.0250\ M \text{ OH}^-$

(b) $\frac{6.00 \times 10^{-3} \text{ mol OH}^- - 5.93 \times 10^{-3} \text{ mol H}^+}{0.0695 \text{ L}} = 1 \times 10^{-3}\ M \text{ OH}^-$

(c) $\frac{6.00 \times 10^{-3} \text{ mol OH}^- - 5.99 \times 10^{-3} \text{ mol H}^+}{0.0699 \text{ L}} = 1 \times 10^{-4}\ M \text{ OH}^-$

(d) equivalence point, mol OH^- = mol H^+

$KClO_4$ does not hydrolyze, so $[H^+] = [OH^-] = 1 \times 10^{-7}\ M$

(e) $\frac{6.02 \times 10^{-3} \text{ mol H}^+ - 6.00 \times 10^{-3} \text{ mol OH}^-}{0.0701 \text{ L}} = 3 \times 10^{-4}\ M \text{ H}^+$

17.13 (a) The quantity of base required to reach the equivalence point is the same in the two titrations.

(b) The pH is higher initially in the titration of a weak acid.

(c) It is higher at the equivalence point in the titration of a weak acid.

(d) The pH in excess base is essentially the same for the two cases.

(e) In titrating a weak acid, one needs an indicator that changes at a higher pH than for the strong acid titration. The choice is more critical because the change in pH close to the equivalence point is smaller for the weak acid titration.

17.14 (a) At the equivalence point, moles HX added = moles B initially present = 0.10 $M \times$ 0.03 L = 0.003 moles HX added.

(b) BH^+

(c) Both K_a for BH^+ and concentration BH^+ determine pH at the equivalence point.

(d) Because the pH at the equivalence point will be less than 7, methyl red would be more appropriate.

17.15 (a) At 0 mL, only weak acid, $HC_2H_3O_2$, is present in solution. Using the acid dissociation equilibrium

	$HC_2H_3O_2(aq)$ ⇌	$H^+(aq)$ +	$C_2H_3O_2^-(aq)$
initial	0.150 M	0	0
equil	0.150-x M	x M	x M

$$K_a = \frac{[H^+][C_2H_3O_2^-]}{[HC_2H_3O_2]} = 1.8 \times 10^{-5} \text{ (Appendix D)}$$

$$1.8 \times 10^{-5} = \frac{x^2}{(0.150-x)} \approx \frac{x^2}{0.150};\ x^2 = 2.7 \times 10^{-6};\ x = [H^+] = 1.6 \times 10^{-3};\ pH = 2.78$$

(b)-(f) Calculate the moles of each component after the acid-base reaction takes place. Moles $HC_2H_3O_2$ originally present = $M \times L$ = 0.150 M × 0.0500 L = 7.50×10^{-3} mol. Moles NaOH added = $M \times L$ = 0.150 M × y mL.

		NaOH(aq)	+ $HC_2H_3O_2(aq)$ →	$Na^+C_2H_3O_2^-(aq)$ + $H_2O(l)$
		(0.150 M × 0.0250 L) =		
(b)	before rx	3.75×10^{-3} mol	7.50×10^{-3} mol	
	after rx	**0**	**3.75×10^{-3} mol**	**3.75×10^{-3} mol**
		(0.150 M × 0.0490 L) =		
(c)	before rx	7.35×10^{-3} mol	7.50×10^{-3} mol	
	after rx	**0**	**0.15×10^{-3} mol**	**7.35×10^{-3} mol**
		(0.150 M × 0.0500 L) =		
(d)	before rx	7.50×10^{-3} mol	7.50×10^{-3} mol	
	after rx	**0**	**0**	**7.50×10^{-3} mol**
		(0.150 M × 0.0510 L) =		
(e)	before rx	7.65×10^{-3} mol	7.50×10^{-3}	
	after rx	**0.15×10^{-3} mol**	**0**	**7.50×10^{-3}**
		(0.150 M × 0.0750 L) =		
(f)	before rx	11.25×10^{-3} mol	7.50×10^{-3} mol	
	after rx	**3.75×10^{-3} mol**	**0**	**7.50×10^{-3} mol**

Calculate the molarity of each species (M = mol/L) and solve the appropriate equilibrium problem in each part.

(b) V_T = 50.0 mL $HC_2H_3O_2$ + 25.0 mL NaOH = 75.0 mL = 0.0750 L

$$[HC_2H_3O_2] = \frac{3.75 \times 10^{-3} \text{ mol}}{0.0750} = 0.0500\ M$$

$$[C_2H_3O_2^-] = \frac{3.75 \times 10^{-3} \text{ mol}}{0.0750 \text{ L}} = 0.0500\ M$$

$$HC_2H_3O_2(aq) \rightleftharpoons H^+ + C_2H_3O_2^-$$

equil. $0.0500 - x\ M$ $\quad xM \quad$ $0.0500 + x\ M$

$$K_a = \frac{[H^+][C_2H_3O_2^-]}{[HC_2H_3O_2]}\ ; \quad [H^+] = \frac{K_a\ [HC_2H_3O_2]}{[C_2H_3O_2^-]}$$

$$[H^+] = \frac{1.8 \times 10^{-5}\ (0.0500 - x)}{(0.0500 + x)} = 1.8 \times 10^{-5}\ M\ H^+\ ; \quad pH = 4.74$$

(c) $$[HC_2H_3O_2] = \frac{0.15 \times 10^{-3} \text{ mol}}{0.0990 \text{ L}} = 1.5 \times 10^{-3}\ M$$

$$[C_2H_3O_2^-] = \frac{7.35 \times 10^{-3} \text{ mol}}{0.0990 \text{ L}} = 0.074\ M$$

$$[H^+] = \frac{1.8 \times 10^{-5}\ (1.5 \times 10^{-3} - x)}{(0.074 + x)} \approx 3.6 \times 10^{-7}\ M\ H^+\ ; \quad pH = 6.44$$

(d) At the equivalence point, only $C_3H_5O_2^-$ is present.

$[C_2H_3O_2^-] = \frac{7.50 \times 10^{-3} \text{ mol}}{0.100 \text{ L}} = 0.0750\ M$ The pertinent equilibrium is the base hydrolysis of $C_2H_3O_2^-$.

	$C_2H_3O_2^-(aq) + H_2O(l) \rightleftharpoons$	$HC_2H_3O_2(aq)$	$+ OH^-(aq)$
initial	$0.0750\ M$	0	0
equil	$0.0750\text{-x}\ M$	x	x

$$K_b = \frac{K_w}{K_a \text{ for } HC_2H_3O_2} = \frac{1.0 \times 10^{-14}}{1.8 \times 10^{-5}} = 5.6 \times 10^{-10} = \frac{[HC_2H_3O_2][OH^-]}{[C_2H_3O_2^-]}$$

$$5.6 \times 10^{-10} = \frac{x^2}{0.0750\text{-x}}\ ;\ x^2 \approx 5.6 \times 10^{-10}\ (0.0750)\ ;\ x = 6.5 \times 10^{-6}\ M\ OH^-$$

$pOH = -\log(6.5 \times 10^{-6}) = 5.19$; $pH = 14.00 - pOH = 8.81$

(e) After the equivalence point, the excess strong base determines the pOH and pH. The $[OH^-]$ from the hydrolysis of $C_2H_3O_2^-$ is small and can be ignored.

$$[OH^-] = \frac{0.15 \times 10^{-3} \text{ mol}}{0.101 \text{ L}} = 1.5 \times 10^{-3}\ M;\ pOH = 2.83\ ;\ pH = 14.0 - 2.83 = 11.17$$

(f) $$[OH^-] = \frac{3.75 \times 10^{-3} \text{ mol}}{0.125 \text{ L}} = 0.0300\ M\ OH^-;\ pOH = 1.52;\ pH = 14.00 - 1.52 = 12.48$$

17.16 (a) Weak base problem: $K_b = 1.8 \times 10^{-5} = \frac{[NH_4^+][OH^-]}{[NH_3]}$

At equilibrium, $[OH^-] = x$, $[NH_3] = (0.100\text{-x})$; $[NH_4^+] = x$

$$1.8 \times 10^{-5} = \frac{x^2}{(0.100\text{-x})} \approx \frac{x^2}{0.100}\ ;\ x = [OH^-] = 1.3 \times 10^{-3}\ M$$

$pH = 14 - 2.87 = 11.13$

(b-f) Calculate mol NH_3 and mol NH_4^+ after the acid-base reaction takes place.
$0.100\ M\ NH_3 \times 0.0600 \text{ L} = 6.00 \times 10^{-3}$ mol NH_3 present initially.

		$NH_3(aq)$ +	$HCl(aq)$ →	$NH_4^+(aq) + Cl^-(aq)$
(b)	before rx	6.00×10^{-3} mol	$(0.150\ M \times 0.0200 \text{ L}) =$ 3.00×10^{-3} mol	0 mol
	after rx	**3.00×10^{-3} mol**	**0 mol**	**3.00×10^{-3} mol**
(c)	before rx	6.00×10^{-3} mol	$(0.150\ M \times 0.0395 \text{ L}) =$ 5.93×10^{-3} mol	0 mol
	after rx	**0.07×10^{-3} mol**	**0 mol**	**5.93×10^{-3} mol**

(d)	before rx	6.00×10^{-3} mol	$(0.150\ M \times 0.0400\ \text{L}) =$ 6.00×10^{-3} mol	0 mol
	after rx	**0 mol**	**0 mol**	**6.00×10^{-3} mol**
(e)	before rx	6.00×10^{-3} mol	$(0.150\ M \times 0.0405\ \text{L}) =$ 6.08×10^{-3} mol	0 mol
	after rx	**0 mol**	**0.08×10^{-3} mol**	**6.00×10^{-3} mol**
(f)	before rx	6.00×10^{-3} mol	$(0.150\ M \times 0.0600\ \text{L}) =$ 9.00×10^{-3} mol	0 mol
	after rx	**0 mol**	**3.00×10^{-3} mol**	**6.00×10^{-3} mol**

(b) Using the acid dissociation equilibrium for NH_4^+ (so that we calculate $[H^+]$ directly), $NH_4^+(aq) \rightleftharpoons H^+(aq) + NH_3(aq)$

$$K_a = \frac{[H^+][NH_3]}{[NH_4^+]} = \frac{K_w}{K_b \text{ for } NH_3} = \frac{1.0 \times 10^{-14}}{1.8 \times 10^{-5}} = 5.6 \times 10^{-10}$$

$$[NH_3] = \frac{3.00 \times 10^{-3} \text{ mol}}{0.0800 \text{ L}} = 0.0375\ M;\ [NH_4^+] = \frac{3.00 \times 10^{-3} \text{ mol}}{0.0800 \text{ L}} = 0.0375\ M$$

$$[H^+] = \frac{5.6 \times 10^{-10}\ [NH_4^+]}{[NH_3]} \approx \frac{5.6 \times 10^{-10}\ (0.0375)}{(0.0375)} = 5.6 \times 10^{-10}\ ;\ \ pH = 9.25$$

(We will assume $[H^+]$ is small compared to $[NH_3]$ and $[NH_4^+]$)

(c) $$[NH_3] = \frac{7.0 \times 10^{-5} \text{ mol}}{0.0995 \text{ L}} = 7 \times 10^{-4}\ M;\ [NH_4^+] = \frac{5.93 \times 10^{-3} \text{ mol}}{0.0995 \text{ L}} = 5.96 \times 10^{-2}\ M$$

$$[H^+] = \frac{5.6 \times 10^{-10}\ (5.96 \times 10^{-2})}{(7 \times 10^{-4})} = 5 \times 10^{-8}\ M;\ \ pH = 7.3$$

(d) At the equivalence point, $[H^+] = [NH_3] = x$

$$[NH_4^+] = \frac{6.00 \times 10^{-3}\ M}{0.100 \text{ L}} = 0.0600\ M$$

$$5.6 \times 10^{-10} \approx \frac{x^2}{0.060}\ ;\ \ x = [H^+] = 5.8 \times 10^{-6}\ ;\ \ pH = 5.24$$

(e) Past the equivalence point, $[H^+]$ from the excess HCl determines the pH.

$$[H^+] = \frac{8 \times 10^{-5} \text{ mol}}{0.105 \text{ L}} = 8 \times 10^{-4}\ M\ ;\ \ pH = 3.1$$

(f) $$[H^+] = \frac{3.00 \times 10^{-3} \text{ mol}}{0.120 \text{ L}} = 0.0250\ M\ ;\ \ pH = 1.602$$

17.17 The volume of 0.200 M HBr required in all cases equals the volume of base and the final volume = $2V_{base}$. The concentration of the salt produced at the equivalence point = $\frac{0.200\ M \times V_{base}}{2V_{base}} = 0.100\ M$

(a) 0.100 *M* NaBr, pH = 7.00

(b) 0.100 *M* $HONH_3^+Br^-$; $HONH_3^+(aq) \rightleftharpoons H^+(aq) + HONH_2$
[equil.] 0.100-x x x

$$K_a = \frac{[H^+][HONH_2]}{[HONH_3^+]} = \frac{K_w}{K_b} = \frac{1.0 \times 10^{-14}}{1.1 \times 10^{-8}} = 9.09 \times 10^{-7}$$

Assume x is small with respect to [salt].

$K_a = x^2/0.100$; x = $[H^+] = 3.02 \times 10^{-4}$ *M*, pH = 3.52

(c) 0.100 *M* $C_6H_5NH_3^+Br^-$. Proceeding as in (b):

$$K_a = \frac{[H^+][C_6H_5NH_2]}{[C_6H_5NH_3^+]} = \frac{K_w}{K_b} = 2.33 \times 10^{-5}$$

$[H^+]^2 = 0.100(2.33 \times 10^{-5})$; $[H^+] = 1.52 \times 10^{-3}$ *M*, pH = 2.82

17.18 The volume of NaOH solution required in all cases is

$$V_{base} = \frac{V_{acid} \times M_{acid}}{M_{base}} = \frac{(0.100)V_{acid}}{(0.080)} = 1.25\ V_{acid}$$

The total volume at the equivalence point is $V_{base} + V_{acid} = 2.25\ V_{acid}$.

The concentration of the salt at the equivalence point is

$$\frac{M_{acid}V_{acid}}{2.25\ V_{acid}} = \frac{0.100}{2.25} = 0.0444\ M$$

(a) 0.044 *M* NaBr, pH = 7.00

(b) 0.044 *M* $Na^+C_3H_5O_3^-$; $C_3H_5O_3^-(aq) + H_2O(l) \rightleftharpoons HC_3H_5O_3(aq) + OH^-(aq)$

$$K_b = \frac{[HC_3H_5O_3][OH^-]}{[C_3H_5O_3^-]} = \frac{K_w}{K_a} = \frac{1.0 \times 10^{-14}}{1.4 \times 10^{-4}} = 7.1 \times 10^{-11}$$

$[HC_3H_5O_3] = [OH^-]$; $[C_3H_5O_3^-] \approx 0.044$

$[OH^-]^2 \approx 0.044(7.1 \times 10^{-11})$; $[OH^-] = 1.8 \times 10^{-6}$, pOH = 5.75; pH = 8.25

(c) $CrO_4^{-2}(aq) + H_2O(l) \rightleftharpoons HCrO_4^-(aq) + OH^-(aq)$

$$K_b = \frac{[HCrO_4^-][OH^-]}{[CrO_4^{2-}]} = \frac{K_w}{K_a} = 3.3 \times 10^{-8}$$

$[OH^-]^2 \approx 0.044(3.3 \times 10^{-8})$; $[OH^-] = 3.8 \times 10^{-5}$, pH = 9.58

Buffers

17.19 $HC_2H_3O_2$ and $NaC_2H_3O_2$ are a weak conjugate acid/conjugate base pair which act as a buffer because undissociated $HC_2H_3O_2$ reacts with added base, while $C_2H_3O_2^-$ combines with added acid, leaving $[H^+]$ relatively unchanged. Although HCl and KCl are a conjugate acid/conjugate base pair, Cl^- is a negligible base. That is, it has no tendency to combine with added acid to form undissociated HCl. Any added acid simply increases $[H^+]$ in an HCl/KCl mixture. In general, the conjugate bases of strong acids are negligible and mixtures of strong acids and their conjugate salts do not act as buffers.

17.20 (a) The pH of a buffer is determined by K_a for the conjugate acid present and the ratio of conjugate base concentration to conjugate acid concentration.

(b) The capacity of a buffer is determined by the absolute concentrations of the conjugate acid and conjugate base present. For example, a buffer which is 1.0 *M* in both $HC_2H_3O_2$ and $C_2H_3O^{2-}$ has a higher capacity than one which is 0.10 *M* in the two components.

17.21 (a) $HC_2H_3O_2(aq) \rightleftharpoons H^+(aq) + C_2H_3O_2^-(aq)$; $K_a = 1.8 \times 10^{-5} = \frac{[H^+][C_2H_3O_2^-]}{[HC_2H_3O_2]}$

$$[HC_2H_3O_2] = \frac{20.0\text{ g } HC_2H_3O_2}{2.00\text{ L soln}} \times \frac{1\text{ mol } HC_2H_3O_2}{60.05\text{ g } HC_2H_3O_2} = 0.167\ M$$

$$[C_2H_3O_2^-] = \frac{20.0\text{ g } NaC_2H_3O_2}{2.00\text{ L soln}} \times \frac{1\text{ mol } NaC_2H_3O_2}{82.04\text{ g } NaC_2H_3O_2} = 0.121\ M$$

$$[H^+] = \frac{K_a[HC_2H_3O_2]}{[C_2H_3O_2^-]} = \frac{1.8 \times 10^{-5}(0.167\text{-}x)}{(0.121+x)} \approx \frac{1.8 \times 10^{-5}(0.167)}{(0.121)}$$

$[H^+] = 2.5 \times 10^{-5}\ M$, pH = 4.60

(b) $Na^+(aq) + C_2H_3O_2^-(aq) + H^+(aq) + Cl^-(aq) \rightarrow HC_2H_3O_2(aq) + Na^+(aq) + Cl^-(aq)$

(c) $HC_2H_3O_2(aq) + Na^+(aq) + OH^-(aq) \rightarrow C_2H_3O_2^-(aq) + H_2O(l) + Na^+(aq)$

17.22 NH_4^+/NH_3 is a basic buffer. Either the hydrolysis of NH_3 or the dissociation of NH_4^+ can be used to determine the pH of the buffer. Using the dissociation of NH_4^+ leads directly to $[H^+]$ and facilitates use of the Henderson-Hasselbach relationship.

(a) $NH_4^+(aq) \rightleftharpoons H^+(aq) + NH_3(aq)$ $K_a = \frac{K_w}{K_b} = \frac{1.0 \times 10^{-14}}{1.8 \times 10^{-5}} = 5.6 \times 10^{-10}$

$$[NH_3] = \frac{5.0 \text{ g } NH_3}{2.50 \text{ L soln}} \times \frac{1 \text{ mol } NH_3}{17.0 \text{ g } NH_3} = 0.12\ M\ NH_3$$

$$[NH_4^+] = \frac{20.0 \text{ g } NH_4Cl}{2.50 \text{ L}} \times \frac{1 \text{ mol } NH_4Cl}{53.50 \text{ g } NH_4Cl} = 0.150\ M\ NH_4^+$$

$$K_a = \frac{[H^+][NH_3]}{[NH_4^+]};\ [H^+] = \frac{K_a[NH_4^+]}{[NH_3]} = \frac{5.6 \times 10^{-10}(0.150-x)}{(0.12+x)} \approx \frac{5.6 \times 10^{-10}(0.150)}{(0.12)}$$

$[H^+] = 7.0 \times 10^{-10}$ *M*, pH = 9.15

(b) $NH_3(aq) + H^+(aq) + NO_3^-(aq) \longrightarrow NH_4^+(aq) + NO_3^-(aq)$

(c) $NH_4^+(aq) + Cl^-(aq) + K^+(aq) + OH^-(aq) \longrightarrow NH_3(aq) + H_2O(l) + Cl^-(aq) + K^+(aq)$

17.23 (a) Complete ionic:

$H^+(aq) + ClO_4^-(aq) + Na^+(aq) + NO_2^-(aq) \rightarrow HNO_2(aq) + Na^+(aq) + ClO_4^-(aq)$

Na^+ and ClO_4^- are spectator ions.

Net ionic: $H^+(aq) + NO_2^-(aq) \rightleftharpoons HNO_2(aq)$

(b) The net ionic equation in part (a) is the reverse of the dissociation of HNO_2.

$$K = \frac{1}{K_a} = \frac{1}{4.5 \times 10^{-4}} = 2.2 \times 10^3$$

(c) For Na^+ and ClO_4^-, this is just a dilution problem.

$M_1V_1 = M_2V_2$; V_2 is 50.0 mL + 50.0 mL = 100.0 mL.

ClO_4^-: $\frac{0.15\ M \times 50.0 \text{ mL}}{100.0 \text{ mL}} = 0.075\ M$; Na^+: $\frac{0.15\ M \times 50.0 \text{ mL}}{100.0 \text{ mL}} = 0.075\ M$

H^+ and NO_2^- react to form HNO_2. Since K >> 1, the reaction essentially goes to completion.

$0.15\ M \times 0.0500 \text{ mL} = 7.5 \times 10^{-3}$ mol H^+

$+ 0.15\ M \times 0.0500 \text{ mL} = 7.5 \times 10^{-3}$ mol NO_2^-

7.5×10^{-3} mol HNO_2

Solve the weak acid problem to determine $[H^+]$, $[NO_2^-]$, $[HNO_2]$ at equilibrium.

$$K_a = \frac{[H^+][NO_2^-]}{[HNO_2]}; \quad [H^+] = [NO_2^-] = x\ M; \quad [HNO_2] = \frac{7.5 \times 10^{-3} \text{ mol}}{0.100 \text{ L}} = (0.075\text{-}x)\ M$$

$$4.5 \times 10^{-4} = \frac{x^2}{(0.075\text{-}x)} \approx \frac{x^2}{0.075}; \quad x = 5.8 \times 10^{-3}\ M$$

$$\frac{[H^+]}{[HNO_2]} \times 100 = \frac{5.8 \times 10^{-3}}{0.075} \times 100 = 7.8\% \text{ dissociation}$$

Using the quadratic formula to determine x: $x^2 + 4.5 \times 10^{-4}\ x - 3.38 \times 10^{-5} = 0$

$x = 5.6 \times 10^{-3}\ M\ H^+$ and NO^{-2}

$$x = \frac{-4.5 \times 10^{-4} + \sqrt{(4.5 \times 10^{-4})^2 - 4(1)(-3.38 \times 10^{-5})}}{2}$$; (Note that $[H^+]$ is not very different from the value obtained using the assumption.)

$[HNO_2] = 0.075 - 0.0056 = 0.069\ M$

In summary:

$[Na^+] = [ClO_4^-] = 0.075\ M$, $[HNO_2] = 0.069\ M$, $[H^+] = [NO_2^-] = 0.0056\ M$.

17.24 (a) Complete ionic:

$$Na^+(aq) + OH^-(aq) + CH_3NH_3^+(aq) + Br^-(aq) \longrightarrow CH_3NH_2(aq) + H_2O(l) + Na^+(aq) + Br^-(aq)$$

Na^+ and Br^- are spectators

Net ionic: $CH_3NH_3^+(aq) + OH^-(aq) \longrightarrow CH_3NH_2(aq) + H_2O(l)$

(b) The net ionic equation in part (a) is the reverse of the hydrolysis of $CH_3NH_2(aq)$.

$$K = \frac{1}{K_b} = \frac{1}{4.4 \times 10^{-4}} = 2.3 \times 10^3$$

(c) For the spectator ions, this is a simple dilution problem.

$M_1V_1 = M_2V_2$; V_2 is 0.250 L + 0.400 L = 0.650 L

$$Na^+: \frac{0.300\ M \times 0.250 \text{ L}}{0.650 \text{ L}} = 0.115\ M\ Na^+; \quad Br^-: \frac{0.100\ M \times 0.400 \text{ L}}{0.650 \text{ L}} = 0.0615\ M\ Br^-$$

$CH_3NH_3^+$ and OH^- react to form CH_3NH_2. Since K >> 1, the reaction essentially goes to completion.

$0.300\ M \times 0.250\ L = 0.0750\ mol\ OH^-$

$+0.100\ M \times 0.400\ L = 0.0400\ mol\ CH_3NH_3^+$

$0.0400\ mol\ CH_3NH_2 + 0.0350\ mol\ OH^-$

$[CH_3NH_3^+] = 0$ (all $CH_3NH_3^+$ is neutralized by NaOH)

$$[CH_3NH_2] = \frac{0.0400\ mol}{0.650\ L} = 0.0615\ M$$

$$[OH^-] = \frac{0.0350\ mol}{0.650\ L} = 0.0538\ M,\ [H^+] = \frac{K_w}{[OH^-]} = \frac{1.00 \times 10^{-14}}{0.0538\ M} = 1.86 \times 10^{-13}\ M$$

17.25 In this problem, $[BrO^-]$ is the unknown.

$pH = 8.80$, $[H^+] = 10^{-8.80} = 1.6 \times 10^{-9}\ M$

$[HBrO] = 0.200 - 1.6 \times 10^{-9} \approx 0.200\ M$

$$K_a = 2 \times 10^{-9} = \frac{1.6 \times 10^{-9}\ [BrO^-]}{0.200};\quad [BrO^-] = 0.25\ M$$

For 1.00 L, 0.25 mol NaBrO are needed.

17.26 $HC_3H_5O_3(aq) \rightleftharpoons H^+(aq) + C_3H_5O_3^-(aq)$

$$[H^+] = \frac{K_a[HC_3H_5O_3]}{[C_3H_5O_3^-]};\quad [H^+] = 10^{-3.90} = 1.26 \times 10^{-4}\ M$$

$[HC_3H_5O_3] = 0.0150\ M$ Calculate $[C_3H_5O_3^-]$

$$[C_3H_5O_3^-] = \frac{K_a[HC_3H_5O_3]}{[H^+]} = \frac{1.4 \times 10^{-4}\ (0.0150)}{1.26 \times 10^{-4}} = 1.7 \times 10^{-2}\ M$$

$$\frac{0.017\ mol\ NaC_3H_5O_3}{1.00\ L} \times \frac{112.1\ g\ NaC_3H_5O_3}{1\ mol\ NaC_3H_5O_3} = 1.9\ g\ NaC_3H_5O_3$$

17.27 (a) $$K_a = \frac{[H^+][C_2H_3O_2^-]}{[HC_2H_3O_2]};\quad [H^+] = \frac{K_a[HC_2H_3O_2]}{[C_2H_3O_2^-]}$$

$$[HC_2H_3O_2] = \frac{1.00\ g\ HC_2H_3O_2}{0.100\ L\ soln} \times \frac{1\ mol\ HC_2H_3O_2}{60.05\ g\ HC_2H_3O_2} = 0.167\ M$$

$$[C_2H_3O_2^-] = \frac{1.50\ g\ NaC_2H_3O_2}{0.100\ L\ soln} \times \frac{1\ mol\ Na_2CH_5O_2}{82.04\ g\ NaC_2H_3O_2} = 0.183\ M$$

$$[H^+] \approx \frac{1.8 \times 10^{-5}\ (0.167)}{(0.183)} = 1.64 \times 10^{-5}\ M;\ pH = 4.78$$

(b)

$C_2H_3O_2^-$	+ HNO_3	→ $HC_2H_3O_2$ + NO_3^-
0.0183 mol	0.001 mol	0.0167 mol
-0.001 mol	-0.001 mol	+0.001 mol
0.0173 mol	0 mol	0.0177 mol

$$[H^+] \approx \frac{1.8 \times 10^{-5}\ (0.0177\ \text{mol}/0.100\ \text{L})}{(0.0173\ \text{mol}/0.100\ \text{L})} = 1.84 \times 10^{-5}\ M;\ \text{pH} = 4.73$$

(c)

$HC_2H_3O_2$	+ KOH	→ $C_2H_3O_2^-$ + H_2O + K^+
0.0167 mol	0.001 mol	0.0183 mol
-0.001 mol	-0.001 mol	+0.001 mol
0.0157 mol	0 mol	0.0193 mol

$$[H^+] \approx \frac{1.8 \times 10^{-5}\ (0.0157\ \text{mol}/0.100\ \text{L})}{(0.0193\ \text{mol}/0.100\ \text{L})} = 1.46 \times 10^{-5}\ M;\ \text{pH} = 4.83$$

17.28 (a) $K_a = \dfrac{[H^+][C_3H_5O_2^-]}{[HC_3H_5O_2]}$; $[H^+] = \dfrac{K_a[HC_3H_5O_2]}{[C_3H_5O_2^-]}$

$$[H^+] = \frac{1.3 \times 10^{-5}\ (0.120-x)}{(0.105+x)} \approx \frac{1.3 \times 10^{-5}\ (0.120)}{0.105} = 1.5 \times 10^{-5}\ M,\ \text{pH} = 4.83$$

(b) $1.50\ M \times 0.010\ \text{L} = 0.015$ mol H^+

$C_3H_5O_2^-$	+ HCl	⟶ $HC_3H_5O_2$ + Cl^-
0.105 mol	0.015 mol	0.120 mol
-0.015 mol	-0.015 mol	+0.015 mol
0.090 mol	0 mol	0.135 mol

$$[H^+] \approx \frac{1.3 \times 10^{-5}\ (0.135)}{(0.090)} = 2.0 \times 10^{-5}\ M,\ \text{pH} = 4.71$$

(c) $0.400\ M \times 0.050\ \text{L} = 0.020$ mol OH^-

$HC_3H_5O_2$	+ OH^-	⟶ $C_3H_5O_2^-$ + H_2O
0.120 mol	0.020 mol	0.105 mol
-0.020 mol	-0.020 mol	+0.020 mol
0.100 mol	0 mol	0.125 mol

$$[H^+] \approx \frac{1.3 \times 10^{-5}\ (0.100)}{(0.125)} = 1.0 \times 10^{-5}\ M,\ \text{pH} = 4.98$$

17.29 $H_2CO_3(aq) \rightleftharpoons H^+(aq) + HCO_3^-(aq)$

$$K_a = \frac{[H^+][HCO_3^-]}{[H_2CO_3]}; \quad \frac{[HCO_3^-]}{[H_2CO_3]} = \frac{K_a}{[H^+]}$$

(a) at pH = 7.4, $[H^+] = 10^{-7.4} = 4.0 \times 10^{-8}$ *M*; $\frac{[HCO_3^-]}{[H_2CO_3]} = \frac{4.3 \times 10^{-7}}{4.0 \times 10^{-8}} = 11$

(b) at pH = 7.4, $[H^+] = 7.9 \times 10^{-8}$ *M*; $\frac{[HCO_3^-]}{[H_2CO_3]} = 5.4$

17.30 $\frac{6.5 \text{ g } NaH_2PO_4}{0.355 \text{ L soln}} \times \frac{1 \text{ mol } NaH_2PO_4}{120 \text{ g } NaH_2PO_4} = 0.15$ *M*

$\frac{8.0 \text{ g } Na_2HPO_4}{0.355 \text{ L soln}} \times \frac{1 \text{ mol } Na_2HPO_4}{142 \text{ g } NaH_2PO_4} = 0.16$ *M*

Using Equation [17.13] to find the pH of the buffer: $pH = 7.21 + \log \frac{0.16}{0.15} = 7.24$

Solubility Equilibria

17.31 (a) The concentration of undissolved solid does not appear in the solubility product expression because it is constant as long as there is solid present. Concentration is a ratio of moles solid to volume of the solid; solids occupy a specific volume not dependent on the solution volume. As the amount (moles) of solid changes, the volume changes proportionally, so that the ratio of moles solid to volume solid is constant.

(b) $K_{sp} = [Ag^+][I^-]$; $Ksp = [Ba^{2+}][CO_3^{2-}]$; $K_{sp} = [Cu^+]^2[S^{2-}]$

$K_{sp} = [Ce^{3+}][F^-]^3$; $K_{sp} = [Ca^{2+}]^3[PO_4^{3-}]^2$

17.32 (a) *Solubility* is the amount (grams, moles) of solute that will dissolve in a certain volume of solution. *Solubility-product constant* is an equilibrium constant, the product of the molar concentrations of all the dissolved ions in solution.

(b) $K_{sp} = [Cu^{2+}][S^{2-}]$; $K_{sp} = [Ni^{2+}][C_2O_4^{2-}]$; $K_{sp} = [Ag^+]^2[SO_4^{2-}]$

$K_{sp} = [Co^{3+}][OH^-]^3$; $K_{sp} = [Fe^{2+}]^3[AsO_4^{3-}]^2$

17.33 (a) $CaF_2(s) \rightleftharpoons Ca^{2+}(aq) + 2F^-(aq)$; $K_{sp} = [Ca^{2+}][F^-]^2$

The molar solubility is the moles of CaF_2 that dissolve in 1.0 L of solution. Each mole of CaF_2 produces 1 mol $Ca^{2+}(aq)$ and 2 mol $F^-(aq)$.

$[Ca^{2+}] = 1.24 \times 10^{-3}$ *M*; $[F^-] = 2 \times 1.24 \times 10^{-3}$ *M* $= 2.48 \times 10^{-3}$ *M*

$K_{sp} = (1.24 \times 10^{-3})(2.48 \times 10^{-3})^2 = 7.63 \times 10^{-9}$

(b) $SrF_2(s) \rightleftharpoons Sr^{2+}(aq) + 2F^-(aq)$; $K_{sp} = [Sr^{2+}][F^-]^2$

Transform the gram solubility to molar solubility.

$$\frac{1.1 \times 10^{-2} \text{ g } SrF_2}{0.100 \text{ L}} \times \frac{1 \text{ mol } SrF_2}{125.6 \text{ g } SrF_2} = 8.8 \times 10^{-4} \text{ mol } SrF_2/\text{L}$$

$[Sr^{2+}] = 8.8 \times 10^{-4}$ *M*; $[F^-] = 2(8.8 \times 10^{-4}$ *M*)

$K_{sp} = (8.8 \times 10^{-4})(2(8.8 \times 10^{-4}))^2 = 2.7 \times 10^{-9}$

(c) $Ba(IO_3)_2(s) \rightleftharpoons Ba^{2+}(aq) + 2IO_3^-(aq)$; $K_{sp} = [Ba^{2+}][IO_3^-]^2$

Since 1 mole of dissolved $Ba(IO_3)_2$ produces 1 mole of Ba^{2+}, the molar solubility of $Ba(IO_3)_2 = [Ba^{2+}]$. Let $x = [Ba^{2+}]$; $[IO_3^-] = 2x$

$K_{sp} = 6.0 \times 10^{-10} = (x)(2x)^2$; $4x^3 = 6.0 \times 10^{-10}$; $x^3 = 1.5 \times 10^{-10}$; $x = 5.3 \times 10^{-4}$ *M*

The molar solubility of $Ba(IO_3)_2$ is 5.3×10^{-4} mol/L.

17.34 (a) $PbBr_2(s) \rightleftharpoons Pb^{2+}(aq) + 2Br^-(aq)$

$K_{sp} = [Pb^{2+}][Br^-]^2$; $[Pb^{2+}] = 1.0 \times 10^{-2}$ *M*, $[Br^-] = 2.0 \times 10^{-2}$ *M*

$K_{sp} = (1.0 \times 10^{-2} M)(2.0 \times 10^{-2} M)^2 = 4.0 \times 10^{-6}$

(b) $AgIO_3(s) \rightleftharpoons Ag^+(aq) + IO_3^-(aq)$

$$K_{sp} = [Ag^+][IO_3^-]; [Ag^+] = [IO_3^-] = \frac{0.0490 \text{ g } AgIO_3}{1.00 \text{ L soln}} \times \frac{1 \text{ mol } AgIO_3}{282.8 \text{ g } AgIO_3} = 1.73 \times 10^{-4} M$$

$K_{sp} = (1.73 \times 10^{-4} M)(1.73 \times 10^{-4} M) = 3.00 \times 10^{-8}$

(c) $Cu(OH)_2(s) \rightleftharpoons Cu^{2+}(aq) + 2OH^-(aq)$; $K_{sp} = [Cu^{2+}][OH^-]^2$

$[Cu^{2+}] = x$, $[OH^-] = 2x$; $K_{sp} = 2.2 \times 10^{-20} = (x)(2x)^2$

$2.2 \times 10^{-20} = 4x^3$; $x = [Cu^{2+}] = 1.76 \times 10^{-7}$ *M*

$$\frac{1.76 \times 10^{-7} \text{ mol } Cu(OH)_2}{1 \text{ L}} \times \frac{97.56 \text{ g } Cu(OH)_2}{1 \text{ mol } Cu(OH)_2} = 1.7 \times 10^{-5} \text{ g } Cu(OH)_2$$

However, $[OH^-]$ from $Cu(OH)_2 = 3.53 \times 10^{-7}$ *M* ; this is similar to $[OH^-]$ from the autoionization of water.

$K_w = [H^+][OH^-]$; $[H^+] = y$, $[OH^-] = (3.53 \times 10^{-7} + y)$

$1.0 \times 10^{-14} = y(3.53 \times 10^{-7} + y)$; $y^2 + 3.53 \times 10^{-7}\, y - 1.0 \times 10^{-14}$

$$y = \frac{-3.53 \times 10^{-7} \pm \sqrt{(3.53 \times 10^{-7})^2 - 4(1)(-1.0 \times 10^{-14})}}{2} ; \quad y = 2.64 \times 10^{-8}$$

$[OH^-]_{total} = 3.53 \times 10^{-7}\, M + 0.264 \times 10^{-7}\, M = 3.79 \times 10^{-7}\, M$

Recalculating $[Cu^{2+}]$ and thus molar solubility of $Cu(OH)_2(s)$:

$2.2 \times 10^{-20} = x(3.79 \times 10^{-7})^2$; $x = 1.53 \times 10^{-7}\, M\, Cu^{2+}$

$$\frac{1.53 \times 10^{-7} \text{ mol } Cu(OH)_2(s)}{1 \text{ L}} \times \frac{97.56 \text{ g } Cu(OH)_2}{1 \text{ mol } Cu(OH)_2} = 1.5 \times 10^{-5} \text{ g } Cu(OH)_2$$

Note that the presence of OH^- as a common ion decreases the water solubility of $Cu(OH)_2$.

17.35 (a) $AgBr(s) \rightleftharpoons Ag^+(aq) + Br^-(aq)$; $K_{sp} = [Ag^+][Br^-] = 5.0 \times 10^{-13}$

molar solubility = $x = [Ag^+] = [Br^-]$; $K_{sp} = x^2$

$x = (5.0 \times 10^{-13})^{1/2}$; $x = 7.1 \times 10^{-7}$ mol AgBr/L

(b) Molar solubility = $x = [Br^-]$; $[Ag^+] = 0.030\, M + x$

$K_{sp} = (0.030 + x)(x) \approx 0.030(x)$

$5.0 \times 10^{-13} = 0.030(x)$; $x = 1.7 \times 10^{-11}$ mol AgBr/L

(c) Molar solubility = $x = [Ag^+]$.

There are two sources of Br^-: NaBr(0.50 *M*) and AgBr(x*M*)

$K_{sp} = (x)(0.50 + x)$; Assuming x is small compared to 0.50 *M*.

$5.0 \times 10^{-13} \approx 0.50(x)$; $x = 1.0 \times 10^{-12}$ mol AgBr/L

17.36 (a) $CaF_2(s) \rightleftharpoons Ca^{2+}(aq) + 2F^-(aq)$; $K_{sp} = [Ca^{2+}][F^-]^2 = 3.9 \times 10^{-11}$

molar solubility = x = $[Ca^{2+}]$; $[F^-] = 2[Ca^{2+}] = 2x$

$K_{sp} = (x)(2x)^2$; $3.9 \times 10^{-11} = 4x^3$, $x = (9.75 \times 10^{-12})^{1/3}$, $x = 2.1 \times 10^{-4}\ M\ Ca^{2+}$

$$\frac{2.1 \times 10^{-4}\ \text{mol}\ CaF_2}{1\ \text{L}} \times \frac{78.1\ \text{g}\ CaF_2}{1\ \text{mol}} = 0.016\ \text{g}\ CaF_2/\text{L}$$

(b) Molar solubility = x = $[Ca^{2+}]$

There are two sources of F^-: KF(0.15 *M*) and CaF_2 (2x *M*).

$K_{sp} = (x)(0.15 + 2x)^2$; Assuming x is small compared to 0.15 *M*,

$3.9 \times 10^{-11} = 0.0225\ x$, $x = 1.7 \times 10^{-9}\ M\ Ca^{2+}$

$$\frac{1.7 \times 10^{-9}\ \text{mol}\ CaF_2}{1\ \text{L}} \times \frac{78.1\ \text{g}\ CaF_2}{1\ \text{mol}} = \frac{1.3 \times 10^{-7}\ \text{g}\ CaF_2}{1\ \text{L}} = \frac{0.13\ \mu\text{g}}{1\ \text{L}} = 0.13\ \text{ppm}\ CaF_2$$

(c) Molar solubility = x, $[F^-] = 2x$, $[Ca^{2+}] = 0.080\ M + x$

$K_{sp} = (0.080 + x)(2x)^2 \approx 0.080\ (2x)^2$; $3.9 \times 10^{-11} = 0.32\ x^2$

$x = (1.22 \times 10^{-10})^{1/2} = 1.1 \times 10^{-5}\ M\ CaF_2$

$$\frac{1.1 \times 10^{-5}\ \text{mol}\ CaF_2}{\text{L}} \times \frac{78.1\ \text{g}\ CaF_2}{1\ \text{mol}} = \frac{8.6 \times 10^{-4}\ \text{g}\ CaF_2}{\text{L}} = \frac{0.86\ \text{mg}\ CaF_2}{\text{L}}$$

17.37 $Cu(OH)_2(s) \rightleftharpoons Cu^{2+}(aq) + 2OH^-(aq)$; $K_{sp} = 2.2 \times 10^{-20}$

Since the $[OH^-]$ is set by the pH of the solution, the solubility of $Cu(OH)_2$ is just $[Cu^{2+}]$.

(a) pH = 7.0, pOH = 14 - pH = 7.0, $[OH^-] = 10^{-pOH} = 1.0 \times 10^{-7}\ M$

$$K_{sp} = 2.2 \times 10^{-20} = [Cu^{2+}](1.0 \times 10^{-7})^2;\ [Cu^{2+}] = \frac{2.2 \times 10^{-20}}{1.0 \times 10^{-14}} = 2.2 \times 10^{-6}\ M$$

(In pure water, $[OH^-]$ from $Cu(OH)_2$ is similar to (OH^-) from the autoionization of water, resulting in a cubic equation for $[Cu^{2+}]$. The solubility of $Cu(OH)_2$ at pH = 7.0 is actually greater than the solubility in pure water.)

(b) pH = 9.0, pOH = 5.0, $[OH^-] = 1.0 \times 10^{-5}$

$$K_{sp} = 2.2 \times 10^{-20} = [Cu^{2+}][1.0 \times 10^{-5}]^2;\ [Cu^{2+}] = \frac{2.2 \times 10^{-20}}{1.0 \times 10^{-10}} = 2.2 \times 10^{-10}\ M$$

(c) pH = 11.0, pOH = 3.0, $[OH^-] = 1.0 \times 10^{-3}$

$$K_{sp} = 2.2 \times 10^{-20} = [Cu^{2+}][1.0 \times 10^{-3}]^2;\ [Cu^{2+}] = \frac{2.2 \times 10^{-20}}{1.0 \times 10^{-6}} = 2.2 \times 10^{-14}\ M$$

17.38 $Mn(OH)_2(s) \rightleftharpoons Mn^{2+}(aq) + 2OH^-(aq);\ K_{sp} = 1.9 \times 10^{-13}$

Since $[OH^-]$ is set by the pH of the solution, the solubility of $Mn(OH)_2$ is just $[Mn^{2+}]$.

(a) pH = 7.0, pOH = 14 - pH = 7.0, $[OH^-] = 10^{-pOH} = 1.0 \times 10^{-7}\ M$

$$K_{sp} = 1.9 \times 10^{-13} = [Mn^{2+}](1.0 \times 10^{-7})^2;\ [Mn^{2+}] = \frac{1.9 \times 10^{-13}}{1.0 \times 10^{-14}} = 19\ M$$

Note that the solubility of $Mn(OH)_2$ in pure water is $3.6 \times 10^{-5}\ M$, and the pH of the resulting solution is 9.9. The relatively low pH of a solution buffered to pH 7.0 actually increases the solubility of $Mn(OH)_2$.

(b) pH = 9.5, pOH = 4.5, $[OH^-] = 3.2 \times 10^{-5}$

$$K_{sp} = 1.9 \times 10^{-13} = [Mn^{2+}]\ (3.2 \times 10^{-5})^2;\ [Mn^{2+}] = \frac{1.9 \times 10^{-13}}{1.0 \times 10^{-9}} = 1.9 \times 10^{-4}\ M$$

(c) pH = 11.8, pOH = 2.2, $[OH^-] = 6.3 \times 10^{-3}$

$$K_{sp} = 1.9 \times 10^{-13} = [Mn^{2+}](6.3 \times 10^{-3})^2;\ [Mn^{2+}] = \frac{1.9 \times 10^{-13}}{4.0 \times 10^{-5}} = 4.8 \times 10^{-9}\ M$$

17.39 Let the molar solubility of CaF_2 = x. $K_{sp} = 3.9 \times 10^{-11}$

BaF_2 = y $K_{sp} = 1.0 \times 10^{-6}$

$[F^-] = 0.010\ M + x + y$; assume x and y are small compared to 0.010.

CaF_2: $K_{sp} = [Ca^{2+}]\,[F^-]^2$; $3.9 \times 10^{-11} \approx (x)\ (0.010)$; $x = 3.9 \times 10^{-9}$ mol CaF_2/L

BaF_2: $K_{sp} = [Ba^{2+}]\,[F^-]^2$; $1.0 \times 10^{-6} \approx (y)\ (0.010)$; $y = 1.0 \times 10^{-4}$ mol BaF_2/L

$$\frac{[Ca^{2+}]}{[Ba^{2+}]} = \frac{3.9 \times 10^{-9}\ M}{1.0 \times 10^{-4}\ M} = 3.9 \times 10^{-5} \text{ mol } Ca^{2+} / 1.0 \text{ mol } Ba^{2+}$$

17.40 $[Ca^{2+}]\,[CO_3^{2-}] = 2.8 \times 10^{-9}$; $[Fe^{2+}]\,[CO_3^{2-}] = 3.2 \times 10^{-11}$

Since $[CO_3^{2-}]$ is the same for both equilibria:

$$[CO_3^{2-}] = \frac{2.8 \times 10^{-9}}{[Ca^{2+}]} = \frac{3.2 \times 10^{-11}}{[Fe^{2+}]};\ \text{rearranging}\ \frac{[Ca^{2+}]}{[Fe^{2+}]} = \frac{2.8 \times 10^{-9}}{3.2 \times 10^{-11}} = 88$$

17.41 $K_{sp} = [Ba^{2+}]\,[MnO_4^-]^2 = 2.5 \times 10^{-10}$

$[MnO_4^-]^2 = 2.5 \times 10^{-10}/2.0 \times 10^{-8} = 0.11\ M$

17.42 (a) $Ce(IO_3)_3(s) \rightleftharpoons Ce^{3+}(aq) + 3IO_3^-(aq)$; $K_{sp} = [Ce^{3+}][IO_3^-]^3 = 3.2 \times 10^{-10}$

Let $[Ce^{3+}] = x$; then $[IO_3^-] = 3x$; $(x)(3x)^3 = 3.2 \times 10^{-10}$

$x = 1.8 \times 10^{-3}$ *M* = molar solubility of $Ce(IO_3)_3$ in water

(b) If $[Ce^{3+}] = 1.8 \times 10^{-4}$ *M*, let $[IO_3^-] = y$; $(1.8 \times 10^{-4})(y)^3 = 3.2 \times 10^{-10}$

$y = 1.2 \times 10^{-2}$ *M* IO_3^-

This is the total concentration of IO_3^- in solution. Of this, a part comes from $Ce(IO_3)_3$. The contribution from this source is $3(1.8 \times 10^{-4}) = 5.4 \times 10^{-4}$ *M*. Thus, the $NaIO_3$ concentration would need to be $(1.2 \times 10^{-2} - 5.4 \times 10^{-4})$ *M* ~ 1.2×10^{-2} *M*.

Precipitation; Dissolution

17.43 Precipitation conditions: will Q (see Chapter 15) exceed K_{sp} for the compound?

(a) In base, Mn^{2+} can form $Mn(OH)_2(s)$.

$Mn(OH)_2(s) \rightleftharpoons Mn^{2+}(aq) + 2OH^-(aq)$; $K_{sp} = [Mn^{2+}][OH^-]^2$

$Q = [Mn^{2+}]_i[OH^-]_i^2$; $[Mn^{2+}]_i = 0.050$ *M* ; pOH = 6; $[OH^-] = 10^{-6} = 1 \times 10^{-6}$ *M*.

$Q = (0.050)(1.0 \times 10^{-6})^2 = 5.0 \times 10^{-14}$. $K_{sp} = 1.9 \times 10^{-13}$ (Appendix D)

$Q < K_{sp}$, no $Mn(OH)_2$ precipitates.

(b) $Ag_2SO_4(s) \rightleftharpoons 2Ag^+(aq) + SO_4^{2-}(aq)$; $K_{sp} = [Ag^+]^2[SO_4^{2-}]$

$$[Ag^+] = \frac{0.010\ M \times 100\ mL}{120\ mL} = 8.3 \times 10^{-3}\ M$$

$$[SO_4^{2-}] = \frac{0.050\ M \times 20\ mL}{120\ mL} = 8.3 \times 10^{-3}\ M$$ (assuming complete dissociation of H_2SO_4 upon reaction with Ag^+)

$Q = (8.3 \times 10^{-3})^2(8.3 \times 10^{-3}) = 5.8 \times 10^{-7}$; $K_{sp} = 1.4 \times 10^{-5}$

$Q < K_{sp}$, no Ag_2SO_4 precipitates.

Considering the two interacting equilibria:

$Ag_2SO_4(s) \rightleftharpoons 2Ag^+(aq) + SO_4^{2-}(aq)$ $K_{sp} = 1.4 \times 10^{-5}$

$H^+(aq) + SO_4^{2-}(aq) \rightleftharpoons HSO_4^-(aq)$ $K = 1/1.2 \times 10^{-2}$

$Ag_2SO_4(s) + H^+(aq) \rightleftharpoons 2Ag^+(aq) + HSO_4^-(aq)$ $K = 1.2 \times 10^{-3}$

Upon dilution, $$[H_2SO_4] = \frac{0.050\ M\ (20\ mL)}{120\ mL} = 0.0083\ M$$

Assuming that the first dissociation step is complete:

	$HSO_4^-(aq)$ ⇌	$H^+(aq)$	+ $SO_4^{2-}(aq)$
initial	0.0083 *M*	0.0083 *M*	0 *M*
equil.	(0.0083-x) *M*	(0.0083+x) *M*	x *M*

$$1.2 \times 10^{-2} = \frac{(0.0083+x)x}{(0.0083-x)}\ ;\quad x^2 + 0.0203x - 1.0 \times 10^{-4} = 0$$

$x = [SO_4^{2-}] = 0.0041\ M$; $[H^+] = 0.0124\ M$; $[HSO_4^-] = 0.0042\ M$

$$K = \frac{[Ag^+]^2[HSO_4^-]}{[H^+]}\ ;\quad Q = \frac{(0.0083)^2(0.0042)}{0.0124} = 2.3 \times 10^{-5}$$

$Q(2.3 \times 10^{-5}) < K\ (1.2 \times 10^{-3})$, so no Ag_2SO_4 precipitates. The result is the same as assuming complete dissociation of H_2SO_4.

17.44 (a) $Co(OH)_2(s) \rightleftharpoons Co^{2+}(aq) + 2OH^-(aq)$; $K_{sp} = [Co^{2+}][OH^-]^2 = 1.6 \times 10^{-15}$

$pH = 8.5$; $pOH = 14 - 8.5 = 5.5$; $[OH^-] = 10^{-5.5} = 3 \times 10^{-6}\ M$

$Q = (0.020)(3 \times 10^{-6})^2 = 2 \times 10^{-13}$; $Q > K_{sp}$, $Co(OH)_2$ will precipitate

(b) $AgIO_3(s) \rightleftharpoons Ag^+(aq) + IO_3^-(aq)$; $K_{sp} = [Ag^+][IO_3^-] = 3.0 \times 10^{-8}$

$$[Ag^+] = \frac{0.010\ M\ Ag^+ \times 0.100\ L}{0.110\ L} = 9.1 \times 10^{-3}\ M$$

$$[IO_3^-] = \frac{0.015\ M\ IO_3^- \times 0.010\ L}{0.110\ L} = 1.4 \times 10^{-3}\ M$$

$Q = (9.1 \times 10^{-3})(1.4 \times 10^{-3}) = 1.3 \times 10^{-5}$; $Q > K_{sp}$, $AgIO_3$ will precipitate

17.45 $Ni(OH)_2(s) \rightleftharpoons Ni^{2+}(aq) + 2OH^-(aq)$; $K_{sp} = [Ni^{2+}][OH^-]^2 = 1.6 \times 10^{-14}$

At equilibrium, $[Ni^{2+}][OH^-]^2 = 1.6 \times 10^{-14}$. Change $[Ni^{2+}]$ to mol/L and solve for $[OH^-]$.

$$\frac{1.0\ \mu g\ Ni^{2+}}{1.0\ L} \times \frac{1 \times 10^{-6}\ g}{1\ \mu g} \times \frac{1\ mol\ Ni^{2+}}{58.7\ g\ Ni^{2+}} = 1.7 \times 10^{-8}\ M\ Ni^{2+}$$

$1.6 \times 10^{-14} = (1.7 \times 10^{-8})[OH^-]^2$; $[OH^-]^2 = 9.4 \times 10^{-7}$; $[OH^-] = 9.7 \times 10^{-4}\ M$

$pOH = 3.01$; $pH = 14.0 - 3.01 = 10.99$

17.46 $AgCl(s) \rightleftharpoons Ag^+(aq) + Cl^-(aq)$; $K_{sp} = [Ag^+][Cl^-] = 1.8 \times 10^{-10}$

$$[Ag^+] = \frac{0.10\ M \times 0.2\ mL}{50\ mL} = 4 \times 10^{-4}\ M\ ;\ [Cl^-] = \frac{1.8 \times 10^{-10}}{4 \times 10^{-4}\ M} = 4.5 \times 10^{-7}\ M$$

$$\frac{4.5 \times 10^{-7}\ mol\ Cl^-}{1\ L} \times \frac{35.45\ g\ Cl^-}{1\ mol\ Cl^-} \times 0.050\ L = 8 \times 10^{-7}\ g\ Cl^-$$

17.47 If the anion in the slightly soluble salt is the conjugate base of a strong acid, there will be no reaction.

(a) $MnS(s) + 2H^+(aq) \rightarrow H_2S(aq) + Mn^{2+}(aq)$

(b) $PbF_2(s) + 2H^+(aq) \rightarrow 2HF(aq) + Pb^{2+}(aq)$

(c) $AuCl_3(s) + H^+(aq) \rightarrow$ no reaction

(d) $Hg_2C_2O_4(s) + 2H^+(aq) \rightarrow H_2C_2O_4(aq) + Hg_2^{2+}(aq)$

(e) $CuBr(s) + H^+(aq) \rightarrow$ no reaction

17.48 If the anion of the salt is the conjugate base of a weak acid, it will combine with H^+, reducing the concentration of the free anion in solution, thereby causing more salt to dissolve.

More soluble in acid: (a) $ZnCO_3$ (b) LaF_3 (d) AgCN (e) $Ba_3(PO_4)_2$

17.49 Calculate $[I^-]$ needed to initiate precipitation of each ion. The cation which requires lower $[I^-]$ will precipitate first.

$$Ag^+ : K_{sp} = [Ag^+][I^-]\ ;\ 8.3 \times 10^{-17} = (2.0 \times 10^{-4})[I^-]\ ;\ [I^-] = \frac{8.3 \times 10^{-17}}{2.0 \times 10^{-4}} = 4.2 \times 10^{-13}\ M\ I^-$$

$$Pb^{2+} : K_{sp} = [Pb^{2+}][I^-]^2\ ;\ 1.4 \times 10^{-8} = (1.5 \times 10^{-3})[I^-]^2\ ;\ [I^-] = \left(\frac{1.4 \times 10^{-8}}{1.5 \times 10^{-3}}\right)^{1/2} = 3.1 \times 10^{-3}\ M\ I^-$$

AgI will precipitate first, at $[I^-] = 4.2 \times 10^{-13}\ M$.

17.50 (a) $BaSO_4$: $K_{sp} = [Ba^{2+}][SO_4^{2-}] = 1.1 \times 10^{-10}$ Precipitation will begin when $Q = K_{sp}$.

$1.1 \times 10^{-10} = (0.010)[SO_4^{2-}]$; $[SO_4^{2-}] = 1.1 \times 10^{-8}$

$SrSO_4$: $K_{sp} = [Sr^{2+}][SO_4^{2-}] = 2.8 \times 10^{-7}$

$2.8 \times 10^{-7} = (0.010)[SO_4^{2-}]$; $[SO_4^{2-}] = 2.8 \times 10^{-5}$

The $[SO_4^{2-}]$ necessary to begin precipitation is the smaller of the two values,

$1.1 \times 10^{-8}\ M\ SO_4^{2-}$.

(b) Ba^{2+} precipitates first, because it requires the smaller $[SO_4^{2-}]$.

(c) Sr^{2+} will begin to precipitate when $[SO_4^{2-}]$ reaches 2.8×10^{-5} *M*.

17.51 First solve for $[S^{2-}]$. From equation 17.25,
$7 \times 10^{-22} = [H^+]^2[S^{2-}]$; at pH = 1.80, $[H^+] = 10^{-1.80} = 0.0158$ *M*

$$[S^{2-}] = \frac{7 \times 10^{-22}}{(0.0158)^2} = 3 \times 10^{-18}\ M$$

$$[Pb^{2+}][S^{2-}] = 8.0 \times 10^{-28};\ [Pb^{2+}] = \frac{8.0 \times 10^{-28}}{3 \times 10^{-18}} = 3 \times 10^{-10}\ M$$

17.52 First solve for $[S^{2-}]$. From equation 17.25,

$[S^{2-}] = 7 \times 10^{-22}/[H^+]^2$; at pH = 2.4, $[H^+] = 4.0 \times 10^{-3}$

$[S^{2-}] = 7 \times 10^{-22}/(4.0 \times 10^{-3})^2 = 4 \times 10^{-17}$

$$[Zn^{2+}][S^{2-}] = 1.1 \times 10^{-21};\ [Zn^{2+}] = \frac{1.1 \times 10^{-21}}{4 \times 10^{-17}} = 3 \times 10^{-5}\ M$$

17.53 (a) $Zn(OH)_2(s) + 2H^+(aq) \rightarrow Zn^{2+}(aq) + 2H_2O(l)$

OH^- is the Lewis base, H^+ is the Lewis acid.

(b) $Cd(CN)_2(s) + 2CN^-(aq) \rightarrow [Cd(CN)_4]^{2-}(aq)$

Cd^{2+} is the Lewis acid, CN^- is the Lewis base.

(c) $CrF_3(s) + 3OH^-(aq) \rightarrow Cr(OH)_3(s) + 3F^-(aq)$

Cr^{3+} is the Lewis acid, OH^- is the Lewis base

17.54 (a) $Cu^{2+}(aq) + 4CN^-(aq) \rightarrow [Cu(CN)_4]^{2-}(aq)$

Cu^{2+} is the Lewis acid, CN^- is the Lewis base

(b) $AgCl(s) + S_2O_3^{2-} \rightarrow [Ag(S_2O_3)_2]^{3-}(aq) + Cl^-(aq)$

Ag^+ is the Lewis acid, $S_2O_3^{2-}$ is the Lewis base

(c) $Cu(OH)_2(s) + 4NH_3(aq) \rightarrow [Cu(NH_3)_4]^{2+}(aq) + 2OH^-(aq)$

Cu^{2+} is the Lewis acid, NH_3 is the Lewis base

17.55 The equilibrium of importance is

$$Cu^{2+}(aq) + 4NH_3(aq) \rightleftharpoons Cu(NH_3)_4^{2+}(aq) \quad K_f = \frac{[Cu(NH_3)_4^{2+}]}{[Cu^{2+}][NH_3]^4} = 5 \times 10^{12}$$

Assuming that nearly all the Cu^{2+} is in the form $Cu(NH_3)_4^{2+}$,

$[Cu(NH_3)_4^{2+}] = 1 \times 10^{-3}$ *M*; $[Cu^{2+}] = x$; $[NH_3] = 0.10$ *M*

$$5 \times 10^{12} = \frac{(1 \times 10^{-3})}{x(0.10)^4}; \quad x = 2 \times 10^{-12}\ M = [Cu^{2+}]$$

17.56 $NiC_2O_4(s) \rightleftharpoons Ni^{2+}(aq) + C_2O_4^{2-}(aq); \quad K_{sp} = [Ni^{2+}][C_2O_4^{2-}] = 4 \times 10^{-10}$

When the salt has just dissolved, $[C_2O_4^{2-}]$ will be 0.020 M. Thus $[Ni^{2+}]$ must be less than $4 \times 10^{-10}/0.020 = 2 \times 10^{-8}$ M. To achieve this low $[Ni^{2+}]$ we must complex the Ni^{2+} ion with NH_3: $Ni^{2+}(aq) + 6NH_3(aq) \rightleftharpoons Ni(NH_3)_6^{2+}(aq)$. Essentially all Ni(II) is in the form of the complex, so $[Ni(NH_3)_6^{2+}] = 0.020$.

$$K_f = \frac{[Ni(NH_3)_6^{2+}]}{[Ni^{2+}][NH_3]^6} = \frac{(0.020)}{(2 \times 10^{-8})[NH_3]^6} = 5.5 \times 10^8; [NH_3]^6 = 1.8 \times 10^{-3}; [NH_3] = 0.35\ M$$

17.57

$$AgI(s) \rightleftharpoons Ag^+(aq) + I^-(aq)$$
$$Ag^+(aq) + 2CN^-(aq) \rightleftharpoons Ag(CN)_2^-(aq)$$

$$AgI(s) + 2CN^-(aq) \rightleftharpoons Ag(CN)_2^-(aq) + I^-(aq)$$

$$K = K_{sp} \times K_f = [Ag^+][I^-] \times \frac{[Ag(CN)_2^-]}{[Ag^+][CN^-]^2} = (8.3 \times 10^{-17})(1 \times 10^{21}) = 8 \times 10^4$$

17.58

$Ag_2S(s) \rightleftharpoons 2Ag^+(aq) + S^{2-}(aq)$	K_{sp}
$S^{2-}(aq) + 2H^+(aq) \rightleftharpoons H_2S(aq)$	$1/(K_{a1} \times K_{a2})$
$2[Ag^+(aq) + 2Cl^-(aq) \rightleftharpoons AgCl_2^-(aq)]$	K_f^2

Add: $Ag_2S(s) + 2H^+(aq) + 4Cl^-(aq) \rightarrow 2AgCl_2^-(aq) + H_2S(aq)$

$$K = \frac{K_{sp} \times K_f^2}{K_{a1} \times K_{a2}} = \frac{(6.3 \times 10^{-50})(1.1 \times 10^5)^2}{7.4 \times 10^{-21}} = 1.0 \times 10^{-19}$$

Qualitative Analysis

17.59 The first two experiments eliminate Group 1 and 2 ions (Figure 17.21). The fact that no insoluble carbonates form in the filtrate from the third experiment rules out Group 4 ions. The ions which might be in the sample are those of Group 3, that is, Al^{3+}, Fe^{2+}, Zn^{2+}, Cr^{3+}, Ni^{2+}, Co^{2+}, or Mn^{2+}, and those of Group 5, NH_4^+, Na^+ or K^+.

17.60 Initial solubility in water rules out CdS and HgO. Formation of a precipitate on addition of HCl indicates the presence of $Pb(NO_3)_2$ (formation of $PbCl_2$). Formation of a precipitate on addition of H_2S at pH 1 probably indicates $Cd(NO_3)_2$ (formation of CdS). (This test can be misleading because enough Pb^{2+} can remain in solution after filtering $PbCl_2$ to lead to visible precipitation of PbS.) Absence of a precipitate on addition of H_2S at pH 8 indicates that $ZnSO_4$ is not present. The yellow flame test indicates presence of Na^+. In summary, $Pb(NO_3)_2$ and Na_2SO_4 are definitely present, $Cd(NO_3)_2$ is probably present, and CdS, HgO and $ZnSO_4$ are definitely absent.

17.61 (a) Make the solution acidic using 0.5 M HCl; saturate with H_2S. CdS will precipitate, ZnS will not.

(b) Add excess base; $Fe(OH)_3(s)$ precipitates, but Cr^{3+} forms the soluble complex $Cr(OH)_4^-$.

(c) Add $(NH_4)_2HPO_4$; Mg^{2+} precipitates as $MgNH_4PO_4$, K^+ remains soluble.

(d) Add 6M HCl, precipitate Ag^+ as AgCl(s).

17.62 (a) Make the solution slightly basic and saturate with H_2S; CdS will precipitate, Na^+ remains in solution.

(b) Make the solution acidic, saturate with H_2S; CuS will precipitate, Mg^{2+} remains in solution.

(c) Add HCl, $PbCl_2$ precipitates. (It is best to carry out the reaction in an ice-water bath to reduce the solubility of $PbCl_2$.)

(d) Add dilute HCl; AgCl precipitates, Hg^{2+} remains in solution.

17.63 (a) Because phosphoric acid is a weak acid, the concentration of free $PO_4^{3-}(aq)$ in an aqueous phosphate solution is low except in strongly basic media. In less basic media the solubility product of the phosphates that one wishes to precipitate is not exceeded.

(b) K_{sp} for those cations in Group 3 is much larger. Thus, to exceed K_{sp} a higher $[S^{2-}]$ is required. This is achieved by making the solution more basic.

(c) They should all redissolve in strongly acidic solution, e.g., in 12 M HCl (all the chlorides of Group 3 metals are soluble).

17.64 The addition of $(NH_4)_2HPO_4$ could result in precipitation of salts from metal ions of the other groups. The $(NH_4)_2HPO_4$ will render the solution basic, so metal hydroxides could form as well as insoluble phosphates. It is essential to separate the metal ions of a group from other metal ions before carrying out the specific tests for that group.

Additional Exercises

17.65 The equilibrium of interest is

$$HC_5H_3O_3(aq) \rightleftharpoons H^+(aq) + C_5H_3O_3^-(aq);\; K_a = 6.76 \times 10^{-4} = \frac{[H^+][C_5H_3O_3^-]}{[HC_5H_3O_3]}$$

Begin by calculating $[HC_5H_3O_3]$ and $[C_5H_3O_3^-]$ for each case.

(a) $$\frac{35.0 \text{ g } HC_5H_3O_3}{0.250 \text{ L soln}} \times \frac{1 \text{ mol } HC_5H_3O_3}{112.1 \text{ g } HC_5H_3O_3} = 1.25\ M\ HC_5H_3O_3$$

$$\frac{30.0 \text{ g } NaC_5H_3O_3}{0.250 \text{ L soln}} \times \frac{1 \text{ mol } NaC_5H_3O_3}{134.1 \text{ g } NaC_5H_3O_3} = 0.895\ M\ C_5H_3O_3^-$$

$$[H^+] = \frac{K_a[HC_5H_3O_3]}{[C_5H_3O_3^-]} = \frac{6.76 \times 10^{-4}\ (1.25 - x)}{(0.895 + x)} \approx \frac{6.76 \times 10^{-4}\ (1.25)}{(0.895)}$$

$[H^+] = 9.44 \times 10^{-4}\ M$, pH = 3.02

(b) For dilution, $M_1V_1 = M_2V_2$

$$[HC_5H_3O_3] = \frac{0.250\ M \times 30.0 \text{ mL}}{125 \text{ mL}} = 0.0600\ M$$

$$[C_5H_3O_3^-] = \frac{0.220\ M \times 20.0 \text{ mL}}{125 \text{ mL}} = 0.0352\ M$$

$$[H^+] \approx \frac{6.76 \times 10^{-4}\ (0.0600)}{0.0352} = 1.15 \times 10^{-3}\ M,\ \text{pH} = 2.938$$

(yes, $[H^+]$ is < 5% of 0.0352 M

(c) $0.0850\ M \times 0.500 \text{ L} = 0.0425 \text{ mol } HC_5H_3O_3$

$1.65\ M \times 0.0500 \text{ L} = 0.0825 \text{ mol NaOH}$

	$HC_5H_3O_3$	+ NaOH	→ $NaC_5H_3O_3$	+ H_2O
initial	0.0425 mol	0.0825 mol		
reaction	-0.0425 mol	-0.0425 mol	+ 0.0425 mol	
after	0 mol	0.0400 mol	0.0425 mol	

The strong base NaOH dominates the pH; the contribution of $C_5H_3O_3^-$ is negligible. This combination would be "after the equivalence point" of a titration. The total volume is 0.550 L.

$$[OH^-] = \frac{0.0400 \text{ mol}}{0.550 \text{ L}} = 0.0727\ M;\ \text{pOH} = 1.138,\ \text{pH} = 12.862$$

17.66 From Equation [17.13] we see that when the acid and base forms are present in equal concentrations the pH = pK_a. Thus $pK_a = 7.80$.

17.67 $K_a = \frac{[H^+][In^-]}{[HIn]}$; at pH = 4.68, [HIn] = [In^-] ; [H^+] = K_a; pH = pK_a = 4.68

17.68 (a) For a monoprotic acid (one H^+ per mole of acid), at the equivalence point moles OH^- added = moles H^+ originally present

$$M_B \times L_B = \text{g acid/molar mass}$$

$$M = \frac{\text{g acid}}{M_B \times L_B} = \frac{0.1355 \text{ g}}{0.0950\ M \times 0.0193 \text{ L}} = 73.9 \text{ g/mol}$$

(b) $$\text{initial mol HA} = \frac{0.1355 \text{ g}}{73.9 \text{ g/mol}} = 1.83 \times 10^{-3} \text{ mol HA}$$

mol OH^- added to pH 5.10 = 0.0950 $M \times 0.012$ L = 1.14×10^{-3} mol OH^-

	HA(aq)	+ NaOH(aq)	→ NaA(aq) + $H_2O(l)$
before rx	1.83×10^{-3} mol	1.14×10^{-3} mol	0
change	-1.14×10^{-3} mol	-1.14×10^{-3} mol	$+1.14 \times 10^{-3}$ mol
after rx	0.69×10^{-3} mol	0	1.14×10^{-3} mol

$$[HA] = \frac{0.69 \times 10^{-3} \text{ mol}}{0.037 \text{ L}} = 0.0186\ M\ ; \quad [A^-] = \frac{1.14 \times 10^{-3} \text{ mol}}{0.037 \text{ L}} = 0.0308\ M$$

$$[H^+] = 10^{-5.10} = 7.94 \times 10^{-6}\ M$$

The mixture after reaction (a buffer) can be described by the acid dissociation equilibrium

	HA(aq)	⇌ H^+(aq)	+ A^-(aq)
initial	0.0186 M	0	0.0308 M
equil.	$(0.0186 - 7.94 \times 10^{-6}\ M)$	$7.94 \times 10^{-6}\ M$	$(0.0308 + 7.94 \times 10^{-6})\ M$

$$K_a = \frac{[H^+][A^-]}{[HA]} \approx \frac{(7.94 \times 10^{-6})(0.0308)}{(0.0186)} = 1.3 \times 10^{-5}$$

(Although we have carried 3 figures through the calculation to avoid rounding errors, the data indicate an answer with 2 significant figures.)

17.69 (a) $\frac{0.4885 \text{ g KHP}}{0.100 \text{ L}} \times \frac{1 \text{ mol KHP}}{204.2 \text{ g KHP}} = 0.02392\ M\ P^{2-}$ at the equivalence point

The pH at the equivalence point is determined by the hydrolysis of P^{2-}.

$P^{2-}(aq) + H_2O(l) \rightleftharpoons HP^-(aq) + OH^-(aq)$

$$K_b = \frac{[HP^-][OH^-]}{[P^{2-}]} = \frac{K_w}{K_a \text{ for } HP^-} = \frac{1.0 \times 10^{-14}}{3.1 \times 10^{-6}} = 3.2 \times 10^{-9}$$

$$3.2 \times 10^{-9} = \frac{x^2}{(0.02392 - x)} \approx \frac{x^2}{0.02392}; \ X = [OH^-] = 8.8 \times 10^{-6}\ M$$

pH = 14 - 5.06 = 8.94. From Figure 16.7, either phenolphthalein (pH 8.2-10.0) or thymol blue (pH 8.0-9.6) could be used to detect the equivalence point. Phenolphthalein is usually the indicator of choice because the colorless to pink change is easier to see.

(b) $0.4885 \text{ g KHP} \times \frac{1 \text{ mol KHP}}{204.2 \text{ g KHP}} \times \frac{1 \text{ mol NaOH}}{1 \text{ mol KHP}} \times \frac{1}{0.03855 \text{ L NaOH}}$

$= 0.06206\ M$ NaOH

17.70

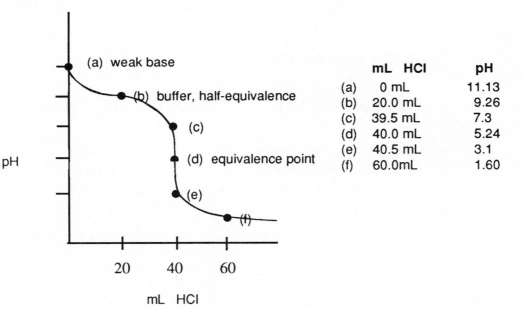

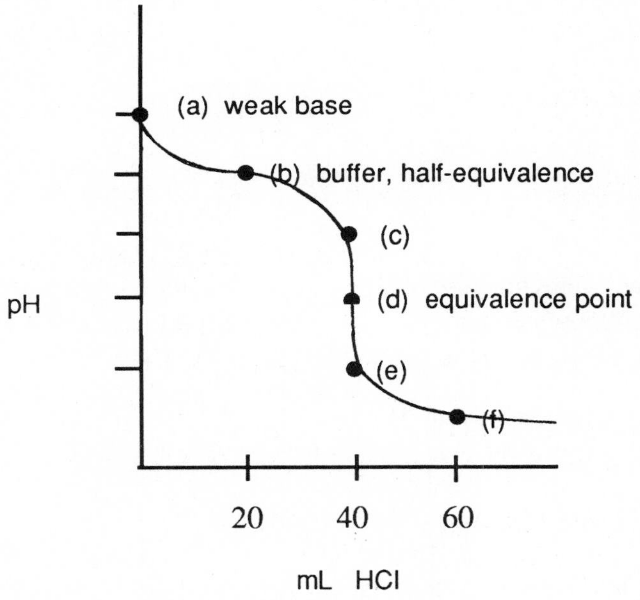

	mL HCl	pH
(a)	0 mL	11.13
(b)	20.0 mL	9.26
(c)	39.5 mL	7.3
(d)	40.0 mL	5.24
(e)	40.5 mL	3.1
(f)	60.0mL	1.60

17.71 (a) Initially, the solution is a 0.100 M in SO_3^{2-}.
$SO_3^{2-}(aq) + H_2O(l) \rightleftharpoons HSO_3^-(aq) + OH^-$
Proceeding in the ususal way for a weak base, calculate pH = 10.10.

(b) It will require 50.00 mL of 0.100 M HCl to reach the first equivalence point, at which point HSO_3^- is the predominant species.

(c) An additional 50.00 mL are required to react with this HSO_3^- to form H_2SO_3. At the second equivalence point there is a 0.0500 *M* solution of H_2SO_3. By the usual procedure for a weak acid:

$$H_2SO_3(aq) \rightleftharpoons H^+(aq) + HSO_3^-(aq)$$

$$K_a = \frac{[H^+][HSO_3^-]}{[H_2SO_3]} = 1.7 \times 10^{-2}\ ; \quad \frac{(x)^2}{(0.050-x)} = 1.7 \times 10^{-2}$$

Use the full quadratic form to solve for $x = [H^+] = 2.2 \times 10^{-2}$; pH = 1.66.

17.72 (a) $HA(aq) + B(aq) \rightleftharpoons HB^+(aq) + A^-(aq) \quad K_c = \frac{[HB^+][A^-]}{[HA][B]}$

(b) Note that the solution is slightly basic because B is a stronger base than HA is an acid. (Or, equivalently, that A^- is a stronger base than HB^+ is an acid.) Thus, a little of the A^- is used up in reaction: $A^-(aq) + H_2O(l) \rightleftharpoons HA(aq) + OH^-(aq)$. Since pH is not very far from neutral, it is reasonable to assume that the reaction in part (a) has gone far to the right, and that $[A^-] \approx [HB^+]$ and $[HA] \approx [B]$. Then

$$K_a = \frac{[A^-][H^+]}{[HA]} = 5.0 \times 10^{-6}\ ; \quad \text{when pH} = 8.8,\ [H^+] = 1.6 \times 10^{-9}\ M$$

$$\frac{[A^-]}{[HA]} = 5.0 \times 10^{-6}/1.6 \times 10^{-9} = 3.1 \times 10^3$$

From the assumptions above, $\frac{[A^-]}{[HA]} \approx \frac{[HB^+]}{[B]}$, so $K_c \approx \frac{[A^-]^2}{[HA]^2} = 9.8 \times 10^6$

(c) K_b for the reaction $B(aq) + H_2O(l) \rightleftharpoons BH^+(aq) + OH^-(aq)$ can be calculated by noting that the equilibrium constant for the reaction in part (a) can be written as $K_c = K_a(HA)K_b(B)/K_w$. (You should prove this to yourself.) Then,

$$K_b(B) = \frac{K_c \times K_w}{K_a} = \frac{(9.8 \times 10^6)(1.0 \times 10^{-14})}{5.0 \times 10^{-6}} = 2.0 \times 10^{-2}$$

$K_b(B)$ is larger than $K_a(A)$, as it must be if the solution is basic.

17.73 If the buffer contains Na_2CO_3 and $NaHCO_3$, the pertinent equilibrium is:

$$HCO_3^-(aq) \rightleftharpoons H^+(aq) + CO_3^{2-}(aq)$$

H^+ will react with the conjugate base CO_3^{2-}: $H^+(aq) + CO_3^{2-}(aq) \rightarrow HCO_3^-(aq)$

OH^- will react with the conjugate acid HCO_3^-: $OH^-(aq) + HCO_3^-(aq) \rightarrow CO_3^{2-}(aq) + H_2O(l)$

17.74 The reaction involved is $HA(aq) + OH^-(aq) \rightleftharpoons A^-(aq) + H_2O(l)$. We thus have 0.080 mol A^- and 0.12 mol HA in a total volume of 1 L, so the "initial" molarities of A^- and HA are 0.080 M and 0.12 M, respectively. The weak acid equilibrium of interest is

$$HA(aq) \rightleftharpoons H^+(aq) + A^-(aq).$$

(a) $K_a = \frac{[H^+][A^-]}{[HA]}$; $[H^+] = 10^{-4.80} = 1.6 \times 10^{-5}$ *M*

Assuming $[H^+]$ is small compared to [HA] and $[A^-]$,

$K_a \approx \frac{(1.6 \times 10^{-5})(0.080)}{(0.12)} = 1.1 \times 10^{-5}$, $pK_a = 4.98$

(b) At pH = 5.00, $[H^+] = 1.0 \times 10^{-5}$ *M*. Let b = extra moles NaOH.

$[HA] = 0.12 - b$, $[A^-] = 0.080 + b$

$1.1 \times 10^{-5} \approx \frac{(1.0 \times 10^{-5})(0.080 + b)}{0.12 - b}$; $2.1 \times 10^{-5}\ b = 5.0 \times 10^{-7}$;

b = 0.024 mol NaOH

17.75 (a) $K_a = \frac{[H^+][CHO_2^-]}{[HCHO_2]}$; $[H^+] = \frac{K_a[HCHO_2]}{[CHO_2^-]}$

Buffer A: $[HCHO_2] = [CHO_2^-] = \frac{1.00 \text{ mol}}{1.00 \text{ L}} = 1.00$ *M*

$[H^+] = \frac{1.8 \times 10^{-4}\ (1.00\ M)}{(1.00\ M)} = 1.8 \times 10^{-4}$ *M*, pH = 3.74

Buffer B: $[HCHO_2] = [CHO_2^-] = \frac{0.010 \text{ mol}}{1.00 \text{ L}} = 0.010$ *M*

$[H^+] = \frac{1.8\ .\ 10^{-4}(0.010\ M)}{(0.010\ M)} = 1.8 \times 10^{-4}$ *M*, pH = 3.74

The pH of a buffer is determined by the identity of the conjugate acid/conjugate base pair (that is, the relevant K_a value) and the *ratio of concentrations* of the conjugate acid and conjugate base. The absolute concentrations of the components is not relevant. The pH values of the two buffers are equal because they both contain $HCHO_2$ and $NaCHO_2$ and the $[HCHO_2]/[CHO_2^-]$ *ratio* is the same in both solutions.

(b) Buffer capacity is determined by the absolute amount of conjugate acid and conjugate base available to absorb strong acid (H^+) or strong base (OH^-) that is added to the buffer. Buffer A has the greater capacity because it contains the greater absolute concentrations of $HCHO_2$ and CHO_2^-.

(c) Buffer A: $CHO_2^- + HCl \longrightarrow HCHO_2 + Cl^-$

1.00 mol	0.001 mol	1.00 mol
0.999 mol	0	1.001 mol

$$[H^+] = \frac{1.8 \times 10^{-4}(1.001)}{(0.999)} = 1.8 \times 10^{-4}\ M, \quad pH = 3.74$$

(In a buffer calculation, volumes cancel and we can substitute moles directly into the K_a expression.)

Buffer B: $CHO_2^- + HCl \longrightarrow HCHO_2 + Cl^-$

0.010 mol	0.001 mol	0.010 mol
0.009 mol	0	0.011 mol

$$[H^+] = \frac{1.8 \times 10^{-4}(0.011)}{(0.009)} = 2.2 \times 10^{-4}\ M, \quad pH = 3.66$$

(d) Buffer A: $1.00\ M\ HCl \times 0.010\ L = 0.010$ mol H^+ added

mol $HCHO_2 = 1.00 + 0.010 = 1.01$ mol

mol $CHO_2^- = 1.00 - 0.010 = 0.99$ mol

$$[H^+] = \frac{1.8 \times 10^{-4}(1.01)}{(0.99)} = 1.8 \times 10^{-4}\ M, \quad pH = 3.74$$

Buffer B: mol $HCHO_2 = 0.010 + 0.010 = 0.020$ mol

mol $CHO_2^- = 0.010 - 0.010 = 0.000$ mol

The solution is no longer a buffer; the only source of CHO_2^- is the dissociation of $HCHO_2$.

$$K_a = \frac{[H^+][CHO_2^-]}{[HCHO_2]} = \frac{x^2}{(0.020\text{-}x)\ M}$$

The extent of dissociation is greater than 5%; from the quadratic formula, $x = [H^+] = 1.8 \times 10^{-3}$, pH = 2.74.

(e) Adding 10 mL of 1.00 *M* HCl to buffer B exceeded its capacity, while the pH of buffer A was unaffected. This is quantitative confirmation that buffer A has a significantly greater capacity than buffer B. In fact, 1.0 L of 1.0 *M* HCl would be required to exceed the capacity of buffer A. Buffer A, with 100 times more $HCHO_2$ and CHO_2^- has 100 times the capacity of buffer B.

17.76 The pH of a buffer is centered around pK_a for its conjugate acid. For the bases in Table D.2, pK_a for the conjugate acids = 14 - pK_b. 14 - pk_b = 10.6; pK_b = 3.4, $K_b = 10^{-3.4} = 4 \times 10^{-4}$. Select two bases with K_b values near 4×10^{-4}.

Methylamine, dimethylamine and ethylamine have K_b values closest to 4×10^{-4}, and ammonia and trimethylamine would probably also work. We will select methylamine and dimethylamine.

In general, $BH^+(aq) \rightleftharpoons B(aq) + H^+(aq)$

$$K_a = \frac{[B][H^+]}{[BH^+]}\ ;\ [H^+] = \frac{K_a[BH^+]}{[B]}\ ;\ [H^+] = 10^{-10.6} = 2.5 \times 10^{-11}\ M$$

$$\text{For methylamine, } K_a = \frac{1.0 \times 10^{-14}}{4.4 \times 10^{-4}} = 2.3 \times 10^{-11}\ ;\ \frac{[BH^+]}{[B]} = \frac{[H^+]}{K_a} = \frac{2.5 \times 10^{-11}}{2.3 \times 10^{-11}} = 1.1$$

The ratio of $[CH_3NH_3^+]$ to $[CH_3NH_2]$ is 1.1 to 1.

$$\text{For dimethylamine, } K_a = \frac{1.0 \times 10^{-14}}{5.4 \times 10^{-4}} = 1.8 \times 10^{-11}\ ;\ \frac{[BH^+]}{[B]} = \frac{[H^+]}{K_a} = \frac{2.5 \times 10^{-11}}{1.8 \times 10^{-11}} = 1.4$$

The ratio of $[(CH_3)_2NH_2^+]$ to $[(CH_3)_2NH]$ is 1.4 to 1. (The stronger base requires more of its conjugate acid to achieve a buffer of the same pH.)

17.77 Assume that H_3PO_4 will react with NaOH in a stepwise fashion. (This is not unreasonable, since the three K_a values for H_3PO_4 are significantly different.)

	$H_3PO_4(aq)$ +	$NaOH(aq)$ →	$H_2PO_4^-(aq)$ + $Na^+(aq)$ + $H_2O(l)$
before	0.20 mol	0.30 mol	0 mol
after	0 mol	0.10 mol	0.20 mol

	$H_2PO_4^-(aq)$ +	$NaOH(aq)$ →	$HPO_4^-(aq)$ + $Na^+(aq)$ + $H_2O(l)$
before	0.20 mol	0.10 mol	0.25 mol
after	0.10 mol	0	0.35 mol

Thus, after all NaOH has reacted, the resulting 1.00 L solution is a buffer containing 0.10 mol $H_2PO_4^-$ and 0.35 mol HPO_4^-. $H_2PO_4^-(aq) \rightleftharpoons H^+(aq) + HPO_4^-(aq)$

$$K_a = 6.2 \times 10^{-8} = \frac{[HPO_4^-][H^+]}{[H_2PO_4^-]}\ ;\ [H^+] = \frac{6.2 \times 10^{-8}\ (0.10\ M)}{(0.35\ M)} = 1.8 \times 10^{-8}\ M;\quad pH = 7.75$$

17.78 The pH of a buffer system is centered around pK_a for the conjugate acid component. For a diprotic acid, two conjugate acid/conjugate base pairs are possible:

$H_2X(aq) \rightleftharpoons H^+(aq) + HX^-(aq)\ ;\ K_{a1} = 2.0 \times 10^{-2},\ pK_{a1} = 1.70$

$HX^-(aq) \rightleftharpoons H^+(aq) + X^{2-}(aq)\ ;\ K_{a2} = 5.0 \times 10^{-7}\ ;\ pK_{a2} = 6.30$

Clearly HX^-/X^{2-} is the more appropriate combination for preparing a buffer with pH = 6.50. The $[H^+]$ in this buffer = $10^{-6.50} = 3.16 \times 10^{-7}\ M$. Using the K_{a2} expression to calculate the $[X^{2-}]$ / [HX] ratio:

$$K_{a2} = \frac{[H^+][X^{2-}]}{[HX^-]}\ ;\ \frac{K_{a2}}{[H^+]} = \frac{[X^{2-}]}{[HX]} = \frac{5.0 \times 10^{-7}}{3.16 \times 10^{-7}} = 1.58$$

Since X^{2-} and HX^- are present in the same solution, the ratio of concentrations is also a ratio of moles.

$$\frac{[X^{2-}]}{[HX^-]} = \left(\frac{\text{mol } X^{2-}/\text{L soln}}{\text{mol } HX^-/\text{L soln}}\right) = \frac{\text{mol } X^{2-}}{\text{mol } HX^-} = 1.58 \;;\quad \text{mol } X^{2-} = (1.58) \text{ mol } HX^-$$

In the 1.0 L of 1.0 *M* H_2X, there is 1.0 mol of X^{-2} containing material.

Thus, mol HX^- + 1.58 (mol HX^-) = 1.0 mol. 2.58 (mol HX^-) = 1.0 ;
mol HX^- = 1.0/2.58 = 0.39 mol HX^-; mol X^{2-} = 1.0 - 0.39 = 0.61 mol X^{2-}.

Thus, enough 1.0 *M* NaOH must be added to produce 0.39 mol HX^- and 0.61 mol X^{2-}.

Considering the neutralization in a step-wise fashion (see discussion of titrations of polyprotic acids in Section 17.2),

	H_2X	+ NaOH	→ HX^-	+ H_2O
before	1.0 mol	1 mol	0	
after	0	0	1.0 mol	

	HX^-	+ NaOH	→ X^{2-}	+ H_2O
before	1.0	0.61		
charge	-0.61	-0.61	+0.61	
after	0.39	0	0.61	

Starting with 1.0 mol of H_2X, 1.0 mol of NaOH is added to completely convert it to 1.0 mol of HX^-. Of that 1.0 mol of HX^-, 0.61 mol must be converted to 0.61 mol X^{2-}. The total moles of NaOH added is (1.00 + 0.61) = 1.61 mol NaOH.

$$\text{L NaOH} = \frac{\text{mol NaOH}}{M \text{ NaOH}} = \frac{1.61 \text{ mol}}{1.0\ M} = 1.6 \text{ L of } 1.0\ M \text{ NaOH}$$

17.79 The simplest way to obtain the desired buffer mixture would be to titrate the H_3PO_4 solution with NaOH, using a pH meter, until the pH had risen to 7.20. From the values of K_a for phosphoric acid, we can guess that the equilibrium of importance will be the second acid dissociation.

$$H_2PO_4^-(aq) \rightleftharpoons HPO_4^{2-}(aq) + H^+(aq);\ K_a = 6.2 \times 10^{-8},\ pK_a = 7.21$$

Using Equation [17.13]

$$7.20 = 7.21 + \log\frac{[HPO_4^{2-}]}{[H_2PO_4^-]} \;;\quad \frac{[HPO_4^{2-}]}{[H_2PO_4^-]} = 0.97$$

That is, at pH 7.20 we have added enough 1 *M* NaOH to the solution to have produced approximately equal concentrations of the two anions $H_2PO_4^-$ and HPO_4^{2-}. If we had begun with 100 mL of the 1 *M* H_3PO_4 solution we would have added about 150 mL of 1 M NaOH, so the total volume of solution would be 250 mL. The total moles phosphate present = 1.0 M × 0.1 L = 0.1 mol, so mol $H_2PO_4^-$ = mol HPO_4^{2-} = 0.05 mol. The concentration of each component is 0.05 mol/0.25 L = 0.20 *M*.

17.80 $\frac{0.30 \text{ mol } HC_2H_3O_2}{1 \text{ L soln}} \times 0.750 \text{ L} = 0.22 \text{ mol } HC_2H_3O_2$

$$0.22 \text{ mol } HC_2H_3O_2 \times \frac{60 \text{ g } HC_2H_3O_2}{1 \text{ mol } HC_2H_3O_2} \times \frac{1 \text{ g gl acetic acid}}{0.99 \text{ g } HC_2H_3O_2}$$

$$\times \frac{1.00 \text{ mL gl acetic acid}}{1.05 \text{ g gl acetic acid}} = 13 \text{ mL glacial acetic acid}$$

At pH 4.50, $[H^+] = 10^{-4.50} = 3.2 \times 10^{-5}$ *M*; this is small compared to 0.30 M $HC_2H_3O_2$.

$$K_a = \frac{(3.2 \times 10^{-5})[C_2H_3O_2^-]}{0.30} = 1.8 \times 10^{-5};\ [C_2H_3O_2^-] = 0.17\ M$$

$$\frac{0.17 \text{ mol } NaC_2H_3O_2}{1 \text{ L soln}} \times 0.750 \text{ L} \times \frac{82 \text{ g } NaC_2H_3O_2}{1 \text{ mol } NaC_2H_3O_2} = 10 \text{ g } NaC_2H_3O_2$$

17.81 $C_3H_5O_3^-$ will be formed by reaction of $HC_3H_5O_3$ with NaOH.

$0.1000\ M \times 0.05000 \text{ L} = 5.000 \times 10^{-3}$ mol $HC_3H_5O_3$; b = mol NaOH needed

	$HC_3H_5O_3$	+ NaOH	$\rightarrow C_3H_5O_3^-$	+ H_2O + Na^+
initial	5.000×10^{-3}	b mol		
rx	-b mol	-b mol	+ b mol	
after rx	5.000×10^{-3} - b mol	0	b mol	

$$K_a = \frac{[H^+][C_3H_5O_3^-]}{[HC_3H_5O_3]};\ K_a = 1.4 \times 10^{-4};\ [H^+] = 10^{-pH} = 10^{-3.50} = 3.16 \times 10^{-4}\ M$$

Since solution volume is the same for $HC_3H_5O_3$ and $C_3H_5O_3^-$, we can use moles in the equation for $[H^+]$.

$$K_a = 1.4 \times 10^{-4} = \frac{3.16 \times 10^{-4}\ (b)}{(5.000 \times 10^{-3} - b)}$$

$0.4427\ (5.000 \times 10^{-3} - b) = b,\ 2.214 \times 10^{-3} = 1.4427b,\ b = 1.5 \times 10^{-3}$ mol OH^-

(The precision of K_a dictates that the result has 2 sig. figs.)

Substituting this result into the K_a expression gives $[H^+] = 3.27 \times 10^{-4}$.

Using 1.53×10^{-3} mol OH^- (3 sig. figs) gives, $[H^+] = 3.16 \times 10^{-4}$, a more reassuring cross check.)

Calculate volume NaOH required from *M* = mol/L.

$$1.5 \times 10^{-3} \text{ mol } OH^- \times \frac{1 \text{ L}}{1.000 \text{ mol}} \times \frac{1\ \mu L}{1 \times 10^{-6} \text{ L}} = 1.5 \times 10^3\ \mu L\ (1.5 \text{ mL})$$

17.82 $K_{sp} = 1.9 \times 10^{-13} = [Mn^{2+}][OH^-]^2$; $x = [Mn^{2+}]$, $2x = [OH^-]$

$1.9 \times 10^{-13} = (x)(2x)^2$; $4x^3 = 1.9 \times 10^{-13}$, $x^3 = 4.75 \times 10^{-14}$, $x = 3.6 \times 10^{-5}$

$[OH^-] = 2x = 7.2 \times 10^{-5}$, pOH = 4.14, pH = 9.86

(OH^- from the autoionization of H_2O is small compared to 7.2×10^{-5}.)

17.83 (a) CdS: 8.0×10^{-27}; CuS: 6.3×10^{-36}
CdS has greater molar solubility.

(b) $PbCO_3$: 7.4×10^{-14}; $BaCrO_4$: 1.2×10^{-10}
$BaCrO_4$ has greater molar solubility.

(c) $Hg(OH)_2$: 3.0×10^{-26} $Fe(OH)_3$: 4×10^{-38}

Since the stoichiometry of the two complexes is not the same, K_{sp} values can't be compared directly; molar solubilities must be calculated from K_{sp} values. For both compounds, $[OH^-]$ from the autoionization of water is large compared to $[OH^-]$ from dissociation of the salt.

$Hg(OH)_2$: $K_{sp} = 3.0 \times 10^{-26} = [Hg^{2+}][OH^-]^2$, $[Hg^{2+}] = x$, $[OH^-] = 1.0 \times 10^{-7}$

$3.0 \times 10^{-26} = (x)(1.0 \times 10^{-7})^2$, $x = 3.0 \times 10^{-12}$ *M*

$Fe(OH)_3$: $K_{sp} = 4 \times 10^{-38} = [Fe^{3+}][OH^-]^3$, $[Fe^{3+}] = x$, $[OH^-] = 1.0 \times 10^{-7}$

$4 \times 10^{-38} = (x)(1.0 \times 10^{-7})^3$, $x = 4 \times 10^{-17}$ *M*

$Hg(OH)_2$ has greater molar solubility than $Fe(OH)_3$.

(d) Again, molar solubilities must be calculated for comparison.

$Ba_3(PO_4)_2$: $K_{sp} = 3.4 \times 10^{-23} = [Ba^{2+}]^3[PO_4^{3-}]^2$

x = molar solubility, $[Ba^{2+}] = 3x$, $[PO_4^{3-}] = 2x$

$3.4 \times 10^{-23} = (3x)^3(2x)^2 = 36x^5$, $x = 1.6 \times 10^{-5}$ *M*

NiC_2O_4: $K_{sp} = 4 \times 10^{-10} = [Ni^{2+}][C_2O_4^{2-}]$, $x = [Ni^{2+}] = [C_2O_4^{2-}]$

$4 \times 10^{-10} = x^2$, $x = 2 \times 10^{-5}$ *M*

NiC_2O_4 has slightly greater molar solubility than $Ba_3(PO_4)_2$, but the solubilities are much closer than one might expect from a cursory inspection of the K_{sp} values.

17.84 $PbSO_4(s) \rightleftharpoons Pb^{2+}(aq) + SO_4^{2-}(aq)$; $K_{sp} = 1.6 \times 10^{-8} = [Pb^{2+}][SO_4^{2-}]$

$SrSO_4(s) \rightleftharpoons Sr^{2+}(aq) + SO_4^{2-}(aq)$; $K_{sp} = 2.8 \times 10^{-7} = [Sr^{2+}][SO_4^{2-}]$

Let $x = [Pb^{2+}]$, $y = [Sr^{2+}]$, $x + y = [SO_4^{2-}]$

$$\frac{x(x+y)}{y(x+y)} = \frac{1.6 \times 10^{-8}}{2.8 \times 10^{-7}}; \ \frac{x}{y} = 0.057; \ x = 0.057\,y$$

$y(0.057\,y + y) = 2.8 \times 10^{-7}$; $1.057\,y^2 = 2.8 \times 10^{-7}$; $y = 5.1 \times 10^{-4}$

$x = 0.057\,y$; $x = 0.057(5.1 \times 10^{-4}) = 2.9 \times 10^{-5}$

$[Pb^{2+}] = 2.9 \times 10^{-5}\ M$, $[Sr^{2+}] = 5.1 \times 10^{-4}\ M$, $[SO_4^{2-}] = 5.4 \times 10^{-4}\ M$

17.85 To determine precipitation conditions, we must know K_{sp} for $CaF_2(s)$ and calculate Q under the specified conditions. $K_{sp} = 3.9 \times 10^{-11} = [Ca^{2+}][F^-]^2$

$[Ca^{2+}]$ and $[F^-]$: The term 1 ppb means 1 part per billion or 1 g solute per billion g solution. Assuming that the density of this very diulte solution is the density of water:

$$1 \text{ ppb} = \frac{1 \text{ g solute}}{1 \times 10^9 \text{ g solution}} \times \frac{1 \text{ g solution}}{1 \text{ mL solution}} \times \frac{1 \times 10^3 \text{ mL}}{1 \text{ L}} = \frac{1 \times 10^{-6} \text{ g solute}}{1 \text{ L solution}}$$

$$\frac{1 \times 10^{-6} \text{ g solute}}{1 \text{ L solution}} \times \frac{1\ \mu g}{1 \times 10^{-6} \text{ g}} = \frac{1\ \mu g}{1 \text{ L}}$$

$$8 \text{ ppb } Ca^{2+} \times \frac{1\ \mu g}{L} = \frac{8\ \mu g\ Ca^{2+}}{1 \text{ L}} = \frac{8 \times 10^{-6} \text{ g } Ca^{2+}}{1 \text{ L}} \times \frac{1 \text{ mol } Ca^{2+}}{40 \text{ g}} = 2 \times 10^{-7}\ M\ Ca^{2+}$$

$$1 \text{ ppb } F^- \times \frac{1\ \mu g}{L} = \frac{1\ \mu g\ F^-}{1 \text{ L}} = \frac{1 \times 10^{-6} \text{ g } F^-}{1 \text{ L}} \times \frac{1 \text{ mol } F^-}{19.0 \text{ g}} = 5 \times 10^{-8}\ M\ F^-$$

$Q = [Ca^{2+}][F^-]^2 = (2 \times 10^{-7})(5 \times 10^{-8})^2 = 5 \times 10^{-22}$

$5 \times 10^{-22} < 3.9 \times 10^{-11}$, $Q < K_{sp}$, no CaF_2 will precipitate

17.86 $MgC_2O_4(s) \rightleftharpoons Mg^{2+}(aq) + C_2O_4^{2-}(aq)$

$K_{sp} = [Mg^{2+}][C_2O_4^{2-}] = 8.6 \times 10^{-5}$

If $[Mg^{2+}]$ is to be $3.0 \times 10^{-2}\ M$, $[C_2O_4^{2-}] = 8.6 \times 10^{-5}/3.0 \times 10^{-2} = 2.9 \times 10^{-3}\ M$

The oxalate ion undergoes hydrolysis:

$C_2O_4^{2-}(aq) + H_2O(l) \rightleftharpoons HC_2O_4^-(aq) + OH^-(aq)$

$$K_b = \frac{[HC_2O_4^-][OH^-]}{[C_2O_4^{2-}]} = 1 \times 10^{-14}/6.4 \times 10^{-5} = 1.6 \times 10^{-10}$$

If $[Mg^{2+}] = 3.0 \times 10^{-2} M$, and $[C_2O_4^{2-}]$ is 2.9×10^{-3} *M*, then

$[HC_2O_4^-] = (3.0 \times 10^{-2} - 2.9 \times 10^{-3})\ M = 2.7 \times 10^{-2}$ *M*, then

$$[OH^-] = 1.6 \times 10^{-10} \frac{[C_2O_4^{2-}]}{[HC_2O_4^-]} = 1.6 \times 10^{-10} \frac{(2.9 \times 10^{-3})}{(2.7 \times 10^{-2})} = 1.7 \times 10^{-11};\quad pH = 3.24$$

17.87 $CuCO_3(s) \rightleftharpoons Cu^{2+}(aq) + CO_3^{2-}(aq)$

The carbonate ion undergoes hydrolysis: $CO_3^{2-} + H_2O(l) \rightleftharpoons HCO_3^-(aq) + OH^-(aq)$

$$K_b = \frac{[HCO_3^-][OH^-]}{[CO_3^{2-}]} = 1.8 \times 10^{-4}$$

When pH = 7.5, $[OH^-] = 3.2 \times 10^{-7}$ *M*. Then, $\frac{[HCO_3^-]}{[CO_3^{2-}]} = 1.8 \times 10^{-4}/3.2 \times 10^{-7} = 5.6 \times 10^2$

This means that only one in 560 of the CO_3^{2-} ions produced when $CuCO_3$ dissolves remain as CO_3^{2-}. The others form HCO_3^-. But a free Cu^{2+} ion is formed for each total CO_3^{2-} or HCO_3^- ion. Thus, $[Cu^{2+}] = 560[CO_3^{2-}]$; $(x)(560x) = K_{sp} = 1.4 \times 10^{-10}$; $x = 5.0 \times 10^{-7}$ *M* Solubility = $[Cu^{2+}] = 560x = 2.8 \times 10^{-4}$ *M*.

17.88 $K_{sp} = [Pb^{2+}][Cl^-]^2 = 1.6 \times 10^{-5}$. When $[Cl^-] = 0.10$ *M*, $[Pb^{2+}] = 1.6 \times 10^{-3}$ *M*. Using Equation [17.25], we find that when pH = 1 (that is, $[H^+] = 0.10$ *M*), $[S^{2-}] = 7 \times 10^{-20}$ *M*. K_{sp} for PbS = $[Pb^{2+}][S^{2-}] = 8.0 \times 10^{-28}$. $Q = (1.6 \times 10^{-3})(7 \times 10^{-20}) = 1 \times 10^{-22}$. $Q > K_{sp}$ and PbS will precipitate from the solution. Normally, rather concentrated Cl^- solutions are used to precipitate Pb^{2+} and the other metal ions of Group 1, Figure 17.21. Even though K_{sp} may be exceeded later when H_2S is added, the quantity of Pb^{2+} present is so slight that a precipitate would not usually be noticeable.

17.89 The student erred in failing to take account of the hydrolysis of the AsO_4^{3-} ion. If there were no hydrolysis $[Mg^{2+}]$ would indeed be 1.5 times that of $[AsO_4^{3-}]$ However, as the reaction $AsO_4^{3-}(aq) + H_2O(l) \rightleftharpoons HAsO_4^{2-}(aq) + OH^-(aq)$ proceeds, the ion product $[Mg^{2+}]^3[AsO_4^{3-}]^2$ falls below the value for K_{sp}. More $Mg_3(AsO_4)_2$ dissolves, more hydrolysis occurs, and so on, until an equilibrium is reached. At this point $[Mg^{2+}]$ is much larger than $[AsO_4^{3-}]$. However, it is exactly 1.5 times the total concentration of all arsenic-containing species. That is,

$$[Mg^{2+}] = 1.5\,([AsO_4^{3-}] + [HAsO_4^{2-}] + [H_2AsO_4^-] + [H_3AsO_4])$$

17.90 At pH = 2.0, $[S^{2-}] = (7 \times 10^{-22})/(1 \times 10^{-2})^2 = 7 \times 10^{-18}$ *M*

For HgS, $Q = (7 \times 10^{-18})(0.020) = 1.4 \times 10^{-19} > K_{sp}$

For NiS, $Q = (7 \times 10^{-18})(0.020) = 1.4 \times 10^{-19} < K_{sp}$

For PbS, $Q = (7 \times 10^{-18})(0.020) = 1.4 \times 10^{-19} > K_{sp}$

Thus, we expect HgS and PbS to precipitate, but not NiS.

17.91 $[Hg^{2+}][S^{2-}] = 4 \times 10^{-53}$; $[S^{2-}] = \frac{4 \times 10^{-53}}{0.010\ M} = 4 \times 10^{-53}\ M$

$[S^{2-}]$ in excess of 4×10^{-53} *M* will cause HgS to precipitate.

$[Fe^{2+}][S^{2-}] = 6.3 \times 10^{-18}$; $[S^{2-}] = \frac{6.3 \times 10^{-18}}{0.010} = 6.3 \times 10^{-16}\ M$

$[S^{2-}]$ in excess of 6.3×10^{-16} *M* will cause FeS to precipitate. If the pH is adjusted so that $4 \times 10^{-53}\ M < [S^{2-}] < 6.3 \times 10^{-16}\ M$, HgS will precipitate and FeS will not.

According to Equation [17.25], $[H^+]^2[S^{2-}] = 7 \times 10^{-22}$

$[H^+]^2 = \frac{7 \times 10^{-22}}{4 \times 10^{-53}}$, $[H^+] = 4 \times 10^{15}$, pH is < 0

$[H^+]^2 = \frac{7 \times 10^{-22}}{6.3 \times 10^{-16}}$, $[H^+] = 1 \times 10^{-3}$, pH = 3.0

If 0 < pH < 3.0, HgS will selectively precipitate.

17.92 (a) $K_{sp} = [Fe^{3+}][OH^-]^3 = 4 \times 10^{-38}$, $[Fe^{3+}] = x$, $[OH^-] = 1 \times 10^{-10}$

$(x)(1 \times 10^{-10})^3 = 4 \times 10^{-38}$, $x = 4 \times 10^{-8}\ M$

$\frac{4 \times 10^{-8} \text{ mol } Fe(OH)_3}{1 \text{ L}} \times \frac{107 \text{ g } Fe(OH)_3}{1 \text{ mol } Fe(OH)_3} = \frac{4 \times 10^{-6} \text{ g } Fe(OH)_3}{\text{L}}$

(b) $[Fe^{3+}] = x$, $[OH^-] = 1 \times 10^{-4}$

$(x)(1 \times 10^{-4})^3 = 4 \times 10^{-38}$, $x = 4 \times 10^{-26}\ M$

$\frac{4 \times 10^{-26} \text{ mol } Fe(OH)_3}{1 \text{ L}} \times \frac{107 \text{ g } Fe(OH)_3}{1 \text{ mol } Fe(OH)_3} = \frac{4 \times 10^{-24} \text{ g } Fe(OH)_3}{\text{L}}$

$Fe(OH)_3$ is much more soluble at the lower pH value.

17.93 (a) $K_{sp} = [Cr^{3+}][OH^-]^3 = 6.3 \times 10^{-31}$, $[Cr^{3+}] = x$, $[OH^-] = 1 \times 10^{-7}$

$(x)(1 \times 10^{-7})^3 = 6.3 \times 10^{-31}$, $x = 6.3 \times 10^{-10}\ M$

(b) In pure water, $[Cr^{3+}] = x$, $[OH^-] = 3x$

$x(3x)^3 = 6.3 \times 10^{-31}$, $27x^4 = 6.3 \times 10^{-31}$, $x = [Cr^{3+}] = 1.2 \times 10^{-8}\ M$

$[OH^-] = 3.6 \times 10^{-8}\ M$

But this is on the order of magnitude of the contribution from H_2O.

$[OH^-] = 1.0 \times 10^{-7} + 3x$; $(x)(1.0 \times 10^{-7} + 3x)^3 = 6 \times 10^{-31}$

Since this is a high order equation, try successive approximation.

Begin with $x = 1.0 \times 10^{-8}$

$(1.0 \times 10^{-8})(1.0 \times 10^{-7} + 0.3 \times 10^{-7})^3 = (1.0 \times 10^{-8})(1.3 \times 10^{-7})^3 = 2.2 \times 10^{-29}$

Try 1.0×10^{-9} $(1.0 \times 10^{-9})(1.03 \times 10^{-7})^3 = 1.1 \times 10^{-30}$

Try 6.0×10^{-10} $(6.0 \times 10^{-10})(1.018 \times 10^{-7}) = 6.3 \times 10^{-31}$

The $[OH^-]$ from the autoionization of water, 1.0×10^{-7}, decreases the solubility of $Cr(OH)_3$ enough so the $[OH^-]$ from the dissociation of the solid is small compared to 1.0×10^{-7}. The solubility in pure water is essentially the same as the solubility in a solution buffered to pH 7.

17.94 The two competing equilibria are

$$Au^+(aq) + 2CN^-(aq) \rightleftharpoons Au(CN)_2^-(aq) \qquad K_f = 2.0 \times 10^{38}$$

$$AuCl(s) \rightleftharpoons Au^+(aq) + Cl^-(aq) \qquad K_{sp} = 2.0 \times 10^{-13}$$

$$AuCl(s) + 2CN^-(aq) \rightleftharpoons Au(CN)_2^-(aq) + Cl^-(aq) \qquad K = K_f \times K_{sp} = 4.0 \times 10^{25}$$

K is large and the reaction essentially goes to completion.

$$0.080\ M \times 1.0\ L = 0.080\ mol\ CN^- \times \frac{1\ mol\ AuCl}{2\ mol\ CN^-} = 0.040\ mol\ AuCl(s)$$

17.95

$$Zn(OH)_2(s) \rightleftharpoons Zn^{2+}(aq) + 2OH^-(aq) \qquad K_{sp} = 1.2 \times 10^{-17}$$

$$Zn^{2+}(aq) + 4OH^-(aq) \rightleftharpoons Zn(OH)_4^{2-}(aq) \qquad K_f = 4.6 \times 10^{17}$$

$$Zn(OH)_2(s) + 2OH^-(aq) \rightleftharpoons Zn(OH)_4^{2-}(aq) \qquad K = K_{sp} \times K_f = 5.5$$

$$K = 5.5 = \frac{[Zn(OH)_4^{2-}]}{[OH^-]^2}$$

If 0.010 mol $Zn(OH)_2$ dissolves, 0.010 mol $Zn(OH)_4^{2-}$ should be present at equilibrium.

$[OH^-]^2 = \frac{(0.010)}{5.5}$; $[OH^-] = 0.043\ M$ $[OH^-] \geq 0.043\ M$ or $pH \geq 12.63$

CHAPTER 18

Chemistry of the Enivronment

Earth's Atmosphere

18.1 (a) The temperature profile of the atmosphere (Figure 18.1) is the basis of its division into regions. The center of each peak or trough in the temperature profile corresponds to a new region.

(b) Troposphere, 0-12 km; stratosphere, 12-50 km; mesosphere, 50-85 km; thermosphere, 85-110 km.

18.2 (a) The troposphere is the 12 km layer of the atmosphere immediately above the earth's surface. The tropopause is the boundary between the troposphere and the stratosphere. In the troposhere, temperature decreases with increasing altitude while in the stratosphere, it increases with increasing altitude. The tropopause is the dividing point, the first minimum in the temperature profile of the atmosphere.

(b) The pressure of the atmosphere is greater in the mesosphere than in the thermosphere. Pressure decreases with increasing altitude throughout the atmosphere, and the mesosphere is at lower altitude than the thermosphere.

18.3 $P_{Ar} = \chi_{Ar} \cdot P_{atm}$; $P_{Ar} = 0.00934$ (765 torr) = 7.15 torr

$P_{Ne} = \chi_{Ne} \cdot P_{atm}$; $P_{Ne} = 0.00001818$(765 torr) = 0.0139 torr

18.4 $P_{Xe} = \chi_{Xe} \cdot P_{atm}$; $P_{Xe} = 8.7 \times 10^{-8}(0.94 \text{ atm}) = 8.2 \times 10^{-8}$ atm

$$n_{Xe} = \frac{P_{Xe} V}{RT} = \frac{8.2 \times 10^{-8} \text{ atm} \times 1.0 \text{ L}}{300 \text{ K}} \times \frac{\text{K} \cdot \text{mol}}{0.0821 \text{ L} \cdot \text{atm}} = 3.3 \times 10^{-9} \text{ mol Xe}$$

$$3.3 \times 10^{-9} \text{ mol Xe} \times \frac{6.022 \times 10^{23} \text{ atoms}}{1 \text{ mol}} = 2.0 \times 10^{15} \text{ Xe atoms}$$

18.5 $\frac{2.10 \times 10^3 \text{ J}}{1 \text{ mol}} \times \frac{1 \text{ mol}}{6.02 \times 10^{23} \text{ molecules}} = 3.49 \times 10^{-19}$ J/molecule

$\lambda = c/\nu$ We also have that $E = h\nu$, so $\nu = E/h$. Thus,

$$\lambda = \frac{hc}{E} = \frac{(6.63 \times 10^{-34} \text{ J-sec})(3.00 \times 10^8 \text{ m/sec})}{3.49 \times 10^{-19} \text{ J}} = 5.69 \times 10^{-7} \text{ m} = 569 \text{ nm}$$

18.6 $\frac{339 \times 10^3 \text{ J}}{1 \text{ mol}} \times \frac{1 \text{ mol}}{6.02 \times 10^{23} \text{ molecules}} = 5.63 \times 10^{-19}$ J/molecule

$$\lambda = \frac{hc}{E} = \frac{(6.63 \times 10^{-34} \text{ J} \cdot \text{sec})(3.00 \times 10^8 \text{ m/sec})}{5.63 \times 10^{-19} \text{ J}} = 353 \text{ nm}$$

$$\frac{293 \times 10^3 \text{ J}}{1 \text{ mol}} \times \frac{1 \text{ mol}}{6.02 \times 10^{23} \text{ molecules}} = 4.87 \times 10^{-19} \text{ J/molecule}$$

$$\lambda = \frac{(6.63 \times 10^{-34} \text{ J} \cdot \text{sec})(3.00 \times 10^8 \text{ m/sec})}{4.87 \times 10^{-19} \text{ J}} = 408 \text{ nm}$$

Photons of wavelengths longer than 408 nm cannot cause rupture of the C-Cl bond in either CF_3Cl or CCl_4. Photons with wavelengths between 408 and 353 nm can cause C-Cl bond rupture in CCl_4, but not in CF_3Cl.

18.7 The bond dissociation energy of N_2, 947 kJ/mol, is much higher than that of O_2, 495 kJ/mol. Photons with a wavelength short enough to photodissociate N_2 are not as abundant as the ultraviolet photons which lead to photodissociation of O_2. Also, N_2 does not absorb these photons as readily as O_2 so even if a short-wavelength photon is available, it may not be absorbed by an N_2 molecule.

18.8 Photoionization of O_2 requires 1205 kJ/mol. Photodissociation requires only 495 kJ/mol (Table 8.4). At lower elevations solar radiation with wavelengths corresponding to 1205 kJ/mol or shorter has already been absorbed, while the longer wavelength radiation has passed through relatively well. Below 90 km, the increased concentration of O_2 and the availability of longer wavelength radiation cause the photodissociation process to dominate.

Chemistry of the Stratosphere

18.9 (a) Oygen atoms exist longer at 120 km because there are fewer particles (atoms and molecules) at this altitude and thus fewer collisions and subsequent reactions that consume O atoms.

(b) Ozone is the primary absorber of high energy ultraviolet radiation in the 200-310 nm range. If this radiation were not absorbed in the stratosphere, plants and animals at the earth's surface would be seriously and adversely affected.

18.10 (a) The highest rate of ozone, O_3, formation occurs at about 50 km, near the stratopause. The formation of ozone is an exothermic process as M* carries excess energy away from the O_3 molecule. The heat energy from the formation of O_3 causes the temperature to be higher near the stratopause than the lower altitude tropopause.

(b) The first step in the formation of O_3 is the photodissociation of O_2 to form two O atoms. Then, an O atom and an O_2 molecule collide to form O_3^*, a species with excess energy. If no other collisions occur, O_3^* spontaneously decomposes. If a carrier molecule such as N_2 or O_2 collides with O_3^* and removes the excess energy, O_3 is formed. It is the energy carried by M* that contributes to the temperature maximum at 50 km altitude.

18.11 $NO(g) + O_3(g) \rightarrow NO_2(g) + O_2(g)$ ΔH = 33.8 kJ - 142.3 kJ - 90.4 kJ = -198.9 kJ
$NO_2(g) + O(g) \rightarrow O_2(g) + NO(g)$ ΔH = 90.4 kJ - 247.5 kJ - 33.8 kJ = -190.9 kJ

18.12 For reaction 1: ΔH = 102 - 101 - (142) = -141 kJ/mol
For reaction 2: ΔH = 101 - 102 - (247) = -248 kJ/mol

Because both reactions are distinctly exothermic it is possible that the ClO-ClO_2 pair could be a catalyst for the decomposition of ozone.

18.13 (a) HCl(g), $ClONO_2(g)$

(b) Neither HCl nor $ClONO_2$ react directly with ozone. The chlorine that is present in the "chlorine reservoir" does not participate in the destruction of ozone. Thus, the larger the "chlorine reservoir," the slower the rate of ozone depletion.

18.14 In the presence of polar stratospheric clouds during the Antarctic winter, HCl and $ClONO_2$ react to form Cl_2 (Equation 18.13):

$$HCl(g) + ClONO_2(g) \rightarrow Cl_2(g) + HNO_3(g)$$

Cl_2 remains in the stratosphere until the season changes and the appropriate wavelength of radiation is available to photodissociate Cl_2:

$$Cl_2(g) \xrightarrow{h\nu} 2Cl(g)$$

Free chlorine atoms then react with O_3 to form O_2 and ClO, which goes on to regenerate Cl atoms in a cycle that may destroy thousands of O_3 molecules before the Cl atoms become part of the chlorine reservoir again. This theory agrees with the seasonal variations in ozone concentration.

Chemistry of the Troposphere

18.15 (a) CO binds with hemoglobin in the blood to block O_2 transport to the cells; people with CO poisoning suffocate from lack of O_2.

(b) SO_2 is corrosive to lung tissue and contributes to higher levels of respiratory disease and shorter life expectancy, especially for people with other respiratory problems such as asthma. It also is a major source of acid rain, which damages forests and wildlife in natural waters.

(c) O_3 is extremely reactive and toxic because of its ability to form free radicals upon reaction with organic molecules in the body. It is particularly dangerous for asthma suffers, exercisers and the elderly. O_3 can also react with organic compounds in polluted air to form peroxyacylnitrates, which cause eye irritation and breathing difficulties.

18.16 $2CO + O_2 \rightarrow 2CO_2$ $\qquad 2SO_2 + O_2 \rightarrow 2SO_3$

$CO_2 + H_2O \rightarrow H_2CO_3$ $\qquad SO_3 + H_2O \rightarrow H_2SO_4$

Oxidation of CO to CO_2 and SO_2 to SO_3 produces gases which readily dissolve in atmospheric moisture to form acid rain. CO_2 and SO_3 are actually more dangerous to the environment because of their potential to form acid rain.

18.17 CO in unpolluted air is typically 0.05 ppm, whereas in urban air CO is about 10 ppm. A major source is automobile exhaust. SO_2 is less than 0.01 ppm in unpolluted air and on the order of 0.08 ppm in urban air. A major source is coal and oil-burning power plants, but there is also some SO_2 in auto exhausts. NO is about 0.01 ppm in unpolluted air and about 0.05 ppm in urban air. It comes mainly from auto exhausts.

18.18 (a) Methane, CH_4, arises from decomposition of organic matter by certain microorganisms; it also escapes from underground gas deposits.

(b) SO_2 is released in volcanic gases, and also is produced by bacterial action on decomposing vegetable and animal matter.

(c) Nitric oxide, NO, results from oxidation of decomposing organic matter, and is formed in lightning flashes.

(d) CO is a possible product of some vegetable matter decay.

18.19 Ozone partial pressure is $(0.31 \times 10^{-6})(745 \text{ torr}) = 2.3 \times 10^{-4}$ torr.

$$\frac{n}{V} = \frac{2.3 \times 10^{-4} \text{ torr}}{289 \text{ K}} \times \frac{1 \text{ atm}}{760 \text{ torr}} \times \frac{\text{K} \cdot \text{mol}}{0.0821 \text{ L} \cdot \text{atm}} \times \frac{10^3 \text{ L}}{1 \text{ m}^3}$$

$$\times \frac{6.02 \times 10^{23} \text{ molecules}}{1 \text{ mol}} = 7.7 \times 10^{18} \text{ molecules/m}^3$$

18.20 Partial pressure is $(0.45 \times 10^{-6})(730 \text{ torr}) = 3.3 \times 10^{-4}$ torr

$$\frac{n}{V} = \frac{3.3 \times 10^{-4} \text{ torr}}{291 \text{ K}} \times \frac{1 \text{ atm}}{760 \text{ torr}} \times \frac{\text{K} \cdot \text{mol}}{0.0821 \text{ L} \cdot \text{atm}} \times \frac{10^3 \text{ L}}{1 \text{ m}^3}$$

$$\times \frac{6.02 \times 10^{23} \text{ molecules}}{1 \text{ mol}} = 1.1 \times 10^{19} \text{ molecules/m}^3$$

18.21 All oxides of nonmetals produce acid solutions when dissolved in water. Sulfur oxides are produced naturally during volcanic eruptions and carbon oxides are products of combustion and metabolism. These dissolved gases cause rainwater to be naturally acidic.

18.22 (a) $H_2SO_4(aq) + Fe(s) \rightarrow FeSO_4(aq) + H_2(g)$

(The actual product may be red rust, Fe_2O_3, formed by the further oxidation of Fe^{2+} by $O_2(g)$.)

(b) $H_2SO_4(aq) + CaCO_3(s) \rightarrow CaSO_4(s) + H_2O(l) + CO_2(g)$

18.23 The reactions involved are: $CaCO_3(s) \rightarrow CaO(s) + CO_2(g)$;
$S(s) + O_2(g) \rightarrow SO_2(g)$; $CaO(s) + SO_2(s) \rightarrow CaSO_3(s)$

$$1 \text{ ton oil} \times \frac{0.027 \text{ ton S}}{1 \text{ ton oil}} \times \frac{2000 \text{ lb}}{1 \text{ ton}} \times \frac{454 \text{ g}}{1 \text{ lb}} \times \frac{1 \text{ mol S}}{32 \text{ g S}} \times \frac{1 \text{ mol } SO_2}{1 \text{ mol S}}$$

$$\times \frac{1 \text{ mol CaO}}{0.30 \text{ mol } SO_2} \times \frac{1 \text{ mol } CaCO_3}{1 \text{ mol CaO}} \times \frac{100 \text{ g } CaCO_3}{1 \text{ mol } CaCO_3} = 2.6 \times 10^5 \text{ g } CaCO_3$$

18.24 (a)

$$8{,}376{,}726 \text{ tons coal} \times \frac{83 \text{ ton C}}{100 \text{ ton coal}} \times \frac{44.01 \text{ ton } CO_2}{12.01 \text{ ton C}} = 2.5 \times 10^7 \text{ ton } CO_2$$

$$8{,}376{,}726 \text{ tons coal} \times \frac{2.5 \text{ ton S}}{100 \text{ ton coal}} \times \frac{64.06 \text{ ton } SO_2}{32.06 \text{ ton S}} = 4.2 \times 10^5 \text{ ton } SO_2$$

(b) $CaO(s) + SO_2(g) \rightarrow CaSO_3(s)$

$$4.2 \times 10^5 \text{ ton } SO_2 \times \frac{55 \text{ ton } SO_2 \text{ removed}}{100 \text{ ton } SO_2 \text{ produced}} \times \frac{120.14 \text{ ton } CaSO_3}{64.06 \text{ ton } SO_2}$$

$$= 4.3 \times 10^5 \text{ ton } CaSO_3$$

18.25 (a) Visible (Figure 6.3)

(b)

$$E_{photon} = \frac{hc}{\lambda} = \frac{6.626 \times 10^{-34} \text{ J} \cdot \text{s} \times 3.00 \times 10^8 \text{ m/s}}{420 \times 10^{-9} \; M} = \frac{4.73 \times 10^{-19} \text{ J}}{1 \text{ photon}}$$

$$\frac{4.73 \times 10^{-19} \text{ J}}{1 \text{ photon}} \times \frac{6.022 \times 10^{23} \text{ photons}}{1 \text{ mol}} \times \frac{1 \text{ kJ}}{1000 \text{ J}} = \frac{285 \text{ kJ}}{\text{mol}}$$

18.26 (a) Ultraviolet (Figure 6.3)

(b) $E_{photon} = \frac{hc}{\lambda} = \frac{6.626 \times 10^{-34} \text{ J}\cdot\text{s} \times 3.00 \times 10^{8} \text{ m/s}}{335 \times 10^{-9}\ m} = \frac{5.93 \times 10^{-19} \text{ J}}{1 \text{ photon}}$

$$\frac{5.93 \times 10^{-19} \text{ J}}{1 \text{ photon}} \times \frac{6.022 \times 10^{23} \text{ photons}}{1 \text{ mol}} \times \frac{1 \text{ kJ}}{1000 \text{ J}} = \frac{357 \text{ kJ}}{1 \text{ mol}}$$

(c) The average C-H bond energy from Table 8.4 is 413 kJ/mol. The energy calculated in part (b), 357 kJ/mol, is the energy required to break 1 mol of C-H bonds in formaldehyde, CH_2O. The C-H bond energy in CH_2O must be less than the "average" C-H bond energy.

18.27 From the composition of air at sea-level (Table 18.1) calculate the partial pressures of $N_2(g)$ and $O_2(g)$ in the original sample.

$P_x = \chi_x P_T$; P_{N_2} = 0.78084 (1.0 atm) = 0.78 atm; P_{O_2} =0.20948 (1.0 atm) = 0.21 atm

	$N_2(g)$	+ $O_2(g)$	$\rightleftharpoons$ $2NO(g)$
inital	0.78 atm	0.21 atm	0
charge	-x	-x	+2x
equil.	(0.78-x) atm	(0.21-x) atm	2x atm

$$K_p = \frac{P^2_{NO}}{P_{N_2} \times P_{O_2}} = \frac{(2x)^2}{(0.78\text{-}x)(0.21\text{-}x)} = \frac{4x^2}{0.164\text{-}0.99x + x^2} = 0.05$$

$0.05\ (0.164\text{-}\ \text{-}\ 0.99x + x^2 = 4x^2; \quad 0 = 3.95x^2 + 0.05x - 0.0082$

Using the quadratic formula, $x = \frac{-b \pm \sqrt{b^2 - 4ac}}{2a} = \frac{-0.05 \pm \sqrt{(0.05)^2 - 4(3.95)\ (-0.0082)}}{2(3.95)}$

$$x = \frac{-0.05 \pm \sqrt{0.0025 + 0.1296}}{7.90} = \frac{-0.05 \pm 0.363}{7.90}$$

The negative result is meaningless; x = 0.04 atm; P_{NO} = 2x = 0.08 atm

Assuming that the total pressure of the gaseous mixture at equilibrium is still 1.0 atm,

$x_{NO} = P_{NO}/P_T$ = 0.08 atm/1.0 atm = 0.08

ppm for gases = $x \times 10^6$ (see Section 18.1)

$ppm_{CO} = 0.08 \times 10^6 = 8 \times 10^4$ ppm

18.28 Initial pressures: P_{N_2} = 0.78 (1.50 atm) = 1.2 atm; P_{O_2} = 0.3 atm

$$P_{NO} = \frac{3200 \text{ atm}}{1 \times 10^6 \text{ atm}} \times 1.5 \text{ atm} = 4.8 \times 10^{-3} \text{ atm}$$

Calculate Q to determine which direction the reaction will proceed.

$$Q = \frac{P^2_{NO}}{P_{N_2} \times P_{O_2}} = \frac{(4.8 \times 10^{-3})^2}{(1.2)\,(0.3)} = 6.4 \times 10^{-5},\ K_p = 1.0 \times 10^{-5}$$

$Q > K_p$, the reaction proceeds to the left.

	$N_2(g)$ +	$O_2(g)$ ⇌	$2NO(g)$
initial	1.2 atm	0.3 atm	4.8×10^{-3} atm
change	+x	+x	-2x
equil.	(1.2 + x)atm	(0.3 + x)atm	$(4.8 \times 10^{-3} - 2x)$atm

$$K_p = 1.0 \times 10^{-5} = \frac{(4.8 \times 10^{-3} - 2x)^2}{(1.2 + x)\,(0.3 + x)}$$

Assume x is small compared to 1.2 and 0.3 (but not 4.8×10^{-3}).

P_{NO} at equilibrium = $(4.8 \times 10^{-3} - 2x) = y$

$$K_p = 1.0 \times 10^{-5} = \frac{y^2}{(1.2)\,(0.3)}\ ;\ y^2 = 3.6 \times 10^{-6}\ ;\ y = 1.9 \times 10^{-3} \text{ atm}$$

$$\text{Concentration, } M = \frac{\text{mol}}{\text{L}} = \frac{n}{V} = \frac{P}{RT}\ ;$$

$$[N_2] = \frac{1.2 \text{ atm}}{1173 \text{ K}} \times \frac{\text{K} \bullet \text{mol}}{0.0821 \text{ L} \bullet \text{atm}} = 1.2 \times 10^{-2}\ M$$

$$[O_2] = \frac{0.3 \text{ atm}}{1173 \text{ K}} \times \frac{\text{K} \bullet \text{mol}}{0.0821 \text{ L} \bullet \text{atm}} = 3 \times 10^{-2}\ M$$

$$[NO] = \frac{1.9 \times 10^{-3} \text{ atm}}{1173 \text{ K}} \times \frac{\text{K} \bullet \text{mol}}{0.0821 \text{ L} \bullet \text{atm}} = 2.0 \times 10^{-5}\ M \text{ (or } 1.3 \times 10^3 \text{ ppm)}$$

Note that the effect of attaining equilibrium is to reduce the concentration of NO, without significantly changing the concentrations of N_2 and O_2.

The World Ocean

18.29 A salinity of 5 denotes that there are 5 g of dry salt per kg of water.

$$\frac{5 \text{ g NaCl}}{1 \text{ kg soln}} \times \frac{1 \text{ kg soln}}{1 \text{ L soln}} \times \frac{1 \text{ mol NaCl}}{58 \text{ g NaCl}} \times \frac{1 \text{ mol Na}^+}{1 \text{ mol NaCl}} = 0.09\ M \text{ Na}^+$$

18.30 If the phosphorous is present as phosphate, there is a 1:1 ratio between the molarity of phosphorus and molarity of phosphate. Thus, we can calculate the molarity based on the given mass of P.

$$\frac{0.07 \text{ g P}}{10^6 \text{ g H}_2\text{O}} \times \frac{1 \text{ mol P}}{31 \text{ g P}} \times \frac{1 \text{ mol PO}_4^{3-}}{1 \text{ mol P}} \times \frac{10^3 \text{ g H}_2\text{O}}{1 \text{ L H}_2\text{O}} = \frac{2.3 \times 10^{-6} \text{ mol PO}_4^{3-}}{1 \text{ L}}$$

18.31 $1 \times 10^{11}\ \text{g Br} \times \frac{1 \times 10^3\ \text{g H}_2\text{O}}{0.067\ \text{g Br}} \times \frac{1\ \text{L H}_2\text{O}}{1 \times 10^3\ \text{g H}_2\text{O}} = 1.5 \times 10^{12}\ \text{L H}_2\text{O}$

Because the process is only 10% efficient, ten times this much, or 1.5×10^{13} L H_2O, must be processed.

18.32 $4.0 \times 10^7\ \text{g Mg(OH)}_2 \times \frac{1\ \text{mol Mg(OH)}_2}{58\ \text{g Mg(OH)}_2} \times \frac{1\ \text{mol CaO}}{1\ \text{mol Mg(OH)}_2} \times \frac{56\ \text{g CaO}}{1\ \text{mol CaO}}$

$= 3.9 \times 10^7$ g CaO

Fresh Water

18.33 Decomposition of dissolved organic matter by aerobic bacteria uses O_2 dissolved in the water. If the level of dissolved O_2 is lower than when the water is saturated with air, it is an indication that the water is polluted with organic matter and other materials which require O_2 for their decomposition.

18.34 Lack of sufficient O_2 to sustain aerobic decay leads to growth of anaerobic bacteria. These organisms are capable of processing organic matter without O_2. The anaerobic decomposition products are less oxidized than the products of aerobic decay. Instead of CO_2 or organic acids, methane, CH_4, might be produced. Other products could include H_2S, PH_3 and NH_3.

18.35 $1.0\ \text{g C}_{18}\text{H}_{29}\text{O}_3\text{S}^- \times \frac{1\ \text{mol C}_{18}\text{H}_{29}\text{O}_3\text{S}^-}{325\ \text{g C}_{18}\text{H}_{29}\text{O}_3\text{S}^-} \times \frac{51\ \text{mol O}_2}{2\ \text{mol C}_{18}\text{H}_{29}\text{O}_3\text{S}^-} \times \frac{32\ \text{g O}_2}{1\ \text{mol O}_2} = 2.5\ \text{g O}_2$

Notice that the mass of O_2 required is 2.5 times greater than the mass of biodegradable material.

18.36 $50{,}000\ \text{persons} \times \frac{59\ \text{g O}_2}{1\ \text{person}} \times \frac{1 \times 10^6\ \text{g H}_2\text{O}}{9\ \text{g O}_2} \times \frac{1\ \text{L H}_2\text{O}}{1 \times 10^3\ \text{g H}_2\text{O}} = 3.3 \times 10^8\ \text{L H}_2\text{O}$

18.37 $Ca^{2+}(aq) + 2HCO_3^-(aq) \rightarrow CaCO_3(s) + CO_2(g) + H_2O(l)$

18.38 $Mg^{2+}(aq) + Ca(OH)_2(s) \rightarrow Mg(OH)_2(s) + Ca^{2+}(aq)$

The excess $Ca^{2+}(aq)$ is removed as $CaCO_3$ (See solution 18.37) by naturally occurring bicarbonate or added Na_2CO_3.

18.39 $Ca(OH)_2$ is added to remove Ca^{2+} as $CaCO_3(s)$, and Na_2CO_3 removes the remaining Ca^{2+}. $Ca^{2+}(aq) + 2HCO_3^-(aq) + [Ca^{2+}(aq) + 2OH^-(aq)] \rightarrow 2CaCO_3(s) + 2H_2O(l)$.
One mole $Ca(OH)_2$ is needed for each 2 moles of $HCO_3^-(aq)$ present.

If there are 7.0×10^{-4} mol $HCO_3^-(aq)$ per liter, we must add 3.5×10^{-4} mol $Ca(OH)_2$ per liter, or a total of 0.35 mol $Ca(OH)_2$ for 10^3 L. This reaction removes 3.5×10^{-4} mol of the original Ca^{2+} from each liter of solution, leaving 1.5×10^{-4} M $Ca^{2+}(aq)$. To remove this $Ca^{2+}(aq)$, we add 1.5×10^{-4} mol Na_2CO_3 per liter, or a total of 0.15 mol Na_2CO_3, forming $CaCO_3(s)$.

18.40 $Ca(OH)_2$ is added to remove Ca^{2+} as $CaCO_3(s)$, and Na_2CO_3 removes the remaining Ca^{2+}. $Ca^{2+}(aq) + 2HCO_3^-(aq) + [Ca^{2+}(aq) + 2OH^-(aq)] \rightarrow 2CaCO_3(s) + 2H_2O(l)$.

One mole $Ca(OH)_2$ is needed for each 2 moles of $HCO_3^-(aq)$ present.

$$5.0 \times 10^7 \text{ L } H_2O \times \frac{1.7 \times 10^{-3} \text{ mol } HCO_3^-}{1 \text{ L } H_2O} \times \frac{1 \text{ mol } Ca(OH)_2}{2 \text{ mol } HCO_3^-} \times \frac{74 \text{ g } Ca(OH)_2}{1 \text{ mol } Ca(OH)_2}$$

$$= 3.1 \times 10^6 \text{ g } Ca(OH)_2$$

This operation reduces the Ca^{2+} concentration from 5.7×10^{-3} M to $(5.7 \times 10^{-3} - 1.7 \times 10^{-3})$ $M = 4.0 \times 10^{-3}$ M. Next we must add sufficient Na_2CO_3 to further reduce $[Ca^{2+}]$ to 1.1×10^{-3} M. We thus need to reduce $[Ca^{2+}]$ by $(4.0 \times 10^{-3} - 1.1 \times 10^{-3})$ M = 2.9×10^{-3} M. $Ca^{2+}(aq) + CO_3^{2-}(aq) \rightarrow CaCO_3(s)$.

$$5.0 \times 10^7 \text{ L } H_2O \times \frac{2.9 \times 10^{-3} \text{ mol } Ca^{2+}}{1 \text{ L } H_2O} \times \frac{1 \text{ mol } Na_2CO_3}{1 \text{ mol } Ca^{2+}} \times \frac{106 \text{ g } Na_2CO_3}{1 \text{ mol } Na_2CO_3}$$

$$= 1.5 \times 10^7 \text{ g } Na_2CO_3$$

18.41 (a) $Al(OH)_3(s) \rightleftharpoons Al^{3+}(aq) + 3OH^-(aq) \quad K_{sp} = 3.7 \times 10^{-15} = [Al^{3+}][OH^-]^3$

This is a precipitation conditions problem. At what $[OH^-]$ (we can get pH from $[OH^-]$) will $Q = 3.7 \times 10^{-15}$, the requirement for the onset of precipitation?

$Q = 3.7 \times 10^{-15} = [Al^{3+}][OH^-]^3$. Find the molar concentration of $Al_2(SO_4)_3$ and thus $[Al^{3+}]$.

$$\frac{2.0 \text{ lb } Al_2(SO_4)_3}{1000 \text{ gal } H_2O} \times \frac{454 \text{ g}}{1 \text{ lb}} \times \frac{1 \text{ mol } Al_2(SO_4)_3}{342 \text{ g } Al_2(SO_4)_3} \times \frac{1 \text{ gal}}{4 \text{ qt}} \times \frac{1 \text{ qt}}{0.946 \text{ L}}$$

$$= 7.0 \times 10^{-4} \, M \, Al_2(SO_4)_3 = 1.4 \times 10^{-3} \, M \, Al^{3+}$$

$Q = 3.7 \times 10^{-15} = (1.4 \times 10^{-3})[OH^-]^3$; $[OH^-]^3 = 2.64 \times 10^{-12}$

$[OH^-] = 1.4 \times 10^{-4}$; pOH = 3.86 ; pH = 14 - 3.86 = 10.14

(b) $CaO(s) + H_2O(l) \rightarrow Ca^{2+}(aq) + 2OH^-(aq)$; $[OH^-] = 1.4 \times 10^{-4}$ mol/L

$$\text{mol OH}^- = \frac{1.4 \times 10^{-4}\ \text{mol}}{1\ \text{L}} \times 1000\ \text{gal} \times \frac{4\ \text{qt}}{1\ \text{gal}} \times \frac{0.946\ \text{L}}{1\ \text{qt}} = 0.53\ \text{mol OH}^-$$

$$0.53\ \text{mol OH}^- \times \frac{1\ \text{mol CaO}}{2\ \text{mol OH}^-} \times \frac{56.1\ \text{g CaO}}{1\ \text{mol CaO}} \times \frac{1\ \text{lb}}{454\ \text{g}} = 0.033\ \text{lb}$$

18.42 (a) CO_3^{2-} is a relatively strong Bronsted base and produces OH^- in aqueous solution according to the hydrolysis reaction:

$$CO_3^{2-}(aq) + H_2O(l) \rightleftharpoons HCO_3^-(aq) + OH^-(aq),\ K_b = 1.8 \times 10^{-4}$$

If enough $OH^-(aq)$ is produced to exceed K_{sp} for $Mg(OH)_2$, the solid will precipitate.

(b) $$\frac{150\ \text{mg Mg}^{2+}}{1\ \text{kg soln}} \times \frac{1\ \text{g Mg}^{2+}}{1000\ \text{mg Mg}^{2+}} \times \frac{1\ \text{kg soln}}{1\ \text{L soln}} \times \frac{1\ \text{mol Mg}^{2+}}{24.3\ \text{g Mg}^{2+}} = 6.2 \times 10^{-3}\ M\ \text{Mg}^{2+}$$

$$\frac{5.0\ \text{g Na}_2\text{CO}_3}{1.0\ \text{L soln}} \times \frac{1\ \text{mol CO}_3^{2-}}{106.0\ \text{g Na}_2\text{CO}_3} = 0.047\ M\ \text{CO}_3^{2-}$$

$$K_b = 1.8 \times 10^{-4} = \frac{[HCO_3^-][OH^-]}{[CO_3^{2-}]} \approx \frac{x^2}{0.047};\ x = [OH^-] = 2.9 \times 10^{-3}\ M$$

(This represents 6.2% hydrolysis, but the result will not be significantly different using the quadratic formula.)

$$Q = [Mg^{2+}][OH^-]^2 = (6.2 \times 10^{-3})(2.9 \times 10^{-3})^2 = 5.2 \times 10^{-8}$$

K_{sp} for $Mg(OH)_2 = 1.8 \times 10^{-11}$; $Q > K_{sp}$, so $Mg(OH)_2$ will precipitate.

Additional Exercises

18.43 (a) *Acid rain* is rain with a larger $[H^+]$ and thus a lower pH than expected. The additional H^+ is produced by the dissolution of sulfur and nitrogen oxides in rain droplets to form sulfuric and nitric acid.

(b) The *greenhouse effect* is warming of the atmosphere caused by heat trapping gases such as $CO_2(g)$ and $H_2O(g)$. That is, these gases absorb infrared or "heat" radiation emitted from the earth's surface and serve to maintain a relatively constant temperature on the surface. A significant increase in the amount of atmospheric CO_2 (from burning fossel fuels and other sources) could cause a corresponding increase in the average surface temperature and drastically change the global climate.

(c) *Photochemical smog* is an unpleasant collection of atmospheric pollutants initiated by photochemical dissociation of NO_2 to form NO and O atoms. The major components are NO(g), $NO_2(g)$, CO(g) and unburned hydrocarbons, all produced by automobile engines, and $O_3(g)$, ozone.

(d) The *ozone hole* is a region of depleted O_3 in the stratosphere over Antarctica caused by reactions between O_3 and Cl atoms originating from chlorofluorocarbons (CFC's), CF_xCl_{4-x}. Depletion of the ozone layer would allow damaging ultraviolet radiation disruptive to the plant and animal life in our ecosystem to reach earth.

18.44 Avg. mol wt. at the surface = 40.0(0.20) + 16.0(0.35) + 32.0(0.45) = 28.0 g/mol

Next, calculate the percentage composition at 200 km. The fractions can be "normalized" by saying that the 0.45 fraction of O_2 is converted into two 0.45 fractions of O atoms, then dividing by the total fractions, 0.20 + 0.35 + 0.45 + 0.45 = 1.45:

$$\text{Avg. mol. wt.} = \frac{40.0(0.20) + 16.0(0.35) + 16.0(0.90)}{1.45} = 19.3 \text{ g/mol}$$

18.45 The ionization energies of metal atoms are much lower than those of any of the atomic or molecular ions present at 120 km; that is, O_2^+, O^+, NO^+. Thus, any of these ions react with metal atoms as illustrated for NO^+: $M + NO^+ \rightarrow M^+ + NO$.

18.46 The answer lies in Figure 18.2. Atmospheric pressure drops rapidly with increasing elevation. Thus at 12 km the pressure is only about 150 torr, and it declines to around 1 torr at 50 km. The number of molecules per unit volume is thus much lower on average in the stratosphere.

18.47 (a) A rate constant of $M^{-1}s^{-1}$ is indicative of a reaction that is second order overall. For the reaction given, the rate law is probably rate = $k[O][O_3]$. (Although rate = $[O]^2$ or $k[O_3]^2$ are possibilities, it is difficult to envision a mechanism consistent with either one that would result in two molecules of O_2 being produced.)

(b) Yes. Most atmospheric processes are initiated by collision. One could imagine an activated complex of four O atoms collapsing to form two O_2 molecules. Also, the rate constant is large, which is less likely for a multistep process. The reaction is analogous to the destruction of O_3 by Cl atoms (Equation 18.7), which is also second order with a large rate constant.

(c) According to the Arrhenius equation, $k = Ae^{-E_a/RT}$. Thus, the larger the value of k, the smaller the activation energy, E_a. The value of the rate constant for this reaction is large, so the activation energy is small.

(d) $\Delta H_f^\circ = 2\Delta H_f^\circ O_2(g) - \Delta H_f^\circ O(g) - \Delta H_f^\circ O_3(g)$

$\Delta H_f^\circ = 0 - 247.5 \text{ kJ} - 142.3 \text{ kJ} = -389.8 \text{ kJ}$

The reaction is exothermic, so energy is released; the reaction would raise the temperature of the stratosphere.

18.48 (a) The production of Cl atoms in the stratosphere is the result of the photodissociation of a C-Cl bond in the chlorofluorocarbon molecule.

$$CF_2Cl_2(g) \xrightarrow{h\nu} CF_2Cl(g) + Cl(g)$$

According to Table 8.4, the bond dissociation energy of a C-Br bond is 276 kJ/mol, while the value for a C-Cl bond is 328 kJ/mol. Photodissociation of $CBrF_3$ to form Br atoms requires less energy than the production of Cl atoms and should occur readily in the stratosphere.

(b) $CBrF_3(g) \xrightarrow{h\nu} CF_3(g) + Br(g)$

$Br(g) + O_3(g) \longrightarrow BrO(g) + O_2(g)$

Also, under certain conditions:

$BrO(g) + BrO(g) \longrightarrow Br_2O_2(g)$

$Br_2O_2(g) + h\nu \longrightarrow O_2(g) + 2Br(g)$

18.49 (a) HNO_3 is a major component in acid rain.

(b) While it removes CO, the reaction produces NO_2. The photodissociation of NO_2 to form O atoms is the first step in the formation of tropospheric ozone and photochemical smog.

(c) Again, NO_2 is the initiator of photochemical smog. Also, methoxyl radical, OCH_3, is a reactive species capable of initiating other undesireable reactions.

18.50 (a) Anthracite coal would be more expensive than bituminous coal because the supply is lower and the demand is higher. Also, since anthracite fuel requires less treatment of the combustion products, it might be cost effective even at a higher purchase price.

(b) Burning bituminous coal releases $SO_2(g)$ into the atmosphere where it can be oxidized to $SO_3(g)$. In the troposphere, these gases can be carried from their point of origin to other parts of the country or the world. The gases dissolve in atmospheric water droplets to produce H_2SO_3 and H_2SO_4. When the water droplets collect, acid rain falls at a point far from the power plant which released the combustion products.

18.51 The formation of NO(g) from $N_2(g)$ and $O_2(g)$ is endothermic:

$$1/2N_2(g) + 1/2O_2(g) \rightleftharpoons NO(g) \quad \Delta H^\circ = 90.37 \text{ kJ}$$

The production of $NO_2(g)$ from NO(g) and $O_2(g)$ is exothermic:

$$NO(g) + 1/2O_2(g) \rightleftharpoons NO_2(g) \quad \Delta H^\circ = -56.53 \text{ kJ}$$

At the high temperature of combustion reactions, the endothermic production of NO(g) is favored.

18.52 From Section 18.4:

$$N_2(g) + O_2(g) \rightleftharpoons 2NO(g) \qquad \Delta H = +180.8 \text{ kJ} \quad (1)$$

$$2NO(g) + O_2(g) \rightleftharpoons 2NO_2(g) \qquad \Delta H = -113.1 \text{ kJ} \quad (2)$$

In an endothermic reaction, heat is a reactant. As the temperature of the reaction increases, the addition of heat favors formation of products and the value of K increases. The reverse is true for exothermic reactions; as temperature increases, the value of K decreases. Thus, K for reaction (1), which is endothermic, increases with increasing temperature and K for reaction (2), which is exothermic, decreases with increasing temperature.

18.53 $$P_{CO} = \frac{10.5 \text{ torr}}{1 \times 10^6 \text{ torr}} \times 745 \text{ torr} \times \frac{1 \text{ atm}}{760 \text{ torr}} = 1.03 \times 10^{-5} \text{ atm}$$

$$\text{Volume of room} = (12 \text{ ft} \times 14 \text{ ft} \times 8 \text{ ft}) \times \frac{12^3 \text{ in}^3}{1 \text{ ft}^3} \times \frac{2.54^3 \text{ cm}^3}{1 \text{ in}^3} \times \frac{1 \text{ L}}{1000 \text{ cm}^3} = 3.8 \times 10^4 \text{ L}$$

$$n = \frac{PV}{RT}\ ;\ 1.03 \times 10^{-5} \text{ atm} \times \frac{\text{K} \cdot \text{mol}}{0.0821 \text{ L} \cdot \text{atm}} \times \frac{3.8 \times 10^4 \text{ L}}{291 \text{ K}} \times \frac{6.02 \times 10^{23} \text{ CO molecules}}{1 \text{ mol CO}}$$

$$= 9.9 \times 10^{21} \text{ CO molecules}$$

18.54 Oxygen is present in the atmosphere to the extent of 209,000 parts per million. If CO binds 210 times more effectively than O_2, then the effective concentration of CO is 210 × 86 ppm = 18,000 ppm. The fraction of carboxyhemoglobin in the blood leaving the lungs is thus $\frac{18{,}000}{18{,}000 + 209{,}000} = 0.079$. Thus, 7.9 percent of the blood is in the form of carboxyhemoglobin, 92.1 percent as the O_2-bound oxyhemoglobin.

18.55 (a) $CH_4(g) + 2O_2(g) \rightarrow CO_2(g) + 2H_2O(g)$

(b) $2CH_4(g) + 3O_2(g) \rightarrow 2CO(g) + 4H_2O(g)$

(c) vol CH_4 → vol O_2 → volume air ($\chi_{O_2} = 0.20948$)

Equal volumes of gases at the same temperature and pressure contain equal numbers of moles (Avogadro's law). If 2 moles of O_2 are required for 1 mole of CH_4, 2.0 L of pure O_2 are needed to burn 1.0 L of CH_4.

$$\text{vol } O_2 = \chi_{O_2} \times \text{vol}_{air}\ ;\ \text{vol}_{air} = \frac{\text{vol } O_2}{\chi_{O_2}} = \frac{2.0 \text{ L}}{0.20948} = 9.5 \text{ L air}$$

(d) Since incomplete combustion requires less O_2 per mole of CH_4, running a home furnace with too little O_2 or too much CH_4 decreases the O_2:CH_4 ratio and encourages the production of toxic CO(g) rather than CO_2(g).

18.56 (a) According to Section 13.4, the solubility of gases in water decreases with increasing temperature. Thus, the solubility of $CO_2(g)$ in the ocean would decrease if the temperature of the ocean increased.

(b) If the solubility of $CO_2(g)$ in the ocean decreased because of global warming, more $CO_2(g)$ would be released into the atmosphere, perpetuating a cycle of increasing temperature and concomittant release of $CO_2(g)$ from the ocean.

18.57 (a) According to Table 18.1, the mole fraction of CO_2 in air is 0.000355.

$$P_{CO_2} = \chi_{CO_2} \cdot P_{atm} = 0.000355\ (1.00\ \text{atm}) = 3.55 \times 10^{-4}\ \text{atm}$$

$$C_{CO_2} = kP_{CO_2} = 3.1 \times 10^{-2}\ M/\text{atm} \times 3.55 \times 10^{-4}\ \text{atm} = 1.1 \times 10^{-5}\ M$$

(b) H_2CO_3 is a weak acid, so the $[H^+]$ is regulated by the equilibria:

$$H_2CO_3(aq) \rightleftharpoons H^+(aq) + HCO_3^-(aq) \qquad K_{a1} = 4.3 \times 10^{-7}$$

$$HCO_3^-\ (aq) \rightleftharpoons H^+(aq) + CO_3^{2-}(aq) \qquad K_{a2} = 5.6 \times 10^{-11}$$

Since the value of K_{a2} is small compared to K_{a1}, we will assume that most of the $H^+(aq)$ is produced by the first dissociation.

$$K_{a1} = 4.3 \times 10^{-7} = \frac{[H^+][HCO_3^-]}{[H_2CO_3]}\ ;\ [H^+] = [HCO_3^-] = x,\ [H_2CO_3] = 1.1 \times 10^{-5} - x$$

Since K_{a1} and $[H_2CO_3]$ have similar values, we cannot assume x is small compared to 1.0×10^{-5}.

$$4.3 \times 10^{-7} = \frac{x^2}{(1.1 \times 10^{-5} - x)}\ ;\ 4.7 \times 10^{-12} - 4.3 \times 10^{-7}\ x = x^2$$

$$0 = x^2 + 4.3 \times 10^{-7} - 4.7 \times 10^{-12}\ ;$$

$$x = \frac{-4.3 \times 10^{-7} \pm \sqrt{(4.3 \times 10^{-7})^2 - 4(1)(-4.7 \times 10^{-12})}}{2(1)}$$

$$x = \frac{-4.3 \times 10^{-7} \pm \sqrt{1.85 \times 10^{-13} + 1.88 \times 10^{-11}}}{2} = \frac{-4.3 \times 10^{-7} \pm 4.36 \times 10^{-6}}{2}$$

The negative result is meaningless; $x = 2.0 \times 10^{-6}\ M\ H^+$; pH = 5.71

Since this $[H^+]$ is quite small, the $[H^+]$ from the autoionization of water might be significant. Calculation shows that for $[H^+] = 2.0 \times 10^{-6}\ M$, from H_2CO_3, $[H^+]$ from $H_2O = 5.2 \times 10^{-9}\ M$, which we can ignore.

18.58 (a) $NO(g) + h\nu \rightarrow N(g) + O(g)$

(b) $NO(g) + h\nu \rightarrow NO^+(g) + e^-$

(c) $NO(g) + O_3(g) \rightarrow NO_2(g) + O_2(g)$

(d) $3NO_2(g) + H_2O(l) \rightarrow 2HNO_3(aq) + NO(g)$

18.59 Osmotic pressure is determined by total concentration of dissolved particles. From Table 18.5, M_T = 1.13. The minimum pressure needed to initiate reverse osmosis is $\pi = MRT$, Equation [13.13]. Assuming 298 K,

$$\pi = MRT = \frac{1.13 \text{ mol}}{1 \text{ L}} \times 298 \text{ K} \times \frac{0.0821 \text{ L} \cdot \text{atm}}{\text{K} \cdot \text{mol}} = 27.6 \text{ atm}$$

18.60 (a) In Figure 18.18, the aeration step speeds oxidation of any remaining dissolved organic matter by aerobic bacteria and ensures an adequate concentration of dissolved O_2 (to prevent build-up of anaerobic bacteria) in the processed water supplied to users.

(b) Cl_2 is used as an antibacterial agent in water treatment because it is more convenient than O_3. Cl_2 can be stored and transported as a liquid, whereas O_3 must be generated on-site as a gas.

(c) Boiling the water (100° C) would kill any *cryptosporidium* bacteria present. This measure should have been possible for all water destined for human consumption or hygiene.

18.61 (a) $Cl_2(g) + H_2O(l) \rightarrow HOCl(aq) + H^+(aq) + Cl^-(aq)$

(b) $Ca^{2+}(aq) + 2HCO_3^-(aq) \xrightarrow{\Delta} CaCO_3(s) + CO_2(g) + H_2O(l)$

18.62 Because NO has an odd electron, like Cl(g), it could act as a catalyst for decomposition of ozone in the stratosphere. The increased destruction of ozone by NO would result in less absorption of short wavelength UV radiation now being screened out primarily by the ozone. Radiation of this wavelength is known to be harmful to humans; it causes skin cancer. There is evidence that many plants don't tolerate it very well either, though more research is needed to test this idea.

In Chapter 22 the oxidation of NO to NO_2 by oxygen is described. On dissolving in water, NO_2 disproportionates into $NO_3^-(aq)$ and NO(g). Thus, over time the NO in the troposphere will be converted into NO_3^- which is in turn incorporated into soils.

CHAPTER 19

Chemical Thermodynamics

Spontaneity and Entropy

19.1 Spontaneous: a, d, e; nonspontaneous: b, c

19.2 Spontaneous: b, c, d; nonspontaneous: a, e

19.3 Berthelot's suggestion is incorrect. Some spontaneous processes that are nonexothermic are expansion of certain pressurized gases, dissolving of one liquid in another, and dissolving of many salts in water.

19.4 (a) At or below 0°C (b) above 0°C

(c) An increase in the entropy of the universe always accompanies a spontaneous process. When water freezes, ΔS_{sys} decreases, but because the reaction is exothermic, ΔS_{surr} increases. Below 0°C, the increase in ΔS_{surr} predominates, and the process is spontaneous. Above 0°C, the increase in ΔS_{surr} does not compensate for the decrease in ΔS_{sys}, and the process is not spontaneous.

19.5 For a process to be spontaneous, $\Delta S_{universe}$ must be positive. Thus, there must be an increase in the entropy or the surroundings, and it must be greater than the decrease in entropy of the system.

19.6 Yes. In using the second law, both the system and the surroundings must be considered. The increased organization within the cell is only the change in the system. There must be a greater decrease in organization of the surroundings.

19.7 S increases in a, b and c; S decreases in d.

19.8 More disorder is associated with forming a gas because of the large volume expansion and increased motional freedom associated with a gas.

19.9 (a) $O_2(g)$ at 0.5 atm (larger volume and more motional freedom)

(b) $Br_2(g)$ (gases have higher entropy due primarily to much larger volume)

(c) 1 mol of $N_2(g)$ in 22.4 L (larger volume provides more motional freedom)

(d) $CO_2(g)$ (more motional freedom)

19.10 (a) 1 mol of $Cl_2(g)$ at 373 K and 1 atm (higher temperature, greater molecular speeds, more overall motion)

(b) 1 mol $H_2O(g)$ at 100°C, 1 atm (larger volume occupied by $H_2O(g)$)

(c) 2 mol of O(g) at 300 K, 60.0 L (more total particles, larger volume in which to move)

(d) 1 mol of $KNO_3(aq)$ at 40°C (more motional freedom in aqueous solution)

19.11 (a) S° for $CCl_4(l)$ is 214.4 J/K • mol and for $CCl_4(g)$ is 309.4 J/K • mol. In general, the gas phase of a substance has a larger S° than the liquid phase because of the greater volume and motional freedom of the gas molecules.

(b) S° for C(diamond) is 2.43 J/K • mol and for C(graphite) is 5.69 J/K • mol. Diamond is a network covalent solid with each C atom tetrahedrally bound to four other C atoms. Graphite consists of sheets of fused planar 6-membered rings with each C atom bound in a trigonal planar arrangement to three other C atoms. The internal entropy in graphite is greater because there is motional freedom between the planar sheets of C atoms while there is very little vibrational freedom within the diamond lattice.

(c) S° for 1 mol of $N_2O_4(g)$ is 304.3 J/K and for 2 mol $NO_2(g)$ is 480.9 J/K. Although N_2O_4 molecules have more ways to store energy than NO_2 molecules, 2 mol of $NO_2(g)$ occupy a larger volume and have greater motional freedom than 1 mol $N_2O_4(g)$.

(d) S° for 1 mol of $KClO_3(s)$ is 143.0 J/K and for 1 mol $KClO_3(aq)$ is 265.7. Ions in solution have greater motional freedom than ions in a solid lattice.

19.12 (a) Si(g), 167.8 J/K • mol; Si(s), 18.7 J/K • mol. Atoms in the gas state have more motional freedom than atoms in the solid state.

(b) 1 mol $C_6H_{12}O_6(s)$, 212.1 J/K; 6 mol C(graphite) + 6 mol $H_2O(l)$ = 6(5.59) + 6(69.96) = 453.9 J/K. The second member of the pair has more total particles and half of them (6 mol $H_2O(l)$) are in the liquid state, for greater total motional freedom. However, note that 1 mol $C_6H_{12}O_6(s)$ has considerably greater entropy than 6 mol C(graphite) because of the many ways to store energy in the chemical bonds in $C_6H_{12}O_6$ and the weak intermolecular forces in $C_6H_{12}O_6$ relative to the strong covalent bonds in graphite sheets.

(c) 2 mol $P_2(g)$, 2(218.1) = 436.2 J/K; 1 mol $P_4(g)$ = 280 J/K. More particles have greater motional freedom.

(d) NaOH(s), 64.46 J/K • mol; NaOH(aq), 49.8 J/K • mol. It is unusual that an ionic compound in the solid state has a lower standard entropy than the compound in an aqueous solution. However, because of the strong hydrogen bonding interactions between OH^- ions and H_2O molecules, the disorder and S° of the aqueous solution is much smaller than expected.

19.13 (a) ΔS negative (increased order) (b) ΔS positive (increased disorder)
(c) ΔS negative (increased order) (d) ΔS negative (increased order)

19.14 (a) ΔS positive (moles of gas increase)
(b) ΔS negative (increased order)
(c) ΔS positive (gas produced, increased disorder)
(d) ΔS negative (fewer moles of gas, increased order)

19.15 (a) $\Delta S^\circ = 2S^\circ\ HF(g) + S^\circ\ Br_2(g) - 2S^\circ\ HBr(g) - S^\circ\ F_2(g)$
$\Delta S^\circ = 2(173.5) + 245.3 - 2(198.5) - 202.7 = -7.4$ J/K

(b) $\Delta S^\circ = 2S^\circ\ NO_2(g) - 2S^\circ\ NO(g) - S^\circ\ O_2(g)$
$\Delta S^\circ = 2(240.4) - 2(210.6) - 205.0 = -145.3$ J/K

(c) $\Delta S^\circ = 4S^\circ\ H_2O(g) + 2S^\circ\ CO_2(g) - 3S^\circ\ O_2(g) - 2S^\circ\ CH_3OH(g)$
$\Delta S^\circ = 4(188.7) + 2(213.6) - 3(205.0) - 2(237.6) = +91.8$ J/K

(d) $\Delta S^\circ = 2S^\circ\ Fe_2O_3(s) - 4S^\circ\ FeO(s) - S^\circ\ O_2(g)$
$\Delta S^\circ = 2(90.0) - 4(60.8) - (205.0) = -268.2$ J/K

In (a) ΔS° is near zero because the number of moles (and the kinds of molecules) are the same on both sides of the equation. The reason for the slightly negative value of ΔS° cannot be determined from the equation alone. In (b) ΔS° is negative because the number of moles of gas is lower in the products. It is positive in (c) because there are six moles of gas among the products, five among the reactants. It is negative in (d) because a gaseous reactant is converted to a solid product.

19.16 (a) $\Delta S^\circ = 3S^\circ\ O_2(g) - 2S^\circ\ O_3(g)$
$\Delta S^\circ = 3(205.0) - 2(237.6) = +139.8$ J/K

(b) $\Delta S^\circ = S^\circ\ SiO_2(s) + 4S^\circ\ HCl(g) - S^\circ\ SiCl_4(l) - 2S^\circ\ H_2O(l)$
$\Delta S^\circ = 41.84 + 4(186.69) - 239.3 - 2(69.96) = +409.4$ J/K

(c) $\Delta S^\circ = S^\circ\ C_6H_6(g) - 3S^\circ\ C_2H_2(g)$
$\Delta S^\circ = 269.2 - 3(200.8) = -333.2$ J/K

(d) $\Delta S^\circ = 2S^\circ\ HCl(g) - S^\circ\ H_2(g) - S^\circ\ Cl_2(g)$
$\Delta S^\circ = 2(186.69) - [130.58 + 222.96] = +19.84$ J/K

Gibbs Free Energy

19.17 (a) $\Delta G = \Delta H - T\Delta S$

(b) There is no relationship between ΔG and rate of reaction. A spontaneous reaction, one with a $-\Delta G$, may occur at a very slow rate. For example: $2H_2(g) + O_2(g) \rightarrow 2H_2O(g)$, $\Delta G = -457$ kJ is very slow if not initiated by a spark.

19.18 (a) When $\Delta G = 0$, the system is at equilibrium.

(b) The standard free energy change, ΔG°, represents the free energy change for the process when all reactants and products are in their standard states. When any or all reactants or products are not in their standard states, the free energy is represented simply as ΔG. The value for ΔG thus depends on the specific states of all reactants and products.

19.19 $\Delta G^\circ = \Delta H^\circ - T\Delta S^\circ = -1.06 \times 10^5$ J - 298 K(58 J/K) = -1.23×10^5 J
The process is highly spontaneous. One does not expect that $H_2O_2(g)$ would be stable at 298 K.

19.20 (a) $\Delta H^\circ = -1216.3 - [-553.5 - 393.5] = -269.3$ kJ
$\Delta G^\circ = -1137.6 - [-525.1 - 394.4] = -218.1$ kJ
$\Delta S^\circ = 112.1 - [70.42 + 213.6] = -171.9$ J/K
$\Delta G^\circ = -269.3$ kJ $- 298(-171.9)$ J $= -218.1$ kJ

(b) $\Delta H^\circ = 2(-435.9) + 3(0) - 2(-391.2) = -89.4$ kJ
$\Delta G^\circ = 2(-408.3) + 3(0) - 2(-289.9) = -236.8$ kJ
$\Delta S^\circ = 2(82.7) + 3(205.0) - 2(143.0) = 494.4$ J/K
$\Delta G^\circ = -89.4$ kJ $- 298(494.4)$ J $= -236.7$ kJ

(c) $\Delta H^\circ = 4(-241.8) + 2(-393.5) - [3(0) + 2(-238.6)] = -1277$ kJ
$\Delta G^\circ = 4(-228.6) + 2(-394.4) - [3(0) + 2(-166.20) = -1371$ kJ
$\Delta S^\circ = 4(188.7) + 2(213.6) - [3(205.0) + 2(126.8)] = 313.4$ J/K
$\Delta G^\circ = -1277$ kJ $- 298(313.4)$ J $= -1370$ kJ

(d) $\Delta H^\circ = 0 + 90.37 - [127.7 + 52.6] = -89.9$ kJ
$\Delta G^\circ = 0 + 86.71 - [105.7 + 66.3] = -85.3$ kJ
$\Delta S^\circ = 223.0 + 210.6 - [165.2 + 264] = +4.4$ J/K
$\Delta G^\circ = -89.9$ kJ $- 298(4.4)$ J $= -91.2$ kJ

(The discrepancy in ΔG° values is due to experimental uncertainties in the tabulated thermodynamic properties.)

19.21 (a) $\Delta G^\circ = 2\Delta G^\circ$ HCl(g) - [ΔG° $H_2(g)$ + ΔG° $Cl_2(g)$]
= 2(-95.27 kJ) - 0 - 0 = -190.5 kJ, spontaneous

(b) $\Delta G^\circ = \Delta G^\circ$ MgO(s) + $2\Delta G^\circ$ HCl(g) - [ΔG° $MgCl_2(s)$ + ΔG° $H_2O(l)$]
= -569.6 + 2(-95.27) - [-592.1 + (-236.81)] = +68.8 kJ, nonspontaneous

(c) $\Delta G^\circ = \Delta G^\circ\ N_2H_4(g) + \Delta G^\circ\ H_2(g) - 2\Delta G^\circ\ NH_3(g)$
$= 159.4 + 0 - 2(-16.66) = +192.7$ kJ, nonspontaneous

(d) $\Delta G^\circ = 2\Delta G^\circ\ NO(g) + \Delta G^\circ\ Cl_2(g) - 2\Delta G^\circ\ NOCl(g)$
$= 2(86.71) + 0 - 2(66.3) = +40.8$ kJ, nonspontaneous

19.22 (a) $\Delta G^\circ = 2\Delta G^\circ\ SO_3(g) - 2\Delta G^\circ\ SO_2(g) - \Delta G^\circ\ O_2(g)$
$= 2(-370.4) - 2(-300.4) - 0 = -140.0$ kJ, spontaneous

(b) $\Delta G^\circ = 3\Delta G^\circ\ NO(g) - \Delta G^\circ\ NO_2(g) - \Delta G^\circ\ N_2O(g)$
$= 3(86.71) - (51.84) - (103.59) = +104.70$ kJ, nonspontaneous

(c) $\Delta G^\circ = 4\Delta G^\circ\ FeCl_3(s) + 3\Delta G^\circ\ O_2(g) - 6\Delta G^\circ\ Cl_2(g) - 2\Delta G^\circ\ Fe_2O_3(s)$
$= 4(-334) + 3(0) - 6(0) -2(-740.98) = +146$ kJ, nonspontaneous

(d) $\Delta G^\circ = \Delta G^\circ\ S(s) + 2\Delta G^\circ\ H_2O(g) - \Delta G^\circ\ SO_2(g) - 2\Delta G^\circ\ H_2(g)$
$= 0 + 2(-228.61) - (-300.4) - 2(0) = -156.8$ kJ, spontaneous

19.23 (a) ΔG is negative at low temperatures, positive at high temperature. That is, the reaction proceeds in the forward direction spontaneously at lower temperatures but spontaneously reverses at higher temperatures.

(b) ΔG is positive at all temperatures. The reaction is nonspontaneous in the forward direction at all temperatures.

(c) ΔG is positive at low temperatures, negative at high temperatures. That is, the reaction will proceed spontaneously in the forward direction at high temperature.

(d) ΔG is positive at lower temperatures, negative at higher temperatures. From our knowledge of the stability of the water molecule, we expect that it will require a very high temperature to dissociate the H_2O molecule.

19.24 $\Delta G^\circ = \Delta H^\circ - T\Delta S^\circ$

(a) $\Delta G^\circ = -844$ kJ $- 298$ K$(-0.165$ kJ/K$) = -795$ kJ, spontaneous

(b) $\Delta G^\circ = +572$ kJ $- 298$ K$(0.179$ kJ/K$) = +519$ kJ, nonspontaneous

To be spontaneous, ΔG must be negative ($\Delta G < 0$).

Thus, $\Delta H^\circ - T\Delta S^\circ < 0$; $\Delta H^\circ < T\Delta S^\circ$; $T > \Delta H^\circ/\Delta S^\circ$; $T > \dfrac{572 \text{ kJ}}{0.179 \text{ kJ/K}} = 3{,}200$ K

19.25 (a) If the reaction is nonspontaneous, ΔG is positive and ΔH must be positive (endothermic) and have a greater magnitude than $T\Delta S$.

(b) The minimum value of ΔH is greater than $T\Delta S$, 298 K (121 J/K) = 36.1 kJ.
$\Delta H > +36.1$ kJ

19.26 $\Delta G < 0$ at 85°C (358 K); $\Delta H = +34$ kJ
$\Delta H - T\Delta S < 0$; $T\Delta S > \Delta H$; $\Delta S > \Delta H/T$
$\Delta S > +34$ kJ/358 K, $\Delta S > 0.095$ kJ/K or 95 J/K

19.27 ΔG is negative when $T\Delta S > \Delta H$ or $T > \Delta H/\Delta S$.

$$T > \frac{\Delta H}{\Delta S} > \frac{+178 \text{ kJ}}{0.160 \text{ kJ/K}} > 1112 \text{ K.}$$

The reaction will be spontaneous above 1112 K or 839°C.

19.28 $T > \Delta H/\Delta S$ (see Exercise 19.27)

$\Delta H^\circ = \Delta H^\circ\ CH_4(g) + \Delta H^\circ\ CO_2(g) - \Delta H^\circ\ CH_3COOH(l)$
$= -74.8 + (-393.5) - (-487.0) = +18.7$ kJ

$\Delta S^\circ = S^\circ\ CH_4(g) + S^\circ\ CO_2(g) - S^\circ\ CH_3COOH(l) = +186.3 + 213.6 - 159.8 = +240.1$ J/K

$$T > \frac{18.7\ \text{kJ}}{0.2401\ \text{kJ/K}} = 77.9\ \text{K}$$

The reaction is spontaneous above 77.9 K (-195°C).

19.29 (a) $\Delta H^\circ = 2(144.3) - 58.9 = 229.7$ kJ; $\Delta S^\circ = 2(218.1) - 280 = 156$ J/K
$\Delta G = \Delta H - T\Delta S$. ΔG will be positive at low temperature and negative at considerably higher ones. That is, the reaction becomes spontaneous at higher temperatures.

(b) $\Delta G^\circ(900\ K) = 229.7 - 900(0.156) = 89$ kJ
(No more than two significant figures is justified.)

19.30 (a) $\Delta H^\circ = (-296.9 - 241.8) - [0 + (-395.2)] = -143.5$ kJ
$\Delta S^\circ = (248.5 + 188.7) - [130.6 + 256.2] = 50.4$ J/K
Because the same numbers of moles of gas are involved in both reactants and products, ΔS° is not large, and ΔG° should, therefore, not change much with temperature.

(b) $\Delta G^\circ\ (600\ K) = 143.5 - 600(0.0504) = -173.7$ kJ
This is to be compared with $G^\circ = -158.5$ kJ at 298 K.

19.31 (a) $CH_4(g) + 2O_2(g) \rightarrow CO_2(g) + 2H_2O(l)$
$\Delta H^\circ = 2\Delta H^\circ\ H_2O(l) + \Delta H^\circ\ CO_2(g) - \Delta H^\circ\ CH_4(g) - 2\Delta H^\circ\ O_2(g)$
$$= 2(-285.85) - 393.5 - (-74.8 + 0) = \frac{-890.4\ \text{kJ produced}}{1\ \text{mol}\ CH_4\ \text{burned}}$$

(b) $w_{max} = \Delta G^\circ = 2\Delta G^\circ H_2O(l) + \Delta G^\circ\ CO_2(g) - \Delta G^\circ\ CH_4 - 2\Delta G^\circ\ O_2(g)$
$= 2(-236.81) - 394.4 + 50.8 = -817.2$ kJ

The negative sign indicates that the system does work on the surroundings; the system can accomplish a maximum of 817.2 kJ of work on its surroundings.

19.32 (a) $C_2H_2(g) + 5/2\ O_2(g) \rightarrow 2CO_2(g) + H_2O(l)$
$\Delta H^\circ = 2\Delta H^\circ\ CO_2(g) + \Delta H^\circ\ H_2O(l) - \Delta H^\circ\ C_2H_2(g) - \Delta H^\circ\ O_2(g)$

$$= 2(-393.5) - 285.85 - 226.7 = \frac{-1299.6\ \text{kJ produced}}{1\ \text{mol}\ C_2H_2\ \text{burned}}$$

(b) $w_{max} = \Delta G^\circ = 2\Delta G^\circ\ CO_2(g) + \Delta G^\circ\ H_2O(l) - \Delta G^\circ\ C_2H_2(g) - \Delta H^\circ\ O_2(g)$
$= 2(-394.4) - 236.81 - 209.2 = -1234.8$ kJ

Free Energy and Equilibrium

19.33 Use $\Delta G = \Delta G^o + RT \ln Q$, where Q is the reaction quotient. When the numerator in Q increases, ln Q becomes more positive, in turn ΔG is more positive. When the denominator in Q becomes larger, Q decreases, ln Q becomes smaller, or more negative, and ΔG becomes smaller or more negative. (a) ΔG becomes more positive; (b) ΔG becomes more negative; (c) ΔG becomes more positive

19.34 Consider the relationship $\Delta G = \Delta G^o + RT \ln Q$ where Q is the reaction quotient.

(a) $H_2(g)$ appears in the denominator of Q for this reaction. An increase in pressure of H_2 decreases Q and ΔG becomes smaller or more negative. Increasing the concentration of a reactant increases the tendency for a reaction to occur.

(b) $H_2(g)$ appears in the numerator of Q for this reaction. Increasing the pressure of H_2 increases Q and ΔG becomes more positive. Increasing the concentration of a product decreases the tendency for the reaction to occur.

19.35 (a) $\Delta G^o = \Delta G^o\ N_2O_4(g) - 2\Delta G^o\ NO_2(g) = 98.28 - 2(51.84) = -5.40$ kJ

(b) $\Delta G = \Delta G^o + RT \ln \dfrac{P_{N_2O_4}}{P^2_{NO_2}}$

$$= -5.40 \text{ kJ} \times \frac{8.314 \text{ J}}{1 \text{ K} \cdot \text{mol}} \times 298 \ln \frac{0.10}{(2.00)^2} = -14.54 \text{ kJ}$$

19.36 (a) $\Delta G^o = 2\Delta G^o\ CO_2(g) - 2\Delta G^o\ CO(g) - \Delta G^o\ O_2(g)$

$\Delta G^o = 2(-394.4) - 2(-137.3) - 0 = -514.2$ kJ

(b) $\Delta G = \Delta G^o + RT \ln \dfrac{P^2_{CO_2}}{P_{O_2} \times P^2_{CO}}$

$$= -514.2 \text{ kJ} \times \frac{8.314 \text{ J}}{1 \text{ K} \cdot \text{mol}} \times 298 \text{ K} \times \ln \frac{(0.10)^2}{300(6.0)^2}$$

$$= -514.2 \text{ kJ} - 34.4 \text{ kJ} = -548.6 \text{ kJ}$$

19.37 For gas phase reaction, $\Delta G^o = -RT \ln K_p$ [Equation 19.14]. Calculate ΔG^o from the data in Appendix C and use it to calculate K_p.

(a) $\Delta G^o = -190.5$ kJ (See Exercise 19.21(a))

$$-190.5 \text{ kJ} = \frac{-8.314 \times 10^{-3} \text{ kJ}}{\text{K} \cdot \text{mol}} \times 298 \text{ K} \ln K_p$$

$$\ln K_p = \frac{-190.5 \text{ kJ}}{-2.478 \text{ kJ}} = +76.88\ ; \quad K_p = e^{76.88} = 2.4 \times 10^{33}$$

(b) $\Delta G^o = \Delta G^o\ C_6H_6(g) - 3\Delta G^o\ C_2H_2(g) = 129.7 - 3(209.2) = -497.9$ kJ

$$\ln K_p = \frac{\Delta G^o}{-RT} = \frac{-497.9 \text{ kJ}}{-2.478 \text{ kJ}} = 200.9\ ; \quad K_p = e^{200.9} = 1.8 \times 10^{87}$$

(c) $\Delta G^\circ = 3\Delta G^\circ\ NO(g) - \Delta G^\circ\ N_2O(g) - \Delta G^\circ\ NO_2(g)$
$= 3(86.71) - (103.59) - (51.84) = +104.70$ kJ

$$\ln K_p = \frac{\Delta G^\circ}{-RT} = \frac{104.70\ kJ}{-2.4777} = -42.257\ ;\quad K_p = e^{-42.257} = 4.45 \times 10^{-19}$$

19.38 Following the same procedure as Exercise 19.37:

(a) $\Delta G^\circ = \Delta G^\circ\ NaOH(s) + \Delta G^\circ\ CO_2(g) - \Delta G^\circ\ NaHCO_3(s)$
$= -379.5 + (-394.4) - (-851.8) = +77.9$ kJ

$$\ln K_p = \frac{\Delta G^\circ}{-RT} = \frac{+77.9\ kJ}{-2.48} = -31.4\ ;\quad K_p = e^{-31.4} = 2 \times 10^{-14}$$

$K_p = P_{CO_2} = 2 \times 10^{-14}$

(b) $\Delta G^\circ = 2\Delta G^\circ\ HCl(g) + \Delta G^\circ\ Br_2(g) - 2\Delta G^\circ\ HBr(g) - \Delta G^\circ\ Cl_2(g)$
$= 2(-95.27) + 3.14 - 2(-53.22) - 0 = -80.96$ kJ

$$\ln K_p = \frac{-80.96}{-2.478} = +32.67\ ;\quad K_p = e^{32.67} = 1.5 \times 10^{14}$$

$$K_p = \frac{P^2_{HCl} \times P_{Br_2}}{P^2_{HBr} \times P_{Cl_2}} = 1.5 \times 10^{14}$$

(c) From Exercise 19.22(a), ΔG° at 298 K = -140.0 kJ.

$$\ln K_p = \frac{\Delta G^\circ}{-RT} = \frac{-140.0}{-2.478} = 56.50;\quad K_p = e^{56.50} = 3.4 \times 10^{24}$$

$$K_p = \frac{P^2_{SO_3}}{P^2_{SO_2} \times P_{O_2}} = 3.4 \times 10^{24}$$

19.39 $K_p = P_{CO_2}$. Calculate ΔG° at the two temperatures using $\Delta G^\circ = \Delta H^\circ - T\Delta S^\circ$ and then calculate K_p and P_{CO_2}.

$\Delta H^\circ = \Delta H^\circ_f\ CaO(s) + \Delta H^\circ_f\ CO_2(g) - \Delta H^\circ_f\ CaCO_3(s)$
$= -635.5 + -395.5 - (-1207.1) = +176.1$ kJ

$\Delta S^\circ = S^\circ\ CaO(s) + S^\circ\ CO_2(g) - S^\circ\ CaCO_3(s)$
$= 39.75 + 213.6 - 92.88 = +160.47$ J/K = 0.1605 kJ/K

(a) ΔG at 298 K = 176.1 kJ - 298 K$\left(\frac{0.1605\ kJ}{K}\right)$ = 128.3 kJ

$$\ln K_p = \frac{\Delta G^\circ}{-RT} = \frac{128.3}{-2.478} = -51.78$$

$K_p = e^{-51.78} = 3.3 \times 10^{-23}$; $P_{CO_2} = 3.3 \times 10^{-23}$ atm

(b) ΔG at 800 K = 176.1 kJ - 800 K$\left(\frac{0.1605\ kJ}{K}\right)$ = +47.7 kJ

$$\ln K_p = \frac{\Delta G^\circ}{-RT} = \frac{47.7}{-2.48} = -19.2$$

$K_p = e^{-19.2} = 4.6 \times 10^{-9}$; $P_{CO_2} = 4.6 \times 10^{-9}$ atm

19.40 (a) $\Delta G^\circ = \Delta G^\circ\ PbO(s) + \Delta G^\circ\ CO_2(g) - \Delta G^\circ\ PbCO_3(s)$

$= -187.9 - 394.4 - (-625.5) = 43.2\ kJ$

$$\ln K_p = \frac{\Delta G^\circ}{-RT} = \frac{43.2 \times 10^3\ J}{-8.314\ J/K \cdot mol\ (298\ K)} = -7.57$$

$K_p = P_{CO_2} = 10^{-7.57} = 2.7 \times 10^{-8}$

(b) $\Delta H^\circ = \Delta H^\circ\ PbO(s) + \Delta H^\circ\ CO_2(g) - \Delta H^\circ\ PbCO_3(s)$

$= -217.3 - 393.5 + 699.1 = 88.3\ kJ$

$\Delta S^\circ = S^\circ\ PbO(s) + S^\circ\ CO_2(g) - S^\circ\ PbCO_3(s)$

$= 68.70 + 213.6 - 131.0 = 151.3\ J/K$ or $0.1513\ kJ/K$

$\Delta G^\circ = \Delta H^\circ - T\Delta S^\circ = 88.3\ kJ - 773\ K\ (0.1513\ kJ) = -28.65\ kJ$

$$\ln K_p = \frac{-28.65 \times 10^3\ J}{-8.314\ J/K \cdot mol\ (773\ K)} = 1.94$$

$K_p = P_{CO_2} = 10^{1.94} = 86\ atm$

19.41 (a) $\Delta G^\circ = -RT \ln K_a = -(8.314)(298) \log (6.5 \times 10^{-5}) = +23.9\ kJ$

(b) When the system is at equilibrium ΔG is, by definition, equal to zero.

(c) $\Delta G = \Delta G^\circ + RT \ln Q$

$$= +23.9 \times 10^3 + (8.314)(298) \times \ln \frac{(3.0 \times 10^{-3})(2.0 \times 10^{-5})}{(0.10)} = -11.6\ kJ$$

The fact that ΔG is negative tells us that this particular mixture will move spontaneously in the direction of further ionization.

19.42 (a) $\Delta G^\circ = -RT \ln K_b = -\frac{8.314\ J}{1\ K \cdot mol}(298) \ln 1.8 \times 10^{-5}$

$= 2.70 \times 10^4\ J$ or $27.0\ kJ$

(b) By definition $\Delta G = 0$ at equilibrium

(c) $\Delta G = \Delta G^\circ + RT \ln ([NH_4^+][OH^-]/[NH_3])$

$$\Delta G = 27.0\ kJ + \frac{8.314\ J}{1\ K \cdot mol}(298) \times \ln \frac{(0.10)(0.05)}{0.10}$$

$\Delta G = 27.0\ kJ - 7.4\ kJ = +19.6\ kJ$

Additional Exercises

19.43 (a) False. The essential question is whether the reaction proceeds far to the right before arriving at equilibrium. The position of equilibrium, which is the essential aspect, is dependent on not only ΔH but the entropy change as well.

(b) True

(c) False. Spontaneity relates to the position of equilibrium in a process, not to the rate at which that equilibrium is approached.

(d) True

(e) False. Such a process might be spontaneous, but would not necessarily be so. Spontaneous processes are those that are exothermic and/or that lead to increased disorder in the system.

19.44

Process	ΔH	ΔS
(a)	+	+
(b)	-	-
(c)	+	+
(d)	+	+
(e)	-	+

19.45 There is no inconsistency. The second law states that in any spontaneous process there is an increase in the entropy of the universe. While there may be a decrease in entropy of the system, as in the present case, this decrease is more than offset by an increase in entropy of the surroundings.

19.46

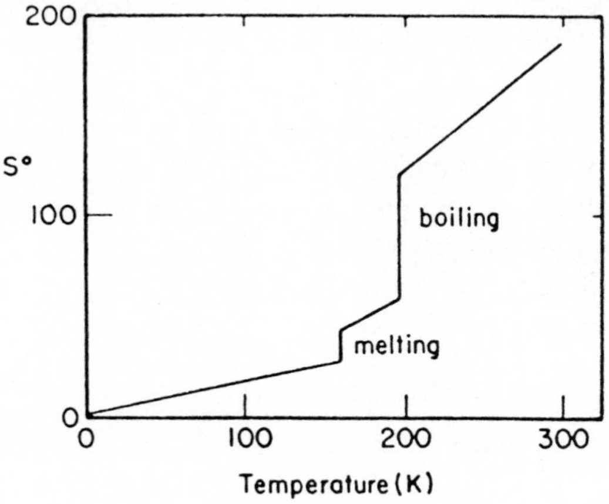

19.47 (a) (i) $2RbCl(s) + 3O_2(g) \rightarrow 2RbClO_3(s)$

$\Delta H^\circ = 2\Delta H_f^\circ\ RbClO_3(s) - 3\Delta H_f^\circ\ O_2(g) - 2\Delta H_f^\circ\ RbCl(s)$

$= 2(-392.4) - 3(0) - 2(-430.5) = +76.2$ kJ

$\Delta S^\circ = 2(152) - 3(205.0) - 2(92) = -495$ J/K $= -0.495$ kJ/K

$\Delta G^\circ = 2(-292.0) - 3(0) - 2(-412.0) = +240.0$ kJ

(ii) $2CH_4(g) \rightarrow C_2H_6(g) + H_2(g)$

$\Delta H^\circ = \Delta H_f^\circ\ C_2H_6(g) + \Delta H_f^\circ\ H_2(g) - 2\Delta H_f^\circ\ CH_4(g)$

$= -84.68 + 0 - 2(-74.8) = 64.9$ kJ

$\Delta S^\circ = 229.5 + 130.58 - 2(186.3) = -12.5$ J/K $= -0.0125$ kJ/K

$\Delta G^\circ = -32.89 + 0 - 2(-50.8) = +68.71$ kJ

(iii) $C_2H_2(g) + 4Cl_2(g) \rightarrow 2CCl_4(l) + H_2(g)$

$\Delta H^\circ = 2\Delta H_f^\circ\ CCl_4(l) + \Delta H_f^\circ\ H_2(g) - \Delta H_f^\circ\ C_2H_2(g) - 4\Delta H_f^\circ\ Cl_2(g)$

$= 2(-139.3) + 0 - (226.7) - 4(0) = -505.3$ kJ

$\Delta S^\circ = 2(214.4) + 130.58 - (200.8) - 4(222.96) = -533.3$ J/K $= -0.5333$ kJ/K

$\Delta G^\circ = 2(-68.6) + 0 - (209.2) - 4(0) = -346.4$ kJ

(b) (i) ΔG° is (+), nonspontaneous

(ii) ΔG° is (+), nonspontaneous

(iii) ΔG° is (-), spontaneous

(c) In each case the manner in which free energy change varies with temperature depends mainly on ΔS: $\Delta G = \Delta H - T\Delta S$. When ΔS is substantially positive, ΔG becomes more negative as temperature increases. When ΔS is substantially negative, ΔG becomes more positive as temperature increases.

(i) ΔS° is negative, ΔG° becomes more positive with increasing temperature.

(ii) ΔS° is small and negative, so ΔG° becomes slightly more positive with increasing temperature, but the effect is not large.

(iii) ΔS° is negative, ΔG° becomes more positive with increasing temperature. (The reaction will become nonspontaneous at some temperature.)

19.48 (a) When an equilibrium exists between the gaseous and liquid states, $A(g) \rightarrow A(l)$, ΔG for the process is zero. Under these conditions $\Delta G = \Delta H - T\Delta S = 0$. Thus $\Delta S = \Delta H/T$. Because the equilibrium between gas and liquid applies when the pressure of gas is 1 atm, both gas and liquid phases are present under standard conditions at the temperature in question. Thus $\Delta S = \Delta S^\circ$. The values of ΔS° are thus calculated as follows:

acetone: $\Delta S^\circ = 30.3 \times 10^3/329 = 92.1$ J/mol • K

benzene: $\Delta S^\circ = 30.7 \times 10^3/353 = 87.0$ J/mol • K

ammonia: $\Delta S^\circ = 23.4 \times 10^3/240 = 97.5$ J/mol • K

water: $\Delta S^\circ = 40.6 \times 10^3/373 = 109$ J/mol • K

(b) Note that ΔS° is significantly larger for NH_3 and H_2O, the two substances for which hydrogen bonding exists in the liquid. Furthermore, it is larger for H_2O than for NH_3, and we know that hydrogen bonding is more extensive in H_2O than in NH_3 (note the much higher boiling point for H_2O). The process of vaporization requires that the intermolecular hydrogen bonds be broken to a large degree. This process results in an increased randomness or disorder of the system; ΔS is thus more positive.

19.49 $\Delta G = \Delta G^\circ + RT \ln Q$

(a) $Q = \dfrac{P^2_{NH_3}}{P_{N_2} \times P^3_{H_2}} = \dfrac{(4.7)^2}{(8.5)(2.4)^3} = 0.19$

$\Delta G^\circ = 2\Delta G^\circ\ NH_3(g) - \Delta G^\circ\ N_2(g) - 3\Delta G^\circ\ H_2(g)$

$= 2(-16.66) - 0 - 3(0) = -33.32$ kJ

$\Delta G = -33.32\ \text{kJ} + \dfrac{8.314 \times 10^{-3}\ \text{kJ}}{\text{K} \cdot \text{mol}} \times 298\ \text{K} \times \ln(0.19)$

$\Delta G = -33.32 + 2.478 \ln(0.19) = -37.5$ kJ

(b) $Q = \dfrac{P^3_{N_2} \times P^4_{H_2O}}{P^2_{N_2H_4} \times P^2_{NO_2}} = \dfrac{(2.5)^3(1.2)^4}{(1.0 \times 10^{-3})^2(1.0 \times 10^{-3})^2} = 3.2 \times 10^{13}$

$\Delta G^\circ = 3\Delta G^\circ_f\ N_2(g) + 4\Delta G^\circ_f\ H_2O(g) - 2\Delta G^\circ_f\ N_2H_4(g) - 2\Delta G^\circ_f\ NO_2(g)$

$= 0 + 4(-228.61) - 2(159.4) - 2(51.84) = -1336.9$ kJ

$\Delta G = -1336.9\ \text{kJ} + 2.478 \ln 3.2 \times 10^{13} = -1259.8$ kJ

(c) $Q = \dfrac{P_{N_2} \times P^2_{H_2}}{P_{N_2H_4}} = \dfrac{(1 \times 10^{-3})(2 \times 10^{-4})^2}{6.0} = 6.7 \times 10^{-12} \rightarrow 7 \times 10^{-12}$

$\Delta G^\circ = \Delta G^\circ_f\ N_2(g) + 2\Delta G^\circ_f\ H_2(g) - \Delta G^\circ_f\ N_2H_4$

$= 0 + 2(0) - 159.4 = -159.4$ kJ

$\Delta G = -159.4\ \text{kJ} + 2.478 \ln 7 \times 10^{-12} = -223.0$ kJ

19.50 $\Delta G^\circ = -RT \ln K_p$

(a) $\Delta G^\circ = \dfrac{-8.314\ \text{J}}{1\ \text{K}} \times 718\ \text{K} \times \ln 50.2 = -23.4$ kJ

(b) $\Delta G^\circ = \dfrac{-8.314\ \text{J}}{1\ \text{K}} \times 1073\ \text{K} \times \ln 9.1 \times 10^2 = -60.8$ kJ

19.51

Reaction	Sign of ΔH°	Sign of ΔS°	K > 1?	Variation in K as Temp. Increases
(i)	-	-	yes	decrease
(ii)	+	+	no	increase
(iii)	+	+	no	increase
(iv)	+	+	no	increase

Note that positive ΔH° leads to a smaller value of K, while positive ΔS° increases the value of K, particularly at high temperatures.

19.52 (a) To obtain ΔH° from the equilibrium constant data, graph ln K at various temperatures vs. 1/T, being sure to employ absolute temperature. The slope of the linear relationship that should result is $\Delta H^\circ/R$; thus, ΔH° is easily calculated.

(b) Use $\Delta G^o = \Delta H^o - T\Delta S^o$ and $\Delta G^o = -RT \ln K$. Substituting this second expression into the first, we obtain

$$-RT \ln K = \Delta H^o - T\Delta S^o \;; \quad \ln K = \frac{-\Delta H^o}{RT} - \frac{-\Delta S^o}{R}$$

Thus, the constant in the equation given in the exercise is $\Delta S^o/R$.

19.53 (a) $\Delta G^o = -RT \ln K_p$ (Equation 19.14); $\ln K_p = -\Delta G^o/RT$

Use $\Delta G^o = \Delta H^o - T\Delta S^o$ to get ΔG^o at the two temperatures. Calculate ΔH^o and ΔS^o using data in Appendix C.

$2CH_4(g) \rightarrow C_2H_6(g) + H_2(g)$

$\Delta H^o = \Delta H_f^o\ C_2H_6(g) + \Delta H_f^o\ H_2(g) - 2\Delta H_f^o\ CH_4(g) = -84.68 + 0 - 2(-74.8) = +64.9$ kJ

$\Delta S^o = S^o\ C_2H_6(g) + S^o\ H_2(g) - 2S^o\ CH_4(g) = 229.5 + 130.58 - 2(186.3) = -12.5$ J/K

at 298 K, $\Delta G = 64.9 \text{ kJ} - 298 \text{ K}(-12.5 \times 10^{-3} \text{ kJ/K}) = 68.6$ kJ

$$\ln K_p = \frac{68.6 \text{ kJ}}{-(8.314 \times 10^{-3} \text{ kJ/K})(298 \text{ K})} = -27.7, \ K_p = e^{-27.7} = 9.4 \times 10^{-13}$$

at 773 K, $\Delta G = 64.9 \text{ kJ} - 773 \text{ K}(-12.5 \times 10^{-3} \text{ J/K}) = 74.6$ kJ

$$\ln K_p = \frac{74.6 \text{ kJ}}{-(8.314 \times 10^{-3} \text{ kJ/K})(773 \text{ K})} = -11.6, \ K_p = e^{-11.6} = 9.1 \times 10^{-6}$$

Because the reaction is endothermic, the value of K_p increases with an increase in temperature.

$2CH_4(g) + 1/2\ O_2(g) \rightarrow C_2H_6(g) + H_2O(g)$

$\Delta H^o = \Delta H_f^o\ C_2H_6(g) + \Delta H_f^o\ H_2O(g) - 2\Delta H_f^o\ CH_4(g) - 1/2\ \Delta H_f^o O_2(g)$

$= -84.68 + (-241.8) - 2(-74.8) - 1/2\ (0) = -176.9$ kJ

$\Delta S^o = S^o\ C_2H_6(g) + S^o\ H_2O(g) - 2S^o\ CH_4(g) - 1/2\ S^o\ O_2(g)$

$= 229.5 + 188.7 - 2(186.3) - 1/2\ (205.0) = -56.9$ J/K

at 298 K, $\Delta G = -176.9 \text{ KJ} - 298 \text{ K}(-56.9 \times 10^{-3} \text{ kJ/K}) = -159.9$ kJ

$$\ln K_p = \frac{-159.9 \text{ kJ}}{-(8.314 \times 10^{-3} \text{ kJ/K})(298 \text{ K})} = +64.54; \ \ K_p = e^{64.54} = 1.1 \times 10^{28}$$

at 773 K, $\Delta G = -176.9 \text{ kJ} - 773 \text{ K}(-56.9 \times 10^{-3} \text{ kJ/K}) = -132.9 \text{ kJ}$

$$\ln K_p = \frac{-132.9 \text{ kJ}}{-(8.314 \times 10^{-3} \text{ kJ/K})(773 \text{ K})} = +20.68; \quad K_p = e^{20.68} = 9.6 \times 10^8$$

Because this reaction is exothermic, the value of K_p decreases with increasing temperature.

(b) The difference in ΔG° for the two reactions is primarily enthalpic; the first reaction is endothermic and the second exothermic. Both reactions have $-\Delta S^\circ$, which inhibits spontaneity.

(c) This is an example of coupling a useful but nonspontaneous reaction with a spontaneous one to spontaneously produce a desired product.

$$2CH_4(g) \rightarrow C_2H_6(g) + H_2(g) \qquad \Delta G^\circ_{298} = +64.9 \text{ kJ, nonspontaneous}$$
$$H_2(g) + 1/2\, O_2(g) \rightarrow H_2O(g) \qquad \Delta G^\circ_{298} = -228.61 \text{ kJ, spontaneous}$$

$$2CH_4(g) + 1/2\, O_2(g) \rightarrow C_2H_6(g) + H_2O(g) \qquad \Delta G^\circ_{298} = -163.7\text{, spontaneous}$$

(ΔG° for the first reaction calculated using ΔG°_f values is 68.7 kJ; this difference results in the difference in ΔG°_{298} values calculated in parts (a) and (c) for the rection in the presence of O_2.)

(d) $CH_4(g) + 2O_2(g) \rightarrow CO_2(g) + 2H_2O(g)$

19.54 (a) The equilibrium of interest here can be written as:

$$K^+(\text{plasma}) \rightleftharpoons K^+(\text{muscle})$$

Since an aqueous solution is involved in both cases, assume that the equilibrium constant for the above process is exactly 1, that is, $\Delta G^\circ = 0$. However, ΔG is not zero because the concentrations are not the same on both sides of the membrane. Use equation [19.14] to calculate ΔG:

$$\Delta G = \Delta G^\circ + RT \ln \frac{[K^+ \text{ (muscle)}]}{[K^+ \text{ (plasma)}]}$$

$$= 0 + (8.314)(310) \ln \frac{(0.15)}{(5.0 \times 10^{-3})} = 8.77 \text{ kJ}$$

(b) Note that ΔG is positive. This means that work must be done on the system (blood plasma plus muscle cells) to move the K^+ ions "uphill," as it were. The minimum amount of work possible is given by the value for ΔG. This value represents the minimum amount of work required to transfer one mole of K^+ ions from the blood plasma at 5×10^{-3} M to muscle cell fluids at 0.15 M, assuming constancy of concentrations. In practice, a larger than minimum amount of work is required.

19.55 (a) $K = \dfrac{\chi_{CH_3COOH}}{\chi_{CH_3OH}\, P_{CO}}$

$\Delta G^\circ = -RT \ln K;\ \ln K_p = -\Delta G/RT$

$\Delta G^\circ = \Delta G_f^\circ\ CH_3COOH(l) - \Delta G_f^\circ\ CH_3OH(l) - \Delta G_f^\circ\ CO(g)$

$= -392.4 - (-166.23) - (-137.2) = -89.0$ kJ

$$\ln K_p = \frac{-89.0 \text{ kJ}}{(-8.314 \times 10^{-3} \text{ kJ/K})(298 \text{ K})} = 35.9;\ K_p = e^{35.9} = 4 \times 10^{15}$$

(b) $\Delta H^\circ = \Delta H_f^\circ\ CH_3COOH(l) - \Delta H_f^\circ\ CH_3OH(l) - \Delta H_f^\circ\ CO(g)$

$= -487.0 - (-238.6) - (-110.5) = -137.9$ kJ

The reaction is exothermic, so the value of K will decrease with increasing temperature, and the mole fraction of CH_3COOH will also decrease. Elevated temperatures must be used to increase the speed of the reaction. Thermodynamics cannot predict the rate at which a reaction reaches equilibrium.

(c) $\Delta G^\circ = -RT \ln K;\ K = 1, \ln K = 0, \Delta G^\circ = 0$

$\Delta G^\circ = \Delta H - T\Delta S$; when $\Delta G^\circ = 0$, $\Delta H^\circ = T\Delta S^\circ$

$\Delta S^\circ = S^\circ\ CH_3COOH(l) - S^\circ\ CH_3OH(l) - S^\circ\ CO(g)$

$= 159.8 - 126.8 - 197.9 = -164.9$ J/K $= -0.1649$ kJ/K

-137.9 kJ $= T(-0.1649$ kJ/K$)$, $T = 836.3$ K

The equilibrium favors products up to 836 K or 563° C, so elevated temperatures to increase the rate of reaction can be safely employed.

19.56 (a) First calculate ΔG° for each reaction:

For $C_6H_{12}O_6(s) + 6O_2(g) \rightleftharpoons 6CO_2(g) + 6H_2O(l)$ (A)

$\Delta G^\circ = 6(-236.8) + 6(-394.4) - (-912) = -2875$ kJ

For $C_6H_{12}O_6(s) \rightleftharpoons 2C_2H_5OH(l) + 2CO_2(g)$ (B)

$\Delta G^\circ = 2(-394.4) + 2(-174.8) - (-912) = -226$ kJ

For (A), $\ln K = 2875 \times 10^3/(8.314)(298) = 1160$; $K = 3 \times 10^{504}$

For (B), $\ln K = 226 \times 10^3/(8.314)(298) = 91.22$; $K = 4 \times 10^{39}$

(b) Both these values for K are unimaginably large. However, K for reaction (A) is larger, because ΔG° is more negative. The magnitude of the work that can be accomplished by coupling a reaction to its surroundings is measured by ΔG. According to the calculations above, considerably more work can in principle be obtained from reaction (A), because ΔG° is more negative.

19.57 (a) ΔH for the process is zero if N_2 behaves as an ideal gas. The process can be written as $N_2(1\ atm) \rightarrow N_2(0.5\ atm)$.

(b) The gas can be returned to its original condition by exerting a force on the piston to push the gas back into flask A. Again ΔH is zero.

(c) The overall changes in ΔG and ΔH for the system are zero, as must be the case if the system is returned to its original condition, because both enthalpy and free energy are state functions.

(d) In the expansion, the system does no work on its surroundings. In the compression phase, the surroundings must do work on the system. This work corresponds to the force exerted on the piston as it moves through chamber B, compressing the gas.

(e) The work done by the surroundings on the system in the compression of the gas results in an increase in entropy of the surroundings. Thus the overall entropy of the universe increases.

19.58 (a) $\Delta G^\circ = 3\Delta G_f^\circ\ S(s) + 2\Delta G_f^\circ\ H_2O(g) - \Delta G_f^\circ\ SO_2(g) - 2\Delta G_f^\circ\ H_2S(g)$
$= 3(0) + 2(-228.61) - (-300.4) - 2(-33.01) = -90.8\ kJ$

$$\ln K = \frac{\Delta G^\circ}{-RT} = \frac{-90.8\ kJ}{-(8.314 \times 10^{-3}\ kJ/K)(298\ K)} = +36.65;\quad K = e^{36.65} = 8.2 \times 10^{15}$$

(b) The reaction is highly spontaneous at 298 K and feasible in principle. However, use of $H_2S(g)$ produces a severe safety hazard for workers and the surrounding community.

(c) $$P_{H_2O} = \frac{25\ torr}{760\ torr/atm} = 0.033\ atm$$

$$K = \frac{P^2_{H_2O}}{P_{SO_2} \times P^2_{H_2S}};\quad P_{SO_2} = P_{H_2S} = x\ atm$$

$$K = 8.2 \times 10^{15} = \frac{(0.033)^2}{x(x)^2};\quad x^3 = \frac{(0.033)^2}{8.2 \times 10^{15}}$$

$x = 5.1 \times 10^{-7}$ atm

(d) $\Delta H^\circ = 3\Delta H_f^\circ\ S(s) + 2\Delta H_f^\circ\ H_2O(g) - \Delta H_f^\circ\ SO_2(g) - 2\Delta H_f^\circ\ H_2S(g)$
$= 3(0) + 2(-241.8) - (-296.9) - 2(-20.17) = -146.4\ kJ$
$\Delta S^\circ = 3S^\circ\ S(s) + 2S^\circ\ H_2O(g) - S^\circ\ SO_2(g) - 2S^\circ\ H_2S(g)$
$= 3(31.88) + 2(188.7) - 248.5 - 2(205.6) = -186.7\ J/K$

The reaction is exothermic ($-\Delta H$) and has a negative ΔS°, so the value of K will decrease and the reaction will become nonspontaneous at some higher temperature. The process will be less effective at elevated temperatures.

19.59 ΔG° for the metabolism of glucose is:

$6\Delta G^\circ\ CO_2(g) + 6\Delta G^\circ\ H_2O(l) - \Delta G^\circ\ C_6H_{12}O_6(s) - 6\Delta G^\circ\ O_2(g)$

$\Delta G^\circ = 6(-394.4) + 6(-236.81) - (-910.4\ kJ + 0) = -2876.9\ kJ$

$\text{moles ATP} = -2876.9\ kJ \times \frac{1\ \text{mol ATP}}{-30.5\ kJ} = 94.3\ \text{mol ATP/mol glucose}$

19.60 (a) $K_p = \frac{P^2_{NO_2}}{P_{N_2O_4}}$

Assume equal amounts means equal numbers of moles. For gases, $P = n(RT/V)$. In an equilibrium mixture, RT/V is a constant, so moles of gas is directly proportional to partial pressure. Gases with equal partial pressures will have equal moles of gas present. The condition $P_{NO_2} = P_{N_2O_4}$ leads to the expression $K_p = P_{NO_2}$. Thus, for any value of P_{NO_2} (or P_T), the condition of equal moles of the two gases can be achieved at some temperature.

Assume $P_{NO_2} = P_{N_2O_4} = 1.0$ atm.

$K_p = \frac{(1.0)^2}{1.0} = 1.0;\ \ \ln K_p = 0;\ \ \Delta G^\circ = 0 = \Delta H^\circ - T\Delta S^\circ\ ;\ \ T = \frac{\Delta H^\circ}{\Delta S^\circ}$

$\Delta H^\circ = 2\Delta H_f^\circ NO_2(g) - \Delta H_f^\circ N_2O_4(g) = 2(33.84) - 9.66 = +58.02\ kJ$

$\Delta S^\circ = 2S^\circ\ NO_2(g) - S^\circ\ N_2O_4(g) = 2(240.45) - 304.3 = 0.1766\ kJ/K$

$T = \frac{58.02\ kJ}{0.1766\ kJ/K} = 328.5\ K \text{ or } 55.5^\circ C$

(b) $P_T = 1.00$ atm; $P_{N_2O_4} = x$, $P_{NO_2} = 2x$; $x + 2x = 1.00$ atm

$x = P_{N_2O_4} = 0.33$ atm; $P_{NO_2} = 0.67$ atm

$K_p = \frac{(0.67)^2}{0.33} = 1.36;\ \ \Delta G^\circ = -RT \ln K = \Delta H^\circ - T\Delta S^\circ$

$-(8.314 \times 10^{-3}\ kJ/K)(\ln 1.36)\ T = 58.02\ kJ - (0.1766\ kJ/K)\ T$

$(-0.0026\ kJ/K)\ T + (0.1766\ kJ/K)\ T = 58.02\ kJ$

$(0.1740\ kJ/K)\ T = 58.02\ kJ;\ \ T = 333.4\ K$

(c) $P_T = 10.00$ atm; $x + 2x = 10.00$ atm

$x = P_{N_2O_4} = 3.33$ atm; $P_{NO_2} = 6.67$ atm

$K_p = \frac{(6.67)^2}{3.33} = 13.4\ ;\ \ -RT \ln K = \Delta H^\circ - T\Delta S^\circ$

$-(8.314 \times 10^{-3}\ kJ/K)(\ln 13.4)\ T = 58.02\ kJ - (0.1766\ kJ/K)\ T$

$-(0.0216\ kJ/K)T + (0.1766\ kJ/K)T = 58.02\ kJ$

$(0.1550\ kJ/K)T = 58.02\ kJ;\ \ \ \ T = 374.2\ K$

(d) The reaction is endothermic, so an increase in the value of K_p as calculated in parts (b) and (c) should be accompanied by an increase in T.

19.61 (a) $S = k \ln W$ (Equation 19.7), $k = \frac{R}{N}$, $W \alpha V^n$

$\Delta S = S_2 - S_1$; $S_1 = k \ln W_1$, $S_2 = k \ln W_2$

$\Delta S = k \ln W_2 - k \ln W_1$; $W_2 = cV_2^N$; $W_1 = cV_1^N$

(The number of particles in both cases is 6.02×10^{23}, or N.)

$\Delta S = k \ln cV_2^N - k \ln cV_1^N$; $\ln a^b = b \ln a$

$\Delta S = k\,N \ln cV_2 - k\,N \ln cV_1$; $\ln a - \ln b = \ln (a/b)$

$$\Delta S = k\,N \ln \left(\frac{cV_2}{cV_1}\right) = k\,N \ln \left(\frac{V_2}{V_1}\right); \quad R = kN$$

$$\Delta S = R \ln \left(\frac{V_2}{V_1}\right)$$

(b) In Figure 19.12, if the two flasks have equal volume x, the expanded volume, V_2 = 2x and the initial volume V_1 = x.

$$\Delta S = \frac{8.314\ J}{K \cdot mol} \ln \left(\frac{2x}{x}\right) = \frac{8.314\ J}{K \cdot mol} \ln 2 = 5.763\ J/K$$

(c) $\Delta G = \Delta H - T\Delta S = 0 - 298\ K(5.763\ J/K) = -1.717\ kJ$

The expansion is spontaneous at 298 K.

CHAPTER 20

Electrochemistry

Oxidation-Reduction Reactions

20.1 (a) I is reduced from +5 to 0; C is oxidized from +2 to +4.
(b) Hg is reduced from +2 to 0; N is oxidized from -2 to 0.
(c) N is reduced from +5 to +2; S is oxidized from -2 to 0.
(d) Cl is reduced from +4 to +3; O is oxidized from -1 to 0.

20.2 (a) No oxidation-reduction.
(b) N is both oxidized and reduced; it is reduced from +4 to +2 and oxidized from +4 to +5.
(c) No oxidation-reduction.
(d) S is reduced from +6 to +4; Br is oxidized from -1 to 0.

20.3 (a) $2N_2H_4(g) + N_2O_4(g) \rightarrow 3N_2(g) + 4H_2O(l)$
(b) N_2O_4 serves as the oxidizing agent; it is itself reduced. N_2H_4 serves as the reducing agent; it is itself oxidized.

20.4 (a) $2PbS(s) + 3O_2(g) \xrightarrow{\Delta} 2PbO(s) + 2SO_2(g)$
(b) O_2 is the oxidant, it undergoes reduction from the 0 to the -2 state. S is the reductant; it is oxidized from the -2 to the +4 state. The oxidation state of Pb is unchanged in the reaction.

20.5 (a) $H_2O_2(aq) + 2H^+(aq) \rightarrow 2H_2O(l)$

$H_2O_2(aq) + 2H^+(aq) + 2e^- \rightarrow 2H_2O(l)$ reduction

(b) $Cl_2(g) + 6H_2O(l) \rightarrow 2ClO_3^-(aq) + 12H^+(aq)$

$Cl_2(g) + 6H_2O(l) \rightarrow 2ClO_3^-(aq) + 12H^+(aq) + 10\ e^-$ oxidation

(c) $2OH^-(aq) \rightarrow O_2(g) + 2H^+(aq)$

$+2OH^-(aq) \qquad + 2OH^-(aq)$

$4\ OH^-(aq) \rightarrow O_2(g) + 2H_2O(l)$

$4\ OH^-(aq) \rightarrow O_2(g) + 2H_2O(l) + 4e^-$ oxidation

(d) $CrO_4^{2-}(aq) + 5H^+(aq) \rightarrow Cr(OH)_3(s) + H_2O(l)$

$+ 5OH^-(aq) \qquad + 5OH^-(aq)$

$CrO_4^{2-}(aq) + 5H_2O(l) \rightarrow Cr(OH)_3(s) + H_2O(l) + 5OH^-(aq)$

$CrO_4^{2-}(aq) + 4H_2O(l) + 3e^- \rightarrow Cr(OH)_3(s) + 5OH^-(aq)$ reduction

20.6 (a) $NO_3^-(aq) + 4H^+(aq) + 3e^- \rightarrow NO(g) + 2H_2O(l)$ reduction

(b) $H_2O_2(aq) \rightarrow O_2(g) + 2H^+(aq) + 2e^-$ oxidation

(c) $SO_3^{2-}(aq) + H_2O(l) + 2OH^- \rightarrow SO_4^{2-}(aq) + 2H^+(aq) + 2OH^-(aq)$

$SO_3^{2-}(aq) + 2OH^-(aq) \rightarrow SO_4^{2-}(aq) + H_2O(l) + 2\ e^-$ oxidation

(d) $2OH^-(aq) + 2H^+(aq) + ClO^-(aq) + 2e^- \rightarrow Cl^-(aq) + H_2O(l) + 2OH^-(aq)$

$ClO^-(aq) + H_2O(l) + 2e^- \rightarrow Cl^-(aq) + 2OH^-(aq)$ reduction

20.7 (a) $Pb(OH)_4^{2-}(aq) + ClO^-(aq) \rightarrow PbO_2(s) + Cl^-(aq) + 2OH^-(aq) + H_2O(l)$

(b) The half reactions are:

$3H_2O(l) + Tl_2O_3(s) + 4e^- \rightarrow 2TlOH(s) + 4OH^-(aq)$

$2[2NH_2OH(aq) + 2OH^-(aq) \rightarrow N_2(g) + 4H_2O(l) + 2e^-]$

Net: $Tl_2O_3(s) + 4NH_2OH(aq) \rightarrow 2TlOH(s) + 2N_2(g) + 5H_2O(l)$

(c) $2[Cr_2O_7^{2-}(aq) + 14H^+(aq) + 6e^- \rightarrow 2Cr^{3+}(aq) + 7H_2O(l)]$

$3[CH_3OH(aq) + H_2O(l) \rightarrow HCO_2H(aq) + 4H^+(aq) + 4e^-]$

Net: $2Cr_2O_7^{2-}(aq) + 3CH_3OH(aq) + 16H^+(aq) \rightarrow 4Cr^{3+}(aq) + 3HCO_2H(aq) + 11\ H_2O(l)$

(d) $2[MnO_4^-(aq) + 8H^+(aq) + 5e^- \rightarrow Mn^{2+}(aq) + 4H_2O(l)]$

$5[2Cl^-(aq) \rightarrow Cl_2(g) + 2e^-]$

Net: $2MnO_4^-(aq) + 10Cl^-(aq) + 16H^+(aq) \rightarrow 2Mn^{2+}(aq) + 5Cl_2(g) + 8H_2O(l)$

(e) $H_2O_2(aq) + 2e^- \rightarrow O_2(g) + 2H^+(aq)$

Since the reaction is in base, the H^+ can be "neutralized" by adding $2OH^-$ to each side of the equation to give $H_2O_2(aq) + 2OH^-(aq) + 2e^- \rightarrow O_2(g) + 2H_2O(l)$.
The other half reaction is $2[ClO_2(aq) + e^- \rightarrow ClO_2^-(aq)]$.

Net: $H_2O_2(aq) + 2ClO_2(aq) + 2OH^-(aq) \rightarrow O_2(g) + 2ClO_2^-(aq) + 2H_2O(l)$

(f) $3[NO_2^-(aq) + H_2O(l) \rightarrow NO_3^-(aq) + 2H^+(aq) + 2\,e^-]$

$Cr_2O_7^{2-}(aq) + 14H^+(aq) + 6\,e^- \rightarrow 2Cr^{3+}(aq) + 7H_2O(l)$

Net: $3NO_2^-(aq) + Cr_2O_7^{2-}(aq) + 8H^+(aq) \rightarrow 3NO_3^-(aq) + 2Cr^{3+}(aq) + 4H_2O(l)$

20.8 (a) $Cr_2O_7^{2-}(aq) + I^-(aq) + 8H^+(aq) \rightarrow 2Cr^{3+}(aq) + IO_3^-(aq) + 4H_2O(l)$

(b) $4MnO_4^-(aq) + 5CH_3OH(aq) + 12H^+(aq) \rightarrow 4Mn^{2+}(aq) + 5HCO_2H(aq) + 11H_2O(l)$

(c) $4As(s) + 3ClO_3^-(aq) + 3H^+(aq) + 6H_2O(l) \rightarrow 4H_3AsO_3(aq) + 3HClO(aq)$

(d) $As_2O_3(s) + 2NO_3^-(aq) + 2H_2O(l) + 2H^+(aq) \rightarrow 2H_3AsO_4(aq) + N_2O_3(aq)$

(e) $2MnO_4^-(aq) + Br^-(aq) + H_2O(l) \rightarrow 2MnO_2(s) + BrO_3^-(aq) + 2OH^-(aq)$

(f) $4H_2O_2(aq) + Cl_2O_7(aq) + 2OH^-(aq) \rightarrow 2ClO_2^-(aq) + 4O_2(g) + 5H_2O(l)$

Voltaic Cells; Emf

20.9 The salt bridge provides a mechanism by which ions not directly involved in the redox reaction can migrate into the anode and cathode compartments to maintain charge neutrality of the solutions. Ionic conduction within the cell, through the salt bridge, completes the cell circuit.

20.10 In the cathode compartment of the voltaic cell in Figure 20.5, Cu^{2+} cations are reduced to Cu° atoms, decreasing the number of positively charged particles in the compartment. Na^+ cations are drawn into the compartment to maintain charge balance as Cu^{2+} ions are removed.

20.11 (a) $Zn(s) \rightarrow Zn^{2+}(aq) + 2e^-$; $Ni^{2+}(aq) + 2e^- \rightarrow Ni(s)$

(b) Zn(s) is the anode; Ni(s) is the cathode.

(c) Zn(s) is negative (-); Ni(s) is positive(+).

(d) Electrons flow from the Zn(-) electrode toward the Ni(+) electrode.

(e) Cations migrate toward the Ni(s) cathode; anions migrate toward the Zn(s) anode.

20.12 (a) $Ag^+(aq) + 1e^- \rightarrow Ag^\circ(s)$; $Ni(s) \rightarrow Ni^{2+}(aq) + 2e^-$

(b) Ni(s) is the anode, Ag(s) is the cathode.

(c) Ni(s) is negative, Ag(s) is positive.

(d) Electrons flow from the Ni(-) electrode to the Ag(+) electrode.

(e) Cations migrate toward the Ag(s) cathode, anions migrate toward the Ni(s) anode.

20.13 (a) The two half-reactions are:

$Tl^{3+}(aq) + 2e^- \rightarrow Tl^+(aq)$ $\quad E^\circ_{red} = ?$

$2(Cr^{2+}(aq) \rightarrow Cr^{3+}(aq) + e^-)$ $\quad E^\circ_{ox} = 0.41$

$E^\circ = E^\circ_{red} + E^\circ_{ox} = 1.19\text{ V};\ E^\circ_{red} = E^\circ - E^\circ_{ox} = 1.19\text{ V} - 0.41\text{ V} = 0.78\text{ V}$

(b) "Standard Conditions" refers to solutions that are 1 *M* in concentration with all gases at 1 atm pressure.

(c)

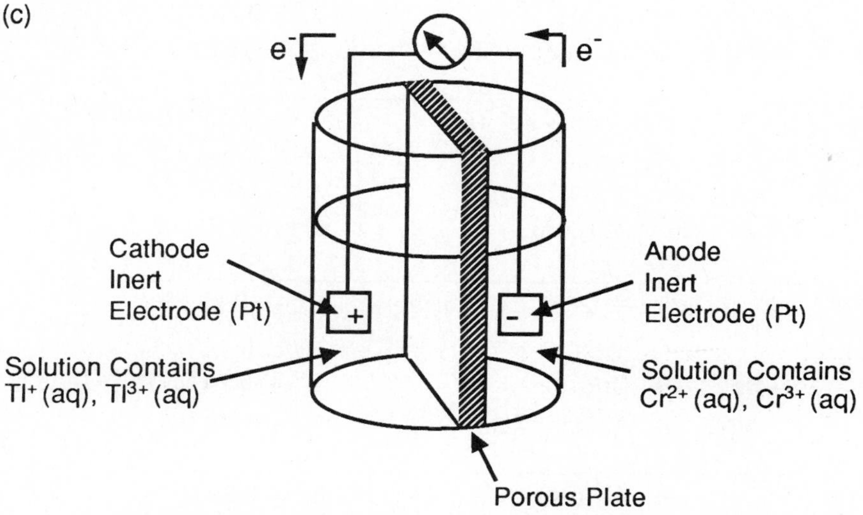

Note that because $Cr^{2+}(aq)$ is readily oxidized, it would be necessary to keep oxygen out of the right-hand cell compartment.

20.14 (a) $PdCl_4^{2-}(aq) + 2e^- \rightarrow Pd(s) + 4Cl^-$ $\quad E^\circ = ?$

$Cd(s) \rightarrow Cd^{2+}(aq) + 2e^-$ $\quad E^\circ = 0.403\text{ V}$

Note that E° for oxidation of Cd is opposite in sign to E° for the reduction of Cd^{2+}.

(b) $E^\circ = E^\circ_{red} + E^\circ_{ox};\ 1.03\text{ V} = E^\circ_{red} + 0.403\text{ V};\ E^\circ_{red} = 0.63\text{ V}$

(c)

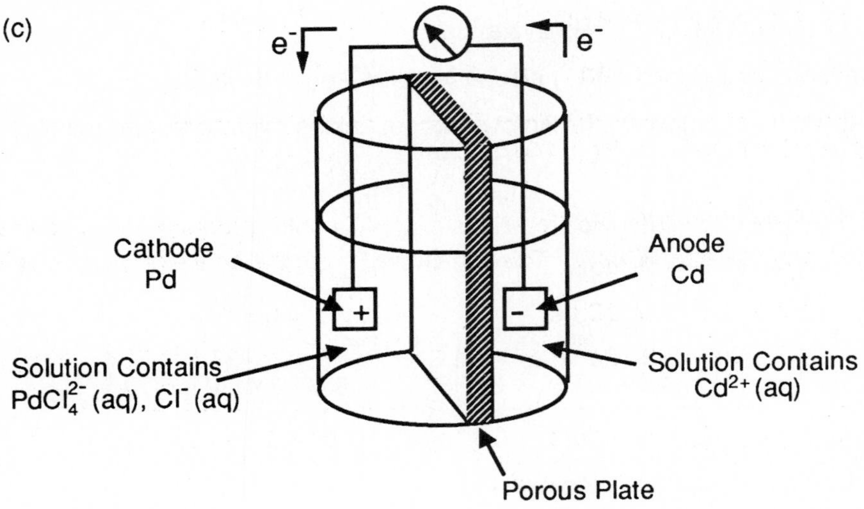

20.15 (a)

$$Cu^{2+}(aq) + 2e^- \rightarrow Cu(s) \qquad E^\circ = 0.337\ V$$
$$Fe(s) \rightarrow Fe^{2+}(aq) + 2e^- \qquad E^\circ = 0.440\ V$$

$$Cu^{2+}(aq) + Fe(s) \rightarrow Cu(s) + Fe^{2+}(aq) \qquad E^\circ = 0.777\ V$$

(b)

$$2[Fe^{3+}(aq) + 1e^- \rightarrow Fe^{2+}(aq)] \qquad E^\circ = 0.771\ V$$
$$2I^-(aq) + 2e^- \rightarrow I_2(aq) \qquad E^\circ = -0.536\ V$$

$$2Fe^{3+}(aq) + 2I^-(aq) \rightarrow 2Fe^{2+}(aq) + I_2(aq) \qquad E^\circ = 0.235\ V$$

2.16 (a)

$$Cr^{3+}(aq) + 3e^- \rightarrow Cr(s) \qquad E^\circ = -0.74\ V$$
$$Al(s) \rightarrow Al^{3+}(aq) + 3e^- \qquad E^\circ = 1.66\ V$$

$$Al(s) + Cr^{3+}(aq) \rightarrow Al^{3+}(aq) + Cr(s) \qquad E^\circ = 0.92\ V$$

(b)

$$Cl_2(g) + 2e^- \rightarrow 2Cl^-(aq) \qquad E^\circ = 1.359\ V$$
$$2[Fe^{2+}(aq) \rightarrow Fe^{3+}(aq) + 1e^-] \qquad E^\circ = -0.771\ V$$

$$2Fe^{2+}(aq) + Cl_2(g) \rightarrow 2Fe^{3+}(aq) + 2Cl^-(aq) \qquad E^\circ = 0.588\ V$$

20.17 (a)

$$2[Au(s) + 4Br^-(aq) \rightarrow AuBr_4^-(aq) + 3e^-] \qquad E^\circ = 0.858\ V$$
$$3[2e^- + IO^-(aq) + H_2O(l) \rightarrow I^-(aq) + 2OH^-(aq)] \qquad E^\circ = 0.49\ V$$

$$2Au(s) + 8Br^-(aq) + 3IO^-(aq) + 3H_2O(l) \rightarrow 2AuBr_4^-(aq) + 3I^-(aq) + 6OH^-(aq)$$
$$E^\circ = 0.858 + 0.49 = 1.35\ V$$

(b)

$$2Eu^{2+}(aq) + Sn^{2+}(aq) \rightarrow 2Eu^{3+}(aq) + Sn(s)$$
$$E^\circ = 0.43 - 0.14 = 0.29\ V$$

20.18 (a)

$$Hg^{2+}(aq) + 2\,e^- \rightarrow Hg(l) \qquad E^\circ = 0.854\ V$$
$$Mn(s) \rightarrow Mn^{2+}(aq) + 2\,e^- \qquad E^\circ = 1.18\ V$$

$$Hg^{2+}(aq) + Mn(s) \rightarrow Hg(l) + Mn^{2+}(aq) \qquad E^\circ = 2.03\ V$$

(b) The reverse of the two reactions above would produce the smallest (most negative) cell emf, -2.03 V. The combination with the smallest positive cell emf is

$$2Cu^+(aq) + 2\,e^- \rightarrow 2Cu(s) \qquad E^\circ = 0.521\ V$$
$$Co(s) \rightarrow Co^{2+}(aq) + 2\,e^- \qquad E^\circ = 0.277\ V$$

$$2Cu^+(aq) + Co(s) \rightarrow 2Cu(s) + Co^{2+}(aq) \qquad E^\circ = 0.798\ V$$

Ease of Oxidation or Reduction; Spontaneity

20.19 The reduction half-reactions are:

$$Fe^{3+}(aq) + 1e^- \rightarrow Fe^{2+}(aq) \qquad E^\circ = 0.771\ V$$
$$H_2O_2(aq) + 2H^+(aq) + 2e^- \rightarrow 2H_2O(l) \qquad E^\circ = 1.776\ V$$

(a) Because it has the larger E°_{red}, the second half-reaction occurs more readily.

(b) Since the second half-reaction occurs more readily, it is the reduction half-reaction and occurs at the cathode of the complete cell.

(c) $Fe^{2+}(aq) \rightarrow Fe^{3+}(aq) + 1e^-$

(d) $E^\circ_{cell} = 1.776\ V - 0.771\ V = 1.005\ V$

20.20 (a) The half-reactions are:

$$2H^+(aq) + 2e^- \rightarrow H_2(g) \qquad E^\circ = 0.00\ V$$
$$Al(s) \rightarrow Al^{3+}(aq) + 3e^- \qquad E^\circ = 1.66\ V$$

Because it has the larger E°_{red} (0.00 V vs. -1.66 V) the first half reaction is the reduction half reaction in a voltaic cell. The standard hydrogen electrode (SHE) is the cathode and the Al strip is the anode.

(b) The Al strip will lose mass as the reaction proceeds, because Al(s) is transformed to $Al^{3+}(aq)$.

(c) $E^\circ_{cell} = 0.00\ V - (-1.66\ V) = 1.66\ V$

20.21 The reduction half-reactions are:

$$Cu^{2+}(aq) + 2e^- \rightarrow Cu(s) \qquad E^\circ = 0.337\ V$$
$$Sn^{2+}(aq) + 2e^- \rightarrow Sn(s) \qquad E^\circ = -0.136\ V$$

(a) It is evident that Cu^{2+} is more readily reduced. Therefore, Cu serves as the cathode, Sn as the anode.

(b) The copper electrode gains mass as Cu is plated out, the Sn electrode loses mass as Sn is oxidized.

(c) The overall cell reaction is $Cu^{2+}(aq) + Sn(s) \rightarrow Cu(s) + Sn^{2+}(aq)$

(d) $E^o = 0.337 + 0.136 = 0.473$ V

20.22 (a) The two half-reactions are:

$Pb^{+2} + 2e^- \rightarrow Pb^o \qquad E^o = -0.126$ V

$Cl_2 + 2e^- \rightarrow 2Cl^- \qquad E^o = 1.359$ V

Because E^o for the reduction of Cl_2 is higher, the reduction of Cl_2 occurs at the Pt cathode. The Pb electrode is the anode.

(b) Neither elecrode gains mass, because the Cl^- ions produced at the cathode are not deposited on the Pt, and the Pb anode loses mass as $Pb^{2+}(aq)$ is produced.

(c) $Cl_2(g) + Pb(s) \rightarrow Pb^{2+}(aq) + 2Cl^-(aq)$

(d) $E^o = 1.359$ V $+ 0.126$ V $= 1.485$ V

20.23 (a) Arranged in order of increasing strength as oxidizing agents (and reduction potential):

$Cu^{2+}(aq) < O_2(g) < Cr_2O_7^{2-}(aq) < Cl_2(g) < H_2O_2(aq)$

(b) Arranged in order of increasing strength as reducing agents (and oxidation potential):

$H_2O_2(aq) < I^-(aq) < Sn^{2+}(aq) < Zn(s) < Al(s)$

20.24 (a) The strongest oxidizing agent is the species most readily reduced, as evidenced by a large, positive reduction potential. That species is Ce^{4+}. The weakest oxidizing agent is the species that least readily accepts an electron. We expect that it will be very difficult to reduce Zn(s); indeed, Zn(s) acts as a comparatively strong reducing agent. No potential is listed for reduction of Zn(s), but we can safely assume that it is less readily reduced than any of the other species present.

(b) The strongest reducing agent is the species most easily oxidized (the largest positive oxidation potential). Zn, $E^o_{ox} = 0.76$ V, is the strongest reducing agent and F^-, $E^o_{ox} = -2.87$ V, is the weakest.

20.25 Any of the reduced species in Table 20.1 for which the oxidation potential is greater than 0.43 V (the reverse of any half-reaction with a reduction potential more negative than -0.43 V) will reduce Eu^{3+} to Eu^{2+}. These include Zn(s), $H_2(g)$, etc. Fe(s) is questionable.

20.26 Any oxidized species with a reduction potential greater than 0.59 V will oxidize RuO_4^{2-} to RuO_4^-. MnO_4^- is on the borderline, but any species above it on Table 20.1 would be strong enough.

20.27 The criterion for spontaneity is the value of E° for each reaction. If E° is positive, the process is spontaneous; if negative, the process is spontaneous in the reverse direction. To determine E° we must sum E° values for the two half-reactions, taking care to reverse the sign of E° when we reverse the direction of reaction.

(a)
$$Hg_2^{2+}(aq) + 2e^- \rightarrow 2Hg(l) \qquad E^\circ = 0.789\ V$$
$$Zn(s) \rightarrow Zn^{2+}(aq) + 2e^- \qquad E^\circ = 0.763\ V$$
$$Hg_2^{2+}(aq) + Zn(s) \rightarrow 2Hg(l) + Zn^{2+}(aq) \qquad E^\circ = 1.552\ V\text{, spontaneous}$$

(b)
$$I_2(aq) + 2e^- \rightarrow 2I^-(aq) \qquad E^\circ = 0.536\ V$$
$$2Ag(s) + 2I^-(aq) \rightarrow 2AgI(s) + 2e^- \qquad E^\circ = 0.151\ V$$
$$2Ag(s) + I_2(aq) \rightarrow 2AgI(s) \qquad 0.687\ V\text{, spontaneous}$$

(c)
$$2CO_2(g) + 2H^+(aq) + 2e^- \rightarrow H_2C_2O_4(aq) \qquad E^\circ = -0.49\ V$$
$$Cl_2(g) + 2H_2O(l) \rightarrow 2HClO(aq) + 2H^+(aq) + 2e^- \qquad E^\circ = -1.63\ V$$
$$2CO_2(g) + Cl_2(g) + 2H_2O(l) \rightarrow H_2C_2O_4 + 2HClO \qquad E^\circ = -2.12\ V\text{, nonspontaneous}$$

20.28 (a)
$$2I^-(aq) \rightarrow I_2(s) + 2e^- \qquad E^\circ = -0.536\ V$$
$$Hg_2^{2+}(aq) + 2e^- \rightarrow 2Hg(l) \qquad E^\circ = +0.789\ V$$
$$2I^-(aq) + Hg_2^{2+}(aq) \rightarrow I_2(s) + 2Hg(l) \qquad E^\circ = 0.253\ V\text{, spontaneous}$$

(b)
$$3[Cu^+(aq) \rightarrow Cu^{2+}(aq) + 1\ e^-] \qquad E^\circ = -0.153\ V$$
$$NO_3^-(aq) + 4H^+(aq) + 3e^- \rightarrow NO(g) + H_2O(l) \qquad E^\circ = 0.96\ V$$
$$3Cu^+(aq) + NO_3^-(aq) + 4H^+(aq) \rightarrow 3Cu^{2+}(aq) + NO(g) + 2H_2O(l) \qquad E^\circ = +0.81\ V\text{, spontaneous}$$

(c)
$$2[Cr(OH)_3(s) + 5OH^-(aq) \rightarrow CrO_4^{2-}(aq) + 4H_2O(l) + 3e^-] \qquad E^\circ = 0.13\ V$$
$$3[ClO^-(aq) + H_2O(l) + 2e^- \rightarrow Cl^-(aq) + 2OH^-(aq)] \qquad E^\circ = 0.89\ V$$
$$2Cr(OH)_3(s) + 3ClO^-(aq) + 4OH^-(aq) \rightarrow 2CrO_4^{2-}(aq) + 3Cl^-(aq) + 5H_2O(l) \qquad E^\circ = +1.02\ V\text{ spontaneous}$$

20.29 $\Delta G^\circ = -nFE^\circ$

(a) $\Delta G^\circ = -2\text{ mol e}^- \times 96{,}500\ \dfrac{J}{V\bullet \text{mol e}^-} \times \dfrac{1\ kJ}{1000\ J} \times 1.552\ V = -299.5\ kJ$

(b) $\Delta G^\circ = -2(96.5)(0.687) = -133\ kJ$

(c) $\Delta G^\circ = -2(96.5)(-212) = +409\ kJ$

20.30 (a) $2Fe^{2+}(aq) + S_2O_6^{2-}(aq) + 4H^+(aq) \rightarrow 2Fe^{3+}(aq) + 2H_2SO_3(aq)$

$E° = 0.60\ V - 0.771\ V = -0.17\ V$

$2Fe^{2+}(aq) + N_2O(aq) + 2H^+(aq) \rightarrow 2Fe^{3+}(aq) + N_2(g) + H_2O(l)$

$E° = -1.77\ V - 0.771\ V = -2.54\ V$

$Fe^{2+}(aq) + VO_2^+(aq) + 2H^+(aq) \rightarrow Fe^{3+}(aq) + VO^{2+}(aq) + H_2O(l)$

$E° = 1.00\ V - 0.771\ V = +0.23\ V$

(b) $\Delta G° = -nFE°$. For the first reaction,

$\Delta G° = -2\ mol \times \frac{96,500\ J}{1\ V \cdot mol} \times -0.17\ V = 3.3 \times 10^5\ J$ or 33 kJ

For the second reaction, $\Delta G° = -2(96,500)(-2.54) = 4.90 \times 10^3$ kJ

For the third reaction, $\Delta G° = -1(96,500)(0.23) = -22$ kJ

Nernst Equation; K

20.31 $Al(s) + 3Ag^+(aq) \rightarrow Al^{3+}(aq) + 3Ag(s)$; $E = E° - \frac{0.0592}{n} \log Q$; $Q = \frac{[Al^{3+}]}{[Ag^+]^3}$

Any change that causes the reaction to be less spontaneous (that causes Q to increase and ultimately shifts the equilibrium to the left) will result in a less positive value for E°.

(a) Decreases E by increasing $[Al^{3+}]$ on the right side of the equation.

(b) No effect; the "concentrations" of pure solids and liquids do not influence the value of K for a heterogeneous equilibrium.

(c) Decreases E; adding water decreases the concentration of Ag^+, which increases Q.

(d) No effect; the concentration of Ag^+ and the value of Q are unchanged.

20.32 $Zn(s) + 2H^+(aq) \rightarrow Zn^{2+}(aq) + H_2(g)$; $E = E° - \frac{0.0592}{n} \log \frac{[Zn^{2+}]\, P_{H_2}}{[H^+]^2}$

(a) P_{H_2} decreases, Q decreases, E increases

(b) No effect (Zn(s) does not appear in Q expression.)

(c) $[H^+]$ increases, Q decreases, E increases

(d) No effect ($NaNO_3$ does not participate in the reaction.)

20.33 (a)

$3[Mn^{2+}(aq) + 2e^- \rightarrow Mn(s)]$ $E° = -1.18\ V$

$2[Al(s) \rightarrow Al^{3+}(aq) + 3e^-]$ $E° = 1.66\ V$

$3Mn^{2+}(aq) + 2Al(s) \rightarrow 3Mn(s) + 2Al^{3+}(aq)$ $E° = 0.48\ V$

(b) $E = E° - \frac{0.0592}{n} \log \frac{[Al^{3+}]^2}{[Mn^{2+}]^3}$; n = 6

$E = 0.48 - \frac{0.0592}{6} \log \frac{(1.5)^2}{(0.10)^3} = 0.48 - \frac{0.592}{6} \log (2.25 \times 10^3)$

$E = 0.48 - \frac{0.0592}{6} (3.35) = 0.48 - 0.033 = 0.45 \text{ V}$

20.34 (a)

$3[Ce^{4+}(aq) + 1e^- \rightarrow Ce3^{+}(aq)]$ $E° = 1.61 \text{ V}$

$Cr(s) \rightarrow Cr^{3+}(aq) + 3e^-$ $E° = 0.74 \text{ V}$

$3Ce^{4+}(aq) + Cr(s) \rightarrow 3Ce^{3+}(aq) + Cr^{3+}(aq)$ $E° = 2.35 \text{ V}$

(b) $E = E° - \frac{0.0592}{n} \log \frac{[Ce^{3+}]^3 [Cr^{3+}]}{[Ce^{4+}]^3}$; n = 3

$E = 2.35 - \frac{0.0592}{3} \log \frac{(0.010)^3 (0.010)}{(1.5)^3} = 2.35 - \frac{0.0592}{3} \log (3.0 \times 10^{-9})$

$E = 2.35 - \frac{0.0592 (-8.53)}{3} = 2.35 + 0.168 = 2.52 \text{ V}$

20.35 (a)

$4[Fe^{2+}(aq) \rightarrow Fe^{3+}(aq) + 1e^-]$ $E° = -0.771 \text{ V}$

$O_2(g) + 4H^+(aq) + 4e^- \rightarrow 2H_2O(l)$ $E° = 1.23 \text{ V}$

$4Fe^{2+}(aq) + O_2(g) + 4H^+(aq) \rightarrow 4Fe^{3+}(aq) + 2H_2O(l)$ $E° = 0.46 \text{ V}$

(b) $E = E° - \frac{0.0592}{n} \log \frac{[Fe^{3+}]^4}{[Fe^{2+}]^4 [H^+]^4 P_{O_2}}$; n = 4, $[H^+] = 1.00 \times 10^{-3}$ *M*

$E = 0.46 \text{ V} - \frac{0.0592}{4} \log \frac{(1.0 \times 10^{-3})^4}{(2.0)^4 (1.0 \times 10^{-3})^4 (0.50)} = 0.46 - \frac{0.0592}{4} \log(0.125)$

$E = 0.46 - \frac{0.0592}{4} (-0.903) = 0.46 + 0.0134 = 0.47 \text{ V}$

30.36 (a)

$2[Fe^{3+}(aq) + 1e^- \rightarrow Fe^{2+}(aq)]$ $E° = 0.771 \text{ V}$

$H_2(g) \rightarrow 2H^+(aq) + 2e^-$ $E° = 0.000 \text{ V}$

$2Fe^{3+}(aq) + H_2(g) \rightarrow 2Fe^{2+}(aq) + 2H^+(aq)$ $E° = 0.771 \text{ V}$

(b) $E = E^\circ - \frac{0.0592}{n} \log \frac{[Fe^{2+}]^2[H^+]^2}{[Fe^{3+}]^2\ p_{H_2}}$; $[H^+] = 10^{-pH} = 1.0 \times 10^{-4}$, $n = 2$

$E = 0.771 - \frac{0.0592}{2} \log \frac{(0.010)^2\ (1.0 \times 10^{-4})^2}{(0.50)^2 (0.25)} = 0.771 - \frac{0.0592}{2} \log (1.6 \times 10^{-11})$

$E = 0.771 - \frac{0.0592\ (-10.80)}{2} = 0.771 + 0.320 = 1.091\ V$

20.37 $E = E^\circ - \frac{0.0592}{2} \log \frac{[H_2]\ [Zn^{2+}]}{[H^+]^2}$; $E^\circ = 0.0\ V + 0.763\ V = 0.763\ V$

$0.720 = 0.763 - \frac{0.0592}{2} \times (\log [H_2]\ [Zn^{2+}] - 2 \log [H^+]) = 0.763 - \frac{0.0592}{2} \times (-1.00 - 2 \log[H^+])$

$0.720 = 0.763 + 0.0296 + 0.0592 \log [H^+]$; $\log [H^+] = \frac{0.720 - 0.763 - 0.0296}{0.0592}$

$\log [H^+] = -1.23$; $[H^+] = 0.059\ M$; $pH = 1.23$

20.38 $E^\circ = -0.136\ V + 0.126\ V = -0.010\ V$; $n = 2$

$0.22 = -0.010 - \frac{0.0592}{2} \log \frac{[Pb^{2+}]}{[Sn^{2+}]} = -0.010 - \frac{0.0592}{2} \log \frac{[Pb^{2+}]}{1.00}$

$\log [Pb^{2+}] = \frac{-0.23\ V(2)}{0.0592} = -7.77$; $[Pb^{2+}] = 1.7 \times 10^{-8}$

For $PbSO_4(s)$, $K_{sp} = [Pb^{2+}]\ [SO_4^{2-}] = (1.0)\ (1.7 \times 10^{-8}) = 1.7 \times 10^{-8}$

20.39 (a) $E^\circ = 0.763 + (-0.136) = 0.627\ V$; $n = 2$ ($Sn^{+2} + 2e^- \rightarrow Sn$)

$\log K = \frac{2(0.627)}{0.0592} = 21.2$; $K = 2 \times 10^{21}$

(b) $E^\circ = 0.277 + (0) = 0.277\ V$; $n = 2$ ($2H^+ + 2e^- \rightarrow H_2$)

$\log K = \frac{2(0.277)}{0.0592} = 9.36$; $K = 2.3 \times 10^9$

(c) $E^\circ = -1.065 + 1.51 = 0.44\ V$; $n = 10$ ($2MnO_4^- + 10e^- \rightarrow 2Mn^{+2}$)

$\log K = \frac{10(0.44)}{0.0592} = 74.3 \approx 74$; $K = 1 \times 10^{74}$

20.40 $E^\circ = \frac{0.0592\ V}{n} \log K$; $\log K = \frac{nE^\circ}{0.0592\ V}$

(a) $E^\circ = -0.28\ V + 1.00\ V = 0.72\ V$; $n = 2$ ($Ni^{+2} + 2e^- \rightarrow Ni$)

$\log K = \frac{2(0.72\ V)}{0.0592\ V} = 24$; $K = 1 \times 10^{24}$

(b) $E^\circ = 1.61\ V - 0.32\ V = 1.29\ V$; $n = 3\ (Ce^{4+} + 3e^- \rightarrow 3Ce^{3+})$

$$\log K = \frac{3(1.29)}{0.0592} = 65.5;\ K = 3 \times 10^{65}$$

(c) $E^\circ = 0.36\ V + 0.23\ V = 0.59\ V$; $n = 4\ (4Fe(CN)_6^{3-} + 4e^- \rightarrow 4Fe(CN)_6^{4-})$

$$\log K = \frac{4(0.59)}{0.0592} = 40;\ K = 1 \times 10^{40}$$

20.41 $E^\circ = \dfrac{0.0592}{n} \log K$; $\log K = \dfrac{nE^\circ}{0.0592\ V}$

(a) $\log K = \dfrac{1(0.35\ V)}{0.0592\ V} = 5.91$; $K = 8.2 \times 10^5$

(b) $\log K = \dfrac{2(0.35\ V)}{0.0592\ V} = 11.82$; $K = 6.7 \times 10^{11}$

(c) $\log K = \dfrac{3(0.35\ V)}{0.0592\ V} = 17.74 \quad K = 5.5 \times 10^{17}$

20.42 $E^\circ = \dfrac{0.0592\ V}{n} \log K$; $n = \dfrac{0.0592\ V}{E^\circ} \log K$

$$n = \frac{0.0592\ V}{0.21\ V} \log 1.31 \times 10^7;\quad n = 2$$

Commercial Voltaic Cells

20.43 The overall cell reaction is:

$$Zn(s) + 2NH_4^+(aq) + 2MnO_2(s) \rightarrow Mn_2O_3(s) + 2NH_3(aq) + H_2O(l)$$

$$120\ g\ Zn \times \frac{1\ mol\ Zn}{65.4\ g\ Zn} \times \frac{2\ mol\ MnO_2}{1\ mol\ Zn} \times \frac{86.9\ g\ MnO_2}{1\ mol\ MnO_2} = 319\ g\ MnO_2$$

20.44 The overall cell reaction (Equation [20.19]) is:

$$Pb(s) + PbO_2(s) + 4H^+(aq) + 2SO_4^{2-}(aq) \rightarrow 2PbSO_4(s) + 2H_2O(l)$$

$$600\ g\ Pb \times \frac{1\ mol\ Pb}{207\ g\ Pb} \times \frac{1\ mol\ PbO_2}{1\ mol\ Pb} \times \frac{239\ g\ PbO_2}{1\ mol\ PbO_2} = 693\ g\ PbO_2$$

20.45 (a) In discharge: $Cd(s) + NiO_2(s) + 2H_2O(l) \rightleftarrows Cd(OH)_2(s) + Ni(OH)_2(s)$

In charging, the reverse reaction occurs.

(b) $E^\circ = 0.76\ V + 0.49\ V = 1.25\ V$

20.46 (a) $HgO(s) + Zn(s) \rightarrow Hg(l) + ZnO(s)$

(b) $E^\circ = E^\circ_{red} + E^\circ_{ox}$; $E^\circ_{ox} = 1.35\ V - 0.098\ V = 1.25\ V$

E°_{ox} is different from E° for $Zn(s) \rightarrow Zn^{2+}(aq) + 2e^-$ (+0.76 V) because in the battery the process happens in the presence of base and the product is stabilized as ZnO(s). Stabilization of the product of a half-reaction increases the driving force for the reaction, so E° is more positive.

20.47 E° for the half-cell reaction $Cd(s) \rightarrow Cd^{2+}(aq) + 2e^-$ is +0.40 V, smaller than the value of +0.76 V for the corresponding oxidation of Zn. Thus, the overall cell emf will be reduced.

20.48 The alkali metals Li and Na have much more metallic character than Zn, Cd, Pb or Ni. The oxidation potentials for Li and Na are thus more positive, leading to greater overall cell potentials for the battery. Also, Li and Na have lower density, so greater total energy for a battery could be achieved for a given total mass of material. One disadvantage is that Na and Li are very reactive and the cell reactions are difficult to control.

Electrolysis

20.49 (a) The products are different because in aqueous electrolysis water is reduced in preference to Mg^{2+}.

(b) $MgCl_2(l) \rightarrow Mg(l) + Cl_2(g)$

$2Cl^-(aq) + 2H_2O(l) \rightarrow Cl_2(g) + H_2(g) + 2OH^-(aq)$

The aqueous solution electrolysis is entirely analogous to that for NaCl(aq), Section 20.8.

(c) $Mg^{2+}(aq) + 2e^- \rightarrow Mg(s)$ $E^\circ = -2.37\ V$

$2Cl^-(aq) \rightarrow Cl_2(g) + 2e^-$ $E^\circ = -1.359\ V$

$MgCl_2(aq) \rightarrow Mg(s) + Cl_2(g)$ $E^\circ = -3.73\ V$

$2H_2O(l) + 2e^- \rightarrow H_2(g) + 2OH^-(aq)$ $E^\circ = -0.83\ V$

$2Cl^-(aq) \rightarrow Cl_2(g) + 2e$ $E^\circ = -1.359$

$3Cl^-(aq) + 2H_2O(l) \rightarrow Cl_2(g) + H_2(g) + 2OH^-(aq)$ $E^\circ = -2.19\ V$

20.50 (a) anode: $2Br^-(l) \rightarrow Br_2(l) + 2e^-$

cathode: $Al^{3+}(l) \rightarrow Al(l) + 3e^-$ (Al(s) melts at a lower temperature than $AlBr_3(s)$)

(b) anode: $2Br^-(l) \rightarrow Br_2(l) + 2e^-$

cathode: $2H_2O(l) + 2e^- \rightarrow H_2(g) + 2OH^-(aq)$

The reduction potential for water is less negative than that for $Al^{3+}(aq)$, so water is reduced in preference to Al^{3+}.

(c) The minimum emf values for the two possibilities in aqueous solution are:
Al^{3+} is reduced: -1.66 V - 1.065 V = -2.73 V
H_2O is reduced: -0.83 V - 1.065 V = -1.90 V

20.51

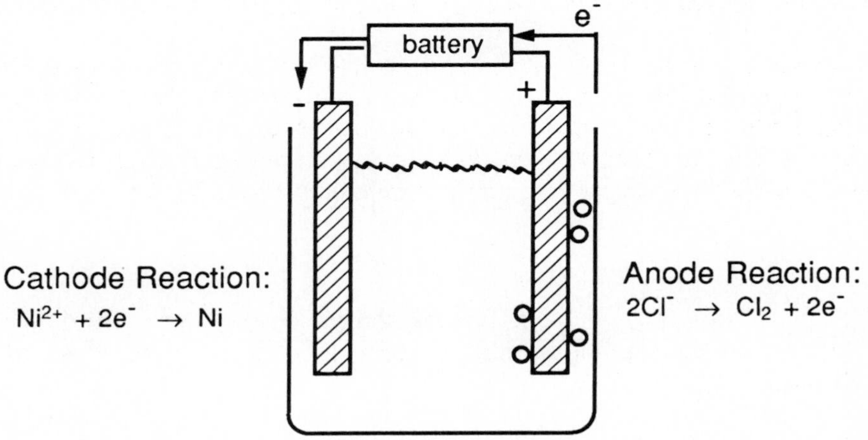

Chlorine is produced in preference to oxidation of water because of a large overvoltage for O_2 formation.

20.52

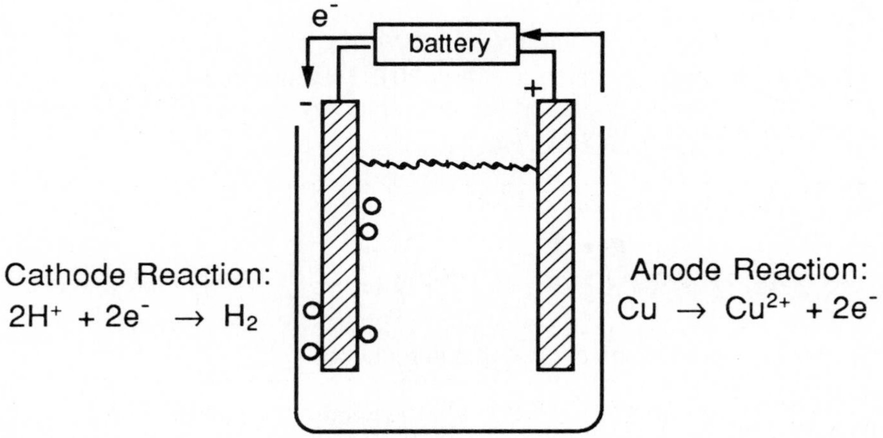

Overall cell reaction: $Cu(s) + 2H^+(aq) \rightarrow Cu^{2+}(aq) + H_2(g)$ $E^o = -0.34$ V
This value of E^o is essentially E_{min}, the minimum voltage required to produce the cell reaction under standard conditions. Note, however, that these standard conditions assume $[H^+] = 1$ *M*, and $[Cu^{2+}] = 1$ *M*. We also ignore the overvoltage, which is likely to be several tenths of a volt.

20.53 (a) Coulombs = amps × s (Equation [20.20]); since this is a $3e^-$ reduction, 1 mol Cr(s) requires 3 Faradays.

$$0.850 \text{ A} \times 24 \text{ hr} \times \frac{60 \text{ min}}{1 \text{ hr}} \times \frac{60 \text{ s}}{1 \text{ min}} \times \frac{1 \text{ C}}{1 \text{ amp} \cdot \text{s}} \times \frac{1 \text{ F}}{96{,}500 \text{ C}} \times \frac{1 \text{ mol Cr(s)}}{3 \text{ F}}$$

$$\times \frac{52.0 \text{ g Cr(s)}}{1 \text{ mol Cr(s)}} = 13 \text{ g Cr(s)}$$

(The time 24 hours has 2 significant figures, so the result has 2 significant figures.)

(b) $$1.65 \text{ g Cr} \times \frac{1 \text{ mol Cr}}{52.0 \text{ g Cr}} \times \frac{3 \text{ F}}{1 \text{ mol Cr}} \times \frac{96{,}500 \text{ C}}{\text{F}} \times \frac{1 \text{ amp} \cdot \text{s}}{1 \text{ C}} \times \frac{1}{2.50 \text{ hr}} \times \frac{1 \text{ hr}}{60 \text{ min}}$$

$$\times \frac{1 \text{ min}}{60 \text{ s}} = 1.02 \text{ amps}$$

20.54 As in Exercise 20.53:

(a) $$2.50 \text{ A} \times 550 \text{ min} \times \frac{60 \text{ s}}{1 \text{ min}} \times \frac{1 \text{ C}}{1 \text{ amp} \cdot \text{s}} \times \frac{1 \text{ F}}{96{,}500 \text{ C}} \times \frac{1 \text{ mol Mg(s)}}{2 \text{ F}} \times \frac{24.3 \text{ g Mg(s)}}{1 \text{ mol Mg(s)}}$$

$$= 10.4 \text{ g Mg}$$

(b) $$2.75 \text{ g Mg} \times \frac{1 \text{ mol Mg}}{24.3 \text{ g Mg}} \times \frac{2 \text{ F}}{1 \text{ mol Mg}} \times \frac{96{,}500 \text{ C}}{\text{F}} \times \frac{1 \text{ amp} \cdot \text{s}}{\text{C}} \times \frac{1}{5.00 \text{ A}}$$

$$= 4.37 \times 10^3 \text{ s}$$

20.55 (a) $$5.50 \text{ amp} \times 100 \text{ min} \times \frac{60 \text{ S}}{1 \text{ min}} \times \frac{1 \text{ C}}{1 \text{ amp} \cdot \text{s}} \times \frac{1 \text{ Faraday}}{96{,}500 \text{ C}} \times \frac{1 \text{ mol } Cl_2}{2 \text{ Faraday}}$$

$$\frac{22.400 \text{ L } Cl_2}{1 \text{ mol } Cl_2} = 3.83 \text{ L } Cl_2$$

(b) From the balanced equation (Section 20.8), we see that 2 mol NaOH are formed per mol Cl_2. Proceeding as in (a), but replacing the last factor by (2 mol NaOH/1 mol Cl_2), we obtain 0.342 mol NaOH.

20.56 The cell reaction is presumed to be:

$$Sn^{2+}(aq) + H_2O(l) \rightarrow Sn(s) + 1/2\ O_2(g) + 2H^+(aq)$$

Calculate the number of moles of Sn^{2+} that is reduced:

$$4.60 \text{ amp} \times 30.0 \text{ min} \times \frac{60 \text{ s}}{1 \text{ min}} \times \frac{1 \text{ C}}{1 \text{ amp} \cdot \text{s}} \times \frac{1 \text{ Faraday}}{96{,}500 \text{ C}} \times \frac{1 \text{ mol } Sn^{2+}}{2 \text{ Faraday}} = 0.0429 \text{ mol } Sn^{2+}$$

Initially there were $\frac{0.600 \text{ mol}}{\text{L}} \times (0.500 \text{ L}) = 0.300$ mol. Thus, following electrolysis, there are 0.300 - 0.043 = 0.257 mol Sn^{2+}.

$[Sn^{2+}] = \frac{0.257 \text{ mol}}{0.500 \text{ L}} = 0.514\ M$

Electrolysis also produces 2(0.0429) = 0.0858 moles $H^+(aq)$. Thus,

$[H^+] = \frac{0.0858 \text{ mol}}{0.500 \text{ L}} = 0.172\ M$

The concentration of SO_4^{2-} remains unchanged.

Electrical Work

20.57 (a) True
(b) True
(c) False; overvoltage requires a higher applied potential than calculated theoretically; more energy than the minimum is expended in producing the cell reaction.

20.58 The free energy change in a process, ΔG, is a measure of the maximum useful work that can be extracted from the process. Applied to a cell reaction this means that $w_{max} = \Delta G$. But G is a state function. Therefore, for a given process such as a cell reaction, there is only one possible value for the free energy change, ΔG, and one possible value for w_{max}. The actual work obtainable is always less than this. For example, the work obtainable from an alkaline cell will be less than that calculated from a knowledge of the cell potential, because some of the energy released in the reaction is used to overcome the internal resistance of the cell.

20.59 $w_{max} = \Delta G = -nFE$. Since the cell is at standard conditions, calculate E^o.

$$2H^+(aq) + 2e^- \rightarrow H_2(g) \qquad E^o = 0.000 \text{ V}$$
$$Cd(s) \rightarrow Cd^{2+}(aq) + 2e^- \qquad E^o = 0.403 \text{ V}$$
$$2H^+(aq) + Cd(s) \rightarrow H_2(g) + Cd^{2+}(aq) \qquad E^o_{cell} = 0.403 \text{ V}$$

w_{max} = -2(96,500) (0.403) = -77.8 kJ/mol Cd(s)

$\frac{-77.8 \text{ kJ}}{\text{mol Cd(s)}} \times 0.780$ mol Cd(s) = -60.7 kJ (The (-) sign means work is done by the cell.)

20.60 At standard conditions, $w_{max} = \Delta G^o = -nFE^o$.

$$I_2(s) + 2e^- \rightarrow 2I^-(aq) \qquad E^o = 0.536 \text{ V}$$
$$Sn(s) \rightarrow Sn^{2+}(aq) + 2e^- \qquad E^o = 0.136 \text{ V}$$
$$I_2(s) + Sn(s) \rightarrow 2I^-(aq) + Sn^{2+}(aq) \qquad E^o = 0.672 \text{ V}$$

w_{max} = -2(96,500) (0.672) = -130 kJ/mol Sn

$\frac{-130 \text{ kJ}}{\text{mol Sn(s)}} \times 0.460$ mol Sn = -59.8 kJ

20.61 (a) $90{,}000 \text{ amp} \times 16 \text{ hr} \times \frac{3600 \text{ s}}{1 \text{ hr}} \times \frac{1 \text{ C}}{1 \text{ amp-s}} \times \frac{1 \text{ Faraday}}{96{,}500 \text{ C}} \times \frac{1 \text{ mol Mg}}{2 \text{ Faraday}}$

$\times \frac{24 \text{ g Mg}}{1 \text{ mol Mg}} (0.50) = 3.2 \times 10^5 \text{ g Mg}$

(b) Coulombs = $9 \times 10^4 \text{ amp} \times 16 \text{ hr} \times \frac{3600 \text{ s}}{1 \text{ hr}} \times \frac{1 \text{ C}}{1 \text{ amp} \cdot \text{s}} = 5.2 \times 10^9 \text{ C}$

kilowatt-hours = $5.2 \times 10^9 \text{ C} \times 4.20 \text{ V} \times \frac{1 \text{ J}}{1 \text{ C} \cdot \text{V}} \times \frac{1 \text{ kWh}}{3.6 \times 10^6 \text{ J}} = 6.1 \times 10^3 \text{ kWh}$

20.62 (a) $66{,}000 \text{ amp} \times 12 \text{ hr} \times \frac{3600 \text{ s}}{1 \text{ hr}} \times \frac{1 \text{ C}}{1 \text{ amp} \cdot \text{s}} \times \frac{1 \text{ F}}{96{,}500 \text{ C}} \times \frac{1 \text{ mol Li}}{1 \text{ F}}$

$\times \frac{6.94 \text{ g Li}}{1 \text{ mol Li}} \times 0.85 = 1.7 \times 10^5 \text{ g Li}$

(b) If the cell is 85% efficient, $\frac{96{,}500 \text{ C}}{\text{F}} \times \frac{1 \text{ F}}{0.85 \text{ mol}} = \frac{1.1 \times 10^5 \text{ C}}{1 \text{ mol Li}}$ required

Energy = $5.50 \text{ V} \times \frac{1.1 \times 10^5 \text{ C}}{1 \text{ mol Li}} \times \frac{1 \text{ J}}{1 \text{ C} \cdot \text{V}} \times \frac{1 \text{ kWh}}{3.6 \times 10^6 \text{ J}} = 0.17 \text{ kWh}$

Corrosion

20.63 No. The potential for oxidation of Co° to Co^{2+}, 0.28 V, is lower than that for oxidation of Fe° to Fe^{2+}, 0.44 V. To afford cathodic protection a metal must be more readily oxidized than iron.

20.64 No. Tin is less readily oxidized than iron; E° for $Sn(s) \rightarrow Sn^{2+} + 2e^-$ is only 0.14 V, as compared with 0.44 V for the corresponding reaction of iron. Tin gives protection by providing a complete cover for iron.

20.65 The pH of an aqueous medium is an important factor in corrosion. Note from Section 20.10 that an increased $[H^+]$ shifts the equilibrium to the right. When $[H^+]$ is depressed by the presence of a base, the reduction of O_2 is less favorable, and corrosion slows down. The added amines serve the function of keeping $[H^+]$ low.

20.66 It is well established that corrosion occurs most readily when the metal surface is in contact with water. Thus, moisture is a requirement for corrosion. Corrosion also occurs more readily in acid solution, because O_2 has a more positive reduction potential in the presence of $H^+(aq)$. SO_2 and its oxidation products dissolve in water to produce acidic solutions, which encourage corrosion. The anodic and cathodic reactions for the corrosion of Ni are:

$$Ni(s) \rightarrow Ni^{2+}(aq) + 2e^- \qquad E^{\circ} = 0.28 \text{ V}$$
$$O_2(g) + 4H^+(aq) + 4e^- \rightarrow 2H_2O(l) \qquad E^{\circ} = 1.23 \text{ V}$$

Nickel(II) oxide, NiO(s), can form by the dry air oxidation of Ni. This NiO coating serves to protect against further corrosion. However, NiO dissolves in acidic solutions such as those produced by SO_2 or SO_3, according to the reaction:

$$NiO(s) + 2H^+(aq) \rightarrow Ni^{2+}(aq) + H_2O(l)$$

This exposes Ni(s) to further wet corrosion.

Additional Exercises

20.67 (a)

$$2HBrO_3(aq) + 10H^+(aq) + 10e^- \rightarrow Br_2(l) + 6H_2O(l)$$
$$10[Fe^{2+}(aq) \rightarrow Fe^{3+}(aq) + 1e^-]$$

$$10Fe^{2+}(aq) + 2HBrO_3(aq) + 10H^+(aq) \rightarrow 10Fe^{3+}(aq) + Br_2(l) + 6H_2O(l)$$

(b)

$$2[NO_3^-(aq) + 4H^+(aq) + 3e^- \rightarrow NO(g) + 2H_2O(l)]$$
$$3[Cu(s) \rightarrow Cu^{2+}(aq) + 2e^-]$$

$$3Cu(s) + 2NO_3^-(aq) + 8H^+(aq) \rightarrow 3Cu^{2+}(aq) + 2NO(g) + 4H_2O(l)$$

(c)

$$3IO_3^-(aq) + 18H^+(aq) + 16e^- \rightarrow I_3^-(aq) + 9H_2O(l)$$
$$+ 18OH^-(aq) \qquad + 18OH^-(aq)$$

$$3IO_3^-(aq) + 9H_2O(l) + 16e^- \rightarrow I_3^-(aq) + 18OH^-(aq)$$
$$8[3I^-(aq) \rightarrow I_3^-(aq) + 2e^-]$$

$$1/3[3IO_3^-(aq) + 24I^-(aq) + 9H_2O(l) \rightarrow 9I_3^-(aq) + 18OH^-(aq)]$$
$$IO_3^-(aq) + 8I^-(aq) + 3H_2O(l) \rightarrow 3I_3^-(aq) + 6OH^-(aq)$$

20.68 (a)

$$MnO_4^{2-}(aq) + 4H^+(aq) + 2e^- \rightarrow MnO_2(s) + 2H_2O(l)$$
$$2[MnO_4^{2-}(aq) \rightarrow MnO_4^-(aq) + 1e^-]$$

$$3MnO_4^{2-}(aq) + 4H^+(aq) \rightarrow 2MnO_4^-(aq) + MnO_2(s) + 2H_2O(l)$$

(b)

$$H_2SO_3(aq) + 4H^+(aq) + 4e^- \rightarrow S(s) + 3H_2O(l)$$
$$2[H_2SO_3(aq) + H_2O \rightarrow HSO_4^-(aq) + 3H^+(aq) + 2e^-]$$

$$3H_2SO_3(aq) \rightarrow S(s) + 2HSO_4^-(aq) + 2H^+(aq) + H_2O(l)$$

(c)

$$Cl_2(aq) + 2H_2O(l) \rightarrow 2ClO^-(aq) + 4H^+(aq) + 2e^-$$
$$+ 4OH^-(aq) \qquad + 4OH^-(aq)$$

$$Cl_2(aq) + 2e^- \rightarrow 2Cl^-(aq)$$

$$1/2[2Cl_2(aq) + 4OH^-(aq) \rightarrow 2Cl^-(aq) + 2ClO^-(aq) + 2H_2O(l)]$$
$$Cl_2(aq) + 2OH^-(aq) \rightarrow Cl^-(aq) + ClO^-(aq) + H_2O(l)$$

20.69 (a) Anode reaction: $5Fe^{2+}(aq) \rightarrow 5Fe^{3+}(aq) + 5e^-$.
Cathode reaction: $8H^+(aq) + MnO_4^-(aq) + 5e^- \rightarrow Mn^{2+}(aq) + 4H_2O(l)$.
Electrons move from the Pt electrode in the $Fe^{2+}(aq)$ solution to the Pt electrode in the $MnO_4^-(aq)$ solution. Anions migrate through the salt bridge from the cathode beaker to the anode beaker; cations migrate in the opposite direction. The electrode in the iron-containing beaker has a negative sign, that in the MnO_4^- beaker has a positive sign.

(b) Using Table 20.1, $E^\circ = (-0.77\ V) + 1.51\ V = 0.74\ V$.

20.70 (a)

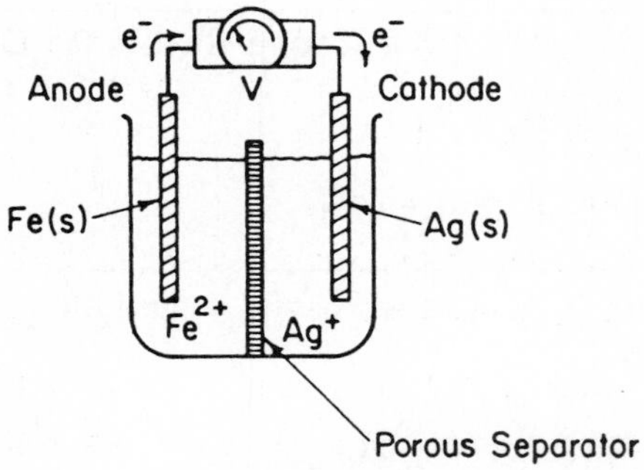

$$Fe(s) \rightarrow Fe^{2+}(aq) + 2e^-$$
$$2Ag^+(aq) + 2e^- \rightarrow 2Ag(s)$$
$$Fe(s) + 2Ag^+(aq) \rightarrow Fe^{2+}(aq) + 2Ag(s)$$

(b)

$$Zn(s) \rightarrow Zn^{2+}(s) + 2e^-$$
$$2H^+(aq) + 2e^- \rightarrow H_2(g)$$
$$Zn(s) + 2H^+(aq) \rightarrow Zn^{2+}(aq) + H_2(g)$$

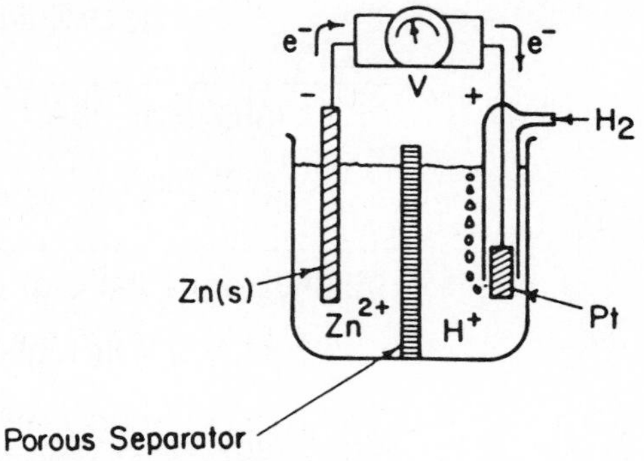

(c) $Cu/Cu^{2+}//ClO_3^-, Cl^-/Pt$ Here, both the oxidized and reduced forms of the cathode solution are in the same phase, so we separate them by a comma, and then indicate an inert electrode.

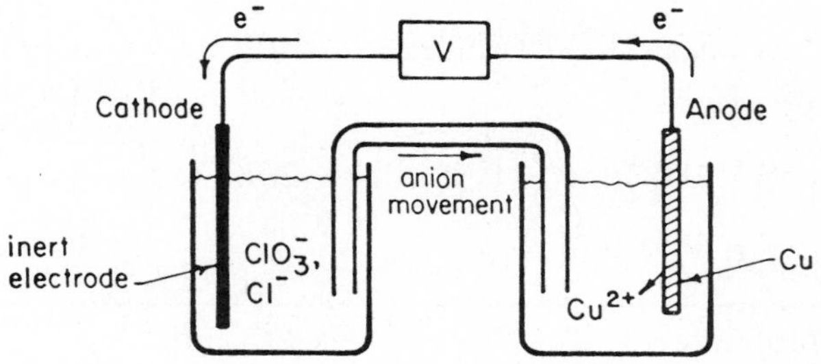

20.71 (a)

$$2[Rh^{3+}(aq) + 3e^- \rightarrow Rh(s)] \qquad E^o = ?$$
$$3[Cd(s) \rightarrow Cd^{2+}(aq) + 2e^-] \qquad E^o = +0.403\ V$$
$$\overline{2Rh^{3+}(aq) + 3Cd(s) \rightarrow 2Rh(s) + 3Cd^{2+}(aq) \qquad E^o = 1.20\ V}$$

(b) Cd(s) is the anode, and Rh(s) is the cathode.

(c) The cell is at standard conditions.

$E^o_{cell} = E^o_{anode} + E^o_{cathode}$; $E^o_{cathode} = E^o_{cell} - E^o_{anode} = 1.20\ V - 0.403 = 0.80\ V$

(d) $\Delta G^o = -nFE^o = -6(96{,}500)\ (1.20) = -695\ kJ$

20.72 (a) Reversal of the first half-reaction, then adding, leads to a large, positive overall cell potential. Zinc is oxidized to Zn^{2+}, silver oxide is reduced to metallic silver. (The silver ion actually undergoes the reduction.)

(b) Electrons move away from the anode ($Zn \rightarrow Zn^{2+} + 2e^-$) during cell operation. Thus, the anode is negatively charged. The cathode is positively charged; it is here that electrons are drawn to reduce the Ag_2O during cell operation.

(c)

$$Zn(s) \rightarrow Zn^{2+}(aq) + 2e^- \qquad E^o = 0.763\ V$$
$$Ag_2O(s) + H_2O(l) \rightarrow 2Ag(s) + 2OH^-(aq) \qquad E^o = 0.344\ V$$
$$\overline{Zn(s) + Ag_2O(s) + H_2O(l) \rightarrow Zn^{2+}(aq) + 2Ag(s) + 2OH^-(aq) \qquad E^o = 1.107\ V}$$

20.73 We need in each case to determine whether E^o is positive (spontaneous) or negative (nonspontaneous).

(a) $E^o = 0.672$ V, spontaneous (b) $E^o = -0.82$ V, nonspontaneous

(c) $E^o = 0.55$ V spontaneous (d) $E^o = -0.98$ V, nonspontaneous

20.74 Let us first balance the equation:

$4CyFe^{2+}(aq) + O_2(g) + 4H^+(aq) \rightarrow CyFe^{3+}(aq) + 2H_2O(l)$; $E = +0.60$ V; $n = 4$

(a) From Equation [20.11] we can relate ΔG for the process under the conditions specified for the measured potential E:

$$\Delta G = -nFE = -(4\ mol\ e^-) \times \frac{96{,}500\ J}{1\ V \cdot mol\ e^-}\ (0.60\ V) = -232\ kJ$$

(b) The number of ATP molecules synthesized per O_2 molecule is given by:

$$\frac{232\ kJ}{O_2\ molecule} \times \frac{1\ ATP\ formed}{37.7\ kJ} = \text{approximately } \frac{6\ ATP}{O_2}$$

20.75 The cell half-reactions are:

$$Cr(s) \rightarrow Cr^{3+}(aq)\ (Conc\ 1) + 3e^-$$
$$Cr^{3+}(aq)\ (Conc\ 2) + 3e^- \rightarrow Cr(s)$$
$$\overline{Cr^{3+}(aq)\ (Conc\ 2) \rightarrow Cr^{3+}(aq)(Conc\ 1)}$$

$E = 0 - \frac{0.0592}{3} \log \frac{[Cr^{3+}]_1}{[Cr^{3+}]_2}$; ($E^o = 0$) . Let $[Cr^{3+}]_1 = 0.040$ M, $[Cr^{3+}]_2 = 2.0$ M

$$E = -\frac{0.0592}{3} \log \frac{0.040}{2.0} = +0.034 \text{ V}$$

The fact that E is positive means that the cell reaction is spontaneous as written: Cr^{3+} (0.040 *M*) $\rightarrow$ Cr^{3+}(2.0 *M*). This means that the cathode is in the compartment containing Cr^{3+}(2.0 *M*); the anode is in the compartment containing the dilute (0.040 *M*) solution. For concentration cells in general, oxidation occurs in the compartment with the dilute solution and reduction in the compartment with the concentrated solution, so that the concentrations of cations in the two compartments become more similar.

20.76 (a) The reduction potential for $O_2(g)$ in the presence of acid is 1.23 V. $O_2(g)$ cannot oxidize Au(s) to $Au^+(aq)$ or $Au^{3+}(aq)$, even in the presence of acid.

(b) The substances need a reduction potential greater than 1.50 V. These include $Co^{3+}(aq)$, $F_2(g)$, $H_2O_2(aq)$ and $O_3(g)$. Marginal oxidizing agents (those with reduction potential near 1.50 V) from Appendix E are $BrO_3^-(aq)$, $Ce^{4+}(aq)$, HClO(aq), $MnO_4^-(aq)$ and $PbO_2(s)$.

(c)

$3[Au^+(aq) + 1e^- \rightarrow Au(s)]$	$E^o = 1.69$ V
$Au(s) \rightarrow Au^{3+}(aq) + 3e^-$	$E^o = -1.50$ V
$3Au^+(aq) \rightarrow 2Au(s) + Au^{3+}(aq)$	$E^o = +0.19$ V

If E^o is positive, ΔG^o is negative and the disproportionation of $Au^+(aq)$ is spontaneous.

(d) Since the Au^+ spontaneously disproportionates, the product of the reaction is AuF_3.

$3[F_2(g) + 2e^- \rightarrow 2F^-(aq)]$	$E^o = 2.87$ V
$2[Au(s) \rightarrow Au^{3+}(aq) + 3e^-]$	$E^o = -1.50$ V
$2Au(s) + 3F_2(g) \rightarrow 2AuF_3(aq)$	$E^o = 1.37$ V

20.77 (a)

$I_2(s) + 2e^- \rightarrow 2I^-(aq)$	$E^o = 0.536$ V
$2[Cu(s) \rightarrow Cu^+(aq) + 1\ e^-]$	$E^o = -0.521$ V
$I_2(s) + 2Cu(s) \rightarrow 2Cu^+(aq) + 2I^-(aq)$	$E^o = 0.015$ V

$$E = E^o - \frac{0.0592}{n} \log Q = 0.015 - \frac{0.0592}{2} \log [Cu^+]^2[I^-]^2$$

$$E = +0.015 - \frac{0.0592}{2} \log (2.1)^2(3.2)^2 = +0.015 - 0.049 = -0.034 \text{ V}$$

(b) Since the cell potential is negative at these concentration conditions, the cell would be spontaneous in the opposite direction and the inert electrode in the I_2/I^- compartment would be the anode; Cu(s) would be the cathode.

(c) No. At standard conditions the cell reaction is as written in part (a) and Cu(s) is the anode.

(d) $E = 0, E^\circ = \frac{0.0592}{2} \log (1.4)^2 [I^-]^2$

$$\frac{2(0.015)}{0.0592} = \log (1.4)^2 + 2 \log [I]^- \;;\; 0.107 = \log [I^-]\;;\; [I^-] = 10^{0.107} = 1.28\ M\ I^-$$

20.78

$AgSCN(s) + e^- \rightarrow Ag(s) + SCN^-(aq)$		$E^\circ = 0.0895$ V
$Ag(s) \rightarrow Ag^+(aq) + e^-$		$E^\circ = -0.799$ V
$AgSCN(s) \rightarrow Ag^+(aq) + SCN^-(aq)$		$E^\circ = -0.710$ V

$$E^\circ = \frac{0.0592}{n} \log K_{sp}\;;\; \log K_{sp} = \frac{(-0.710)(1)}{0.0592} = -11.993$$

$K_{sp} = 10^{-11.993} = 1.0 \times 10^{-12}$

(Intermediate rounding of E° and log K_{sp} lead to slight variations in the calculated value of K_{sp}.)

20.79 The reaction can be written as a sum of the steps:

$Pb^{2+}(aq) + 2e^- \rightarrow Pb(s)$	$E^\circ = -0.126$ V
$PbS(s) \rightarrow Pb^{2+}(aq) + S^{2-}(aq)$	$E^\circ =$?
$PbS(s) + 2e^- \rightarrow Pb(s) + S^{2-}(aq)$	$E^\circ =$?

"E°" for the second step can be calculated from K_{sp}.

$$E^\circ = \frac{0.0592}{n} \log K_{sp} = \frac{0.0592}{2} \log (8.0 \times 10^{-28}) = \frac{0.0592}{2}(-27.10) = -0.802 \text{ V}$$

E° for the half-reaction = -0.126 V + (-0.802 V) = -0.928 V

Calculating an imaginary E° for a nonredox process like step 2 may be a disturbing idea. Alternatively, one could calculate K for step 1 (= 5.4×10^{-5}), K for the reaction in question ($K = K_1 \times K_{sp} = 4.4 \times 10^{-32}$) and then E° for the half-reaction. The result is the same.

20.80 $Cu^{+}(aq) + 1e^{-} \rightarrow Cu^{2+}(aq)$ $E^{o} = -0.153$ V

$Cu^{+}(aq) \rightarrow Cu^{o}(s) + 1e^{-}$ $E^{o} = 0.521$ V

$2Cu^{+}(aq) \rightarrow Cu^{o}(s) + Cu^{2+}(aq)$ $E^{o} = 0.368$ V

$$E^{o} = \frac{0.0592}{n}\log K;\quad \log K = \frac{nE^{o}}{0.0592} = \frac{1(0.368)}{0.0592} = 6.22$$

$K = 10^{6.22} = 1.7 \times 10^{6}$

20.81 (a) Total volume of Cr = 2.3×10^{-4} m $\times$ 0.32 m^2 = 7.4×10^{-5} m^3

$$\text{mol Cr} = 7.4 \times 10^{-5}\ \text{m}^3\ \text{Cr} \times \frac{100^3\ \text{cm}^3}{1\ \text{m}^3} \times \frac{7.20\ \text{g Cr}}{1\ \text{cm}^3} \times \frac{1\ \text{mol Cr}}{52.0\ \text{g Cr}} = 10\ \text{mol Cr}$$

The electrode reaction is:

$CrO_4^{2-}(aq) + 4H_2O(l) + 6e^{-} \rightarrow Cr(s) + 8OH^{-}(aq)$

$$\text{Coulombs requires} = 10\ \text{mol Cr} \times \frac{6\ \text{Faraday}}{1\ \text{mol Cr}} \times \frac{96{,}500\ \text{C}}{1\ \text{Faraday}} = 5.9 \times 10^{6}\ \text{C}$$

(b) $$5.9 \times 10^{6}\ \text{C} \times \frac{1\ \text{amp}\cdot\text{s}}{1\ \text{C}} \times \frac{1}{6.0\ \text{s}} = 9.8 \times 10^{5}\ \text{amp}$$

20.82 $$3.20\ \text{amp} \times 40\ \text{min} \times \frac{60\ \text{s}}{1\ \text{min}} \times \frac{1\ \text{C}}{1\ \text{amp}\cdot\text{s}} \times \frac{1\ \text{Faraday}}{96{,}500\ \text{C}} \times \frac{1\ \text{mol e}^-}{1\ \text{Faraday}} = 0.079\ \text{mol e}^-$$

$$\frac{4.57\ \text{g In}}{0.079\ \text{mol e}^-} \times \frac{1\ \text{mol In}}{115\ \text{g In}} = \frac{0.50\ \text{mol In}}{1\ \text{mol e}^-}$$

This result tells us that the In must be in the +2 oxidation state in the molten halide.

20.83 (a) $$1\ \text{lb Al} \times \frac{453.6\ \text{g}}{1\ \text{lb}} \times \frac{1\ \text{mol Al}}{26.98\ \text{g Al}} \times \frac{3\ \text{F}}{1\ \text{mol Al}} \times \frac{96{,}500\ \text{C}}{\text{F}} \times \frac{1\ \text{amp}\cdot\text{s}}{\text{C}} \times \frac{1}{11.2\ \text{A}}$$

$$\times \frac{1\ \text{min}}{60\ \text{s}} \times \frac{1\ \text{hr}}{60\ \text{min}} = 1.2 \times 10^{2}\ \text{hr}$$

(b) $$\text{Coulombs} = 453.6\ \text{g Al} \times \frac{1\ \text{mol Al}}{26.98\ \text{g Al}} \times \frac{3\ \text{F}}{1\ \text{mol Al}} \times \frac{96{,}500\ \text{C}}{\text{F}} = 4.867 \times 10^{6}\ \text{C}$$

$$\text{kWh} = 4.867 \times 10^{6}\ \text{C} \times 6.0\ \text{V} \times \frac{1\ \text{J}}{\text{C}\cdot\text{V}} \times \frac{1\ \text{kWh}}{3.6 \times 10^{6}\ \text{J}} = 8.1\ \text{kWh}$$

This is the amount of electrical power expended if the cell is 100% efficient. 8.2 kWh/0.40 = **20 kWh** required to produce 1 lb of Al.

20.84 In the electrolysis of aqueous Na_2SO_4 solution, the electrode reactions are:

$2H_2O(l) + 2e^- \rightarrow H_2(g) + 2OH^-(aq)$

$2H_2O(l) \rightarrow 4H^+(aq) + O_2(g) + 4e^-$

The first reaction, occurring at the cathode, produces $OH^-(aq)$; thus, litmus turns blue. The second reaction, occuring at the anode, produces $H^+(aq)$; thus, litmus turns red.

20.85 $$24.0 \text{ hr} \times 20.0 \text{ amp} \times \frac{3600 \text{ s}}{1 \text{ hr}} \times \frac{1 \text{ mol e}^-}{96{,}500 \text{ amp} \cdot \text{s}} \times \frac{4 \text{ mol NaBO}_3}{8 \text{ mol e}^-}$$

$$\times \frac{81.8 \text{ g NaBO}_3}{1 \text{ mo NaBO}_3} = 732 \text{ g NaBO}_3$$

20.86 (a) $Ni^{2+}(aq) + 2e^- \rightarrow Ni(s) \quad E^\circ = -0.28 \text{ V}$

$2Br^-(aq) \rightarrow Br_2(l) + 2e^- \quad E^\circ = -1.065 \text{ V}$

$E_{min} = 1.065 + 0.28 = 1.34 \text{ V}$

(b) A larger voltage than the calculated E_{min} is required to overcome cell resistance, if the reaction is to proceed at a reasonable rate. There may also be an overvoltage for reduction of $Br^-(aq)$ at Pt, although this should not be large.

20.87 (a) The work obtainable is given by the product of the voltage, which has units of J/C, times the number of Coulombs of electricity produced:

$$w_{max} = 300 \text{ amp} \cdot \text{hr} \times \frac{3600 \text{ s}}{1 \text{ hr}} \times \frac{1 \text{ C}}{1 \text{ amp} \cdot \text{s}} \times \frac{6 \text{ J}}{1 \text{ C}} \times \frac{1 \text{ kWh}}{3.6 \times 10^6 \text{ J}} = 1.8 \text{ kWh} \simeq 2 \text{ kWh}$$

(b) This maximum amount of work is never realized because some of the electrical energy is dissipated in overcoming the internal resistance of the battery; because the cell voltage does not remain constant as the reaction proceeds; because the systems to which the electrical energy is delivered are not capable of completely converting electrical energy into work.

20.88 (a) The major reason "deep discharge" causes battery failure is loss of materials from the electrode surfaces, during either discharge or charging. When PbO_2 is almost completely removed from the cathode during discharge, as it is converted to $PbSO_4$ there may be some flaking off of $PbSO_4$. This solid settles in the bottom of the cell, and is no longer able to take part in the cell reaction. It may eventually cause shorting between the plates.

(b) In an auto system, a running engine powers a generator and voltage regulator which together keep the battery fully charged, or nearly so. Whatever charge is used to start the engine is replaced while the auto is running. Except when lights are left on accidentally, or the battery is run down in cold weather starting, it carries nearly a full charge.

20.89 The ship's hull should be made negative. By keeping an excess of electrons in the metal of the ship, the tendency for iron to undergo oxidation, with release of electrons, is reduced. The ship, as a negatively charged "electrode," becomes the site of reduction, rather than oxidation, in an electrolytic process.

20.90 (a) $7 \times 10^8 \text{ mol } H_2 \times \frac{2 \text{ F}}{1 \text{ mol } H_2} \times \frac{96{,}500 \text{ C}}{1 \text{ F}} = 1.4 \times 10^{14} \text{ C}$

(b)

$$\begin{array}{ll} 2H_2O(l) \rightarrow O_2(g) + 4H^+(aq) + 4e^- & E^\circ = -1.23 \text{ V} \\ 2[H^+(aq) + 2e^- \rightarrow H_2(g)] & E^\circ = 0 \text{ V} \\ \hline 2H_2O(l) \rightarrow O_2(g) + 2H_2(g) & E^\circ = -1.23 \text{ V} \end{array}$$

$$E = E^\circ - \frac{0.0592}{4} \log [O_2][H_2]^2 = -1.23 \text{ V} - \frac{0.0592}{4} \log (300)^3$$

$$= -1.23 \text{ V} - 0.11 \text{ V} = -1.34 \text{ V}$$

(c) $\text{Energy} = nFE = 2(7 \times 10^8 \text{ mol})(1.34 \text{ V})(96{,}500 \frac{1}{\text{V} \cdot \text{mol}}) = 2 \times 10^{14} \text{ J}$

(d) $2 \times 10^{14} \text{ J} \times \frac{1 \text{ kWh}}{3.6 \times 10^6 \text{ J}} \times \frac{\$0.23}{\text{kWh}} = \$1.3 \times 10^7$

It will cost more than ten million dollars for the electricity alone!

20.91 $1.62 \text{ g Zn} \times \frac{1 \text{ mol Zn}}{65.4 \text{ g Zn}} \times \frac{2 \text{ mol } e^-}{1 \text{ mol Zn}} \times \frac{96{,}500 \text{ C}}{1 \text{ mol } e^-} \times \frac{1}{3.5 \times 10^{-3}} = 1.37 \times 10^6 \text{ C}$

CHAPTER **21**

Nuclear Chemistry

Radioactivity

21.1 p = protons, n = neutrons, e = electrons; # of protons = atomic number; # of neutrons = mass number - atomic number

(a) $^{56}_{26}Fe$ - 26p, 30n (b) ^{88}Sr - 38p, 50n (c) ^{31}P - 15p, 16n

21.2 (a) $^{75}_{33}As$ - 33p, 42n (b) ^{101}Ru - 44p, 57n (c) ^{85}Rb - 37p, 48n

21.3 (a) $^{0}_{-1}e$ (b) $^{0}_{1}e$ (c) $^{1}_{0}n$

21.4 (a) $^{1}_{1}p$ (b) $^{4}_{2}He$ (c) $^{0}_{-1}e$ or $^{0}_{-1}\beta$

21.5 (a) $^{131}_{53}I \rightarrow ^{131}_{54}Xe + ^{0}_{-1}e$

(b) $^{230}_{90}Th \rightarrow ^{226}_{88}Ra + ^{4}_{2}He$; α emission is the only decay process that reduces the atomic number by two.

(c) $^{181}_{74}W + ^{0}_{-1}e$ (orbital electron) $\rightarrow ^{181}_{73}Ta$

(d) $^{13}_{7}N \rightarrow ^{13}_{6}C + ^{0}_{1}e$

21.6 (a) $^{214}_{83}Bi \rightarrow ^{214}_{84}Po + ^{0}_{-1}e$ (b) $^{195}_{79}Au + ^{0}_{-1}e$ (orbital electron) $\rightarrow ^{195}_{78}Pt$

(c) $^{38}_{19}K \rightarrow ^{38}_{18}Ar + ^{0}_{1}e$ (d) $^{242}_{94}Pu + ^{238}_{92}U + ^{4}_{2}He$

21.7 The total mass number change is (235-207) = 28. Since each α particle accompanies a change of -4 in mass number, whereas emission of a β particle does not correspond to a mass change, there are 7 α particle emissions. The change in atomic number in the series is 10. Each α particle results in an atomic number lower by two. The 7 α particle emissions alone would cause a decrease of 14 in atomic number. Each β particle emission raises the atomic number by one. To obtain the observed lowering of 10 in the series, there must be 4 β particle emissions.

21.8 This decay series represents a change of (237-209 =) 28 mass units. Since only alpha emissions change the nuclear mass, and each changes the mass by four, there must be a total of 7 α emissions. Each alpha emission causes a decrease of two in atomic number. Therefore, the 7 alpha emissions, by themselves, should have caused a decrease in atomic number of 14. The series as a whole involves a decrease of 10 in atomic number. Thus, there must be a total of 4 β emissions, each of which increase atomic number by one.

Nuclear Stability

21.9 (a) ${}^{8}_{5}B$ - low neutron/proton ratio, positron emission

(b) ${}^{68}_{29}Cu$ - high neutron/proton ratio, beta emission

(c) ${}^{241}_{93}Np$ - high neutron/proton ratio, beta emission (Even though ${}^{241}Np$ has an atomic number $\geq$ 84, the most common decay pathway for nuclides with neutron/proton ratios higher than the isotope listed on the periodic chart is beta decay.)

(d) ${}^{39}_{17}Cl$ - high neutron/proton ratio, beta emission

21.10 (a) ${}^{66}_{32}Ge$ - lower neutron/proton ratio, positron emission or electron capture

(b) ${}^{105}_{45}Rh$ - high neutron/proton ratio, beta emission

(c) ${}^{137}_{53}I$ - high neturon/proton ratio, beta emission

(d) ${}^{133}_{58}Ce$ - low neutron/proton ratio, electron capture or positron emission

21.11 (a) It is on the edge -- slightly low neutron/proton ratio; could be a positron emitter or undergo orbital electron capture.
(b) No -- low neutron/proton ratio; should be a positron emitter, or possibly undergo orbital electron capture.
(c) No -- high neutron/proton ratio; should be a beta emitter.
(d) No -- high atomic number; it should be an alpha emitter.

21.12 (a) It is on the edge - slightly low neutron/proton ratio; could be a positron emitter or undergo orbital electron capture.
(b) No -- somewhat high neutron/proton ratio; beta emitter.
(c) Yes.
(d) No -- high atomic number; alpha emitter.

21.13 Use the criteria listed in Table 21.3.

(a) Stable: ${}^{39}_{19}K$ odd proton, even neutron more abundant than odd proton, odd neturon; 20 neutrons is a magic number.

(b) Stable: ${}^{209}_{83}Bi$ odd proton, even neutron more abundant than odd proton, odd neutron; 126 neutrons is a magic number.

(c) Stable: ${}^{25}_{12}Mg$ even though ${}^{24}_{10}Ne$ is an even proton, even neutron nuclide, it has a very high neutron/proton ratio and lies outside the band of stability.

21.14 Use the criteria listed in Table 21.3.

(a) ${}^{19}_{9}F$ odd proton, even neutron more abundant

(b) ${}^{80}_{34}Se$ even-even more abundant

(c) ${}^{56}_{26}Fe$ even-even more abundant

(d) ${}^{118}_{50}Sn$ even-even more abundant than odd-odd

21.15 ${}^{4}_{2}He$, ${}^{16}_{8}O$, ${}^{40}_{20}Ca$

21.16 ${}^{40}_{20}Ar$ - 18p, 22n, ${}^{208}_{82}Pb$ - 82p, 126n
Pb has closed shells of protons and neutrons, which leads to nuclear stability. Ar has a closed shell of electrons, which limits its chemical reactivity.

21.17 The criterion employed in judging whether the nucleus is likely to be radioactive is the position of the nucleus on the plot shown in Figure 21.2. If the neutron/proton ratio is too high or low, or if the atomic number exceeds 83, the nucleus will be radioactive. Radioactive: (b) low neutron/proton ratio; (c) low neutron/proton ratio; (e) high atomic number; stable: (a) and (d).

21.18 Radioactive:

${}^{14}_{8}O$, ${}^{115}_{52}Te$ - low neutron/proton ratio; ${}^{208}_{84}Po$ - atomic number $\geq$ 84

${}^{84}_{34}Se$ - even though this is an even proton, even neutron nuclide with a magic number of neutrons, the neutron/proton ratio is so high that the nuclide is radioacitve.

Stable: ${}^{32}_{16}S$, ${}^{78}_{34}Se$ - even proton, even neutron, stable neutron, proton ratio

Nuclear Transmutations

21.19 Protons and alpha particles are positively charged and must be moving very fast to overcome electrostatic forces which would repel them from the target nucleus. Neutrons are electrically neutral and not repelled by the nucleus.

21.20 A major difference is that the charge on the nitrogen nucleus, +7, is much smaller than on the gold nucleus, +79. Thus, the alpha particle could more easily penetrate the coulomb barrier (that is, the repulsive energy barrier due to like charges) to make contact with the nitrogen nucleus than the gold nucleus. Rutherford used alpha particles that were being emitted from some radioactive source. He did not have access to machines that can accelerate particles to very high energy. It would be necessary to do just that to observe reaction of an alpha particle with a gold nucleus.

21.21 (a) ${}^{32}_{16}S + {}^{1}_{0}n \rightarrow {}^{1}_{1}H + {}^{32}_{15}P$ (b) ${}^{7}_{4}Be + {}^{0}_{-1}e$ (orbital electron) $\rightarrow {}^{7}_{3}Li$

(c) ${}^{187}_{75}Re \rightarrow {}^{187}_{76}Os + {}^{0}_{-1}e$ (d) ${}^{98}_{42}Mo + {}^{2}_{1}H \rightarrow {}^{1}_{0}n + {}^{99}_{43}Tc$

(e) ${}^{235}_{92}U + {}^{1}_{0}n \rightarrow {}^{135}_{54}Xe + {}^{99}_{38}Sr + 2{}^{1}_{0}n$

21.22 (a) ${}^{252}_{98}Cf + {}^{10}_{5}B \rightarrow 3\,{}^{1}_{0}n + {}^{259}_{103}Lw$ (b) ${}^{2}_{1}H + {}^{3}_{2}He \rightarrow {}^{4}_{2}He + {}^{1}_{1}H$

(c) ${}^{1}_{1}H + {}^{11}_{5}B \rightarrow 3\,{}^{4}_{2}He$ (d) ${}^{122}_{53}I \rightarrow {}^{122}_{54}Xe + {}^{0}_{-1}e$

(e) ${}^{59}_{26}Fe \rightarrow {}^{0}_{-1}e + {}^{59}_{27}Co$

21.23 (a) ${}^{238}_{92}U + {}^{1}_{0}n \rightarrow {}^{239}_{92}U + {}^{0}_{0}\gamma$ (b) ${}^{14}_{7}N + {}^{1}_{1}H \rightarrow {}^{11}_{6}C + {}^{4}_{2}He$

(c) ${}^{18}_{8}O + {}^{1}_{0}n \rightarrow {}^{19}_{9}F + {}^{0}_{-1}e$

21.24 (a) ${}^{14}_{7}N + {}^{1}_{1}H \rightarrow {}^{11}_{6}C + {}^{4}_{2}He$ (b) ${}^{14}_{7}N + {}^{4}_{2}He \rightarrow {}^{17}_{8}O + {}^{1}_{1}H$

(c) ${}^{59}_{26}Fe + {}^{4}_{2}He \rightarrow {}^{63}_{29}Cu + {}^{0}_{-1}e$

Half-life; Dating

21.25 The suggestion is not reasonable. The energies of nuclear states are very large relative to ordinary temperatures. Thus, merely changing the temperature by less than 100 K would not be expected to significantly affect the behavior of nuclei with regard to nuclear decay rates.

21.26 Chemical reactions do not affect the character of atomic nuclei. The energy changes involved in chemical reactions are much too small to allow us to alter nuclear properties via chemical processes. Therefore, the nuclei that are formed in a nuclear reaction will continue to emit radioactivity regardless of any chemical changes we bring to bear. However, we can hope to use chemical means to separate radioactive substances, or remove them from foods or a portion of the environment.

21.27 ${}^{68}_{31}Ga \rightarrow {}^{68}_{30}Zn + {}^{0}_{1}e$ See sample Exercise 21.8.

Calculate k. $k = 0.693/t_{1/2} = 0.693/68.3 \text{ min} = 1.01 \times 10^{-2} \text{ min}^{-1}$

$$\ln\frac{N_t}{N_o} = -kt = \frac{-1.01 \times 10^{-2}}{\text{min}} \times 683 \text{ min} = -6.93$$

$$\frac{N_t}{N_o} = 9.78 \times 10^{-4}\ ;\ N_t = (9.78 \times 10^{-4})(1.00 \times 10^{-2} \text{ g}) = 9.78 \times 10^{-6} \text{ g}\ \ {}^{68}\text{Ga remain.}$$

Alternately, 683 min is exactly 10 half-lives. The amount of ${}^{68}Ga$ remaining is $1.00 \times 10^{-2}\text{ g}(1/2)^{10} = 9.77 \times 10^{-6}\text{ g } {}^{68}Ga$. (The two answers differ due to rounding differences.)

21.28 After 12.3 yr, one half-life, there are (1/2)48.0 = 24.0 mg. 49.2 yr are exactly four half-lives. There are then $(48.0)(1/2)^4 = 3.0$ mg tritium remaining.

21.29 Using Equation [21.19],

$$k = \frac{-1}{t} \times \ln\frac{N_t}{N_o} = \frac{-1}{1\ \text{yr}} \times \ln\frac{2921}{3012} = 0.0307\ \text{yr}^{-1}$$

Using Equation [21.20],

$$t_{1/2} = 0.693/k = 0.693/(0.0307\ \text{yr}^{-1}) = 22.6\ \text{yr}$$

21.30 $$k = \frac{-1}{t} \ln \frac{N_t}{N_o} = \frac{-1}{350\ \text{min}} \ln \frac{1250}{8540} = 5.49 \times 10^{-3}\ \text{min}^{-1}$$

$$t_{1/2} = 0.693/k = 0.693/5.49 \times 10^{-3}\ \text{min}^{-1} = 126\ \text{min}$$

21.31 Assume that no depletion of iodide from the water due to plant uptake has occurred. Then the activity after 32 days would be:

$$k = 0.693/t_{1/2} = 0.693/8.1\ \text{d} = 0.086\ \text{d}^{-1}$$

$$\ln \frac{N_t}{N_o} = -(0.086\ \text{d}^{-1})(32.\text{d}) = -2.75; \quad \frac{N_t}{N_o} = 0.0638$$

We thus expect $N_t = 0.064(89) = 5.7$ counts/min. This is just the observed activity; we can assume that the plants did not absorb iodide, because that would have resulted in a lower level of remaining activity.

21.32 $$\ln \frac{N_t}{N_o} = -kt \text{ ; solve for } k$$

$$k = -\ln \frac{N_t}{N_o} \times \frac{1}{t} = -\ln \frac{3130}{4600} \times \frac{1}{30.0} = 0.0128\ \text{d}^{-1} \text{ ; } \quad t_{1/2} = 0.693/0.0128\ \text{d}^{-1} = 54.1\ \text{d}$$

21.33 $${}^{226}_{88}\text{Ra} \rightarrow {}^{222}_{86}\text{Rn} + {}^{4}_{2}\text{He}$$

1 α particle is produced for each ^{226}Ra that decays. Calculate the mass of ^{226}Ra remaining after 1.0 min, calculate by subtraction the mass that has decayed, and use Avogadro's number to get the number of ${}^{4}_{2}$He particles.

Calculate k in min^{-1}. $$1622\ \text{yr} \times \frac{365\ \text{d}}{1\ \text{yr}} \times \frac{24\ \text{hr}}{1\ \text{d}} \times \frac{60\ \text{min}}{1\ \text{hr}} = 8.525 \times 10^8\ \text{min}$$

$$k = \frac{0.693}{t_{1/2}} = \frac{0.693}{8.525 \times 10^8\ \text{min}} = 8.129 \times 10^{-10}\ \text{min}^{-1}$$

$$\ln\frac{N_t}{N_o} = -kt = -(8.129 \times 10^{-10}\ \text{min}^{-1})(1.0\ \text{min}) = -8.129 \times 10^{-10}$$

$\frac{N_t}{N_o} = (1.000 - 8.0 \times 10^{-10})$; (don't round here!)

$N_t = 5.0 \times 10^{-3}$ g $(1.00 - 8.0 \times 10^{-10})$. The amount that decays is $N_o - N_t$:

$$5.0 \times 10^{-3} \text{ g} - [5.0 \times 10^{-3}(1.00 - 8.0 \times 10^{-10})] = 5.0 \times 10^{-3} \text{ g } (8.0 \times 10^{-10})$$

$$[N_o - N_t] = 4.0 \times 10^{-12} \text{ g Ra} \times \frac{1 \text{ mol Ra}}{226 \text{ g Ra}} \times \frac{6.022 \times 10^{23} \text{ Ra atoms}}{1 \text{ mol Ra}} \times \frac{1\ {}^4_2\text{He}}{1 \text{ Ra atom}}$$

$$= 1.1 \times 10^{10}\ \alpha \text{ particles emitted in 1 min}$$

21.34 Proceeding as in Exercise 21.33, calculate k in s^{-1}.

$$5.26 \text{ yr} \times \frac{365 \text{ d}}{1 \text{ yr}} \times \frac{24 \text{ hr}}{1 \text{ d}} \times \frac{3600 \text{ sec}}{1 \text{ hr}} = 1.66 \times 10^8 \text{ s}$$

$$k = \frac{0.693}{t_{1/2}} = \frac{0.693}{1.66 \times 10^8} = 4.18 \times 10^{-9} \text{ s}^{-1}$$

$$\ln \frac{N_t}{N_o} = -kt = -(4.18 \times 10^{-9} \text{ s}^{-1})(10.0 \text{ s}) = -4.18 \times 10^{-8}$$

$$\frac{N_t}{N_o} = e^{-4.18 \times 10^{-8}} = (1.000 - 4.2 \times 10^{-8})\ ;\ N_t = 650 \times 10^{-6} \text{ g } (1.000 - 4.2 \times 10^{-8})$$

The amount that decays is $N_o - N_t$:

$$650 \times 10^{-6} \text{ g} - [650 \times 10^{-6} \text{ g } (1.000 - 4.2 \times 10^{-8})] = 650 \times 10^{-6} \text{ g } (4.2 \times 10^{-8})$$

$$N_o - N_t = 2.73 \times 10^{-11} \text{ g Co} \times \frac{1 \text{ mol Co}}{60 \text{ g Co}} \times \frac{6.022 \times 10^{23} \text{ Co atoms}}{1 \text{ mol Co}} \times \frac{1\ \beta}{1 \text{ Ra atom}}$$

$$= 2.74 \times 10^{11}\ \beta \text{ particles emitted in 10.0 seconds}$$

21.35 $t = \frac{-1}{k} \ln \frac{N_t}{N_o}$; $k = 0.693/5.73 \times 10^3 \text{ yr} = 1.21 \times 10^{-4} \text{ yr}^{-1}$

$$t = \frac{-1}{1.21 \times 10^{-4} \text{ yr}^{-1}} \ln \frac{25.8}{31.7} = 1.70 \times 10^3 \text{ yr}$$

21.36 $t_{1/2} = 5.73 \times 10^3$ yr; $k = 0.693/5.73 \times 10^3 \text{ yr} = 1.21 \times 10^{-4} \text{ yr}^{-1}$

$$t = \frac{-1}{k} \times \ln \frac{N_t}{N_o} = \frac{-1}{1.21 \times 10^{-4} \text{ yr}^{-1}} \times \ln \frac{9.6}{18.4} = 5.38 \times 10^3 \text{ yr}$$

21.37 Follow the procedure outlined in Sample Exercise 21.7. The original quantity of ^{238}U is 50.0 mg plus the amount that gave rise to 14.0 mg of ^{206}Pb. This amount is 14.0(238/206) = 16.2 mg.

$k = 0.693/4.5 \times 10^9\ yr = 1.5 \times 10^{-10}\ yr^{-1}$

$$t = \frac{-1}{k} \ln \frac{N_t}{N_o} = \frac{-1}{1.5 \times 10^{-10}\ yr^{-1}} \times \ln \frac{50.0}{66.2} = 1.9 \times 10^9\ yr$$

21.38 $k = 0.693/1.27 \times 10^9\ yr = 5.5 \times 10^{-10}\ yr^{-1}$
If the mass of ^{40}Ar is 3.6 times that of ^{40}K, then the original mass of ^{40}K must have been 3.6 + 1 = 4.6 times that now present.

$$t = \frac{-1}{5.5 \times 10^{-10}\ yr^{-1}} \times \ln \frac{1}{(4.6)} = 2.8 \times 10^9\ yr$$

Energy Changes

21.39 Calculate the binding energy (note that <u>nuclear</u> mass is provided here): 4(1.00728) + 5(1.00866) = 9.07242 amu; binding energy = (9.07242 - 9.00999) = 0.06243 amu. The energy equivalent of this mass can be evaluated using Einstein's equation, $\Delta E = c^2(\Delta m)$.

$$\Delta E = (2.9979 \times 10^8\ m/s)^2 \times 0.06243\ amu \times \frac{1\ g}{6.0221 \times 10^{23}\ amu} \times \frac{1\ kg}{10^3\ g}$$

$$\Delta E = 9.317 \times 10^{-12}\ kg \cdot m^2/s^2 = 9.317 \times 10^{-12}\ J$$

For one mole of ^9Be nuclei, $\Delta E = (9.317 \times 10^{-12}\ J)(6.022 \times 10^{23})$
$= 5.611 \times 10^{12}\ J/mol$

21.40 $\Delta m = 8(1.00728\ amu) + 8(1.00866\ amu) - 15.99052\ amu = 0.13700\ amu$

$$\Delta E = (2.99793 \times 10^8\ m/s)^2 \times 0.13700\ amu \times \frac{1\ g}{6.0221 \times 10^{23}\ amu} \times \frac{1\ kg}{1 \times 10^3\ g}$$

$= 2.0446 \times 10^{-11}\ J/^{19}F$ nucleus absorbed

(The total mass increases, so energy is absorbed.)

$$2.0442 \times 10^{-11} \frac{J}{nucleus} \times \frac{6.0221 \times 10^{23}\ atoms}{mol} = 1.2310 \times 10^{13}\ J/mol\ ^{19}F$$

21.41 Calculate in each case the mass of nucleons plus atomic electrons.

(a) 6(1.00728) + 6(1.00866) + 6(0.00055) = 12.09894 amu; binding energy per nucleon is
(12.09894 - 12.00000)/12 = 8.245×10^{-3} amu

$$\Delta E = 8.25 \times 10^{-3}\ amu \times \frac{1\ g}{6.02 \times 10^{23}\ amu} \times \frac{1\ kg}{10^3\ g} \times \frac{9.00 \times 10^{16}\ m^2}{s^2}$$

$= 1.23 \times 10^{-12}$ J/nucleon

(b) $(59.48853 - 58.9332)/59 = 9.412 \times 10^{-3}$ amu

$$\Delta E = 9.41 \times 10^{-3} \text{ amu} \times \frac{1 \text{ g}}{6.02 \times 10^{23} \text{ amu}} \times \frac{1 \text{ kg}}{10^3 \text{ g}} \times \frac{9.00 \times 10^{16} \text{ m}^2}{\text{s}^2}$$

$= 1.41 \times 10^{-12}$ J/nucleon

(c) $(207.71698 - 205.97447)/206 = 8.4588 \times 10^{-3}$ amu

$$\Delta E = 8.46 \times 10^{-3} \text{ amu} \times \frac{1 \text{ g}}{6.02 \times 10^{23} \text{ amu}} \times \frac{1 \text{ kg}}{10^3 \text{ g}} \times \frac{9.00 \times 10^{16} \text{ m}^2}{\text{s}^2}$$

$= 1.26 \times 10^{-12}$ J/nucleon

21.42 (a) $3(1.00728) + 4(1.00866) + 3(0.00055) = 7.05813$ amu

Total binding energy = $7.05813 - 7.01600 = 0.04213$ amu

$$\Delta E = 0.04213 \text{ amu} \times \frac{1 \text{ g}}{6.02 \times 10^{23} \text{ amu}} \times \frac{1 \text{ kg}}{1 \times 10^3 \text{ g}} \times \frac{9.00 \times 10^{16} \text{ m}^2}{\text{s}^2}$$

$= 6.30 \times 10^{-12}$ J

Binding energy per nucleon = 6.30×10^{-12} J/7 = 9.00×10^{-13} J

(b) $64.52962 - 63.92914 = 0.60048$ amu; total binding energy $\Delta E = 8.98 \times 10^{-11}$ J

Binding energy per nucleon = 8.98×10^{-11} J/64 = 1.40×10^{-12} J.

(c) $233.93442 - 232.0382 = 1.8962$ amu; total binding energy $\Delta E = 2.83 \times 10^{-10}$

Binding energy per nucleon = 2.83×10^{-10} J/232 = 1.22×10^{-12} J

21.43 $$\frac{1.07 \times 10^{16} \text{ kJ}}{1 \text{ min}} \times \frac{60 \text{ min}}{1 \text{ hr}} \times \frac{24 \text{ hr}}{1 \text{ day}} = 1.54 \times 10^{19} \frac{\text{kJ}}{\text{day}} = 1.54 \times 10^{22} \text{ J/day}$$

$$\Delta m = \frac{1.54 \times 10^{22} \text{ kg} \cdot \text{m}^2/\text{s}^2/\text{d}}{(3.00 \times 10^8 \text{ m/s})^2} = 1.71 \times 10^5 \text{ kg/d}$$

Now calculate the mass change in the given nuclear reaction:
$\Delta m = 140.9140 + 91.9218 + 2(1.00867) - 235.0439 = -0.19076$ amu.
Converting from atoms to moles and amu to grams, it requires 235 g ^{235}U to produce energy equivalent to 0.191 g.
Now 0.1% of 1.71×10^5 kg is 1.71×10^2 kg = 1.71×10^5 g. Then,
$$1.71 \times 10^5 \text{ g} \times \frac{235 \text{ g } ^{235}\text{U}}{0.191 \text{ g}} = 2.10 \times 10^8 \text{ g } ^{235}\text{U}$$
This is about 230 tons of ^{235}U. Since a typical reactor fuel contains about three percent ^{235}U, this energy is the equivalent of about 7600 tons of reactor fuel per day.

21.44 (a) $\Delta m = 4.00260 + 1.00867 - 3.01605 - 2.01410 = -0.01888$ amu.

$$\Delta E = 0.01888 \text{ amu} \times \frac{1 \text{ g}}{1 \text{ amu}} \times \frac{1 \text{ kg}}{10^3 \text{ g}} \times (3.00 \times 10^8 \text{ m/sec})^2 = 1.70 \times 10^{12} \text{ J/mol}$$

(b) $\Delta m = 3.01603 + 1.00867 - 2(2.01410) = -3.50 \times 10^{-3}$ amu

$\Delta E = 3.15 \times 10^{11}$ J/mol

(c) $\Delta m = 4.00260 + 1.00782 - 3.01603 - 2.01410 = -1.97 \times 10^{-2}$ amu

$\Delta E = 1.77 \times 10^{12}$ J/mol

21.45 We can use Figure 21.13 to see that the binding energy per nucleon (which gives rise to the mass defect) is greatest for nuclei of mass numbers around 50. Thus (a) ${}^{59}_{27}Co$ should possess the greatest mass defect per nucleon.

21.46 In a fission reactor absorption of neutrons causes a single heavy nucleus to undergo fission, producing two medium mass nuclei. These are seen in Figure 21.13 to have a larger total mass defect than the starting single heavier nucleus, so energy is released.

Effects and Uses of Radioisotopes

21.47 (a) ${}^{235}_{92}U + {}^{1}_{0}n \rightarrow {}^{160}_{62}Sm + {}^{72}_{30}Zn + 4{}^{1}_{0}n$ (b) ${}^{239}_{94}Pu + {}^{1}_{0}n \rightarrow {}^{144}_{58}Ce + {}^{94}_{36}Kr + 2{}^{1}_{0}n$

21.48 (a) ${}^{2}_{1}H + {}^{2}_{1}H \rightarrow {}^{3}_{2}He + {}^{1}_{0}n$ (b) ${}^{233}_{92}U + {}^{1}_{0}n \rightarrow {}^{133}_{51}Sb + {}^{98}_{41}Nb + 3{}^{1}_{0}n$

21.49 The ${}^{59}Fe$ would be incorporated into the diet component, which in turn is fed to the rabbits. After a time blood samples could be removed from the animals, the red blood cells separated, and the radioactivity of the sample measured. If the iron in the dietary compound has been incorporated into blood hemoglobin, the blood cell sample should show beta emission. Samples could be taken at various times to determine the rate of iron uptake, rate of loss of the iron from the blood, and so forth.

21.50 (a) Add ${}^{36}Cl$ to water as a chloride salt. Then dissolve ordinary CCl_3COOH. After a time, distill the volatile materials away from the salt; CCl_3COOH is volatile, and will distill with water. Count radioactivity in the volatile material. If chlorine exchange has occurred, there will be radioactivity.

(b) Prepare a saturated solution of $BaCl_2$ containing a small amount of solid $BaCl_2$. Add to this solution solid $BaCl_2$ containing ${}^{36}Cl$. If the solid-solution equilibrium is dynamic, some of the ${}^{36}Cl$ in the solid will find itself in solution as chloride ion. After allowing some time for equilibrium to become established, filter the solution, count radioactivity in the solution that is separated from the solid. If there were no dynamic equilibrium, the ${}^{36}Cl^-$ would remain in the added solid, since the solution is already saturated before the addition of more solid.

(c) Utilize ${}^{36}Cl$ in soils of various pH values; grow plants for a given period of time. Remove plants, and directly measure radioactivity in samples from stems, leaves and so forth, or reduce the volume of plant sample by some form of digestion and evaporation of solution to give a dry residue that can be counted.

21.51 The extremely high temperature is required to overcome the electrostatic charge repulsions between the nuclei so that they come together to react.

21.52 Fusion reactors are attractive because, unlike fission reactors, they produce little or no radioactive waste. The problem is, no known material can withstand the extremely high temperatures required for fusion; the fusion reaction cannot be contained with current technology.

21.53 (a) *Control rods* control neutron flux so that there are enough neutrons to sustain the chain reaction but not so many that the core overheats.

(b) A *moderator* slows neutrons so that they are more easily captured by fissioning nuclei.

21.54 (a) In a *chain reaction*, one neutron initiates a nuclear transformation that produces more than one neutron. The product neutrons initiate more transformations, so that the reaction is self-sustaining.

(b) *Critical mass* is the mass of fissionable material required to sustain a chain reaction so that only one product neutron is effective at initiating a new transformation.

21.55 1 becquerel = 1 disintegration/second

The curie (ci) is another common unit for the activity of a sample. 1 Ci = 3.7×10^{10} becq

21.56 A *rad* is a measure of the amount of energy delivered to the body. The *rem* measures total biological damage and is the product of energy delivered times an index of the biological effectiveness of the radiation.

Additional Exercises

21.57 $^{210}_{84}\text{Po} \rightarrow\ ^{206}_{82}\text{Pb} +\ ^{4}_{2}\text{He}$

21.58 (1) $^{238}_{92}\text{U} \rightarrow\ ^{234}_{90}\text{Th} +\ ^{4}_{2}\text{He}$

(2) $^{234}_{90}\text{Th} \rightarrow\ ^{234}_{91}\text{Pa} +\ ^{0}_{-1}\text{e}$

(3) $^{234}_{91}\text{Pa} \rightarrow\ ^{234}_{92}\text{U} +\ ^{0}_{-1}\text{e}$

(4) $^{234}_{92}\text{U} \rightarrow\ ^{230}_{90}\text{Th} +\ ^{4}_{2}\text{He}$

(5) $^{230}_{90}\text{Th} \rightarrow\ ^{226}_{88}\text{Ra} +\ ^{4}_{2}\text{He}$

(6) $^{226}_{88}\text{Ra} \rightarrow\ ^{222}_{86}\text{Rn} +\ ^{4}_{2}\text{He}$

(7) $^{222}_{86}\text{Rn} \rightarrow\ ^{218}_{84}\text{Po} +\ ^{4}_{2}\text{He}$

(8) $^{218}_{84}\text{Po} \rightarrow\ ^{214}_{82}\text{Pb} +\ ^{4}_{2}\text{He}$

(9) $^{214}_{82}\text{Pb} \rightarrow\ ^{214}_{83}\text{Bi} +\ ^{0}_{-1}\text{e}$

(10) $^{214}_{83}\text{Bi} \rightarrow\ ^{214}_{84}\text{Po} +\ ^{0}_{-1}\text{e}$

(11) $^{214}_{84}\text{Po} \rightarrow\ ^{210}_{82}\text{Pb} +\ ^{4}_{2}\text{He}$

(12) $^{210}_{82}\text{Pb} \rightarrow\ ^{210}_{83}\text{Bi} +\ ^{0}_{-1}\text{e}$

(13) $^{210}_{83}\text{Bi} \rightarrow\ ^{210}_{84}\text{Po} +\ ^{0}_{-1}\text{e}$

(14) $^{210}_{84}\text{Po} \rightarrow\ ^{206}_{82}\text{Pb} +\ ^{4}_{2}\text{He}$

21.59 $^{1}_{0}\text{n} \rightarrow\ ^{1}_{1}\text{p} +\ ^{0}_{-1}\text{e}$

The other product of neutron decay is an electron.

21.60 $^{2}_{2}He$ would have two protons and no neutrons in the nucleus. Electrostatic repulsion between the two positively charged protons causes this nucleus to be unstable; all nuclei with two or more protons contain neutrons.

21.61 The most massive radionuclides will have the highest neutron/proton ratios. Thus, they are most likely to decay by a process that lowers this ratio, beta emission. The least massive nuclides, on the other hand, will decay by a process that increases the neutron/proton ratio, positron emission or orbital electron capture.

21.62 (a) $^{36}_{17}Cl \rightarrow {}^{36}_{18}Ar + {}^{0}_{-1}e$

(b) According to Table 21.3, nuclei with even numbers of both protons and neutrons, or an even number of one kind of nucleon, are more stable. ^{35}Cl and ^{37}Cl both have an odd number of protons but an even number of neutrons. ^{36}Cl has an odd number of protons and neutrons (17p, 19n), so it is less stable than the other two isotopes. Also, ^{37}Cl has 20 neutrons, a nuclear closed shell.

21.63 (a) $^{6}_{3}Li + {}^{63}_{28}Ni \rightarrow {}^{69}_{31}Ga$ (b) $^{48}_{20}Ca + {}^{248}_{96}Cm \rightarrow {}^{296}_{116}X$

(c) $^{88}_{38}Sr + {}^{84}_{36}Kr \rightarrow {}^{116}_{46}Pd + {}^{56}_{28}Ni$ (d) $^{48}_{20}Ca + {}^{298}_{92}U \rightarrow {}^{70}_{20}Ca + 4\,{}^{1}_{0}n + 2\,{}^{106}_{46}Pd$

21.64 $k = \dfrac{0.693}{t_{1/2}} = \dfrac{0.693}{5.26\text{ yr}} = 0.132\text{ yr}^{-1}$; $\dfrac{N_t}{N_o} = 0.80$

$$\ln\frac{N_t}{N_o} = -kt\;;\quad t = \frac{-1}{k}\ln\frac{N_t}{N_o} = \frac{-1}{0.132\text{ yr}^{-1}}\ln(0.80) = 1.69\text{ yr}$$

1.69 yr = 20.3 mo = 617 d. The source will have to be replaced sometime in July, 1995.

21.65 (a) $\dfrac{96.485\text{ kJ}}{\text{mo}\cdot\text{eV}} \times \dfrac{10^6\text{ eV}}{1\text{ MeV}} \times \dfrac{1\text{ mol}}{6.022\times10^{23}\text{ photons}} = 1.602\times10^{-16}\text{ kJ/1.0 MeV photons}$

(b) As in Exercise 21.33, $k = \dfrac{0.693}{30\text{ yr}} \times \dfrac{1\text{ yr}}{365\text{ d}} \times \dfrac{1\text{ d}}{24\text{ hr}} \times \dfrac{1\text{ hr}}{60\text{ min}} = 4.4\times10^{-8}\text{ min}^{-1}$

$$\ln\frac{N_t}{N_o} = -kt = -(4.4\times10^{-8}\text{ min}^{-1})\times30\text{ min} = -1.3\times10^{-6}$$

$$\frac{N_t}{N_o} = e^{-1.3\times10^{-6}} = (1.00 - 1.3\times10^{-6})\;;\; N_t = 10.0\times10^{-3}\text{ g}\,(1.00 - 1.3\times10^{-6})$$

$$N_o - N_t = 10.0\times10^{-3}\text{ g} - [10.0\times10^{-3}(1.00 - 1.3\times10^{-6})] = 1.3\times10^{-8}\text{ g } {}^{137}CsCl$$

$$1.3\times10^{-8}\text{ g }{}^{137}CsCl \times \frac{1\text{ mol CsCl}}{172.45\text{ g CsCl}} \times \frac{6.022\times10^{23}\text{ photons}}{1\text{ mol Cs}} \times \frac{0.662\text{ MeV}}{1\text{ photon}}$$

$$\times\frac{1.602\times10^{-16}\text{ kJ}}{1\text{ MeV}} = 4.9\times10^{-3}\text{ kJ } (= 4.9\text{ J})$$

(c) $E_\gamma = 0.662 \text{ MeV} \times \frac{1.602 \times 10^{-16} \text{ kJ}}{1 \text{ MeV photon}} = 1.06 \times 10^{-16} \text{ kJ}$

$$E_{uv} = \frac{hc}{\lambda} = \frac{6.626 \times 10^{-34} \text{ J} \cdot \text{s} \times 3.00 \times 10^8 \text{ m/s}}{300 \times 10^{-9} \text{ m}} \times \frac{1 \text{ kJ}}{1000 \text{ J}} = 6.63 \times 10^{-22} \text{ kJ}$$

$$\frac{E_\gamma}{E_{uv}} = \frac{1.06 \times 10^{-16} \text{ kJ}}{6.63 \times 10^{-22} \text{ kJ}} = 1.60 \times 10^5$$

Gamma rays carry orders of magnitude (10^5 - 10^6) greater energy than ultraviolet photons and pose a much greater health risk upon exposure.

21.66 The danger from inhaling plutonium-containing particles arises from the extensive damage that alpha particles can cause in biological tissues. Alpha particles have only a short range; that is, they cause damage only in a few mm of tissue as they move away from the radiation source. Only when alpha-emitting substances are ingested or inhaled does the radiation actually penetrate sensitive tissues and organs. Unfortunately, once plutonium gets inside the body it is not readily excreted, so this long-lived source of radioactivity continues to cause damage over a long period of time.

21.67 Figure (a) shows a graph of activity (distintegrations per minute) vs. time. Choose $t_{1/2}$ as the time at which the activity is half the initial value. Rearrange Equation [21.19] to obtain

$\ln N_t - \ln N_o = -kt$; $\ln N_t = -kt + \ln N_o$; $\log N_t = \frac{-kt}{2.30} + \frac{\log N_o}{2.30}$

The slope of a graph of $\log N_t$ vs. t is therefore = -k/2.30. The slope of the graph, (b), is -4.9×10^{-2}. $k = -2.30(-4.9 \times 10^{-2}) = 0.11$; $t_{1/2} = 0.693/0.11 = 6.3$ hr

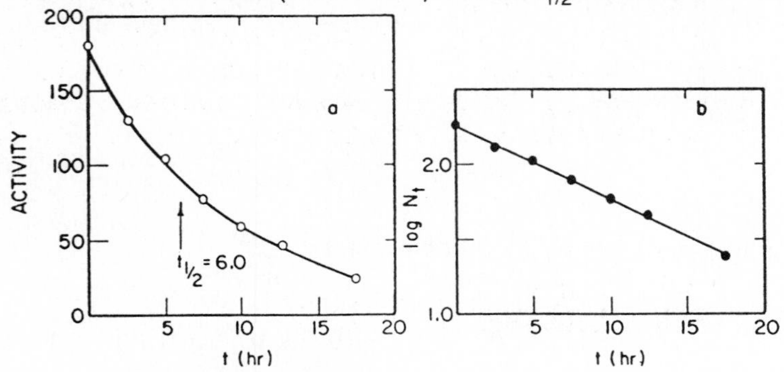

21.68 $1 \times 10^{-6} \text{ curie} \times \frac{3.7 \times 10^{10} \text{ disint/s}}{\text{curie}} = 3.7 \times 10^4 \text{ disint/s}$

rate = 3.7×10^4 nuclei/s = kN

$$k = \frac{0.693}{t_{1/2}} = \frac{0.693}{28.8 \text{ yr}} \times \frac{1 \text{ yr}}{365 \times 24 \times 3600 \text{ sec}} = 7.93 \times 10^{-10}\text{/s}$$

3.7×10^4 nuclei/s = $(7.93 \times 10^{-10}\text{/s})$ N; N = 4.8×10^{13} nuclei

mass ^{90}Sr = $4.8 \times 10^{13} \text{ nuclei} \times \frac{90 \text{ g Sr}}{6.02 \times 10^{23} \text{ nuclei}} = 7.2 \times 10^{-9}$ g Sr

21.69 First calculate k from a knowledge of $t_{1/2}$:

$$k = \frac{0.693}{2.4 \times 10^4 \text{ yr}} \times \frac{1 \text{ yr}}{365 \times 24 \times 3600 \text{ s}} = 9.2 \times 10^{-13} \text{s}^{-1}$$

Now calculate N:

$$N = 0.500 \text{ g Pu} \times \frac{1 \text{ mol Pu}}{239 \text{ g Pu}} \times \frac{6.02 \times 10^{23} \text{ Pu atoms}}{1 \text{ mol Pu}} = 1.3 \times 10^{21} \text{ Pu atoms}$$

rate = $(9.2 \times 10^{-13} \text{ s}^{-1})(1.3 \times 10^{21} \text{ Pu atoms}) = 1.2 \times 10^9$ distin/s

21.70 (a) $1.25 \text{ Ci} \times \frac{3.7 \times 10^{10} \text{ disint/s}}{1 \text{ Ci}} \times \frac{1 \text{ becq}}{1 \text{ disint/s}} = 4.6 \times 10^{10}$ becq

(b) First, calculate k: $k = \frac{0.693}{5.26 \text{ yr}} = 0.132 \text{ yr}^{-1}$

From Equation 21.19, $\ln\frac{N_t}{N_o} = -kt$; $\ln\frac{N_t}{1.25} = 0.132 \text{ yr}^{-1} \times 2 \text{ yr} = -0.264$

$\frac{N_t}{1.25} = 0.768$; N_t 0.960 Ci = 3.6×10^{10} becq

21.71 $k = \frac{0.693}{t_{1/2}} = \frac{0.693}{4.5 \times 10^9 \text{ yr}} \times \frac{1 \text{ yr}}{365 \text{ d}} \times \frac{1 \text{ d}}{24 \text{ hr}} \times \frac{1 \text{ hr}}{3600 \text{ s}} = 4.9 \times 10^{-18} \text{ s}^{-1}$

$$N = 0.0010 \text{ g } {}^{238}\text{U} \times \frac{1 \text{ mol U}}{238 \text{ g U}} \times \frac{6.02 \times 10^{23} \text{ U atoms}}{1 \text{ mol U}} = 2.5 \times 10^{18} \text{ U atoms}$$

$$\text{rate} = kN = \frac{4.9 \times 10^{-18}}{\text{s}} \times 2.5 \times 10^{18} \text{ atoms} = 12 \text{ disint/s}$$

21.72 The health hazard associated with a radioisotope depends on the effective dosage (rems) received by a subject. This not only depends on the rate of decay or half-life of the radioisotope, but the type of decay, the energy associated with the particle or photon emitted and the penetrating power of the radiation. In general, radioisotopes with short half-lives give off a large amount of energy in a short period of time. Given that the subject will be exposed to the radioisotope for a certain amount of time, those with short half-life will deliver a larger dose of radiation. However, the risk of exposure to a long-lived radioisotope is greater for more subjects than the risk from a short-lived isotope.

21.73 First, calculate k:

$$k = \frac{0.693}{12.26 \text{ yr}} \times \frac{1 \text{ yr}}{365 \text{ d}} \times \frac{1 \text{ d}}{24 \text{ hr}} \times \frac{1 \text{ hr}}{3600 \text{ sec}} = 1.79 \times 10^{-9} \text{s}^{-1}$$

From Equation [21.18], $1.50 \times 10^3 \text{ s}^{-1} = (1.79 \times 10^{-9} \text{ s}^{-1})(N)$;

$N = 8.38 \times 10^{11}$. In 26.0 g of water, there are

$$26.0 \text{ g } H_2O \times \frac{1 \text{ mol } H_2O}{18.0 \text{ g } H_2O} \times \frac{6.02 \times 10^{23} \text{ } H_2O}{1 \text{ mol } H_2O} \times \frac{2H}{1 \text{ } H_2O} = 1.74 \times 10^{24} \text{ H atoms}$$

The mole fraction of ${}^3_1H$ atoms in the sample is thus

$8.38 \times 10^{11}/1.74 \times 10^{24} = 4.82 \times 10^{-13}$.

21.74 Determine the wavelengths of the photons by first calculating the energy equivalent of the mass of an electron or positron. (Since two photons are formed by annihilation of two particles of equal mass, we need calculate the energy equivalent of just one particle.)

$$\Delta E = (5.49 \times 10^{-4} \text{ amu})\,(3.0 \times 10^{8} \text{ cm/s})^2 \times \frac{1 \text{ g}}{6.02 \times 10^{23} \text{ amu}} \times \frac{1 \text{ kg}}{10^3 \text{ g}} = 8.2 \times 10^{-14} \text{ J}$$

Also, $\Delta E = h\nu$; $\Delta E = \frac{hc}{\lambda}$; $\lambda = \frac{hc}{\Delta E}$

$$\lambda = \frac{(6.63 \times 10^{-34} \text{ J} \cdot \text{s})\,(3.0 \times 10^{8} \text{ m/s})}{8.2 \times 10^{-14} \text{ J}} = 2.4 \times 10^{-12} \text{ m} = 2.4 \times 10^{-3} \text{ nm}$$

This is a very short wavelength indeed; it lies at the short wavelength end of the range of observed gamma ray wavelengths (see Figure 6.3).

21.75 Because of the relationship $\Delta E = \Delta mc^2$, the mass defect (Δm) is directly related to the binding energy (ΔE) of the nucleus.

^{7}Be - 4p, 3n, ; 4(1.00728) + 3(1.00866) = 7.0551 amu

Total mass defect = 7.0551 - 7.0147 = 0.0404 amu

0.0404 amu/7 nucleons = 5.77×10^{-3} amu/nucleon

$$\Delta E = \Delta m \times c^2 = \frac{5.77 \times 10^{-3} \text{ amu}}{\text{nucleon}} \times \frac{1 \text{ g}}{6.022 \times 10^{23} \text{ amu}} \times \frac{1 \text{ kg}}{1 \times 10^3 \text{ g}} \times \frac{9.00 \times 10^{16} \text{ m}^2}{\text{sec}^2}$$

$$= \frac{5.77 \times 10^{-3} \text{ amu}}{\text{nucleon}} \times \frac{1.495 \times 10^{-10} \text{ J}}{1 \text{ amu}} = 8.63 \times 10^{-13} \text{ J/nucleon}$$

^{9}Be - 4p, 5n; 4(1.00728) + 5(1.00866) = 9.0724 amu

Total mass defect = 9.0724 - 9.0100 = 0.0624 amu

0.0624 amu/9 nucleons = 6.94×10^{-3} amu/nucleon

6.94×10^{-3} amu/nucleon $\times$ 1.495×10^{-10} J/amu = (1.037) $\rightarrow$ 1.04×10^{-12} J/nucleon

^{10}Be - 4p, 6n; 4(1.00728) + 6(1.00866) = 10.0811 amu

Total mass defect = 10.0811 - 10.0113 = 0.0698 amu

0.0698 amu/10 nucleons = 6.98×10^{-3} amu/nucleon

6.98×10^{-3} amu/nucleon $\times$ 1.495×10^{-10} J/amu = (1.044) $\rightarrow$ 1.04×10^{-12} J/nucleon

The binding energies/nucleon for ^{9}Be and ^{10}Be are very similar; that for ^{10}Be is slightly higher.

21.76 $\Delta m = \Delta E/c^2$; $\Delta m = \dfrac{3.9 \times 10^{26}\ \text{kg} \cdot \text{m/s}^2}{(3.0 \times 10^8\ \text{m/s})^2} = 4.3 \times 10^9\ \text{kg}$

This is the mass lost in one second.

21.77 $1000\ \text{M watts} \times \dfrac{1 \times 10^6\ \text{watts}}{1\ \text{M watt}} \times \dfrac{1\ \text{J}}{1\ \text{watt}\cdot\text{s}} \times \dfrac{1\ {}^{235}\text{U atom}}{3 \times 10^{-11}\ \text{J}} \times \dfrac{1\ \text{mol U}}{6.02 \times 10^{23}\ \text{atoms}}$

$\times \dfrac{235\ \text{g U}}{1\ \text{mol}} \times \dfrac{3600\ \text{s}}{1\ \text{hr}} \times \dfrac{24\ \text{hr}}{1\ \text{d}} \times \dfrac{365\ \text{d}}{1\ \text{yr}} \times \dfrac{40}{100}\ \text{(efficiency)} = \dfrac{2 \times 10^5\ \text{g U}}{\text{yr}} = \dfrac{200\ \text{kg U}}{\text{yr}}$

21.78 $2 \times 10^{-12}\ \text{curies} \times \dfrac{3.7 \times 10^{10}\ \text{distint/s}}{1\ \text{curie}} = 7.4 \times 10^{-2}\ \text{disint/s}$

$\dfrac{7.4 \times 10^{-2}\ \text{disint/s}}{75\ \text{kg}} \times \dfrac{8 \times 10^{-13}\ \text{J}}{\text{disint}} \times \dfrac{1\ \text{rad}}{1 \times 10^{-2}\ \text{J/kg}} \times \dfrac{3600\ \text{s}}{\text{hr}} \times \dfrac{24\ \text{hr}}{1\ \text{d}}$

$\times \dfrac{365\ \text{d}}{1\ \text{yr}} = 2.5 \times 10^{-6}\ \text{rad/yr}$

Recall that there are 10 rem/rad for alpha particles.

$\dfrac{2.5 \times 10^{-6}\ \text{rad}}{1\ \text{yr}} \times \dfrac{10\ \text{rem}}{1\ \text{rad}} = 2.5 \times 10^{-5}\ \text{rem/yr}$

CHAPTER 22

Chemistry of Hydrogen, Oxygen Nitrogen, and Carbon

Periodic Trends and Chemical Reactions

22.1 Metals: Sr, Ce, Rh; nonmetals: Se, Kr; semimetal: Sb

22.2 Metals: Zr, In, Cs; nonmetals: Xe, Br; semimetal: Ge

22.3 Metals have high melting points and large thermal and electrical conductivity while nonmetals have much lower melting points and are poor conductors. Metals react with nonmetals to produce high-melting ionic solids; nonmetals react with other nonmetals to form covalent compounds which can be solids, liquids or gases. Oxides of metals produce basic aqueous solutions while oxides of nonmetals produce acidic aqueous solutions.

22.4 The first member of each family of nonmetals can form a maximum of four bonds because only the 2s and three 2p orbitals are available for bonding. Because of their small atomic radii, these elements readily form π bonds. Nonmetals in the third row and beyond can form more than four bonds owing to increased atomic radii and available d orbitals. They do not form stable π bonds because of ineffective overlap of the larger p valence orbitals. Also, because of their large electronegativities, nitrogen, oxygen and fluorine can act as acceptors in hydrogen bonding interactions, while other elements in their families do not.

22.5 Bismuth should be most metallic. It has a metallic luster and a relatively low ionization energy. Bi_2O_3 is soluble in acid but not in base, characteristic of the basic properties of the oxide of a metal rather than the oxide of a nonmetal.

22.6 B is most nonmetallic in Group 3A. Some properties which demonstrate this are its low electrical and thermal conductivity, lack of metallic luster (it is a yellow powder), and the acidity of its oxide.

22.7 (a) N (b) K
(c) K in the gas phase (lowest ionization energy), Li in aqueous solution (most positive E^{o} value)
(d) N has the smallest <u>covalent</u> radius; Ne is difficult to compare because it doesn't form compounds
(e) N

22.8 (a) O (b) I (c) Ba (d) O (e) Cu

22.9 (a) IF_3 (b) B_2O_3 (c) BF_3 (d) H_2O
In each case, the pair of atoms with the greater difference in electronegativity form the more polar bond.

22.10 (a) SnO_2; O is very electronegative so Sn (less electronegative than C) forms the more polar bond with it.
(b) V_2O_3; V is more metallic than P and has a much lower electronegativity.
(c) SiO_2; Si is less electronegative than S.
(d) CCl_4; Cl is more electronegative than Br and has the greater ΔEN with C.

22.11 (a) Nitrogen is too small to accomodate five flourine atoms about it. The P and As atoms are larger. Furthermore, P and As have available 3d and 4d orbitals, respectively, to form hybrid orbitals that can accommodate more than an octet of electrons about the central atom.
(b) Si does not readily form π bonds, which would be necessary to satisfy the octet rule for both atoms in SiO.
(c) A reducing agent is a substance that readily loses electrons. As has a lower electronegativity than N; that is, it more readily gives up electrons to an acceptor and is more easily oxidized.

22.12 (a) Nitrogen is a highly electronegative element. In HNO_3 it is in its highest oxidation state, +5, and thus is more readily reduced than phosphorus, which forms stable P-O bonds.
(b) The difference between the third row element and the second lies in the smaller size of C as compared with Si, and the fact that Si has 3d orbitals available to form an sp^3d^2 hybrid set that can accommodate more than an octet of electrons.
c) Two of the carbon compounds, C_2H_4 and C_2H_2, contain C-C π bonds. Si does not readily form π bonds (to itself or other atoms), so Si_2H_4 and Si_2H_2 are not known as stable compounds.

22.13 (a) $NaNH_2(s) + H_2O(l) \rightarrow NH_3(aq) + Na^+(aq) + OH^-(aq)$
(b) $2C_3H_7OH(l) + 9O_2(g) \rightarrow 6CO_2(g) + 8H_2O(l)$
(c) $NiO(s) + C(s) \rightarrow CO(g) + Ni(s)$ or $2NiO(s) + C(s) \rightarrow CO_2(g) + 2Ni(s)$
(d) $AlP(s) + 3H_2O(l) \rightarrow PH_3(g) + Al(OH)_3(s)$
(e) $Na_2S(s) + 2HCl(aq) \rightarrow H_2S(g) + 2Na^+(aq) + 2Cl^-(aq)$

22.14 (a) $NaOCH_3(s) + H_2O(l) \rightarrow NaOH(aq) + CH_3OH(aq)$

(b) $Na_2O(s) + 2HC_2H_3O_2(aq) \rightarrow 2NaC_2H_3O_2(aq) + H_2O(l)$

(c) $WO_3(s) + H_2(g) \rightarrow W(s) + 3H_2O(g)$

(d) $4NH_2OH(l) + O_2(g) \rightarrow 6H_2O(l) + 2N_2(g)$

(e) $Al_4C_3(s) + 12H_2O(l) \rightarrow 4Al(OH)_3(s) + 3CH_4(g)$

Hydrogen

22.15 ${}^1_1H$ - protium; ${}^2_1H$ - deuterium; ${}^3_1H$ - tritium

22.16 Tritium is radioactive. ${}^3_1H \rightarrow {}^3_2He + {}^{\ 0}_{-1}e$

22.17 Like other elements in group 1A, hydrogen has only one valence electron. Like other elements in group 7A, hydrogen needs only one electron to complete its valence shell. The most common oxidation number of H is +1, like the group 1A elements; H can also exist in the -1 oxidation state, a state common to the group 7A elements.

22.18 Hydrogen does not have a strictly comparable electronic arrangement. There are no closed shells of electrons underlying the valence shell. When hydrogen completes its valence shell it has only two electrons therein, a characteristic it shares only with He.

22.19 (a) $Mg(s) + 2H^+(aq) \rightarrow Mg^{2+}(aq) + H_2(g)$

(b) $C(s) + H_2O(g) \xrightarrow{1000^\circ C} CO(g) + H_2(g)$

(c) $CH_4(g) + H_2O(g) \xrightarrow{1100^\circ C} CO(g) + 3H_2(g)$

22.20 (a) Electrolysis of brine; reaction of carbon with steam; reaction of methane with steam; byproduct in petroleum refining

(b) synthesis of ammonia; synthesis of methanol; reducing agent; hydrogenation of unsaturated vegetable oils

22.21 (a) Ionic (metal hydride) (b) molecular (nonmetal hydride)

(c) metallic (nonstoichiometric transition metal hydride)

22.22 (a) Molecular (b) ionic (c) metallic

22.23 (a) $NaH(s) + H_2O(l) \rightarrow NaOH(aq) + H_2(g)$

(b) $Fe(s) + H_2SO_4(aq) \rightarrow Fe^{2+}(aq) + H_2(g) + SO_4^{2-}(aq)$

(c) $H_2(g) + Br_2(g) \rightarrow 2HBr(g)$

(d) $Na(l) + H_2(g) \rightarrow 2NaH(s)$

(e) $PbO(s) + H_2(g) \xrightarrow{\Delta} Pb(s) + H_2O(g)$

22.24 (a) $2Al(s) + 6H^+(aq) \rightarrow 2Al^{3+}(aq) + 3H_2(g)$

(b) $Mg(s) + H_2O(g) \rightarrow MgO(s) + H_2(g)$

(c) $MnO_2(s) + H_2(g) \rightarrow MnO(s) + H_2O(g)$

(d) $CaH_2(s) + H_2O(l) \rightarrow Ca(OH)_2(aq) + 2H_2(g)$

22.25 $$1.1 \times 10^{13} \text{ g } C_2H_4 \times \frac{1 \text{ mol } C_2H_4}{28 \text{ g } C_2H_4} \times \frac{1 \text{ mol } H_2}{1 \text{ mol } C_2H_4} \times \frac{2.0 \text{ g } H_2}{1 \text{ mol } H_2} = 7.9 \times 10^{11} \text{ g } H_2$$
$$= 7.9 \times 10^{8} \text{ kg } H_2$$

22.26 (a) $H_2(g) + 1/2\ O_2(g) \rightarrow H_2O(l)$; $\Delta H = -285.8$ kJ

$CH_4(g) + 2O_2(g) \rightarrow CO_2(g) + 2H_2O(l)$

$\Delta H = 2(-285.8) - 393.5 - (-74.8) = -890.3$ kJ

(b) for H_2: $$\frac{-285.8 \text{ kJ}}{1 \text{ mol } H_2} \times \frac{1 \text{ mol } H_2}{2.02 \text{ g } H_2} = \frac{-141.5 \text{ kJ}}{1 \text{ g } H_2}$$

for CH_4: $$\frac{-890.3 \text{ kJ}}{1 \text{ mol } CH_4} \times \frac{1 \text{ mol } CH_4}{16.0 \text{ g } CH_4} = \frac{-55.6 \text{ kJ}}{1 \text{ g } CH_4}$$

(c) Find the number of moles of gas that occupy 1 m^3 at STP:

$$n = \frac{1 \text{ atm} \times 1 \text{ m}^3}{273 \text{ K}} \times \frac{1 \text{ K} \cdot \text{mol}}{0.0821 \text{ L} \cdot \text{atm}} \times \left[\frac{100 \text{ cm}}{1 \text{ m}}\right]^3 \times \frac{1 \text{ L}}{10^3 \text{ cm}^3} = 44.6 \text{ mols}$$

for H_2: $$\frac{-285.8 \text{ kJ}}{1 \text{ mol } H_2} \times \frac{44.6 \text{ mol } H_2}{1 \text{ m}^3 \text{ } H_2} = \frac{1.27 \times 10^4 \text{ kJ}}{1 \text{ m}^3 \text{ } H_2}$$

for CH_4: $$\frac{-890.3 \text{ kJ}}{1 \text{ mol } CH_4} \times \frac{44.6 \text{ mol } CH_4}{1 \text{ cm}^3 \text{ } CH_4} = \frac{3.97 \times 10^4 \text{ kJ}}{1 \text{ cm}^3 \text{ } CH_4}$$

Oxygen

22.27 As an oxidizing agent in steel-making; to bleach pulp and paper; in oxyacetylene torches; in medicine to assist in breathing

22.28 Synthesis of pharmaceuticals, lubricants and other organic compounds where C=C bonds are cleaved; in water treatment

22.29

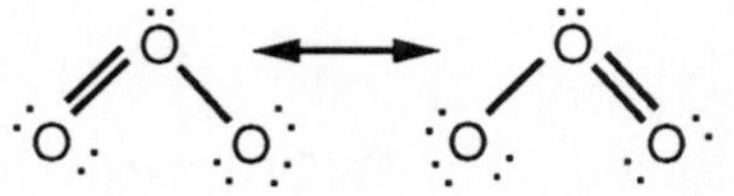

Ozone has two resonance forms (Section 8.7); the molecular structure is bent, with an O-O-O bond angle of approximately 120°. The π bond in ozone is delocalized over the entire molecule; neither individual O-O bond is a full double bond, so the observed O-O distance of 1.28 Å is greater than the 1.21 Å distance in O_2, which has a full O-O double bond.

22.30 Usually by thermal decomposition of $KClO_3$ or H_2O_2:

$$2KClO_3(s) \xrightarrow[\Delta]{MnO_2} 2KCl(s) + 3O_2(g)$$

$$2H_2O_2(aq) \xrightarrow{\Delta} 2H_2O(l) + O_2(g)$$

22.31 (a) $CaO(s) + H_2O(l) \rightarrow Ca^{2+}(aq) + 2OH^-(aq)$
(b) $Al_2O_3(s) + 6H^+(aq) \rightarrow 2Al^{3+}(aq) + 3H_2O(l)$
(c) $Na_2O_2(s) + 2H_2O(l) \rightarrow 2Na^+(aq) + 2OH^-(aq) + H_2O_2(aq)$
(d) $N_2O_3(g) + H_2O(l) \rightarrow 2HNO_2(aq)$
(e) $2KO_2(s) + 2H_2O(l) \rightarrow 2K^+(aq) + 2OH^-(aq) + O_2(g) + H_2O_2(aq)$
(f) $NO(g) + O_3(g) \rightarrow NO_2(g) + O_2(g)$

22.32 (a) $2HgO(s) \xrightarrow{\Delta} Hg(l) + O_2(g)$
(b) $2Cu(NO_3)_2(s) \xrightarrow{\Delta} 2CuO(s) + 4NO_2(g) + O_2(g)$
(c) $PbS(s) + 4O_3(g) \rightarrow PbSO_4(s) + 4O_2(g)$
(d) $2ZnS(s) + 3O_2(g) \rightarrow 2ZnO(s) + 2SO_2(g)$
(e) $2K_2O_2(s) + 2CO_2(g) \rightarrow 2K_2CO_3(s) + O_2(g)$
(f) $2Ag(s) + O_3(g) \rightarrow Ag_2O(s) + O_2(g)$

22.33 (a) Neutral (b) acidic (oxide of a nonmetal) (c) basic (oxide of a metal) (d) amphoteric

22.34 (a) Mn_2O_7 (higher oxidation state of Mn) (b) SnO_2 (higher oxidation state of Sn)
(c) SO_3 (higher oxidation state of S) (d) SO_2 (more nonmetallic character of S)
(e) Ga_2O_3 (more nonmetallic character of Ga) (f) SO_2 (more nonmetallic character of S)

22.35 Assume that the reactions occur in basic solution. The half-reaction for reduction of H_2O_2 is in all cases $H_2O_2(aq) + 2e^- \rightarrow 2OH^-(aq)$.
(a) $H_2O_2(aq) + S^{2-}(aq) \rightarrow 2OH^-(aq) + S(s)$
(b) $SO_2(g) + 2OH^-(aq) + H_2O_2(aq) \rightarrow SO_4^{2-}(aq) + H_2O(l)$
(c) $NO_2^-(aq) + H_2O_2(aq) \rightarrow NO_3^-(aq) + H_2O(l)$
(d) $As_2O_3(s) + 2H_2O_2(aq) + 6OH^-(aq) \rightarrow 2AsO_4^{3-}(aq) + 5H_2O(l)$
(e) This reaction must be occurring in acidic solution, since $Fe(OH)_3$ would form if the solution were basic. The half-reactions are:

$$2H^+(aq) + H_2O_2(aq) + 2e^- \rightarrow 2H_2O(l)$$
$$2[Fe^{2+}(aq) \rightarrow Fe^{3+}(aq) + e^-]$$

$$2Fe^{2+}(aq) + H_2O_2(aq) + 2H^+(aq) \rightarrow 2Fe^{3+}(aq) + 2H_2O(l)$$

CHAPTER 22 Chemistry of Hydrogen, Oxygen Nitrogen and Carbon

Solutions to Exercises

22.36 The half-reaction for oxidation of H_2O_2 in acidic solution is as follows:

$H_2O_2(aq) \rightarrow O_2(g) + 2H^+(aq) + 2e^-$.

(a) $2[MnO_4^-(aq) + 8H^+(aq) + 5e^- \rightarrow Mn^{2+}(aq) + 4H_2O(l)]$

$5[H_2O_2(aq) \rightarrow O_2(g) + 2H^+(aq) + 2e^-]$

$2MnO_4^-(aq) + 5H_2O_2(aq) + 6H^+(aq) \rightarrow 2Mn^{2+}(aq) + 5O_2(g) + 8H_2O(l)$

(b) $H_2O_2(aq) + Cl_2(aq) \rightarrow O_2(g) + 2Cl^-(aq) + 2H^+(aq)$

(c) $H_2O_2(aq) + 2Ce^{4+}(aq) \rightarrow O_2(g) + 2Ce^{3+}(aq) + 2H^+(aq)$

(d) $3H_2O_2(aq) + O_3(g) \rightarrow 3H_2O(l) + 3O_2(g)$

Nitrogen

22.37 Formation of ammonia and fertilizers; as an inert gaseous blanket in manufacturing processes; as a coolant in liquid form.

22.38 Nitrogen fixation is the formation of nitrogen-containing compounds from N_2. N_2 is unreactive because it contains a very strong $N \equiv N$ bond and is nonpolar.

22.39 (a) HNO_2; + 3 (b) N_2H_4; -2 (c) KCN; -3
(d) $NaNO_3$; +5 (e) NH_4Cl; -3 (f) Li_3N; -3

22.40 (a) $NaNO_2$; + 3 (b) NH_3; -3 (c) N_2O; +1
(d) NaCN; -3 (e) HNO_3; +5 (f) NO_2; +4

22.41 (a) 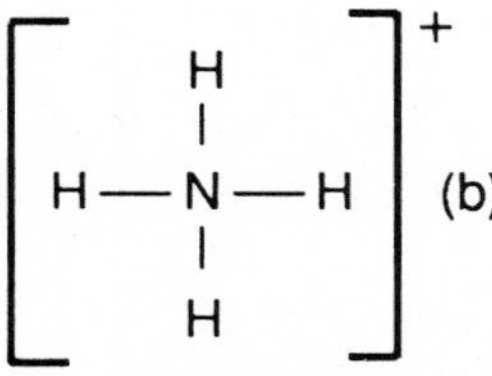

tetrahedral

(b) 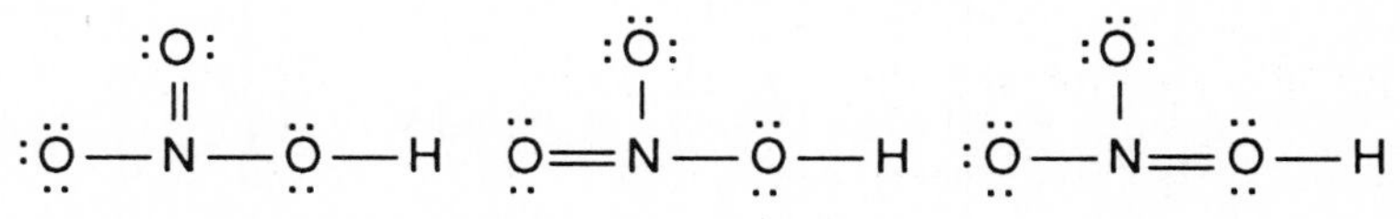

The geometry around nitrogen is trigonal planar, but the hydrogen atom is not required to lie in this plane. The third resonance form makes a much smaller contribution to the structure than the first two.

(c) 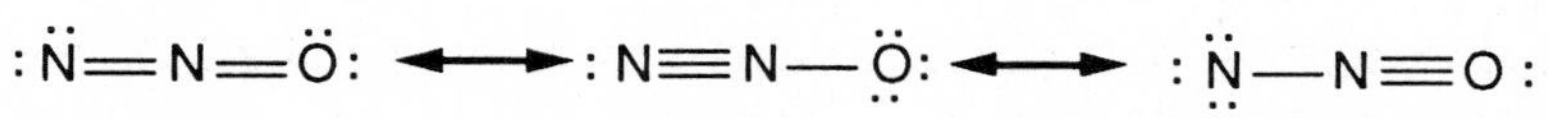

The molecule is linear. Again, the third resonance form makes less contribution to the structure because of the high formal charges involved.

(d) 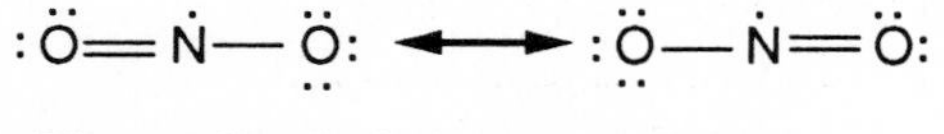

The molecule is bent (nonlinear).

22.42 (a)   :Ö=N—Ö—H

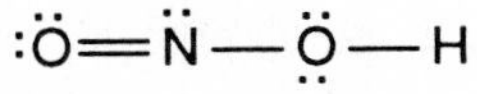

The molecule is bent around the central oxygen and nitrogen atoms; the four atoms need not lie in a plane.

(b) $[:\ddot{N}=N=\ddot{N}:]^-$ ⟷ $[:N\equiv N—\ddot{N}:]^-$ ⟷ $[:\ddot{N}—N\equiv N:]^-$

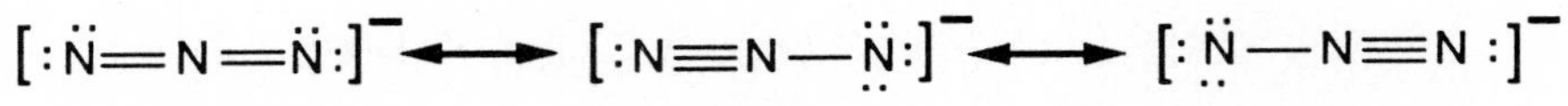

The molecule is linear.

(c)
$$\left[\begin{array}{c} \text{H} \quad \text{H} \\ | \quad\; | \\ \text{H—N—N:} \\ | \quad\; | \\ \text{H} \quad \text{H} \end{array}\right]^+$$

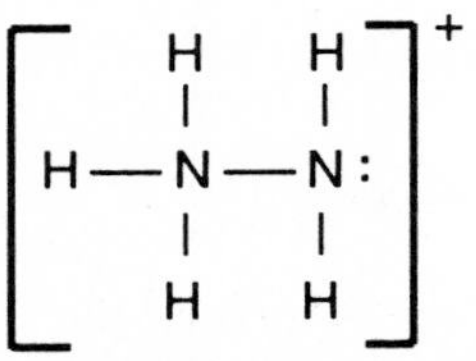

The geometry is tetrahedral around the left nitrogen, trigonal pyramidal around the right.

(d)
$$\left[:\ddot{O}—N\!\!\begin{array}{c}:\ddot{O}:\\ |\\ \end{array}\!\!=\ddot{O}\right]^-$$

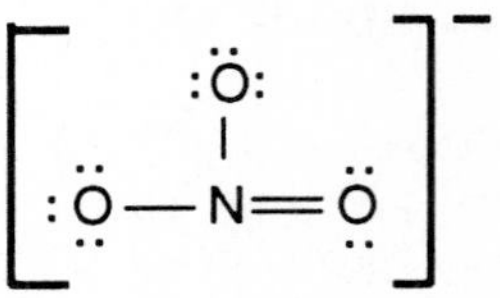

(three equivalent resonance forms) The ion is trigonal planar.

22.43 (a) $Mg_3N_2(s) + 6H_2O(l) \rightarrow 3Mg(OH)_2(s) + 2NH_3(aq)$

(b) $2NO(g) + O_2(g) \rightarrow 2NO_2(g)$

(c) $4NH_3(g) + 3O_2(g) \rightarrow 2N_2(g) + 6H_2O(g)$

(d) $NaNH_2(s) + H_2O(l) \rightarrow Na^+(aq) + OH^-(aq) + NH_3(aq)$

22.44 (a) $N_2O_5(g) + H_2O(l) \rightarrow 2H^+(aq) + 2NO_3^-(aq)$

(b) $Li_3N(s) + 3H_2O(l) \rightarrow NH_3(aq) + 3LiOH(aq)$

(c) $NH_3(aq) + H^+(aq) \rightarrow NH_4^+(aq)$

(d) $N_2H_4(aq) + H^+(aq) \rightarrow N_2H_5^+(aq)$

22.45 (a) $4Zn(s) + 2NO_3^-(aq) + 10H^+(aq) \rightarrow 4Zn^{2+}(aq) + N_2O(g) + 5H_2O(l)$

(b) $4NO_3^-(aq) + S(s) + 4H^+(aq) \rightarrow 4NO_2(g) + SO_2(g) + 2H_2O(l)$

(or $6NO_3^-(aq) + S(s) + 4H^+(aq) \rightarrow 6NO_2(g) + SO_4^{2-}(aq) + 2H_2O(l)$)

(c) $2NO_3^-(aq) + 3SO_2(g) + 2H_2O(l) \rightarrow 2NO(g) + 3SO_4^{2-}(aq) + 4H^+(aq)$

(d) $CO(NH_2)_2(aq) + H_2O(l) \rightarrow 2NH_3(aq) + CO_2(aq)$

22.46 (a) $N_2H_4(g) + 5F_2(g) \rightarrow 2NF_3(g) + 4HF(g)$

(b) $4CrO_4^{2-}(aq) + 3N_2H_4(aq) + 4H_2O(l) \rightarrow 4Cr(OH)_4^-(aq) + 4OH^-(aq) + 3N_2(g)$

(c) $Cu^{2+}(aq) + 2e^- \rightarrow Cu(s)$

$2NH_2OH(aq) \rightarrow N_2(g) + 2H_2O(l) + 2H^+(aq) + 2e^-$

$Cu^{2+}(aq) + 2NH_2OH(aq) \rightarrow Cu(s) + N_2(g) + 2H_2O(l) + 2H^+(aq)$

(d) $2N_3^-(aq) + Cl_2(g) \rightarrow 3N_2(g) + 2Cl^-(aq)$

22.47 (a) $2NO_3^-(aq) + 10e^- + 12H^+(aq) \rightarrow N_2(g) + 6H_2O(l)$ $E^\circ = +1.25$ V

(b) $2NH_4^+(aq) \rightarrow N_2(g) + 8H^+(aq) + 6e^-$ $E^\circ = -0.27$ V

22.48 (a) $NO_3^-(aq) + 4H^+(aq) + 3e^- \rightarrow NO(g) + 2H_2O(l)$ $E^\circ = +0.96$ V

(b) $HNO_2(aq) \rightarrow NO_2(g) + H^+(aq) + 1e^-$ $E^\circ = -1.12$ V

22.49 From Equations [22.12] and [22.13] note that four moles of H_2 are formed from one mole of CH_4. Ammonia synthesis involves the reaction: $N_2(g) + 3H_2(g) \rightleftharpoons NH_3(g)$

$$1.6 \times 10^{10} \text{ kg } NH_3 \times \frac{1000 \text{ g}}{1 \text{ kg}} \times \frac{1 \text{ mol } NH_3}{17.0 \text{ g } NH_3} \times \frac{1 \text{ mol } N_2}{2 \text{ mol } NH_3} \times \frac{3 \text{ mol } H_2}{1 \text{ mol } N_2}$$

$$\times \frac{1 \text{ mol } CH_4}{4 \text{ mol } H_2} \times \frac{16 \text{ g } CH_4}{1 \text{ mol } CH_4} \times \frac{1 \text{ kg}}{1000 \text{ g}} = 5.6 \times 10^9 \text{ kg } CH_4$$

Twice this, or 1.1×10^{10} kg CH_4, are required to produce the NH_3.

22.50 $4NH_3(g) + 5O_2(g) \rightarrow 4NO(g) + 6H_2O(g)$

$\Delta H^\circ = 4(90.37) + 6(-241.8) - 4(-46.19) - 5(0) = -904.56$ kJ

$\Delta G^\circ = 4(86.71) + 6(-228.61) - 4(-16.66) - 5(0) = -958.18$ kJ

$4NH_3 + 3O_2(g) \rightarrow 2N_2(g) + 6H_2O(g)$

$\Delta H^\circ = 2(0) + 6(-241.8) - 4(-46.19) - 3(0) = -1266.04$ kJ

$\Delta G^\circ = 2(0) + 6(-228.61) - 4(-16.66) - 3(0) = -1305.02$ kJ

Carbon

22.51 Fire extinguishers; as coolant (dry ice); manufacture of $Na_2CO_3 \cdot 10H_2O$; carbonation of beverages.

CHAPTER 22 Chemistry of Hydrogen, Oxygen Nitrogen and Carbon

Solutions to Exercises

22.52 Fuel; reducing agent in metallurgy; in formation of organic compounds.

22.53 (a) HCN (b) SiC (c) $CaCO_3$ (d) CaC_2

22.54 (a) H_2CO_3 (b) NaCN (c) $KHCO_3$ (d) C_2H_2

22.55 (a) $[:C{\equiv}N:]^-$ (b) $:C{\equiv}O:$ (c) $[:C{\equiv}C:]^{2-}$

(d) $\ddot{\underset{..}{S}}{=}C{=}\ddot{\underset{..}{S}}$ (e) $\ddot{\underset{..}{O}}{=}C{=}\ddot{\underset{..}{O}}$ (f)

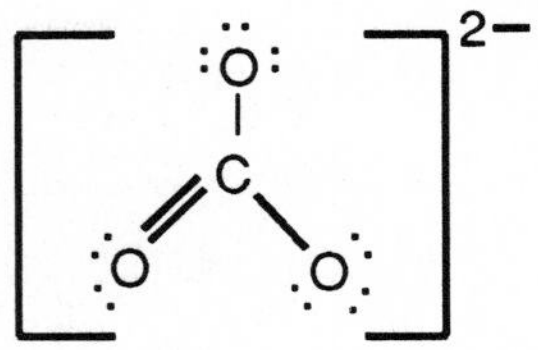

one of three equivalent resonance structures

22.56 (a) The CH_3 carbon is tetrahedral; it employs an sp^3 hybrid orbital set. The other two carbons have a linear geometry about them; they employ an sp hybrid orbital set.

(b) The carbon in CN^- uses an sp hybrid orbital set. The Lewis structure of CN^- is $:C{\equiv}N:^-$ The lone pair and the C-N σ bond electrons occupy the sp hybrid orbitals. The other two p orbitals are employed in π bonding to N.

(c) The carbon in CS_2 has an sp hybrid orbital set, consistent with the linear geometry.

(d) In C_2H_6 each carbon is in a tetrahedral environment of one C-C and three C-H bonds. It employs sp^3 hybrid orbitals.

22.57 (a) $ZnCO_3(s) \xrightarrow{\Delta} ZnO(s) + CO_2(g)$

(b) $BaC_2(s) + 2H_2O(l) \rightarrow Ba^{2+}(aq) + 2OH^-(aq) + C_2H_2(g)$

(c) $C_2H_4(g) + 3O_2(g) \rightarrow 2CO_2(g) + 2H_2O(g)$

(d) $2CH_3OH(l) + 3O_2(g) \rightarrow 2CO_2(g) + 4H_2O(g)$

(e) $NaCN(s) + H^+(aq) \rightarrow Na^+(aq) + HCN(g)$

22.58 (a) $CO_2(g) + OH^-(aq) \rightarrow HCO_3^-(aq)$

(b) $NaHCO_3(s) + H^+(aq) \rightarrow Na^+(aq) + H_2O(l) + CO_2(g)$

(c) $2CaO(s) + 5C(s) \rightarrow 2CaC_2(s) + CO_2(g)$

(d) $C(s) + H_2O(g) \rightarrow H_2(g) + CO(g)$

(e) $CuO(s) + CO(g) \rightarrow Cu(s) + CO_2(g)$

22.59 (a) $2Mg(s) + CO_2(g) \rightarrow 2MgO(s) + C(s)$

(b) $6CO_2(g) + 6H_2O(l) \xrightarrow{h\nu} C_6H_{12}O_6(aq) + 6O_2(g)$

(c) $CO_3^{2-}(aq) + H_2O(l) \rightarrow HCO_3^-(aq) + OH^-(aq)$

22.60 (a) $2CH_4(g) + 2NH_3(g) + 3\,O_2(g) \xrightarrow[\text{cat}]{800^\circ} 2HCN(g) + 6H_2O(g)$

(b) $NaHCO_3(s) + H^+(aq) \rightarrow CO_2(g) + H_2O(l) + Na^+(aq)$

(c) $2BaCO_3(s) + O_2(g) + 2SO_2(g) \rightarrow 2BaSO_4(s) + 2CO_2(g)$

22.61 The equation for the reaction is:

$CaC_2(s) + 2H_2O(l) \rightarrow Ca^{2+}(aq) + 2OH^-(aq) + C_2H_2(g)$

$$10.0\text{ g }CaC_2 \times \frac{1\text{ mol }CaC_2}{64.1\text{ g }CaC_2} \times \frac{1\text{ mol }C_2H_2}{1\text{ mol }CaC_2} = 0.156\text{ mol }C_2H_2$$

$$V = \frac{(0.156)\,(0.0821)\,(300)}{(720/760)} = 4.06\text{ L}$$

22.62 $90.0\text{ g C} \times \frac{1\text{ mol}}{12.0\text{ g C}} = 7.50\text{ mol C}$; $10.0\text{ g H} \times \frac{1\text{ mol H}}{1.01\text{ g H}} = 9.90\text{ mol H}$

The empirical formula is apparently C_3H_4. Using Equation 10.12, $\mathcal{M}$ = dRT/P;
$\mathcal{M}$ = (1.60)(0.0821)(298)/(742/760) = 40.1 g/mol. This corresponds to C_3H_4.

$$0.1443\text{ g }C_3H_4 \times \frac{36\text{ g C}}{40\text{ g }C_3H_4} = 0.130\text{ g C}$$

The mass of Mg is then 0.3052 - 0.130 = 0.175 g Mg.

$$0.175\text{ g Mg} \times \frac{1\text{ mol Mg}}{24.3\text{ g Mg}} = 7.20 \times 10^{-3}\text{ mol Mg}$$

$$0.130\text{ g C} \times \frac{1\text{ mol C}}{12.0\text{ g C}} \times 1.08 \times 10^{-2}\text{ mol C}$$

The ratio of C to Mg is 1.50. Thus, Mg_2C_3.

Additional Exercises

22.63 First find how many moles of gas occupy 1 ft^3 at STP.

$$1\text{ ft}^3 \times \left[\frac{12\text{ in}}{1\text{ ft}}\right]^3 \times \left[\frac{2.54\text{ cm}}{1\text{ in}}\right]^3 \times \frac{1\text{ L}}{10^3\text{ cm}^3} = 28.3\text{ L}$$

$$\text{at STP } n = \frac{(1\text{ atm})\,(28.3\text{ L})}{(0.0821\text{ L}\bullet\text{atm/mol}\bullet\text{K})\,(273\text{ K})} = 1.26\text{ mol/ft}^3$$

(a) $2.0 \times 10^8\text{ kg }H_2 \times \frac{1000\text{ g}}{1\text{ kg}} \times \frac{1\text{ mol }H_2}{2.02\text{ g }H_2} \times \frac{1\text{ ft}^3}{1.26\text{ mol}} = 7.9 \times 10^{10}\text{ ft}^3\ H_2$

(b) $1.1 \times 10^{10}\ \text{kg}\ N_2 \times \frac{1000\ \text{g}}{1\ \text{kg}} \times \frac{1\ \text{mol}\ N_2}{28.0\ \text{g}\ N_2} \times \frac{1\ \text{ft}^3}{1.26\ \text{mol}} = 3.1 \times 10^{11}\ \text{ft}^3\ N_2$

(c) $1.4 \times 10^{10}\ \text{kg}\ O_2 \times \frac{1000\ \text{g}}{1\ \text{kg}} \times \frac{1\ \text{mol}\ O_2}{32.0\ \text{g}\ O_2} \times \frac{1\ \text{ft}^3}{1.26\ \text{mol}} = 3.5 \times 10^{11}\ \text{ft}^3\ O_2$

22.64 (a) $1.00\ \text{kg FeTi} \times \frac{1\ \text{mol FeTi}}{104\ \text{g FeTi}} \times \frac{1000\ \text{g}}{1\ \text{kg}} \times \frac{1\ \text{mol}\ H_2}{1\ \text{mol FeTi}} \times \frac{2.02\ \text{g H}}{1\ \text{mol}\ H_2} = 19.4\ \text{g H}$

(b) $V = \frac{\left[\frac{19.4}{2.02}\right] \times 0.0821 \times 273}{1} = 215\ \text{L}$

22.65 (a) React an ionic nitride with D_2O, e.g.,

$Mg_3N_2(s) + 6D_2O(l) \rightarrow 2ND_3(aq) + 3Mg(OD)_2(s)$

(b) React SO_3 with D_2O: $SO_3(g) + D_2O(l) \rightleftharpoons D_2SO_4(aq)$

(c) React Na_2O with D_2O: $Na_2O(s) + D_2O(l) \rightarrow 2NaOD(aq)$

(d) Dissolve $N_2O_5(g)$ in D_2O: $N_2O_5(g) + D_2O(l) \rightarrow 2DNO_3(aq)$

(e) React CaC_2 with D_2O: $CaC_2(s) + 2D_2O(l) \rightarrow Ca^{2+}(aq) + 2OD^-(aq) + C_2D_2(g)$

(f) Add NaCN to the D_2SO_4 solution prepared in (b):

$NaCN(s) + D^+(aq) \xrightarrow{\Delta} DCN(aq) + Na^+(aq)$

The DCN can be removed as gas from the reaction.

22.66 Tube A - CO_2 (it does not support combustion)
Tube B - H_2 (it is ignited by the splint)
Tube C - O_2 (it supports combustion of the glowing splint)

22.67 Substances that will burn in O_2: SiH_4, CO, Mg.

The others, SiO_2, CO_2 and CaO, have Si, C and Ca in maximum oxidation states, so O_2 cannot act as an oxidizing agent.

22.68 (a) $SO_2(g) + H_2O(l) \rightleftharpoons H_2SO_3(aq)$

(b) $Cl_2O(g) + H_2O(l) \rightleftharpoons 2HClO(aq)$

(c) $Na_2O(s) + H_2O(l) \rightarrow 2Na^{2+}(aq) + 2OH^-(aq)$

(d) $BaC_2(s) + 2H_2O(l) \rightarrow Ba^{2+}(aq) + 2OH^-(aq) + C_2H_2(g)$

(e) $RbO_2(s) + 2H_2O(l) \rightarrow 2Rb^+(aq) + 2OH^-(aq) + O_2(g) + H_2O_2(aq)$

CHAPTER 22 Solutions to Exercises

Chemistry of Hydrogen, Oxygen Nitrogen and Carbon

22.69 (a) $H_2SO_4 - H_2O \rightarrow SO_3$ (b) $2HClO_3 - H_2O \rightarrow Cl_2O_5$

(c) $2HNO_2 - H_2O \rightarrow N_2O_3$ (d) $H_2CO_3 - H_2O \rightarrow CO_2$

(e) $2H_3PO_4 - 3H_2O \rightarrow P_2O_5$

22.70 $CaCO_3(s) \xrightarrow{\Delta} CaO(s) + CO_2(g)$

Electrolyze NaCl(aq): $2Cl^-(aq) + H_2O(l) \rightarrow Cl_2(g) + H_2(g) + 2OH^-(aq)$

$CO_2(g) + 2OH^-(aq) \rightarrow CO_3^{2-}(aq) + H_2O(l)$

22.71 $Ba(s) + O_2(g) \rightarrow BaO_2(s)$

$BaO_2(s) + 2H^+(aq) + SO_4^{2-}(aq) \rightarrow BaSO_4(s) + H_2O_2(aq)$

22.72 First write the balanced equation to give the number of moles of gaseous products per mole of hydrazine.

(A) $(CH_3)_2NNH_2 + 2N_2O_4 \rightarrow 3N_2(g) + 4H_2O(g) + 2CO_2(g)$

(B) $(CH_3)HNNH_2 + 5/4\ N_2O_4 \rightarrow 9/4\ N_2(g) + 3H_2O(g) + CO_2(g)$

In case (A) there are nine moles gas per one mole $(CH_3)_2NNH_2$ plus two moles N_2O_4. The total mass of reactants is 60 + 2(92) = 244 g. Thus, there are

$$\frac{9 \text{ mol gas}}{244 \text{ g reactants}} = \frac{0.0369 \text{ mol gas}}{1 \text{ g reactants}}$$

In case (B) there are 6.25 moles of gaseous product per one mole (CH_3) $HNNH_2$ plus 1.25 moles N_2O_4. The total mass of this amount of reactants is 46.0 + 1.25(92.0) = 161 g.

$$\frac{6.25 \text{ mol gas}}{161 \text{ g reactants}} = \frac{0.0388 \text{ mol gas}}{1 \text{ g reactants}}$$

Thus the methylhydrazine has marginally greater thrust.

22.73 (a) PO_4^{3-}, + 5; NO_3^-, +5

(b) The Lewis structure for NO_4^{3-} would be:

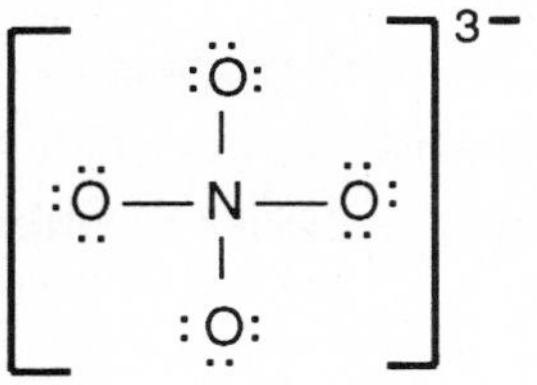

The formal charge on N is +1 and on each O atom is -1. The four electronegative oxygen atoms withdraw electron density, leaving the nitrogen deficient. Since N can form a maximum of four bonds, it cannot form a π bond with one or more of the O atoms to regain electron density, as the P atom in PO_4^{3-} does. Also, the short N-O distance would lead to a tight tetrahedron of O atoms subject to steric repulsion.

22.74 (a) Nitrogen is too small to accommodate five chlorines about it. Further, there are no orbitals of appropriate energy to form a five-orbital hybrid set.

(b) Silicon does not readily form π bonds because of poor overlap of its 3p orbital with the oxygen 2p orbital.

(c) It is impossible to complete an octet of electrons about each atom in P_2O without extensive π bonding, as in N_2O. However, phosphorus does not readily form a π bond with itself or with another element such as O.

(d) An H_3 molecule would require more than two electrons in the orbitals of at least one hydrogen atom. The 1s orbital lies far below the next available orbital in terms of energy. Furthermore, H_3 would possess an odd number of electrons.

22.75 $(NH_2)_2CO$: FW = 60.1

$$\% N = \frac{28.0}{60.1} \times 100 = 46.6\%$$

NH_3: FW = 17.0

$$\% N = \frac{14.0}{17.0} \times 100 = \mathbf{82.3\%}$$

$(NH_4)_2SO_4$: FW = 132.1

$$\% N = \frac{28.0}{132.1} \times 100 = 21.2\%$$

$NaNO_3$: FW = 85.0

$$\% N = \frac{14.0}{85.0} \times 100 = 16.5\ \%$$

22.76 $N_2H_5^+(aq) \rightarrow N_2(g) + 5H^+(aq) + 4e^-$ $E^\circ = +0.23$ V

Reduction of the metal should occur when the electrode potential for reduction of the metal ion is more positive than about -0.15 V. This is the case for Sn^{2+} (marginal), Cu^{2+} and Ag^+.

22.77 $\Delta H^\circ = (-207.3) - 0.5(-285.8) = -64.4$ kJ; $\Delta G^\circ = (-111.3) - 0.5(-236.8) = 7.1$ kJ

Recall that $\Delta G^\circ = -RT \ln K$; $\ln K = \dfrac{-7.1 \times 10^3 \text{ J}}{(8.314 \text{ J/mol} \cdot \text{K})\ 298 \text{ K}} = -2.866$

$$K = 0.057 = \frac{[HNO_3]}{P_{N_2}^{1/2}\,[H_2O]\,P_{O_2}^{5/4}}\ ; \quad P_{N_2} = 0.8 \text{ atm}; \quad P_{O_2} = 0.2 \text{ atm}$$

$[H_2O] = 1$ (nearly pure liquid)

Thus, $[HNO_3] = (0.057)(0.89)(0.13) = 6.8 \times 10^{-3}$ *M*. The process does not lead to a substantial concentration of $HNO_3(aq)$, but it is possible thermodynamically. However, thermodynamics says nothing about the rate of formation of HNO_3, which could be large or small.

22.78 For this reaction, the temperature dependence of the equilibrium constant and the reaction rate are opposing effects. At ambient temperatures, the rate of the reaction is extremely slow, and equilibrium is not established in an economically viable length of time. To establish a reasonable rate, the reaction must be carried out at high temperature.

22.79 $(CH_3)_2N_2H_2(g) + 2N_2O_4(g) \rightarrow 2CO_2(g) + 3N_2(g) + 4H_2O(g)$

$$4.0 \text{ tons } (CH_3)_2N_2H_2 \times \frac{2000 \text{ lb}}{1 \text{ ton}} \times \frac{454 \text{ g}}{1 \text{ lb}} \times \frac{1 \text{ mol } (CH_3)_2N_2H_2}{60.1 \text{ g } (CH_3)_2N_2H_2}$$

$$\times \frac{2 \text{ mol } N_2O_4}{1 \text{ mol } (CH_3)_2N_2H_2} \times \frac{92.0 \text{ g } N_2O_4}{1 \text{ mol } N_2O_4} \times \frac{1 \text{ lb}}{454 \text{ g}} \times \frac{1 \text{ ton}}{2000 \text{ lb}} = 12 \text{ tons } N_2O_4$$

(two significant figures)

22.80 (a) $N_2H_4(g) + O_2(g) \rightarrow N_2(g) + 2H_2O(l)$

(b) $\Delta H^\circ = \Delta H_f^\circ\, N_2(g) + 2\Delta H_f^\circ\, H_2O(l) - \Delta H_f^\circ\, N_2H_4(aq) - \Delta H_f^\circ\, O_2(g)$
$= 0 + 2(-285.85) - 95.40 - 0 = -667.10 \text{ kJ}$

(c) $$\frac{9.1 \text{ g } O_2}{1 \times 10^6 \text{ g } H_2O} \times \frac{1.0 \text{ g } H_2O}{1 \text{ mL } H_2O} \times \frac{1000 \text{ mL}}{1 \text{ L}} \times 3.0 \times 10^4 \text{ L} = 2.7 \times 10^2 \text{ g } O_2$$

$$2.7 \times 10^2 \text{ g } O_2 \times \frac{1 \text{ mol } O_2}{32.0 \text{ g } O_2} \times \frac{1 \text{ mol } N_2H_4}{1 \text{ mol } O_2} \times \frac{32.0 \text{ g } N_2H_4}{1 \text{ mol } N_2H_4} = 2.7 \times 10^2 \text{ g } N_2H_4$$

22.81 (a) $Li_3N(s) + H_2O(l) \rightarrow 3Li^+(aq) + 3OH^-(aq) + NH_3(aq)$

(b) $NH_3(aq) + H_2O(l) \rightleftharpoons NH_4^+(aq) + OH^-(aq)$

(c) $3NO_2(g) + H_2O(l) \rightarrow NO(g) + 2H^+(aq) + 2NO_3^-(aq)$

(d) $2NO_2 \rightleftharpoons N_2O_4(g)$

(e) $4NH_3(g) + 5O_2(g) \rightarrow 4NO(g) + 6H_2O(g)$

(f) $2CO(g) + O_2(g) \rightarrow 2CO_2(g)$

(g) $H_2CO_3(aq) \xrightarrow{\Delta} H_2O(g) + CO_2(g)$

(h) $Ni(s) + CO(g) \rightarrow NiO(s) + C(s)$

(i) $CS_2(g) + O_2(g) \rightarrow CO_2(g) + S_2(g)$

(j) $CaO(s) + SO_2(g) \rightarrow CaSO_3(s)$

(k) $Na(s) + H_2O(l) \rightarrow NaOH(aq) + H_2(g)$

(l) $CH_4(g) + H_2O(g) \xrightarrow{\Delta} CO(g) + 3H_2(g)$

(m) $LiH(s) + H_2O(l) \rightarrow LiOH(aq) + H_2(g)$

(n) $Fe_2O_3(s) + 3H_2(g) \rightarrow 2Fe(s) + 3H_2O(g)$

22.82 (a)

$$2[5e^- + MnO_4^-(aq) + 8H^+(aq) \rightarrow Mn^{2+}(aq) + 4H_2O(l)]$$
$$5[H_2O_2(aq) \rightarrow O_2(g) + 2H^+(aq) + 2e^-]$$
$$2MnO_4^-(aq) + 5H_2O_2(aq) + 6H^+(aq) \rightarrow 2Mn^{2+}(aq) + 5O_2(g) + 8H_2O(l)$$

(b)

$$2[Fe^{2+}(aq) \rightarrow Fe^{3+}(aq) + e^-]$$
$$H_2O_2(aq) + 2H^+(aq) + 2e^- \rightarrow 2H_2O(l)$$
$$2Fe^{2+}(aq) + H_2O_2(aq) + 2H^+(aq) \rightarrow 2Fe^{3+}(aq) + 2H_2O(l)$$

(c)

$$2I^-(aq) \rightarrow I_2(s) + 2e^-$$
$$H_2O_2(aq) + 2H^+(aq) + 2e^- \rightarrow 2H_2O(l)$$
$$2I^-(aq) + H_2O_2(aq) + 2H^+(aq) \rightarrow I_2(s) + 2H_2O(l)$$

(d)

$$MnO_2(s) + 4H^+(aq) + 2e^- \rightarrow Mn^{2+}(aq) + 2H_2O(l)$$
$$H_2O_2(aq) \rightarrow O_2(g) + 2H^+(aq) + 2e^-$$
$$MnO_2(s) + 2H^+(aq) + H_2O_2(aq) \rightarrow Mn^{2+}(aq) + 2H_2O(l) + O_2(g)$$

(e)

$$2I^-(aq) \rightarrow I_2(s) + 2e^-$$
$$O_3(g) + H_2O(l) + 2e^- \rightarrow O_2(g) + 2OH^-(aq)$$
$$2I^-(aq) + O_3(g) + H_2O(l) \rightarrow O_2(g) + I_2(s) + 2OH^-(aq)$$

22.83

	Zinc	Oxygen
melting point	419°	-219°
boiling point	911°	-183°
density	7.1 g/cm^3	1.33×10^{-3} g/cm^3
structure	close-packed	O_2 molecules
ionization energy	906 kJ/mol	1314 kJ/mol
reduction potential	-0.76 V[a]	+1.23 V[b]

(a) for $Zn^{2+} + 2e^- \rightarrow Zn$ (b) for $O_2 + 4H^+ + 4e^- \rightarrow 2H_2O$

Note that oxygen forms a diatomic molecule, and in this form it is low-boiling and low-melting. Zinc, on the other hand, exists in a solid with many nearest neighbors about each zinc. Furthermore, the solid is a good conductor of electricity (Chapter 20). Note finally that whereas Zn is easily oxidized (reverse of reaction a), oxygen is very easily reduced.

22.84 Recall that electronegativity is a measure of the ability of an atom to attract a shared (bonding) pair of electrons to itself. When the charge on an atom increases because electrons are lost, any remaining electrons shared with other elements will be more strongly attracted to the positively charged atom. Thus, electronegativity should increase with increase in oxidation number.

Both $HClO_4$ and $HMnO_4$ are very strong acids (completely ionized) in aqueous solution. The MnO_4^- and ClO_4^- ions are both tetrahedral; both perchlorate and permanganate salts tend to be water-soluble; both ions have the central atom in the +7 oxidation state; both ions are strong oxidizing agents:

$$MnO_4^-(aq) + 8H^+(aq) + 5e^- \rightarrow Mn^{2+}(aq) + 4H_2O(l) \quad E^\circ = +1.51\ V$$

$$ClO_4^-(aq) + 8H^+(aq) + 7e^- \rightarrow 1/2\ Cl_2(aq) + 4H_2O(l) \quad E^\circ = +1.34\ V$$

It is reasonable to conclude that in the high (+7) oxidation state both Mn and Cl will have rather high effective electronegativities. Thus they behave rather similarly. In lower oxidation states, they are much different, because Mn is a metal, and Cl is a nonmetal.

CHAPTER 23

Chemistry of Other Nonmetallic Elements

The Noble Gases and Halogens

23.1 (a) IO_3^-; +5 (b) $HBrO_3$; +5 (c) BrF_3; Br, +3; F, -1
(d) NaOCl; +1 (e) HIO_2; +3 (f) XeO_3; +6

23.2 (a) $CaBr_2$; -1 (b) $HClO_4$; +7 (c) $XeOF_4$; Xe, +6; F, -1 (d) ClO_2^-; +3
(e) HBrO; +1 (f) IF_5; I, +5, F, -1

23.3 (a) potassium chlorate (b) calcium iodate (c) aluminum chloride (d) bromic acid
(e) paraperiodic acid (f) xenon tetrafluoride

23.4 (a) iron(II) perchlorate (b) chlorous acid (c) xenon difluoride (d) iodine pentafluoride
(e) xenon trioxide (f) hydrobromic acid

23.5 (a) linear (b) square-planar (c) trigonal pyramid
(d) octahedral about the central iodine (e) square-planar

23.6 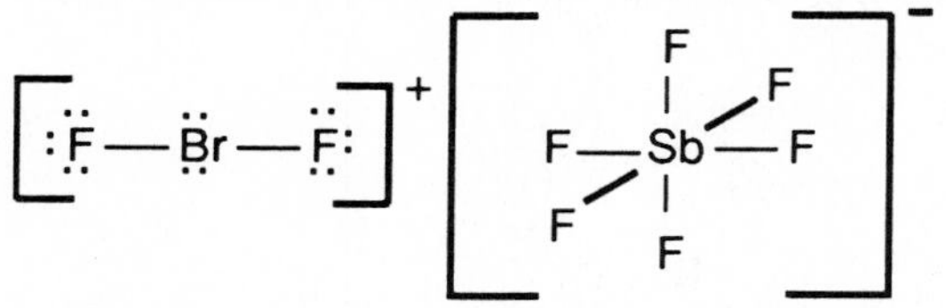

(For clarity, the three unshared electron pairs on each F in the anion are omitted.) The VSEPR model predicts a bent structure for the cation, and an octahedral geometry about Sb for the anion.

23.7 Xenon is larger, and thus can more readily accomodate an expanded octet. More important is the lower ionization energy of xenon; because the valence electrons are a greater average distance from the nucleus, they are more readily promoted to a state in which the Xe atom can form bonds with flourine.

23.8 The noble gases are colorless, odorless, diamagnetic, inert gases. They make up a very small fraction of the atmosphere and do not exist in naturally occuring compounds. Because they are rare and unreactive, they were difficult to detect.

23.9 (a) Van der Waals intermolecular attractive forces increase with increasing numbers of electrons in the atoms.

(b) F_2 reacts with water: $F_2(g) + H_2O(l) \rightarrow 2HF(aq) + O_2(g)$. That is, fluorine is too strong an oxidizing agent to exist in water.

(c) HF has extensive hydrogen bonding.

(d) Oxidizing power is related to electronegativity. Electronegativity decreases in the order given.

23.10 (a) The more electronegative the central atom, the greater the extent to which it withdraws charge from oxygen, in turn making the O-H bond more polar, and enhancing dissociation of H^+.

(b) HF reacts with the silica which is a major component of glass:
$6HF(aq) + SiO_2(s) \rightarrow SiF_6^{2-}(aq) + 2H_2O(l) + 2H^+(aq)$.

(c) Iodide is oxidized by sulfuric acid, as described in Equation 23.10.

(d) The major factor is size; there is not room about Br for the three chlorines plus the two unshared electron pairs that would occupy the bromine valence shell orbitals.

(e) I_2 reacts with I^- to form the triiodide ion, I_3^-.

23.11 Fluorine - to prepare fluorocarbons such as freon 12, CF_2Cl_2, used as a refrigerant and propellant, and teflon, used as a lubricant; chlorine - vinyl chloride, C_2H_3Cl, for plastics, water treatment, and bleach containing ClO^-; bromine - silver bromide used in photographic film; iodine- addition of KI to table salt and as I_2 in the antiseptic tincture of iodine.

23.12 Fluorine: $2KHF_2(l) \rightarrow H_2(g) + F_2(g) + 2KF(l)$ (electrolysis)

Chlorine: $2NaCl(l) \rightarrow 2Na(l) + Cl_2(g)$ (electrolysis)

Bromine: $2Br^-(aq) + Cl_2(g) \rightarrow Br_2(aq) + 2Cl^-(aq)$

Iodine: $2I^-(aq) + Cl_2(g) \rightarrow I_2(s) + 2Cl^-(aq)$

23.13 (a) $PBr_5(l) + 4H_2O(l) \rightarrow H_3PO_4(aq) + 5H^+(aq) + 5Br^-(aq)$

(b) $IF_5(l) + 3H_2O(l) \rightarrow H^+(aq) + IO_3^-(aq) + 5HF(aq)$

(c) $SiBr_4(l) + 4H_2O(l) \rightarrow Si(OH)_4(s) + 4H^+(aq) + 4Br^-(aq)$

(d) $2F_2(g) + 2H_2O(l) \rightarrow 4HF(aq) + O_2(g)$

(e) $2ClO_2(g) + H_2O(l) \rightarrow H^+(aq) + ClO_3^-(aq) + HClO_2(aq)$

(f) $HI(g) \rightarrow H^+(aq) + I^-(aq)$

23.14 (a) $CaF_2(s) + H_2SO_4(l) \rightarrow 2HF(g) + CaSO_4(s)$

(b) $2I^-(aq) + Cl_2(g) \rightarrow I_2(s) + 2Cl^-(aq)$

(c) $Xe(g) + 2F_2(g) \rightarrow XeF_4(s)$

(d) $Ca^{2+}(aq) + 2OH^-(aq) + Cl_2(aq) \rightarrow ClO^-(aq) + Cl^-(aq) + H_2O(l) + Ca^{2+}(aq)$

(e) $S_8(s) + 24F_2(g) \rightarrow 8SF_6(g)$ or $S_8(s) + 24BrF_5(l) \rightarrow 8SF_6(g) + 24BrF_3(l)$

(f) $Cl_2(g) + 2OH^-(aq) \rightarrow ClO^-(aq) + H_2O(l) + Cl^-(aq)$
NaClO (very unstable) can be isolated by cooling a concentrated solution.

23.15 (a) $Br_2(l) + 2OH^-(aq) \rightarrow BrO^-(aq) + Br^-(aq) + H_2O(l)$

(b) $Cl_2(g) + 2Br^-(aq) \rightarrow Br_2(l) + 2Cl^-(aq)$

(c) $Br_2(l) + H_2O_2(aq) \rightarrow 2Br^-(aq) + O_2(g) + 2H^+(aq)$

23.16 (a) $3CaBr_2(s) + 2H_3PO_4(l) \rightarrow Ca_3(PO_4)_2(s) + 6HBr(g)$

(b) $AlBr_3(s) + 3H_2O(l) \rightarrow Al(OH)_3(s) + 3HBr(g)$

(c) $2HF(aq) + CaCO_3(s) \rightarrow CaF_2(s) + H_2O(l) + CO_2(g)$

23.17 (a)

$$2ClO_3^-(aq) + 12H^+(aq) + 10e^- \rightarrow Cl_2(g) + 6H_2O(l)$$
$$10[Fe^{2+}(aq) \rightarrow Fe^{3+}(aq) + e^-]$$

$$2ClO_3^-(aq) + 10Fe^{2+}(aq) + 12H^+(aq) \rightarrow Cl_2(g) + 10Fe^{3+}(aq) + 6H_2O(l)$$

(b) $E^\circ = +1.47 + (-0.771) = +0.70$ V

23.18 (a) $MnO_2(s) + 2Cl^-(aq) + 4H^+(aq) \rightarrow Mn^{2+}(aq) + Cl_2(g) + 2H_2O(l)$

(b) $10Cl^-(aq) + 2MnO_4^-(aq) + 16H^+(aq) \rightarrow 2Mn^{2+}(aq) + 5Cl_2(g) + 8H_2O(l)$

(c) $6Cl^-(aq) + Cr_2O_7^{2-}(aq) + 14H^+(aq) \rightarrow 3Cl_2(g) + 2Cr^{3+}(aq) + 7H_2O(l)$

23.19 From Appendix C, we need only ΔH_f° for F(g), so that we can estimate ΔH for the process: $F_2(g) \rightarrow F(g) + F(g)$; $\Delta H^\circ = +160$ kJ. Then for XeF_2 we have

$XeF_2(g) \rightarrow Xe(g) + F_2(g)$ $\quad -\Delta H_f^\circ = +109$ kJ

$F_2(g) \rightarrow 2F(g)$ $\quad \Delta H^\circ = +160$ kJ

$XeF_2(g) \rightarrow Xe(g) + 2F(g)$ $\quad \Delta H^\circ = 269$ kJ

The average Xe-F bond enthalpy is thus 269/2 = 134 kJ. Similarly,

$XeF_4(g) \rightarrow Xe(g) + 2F_2(g)$ $\quad -\Delta H_f^\circ = +218$ kJ

$2F_2(g) \rightarrow 4F(g)$ $\quad \Delta H^\circ = 320$ kJ

$XeF_4(g) \rightarrow Xe(g) + 4F(g)$ $\quad \Delta H^\circ = 538$ kJ

Average Xe-F bond energy = 538/4 = 134 kJ

$XeF_6(g) \rightarrow Xe(g) + 3F_2(g)$ $-\Delta H_f^o = 298$ kJ
$3F_2(g) \rightarrow 6F(g)$ $\Delta H^o = 480$ kJ

$XeF_6(g) \rightarrow Xe(g) + 6F(g)$ $\Delta H^o = 778$ kJ
Average Xe-F bond energy = 778/6 = 130 kJ

The average bond enthalpies are: XeF_2 - 134 kJ, XeF_4 - 134 kJ, XeF_6 - 130 kJ. They are remarkably constant in the series.

23.20 First calculate the molar solubility of Cl_2 in water.

$$n = \frac{1(0.310\text{ L})}{\frac{0.0821\text{ L}\cdot\text{atm}}{1\text{ mol}\cdot\text{K}} \times 273\text{ K}} = 0.0138\text{ mol Cl}_2\ ;\ M = \frac{0.0138\text{ mol}}{0.100\text{ L}} = 0.138\ M$$

$$K = \frac{[Cl^-][HOCl][H^+]}{[Cl_2]} = 4.7 \times 10^{-4}$$

$[Cl^-] = [HOCl] = [H^+]$. Let this quantity = x. Then,

$$\frac{x^3}{(0.138-x)} = 4.7 \times 10^{-4}$$

Assuming that x is small compared with 0.138:
$x^3 = (0.138)(4.7 \times 10^{-4}) = 6.49 \times 10^{-5}$; $x = 0.0402\ M$

We can correct the denominator using this value, to get a better estimate of x:

$$\frac{x^3}{0.138 - 0.040} = 4.7 \times 10^{-4}\ ;\quad x = 0.0358\ M$$

One more round of approximation gives x = 0.0363 *M*. This is the equilibrium concentration of HClO.

Group 6A

23.21 (a) H_2SeO_3; +4 (b) $KHSO_3$; +4 (c) H_2Te; -2 (d) CS_2; -2 (e) $CaSO_4$; +6 (f) $Na_2S_2O_3$; +2

23.22 (a) SeO_3; +6 (b) H_6TeO_6; +6 (c) $ZnSeO_4$; +6 (d) SF_4; +4 (e) H_2S; -2 (f) H_2SO_3; +4

23.23 (a) Potassium thiosulfate (b) aluminum sulfide (c) sodium hydrogen selenite (d) selenium hexafluoride

23.24 (a) Hydrogen selenide (b) iron pyrite (iron persulfide) (c) sodium hydrogen sulfate (or sodium bisulfate) (d) sodium selenate

23.25 (a) 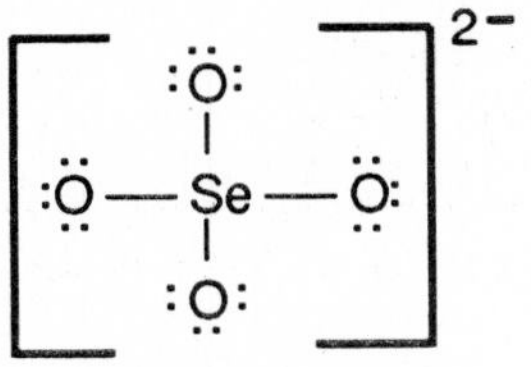
tetrahedral

(b) 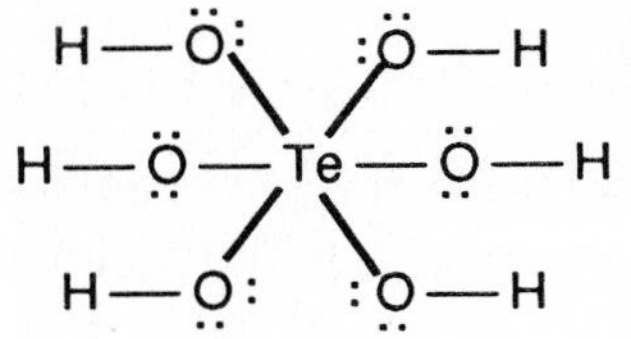
octahedral

(c) 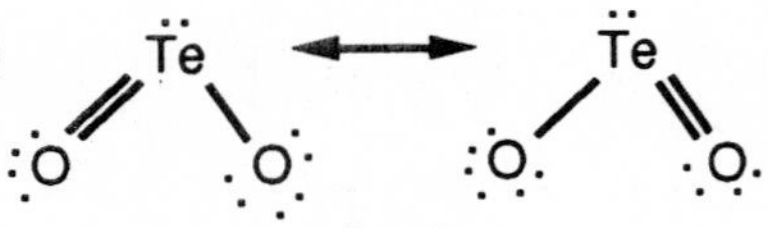
bent

(d) 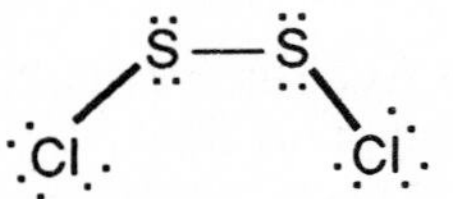
bent (free rotation around S-S bond)

(e) 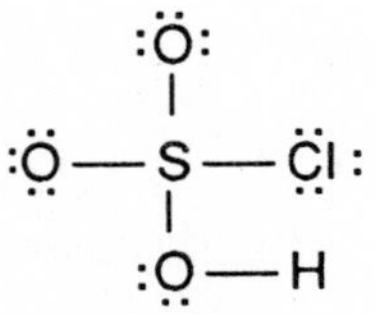
tetrahedral

23.26 SF_4, 34 e^- SF_5^-, 42 e^-

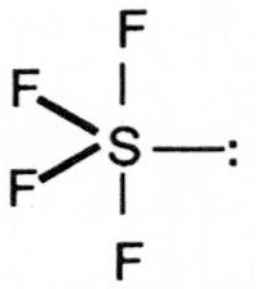
trigonal bipyramid electron pair

see saw molecular geometry

(lone pairs on F atoms omitted for clarity)

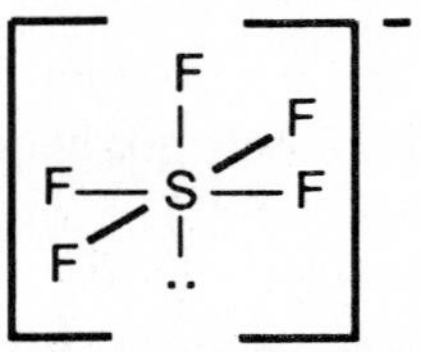
octahedral electron pair geometry

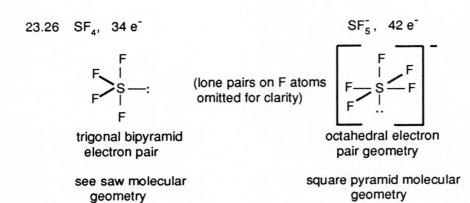

square pyramid molecular geometry

23.27 (a) $SeO_2(s) + H_2O(l) \rightarrow H_2SeO_3(aq) \rightleftharpoons H^+(aq) + HSeO_3^-(aq)$

(b) $ZnS(s) + 2H^+(aq) \rightarrow Zn^{2+}(aq) + H_2S(g)$

(c) $8SO_3^{2-}(aq) + S_8(s) \rightarrow 8S_2O_3^{2-}(aq)$

(d) $Se(s) + 2H_2SO_4(l) \xrightarrow{\Delta} SeO_2(g) + 2SO_2(g) + 2H_2O(g)$

(e) $SO_3(g) + H_2SO_4(l) \rightarrow H_2S_2O_7(l)$

(f) $H_2SeO_3(aq) + H_2O_2(aq) \rightarrow 2H^+(aq) + SeO_4^{2-}(aq) + H_2O(l)$

23.28 (a) $H_2SeO_3(aq) + N_2H_4(aq) \rightarrow Se(s) + N_2(g) + 3H_2O(l)$

(b) $H_6TeO_6(s) \xrightarrow{\Delta} TeO_3(s) + 3H_2O(l)$

(c) $Al_2Se_3(s) + 6H^+(aq) \rightarrow 2Al^{3+}(aq) + 3H_2Se(g)$

(d) $Cl_2(aq) + S_2O_3^{2-}(aq) + H_2O(l) \rightarrow 2Cl^-(aq) + S(s) + SO_4^{2-}(aq) + 2H^+(aq)$

23.29 (a) $2[MnO_4^-(aq) + 8H^+(aq) + 5e^- \rightarrow Mn^{2+}(aq) + 4H_2O(l)]$

$5[H_2SO_3(aq) + H_2O(l) \rightarrow SO_4^{2-}(aq) + 4H^+(aq) + 2e^-]$

$2MnO_4^-(aq) + 5H_2SO_3(aq) \rightarrow 2MnSO_4(s) + 5SO_4^{2-}(aq) + 3H_2O(l) + 4H^+(aq)$

(b) $Cr_2O_7^{2-}(aq) + 14H^+(aq) + 6e^- \rightarrow 2Cr^{3+}(aq) + 7H_2O(l)$

$3[H_2SO_3(aq) + H_2O(l) \rightarrow SO_4^{2-}(aq) + 4H^+(aq) + 2e^-]$

$Cr_2O_7^{2-}(aq) + 3H_2SO_3(aq) + 2H^+(aq) \rightarrow 2Cr^{3+}(aq) + 3SO_4^{2-}(aq) + 4H_2O(l)$

(c) $Hg_2^{2+}(aq) + 2e^- \rightarrow 2Hg(l)$

$H_2SO_3(aq) + H_2O(l) \rightarrow SO_4^{2-}(aq) + 4H^+(aq) + 2e^-$

$Hg_2^{2+}(aq) + H_2SO_3(aq) + H_2O(l) \rightarrow 2Hg(l) + SO_4^{2-}(aq) + 4H^+(aq)$

23.30 The half reaction for oxidation in all these cases is:
$H_2S(aq) \rightarrow S(s) + 2H^+ + 2e^-$. (The product could be written as $S_8(s)$, but this is not necessary. In fact it is not necessarily the case that S_8 would be formed, rather than some other allotropic form of the element.)

(a) $2Fe^{3+}(aq) + H_2S(aq) \rightarrow 2Fe^{2+}(aq) + S(s) + 2H^+(aq)$

(b) $Br_2(l) + H_2S(aq) \rightarrow 2Br^-(aq) + S(s) + 2H^+(aq)$

(c) $2MnO_4^- + 6H^+(aq) + 5H_2S(aq) \rightarrow 2Mn^{2+}(aq) + 5S(s) + 8H_2O(l)$

(d) $2NO_3^-(aq) + H_2S(aq) + 2H^+(aq) \rightarrow 2NO_2(aq) + S(s) + 2H_2O(l)$

23.31 Te is less electronegative than Se or S (see Figure 8.7), so it withdraws less charge from the oxygens, thus making the O-H bonds less polar and less likely to ionize as H^+. Also, tellurium is larger than sulfur or selenium. It is likely that in aqueous solution the tellurium atom is coordinated by additional water molecules. (Indeed, we know that $Te(OH)_6$ can be formed.) The interaction of water molecules with the central tellurium adds electron density that further reduces the withdrawal of charge from the oxygens of H_2TeO_4.

23.32 (a) SF_6 has an octahedral arrangement of fluorines about the central sulfur. There is very little room for a base (an electron pair donor) to approach the central sulfur. By contrast, in SF_4 there are four fluorines and one lone pair about the central atom. A Lewis base can find its way into the central sulfur in SF_4 whereas it cannot do so in SF_6. Further, since sulfur uses an sp^3d^2 hybrid set in SF_6, and an sp^3d set in SF_4, the second 3d orbital in SF_4 can be used as the acceptor orbital as the base approaches.

(b) Lewis acid attack at SF_4 can occur at the unshared electron pair on the central atom. SF_4 could be oxidized, for example, by addition of F_2 to make SF_6.

(c) This can occur because in SF_4 there is an unshared electron pair that can be activated to form two new bonds, as in going from SF_4 to SF_6. SF_6 cannot be further oxidized because all valence electrons are in use.

Group 5A

23.33 (a) H_3PO_4; +5 (b) H_3AsO_3; +3 (c) Sb_2S_3; +3 (d) $Ca(H_2PO_4)_2$; +5 (e) K_3P; -3

23.34 (a) H_3PO_3; +3 (b) $H_4P_2O_7$; +5 (c) $SbCl_3$; +3 (d) $Mg_3(AsO_4)_2$; +5 (e) P_2O_6; +5

23.35 (a) sodium phosphide (b) arsenic acid (c) tetraphosphorus decaoxide
(d) arsenic pentafluoride

23.36 (a) potassium arsenide (b) phosphorus tribromide (c) antimony trioxide
(d) sodium dihydrogen arsenate

23.37 (a)

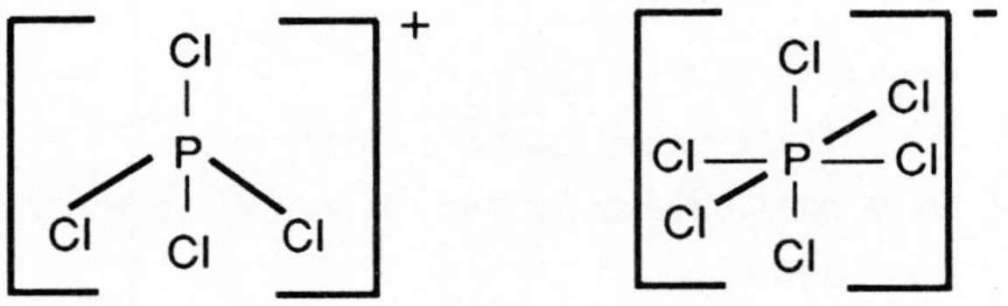

(Each terminal chlorine has three unshared or lone pairs, not shown for clarity.)

(b) In PCl_4^+ the phosphorus employs an sp^3 hybrid set. In PCl_6^- the phosphorus employs a sp^3d^2 hybrid set.

(c) The ionic form is stabilized in the solid state by the lattice energy that is gained by forming the ions. This is a case of spontaneous ion formation between two identical species, driven by the increased stability of the ionic lattice compared with the covalent lattice.

23.38

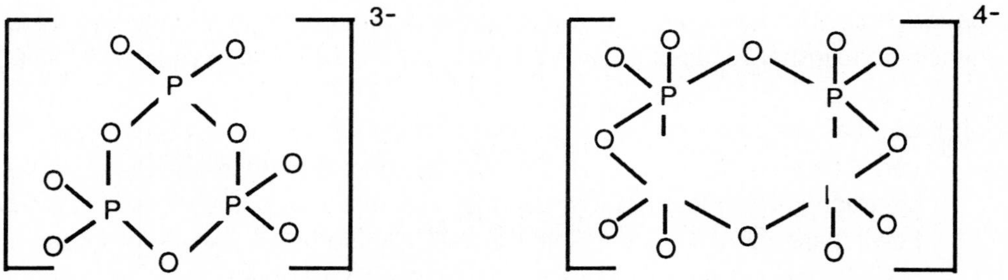

(unshared pairs on all oxygens to give octets; geometry around each P is approximately tetrahedral)

(unshared pairs on all oxygens to give octets; geometry about each P is approximately tetrahedral)

23.39 (a) Only two of the hydrogens in H_3PO_3 are bound to oxygen. The third is attached directly to phosphorus, and not readily ionized, because the H-P bond is not very polar.

(b) The smaller, more electronegative nitrogen withdraws more electron density from the O-H bond, making it more polar and more likely to ionize.

(c) Phosphate rock consists of $Ca_3(PO_4)_2$, which is only slightly soluble in water. The phosphorus is unavailable for plant use.

(d) N_2 can form stable π bonds to complete the octet of both N atoms. Because phosphorus atoms are larger than nitrogen atoms, they do not form stable π bonds with themselves and must form σ bonds with several other phosphorus atoms (producing P_4 tetrahedra or sheet structures) to complete their octets.

(e) In solution Na_3PO_4 is completely dissociated into Na^+ and PO_4^{3-}. PO_4^{3-}, the conjugate base of the very weak acid HPO_4^{2-}, has a K_b of 2.4×10^{-2} and produces a considerable amount of OH^- by hydrolysis of H_2O.

23.40 (a) Phosphorus is a larger atom and can more easily accommodate five surrounding atoms and an expanded octet of electrons than nitrogen can. Also, P has energetically "available" 3d orbitals which participate in the bonding, but nitrogen does not.

(b) Only one of the three hydrogens in H_3PO_2 is bonded to oxygen. The other two are bonded directly to phosphorous and are not easily ionized because the P-H bond is not very polar.

(c) PH_3 is a weaker base than H_2O (PH_4^+ is a stronger acid than H_3O^+). Any attempt to add H^+ to PH_3 in the presence of H_2O merely causes protonation of H_2O.

(d) Antimony is less nonmetallic in character than phosphorus. Because it is larger than P it has less attraction for additional electrons (it is a weaker Lewis acid). Thus, the reaction stops before complete hydrolysis has occurred.

(e) White phosphorus consists of P_4 molecules, with P-P-P bond angles of 60°. Each P atom has four VSEPR pairs of electrons, so the predicted electron pair geometry is tetrahedral and the preferred bond angle is 109°. Because of the severely strained bond angles in P_4 molecules, white phosphorus is highly reactive.

23.41 (a) $2Ca_3(PO_4)_2(s) + 6SiO_2(s) + 10C(s) \rightarrow P_4(g) + 6CaSiO_3(l) + 10CO_2(g)$

(b) $3H_2O(l) + PCl_3(l) \rightarrow H_3PO_3(aq) + 3H^+(aq) + 3Cl^-(aq)$

(c) $6Cl_2(g) + P_4(s) \rightarrow 4PCl_3(l)$

(d) $P_4O_{10}(s) + 6H_2O(l) \rightarrow 4H_3PO_4(aq)$

23.42 (a) $4P_4O_{10}(l) + 15CaF_2(s) \rightarrow 6PF_6(l) + 5Ca_3(PO_4)_2(s)$

(b) $As_2O_3(s) + 3H_2O(l) \rightarrow 2H_3AsO_3(aq)$

(c) $2H_3PO_4(l) \xrightarrow{\Delta} H_4P_2O_7(l) + H_2O(g)$

(d) $As(s) + HNO_3(aq) + H_2O(l) \rightarrow H_3AsO_3(aq) + NO(g)$

Group 4A and Boron

23.43 (a) SiO_2; +4 (b) $GeCl_4$; +4 (c) $NaBH_4$; +3 (d) $SnCl_2$; +2 (e) B_2H_6; -3

23.44 (a) H_3BO_3; +3 (b) $SiBr_4$; +4 (c) $PbCl_2$; +2 (d) $Na_2B_4O_7{\bullet}10H_2O$; +3 (e) B_2O_3; +3

23.45 In C_2H_4 and C_2H_2 the carbons are joined by π bonds as well as σ bonds. Silicon does not form Si-Si π bonds of great stability, so Si compounds involving such bonds are unstable relative to other bonding configurations.

23.46 Carbon has a greater tendency toward catenation (formation of bonds with itself) than Si or other elements. Further, it is capable of forming stable multiple bonds with other elements such as N and O. Finally, many key carbon compounds are gaseous, most notably CO_2. Thus, the product of biological oxidation, CO_2, can be easily vented from the system, and circulated in the biosphere.

23.47 (a) Carbon (b) lead (c) silicon

23.48 (a) Carbon (b) lead (c) germanium

23.49 $GeCl_4(g) + 2H_2O(g) \rightarrow GeO_2(s) + 4HCl(g)$
$SiCl_4(g) + 2H_2O(g) \rightarrow SiO_2(s) + 4HCl(g)$

23.50 (a) $GeCl_4(l) + Ge(s) \rightarrow 2GeCl_2(l)$.
(b) In $GeCl_2$ there are six electrons about Ge. According to the VSEPR model the three electron pairs should be in a trigonal plane. One of the electron pairs is unshared, the other two are in the Ge-Cl bonds. The molecule should therefore appear bent, with a Cl-Ge-Cl angle of somewhat less than 120°

23.51 KSi_3AlO_8

23.52 In this mineral the cations other than the Al and Si of the $AlSi_3O_{10}^{7-}$ portion of the structure are found between aluminosilicate sheets, and provide charge balance. The three Mg^{2+} ions in the mica mineral serve the same charge balance role as the two Al^{3+} ions in Muscovite.

23.53 (a) SiO_4^{4-} (b) SiO_3^{2-} (c) SiO_3^{2-}

23.54 (a) Silica structure. One fourth of Si^{4+} replaced by Al^{3+}, with a Na^+ ion for charge balance (a feldspar).
(b) It contains the $AlSi_2O_6^-$ structure; an aluminosilicate based on $Si_3O_6^{2-} \equiv SiO_2^{2-}$. A cyclic or chain aluminosilicate structure might be expected.

(c) The parent structure consists of SiO_4^{2-} tetrahedra, but notice that Zr is an element of Group 4B. Zr^{4+} should substitute readily for Si^{4+}. Thus this compound can be visualized as the silica structure with half the Si^{4+} replaced by Zr^{4+}. Both cations are in tetrahedral environments.

(d) It is a derivative of $Si_2O_5^{2-}$ in which half of the Si^{4+} are replaced by Ti^{4+}. Ca^{2+} is present for charge balance. It should possess a sheet structure.

Additional Exercises

23.55 (a) Tetraboric acid (b) carborundum, or silicon carbide (c) metaphosphoric acid (d) xenon difluoride (e) sodium sulfide (f) potassium chlorate (g) decaborane

23.56 (a) Sulfur is a less electronegative element than oxygen, and can thus be expected to more readily give up its electrons in the course of being oxidized.

(b) Sulfur is more electronegative than selenium, resulting in a greater polarity of the O-H bond.

(c) Astatine is radioactive, and not present in nature to a significant extent.

(d) Red phosphorus consists of sheets of phosphorus atoms.

(e) Xenon is a larger atom than krypton; there is room for six fluorines about Xe, but not about Kr. Further, the ionization energy of electrons from Xe is lower, so bond formation with fluorine occurs more readily.

(f) SF_4 undergoes hydrolysis: $SF_4(g) + 3H_2O(l) \rightarrow H_2SO_3(aq) + 4HF(aq)$.

(g) In aluminosilicate sheets the overall charge is more negative than in silicates because of replacement of Si^{4+} by Al^{3+}. More cations between layers are required for charge neutralization. The increased electrostatic interactions between cations and the aluminosilicate sheets makes for a more rigid structure.

(h) Sulfite ion, SO_3^{2-}, is readily oxidized to sulfate by $O_2(g)$ in the air.

23.57 The XeO_6^{4-} ion can be visualized as an octahedral array of oxygens about xenon. There is a total of 48 valence shell electrons. Each oxygen has an octet consisting of three unshared pairs and one pair shared with the xenon, thus accounting for all 48 electrons. This means that there are six electron pairs in the xenon valence orbitals. We expect a sp^3d^2 hybrid orbital set, consistent with an octahedral geometry.

23.58 $2XeO_3(s) \rightarrow 2Xe(g) + 3O_2(g)$

$$0.654 \text{ g } XeO_3 \times \frac{1 \text{ mol } XeO_3}{179 \text{ g } XeO_3} \times \frac{5 \text{ mol gas}}{2 \text{ mol } XeO_3} = 9.13 \times 10^{-3} \text{ mol gas}$$

$$P = \frac{(9.13 \times 10^{-3} \text{ mol})(0.0821 \text{ L} \cdot \text{atm/mol} \cdot \text{K})(321 \text{ K})}{0.452 \text{ L}} = 0.533 \text{ atm}$$

23.59 (a) $2Na(l) + 2HCl(g) \rightarrow 2NaCl(s) + H_2(g)$

(b) $H_2SO_3(aq) + Br_2(l) + H_2O(l) \rightarrow HSO_4^-(aq) + 2Br^-(aq) + 3H^+(aq)$
Note that two strong acids are formed, $H_2SO_4(aq)$ and HBr(aq).
H_2SO_4 is not volatile, and remains behind when the HBr is distilled.

(c) The half reactions are:

$$12\ OH^-(aq) + Br_2(l) \rightarrow 2BrO_3^-(aq) + 6H_2O(l) + 10e^-$$

$$5[ClO^-(aq) + H_2O(l) + 2e^- \rightarrow Cl^-(aq) + 2OH^-(aq)]$$

$$2OH^-(aq) + Br_2(l) + 5ClO^-(aq) \rightarrow 5Cl^-(aq) + 2BrO_3^-(aq) + H_2O(l)$$

Now take account of the formation of $KBrO_3(s)$:

$$2K^+(aq) + 2BrO_3^-(aq) \rightarrow 2KBrO_3(s)$$

$$2K^+(aq) + 2OH^-(aq) + Br_2(l) + 5ClO_3^-(aq) \rightarrow 5Cl^-(aq) + 2KBrO_3(s) + H_2O(l)$$

(d) The half-reactions are:

$$2BrO_3^-(aq) + 12H^+(aq) + 10e^- \rightarrow Br_2(l) + 6H_2O(l)$$

$$5[H_2SO_3(aq) + H_2O(l) \rightarrow HSO_4^-(aq) + 3H^+(aq) + 2e^-]$$

$$2BrO_3^-(aq) + 5H_2SO_3(aq) \rightarrow Br_2(l) + 5HSO_4^-(aq) + 3H^+(aq) + H_2O(l)$$

(e) $3UCl_4(s) + 4ClF_3(g) \rightarrow 3UF_4(g) + 8Cl_2(g)$

23.60 (a) $(CN)_2(g) + OH^-(aq) \rightarrow CN^-(aq) + CNO^-(aq) + H_2O(l)$
(b) $N \equiv C\text{-}C \equiv N$; linear

23.61 $BrO_3^-(aq) + XeF_2(aq) + H_2O(l) \rightarrow Xe(g) + 2HF(aq) + BrO_4^-(aq)$

23.62 $I_2 < F_2 < Br_2 < Cl_2$. The lower-than-expected value for F_2 can be ascribed to repulsions between the unshared electron pairs on the fluorine atoms at the short distance necessary to give good overlap for F-F bonding.

23.63 P_4, P_4O_6, and P_4O_{10} all contain a tetrahedron of phosphorus atoms with P-P-P angles of approximately 60°. In the acids containing phosphorus in the +5 oxidation state, H_3PO_4, $H_4P_2O_7$, and $(HPO_3)_n$, P is bound to four O atoms with tetrahedral geometry and approximate 109° bond angles around phosphorus.

23.64 $8Fe(s) + S_8(s) \rightarrow 8FeS(s)$
$S_8(s) + 16F_2(g) \rightarrow 8SF_4(g)$ or $S_8(s) + 24F_2(g) \rightarrow 8SF_6(g)$
$S_8(s) + 8O_2(g) \rightarrow 8SO_2(g)$
$S_8(s) + 8H_2(g) \rightarrow 8H_2S(g)$

Sulfur acts as an oxidizing agent in reactions with Fe or H_2 and as a reducing agent in reactions with O_2 or F_2. Incidentally, these reactions are often written using the symbol S rather than S_8 for sulfur.

23.65 $Na_2SO_4(aq) + 4C(s) \rightarrow 4CO(g) + Na_2S(aq)$

23.66 The valence shell electron arrangement controls the upper and lower limits of oxidation state. For group 6 the configuration is ns^2np^4. Thus addition of two electrons makes a closed shell. "Loss" of all six also leaves a closed shell. The limits are -2 to +6.

23.67

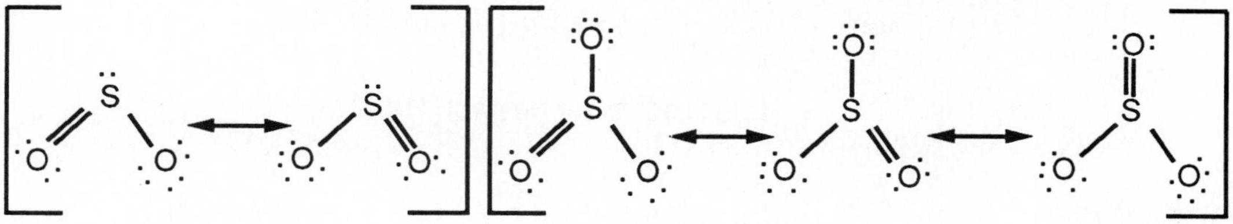

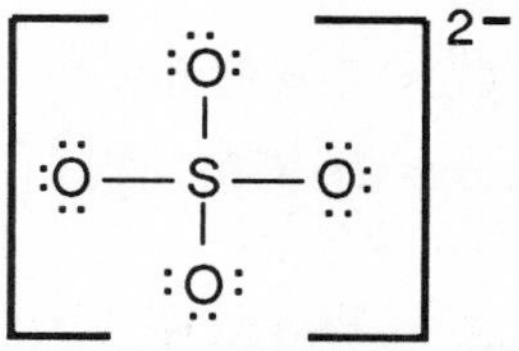

The S-O bond length will be shortest in the structure that has the highest S-O bond order. The bond order in SO_2 can be thought of as 1.5 (average of one single and one double bond); in SO_3, 1.33 (average of two single and one double bonds); in SO_4^{2-}, 1.00. Thus, the S-O bond lengths increase in the order: $SO_2 < SO_3 < SO_4^{2-}$.

23.68 (a) $SO_2(g) + 2H_2S(g) \rightarrow 3S(s) + 2H_2O(g)$. Or, if we assume S_8 is the product, $8SO_2(g) + 16H_2S(g) \rightarrow 3S_8(s) + 16H_2O(g)$.

(b) $$2000 \text{ lb coal} \times \frac{0.035 \text{ lb S}}{1 \text{ lb coal}} \times \frac{454 \text{ g S}}{1 \text{ lb S}} \times \frac{1 \text{ mol S}}{32 \text{ g S}} \times \frac{1 \text{ mol } SO_2}{1 \text{ mol S}} \times \frac{2 \text{ mol } H_2S}{1 \text{ mol } SO_2}$$

$$= 2.0 \times 10^3 \; H_2S; \quad V = \frac{2.0 \times 10^3 \text{ mol } (0.0821 \text{ L} \cdot \text{atm/mol} \cdot \text{K})(300 \text{ K})}{(740/760) \text{ atm}}$$

(c) $$2.0 \times 10^3 \text{ mol } H_2S \times \frac{3 \text{ mol S}}{2 \text{ mol } H_2S} \times \frac{32 \text{ g S}}{1 \text{ mol S}} = 9.6 \times 10^4 \text{ g S}$$

This is about 200 lb S per ton of coal combusted. (However, two-thirds of this comes from the H_2S, which was presumably also obtained from coal.)

23.69 (a) $2H_2Se(g) + O_2(g) \rightarrow 2H_2O(g) + 2Se(s)$

(b) $\Delta G^\circ = 2\Delta G_f^\circ H_2O(g) - 2\Delta G_f^\circ H_2Se(g)$
$= 2(-228.61 \text{ kJ}) - 2(15.9 \text{ kJ}) = -489.0 \text{ kJ}$
$\Delta G^\circ = -RT \ln K$

$$\ln K = \frac{-(489.0 \times 10^3 \text{ J})}{8.314 \text{ J/K} \cdot \text{mol} \times 298 \text{ K}} = 197.4; \quad K = 5 \times 10^{85}$$

23.70

$$
\begin{array}{llll}
S(g) + O_2(g) \rightarrow SO_2(g) & \Delta H = -296.9 \text{ kJ} & (1) \\
SO_2(g) + 1/2\ O_2(g) \rightarrow SO_3(g) & \Delta H = -98.3 \text{ kJ} & (2) \\
SO_3(g) + H_2O(l) \rightarrow H_2SO_4(aq) & \Delta H = -130 \text{ kJ/mol} & (3) \\
\hline
S(g) + 3/2\ O_2(g) + H_2O(l) \rightarrow H_2SO_4(aq) & \Delta H = -525 \text{ kJ} &
\end{array}
$$

$$1 \text{ ton } H_2SO_4 \times \frac{2000 \text{ lb}}{\text{ton}} \times \frac{454 \text{ g}}{1 \text{ lb}} \times \frac{1 \text{ mol } H_2SO_4}{98.1 \text{ g}} \times \frac{-525 \text{ kJ}}{\text{mol } H_2SO_4}$$

$$= -4.86 \times 10^6 \text{ kJ of heat/ton } H_2SO_4$$

23.71 $GeO_2(s) + C(s) \xrightarrow{\Delta} Ge(l) + CO_2(g)$

$Ge(l) + 2Cl_2(g) \rightarrow GeCl_4(l)$

$GeCl_4(l) + 2H_2O(l) \rightarrow GeO_2(s) + 4HCl(g)$

$GeO_2(s) + 2H_2(g) \rightarrow Ge(s) + 2H_2O(l)$

23.72 The fundamental unit is the SiO_4^{4-} tetrahedron. In aluminosilicates some fraction of the tetrahedra have Al^{3+} ions in place of Si^{4+} ions.

23.73

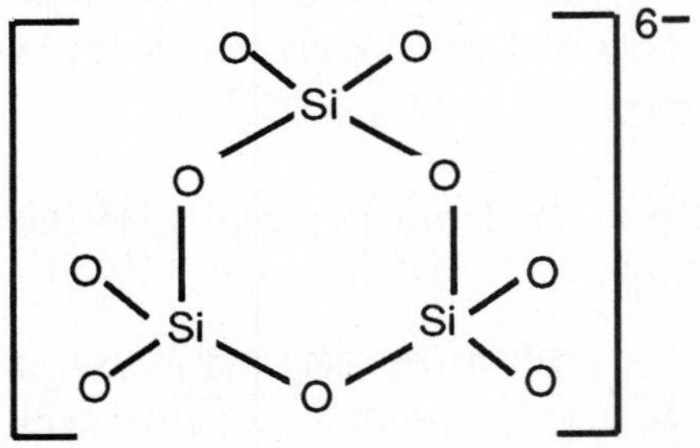

(Note that this ion is isoelectronic with the $P_3O_9^{3-}$ ion, Exercise 23.38.)

(unshared pairs omitted from oxygens)

23.74 The maximum allowable concentration of H_2S is 20 ppm or 20 mol $H_2S/1 \times 10^6$ mol air. The total moles air (29.0 g/mol) in the room is

$$n = \frac{1 \text{ atm} \times 2.7 \times 4.3 \times 4.3 \text{ m}^3 \times \left[\frac{100 \text{ cm}}{1 \text{ m}}\right]^3 \times \frac{1 \text{ L}}{10^3 \text{ cm}^3}}{\frac{0.0821 \text{ L} \cdot \text{atm}}{\text{mol} \cdot \text{K}} \times 298 \text{ K}} = 2.0 \times 10^3 \text{ mol air}$$

$$= 2.0 \times 10^3 \text{ mol air} \times \frac{20 \text{ mol } H_2S}{1 \times 10^6 \text{ mol air}} \times \frac{1 \text{ mol FeS}}{1 \text{ mol } H_2S} \times \frac{88 \text{ g FeS}}{1 \text{ mol FeS}} = 3.6 \text{ g FeS}$$

23.75 The reactions can be written as follows:

	$H_2(g) + X(\text{std state}) \rightarrow H_2X(g)$	ΔH_f^o
	$2H(g) \rightarrow H_2(g)$	$\Delta H_f^o(\text{H-H})$
	$X(g) \rightarrow X(\text{std state})$	ΔH_3
Add:	$2H(g) + X(g) \rightarrow H_2X(g)$	$\Delta H = \Delta H_f^o + \Delta H_f^o(\text{H-H}) + \Delta H_3$

These are all the necessary ΔH values. Thus,

Compound	ΔH
H_2O	ΔH = -242 kJ - 436 kJ - 248 kJ = -926 kJ
H_2S	ΔH = -20 kJ - 436 kJ - 277 kJ = -733 kJ
H_2Se	ΔH = +30 kJ - 436 kJ - 227 kJ = -633 kJ
H_2Te	ΔH = +100 kJ - 436 kJ - 197 kJ = -533 kJ

The average H-X bond energy in each case is just half of ΔH: H_2O, 468 kJ; H_2S, 366 kJ; H_2Se, 316 kJ; H_2Te, 266 kJ. The H-X bond energy decreases steadily in the series. The origin of this effect is probably the increasing size of the orbital from X with which the hydrogen 1s orbital must overlap.

23.76 (a) Use the formula that relates the length of a side of a triangle to the lengths of the other two sides and the opposite angle: $a^2 = b^2 + c^2 - 2bc \cos A$. In this case, a = b = the P-O distance, and A is the P-O-P angle:

For P_4O_6, $a^2 = 2b^2(1 - \cos A) = 2(0.165)^2(1 - \cos 127.5^o)$ a = 0.295 nm
For P_4O_{10}, $a^2 = 2b^2(1 - \cos A) = 2(0.160)^2(1 - \cos 124.5^o)$ a = 0.283 nm

(b) The shorter P-P distance in P_4O_{10} is due both to the sharper angle and to the shorter P-O distance. The latter may be due to the fact that phosphorus has a higher effective charge in P_4O_{10}, so the P atom is slightly contracted as compared with P_4O_6 because of its higher oxidation state.

23.77 The apparent atomic radius calculated for nitrogen in each case is:

NOF: 1.52 - 0.72 = 0.80 Å

NOCl: 1.98 - 1.00 = 0.98 Å

NOBr: 2.14 - 1.15 = 0.99 Å

Note that nitrogen is apparently smaller in the case of NOF. We can see a possible reason for this if we write the dominant and other possible Lewis structures:

The second structure is not expected to be very important, except possibly for fluorine. Because fluorine is a second row element along with nitrogen, we might expect that π bond formation would be more important than for Cl or Br. The double bond character of the second structure suggests that the N-F bond should be shorter than otherwise expected.

23.78 The clay minerals are hydrated aluminosilicates in which there are many H-O-M bonds present. Heating causes condensation reactions to occur. That is, water is expelled:

$$\text{M—O—H} + \text{H—O—M} \xrightarrow{\Delta} \text{M—O—M} + H_2O(g)$$

The loss of water results in cross-links between the clay mineral sheets, resulting in a hard, brittle substance. The final product is usually treated with a surface glaze to reduce the chances of a subsequent reverse reaction.

23.79 BN has the same number of valence electrons per formula unit as carbon. (Three from B, five from N, for an average of four per atom.) To the extent that we can neglect the difference in nuclear charges between B and N, we can therefore think of BN as carbon-like. Indeed, BN takes on the same structural forms as carbon. However, because the B-N bonds are somewhat polar, BN is in fact even harder than diamond.

CHAPTER 24

Metals and Metallurgy

Metallurgy

24.1 The important sources of iron are <u>hematite</u> (Fe_2O_3) and <u>magnetite</u> (Fe_3O_4). The major source of aluminum is <u>bauxite</u> ($Al_2O_3 \cdot xH_2O$). In ores, iron is present as the +3 ion, or in both the +2 and +3 states, as in magnetite. Aluminum is always present in the +3 oxidation state.

2.42 Sphalerite is ZnS; Zn is in the +2 oxidation state.

24.3 (a) $ZnCO_3(s) \xrightarrow{\Delta} ZnO(s) + CO_2(g)$

(b) $MnO(s) + CO(g) \rightarrow Mn(l) + CO_2(g)$

(c) $Al(OH)_3(s) + OH^-(aq) \rightarrow [Al(OH)_4]^-(aq)$

(d) $TiCl_4(g) + K(l) \rightarrow Ti(s) + 4KCl(s)$

(e) $3CaO(l) + P_2O_5(l) \rightarrow Ca_3(PO_4)_2(l)$

24.4 (a) $2PbS(s) + 3O_2(s) \rightarrow 2PbO(s) + 2SO_2(g)$

(b) $PbCO_3(s) \xrightarrow{\Delta} PbO(s) + CO_2(g)$

(c) $WO_3(s) + 3H_2(g) \rightarrow W(s) + 3H_2O(g)$

(d) $ZnO(s) + CO(g) \rightarrow Zn(l) + CO_2(g)$

(e) $ZrC(s) + 4Cl_2(g) \rightarrow ZrCl_4(g) + CCl_4(g)$

24.5 $FeO(s) + H_2(g) \rightarrow Fe(l) + H_2O(g)$

$Fe_2O_3(s) + 3H_2(g) \rightarrow 2Fe(l) + 3H_2O(g)$

$FeO(s) + CO(g) \rightarrow Fe(l) + CO_2(g)$

$Fe_2O_3(s) + 3CO(g) \rightarrow 2Fe(l) + 3CO_2(g)$

24.6 The major reducing agent is CO, formed by partial oxidation of the coke (C) with which the furnace is charged.

$Fe_2O_3(s) + 3CO(g) \rightarrow 2Fe(l) + 3CO_2(g)$

$Fe_3O_4(s) + 4CO(g) \rightarrow 3Fe(l) + 4CO_2(g)$

24.7 (a) $SO_3(g)$

(b) CO(g) provides a reducing environment for the transformation of Pb^{2+} to Pb^0

(c) $PbSO_4(s) \rightarrow PbO(s) + SO_3(g)$

$PbO(s) + CO(g) \rightarrow Pb(s) + CO_2(g)$

24.8 (a) $NiO(s)$, $H_2O(g)$

(b) NiO or $Ni(OH)_2$ might be roasted to Ni^o in a reducing atmosphere such as CO.

(c) $Ni(OH)_2(s) \rightarrow NiO(s) + H_2O(g)$

$NiO(s) + CO(g) \rightarrow Ni(s) + CO_2(g)$

$Ni(OH)_2(s) + CO(g) \rightarrow Ni(s) + CO_2(g) + H_2O(g)$

24.9 The reactions involved are:

$Cu_2S(s) + O_2(g) \rightarrow 2Cu(s) + SO_2(g)$

$FeS(s) + O_2(g) \rightarrow Fe(s) + SO_2(g)$

Do this problem by staying with units of kg, and using the ratios of formula weights to arrive at the desired masses.

$$2.0 \times 10^4 \text{ kg ore} \times \frac{0.32 \text{ kg } Cu_2S}{1 \text{ kg ore}} \times \frac{32 \text{ kg S}}{159 \text{ kg } Cu_2S} \times \frac{64 \text{ kg } SO_2}{32 \text{ kg S}} = 2.6 \times 10^3 \text{ kg } SO_2$$

$$2.0 \times 10^4 \text{ kg ore} \times \frac{0.07 \text{ kg FeS}}{1 \text{ kg ore}} \times \frac{32 \text{ kg S}}{88 \text{ kg FeS}} \times \frac{64 \text{ kg } SO_2}{32 \text{ kg S}} = 1 \times 10^3 \text{ kg } SO_2$$

Total SO_2 is about 3.6×10^3 kg SO_2, or, to one significant figure, 4×10^3 kg SO_2.

24.10 $FeO(s) + SiO_2(l) \rightarrow FeSiO_3(l)$

$$1.2 \times 10^4 \text{ kg ore} \times \frac{0.26 \text{ kg FeO}}{1 \text{ kg ore}} \times \frac{1000 \text{ g FeO}}{1 \text{ kg FeO}} \times \frac{1 \text{ mol FeO}}{72 \text{ g FeO}} \times \frac{1 \text{ mol } SiO_2}{1 \text{ mol FeO}}$$

$$\times \frac{60 \text{ g } SiO_2}{1 \text{ mol } SiO_2} = 2.6 \times 10^6 \text{ g } SiO_2 = 2.6 \times 10^3 \text{ kg } SiO_2$$

24.11 (a) Air serves primarily to oxidize coke to CO, the main reducing agent in the blast furnace. This exothermic reaction also provides heat for the furnace.

$2C(s) + O_2(g) \rightarrow 2CO(g)$ $\Delta H = -110$ kJ

(b) Limestone, $CaCO_3$, is the source of basic oxide for slag formation:

$CaCO_3(s) \xrightarrow{\Delta} CaO(s) + CO_2(g)$; $CaO(l) + SiO_2(l) \rightarrow CaSiO_3(l)$

(c) Coke is the fuel for the blast furnace, and is the source of CO, the major reducing agent in the furnace:

$2C(s) + O_2(g) \rightarrow 2CO(g)$; $CO(g) + Fe_3O_4(s) \rightarrow 4CO_2(g) + 3Fe(l)$

(d) Water acts as a source of hydrogen, and as a means of controlling temperature (see Equation [24.9]). $C(s) + H_2O(g) \rightarrow CO(g) + H_2(g)$ $\Delta H = +113$ kJ

24.12 Recall from the discussion in Chapter 13 that like substances tend to be soluble in one another, whereas unlike substances do not. Molten metal consists of atoms that continue to be bound to one another by metallic bonding, even though the substance is liquid. In a slag, on the other hand, the attractive forces are those between ions. The slag phase is a highly polar, ionic medium, whereas the metallic phase is nonpolar, and the attractive interactions are due to metallic bond formation. There is little driving force for materials with such different characteristics to dissolve in one another.

24.13 Cobalt could be purified by constructing an electrolysis cell in which the crude metal was the anode and a thin sheet of pure cobalt was the cathode. The electrolysis solution is aqueous with a soluble cobalt salt such as $CoSO_4 \cdot 7H_2O$ serving as the electrolyte. (Other soluble salts with anions that do not participate in the cell reactions could be used.) Anode reaction: $Co(s) \rightarrow Co^{2+}(aq) + 2e^-$; cathode reaction: $Co^{2+}(aq) + 2e^- \rightarrow Co(s)$. Although E° for reduction of $Co^{2+}(aq)$ is slightly negative (-0.277 V), we assume that reduction of water or H^+ does not occur because of a large overvoltage.

24.14 $SnO_2(s) + C(s) \xrightarrow{\Delta} Sn(l) + CO_2(g)$

$Sn(s) \rightarrow Sn^{2+}(aq) + 2e^-$ (anode)

$Sn^{2+}(aq) + 2e^- \rightarrow Sn(s)$ (cathode)

24.15 The first equation indicates that one mole Ni^{2+} is formed from passage of two moles of electrons, and the second equation indicates the same thing. Thus, the simple ratio (1 mol Ni^{2+}/2F).

$$66 \text{ A} \times 8.0 \text{ hr} \times \frac{3600 \text{ s}}{1 \text{ hr}} \times \frac{1 \text{ C}}{1 \text{ A} \cdot \text{s}} \times \frac{1 \text{ F}}{96{,}500 \text{ C}} \times \frac{1 \text{ mol Ni}^{2+}}{2 \text{ F}} \times \frac{58.7 \text{ g Ni}^{2+}}{1 \text{ mol Ni}^{2+}} = 580 \text{ g Ni}^{2+}(aq)$$

(two significant figures)

24.16 The half-cell reaction is $Cu^{2+}(aq) + 2e^- \rightarrow Cu(s)$

$$0.80 \times 240 \text{ A} \times 10 \text{ hr} \times \frac{3600 \text{ s}}{1 \text{ hr}} \times \frac{1 \text{ C}}{1 \text{ A} \cdot \text{s}} \times \frac{1 \text{ F}}{96{,}500 \text{ C}} \times \frac{1 \text{ mol Cu}}{2 \text{ F}} \times \frac{63.5 \text{ g Cu}}{1 \text{ mol Cu}}$$

$$= 2.3 \times 10^3 \text{ g Cu}$$

24.17 $\Delta G = \Delta H - T\Delta S$ (assume ΔH and ΔS are standard)

(a) $PbO(s) + CO(g) \rightarrow Pb(s) + CO_2(g)$

$\Delta H^\circ = \Delta H_f^\circ\ CO_2(g) + \Delta H_f^\circ\ Pb(s) - \Delta H_f^\circ\ CO(g) - \Delta H_f^\circ\ PbO(s)$

$= -393.5 - (-110.5) - (-217.3) = -65.7$ kJ

$\Delta S^\circ = S^\circ\ CO_2(g) + S^\circ\ Pb(s) - S^\circ\ CO(g) - S^\circ\ PbO(s)$

$= 213.6 + 68.85 - 197.9 - 68.70 = 15.8$ J/K

$\Delta G = -65.7$ kJ $- 1473$ K $(0.0158$ kJ/K$) = -89.0$ kJ

(b) $Si(s) + 2MnO(s) \rightarrow SiO_2(s) + 2Mn(s)$

$\Delta H^\circ = -910.9 + 0 - 0 -2(-385.2) = -140.5$ kJ

$\Delta S^\circ = 41.84 + 2(32.0) - 18.7 - 2(59.7) = -32.3$ J/K

$\Delta G^\circ = -140.5$ kJ $- 1473$ K$(-0.0323$ kJ/K$) = -92.9$ kJ

(c) $FeO(s) + H_2(g) \rightarrow Fe(s) + H_2O(g)$

$\Delta H^\circ = -241.8 + 0 - 0 - (-271.9) = +30.1$ kJ

$\Delta S^\circ = 27.15 + 188.7 - 60.75 - 130.58 = 24.5$ J/K

$\Delta G = 30.1$ kJ $- 1473$ K $(0.0245$J/ K$) = (30.1 - 36.1)$ kJ $= -6.0$ kJ

24.18 A most important consideration is that molten iron itself not be oxidized too readily. If it were, too much iron oxide would appear in the product. On the other hand, the free energies of formation of the undesirable oxides must be sufficiently large and negative to permit nearly quantitative removal.

Metals and Alloys

24.19 Sodium is metallic; each atom is bonded to many nearest neighbor atoms by metallic bonding involving just one electron per atom, and delocalized over the entire three-dimensional structure. When sodium metal is distorted, each atom continues to have bonding interations with many nearest neighbors. In NaCl the ionic forces are strong, and the arrangement of ions in the solid is very regular. When subjected to physical stress, the three-dimensional lattice tends to cleave along the very regular lattice planes, rather than undergo the large distortions characteristic of metals.

24.20 Test electrical and/or thermal conductivity, which should both be high for a metal. Strike a small sample of Si with a hammer; if it flattens into a sheet, it is a metal. If the sample shatters, it is not a metal.

24.21 In the electron-sea model for metallic bonding, the valence electrons of the silver atoms move about the three-dimensional metallic lattice, while the silver atoms maintain regular lattice positions. Under the influence of an applied potential the electrons can move throughout the structure, giving rise to high electrical conductivity. The mobility of the electrons facilitates the transfer of kinetic energy and leads to high thermal conductivity.

24.22 (a) Ductility (b) In the electron-sea model, electrons are free to move about the lattice and no electron is associated with a particular metal atom. When the copper is shaped into a wire, the positions of the Cu atoms can change without breaking chemical bonds, and the electrons are redistributed to compensate for the change in atomic positions.

24.23 According to the molecular orbital or band theory of metallic bonding, the 4s and 3d atomic orbitals of the fourth row elements are used to form six molecular orbitals, three bonding and three antibonding. The maximum bond order and highest bond strength occurs in metals with six valence electrons (group 6B). Assuming that hardness is directly related to the bond strength between metal atoms (the stronger the bonds, the more difficult it is to distort the solid without disrupting the three-dimensional structure), Cr, with six valence electrons, should have the highest bond strength and greatest hardness in the fourth row.

24.24 The variation in densities reflects shorter metal-metal bond distances. These shorter distances suggest in turn that the extent of metal-metal bonding is increasing in the series. Thus, it would appear that all the valence electrons in these elements (1, 2, 3 and 4, respectively) are involved in metallic bonding.

24.25 According to band theory, an insulator has a completely filled valence band and a large energy gap between the valence band and the nearest empty band; electrons are localized within the lattice. A conductor must have a partially filled energy band; a small excitation will promote electrons to previously empty levels within the band and allow them to move freely throughout the lattice, giving rise to the property of conduction. A semiconductor has a filled valence band, but the gap between the filled and empty bands is small enough so that some electrons from the filled valence band possess enough thermal energy to jump to the empty conduction band. The presence of an impurity may also place an electon in an otherwise empty band (producing an n-type semiconductor), or create a vacancy in an otherwise full band (producing a p-type semiconductor), providing a mechanism for conduction.

24.26 Germanium doped with arsenic (Ge/As) is a better conductor than germanium. Ge, a semiconductor, has a filled valence band but an energetically accessible conduction band. In Ge/As, the additional valence electrons of the doped As atoms cannot fit into the filled valence band and occupy the conduction band. These electrons have access to the vacant orbitals within the conduction bond and serve as carriers of electrical current. Ge/As is an *n-type* semiconductor.

24.27 White tin, with a characteristic metallic structure, is expected to be more metallic in character. The electrical conductivity of the white allotropic form is higher because the valence electrons are shared with 12 nearest neighbors rather than being localized in four bonds to nearest neighbors as in gray tin. The Sn-Sn distance should be longer in white tin; there are only four valence electrons from each atom, and 12 nearest neighbors. The average tin-tin bond order can, therefore, be only about 1/3, whereas in gray tin the bond order is one. (In gray tin the Sn-Sn distance is 2.81 Å in white tin it is 3.02 Å.)

24.28 Each silver atom has many near neighbors with which it interacts. Because each Ag atom contributes only a few valence level electrons to the bonding, there is a comparatively weak bonding interaction with any one other silver atom. As a result of this relatively weak interaction, the Ag-Ag distance is large. The metal itself has quite high melting and boiling points, because each silver atom has a bonding interaction with many neighbors. The total attractive forces experienced by any one Ag atom are very high.

24.29 An alloy contains atoms of more than one element and has the properties of a metal. Solution alloys are homogeneous mixtures with different kinds of atoms dispersed randomly and uniformly. In heterogeneous alloys the components (elements or compounds) are not evenly dispersed and their properties depend not only on composition but methods of preparation. In an intermetallic compound the component elements have interacted to form a compound substance, e.g., Cu_3As; as with more familiar compounds, these are homogeneous and have definite composition and properties.

24.30 Substitutional and interstitial alloys are both solution alloys. In a substitutional alloy, the atoms of the "solute" take positions normally occupied by the "solvent." In an interstitial alloy, the atoms of the "solute" occupy the holes or interstitial positions between "solvent" atoms. Substitutional alloys tend to form when solute and solvent atoms are of comparable size and have similar bonding characteristics.

Transition Metals

24.31 Of the properties listed, (b) the first ionization energy and (c) atomic radius are characteristic of isolated atoms. Electrical conductivity (a) and melting point (d) are properties of the bulk metal.

24.32 Heat of fusion, malleability, hardness, and electrode potential (because the reduction process involves formation of M(s)) all relate to bonding in metals and are thus bulk properties.

24.33 The lanthanide contraction is the name given to the decrease in atomic size due to the build-up in nuclear charge as we move through the lanthanides (elements 58-71) and beyond them. This effect offsets the expected increase in atomic size going from the second to the third transition series. The lanthanide contraction affects size-related properties such as ionization energy, electron affinity and density.

24.34 Zr - $[Kr]5s^2 4d^2$, Z = 40; Hf - $[Kr]6s^2 4f^{14} 5d^2$, Z = 72

Moving down a family of the periodic chart, atomic size increases because the valence electrons are in a higher principle quantum level (and thus further from the nucleus) and are more effectively shielded from the nuclear charge by a larger core electron cloud. However, the build-up in Z that accompanies the filling of the 4f orbitals causes the valence electrons in Hf to experience a much greater relative nuclear charge than those in La, its neighbor to the left. This increase in Z offsets the usual effect of the increase in *n* value of the valence electrons and the radii of Zr and Hf atoms are similar.

24.35 See Figure 24.25 (a) $NbCl_5$ (b) WCl_6 (c) $CoCl_3$ (d) $CdCl_2$ (e) $ReCl_7$

24.36 (a) CrO_3 (b) ZnO (c) HfO_2 (d) OsO_4 (e) Sc_2O_3

24.37 Chromium, $[Ar]4s^1 3d^5$, has six valence-shell electrons, some or all of which can be involved in bonding, leading to multiple stable oxidation states. By contrast, aluminum, $[Ne]3s^2 3p^1$, has only three valence electrons which are all lost or shared during bonding, producing the +3 state exclusively.

24.38 The electron configurations of the two metals are: Zn - $[Ar]\ 3d^{10} 4s^2$; Cu - $[Ar]4s^1 3d^{10}$ (note exception to the normal filling pattern). The +2 state for Zn and the +1 state for Cu follow directly from their electron configurations, having complete n=3 shells. In Cu, the 4s and 3d orbitals are very similar in energy, as evidenced by the fact that the 3d fills before the 4s. Loss of a second electron is easily achieved, giving the +2 state. Cu and Zn are near the end of the first transition-metal series and have relatively high effective nuclear charges, which prevents them from forming +3 ions.

24.39 (a) Cr^{3+} - $[Ar]3d^3$ (b) Au^{3+} - $[Xe]4f^{14}5d^8$ (c) Ru^{2+} - $[Kr]4d^6$

(d) Cu^+ - $[Ar]3d^{10}$ (e) Mn^{4+} - $[Ar]3d^3$ (f) Ir^{3+} - $[Xe]4f^{14}5d^6$

24.40 (a) Fe^{2+} - $[Ar]3d^6$ (b) Sc^{2+} - $[Ar]3d^1$ (c) Ag^+ - $[Kr]4d^{10}$

(d) Mo^{4+} - $[Kr]4d^2$ (e) Nb^{3+} - $[Kr]4d^2$ (f) Rh^{3+} - $[Kr]4d^3$

24.41 Ease of oxidation decreases from left to right across a period (owing to increasing effective nuclear charge); Ti^{2+} should be more easily oxidized than Ni^{2+}.

24.42 The stronger reducing agent is more easily oxidized; Cr^{2+} is the stronger reducing agent (see Exercise 24.41).

24.43 The better oxidizing agent is more easily reduced; it has greater attraction for extra electrons. Ease of reduction decreases going down a family, so MnO_4^- will be a better oxidizing agent than TcO_4^- (Mn and Tc are both in family 7B).

24.44 Ease of reduction decreases going down a family (see Exercise 24.43), so WO_3 will be harder to reduce than CrO_3.

24.45 Au(I) and Au(III); Au(III) would be present in a solution of a strongly oxidizing acid.

24.46
$$2[Cu^+(aq) + 1e^- \rightarrow Cu(s)] \qquad E^\circ = 0.521 \text{ V}$$
$$Cu(s) + 2e^- \rightarrow Cu^{2+}(aq) \qquad E^\circ = -0.337 \text{ V}$$

$$2Cu^+(aq) \rightarrow Cu^{2+}(aq) + Cu(s) \qquad E^\circ = 0.184 \text{ V}$$

According to Equation [20.18], $\log K = \frac{nE^\circ}{0.0592}$; $n = 2$

$\log K = \frac{2(0.184)}{0.0592} = 6.22$; $K = 1.7 \times 10^6$

24.47 Chromate ion, CrO_4^{2-}, is bright yellow. Dichromate, $Cr_2O_7^{2-}$, is orange and more stable in acid solution than CrO_4^{2-} because of the equilibrium
$2CrO_4^{2-}(aq) + 2H^+(aq) \rightleftharpoons Cr_2O_7^{2-}(aq) + H_2O(l)$.

24.48 (Equation [24.27]) Fe^{2+} is a reducing agent which is readily oxidized to Fe^{3+} in the presence of O_2 from air.

24.49 (a) $Fe(s) + 2HCl(aq) \rightarrow FeCl_2(aq) + H_2(g)$

(b) $Fe(s) + 4HNO_3(aq) \rightarrow Fe(NO_3)_3(aq) + NO(g) + 2H_2O(l)$
(See net ionic equation, Equation 24.29) In concentrated nitric acid, the reaction can produce $NO_2(g)$ according to the reaction:
$Fe(s) + 6HNO_3(aq) \rightarrow Fe(NO_3)_3(aq) + 3NO_2(g) + 3H_2O(l)$

24.50 (a) $Fe^{3+}(aq) + NaOH(aq) \rightarrow Fe(OH)_3(s) + 3Na^+(aq)$
(b) $FeCO_3(s) + 2HCl(aq) \rightarrow FeCl_2(aq) + CO_2(g) + H_2O(l)$

24.51 The unpaired electrons in a paramagnetic material cause it to be weakly attracted into a magnetic field. A diamagnetic material, where all electrons are paired, is very weakly repelled by a magnetic field.

24.52 In a ferromagnetic substance, the magnetic moments on sites throughout the lattice interact with one another. That is, they are close enough physically, and the overlap of orbitals is such, that the individual magnetic sites couple to form a much larger magnetic moment throughout the solid. Because of these interactions, the continued existence of the magnetic moment does not require continued application of an external magnetic field. That is, ferromagnetic materials can form "permanent" magnets.

Additional Exercises

24.53 (a) $NiO(s) + 3H^+(aq) \rightarrow Ni^{2+}(aq) + H_2O(l)$

(b) The simple answer is that the solid is subjected to acid hydrolysis:

$CuCo_2S_4(s) + 8H^+(aq) \rightarrow Cu^{2+}(aq) + 2Co^{3+}(aq) + 4H_2S(g)$

However, in the absence of a strong complexing ligand, Co^{3+} is not stable in water. It oxidizes water according to the following reaction:

$4Co^{3+}(aq) + 2H_2O(l) \rightarrow 4Co^{2+}(aq) + O_2(g) + 2H^+(aq)$

(c) $TiO_2(s) + C(s) + 2Cl_2(g) \rightarrow TiCl_4(g) + CO_2(g)$

(d) In this reaction O_2 is reduced and sulfide is oxidized. Writing the sulfur product as S_8, the balanced equation is:

$8ZnS(s) + 4O_2(g) + 16H^+(aq) \rightarrow 8Zn^{2+}(aq) + S_8(s) + 8H_2O(l)$

24.54 (a) $2NH_4VO_3(s) \rightarrow V_2O_5(s) + 2NH_3(g) + H_2O(g)$

(b) $V_2O_5(s) + SO_2(g) \rightarrow 2VO_2(s) + SO_3(g)$

(c) $V_2O_5(s) + 2Mg(s) + 10H^+(aq) \rightarrow 2V^{3+}(aq) + 2Mg^{2+}(aq) + 5H_2O(l)$

(d) $NbCl_5(s) + 3H_2O(l) \rightarrow HNbO_3(s) + 5H^+(aq) + 5Cl^-(aq)$

24.55 $$1\text{ kg ore} \times \frac{660\text{ g }Fe_3O_4}{1\text{ kg ore}} \times \frac{1\text{ mol }Fe_3O_4}{231\text{ g }Fe_3O_4} \times \frac{3\text{ mol Fe}}{1\text{ mol }Fe_3O_4} \times \frac{55.8\text{ g Fe}}{1\text{ mol Fe}}$$

$$\times \frac{1.030\text{ g pig iron}}{1\text{ g Fe}} = 493\text{ g pig iron}$$

Substantial shipping costs could be saved by shipping the pig iron rather than the raw ore. The northern refineries went out of business because the hardwood forests were decimated around the mills, and because reduction with coke became a much better procedure.

24.56 $\Delta G^\circ = -RT\ln K$; $\Delta G^\circ = \Delta H^\circ - T\Delta S^\circ$

Calculate ΔH° and ΔS° using data from Appendix C, assuming ΔH° and ΔS° remain constant with changing temperature. Then calculate ΔG° and K at the two temperatures.

$\Delta H^\circ = 2\Delta H_f\ CO(g) - \Delta H_f^\circ\ C(s) - \Delta H_f^\circ\ CO_2(g)$

$\Delta H^\circ = 2(-110.5) - 0 - (-393.5) = +172.5\text{ kJ}$

$\Delta S^\circ = 2S^\circ\ CO(g) - S^\circ\ C(s) - S^\circ\ CO_2(g)$

$= 2(197.9) - 5.69 - 213.6 = +176.5\text{ J/K} = 0.1765\text{ kJ/K}$

$\Delta G^\circ_{298} = 172.5\text{ kJ} - 298\text{ K }(0.1765\text{ kJ/K}) = +119.9\text{ kJ}$

$$\ln K = \frac{\Delta G^\circ}{-RT} = \frac{119.9\text{ kJ}}{-(8.314 \times 10^{-3}\text{ kJ/K})(298\text{ K})} = -48.39\ ;\quad K = 9.6 \times 10^{-22}$$

$$\Delta G^\circ_{2000} = 172.5\text{ kJ} - 2000\text{ K }(0.1765\text{ kJ/K}) = -180.5\text{ kJ}$$

$$\ln K = \frac{-180.5}{(-8.314 \times 10^{-3}\text{ kJ/K})(2000\text{ K})} = 10.86\text{ ; } K = 5.2 \times 10^4$$

24.57 $97{,}000\text{ A} \times 24\text{ hr} \times \frac{3600\text{ sec}}{1\text{ hr}} \times \frac{1\text{ C}}{1\text{ A}\cdot\text{s}} \times \frac{1\text{F}}{96{,}500\text{ C}} \times \frac{1\text{ mol Mg}}{2\text{ F}} \times \frac{24\text{ g Mg}}{1\text{ mol Mg}}$

$$= 1.0 \times 10^6\text{ g Mg} = 1.0 \times 10^3\text{ kg Mg}$$

24.58 $HfC(s) + 4Cl_2(g) \rightarrow HfCl_4(g) + CCl_4(g)$

$HfCl_4(g) + Na(l) \rightarrow 4NaCl(s) + Hf(s)$

24.59 Calculate the mass of Zn(s) that will be deposited. (We will carry extra figures through the calculation and round the final result.)

$$1.0\ M \times 50\text{ m} \times \frac{(100)^2\text{ cm}^2}{1\text{ m}^2} \times 0.30\text{ mm} \times \frac{1\text{ cm}}{10\text{ mm}} \times \frac{7.1\text{ g}}{\text{cm}^3} = 1.065 \times 10^5\text{ g Zn}$$

(1.1×10^5 g Zn, 2 figures)

$$1.065 \times 10^5\text{ g Zn} \times \frac{1\text{ mol Zn}}{65.38\text{ g Zn}} \times \frac{2\text{ F}}{0.90\text{ mol Zn}} \times \frac{96{,}500\text{ C}}{\text{F}} = 3.493 \times 10^8\text{ C}$$

(3.5×10^8 C)

(2 F/0.90 mol Zn takes the 90% efficiency into account.)

$$3.493 \times 10^8\text{ C} \times 3.5\text{ V} \times \frac{1\text{ J}}{\text{C}\cdot\text{V}} \times \frac{1\text{ kWh}}{3.6 \times 10^6\text{ J}} = 339.6 \rightarrow 3.4 \times 10^2\text{ kWh}$$

$$339.6\text{ kWh} \times \frac{\$0.80}{1\text{ kWh}} = \$27.17 \rightarrow \$27$$

24.60 Because selenium and tellurium are both nonmetals we would expect that the half-cell potential for their oxidation would be relatively negative. (For Se $\rightarrow$ H_2SeO_3, E° is -0.74 V; for Te $\rightarrow$ $TeO_2(s)$, E° is -0.59 V.) Thus, both Se and Te are likely to accumulate as the free elements in the so-called anode slime, along with noble metals that are not oxidized.

24.61 (a) (See Exercise 17.57)

$Ag_2(s) \rightleftharpoons 2Ag^+(aq) + S^{2-}(aq)$	K_{sp}
$2[Ag^+(aq) + 2CN^-(aq) \rightleftharpoons Ag(CN)_2^-]$	K_f^2
$Ag_2S(s) + 4CN^-(aq) \rightleftharpoons 2Ag(CN)_2^-(aq) + S^{2-}(aq)$	

$$K = K_{sp} \times K_f^2 = [Ag^+]^2[S^{2-}] \times \frac{[Ag(CN)_2^-]^2}{[Ag^+]^2[CN^-]^4} = (6.3 \times 10^{-50})(1 \times 10^{21})^2 = 6 \times 10^{-8}$$

(b) The equilibrium constant for the cyanidation of Ag_2S, 6×10^{-8}, is much less than one and favors the presence of reactants rather than products. The process is not practical.

(c) $AgCl(s) \rightleftharpoons Ag^+(aq) + Cl^-(aq)$ $\quad K_{sp}$

$Ag^+(aq) + 2CN^-(aq) \rightleftharpoons Ag(CN)_2^-(aq)$ $\quad K_f$

$AgCl(s) + 2CN^-(aq) \rightleftharpoons Ag(CN)_2^-(aq) + Cl^-(aq)$

$$K = K_{sp} \times K_f = [Ag^+][Cl^-] \times \frac{[Ag(CN)_2^-]}{[Ag^+][CN^-]^2} = (1.8 \times 10^{-10})(1 \times 10^{21}) = 2 \times 10^{11}$$

Since K >> 1 for this process, it is potentially useful for recovering silver from horn silver. However the magnitude of K says nothing about the rate of reaction. The reaction could be slow and require heat, a catalyst or both to be practical.

24.62 The iron accumulates in the solution as $Fe^{2+}(aq)$ from oxidation at the anode. Given that iron can exist in more than one oxidation state and that the presence of air is an important factor, Fe^{2+} is probably being oxidized to $Fe^{3+}(aq)$, Equation [24.27]. Now look at reduction potentials:

$Ni^{2+}(aq) + 2e^- \rightarrow Ni(s)$ $\quad E^\circ = -0.28$ V

$Fe^{3+}(aq) + e^- \rightarrow Fe^{2+}(aq)$ $\quad E^\circ = 0.77$ V

These data indicate that Fe^{3+} will be reduced to Fe^{2+} at the cathode, instead of Ni^{2+} being reduced to Ni(s). As long as air is present to reoxidize the Fe^{2+} to Fe^{3+}, the iron keeps reappearing as Fe^{3+}, using electrical energy that would otherwise be applied to reduction of Ni^{2+}.

24.63 Assuming that SO_2 and N_2 are the nonmetallic products, two half-reactions can be written:

$$5[MoS_2(s) + 7H_2O(l) \rightarrow MoO_3(s) + 2SO_2(g) + 14H^+(aq) + 14e^-]$$

$$7[12H^+(aq) + 2NO_3^-(aq) + 10e^- \rightarrow N_2(g) + 6H_2O(l)]$$

$$5MoS_2(s) + 14H^+(aq) + 14NO_3^-(aq) \rightarrow 5MoO_3(s) + 10SO_2(g) + 7N_2(g) + 7H_2O(l)$$

$$MoO_3(s) + 2NH_3(aq) + H_2O(l) \rightarrow (NH_4)_2MoO_4(s)$$

$$(NH_4)_2MoO_4(s) \rightarrow 2NH_3(g) + H_2O(g) + MoO_3(s)$$

$$MoO_3(s) + 3H_2(g) \rightarrow Mo(s) + 3H_2O(g)$$

24.64 (a) Malleability

(b) $$\frac{1 \text{ oz}}{300 \text{ ft}^2} \times \frac{1 \text{ lb}}{16 \text{ oz}} \times \frac{453.6 \text{ g}}{1 \text{ lb}} \times \frac{1 \text{ ft}^2}{(12 \text{ in})^2} \times \frac{1 \text{ in}^2}{(2.54 \text{ cm})^2} \times \frac{1 \text{ cm}^3}{19.31 \text{ g}} = 5.268 \times 10^{-6} \text{ cm}$$

5.268×10^{-6} cm = $\mathbf{5.268 \times 10^{-5}}$ **mm** = 5.268×10^{-8} m = 52.68 nm

---

24.65 (a) p-type semi-conductor (b) metallic conductor (c) insulator
(d) an n-type semi-conductor (e) metallic conductor (f) insulator

24.66 (a) In a substitutional alloy one element replaces another, atom-for-atom, with no disruption in the lattice structure. In an interstitial alloy a second component fits into the spaces between the major component.

(b) An intermetallic compound is a true chemical substance, with a fixed ratio of numbers of atoms of the different elements present. A heterogeneous alloy is a solid mixture; two or more substances (they may be elements or intermetallic compounds) exist together as an intimate mixture of two mutually insoluble solid phases.

(c) A metal is a substance in which the bonding produces certain key properties, among them, high electrical and thermal conductivity. Metals have delocalized electrons that are free to move throughout the lattice. Insulators possess very low electrical and thermal conductivities. The electrons in insulators are largely confined to localized strong bonds between atoms, or are confined within ions, as in ionic substances that are insulators, such as NaCl.

(d) In Na_2, a gaseous molecule, the two Na atoms are bound together by a localized bond between the two atoms. In metallic Na, each Na atom has many near-neighbors. Its one valence electron is delocalized throughout the three-dimensional lattice.

24.67 Silicon has the diamond structure. As with carbon, the four valence electrons of silicon are completely involved in the four localized bonds to its neighbors. There are thus no electrons free to migrate throughout the solid. Titanium exists as a close-packed lattice; each Ti atom has twelve equivalent nearest neighbors. The valence shell electrons cannot be localized between pairs of atoms; rather they are delocalized and mobile throughout the structure. In terms of the band model described in Figure 24.17, Ti has an incompletely occupied allowed energy band. The origin of the different behaviors with regard to structure has to do with the extent of the orbitals in space. Electrons in Si feel a greater attraction to the nucleus than electrons in Ti, so they are more localized.

24.68 The equilibrium of interest is

$[ZnL_4] \rightleftharpoons Zn^{2+}(aq) + 4L \qquad K = 1/K_f$

Since $Zn(H_2O)_4^{2+}$ is $Zn^{2+}(aq)$, its reduction potential is -0.763 V. As the stability (K_f) of the complexes increases, the $[Zn^{2+}(aq)]$ decreases and the electrode potentials become more negative relative to the reduction of $Zn^{2+}(aq)$ to Zn^0.

24.69 In a ferromagnetic solid, the magnetic centers are coupled such that the spins of all unpaired electrons are parallel. As the temperature of the solid increases, the average kinetic energy of the atoms increases until the energy of motion overcomes the force aligning the electron spins. The substance becomes paramagnetic; it still has unpaired electrons, but their spins are no longer aligned.

24.70 (a) $2NiS(s) + 3O_2(g) \rightarrow 2NiO(s) + 2SO_2(g)$

(b) $2C(s) + O_2(g) \rightarrow 2CO(g)$; $C(s) + H_2O(g) \rightarrow CO(g) + H_2(g)$

$NiO(s) + CO(g) \rightarrow Ni(s) + CO_2(g)$; $NiO(s) + H_2(g) \rightarrow Ni(s) + H_2O(g)$

(c) $Ni(s) + 2HCl(aq) \rightarrow NiCl_2(aq) + H_2(g)$

(d) $NiCl_2(aq) + 2NaOH(aq) \rightarrow Ni(OH)_2(s) + 2NaCl(aq)$

(e) $Ni(OH)_2(s) \rightarrow NiO(s) + H_2O(g)$

CHAPTER 25

Chemistry of Coordination Compounds

Structure and Nomenclature

25.1 (a) coordination number = 4, oxidation number = +2
(b) 6, +3 (c) 6, +3 (d) 5, +4 (e) 2, +1

25.2 (a) coordination number = 6, oxidation number = +4
(b) 6, +2 (c) 4, +2 (d) 6, +3 (e) 4, +2

25.3 (a)

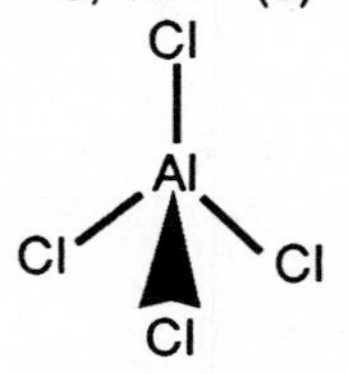

tetrahedral

(b) $[:N \equiv C - Ag - C \equiv N:]^-$

linear

(c)

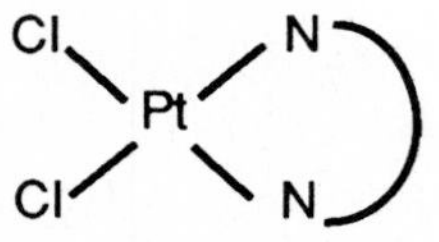

square planar

(d)

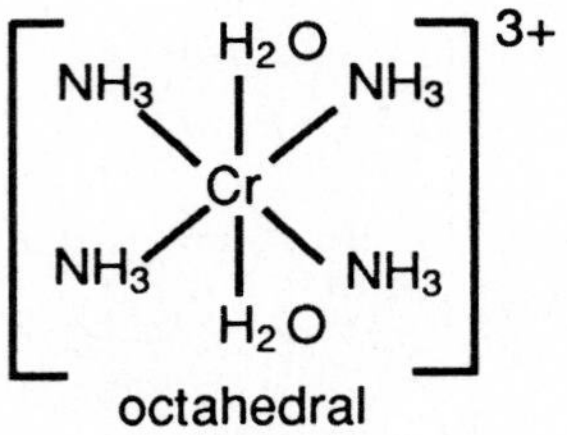

octahedral

25.4 (a)

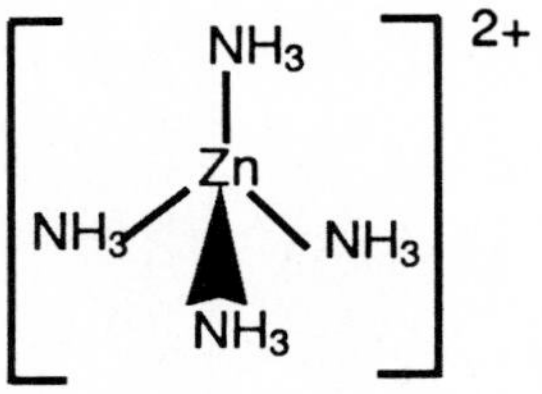

(b)

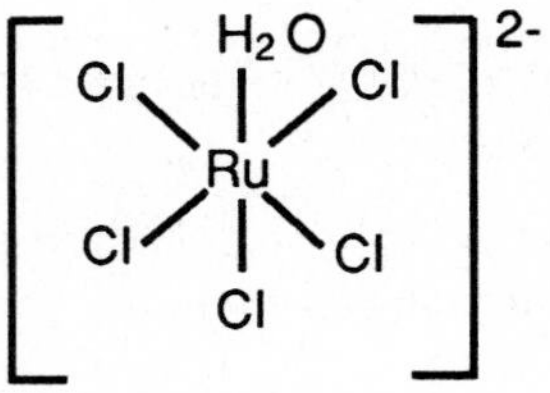

(c) 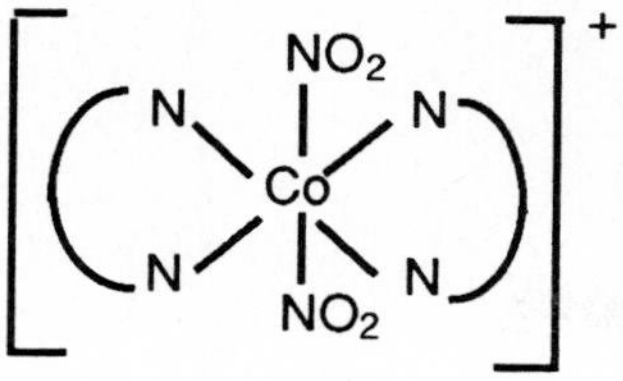

(d) 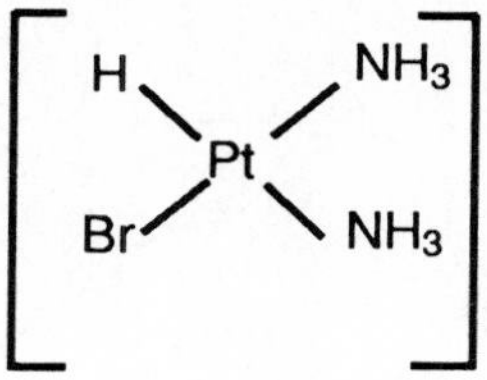

25.5 [25.3] (a) tetrachloroaluminate(III)
(b) dicyanoargentate(I)
(c) dichloroethylenediamineplatinum(II)
(d) trans-tetraamminediaquachromium(III)

[25.4] (a) tetraamminezinc(II)
(b) aquapentachlororuthenate(III)
(c) trans-bis(ethylenediamine)dinitrocobalt(III)
(d) cis-diamminebromohydridoplatinum(II)

25.6 (a) triamminetribromochromium(III)
(b) diamminedichloroethylenediamine cobalt(III) bromide
(c) potassium tris(oxalato)ferrate(III)
(d) cesium dichlorobis(oxalato)chromate(III)
(e) potassium pentachlorothiosulfatoiridate(IV)
(f) bis(ethylenediamine)palladium(II) diamminetetrabromochromate(III)

25.7 (a) $[Cr(NH_3)_6](NO_3)_3$ (b) $[Co(NH_3)_4CO_3]SO_4$ (c) $[Pt(en)_2Cl_2]Br_2$
(d) $K[V(H_2O)_2Br_4]$ (e) $[Zn(en)_2][HgI_4]$

25.8 (a) $[Mn(H_2O)_5Br]SO_4$ (b) $[Ru(bipy)_3](NO_3)_2$
(c) $[Fe(o\text{-}phen)_2Cl_2]ClO_4$ (d) $Na[Co(en)Br_4]$ e) $[Ni(NH_3)_6]_3[Cr(ox)_3]_2$

25.9 (a) ortho-phenanthroline (o-phen) is bidentate
(b) oxalate, $C_2O_4^{2-}$ is bidentate
(c) ethylenediaminetetraacetate, EDTA, is hexadentate
(d) ethylenediamine, en, is bidentate

25.10 (a) 4 (b) 4 (c) 6 (d) 6

Isomerism

25.11 (a)

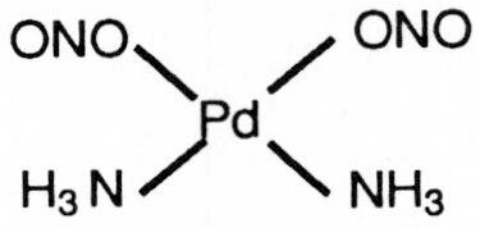

cis

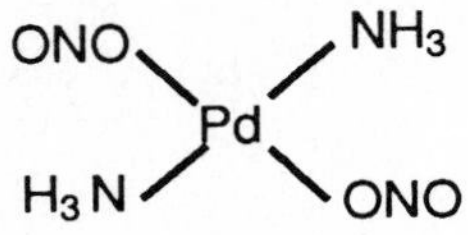

trans

(b) $[Pd(NH_3)_2(ONO)_2]$, $[Pd(NH_3)_2(NO_2)_2]$

(c)

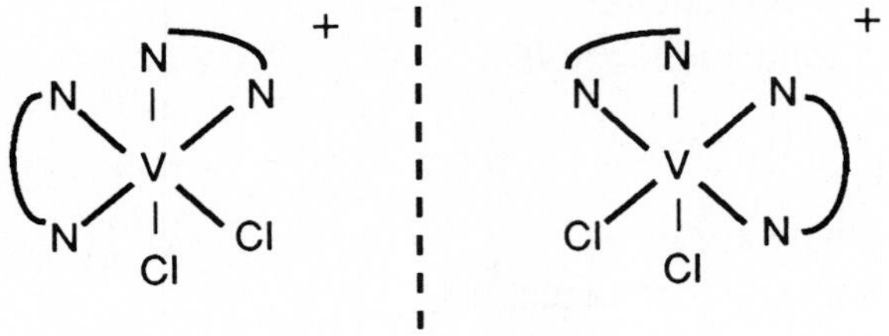

(d) $[Co(NH_3)_4Br_2]Cl$, $[Co(NH_3)_4BrCl]Br$

25.12 (a)

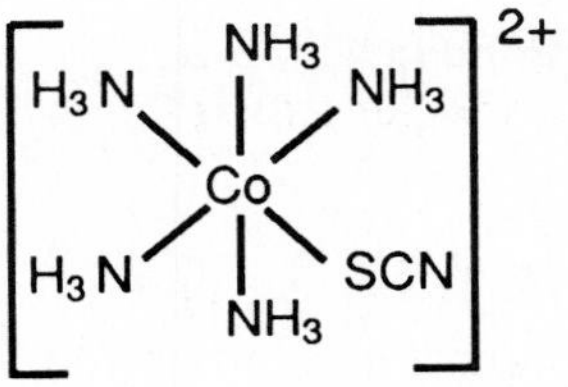

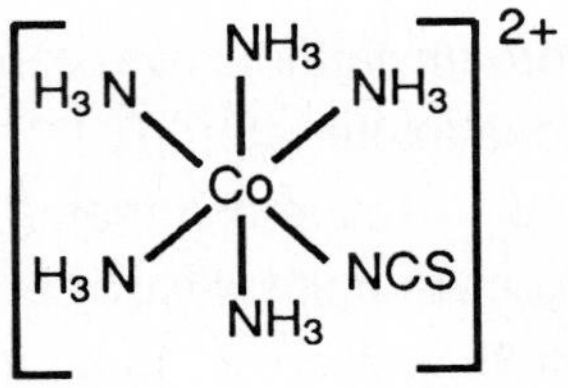

(b)

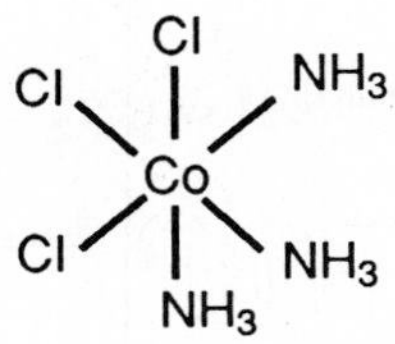

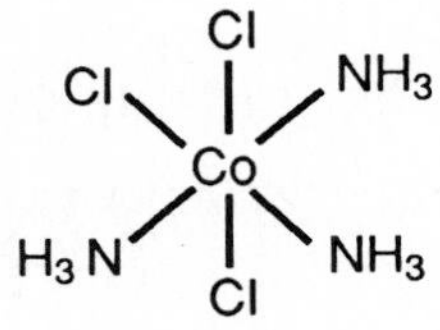

(c)

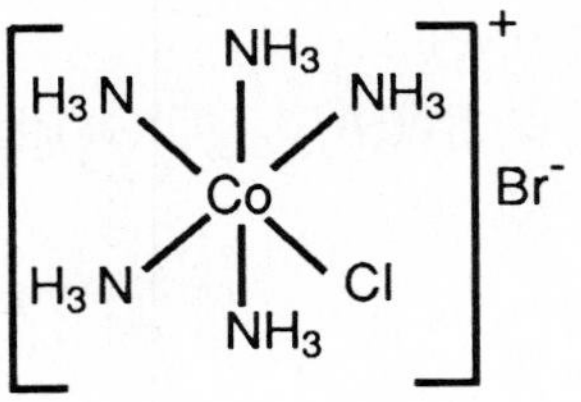

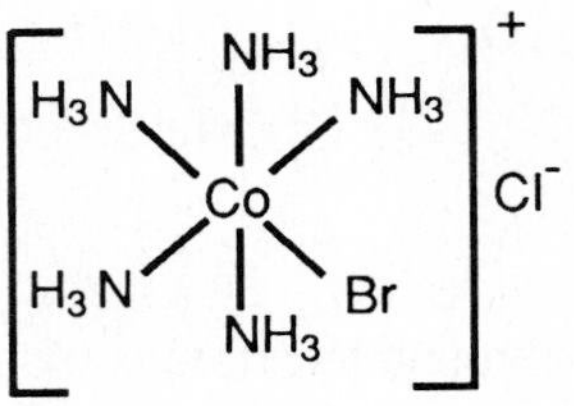

coordination sphere isomerism

25.13

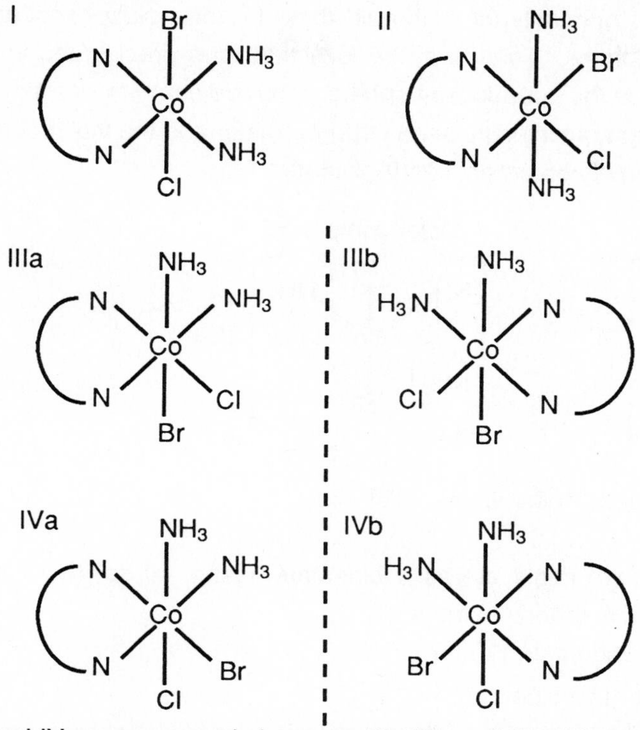

I, II, IIIa and IVa are geometric isomers; IIIa/IIIb and IVa/IVb are the pairs of optical isomers.

25.14 (a) only one: 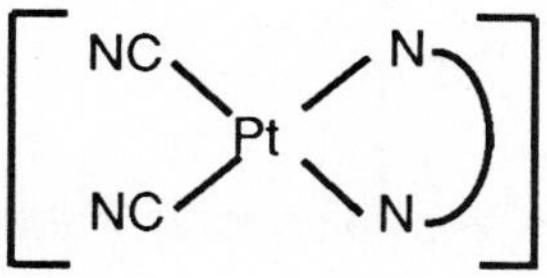 (b) only one: 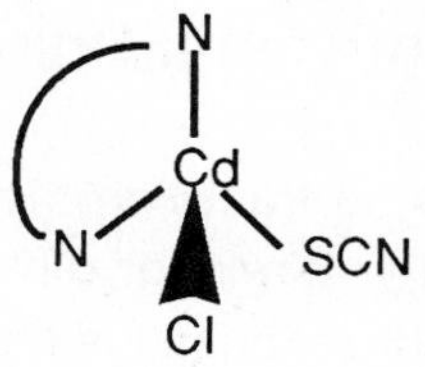

(c) 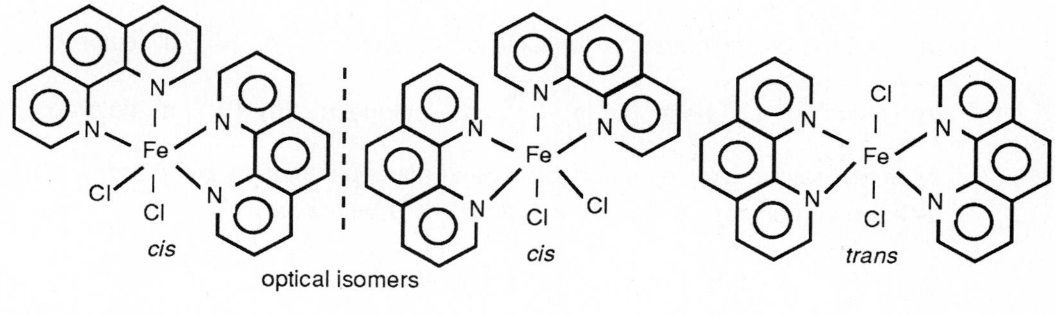

(The 3 isomeric complex ions in part (c) each have a +1 charge.)

(d) 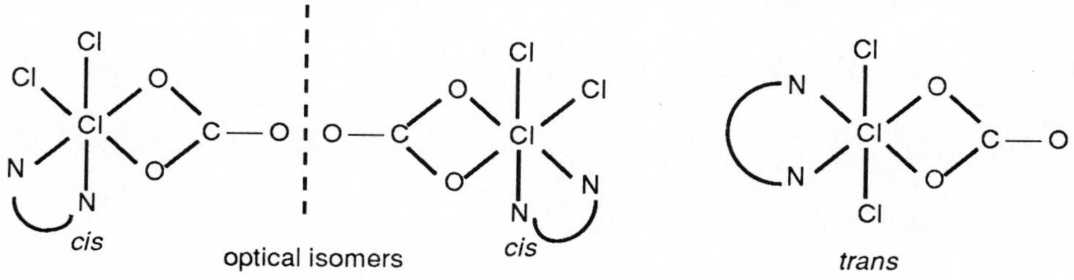

25.15 Cobalt(III) complexes are generally inert; that is, they do not rapidly exchange ligands inside the coordination sphere. Therefore, the ions that form precipitates in these two cases are probably outside the coordination sphere. The red complex can be formulated as $[Co(NH_3)_5SO_4]Br$, pentaamminesulfatocobalt(III) bromide, and the violet compound as $[Co(NH_3)_5Br]SO_4$, pentaamminebromocobalt(III) sulfate.

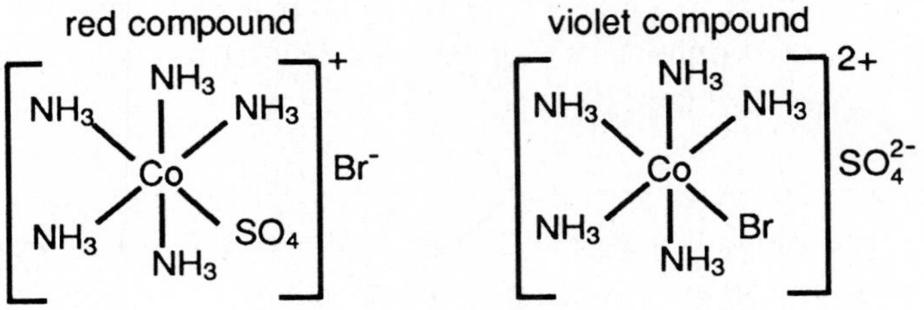

25.16 $[Ce(H_2O)_4(SO_4)_2]$, tetraaquadisulfatocerium (IV)

By analogy to CO_3^{2-}, SO_4^{2-} might act as a bidentate ligand. If so, there are several possibilities for coordination sphere isomers:
$[Ce(H_2O)_4SO_4]SO_4$, (one bidentate SO_4^{2-})
$[Ce(H_2O)_2(SO_4)_2] \cdot 2H_2O$, (two bidentate SO_4^{2-})
$[Ce(H_2O)_3(SO_4)_2] \cdot H_2O$ (This complex has one monodentate and one bidentate SO_4^{2-} ligand and seems less likely than the others.)

Color, Magnetism; Crystal Field Theory

25.17 Color in transition metal compounds arises from electronic transitions between d-orbital energy levels, or d-d transitions. Compounds with d^0 or d^{10} electron configurations are colorless, because d-d transitions are not possible.

(a) Zn^{2+}, d^{10}, colorless; (b) Cr^{4+}, d^2, colored (c) Ni^{2+}, d^8, colored

(d) Al^{3+}, d^0, colorless (e) Cd^{2+}, d^{10}, colorless (f) Fe^{2+}, d^6, colored

25.18 Assume that the simple salts dissociate in aqueous solution and that the metal ions exist as aqua complexes. Color arises from d^1 - d^9 metal ions.

(a) Sc^{3+}, d^0, colorless (b) Ni^{2+}, d^8, colored (c) Sr^{2+}, d^0, colorless

(d) V^{3+}, d^2, colored (e) Rh^{3+}, d^6, colored (f) Pb^{2+}, d^{10}, colorless

25.19 (a) Fe^{4+}, d^4 (b) Co^{2+}, d^7 (c) Ni^{2+}, d^8 (d) Au^{3+}, d^8 (e) Mo^{4+}, d^2

25.20 (a) Ru^{3+}, d^5 (b) Cu^{2+}, d^9 (c) Co^{3+}, d^6 (d) Mo^{5+}, d^1 (e) Re^{3+}, d^4

25.21 Blue to blue-violet (Figure 25.25)

25.22 Approximately 460 nm

25.23 Six ligands in an octahedral arrangement are oriented along the x, y and z axes of the metal. These negatively charged ligands (or the negative end of ligand dipoles) have greater electrostatic repulsion with valence electrons in metal orbitals that also lie along these axes, the d_{z^2}, and $d_{x^2-y^2}$. The d_{xy}, d_{xz} and d_{yz} metal orbitals point between the x, y, and z axes, and electrons in these orbitals experience less repulsion with ligand electrons. Thus, in the presence of an octahedral ligand field, the d_{xy}, d_{xz} and d_{xy} metal orbitals are lower in energy than the $d_{x^2-y^2}$ and d_{z^2}.

25.24 The ligands that possess the greatest ability to interact with the central metal atom cause the largest splitting. The properties of ligands that are important are charge (usually the more negatively charged ligands produce larger splittings) and polarizability, which measures the ability of the ligand to distort its charge distribution as it interacts with the positively charged metal ion.

25.25 Cyanide is a strong field ligand. The d-d electronic transitions occur at relatively high energy, because Δ is large. A yellow color corresponds to absorption of a photon in the violet region of the visible spectrum, between 430 and 400 nm. H_2O is a weaker field ligand than CN^-. The blue or green colors of aqua complexes correspond to absorptions in the region of 620 nm. Clearly, this is a region of lower energy photons than those with characteristic wavelengths in the 430 to 400 nm region. These are very general and imprecise comparisons. Other factors are involved, including whether the complex is high spin or low spin.

25.26 The ions absorb the complement of the color they appear. Green $[Ni(H_2O)_6]^{2+}$ absorbs red light, 650-800 nm. Purple $[Ni(NH_3)_6]^{2+}$ absorbs yellow light, 560-580 nm. Thus, $[Ni(NH_3)_6]^{2+}$ absorbs light with the shorter wavelength. This agrees with the spectrochemical series, which indicates that H_2O will produce a smaller d orbital splitting (Δ) than NH_3. Thus, $[Ni(H_2O)_6]^{2+}$ should absorb light with a smaller energy and longer wavelength.

25.27 (a) $[Kr]5s^1 4d^8$, $[Kr]4d^6$

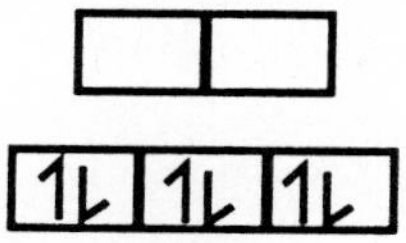

Rh^{3+}

(b) $[Ar]4s^2 3d^5$, $[Ar]3d^4$

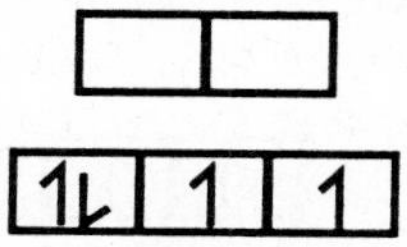

Mn^{3+}

(c) $[Kr]4d^{10}$, $[Kr]4d^7$

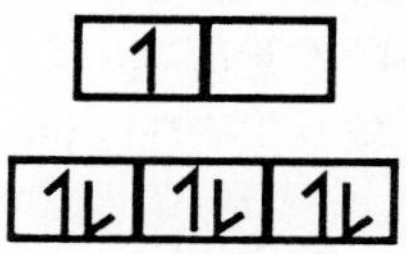

Pd^{3+}

25.28 (a) $[Ar]4s^2 3d^8$, $[Ar]3d^8$

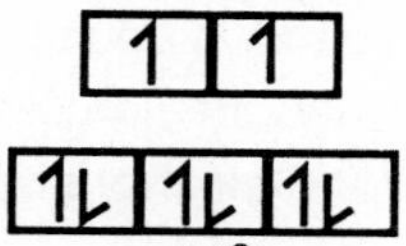

Ni^{2+}

(b) $[Kr]5s^2 4d^5$, $[Kr]4d^5$

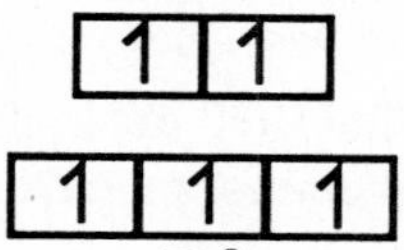

Tc^{2+}

(c) $[Ar]4s^1 3d^5$, $[Ar]4d^4$

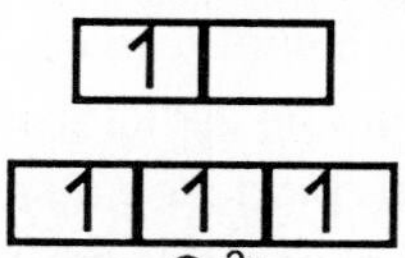

Cr^{2+}

25.29 All complexes in this exercise are six-coordinate octahedral.

(a)

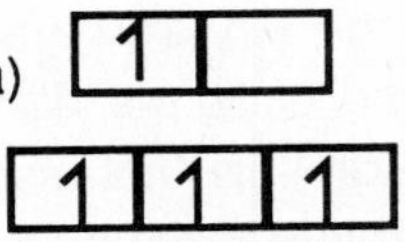

d^4, high spin

(b)

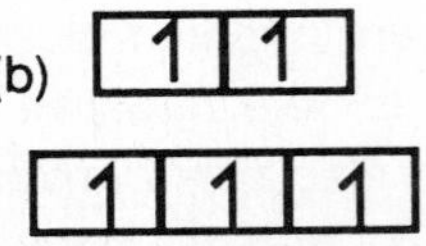

d^5, high spin

(c)

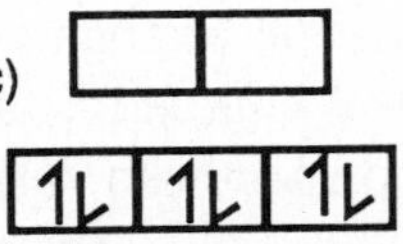

d^6, low spin

(d)

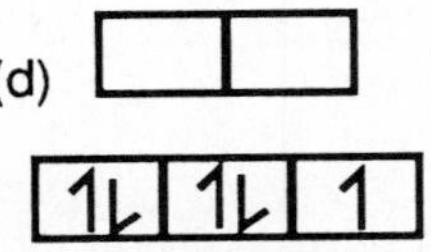

d^5, low spin

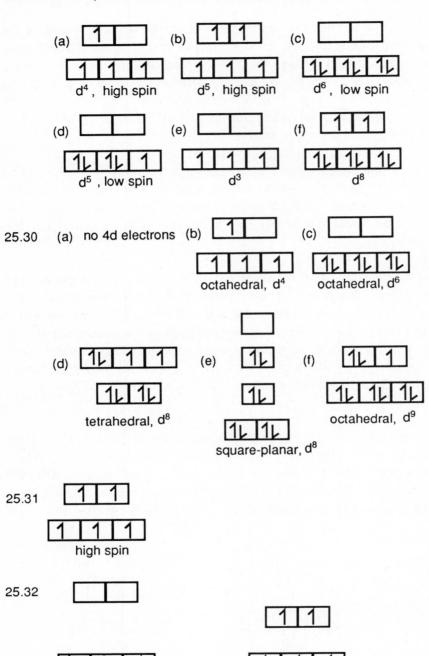

(e)

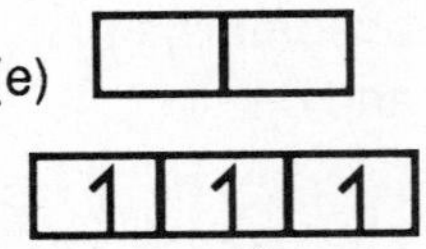

d^3

(f)

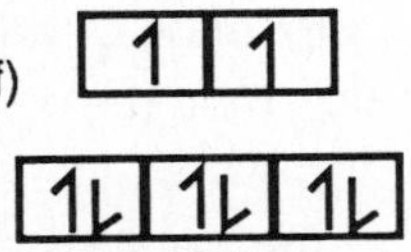

d^8

25.30 (a) no 4d electrons

(b)

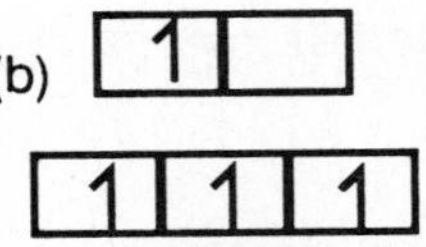

octahedral, d^4

(c)

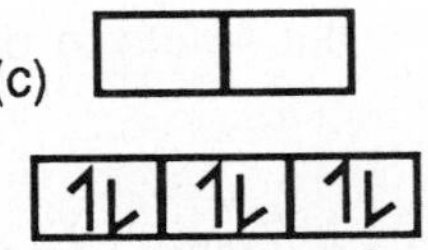

octahedral, d^6

(d)

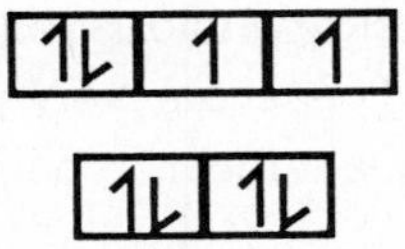

tetrahedral, d^8

(e)

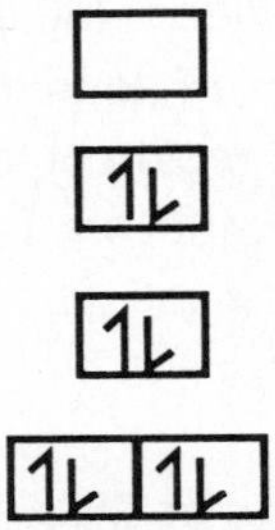

square-planar, d^8

(f)

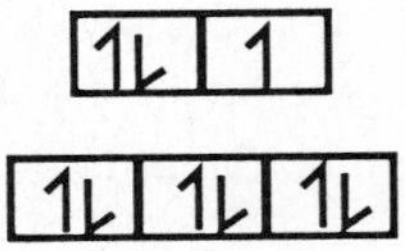

octahedral, d^9

25.31

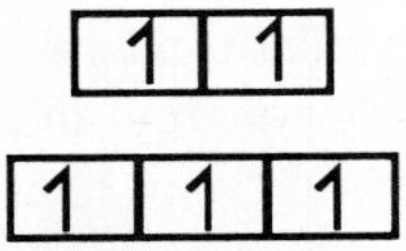

high spin

25.32

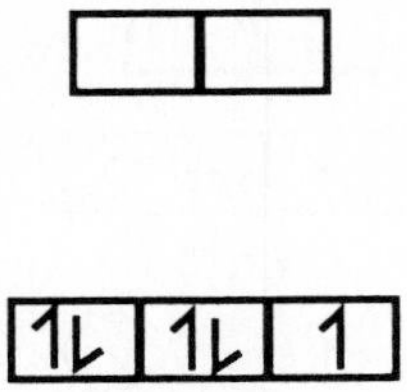

$[Fe(CN)_6]^{3-}$

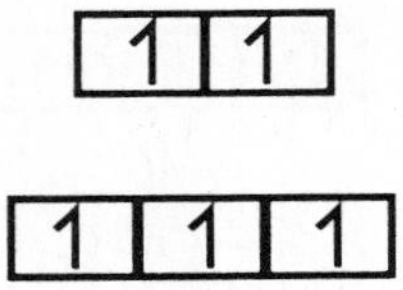

$[Fe(SCN)_6]^{3-}$

Both complexes contain Fe^{3+}, a d^5 ion. CN^-, a strong field ligand, produces such a large Δ that the splitting energy is greater than the pairing energy, and the complex is low spin. SCN^- produces a smaller Δ, so it is energetically favorable for d electrons to be unpaired in the higher energy d orbitals.

Additional Exercises

25.33 (a) $K_2[Ni(en)Cl_4]$; $[Ni(en)(H_2O)_2Cl_2]$
(b) $K_2[Ni(CN)_4]$ $[Zn(H_2O)_4](NO_3)_2$; $[Cu(NH_3)_4]SO_4$
(c) $[CoF_6]^{3-}$, high spin; $[Co(NH_3)_6]^{3+}$ or $[Co(CN)_6]^{3-}$, low spin
(d) thiocyanate, SCN^- or NCS^-; nitrite, NO_2^- or ONO^-.
(e) $[Co(en)_2Cl_2]Cl$; see Exercise 25.14(c) and (d) for other examples.
(f) $[Co(en)_3]Cl_3$, $[Cr(NH_3)_6]Cl_3$

25.34 $[Pt(NH_3)_6]Cl_4$; $[Pt(NH_3)_4Cl_2]Cl_2$; $[Pt(NH_3)_3Cl_3]Cl$; $[Pt(NH_3)_2Cl_4]$; $K[Pt(NH_3)Cl_5]$

25.35 The reaction that occurs produces an ion of higher charge. It would appear from the relative values that the reaction could be

$$[Co(NH_3)_4Br_2]^+(aq) + H_2O(l) \rightarrow [Co(NH_3)_4(H_2O)Br]^{2+}(aq) + Br^-(aq)$$

This reaction would convert the 1:1 electrolyte, $[Co(NH_3)_4Br_2]Br$, to a 1:2 electrolyte, $[Co(NH_3)_3(H_2O)Br]Br_2$. The reaction is not extremely rapid, so we would classify the starting complex as relatively inert, though it does undergo ligand replacement.

25.36 First determine the empirical formula, assuming that the remaining mass of complex is Pd.

	mole	mole ratios
$37.6 \text{ g Br} \times \frac{1 \text{ mol Br}}{79.9 \text{ g Br}}$ =	0.470 mol Br	1.98
$28.3 \text{ g C} \times \frac{1 \text{ mol C}}{12.0 \text{ g C}}$ =	2.36 mol C	9.95
$6.60 \text{ g N} \times \frac{1 \text{ mol N}}{14.0 \text{ g N}}$ =	0.471 mol N	1.99
$2.37 \text{ g H} \times \frac{1 \text{ mol H}}{1.01 \text{ g H}}$ =	2.35 mol H	9.91
$25.13 \text{ g Pd} \times \frac{1 \text{ mol Pd}}{106 \text{ g Pd}}$ =	0.237 mol Pd	1.00

The chemical formula is $Pd(NC_5H_5)_2Br_2$. This should be a square-planar complex of Pd(II), a non-electrolyte. Because the dipole moment is zero, we can infer that it must be the trans-isomer, trans-dibromodipyridinepalladium(II).

CHAPTER 25 Solutions to Exercises
Chemistry of Coordination Compounds

25.37 Determine the empirical formula of the complex, assuming the remaining mass is due to oxygen, and a 100 g sample.

$$10.0 \text{ g Mn} \times \frac{1 \text{ mol Mn}}{54.9 \text{ g Mn}} = 0.182 \text{ mol Mn}; \; 0.182 \div 0.182 = 1$$

$$28.6 \text{ g K} \times \frac{1 \text{ mol K}}{39.1 \text{ g K}} = 0.731 \text{ mol K}; \; 0.731 \div 0.182 = 4$$

$$8.8 \text{ g C} \times \frac{1 \text{ mol C}}{12.0 \text{ g C}} = 0.733 \text{ mol C}; \; 0.733 \div 0.182 = 4$$

$$29.2 \text{ g Br} \times \frac{1 \text{ mol Br}}{79.9 \text{ g Br}} = 0.365 \text{ mol Br}; \; 0.365 \div 0.182 = 2$$

$$23.4 \text{ g O} \times \frac{1 \text{ mol O}}{16.0 \text{ g O}} = 1.46 \text{ mol O}; \; 1.46 \div 0.182 = 8$$

There are 2 C and 4 O per oxalate ion, for a total of two oxalate ligands in the complex. To match the conductivity of $K_4[Fe(CN)_6]$, the oxalate and bromide ions must be in the coordination sphere of the complex anion. Thus, the complex is $K_4[Mn(ox)_2Br_2]$, potassium dibromobis(oxalato)manganate(II).

25.38 (a) 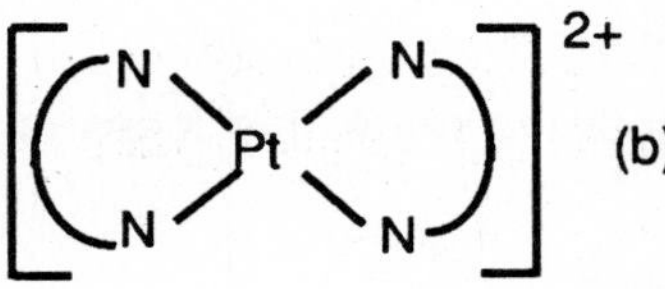

(b) 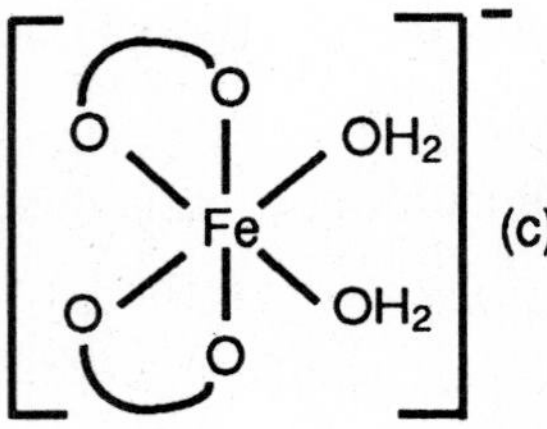

(c) 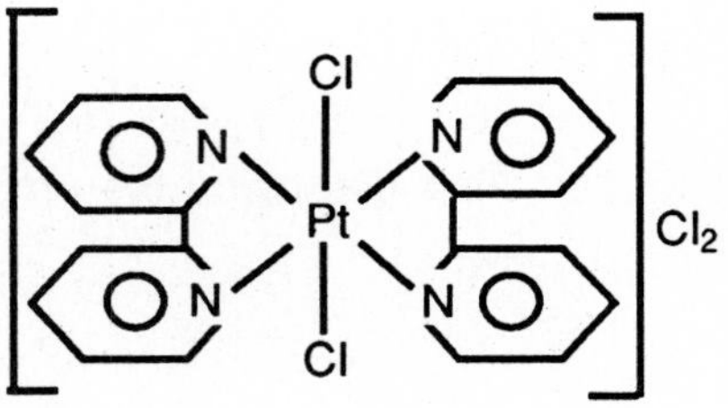

(d) Hg^{2+} is d^{10}, so the complex is probably tetrahedral.

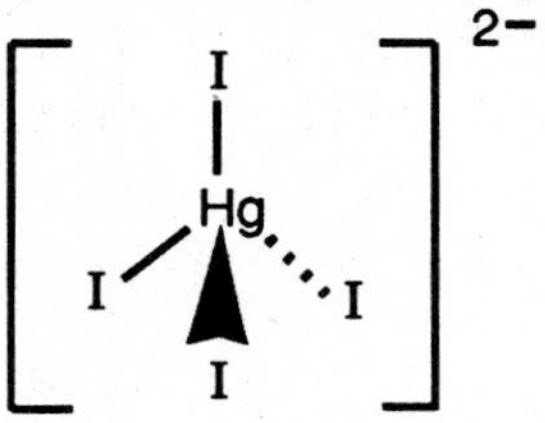

(e) 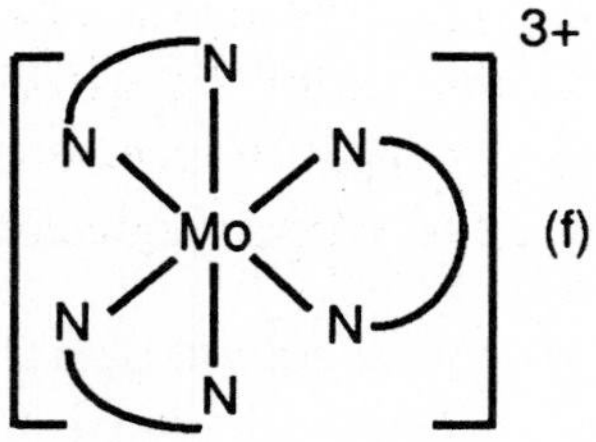

(f) 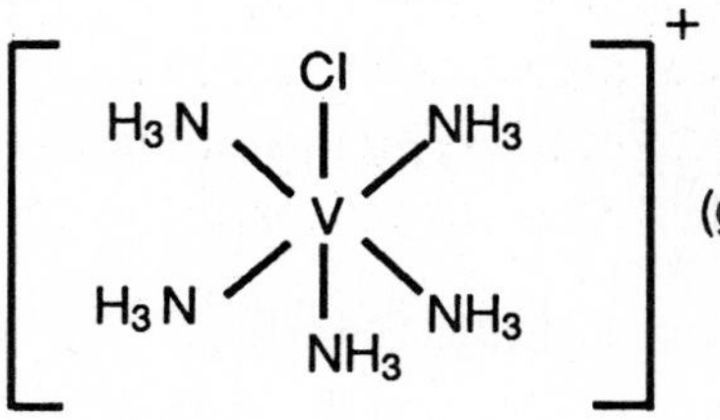

(g) 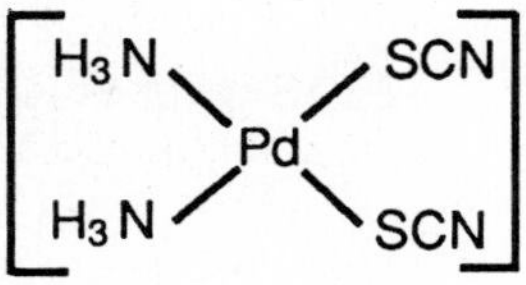

25.39 We will represent the end of the bidentate ligand containing the CF_3 group by a closed circle, the other end by an open circle:

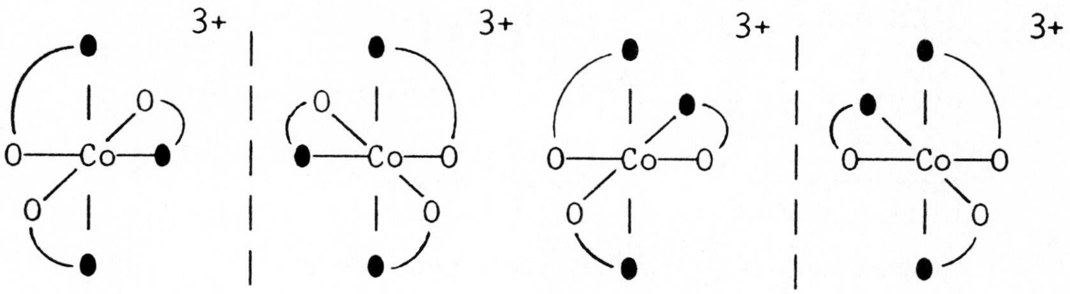

25.40 (a) Only one
(b) Two; (cis or trans arrangement of N and O ends)
(c) Four; two are geometrical, the other two are stereoisomers of each of these; the figure for solution to Exercise 25.39 applies to this problem as well.

25.41 (a) $AgCl(s) + 2NH_3(aq) \rightarrow [Ag(NH_3)_2]^+(aq) + Cl^-(aq)$
(b) $[Cr(en)_2Cl_2]^+(aq) + H_2O(l) \rightarrow [Cr(en)_2(H_2O)Cl]^{2+}(aq) + Cl^-(aq)$
$[Cr(en)_2(H_2O)Cl]^{2+}(aq) + H_2O(l) \rightarrow [Cr(en)_2(H_2O)_2]^{3+}(aq) + Cl^-(aq)$
(c) $Zn(OH)_2(s) + 4NH_3(aq) \rightarrow [Zn(NH_3)_4]^{2+}(aq) + 2OH^-(aq)$
(d) $Co^{2+}(aq) + 4Cl^-(aq) \rightarrow [CoCl_4]^{2-}(aq)$

25.42 The order given is in the direction of increasing Δ in the spectrochemical series. Thus the absorption spectra of $[VF_6]^{3-}$, $[V(NH_3)_6]^{3+}$ and $[V(NO_2)_6]^{3-}$ should exhibit a trend toward absorptions at shorter wavelength (higher energy).

25.43 The $[Co(H_2O)_6]^{2+}$ complex should be rose-colored, or pink. The $CoCl_4^{2-}$ solutions should be green or blue-green. The $CoCl_4^{2-}$ ion is tetrahedral. We know that the d orbital splittings in tetrahedral complexes are only 4/9 as large as for the same ligand in an octahedral complex. Thus, the energy of the transition between the two sets of d orbitals is smaller in the tetrahedral complex, and the absorption is at longer wavelength.

25.44 oxyhemoglobin Fe^{2+} - d^6 deoxyhemoglobin Fe^{2+} - d^6

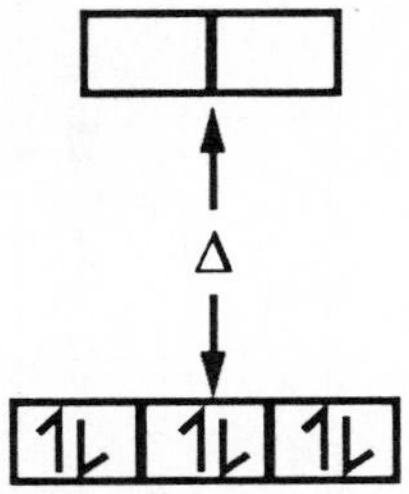

low spin

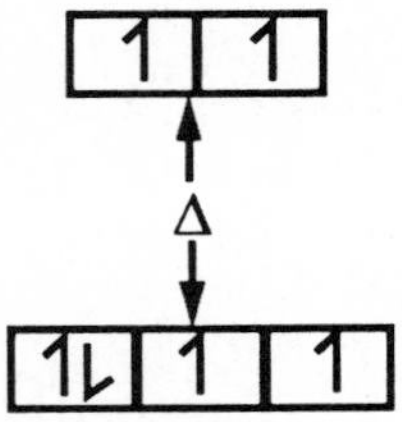

high spin

In general, the crystal field splitting, Δ, is greater in low spin than high spin complexes. The energy, Δ, corresponds to the wavelength of light absorbed by the complex and determines its color. Since oxyhemoglobin absorbs higher energy, shorter wavelength light, longer wavelengths remain and the sample appears red. Deoxyhemoglobin absorbs lower energy (orange-red) light, and the sample appears blue.

25.45 (a) A square-planar complex, d^8

(b) Octahedral, d^8

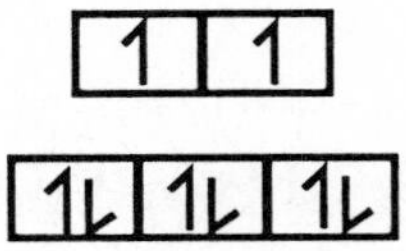

(c) Octahedral, d^4, surely low spin because it is a third-row element

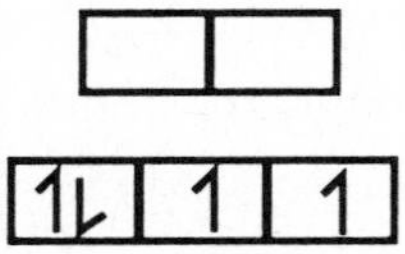

(d) Octahedral, d^5; low spin

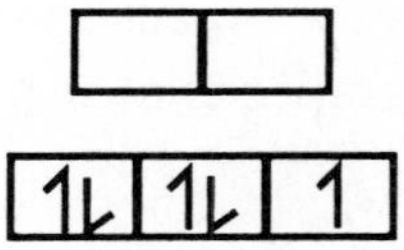

25.46 (a) False. The spin pairing energy is smaller than Δ; it is for this reason that electrons pair up in the lower energy orbitals, in spite of the repulsive energy associated with spin pairing, rather than move to the higher energy orbital, which would cost energy in the amount Δ.

(b) True. Higher metal ion charge causes the ligands to be more strongly attracted, thus producing a larger splitting in the d orbital energies.

(c) False. Square-planar configurations are associated with a d^8 electron configuration (which Ni^{2+} has) and a ligand that produces a strong field, leading to large separations between the energy levels. Cyanide, CN^-, is a much stronger field ligand than Cl^- (see Section 25.7).

25.47 $\frac{182 \times 10^3 \text{ J}}{1 \text{ mol}} \times \frac{1 \text{ mol}}{6.02 \times 10^{23} \text{ molecules}} = 3.02 \times 10^{-9}$ J/photon

$\Delta E = h\nu = 3.02 \times 10^{-19}$ J; $\nu = \Delta E/n$

$\nu = 3.02 \times 10^{-19} \text{ J}/6.63 \times 10^{-34} \text{ J} \bullet \text{s} = 4.56 \times 10^{14} \text{ s}^{-1}$

$\lambda = \frac{3.0 \times 10^8 \text{ m/s}}{4.56 \times 10^{14} \text{ s}^{-1}} = 6.58 \times 10^{-7} \text{ m} = 658 \text{ nm}$

We expect that this complex will absorb in the visible, at around 660 nm. It will thus exhibit a green color (Figure 25.25).

25.48 Longer wavelength absorption corresponds to a smaller splitting of the energies of the d orbitals

(a) $[CoF_6]^{4-}$, because F^- is a weaker field ligand than CN^-.

(b) $[V(H_2O)_6]^{2+}$, because the metal ion of lower charge attracts the ligands less strongly, thus produces a smaller splitting;

(c) $[MnCl_4]^-$, because in a tetrahedral complex the ligand field is much smaller than in an octahedral complex; furthermore, Cl^- is a much weaker field ligand than CN^-.

25.49 Application of pressure would result in shorter metal ion-oxide distances. This would have the effect of increasing the ligand-electron repulsions, and would result in a larger splitting in the d orbital energies. Thus, application of pressure should result in a shift in the absorption to a higher energy (shorter wavelength).

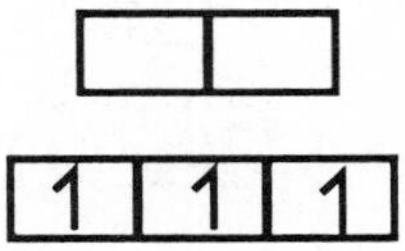

25.50 The d^3 and d^6 electron configurations in octahedral and tetrahedral fields are shown below:

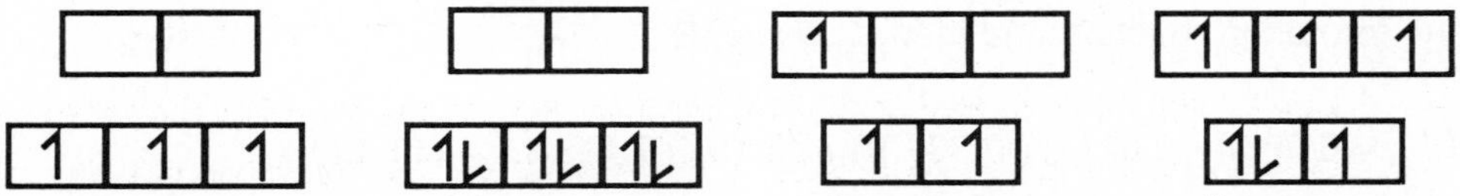

d^3 octahedral　　d^6 octahedral　　d^3 tetrahedral　　d^6 tetrahedral

In the octahedral environment, d^3 and d^6, strong field configurations only have electrons in the d orbitals which are lower in energy than the five degenerate orbitals in the isolated metal atom. In a tetrahedral environment (always weak field, see Section 25.7) d^3 has one electron and d^6 has three electrons in the higher energy d orbitals. Thus the octahedral environment is energetically more favorable for metal ions with d^3 or d^6 configurations.

25.51

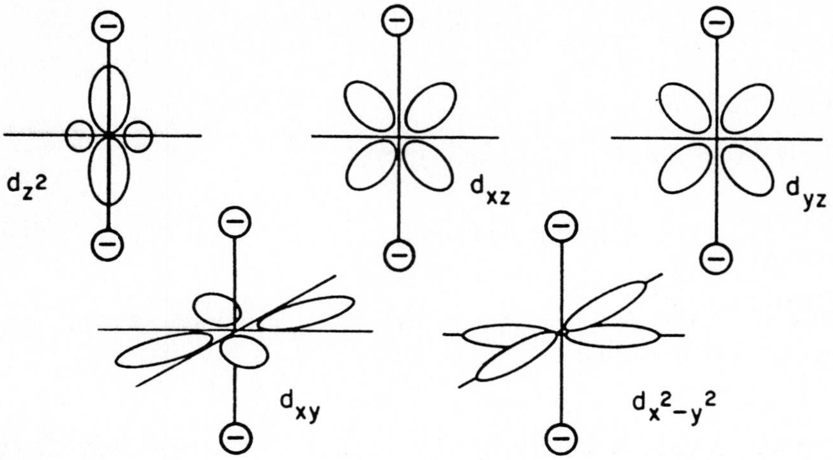

For a d^6 metal ion in a strong ligand field, there would be two unpaired electrons.

25.52 The process can be written:

$$H_2(g) + 2e^- \rightarrow 2H^+(aq) \qquad E^\circ = 0.0\ V$$
$$Cu(s) \rightarrow Cu^{2+}(aq) + 2\ e^- \qquad E^\circ = -0.337\ V$$
$$Cu^{2+}(aq) + 4NH_3(aq) \rightarrow [Cu(NH_3)_4]^{2+}] (aq)$$

$$H_2(g) + Cu(s) + 4NH_3(aq) \rightarrow 2H^+(aq) + [Cu(NH_3)_4]^{2+}(aq)$$

$$E = E^\circ - RT \ln K; \quad K = \frac{[H^+]^2[Cu(NH_3)_4^{2+}]}{P_{H_2}\ [NH_3]^4}$$

P_{H_2} = 1 atm, $[H^+]$ = 1 *M*, $[NH_3]$ = 1 *M*, $[Cu(NH_3)_4]^{2+}$ = 1 *M*, K = 1

$$E = E^\circ - RT \ln(1); \quad E = E^\circ - RT(0); \quad E = E^\circ = 0.08\ V$$

Since we know E° values for two steps and the overall reaction, we can calculate "E°" for the formation reaction and then K_f, using $E^\circ = \frac{0.0592}{n} \log K$ for the step.

$$E_{cell} = 0.08\ V = 0.0\ V + (-0.337\ V) + "E_f^\circ" = +0.417\ V \qquad (0.42\ V)$$

$$"E_f^\circ" = \frac{0.0592}{n} \log K_f\ ; \quad \log K_f = \frac{n(E_f^\circ)}{0.0592} = \frac{2(0.417)}{0.0592} = 14.09$$

$$K_f = 10^{14.09} = 1.2 \times 10^{14}$$

CHAPTER *26*

The Chemistry of Life: Organic and Biological Chemistry

Hydrocarbon Structures and Nomenclature

26.1 (a) $CH_3CH_2CH_2CH_2CH_3$, C_5H_{12}

(b)

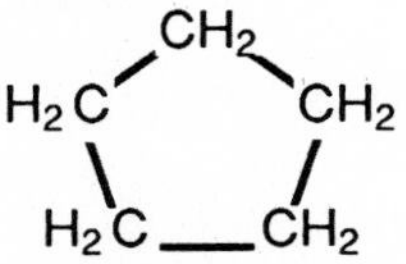

, C_5H_{10}

(c) $CH_2 = CHCH_2CH_2CH_3$, C_5H_{10}

(d) $HC \equiv C\text{-}CH_2CH_2CH_3$, C_5H_8

saturated: (a), (b); unsaturated: (c), (d)

26.2 alkane,

$CH_3 - \overset{\overset{\displaystyle CH_3}{|}}{\underset{\underset{\displaystyle CH_3}{|}}{C}} - CH_2 - CH_3$, C_6H_{14} , saturated

cycloalkane,

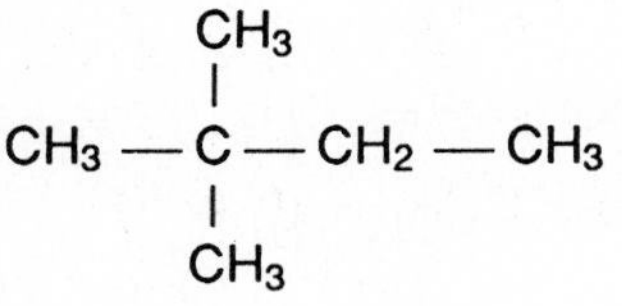

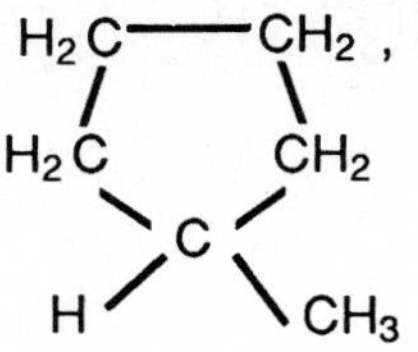

, C_6H_{12} , saturated

alkene, $CH_3 - \underset{\underset{CH_3}{|}}{\overset{\overset{H}{|}}{C}} - CH = CH - CH_3$, C_6H_{12} , unsaturated

alkyne, $CH_3 - CH_2 - C \equiv C - CH_2 - CH_3$, $C_6H_{10,}$ unsaturated

aromtic hydrocarbon, C_6H_6 , unsaturated

26.3 There are 5 isomers. Their carbon skeletons are as follows:

$C - C - C - C - C - C$

n-hexane

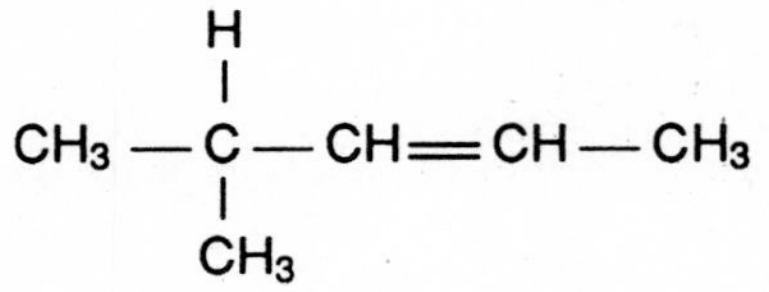

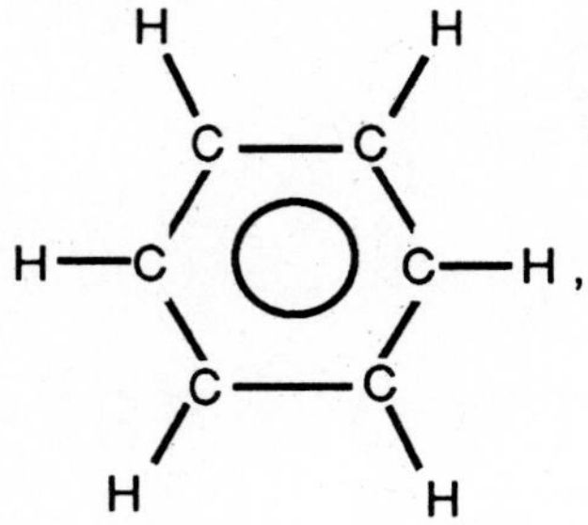

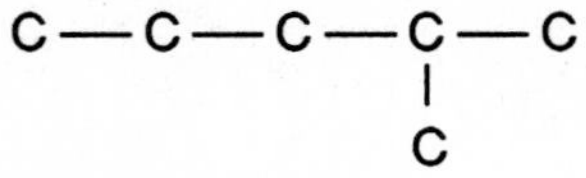

$C - C - C - \underset{\underset{C}{|}}{C} - C$

2-methylpentane

$C - C - \underset{\underset{C}{|}}{C} - C - C$

3-methylpentane

$C - \underset{\underset{C}{|}}{C} - \underset{\underset{C}{|}}{C} - C$

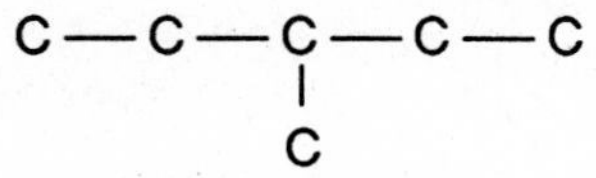

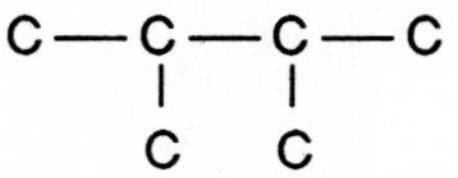

2,3-dimethylbutane

$C - C - \underset{\underset{C}{|}}{\overset{\overset{C}{|}}{C}} - C$

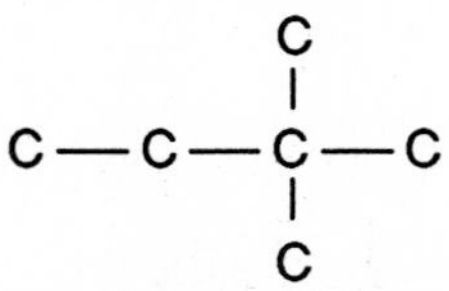

2,2-dimethylbutane

26.4 (a) $CH_3 CH_2 CH_2 C \equiv CH$ $CH_3 CH_2 C \equiv CCH_3$ $CH_3 CH = C = CHCH_3$

$CH_3 \underset{\underset{CH_3}{|}}{C}HC \equiv CH$ $CH_2 = CHCH_2 CH = CH_2$ $CH_2 = CH - \underset{\underset{CH_3}{|}}{C} = CH_2$

$CH_2 = CHCH = CHCH_3$

$CH_2 = C = CHCH_2 CH_3$ $CH_2 = C = \underset{\underset{CH_3}{|}}{C} - CH_3$

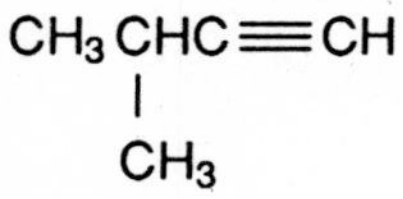

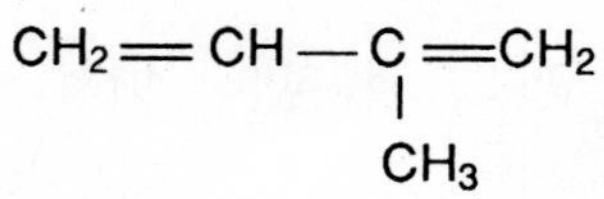

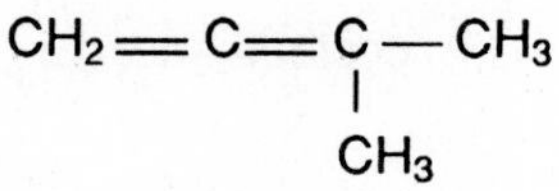

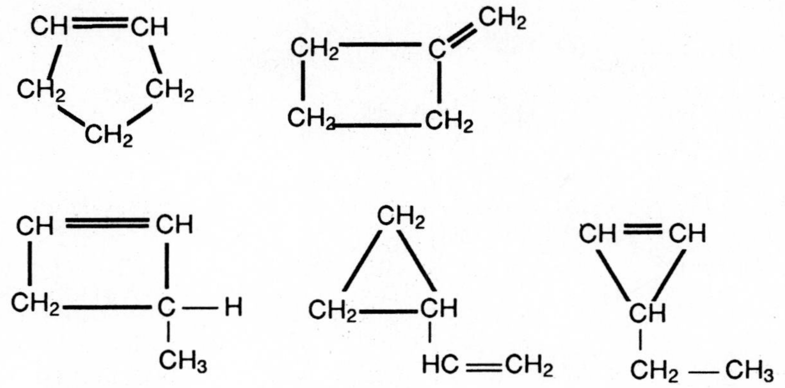

(b) 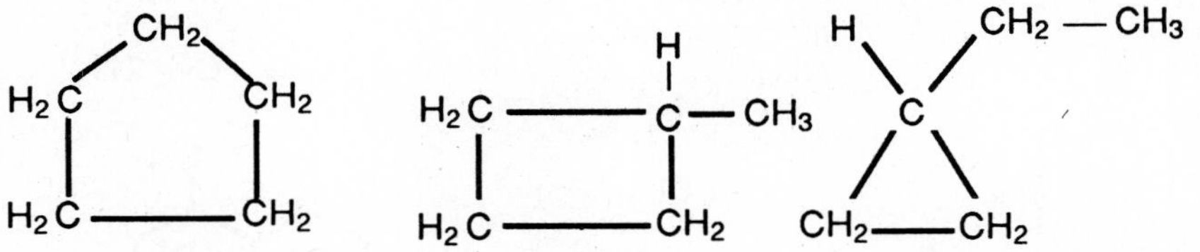

$CH_3CH_2CH{=}CH{-}CH_3$

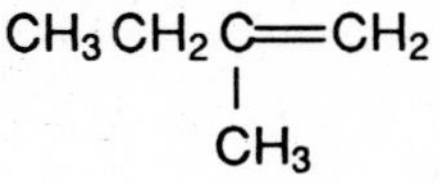

$CH_3CH_2CH_2CH{=}CH_2$

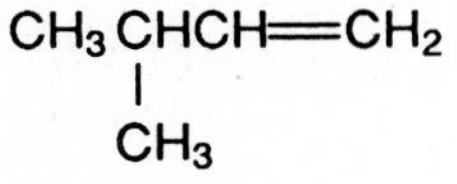

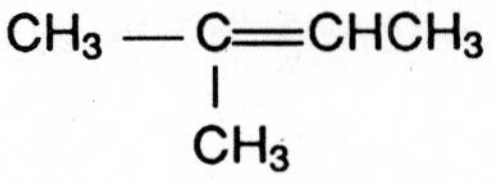

26.5 (a) 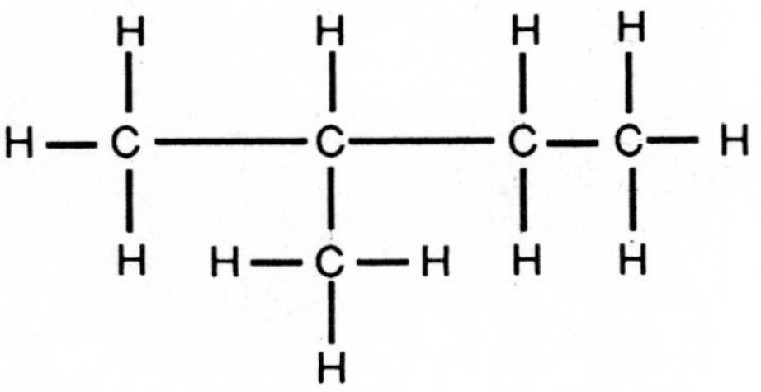 (b) 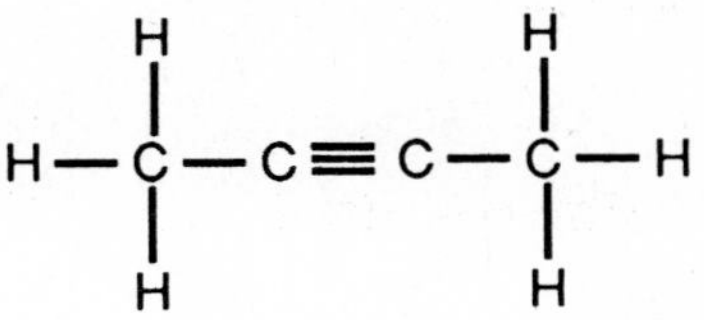

(c) 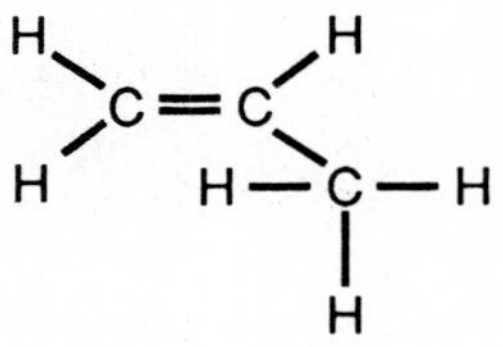

26.6 (a) 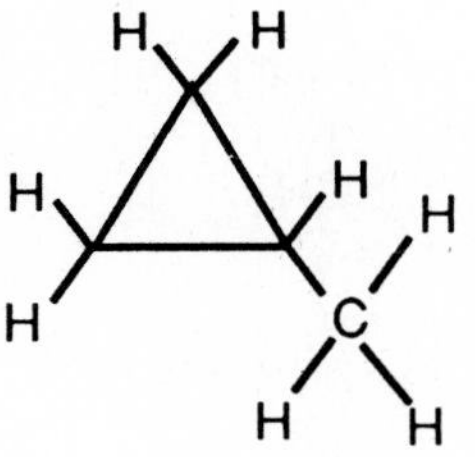 (b) 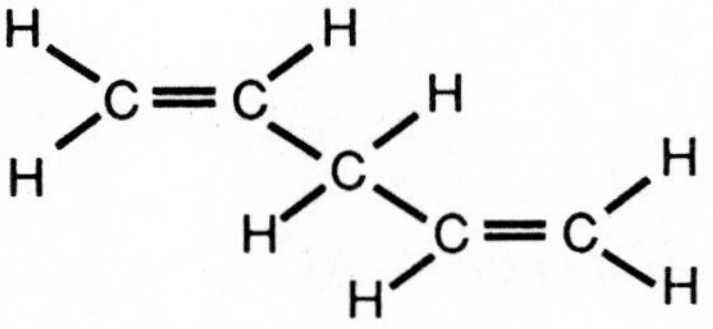 (c) 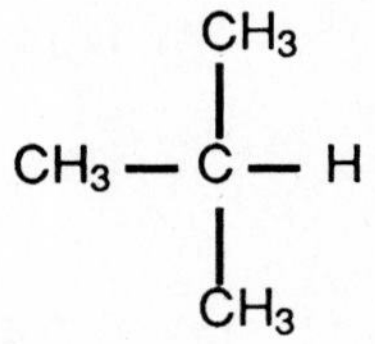

26.7 (a) 109° (b) 120° (c) 180°

26.8 (a) sp^3 (b) sp^2 (c) sp^2 (d) sp

26.9 (a) 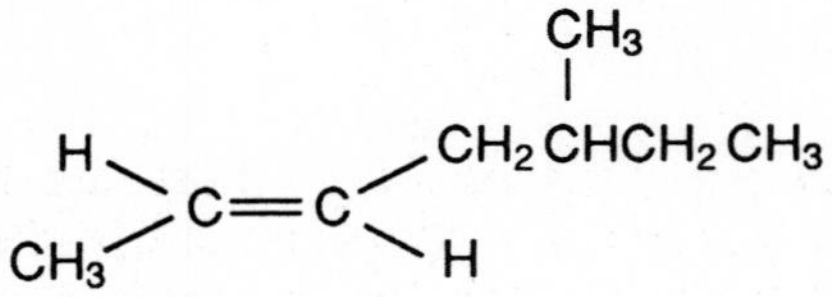

(b) $HC\equiv C-CH_2Cl$

(c) 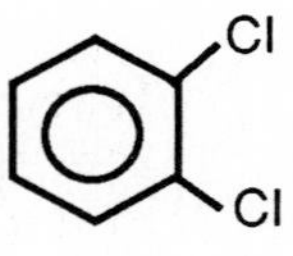

(d) 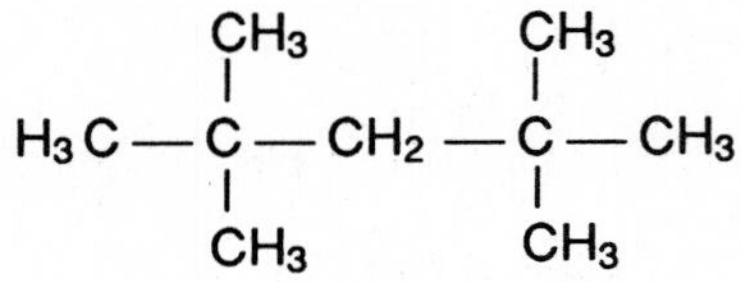

(e) 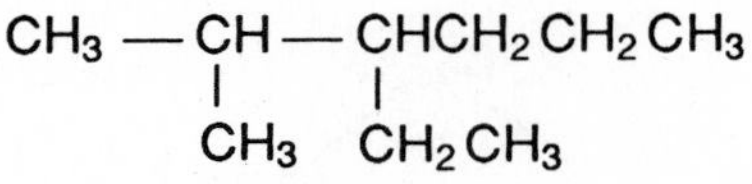

(f) 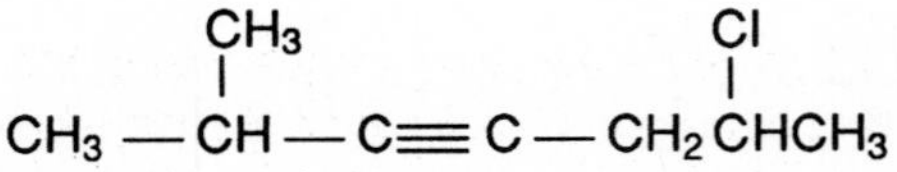

(g) 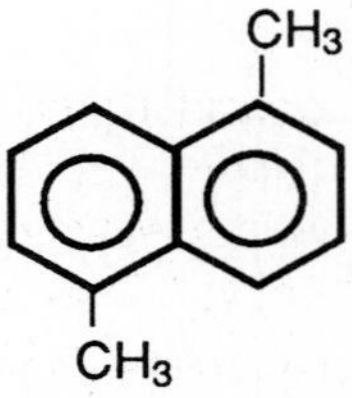

(h) $CH_2{=}CHCH_2CH_2CH_2CH{=}CH_2$

26.10 (a) 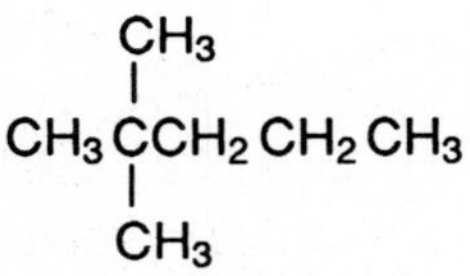

(b) 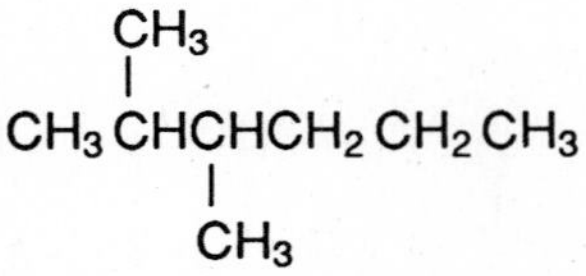

(c) 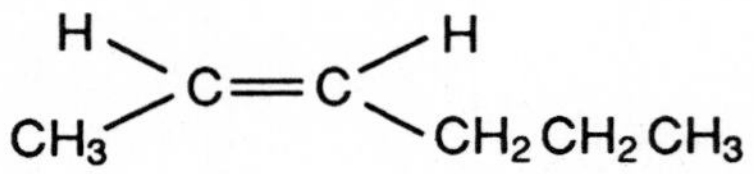

(d) 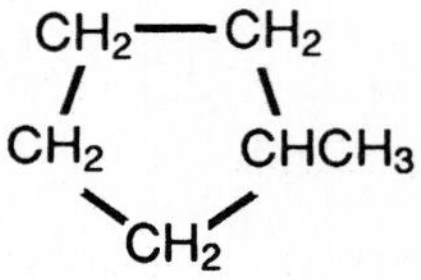

(e) 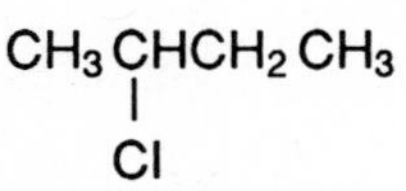

(f) 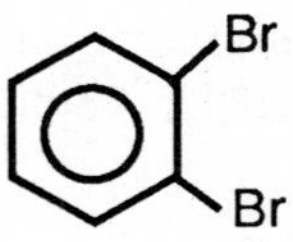

(g) 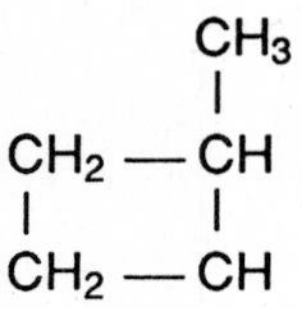

(h) 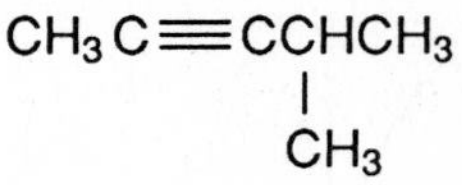

26.11 (a) 3,4-dimethylheptane (b) cis-6-methyl-3-octene (c) para-dibromobenzene
(d) 4,4-dimethyl-1-hexyne (e) methylcyclobutane

CHAPTER 26 — Solutions to Exercises

The Chemistry of Life: Organic and Biological Chemistry

26.12 (a) 2,4-dibromopentane (b) 2,5-octadiene (c) 2-phenyl-3,5-dichlorohexane
(d) 1-chloro-2-cyclopentylpropane (e) trans-dichloroethane

26.13 Butene is an alkene, C_4H_8. There are two possible placements for the double bond:

$$\underset{\text{1-butene}}{CH_2 = CHCH_2CH_3} \quad \text{or} \quad \underset{\text{2-butene}}{CH_3CH = CHCH_3}$$

These two compounds are structural isomers. For 2-butene, there are two different, noninterchangeable ways to construct the carbon skeleton (owing to the absence of free rotation around the double bond). These two compounds are geometric isomers.

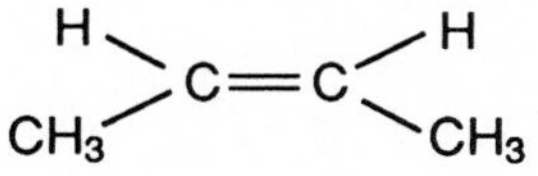

cis-2-butene

or

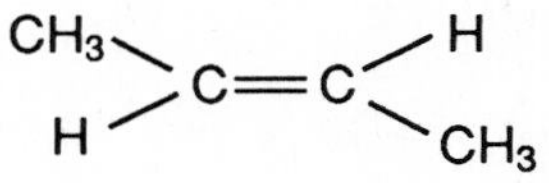

trans-2-butene

26.14 Each doubly bound carbon atom in an alkene has two unique sites for substitution. These sites cannot be interconverted because rotation about the double bond is restricted; geometric isomerism results. In an alkane, carbon forms only single bonds, so the three remaining sites are interchangeable by rotation about the single bond. Although there is also restricted rotation around the triple bond of an alkyne, there is only one additional bonding site on a triply bound carbon, so no isomerism results.

26.15 $C_3H_4Cl_2$

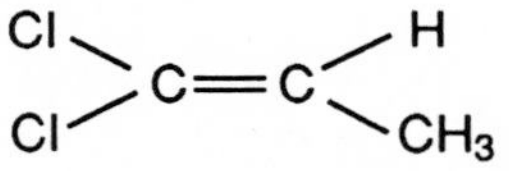

(1,1)-dichloropropene

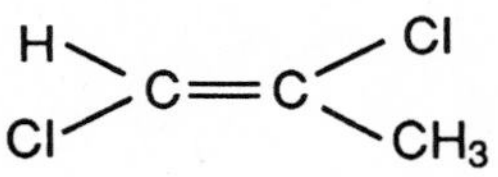

trans-1,2-dichloropropene

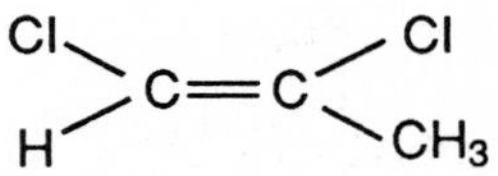

cis-1,2-dichloropropene

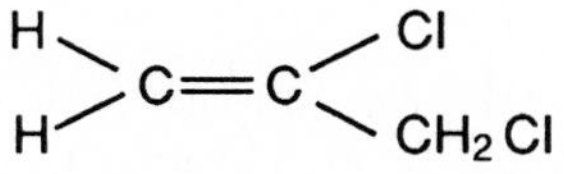

2,3-dichloropropene

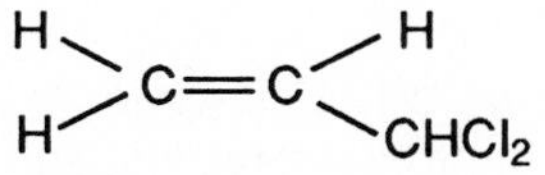

3,3-dichloropropene

The cis and trans compounds are geometric isomers. The others are structural isomers (different placement of the Cl atoms).

26.16 (a) no (b)

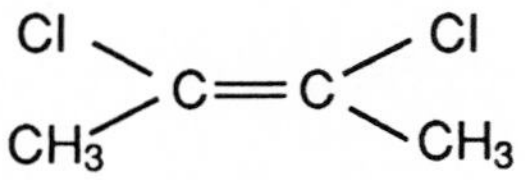

cis

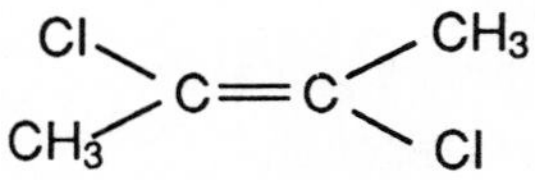

trans

(c) no (d) no

26.17 Assuming that each component retains its effective octane number in the mixture (and this isn't always the case), we obtain: octane number = 0.30(0) + 0.70(100) = 70.

CHAPTER 26 The Chemistry of Life: Organic and Biological Chemistry

Solutions to Exercises

26.18 Octane number can be increased by increasing the fraction of branched-chain alkanes or aromatics, since these have high octane numbers. This can be done by cracking. The octane number also can be increased by adding an anti-knock agent such as tetraethyl lead, $Pb(C_2H_5)_4$, methylcyclopentadienylmanganesetricarbonyl, $CH_3C_5H_4Mn(CO)_3$, or an alcohol, methanol or ethanol.

Reactions of Hydrocarbons

26.19 (a) $CH_3CH_2CH_3(g) + 5O_2(g) \rightarrow 3CO_2(g) + 4H_2O(g)$

(b) $CH_3CH_2C{\equiv}CH + Br_2 \rightarrow CH_3CH_2CBr{=}CHBr$

(c) $C_6H_6 + Br_2 \xrightarrow{FeBr_3} C_6H_5Br + HBr$

or $C_6H_6 + HNO_3 \xrightarrow{H_2SO_4} C_6H_5NO_2 + H_2O$

26.20 (a) $CH_2{=}CHCH_2CH_3 + H_2 \longrightarrow CH_3CH_2CH_2CH_3$

(b) $CH_3CH_2CH{=}CHCH_2CH_3 + H_2O \xrightarrow{H_2SO_4} CH_3CH_2CH(OH)CH_2CH_2CH_3$

(c) 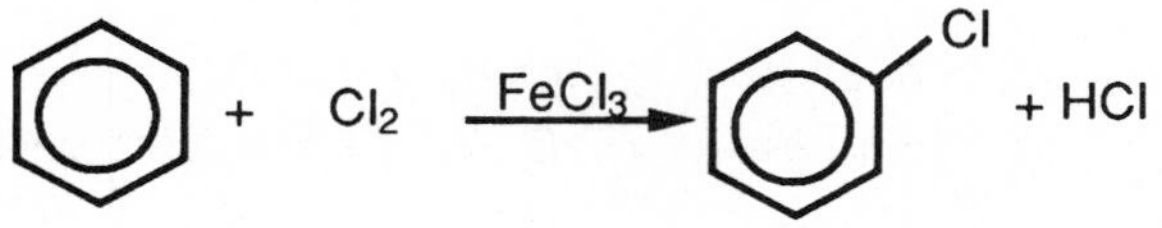

26.21 The small 60° C-C-C angles in the cyclopropane ring cause strain that provides a driving force for reactions that result in ring-opening. There is no comparable strain in the five- or six-membered rings.

26.22 The energy gained by forming two σ bonds (or four σ bonds) more than compensates for the loss of one (or two) π bonds when addition occurs to an alkene (or alkyne). However, in benzene the aromatic ring is especially stabilized by the delocalization of π electrons around the ring. It, therefore, requires a substantial activation energy to cause the loss of this aromatic character. The most common reaction in aromatics is substitution rather than addition, since substitution does not result in loss of aromatic character.

26.23 First form an alkyl halide: $C_2H_4(g) + HBr(g) \rightarrow CH_3CH_2Br(l)$; then carry out a Friedel-Crafts reaction:

$C_6H_6(l) + CH_3CH_2Br(l) \xrightarrow{AlCl_3} C_6H_5CH_2CH_3(l) + HBr(g)$

26.24 Both the ortho and para isomers are formed; they must be separated by distillation or some other technique.

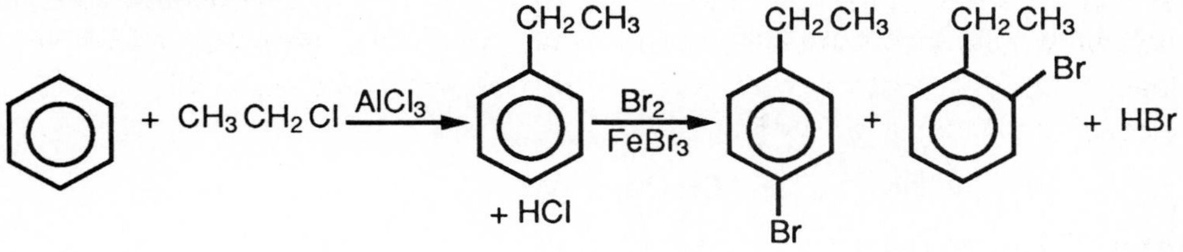

26.25

	ΔH
$C_{10}H_8(l) + 12O_2(g) \rightarrow 10CO_2(g) + 4H_2O(l)$	-5157 kJ
$-[C_{10}H_{18}(l) + 29/2\ O_2(g) \rightarrow 10CO_2(g) + 9H_2O(l)]$	-(-6286) kJ
$C_{10}H_{18}(l) + 5H_2O(l) \rightarrow C_{10}H_{18}(l) + 5/2\ O_2(g)$	+1129 kJ
$5/2\ O_2(g) + 5H_2(g) \rightarrow 5H_2O(l)$	5(-285.8) kJ
$C_{10}H_8(l) + 5H_2(g) \rightarrow C_{10}H_{18}(l)$	-300 kJ

Compare this with the heat of hydrogenation of ethylene:
$C_2H_4(g) + H_2(g) \rightarrow C_2H_6(g)$; $\Delta H = -84.7 - (52.3) = -137$ kJ. This value applies to just one double bond. For five double bonds, we would expect about -685 kJ. The fact that hydrogenation of napthalene yields only -300 kJ indicates that the overall energy of the naphthalene molecule is lower than expected for five isolated double bonds and that there must be some special stability associated with the aromatic system in this molecule.

26.26 Both combustion reactions produce CO_2 and H_2O:

$$C_3H_6(g) + 9/2\ O_2(g) \rightarrow 3CO_2(g) + 3H_2O(l)$$

$$C_5H_{10}(g) + 15/2\ O_2(g) \rightarrow 5CO_2(g) + 5H_2O(l)$$

Thus, we can calculate the $\Delta H_{comb}/CH_2$ group for each compound:

$$\frac{\Delta H_{comb}}{CH_2 \text{ group}} = \frac{2089 \text{ kJ/mol } C_3H_6}{3CH_2 \text{ groups}} = \frac{696.3 \text{ kJ}}{\text{mol } CH_2}\ ; \quad \frac{3317 \text{ kJ/mol } C_5H_{10}}{5CH_2 \text{ groups}} = \frac{663.4 \text{ kJ}}{\text{mol } CH_2}$$

$\Delta H_{comb}/CH_2$ group for cyclopropane is greater because C_3H_6 contains a strained ring. When combustion occurs, the strain is relieved and the stored energy is released during the reaction.

Hydrocarbon Derivates

26.27   (a) ketone   (b) carboxylic acid   (c) alcohol   (d) ester   (e) amide   (f) amine

26.28   (a) $—NH_2$, amine; $—\overset{\overset{\displaystyle O}{\|}}{C}—OH$, carboxylic acid

(b) $CH_2\!=\!CH—$, alkene; $—OH$, alcohol   (c) $—\overset{\overset{\displaystyle O}{\|}}{C}—NH—$, amide

(d) $—\overset{\overset{\displaystyle O}{\|}}{C}—$, ketone   (e) $—\overset{\overset{\displaystyle O}{\|}}{C}—H$, aldehyde

(f) $HC\!\equiv\!C—$, alkyne; $—OH$, alcohol

26.29   propionaldehyde (or propanal): $H—\overset{\overset{\displaystyle H}{|}}{\underset{\underset{\displaystyle H}{|}}{C}}—\overset{\overset{\displaystyle H}{|}}{\underset{\underset{\displaystyle H}{|}}{C}}—\overset{\overset{\displaystyle O}{\|}}{C}\diagdown_{H}$

26.30 $H—\overset{\overset{\displaystyle H}{|}}{\underset{\underset{\displaystyle H}{|}}{C}}—O—\overset{\overset{\displaystyle H}{|}}{\underset{\underset{\displaystyle H}{|}}{C}}—H$

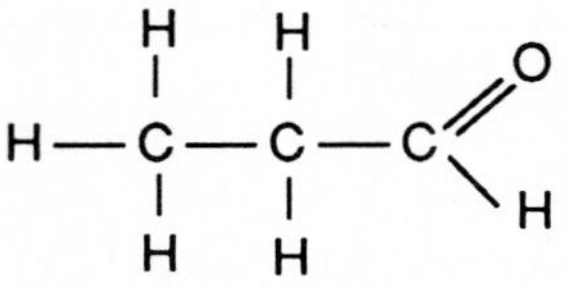

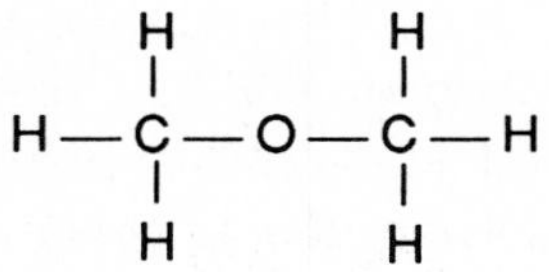

26.31   About each CH_3 carbon, 109°; about the carbonyl carbon, 120° planar.

26.32 $\underset{\underset{\displaystyle sp^2}{\uparrow}}{H—\overset{\overset{\displaystyle sp}{\downarrow}}{C}\equiv\overset{\overset{\displaystyle sp}{\downarrow}}{C}—\overset{\overset{\displaystyle O}{\|}}{C}}—O—\underset{\underset{\displaystyle sp^3}{\uparrow}}{CH_3}$

26.33   (a) $\underset{\underset{\displaystyle CH_3}{|}}{\overset{\overset{\displaystyle CH_3}{|}}{HC}}—O\overset{\overset{\displaystyle O}{\|}}{C}CH_3$ (b) $CH_3CH_2CH_2O—\overset{\overset{\displaystyle O}{\|}}{C}CH_3$ (c) $CH_3CH_2O—\overset{\overset{\displaystyle O}{\|}}{C}H$

2-propylacetate   1-propylacetate   ethylformate

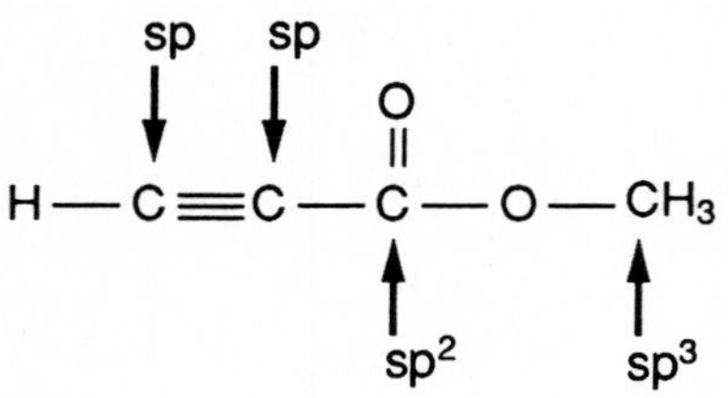

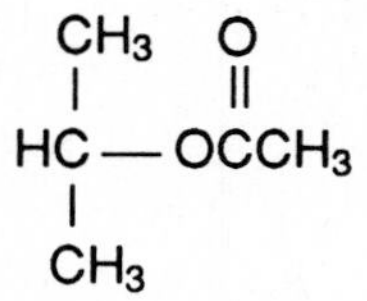

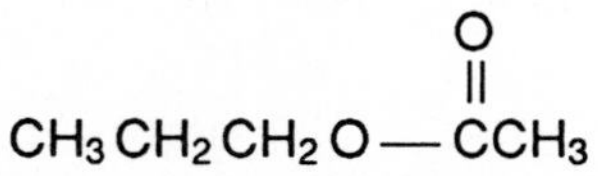

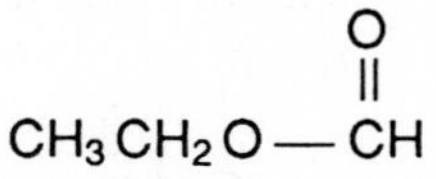

26.34 (a) $CH_3CH_2O-\overset{\overset{\displaystyle O}{\|}}{C}$—⬡

ethylbenzoate

(b) $CH_3\overset{\overset{\displaystyle H}{|}}{N}-\overset{\overset{\displaystyle O}{\|}}{C}CH_2CH_3$

N-methylpropionamide

(c) ⬡$-O-\overset{\overset{\displaystyle O}{\|}}{C}CH_3$

phenylacetate

26.35 $HC\overset{\overset{\displaystyle O}{\|}}{}-O-CH_2CH_3 + NaOH \longrightarrow \left[HC\overset{\diagup O}{\diagdown O} \right]^- + Na^+ + CH_3CH_2OH$

26.36 $CH_3C\overset{\overset{\displaystyle O}{\|}}{}-O-CH_3 + NaOH \longrightarrow \left[CH_3C\overset{\diagup O}{\diagdown O} \right]^- + Na^+ + CH_3OH$

26.37 (a) $CH_3CH_2\overset{\overset{\displaystyle }{|}}{\underset{\underset{\displaystyle OH}{|}}{C}H}CH_3$

(b) $HOCH_2CH_2OH$

(c) $HC\overset{\overset{\displaystyle O}{\|}}{}-OCH_3$

(d) $CH_3CH_2\overset{\overset{\displaystyle O}{\|}}{C}CH_2CH_3$

(e) $CH_3CH_2OCH_2CH_3$

26.38 (a) $Cl-CH_2CH_2\overset{\overset{\displaystyle O}{\|}}{C}H$

(b) $CH_3\overset{\overset{\displaystyle O}{\|}}{C}\underset{\underset{\displaystyle CH_3}{|}}{C}HCH_3$

(c) ⬡ with $\overset{\overset{\displaystyle O}{\|}}{C}H$ and Cl

(d) $CH_3OCH_2\diagdown{}_{C}=C{}^{\diagup CH_3}$ with H substituents

(e) $CH_3CH_2\overset{\overset{\displaystyle O}{\|}}{C}\underset{\underset{\displaystyle C_2H_5}{|}}{N}C_2H_5$

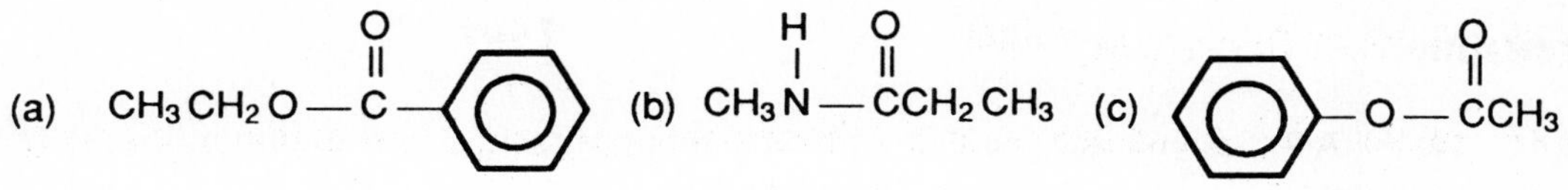

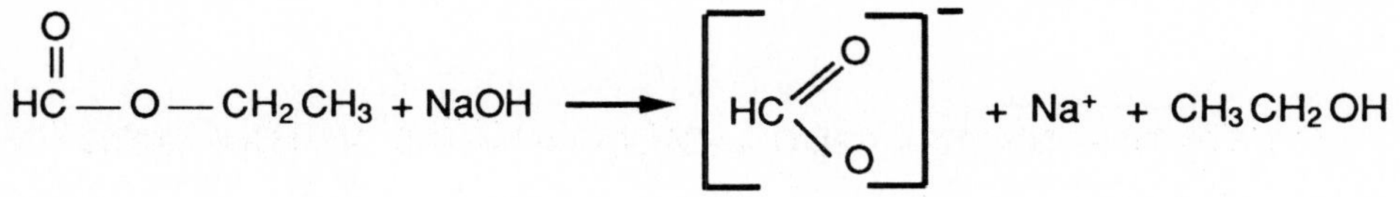

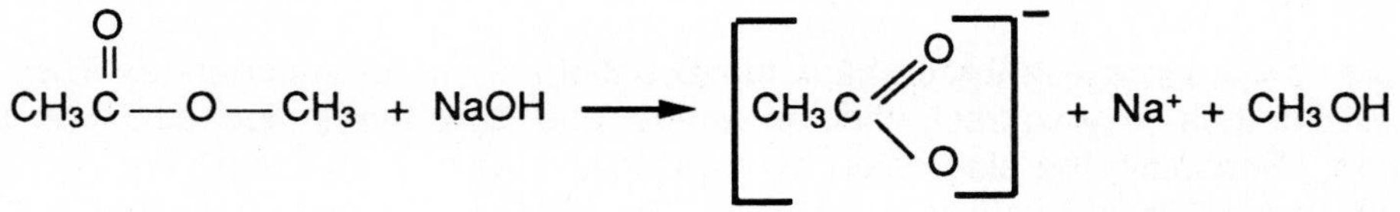

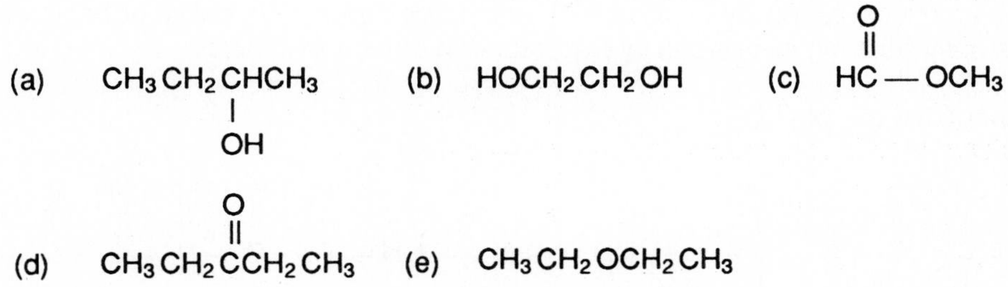

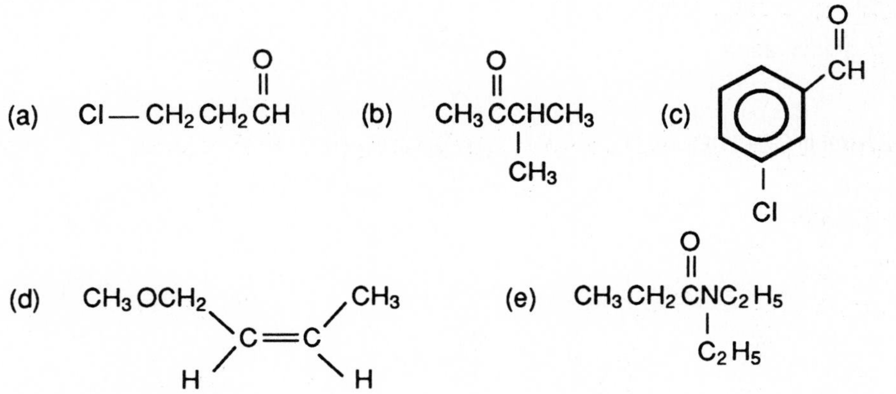

26.39 (a) methanoic acid   (b) butanoic acid   (c) 3-methylpentanoic acid

26.40 (a) $CH_3CH_2\overset{\overset{\displaystyle O}{\|}}{C}-H$

(b) $CH_3CH_2CH_2\overset{\overset{\displaystyle O}{\|}}{C}CH_3$

(c) $CH_3\underset{\underset{\displaystyle CH_3}{|}}{C}H\overset{\overset{\displaystyle O}{\|}}{C}CH_3$

(d) $CH_3CH_2\underset{\underset{\displaystyle CH_3}{|}}{C}H\overset{\overset{\displaystyle O}{\|}}{C}H$

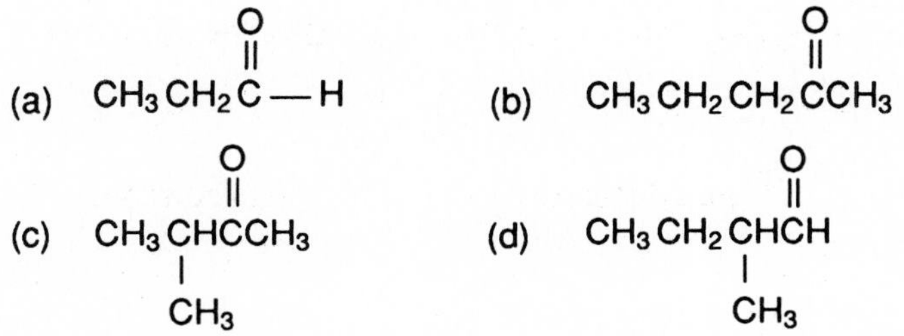

Proteins

26.41 (a) An α-amino acid contains an NH_2 group attached to the carbon that is bound to the carbon of the carboxylic acid function.

(b) In forming a protein amino acids undergo a condensation reaction between the amino group and carboxylic acid:

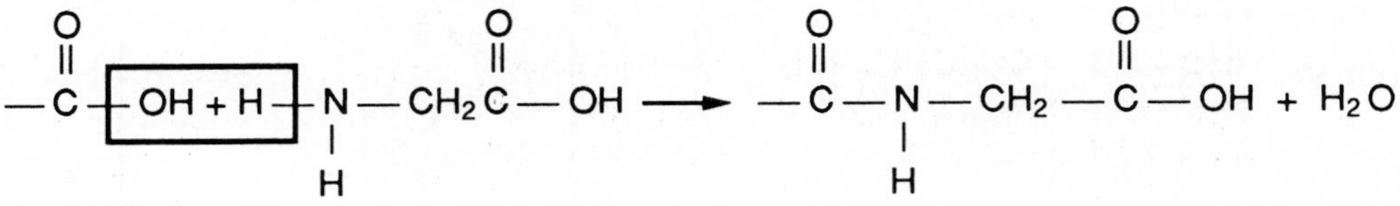

26.42 The side chains possess three characteristics that may be of importance. They may be bulky, and thus impose restraints on where and how the amino acid can undergo reaction. Secondly, the side chain may possess a polar group (e.g., the -OH group in serine), that will in part determine solubility. Finally, the side chain may contain an acidic or basic functional group that will partially determine solubility in acidic or basic medium, and that may become involved in interactions with other amino acids.

26.43 Two dipeptides are possible:

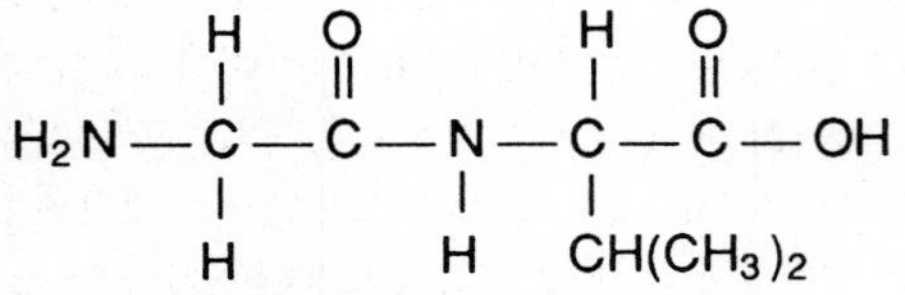

glycylvaline

and

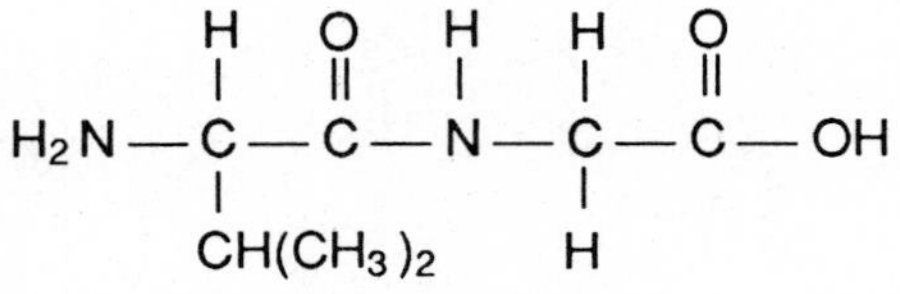

valylglycine

26.44 $H_2NCH(CH_2CO_2H)CO_2H + H_2NCH(CH_2SH)CO_2H \longrightarrow$

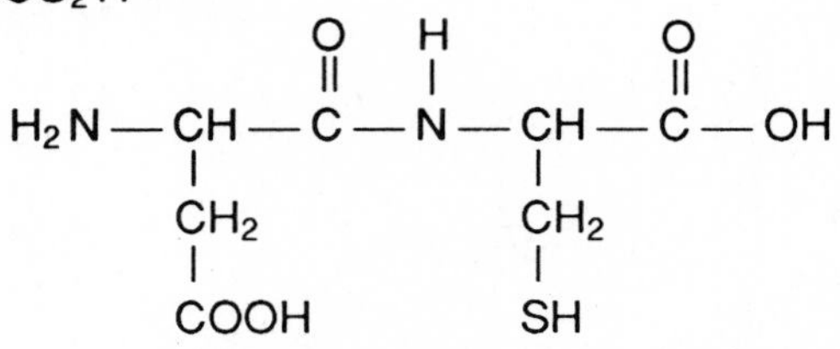

aspartylcysteine

26.45

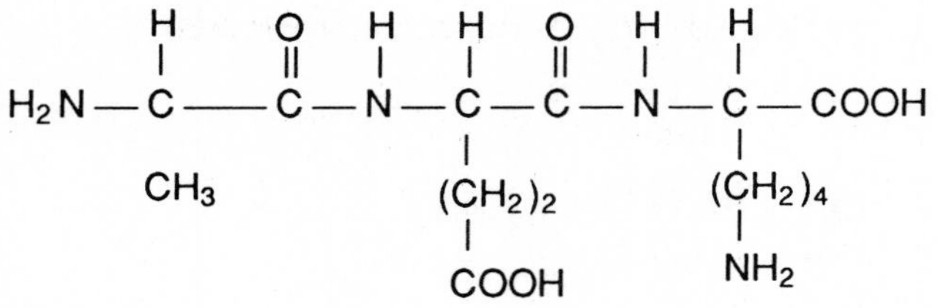

26.46 Alanine, serine

26.47 Eight: ser-ser-ser; ser-ser-phe; ser-phe-ser; phe-ser-ser; ser-phe-phe; phe-ser-phe; phe-phe-ser; phe-phe-phe

27.48 Six: gly-val-ala; gly-ala-val; val-gly-ala; val-ala-gly; ala-gly-val; ala-val-gly

26.49 The primary structure of a protein refers to the sequence of amino acids in the chain. Along any particular section of the protein chain the configuration may be helical, or it may be an open chain, or arranged in some other way. This is called the secondary structure. The overall shape of the protein molecule is determined by the way the segments of the protein chain come together, or pack. The interactions which determine the overall shape are referred to as the tertiary structure.

26.50 It is quite evident from Figure 26.19 that the hydrogen bonds between an NH group along the chain and the unshared electron pairs of a carbonyl group further along are responsible for maintaining the helix. Indeed, the pitch and general shape of the helix are determined by what specific interactions produce a good hydrogen bonding arrangement.

Carbohydrates

26.51 Glucose exists in solution as a cyclic structure in which the aldehyde function on carbon 1 reacts with the OH group of carbon 5 to form what is called a hemiacetal, Figure 26.22. Carbon atom 1 carries an OH group in the hemiacetal form; in α-glucose this OH group is on the opposite side of the ring as the CH_2OH group on carbon atom 5. In the β (beta) form the OH group on carbon 1 is on the same side of the ring as the CH_2OH group on carbon 5.

The condensation products look like this:

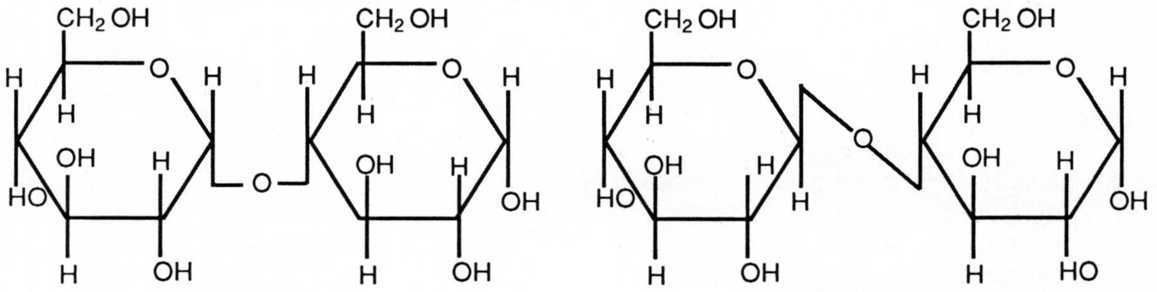

α-linkage β -linkage

26.52 (a) α-form (b) β-form (c) β-form

26.53 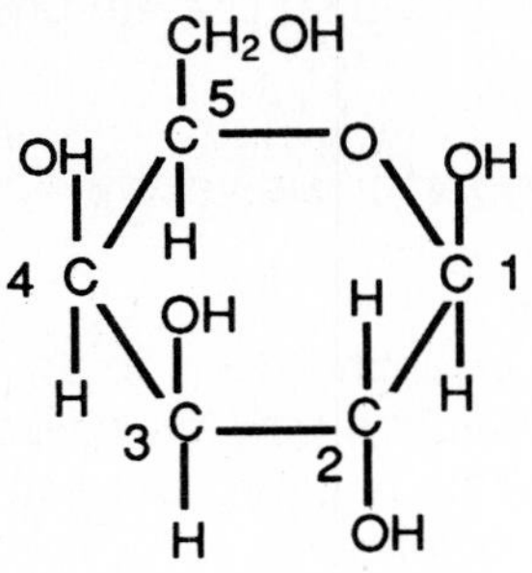

Galactose

The structure is best deduced by comparing galactose with glucose, and inverting the configurations at the appropriate carbon atoms. Recall from Exercise 26.51 that both the β-form (shown here) and the α-form (OH on carbon 1 on the opposite side of ring as the CH_2OH on carbon 5) are possible.

26.54 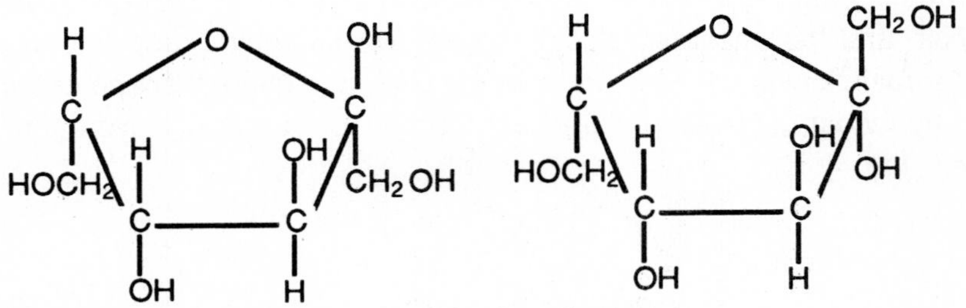

Both the α (left) and β (right) forms are possible.

26.55 In the ring form (see Exercise 26.53), carbon atoms 1, 2, 3, 4 and 5 are chiral, because they carry four different groups on each.

26.56 Numbering the carbons from the top down, atoms 2, 3, 4 and 5 are chiral because each is bound to four different groups.

Nucleic Acids

26.57 A <u>nucleotide</u> is a molecule consisting of a nitrogen-containing aromatic compound, a sugar in the furanose (5-membered) ring form, and a phosphoric acid molecule. The structure of deoxycytidine monosphosphate is shown at right.

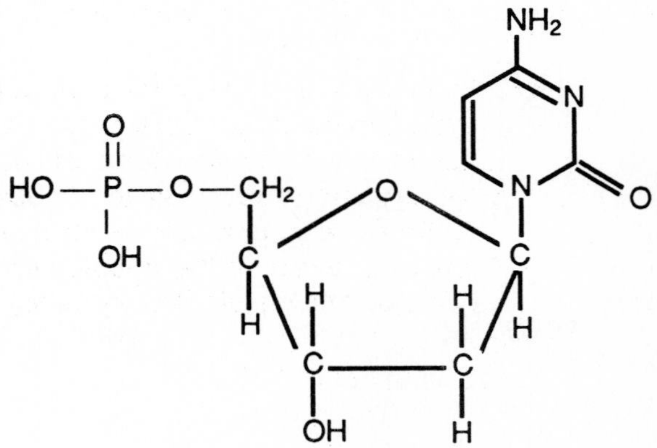

26.58

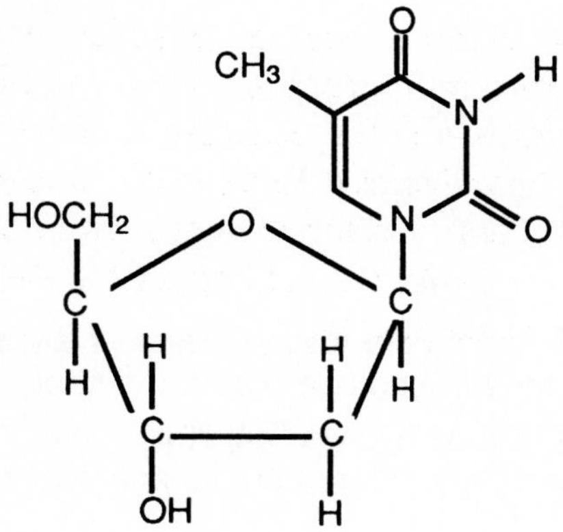

26.59 $C_4H_7CH_2OH + H_3PO_4 \rightarrow C_4H_7CH_2\text{-}O\text{-}PO_3H_2 + H_2O$

26.60

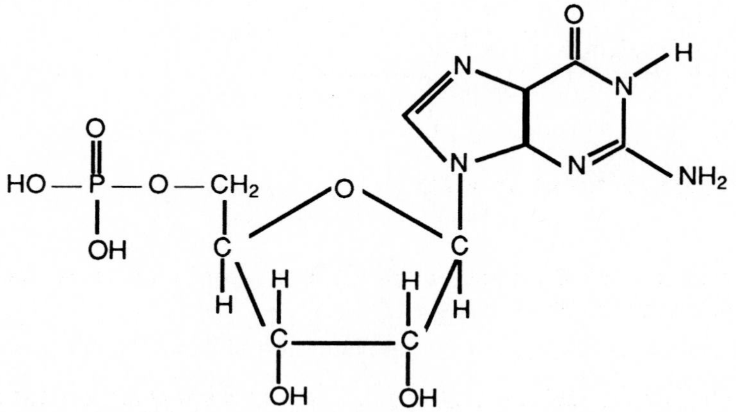

26.61

```
—A—C—T—C—G—A—
 |  |  |  |  |  |
—T—G—A—G—C—T—  ◄—— Complementary strand
```

26.62 In the helical structure for DNA, the strands of the polynucleotides are held together by hydrogen-bonding interactions between particular pairs of bases. It happens that adenine and thymine form an especially effective base pair, and that guanine and cytosine are similarly related. Thus, each adenine has a thymine as its opposite number in the other strand, and each guanine has a cytosine as its opposite number. In the overall analysis of the double strand, total adenine must then equal total thymine, and total guanine equals total cytosine.

Additional Exercises

26.63 $HC \equiv C - CH_2 CH_3$ $CH_3 C \equiv C - CH_3$

$$HC = CH$$
$$| \quad\quad |$$
$$CH_2 - CH_2$$

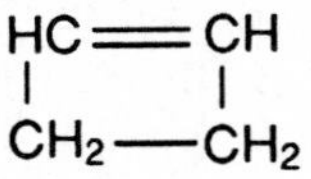

$CH_2 = CH - CH = CH_2$

26.64 No. Hydrocarbons are alkanes in which carbon atoms are involved exclusively in single bonds. There are four bonding electron pairs around each carbon in the chain and the bond angles are 109°. Thus, an unbranched alkane is really a "zig-zag" chain of carbon atoms, with each C atom bound to a maximum of two other C atoms and an approriate number of H atoms.

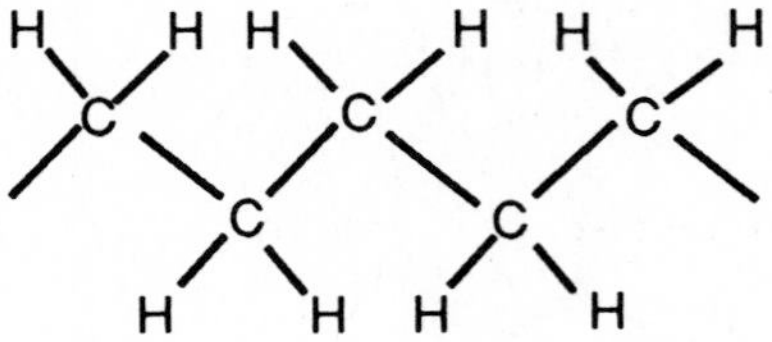

26.65

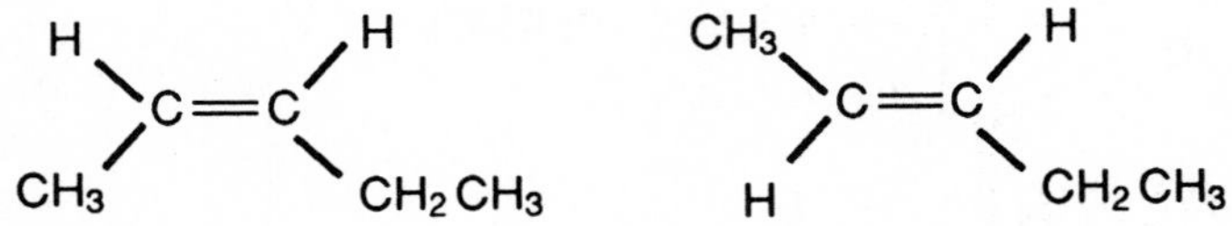

cis   trans

To show complete Lewis structures all C-H bonds should be shown. These have been omitted here to save space. Cyclopentene does not show cis-trans isomerism because the existence of the ring demands that the C-C bonds be cis to one another.

26.66 (a)   2-propanol   (b) dimethyl ether   (c) 3-butene-1-ol
           (d)   3-methyl-3-hexene   (e) 4-methyl-2-pentyne

26.67 $\underset{\displaystyle CH_3 CH_2 CH_2}{\overset{\displaystyle OH}{\overset{\displaystyle |}{}}}$ $\underset{\displaystyle CH_3 CHCH_3}{\overset{\displaystyle OH}{\overset{\displaystyle |}{}}}$ $CH_3 - O - CH_2 CH_3$

26.68 The C-Cl bonds in the trans compound are pointing in exactly opposite directions. Thus, the C-Cl bond dipoles cancel (Section 9.2). This is not the case in the cis compound, as can be seen by writing out the structure:

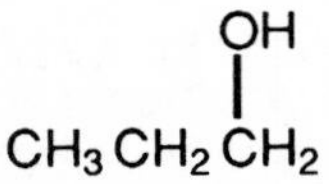

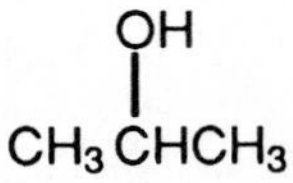

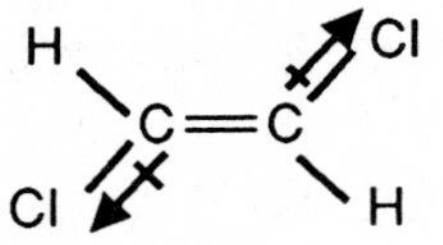

trans

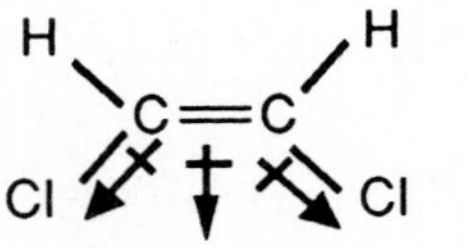

net dipole moment
cis

CHAPTER 26

The Chemistry of Life: Organic and Biological Chemistry

Solutions to Exercises

26.69 Because of the strain in bond angles about the ring, cyclohexyne would not be stable. The alkyne carbons preferentially have a 180° bond angle. However, there are not enough carbons in the ring to make this possible without gross distortions of other bond lengths and angles:

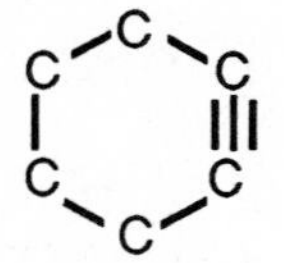

26.70 (a) Ether, C—O—C; alkene, $—CH{=}CH_2$

(b) carboxylic acid, $—\overset{O}{\overset{\|}{C}}—OH$; ester, $CH_3\overset{O}{\overset{\|}{C}}—O—$

(c) ketone, $—\overset{O}{\overset{\|}{C}}—$; alkene, $—CH{=}CH—$; alcohol, $—C—OH$

26.71 (a) $CH_3OCH_2CH_3$ (b) $CH_3CH\overset{O}{\overset{\|}{C}}—H$ (c) $CH_3\overset{O}{\overset{\|}{C}}CH_3$

(d) $CH_3CH_2\overset{O}{\overset{\|}{C}}OH$ (e) $CH_3\overset{O}{\overset{\|}{C}}—OCH_3$

26.72 (a) $CH_3\overset{O}{\overset{\|}{C}}—OH$, C_6H_5OH (b) $C_6H_5\overset{O}{\overset{\|}{C}}—OH$, CH_3OH

26.73 (a) cyclobutane. There is some strain in the four-membered ring because the C-C-C bond angles must be less than the desired 109°.

(b) cyclohexene. The C=C bond is capable of addition reactions, for example with Br_2.

(c) 1-hexene. Whereas the alkene readily undergoes addition reactions, the aromatic hydrocarbon is extremely stable.

(d) 2-hexyne. Alkynes undergo addition reactions even more readily than alkenes.

26.74 The compound is clearly an alcohol. Its slight solubility in water is consistent with the properties expected of a secondary alcohol with a five-carbon chain. The fact that it is oxidized to a ketone rather than to an aldehyde and then to a carboxylic acid tells us that it is a secondary alcohol:

$CH_3\underset{OH}{\underset{|}{C}}HCH_2CH_2CH_3$ $\quad CH_3CH_2\underset{OH}{\underset{|}{C}}HCH_2CH_3$ $\quad CH_3\underset{OH}{\underset{|}{C}}HCH(CH_3)_2$

CHAPTER 26 The Chemistry of Life: Organic and Biological Chemistry

Solutions to Exercises

26.75 $85.7 \text{ g C} \times \frac{1 \text{ mol C}}{12.0 \text{ g C}} = 7.14 \text{ mol C}$

$$14.3 \text{ g H} \times \frac{1 \text{ mol H}}{1.01 \text{ g H}} = 14.2 \text{ mol H}$$

Empirical formula is CH_2. Using Equation 10.12:

$$\mathcal{M} = \frac{(2.21 \text{ g/L})\ (0.0821 \text{ L-atm/mol-K})\ (373 \text{ K})}{(735/760) \text{ atm}} = 70.0 \text{ g/mol}$$

The molecular formula is thus C_5H_{10}. The absence of reaction with aqueous Br_2 indicates that the compound is not an alkene, so the compound is probably the cycloalkane, cyclopentane. To verify this compare the boiling point of the substance with the handbook value for cyclopentane.

26.76 All growth processes, which require synthesis of new cell materials, require energy. For example, in plants transpiration of water requires energy. In animals, the action of the involuntary muscles, such as the heart, maintenance of a uniform body temperature, synthesis of new cells, motions of limbs, and maintenance of correct body fluid compositions all require energy.

26.77 (a) None

(b) The carbon bearing the secondary -OH has four different groups attached, and is thus chiral.

(c) The carbon bearing the $-NH_2$ group and the carbon bearing the CH_3 group are both chiral.

26.78 (a)

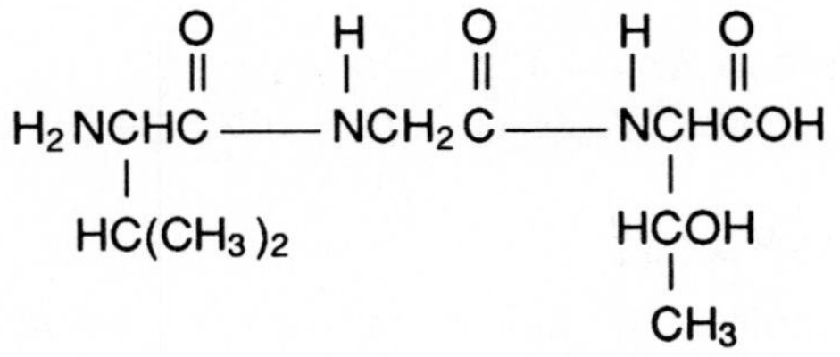

(b)

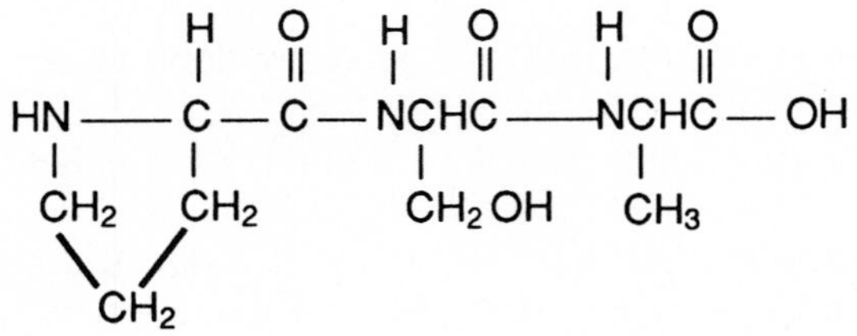

26.79 Glu-cys-gly is the only possible structure.

CHAPTER 26
The Chemistry of Life: Organic and Biological Chemistry

Solutions to Exercises

26.80 The reaction is:

$2NH_2CH_2COOH(aq) \rightarrow NH_2CH_2CONHCH_2COOH(aq) + H_2O(l)$

$\Delta G = (-488) + (-285.8) - 2(-369) = -35.8 \text{ kJ}$

26.81 (a) Sucrose, maltrose, or lactose (see Figure 26.23)

(b) Ribose, and deoxyribose (see Section 26.7)

(c) Glucose (Figure 26.22)

(d) Starch and cellulose (Figures 26.24, 26.25 and 26.26).

26.82 The reaction is approximately:

$\underline{n}C_6H_{12}O_6(aq) \rightarrow -(C_6H_{11}O_5)_{\underline{n}}-(aq) + (n-1)H_2O(l)$

$\Delta G^\circ = n(-662.3) + (n-1)(-285.8) - n(-917.2)$

$= n(-285.8 - 662.3 + 917.2) + 285.8$

$= n(-30.9) + 285.8 \text{ kJ}$

26.83
```
—T—A—T—G—C—A—
 :  :  :  :  :  :
—A—T—A—C—G—T—  ←— Complementary strand
```

26.84 $AMPOH^-(aq) \rightleftharpoons AMPO^{2-}(aq) + H^+(aq)$

$K_a = \dfrac{[AMPO^{2-}][H^+]}{[AMPOH^-]} = 6.2 \times 10^{-8}$. When pH = 7.40, $[H^+] = 4.0 \times 10^{-8}$.

Then $\dfrac{[AMPOH^-]}{[AMPO^{2-}]} = 4.0 \times 10^{-8} / 6.2 \times 10^{-8} = 0.65$